Advanced Structured Materials

Volume 244

Common engineering materials are reaching their limits in many applications, and new developments are required to meet the increasing demands on engineering materials. The performance of materials can be improved by combining different materials to achieve better properties than with a single constituent, or by shaping the material or constituents into a specific structure. The interaction between material and structure can occur at different length scales, such as the micro, meso, or macro scale, and offers potential applications in very different fields.

This book series addresses the fundamental relationships between materials and their structure on overall properties (e.g., mechanical, thermal, chemical, electrical, or magnetic properties, etc.). Experimental data and procedures are presented, as well as methods for modeling structures and materials using numerical and analytical approaches. In addition, the series shows how these materials engineering and design processes are implemented and how new technologies can be used to optimize materials and processes.

Advanced Structured Materials is indexed in Google Scholar and Scopus.

Lakshmipriya Ravindran · Akhila Babu ·
Sreekala Meyyarappallil Sadasivan ·
Miran Mozetič · Sabu Thomas
Editors

Macro, Micro and Nanocomposites from Sustainable Sources

Shrinking Environmental Footprints

Springer

Editors
Lakshmipriya Ravindran
International and Inter University Centre for Nanoscience and Nanotechnology (IIUCNN)
Mahatma Gandhi University
Kottayam, Kerala, India

Sreekala Meyyarappallil Sadasivan
School of Chemical Sciences
Mahatma Gandhi University
Kottayam, Kerala, India

Sabu Thomas
International and Inter University Centre for Nanoscience and Nanotechnology (IIUCNN)
Mahatma Gandhi University
Kottayam, Kerala, India

Akhila Babu
School of Chemical Sciences
Mahatma Gandhi University
Kottayam, Kerala, India

Miran Mozetič
Jozef Stefan Institute
Ljubljana, Slovenia

ISSN 1869-8433 ISSN 1869-8441 (electronic)
Advanced Structured Materials
ISBN 978-981-95-2468-6 ISBN 978-981-95-2469-3 (eBook)
https://doi.org/10.1007/978-981-95-2469-3

This Springer imprint is published by the registered company Springer Nature Singapore Pte Ltd.
The registered company address is: 152 Beach Road, #21-01/04 Gateway East, Singapore 189721, Singapore

Preface

Over the last few years, the international science fraternity saw an unprecedented revolution in material science, and this has happened due to the dire necessity to be sustainable, preserve the environment and be innovative in creating functional materials. The production and use of composites based on biomass, especially at the macro, micro and nano levels, are one of the best perspectives of solutions to the question of ecology, with savings in the use of fossil resources and moving forward with circular economies. *Macro, Micro and Nanocomposites from Sustainable Sources: Shrinking Environmental Footprints* is a handbook that is aimed at collating the latest research on topics together with practical knowledge and emerging trends in sustainable composites design, processing, and application. The emphasis goes through a range of renewable materials, including agricultural wastes and biodegradable polymers, to naturally occurring nanostructures synthesized into high-performance, low-footprint composite systems.

The volume comprehensively covers both fundamental principles and real-world applications from the domain of green synthesis pathways, bio-functional interfacing, and scales of production, life cycle analysis, and diverse application domains such as packaging, construction, automotive, aerospace, military, medicine, energy harvesting and storage, wastewater treatment, and marine technology. Readers will also find valuable insights into cutting-edge areas like self-healing materials, shape memory biopolymers, computational modelling, regulatory frameworks, recycling strategies and market feasibility studies.

The introductory chapters highlight the recent progress and future challenges of sustainable composites, also detailing the relationship between processing methods and material properties, an essential aspect in tailoring sustainable composites for targeted applications. Various applications of green composites like packaging and construction are discussed in depth, with emphasis on the development of lightweight, biodegradable and recyclable materials to replace conventional plastic and cement. This volume also contains chapters dedicated to the applications of biopolymer composites in agriculture and the automotive industry, focusing on their role in enhancing productivity and reducing carbon footprints.

Further, this book explores the contribution of green composites to the energy sector. Chapters are dedicated to energy harvesting and storage, including discussions on piezoelectric materials, biopolymer-based electrodes and integration of sustainable composites in next-generation energy storage devices. The book also features advanced applications designed for high-performance areas like military and aerospace, since durability, lightweight structure and sustainability are of paramount importance.

Other innovative chapters on self-healing and shape memory biomaterials demonstrate the significance of biopolymer systems that can extend product lifespans and reduce maintenance costs. The biomedical and pharmaceutical potential of sustainable composites is also covered. Environmental remediation and sustainability are further addressed with chapters on the use of functional biopolymers for wastewater treatment alongside strategies for improving the recyclability and long-term durability of composites. Additionally, the book presents computational and modelling approaches that enhance our understanding of material behaviour, contributing to the efficient design of new sustainable composites.

The final sections shift towards the broader societal and industrial implications of sustainable composites, including regulatory frameworks, policy considerations, case studies from green industries and economic feasibility studies. These chapters provide essential perspectives for researchers, policymakers and entrepreneurs aiming to implement sustainable material solutions at scale. This handbook draws on the experiences of a diverse group of researchers, engineers and practitioners who are at the forefront of developing innovative solutions that bridge sustainability and material innovation. Every chapter is handpicked, involving a balanced picture of the theoretical grounds, experimental designs, performance evaluation and practical use of the scenarios.

It is our hope that the contents of this handbook will provide a sound guide of reference to the academic researcher, industrialists, engineers and students, all of whom are interested in getting an overview of studying and developing the science and engineering of sustainable composite materials. We all aim at informing, but more importantly, we seek to inspire new ways of creating smaller environmental footprints, employing smart and responsible development of materials. We would like to thank all authors, reviewers and editors whose contribution has ensured that this compilation is realized. May this become part of a contribution to a greener, more sustainable world.

Kottayam, India — Lakshmipriya Ravindran
Kottayam, India — Akhila Babu
Kottayam, India — Sreekala Meyyarappallil Sadasivan
Ljubljana, Slovenia — Miran Mozetič
Kottayam, India — Sabu Thomas

Contents

Chapter 1
State of the Art, Applications, and Future Challenges of Sustainable Macro, Micro, and Nanocomposites

B. Akhila, V. Abhijith, Lakshmipriya Ravindran, Miran Mozetič, Sabu Thomas, and M. S. Sreekala

Abstract The growing necessity for reducing environmental pollution and a shift towards a circular economy have prompted the advancement of sustainable macro, micro, and nanocomposites derived from biopolymers. This chapter provides an overview of biopolymer-based composites, focusing mainly on their various advanced applications. The importance of these composites is highlighted in critical sectors including packaging, construction, agriculture, automotive, biomedical applications, energy harvesting, aerospace, and wastewater treatment. The emerging innovations towards self-healing systems, shape-memory materials, and bio-based sensors are also discussed, addressing the challenges associated with durability, scalability, and economic feasibility of biopolymer-based composites. Thus, the scientific advancements, economic feasibility, and industrial potential of sustainable composites establish a foundation for future research and development in the design of eco-friendly materials for next-generation applications.

Keywords Biopolymers · Packaging · Biomedical · Shape memory · Energy harvesting · Defence

---

B. Akhila · V. Abhijith · M. S. Sreekala (✉)
School of Chemical Sciences, Mahatma Gandhi University, Kottayam, Kerala, India
e-mail: sreekalams@mgu.ac.in

L. Ravindran (✉) · S. Thomas · M. S. Sreekala
International and Inter-University Centre for Nanoscience and Nanotechnology, Mahatma Gandhi University, Kottayam, Kerala, India
e-mail: lakshmipriyaravindran@mgu.ac.in

S. Thomas · M. S. Sreekala
School of Nanoscience and Nanotechnology, Mahatma Gandhi University, Kottayam, Kerala, India

M. Mozetič
Jozef Stefan Institute, Ljubljana, Slovenia

L. Ravindran et al. (eds.), *Macro, Micro and Nanocomposites from Sustainable Sources*, Advanced Structured Materials 244, https://doi.org/10.1007/978-981-95-2469-3_1

1.1 Introduction

Biopolymer-based nanocomposites are a new class of materials that consist of biodegradable polymers and nanoscale fillers to produce lightweight, strong, and functional systems. The materials provide a greener option to conventional plastics with advantages like biodegradability, renewability, and low toxicity. Compared to conventional composite materials, the materials are sourced from widespread biological resources and are increasingly gaining attention in research and industry due to their sustainable life cycle and reduced environmental footprint. Their chemically tuneable structure and capability to establish strong interfacial connections with nanofillers render them highly suitable for multifunctional applications that highlight performance with sustainability (Xiong et al. 2018).

Biopolymer composites with cellulose, chitosan, chitin, starch, silk, collagen, and alginate, and synthetic biopolymers like PLA and PHA, with nanomaterials like graphene, carbon nanotubes, metal oxides, nanocellulose, and clays, can cause remarkable improvement in various properties. Nanofillers improve the strength of the polymer matrix as well as offer additional functionalities such as mechanical strength, electrical conductivity, flame retardation, and antimicrobial activity. The large surface area and high aspect ratio of nanomaterials enable effective stress transfer and enhance the overall performance of the composite. Blending bio-based matrices with nanostructured fillers enables the generation of high-performance materials for advanced technological and sustainable applications (Díez-Pascual 2019; Cherian et al. 2025).

Applications for nanocomposites are found in biomedicine in wound healing scaffolds, implants, and drug delivery; in energy in batteries, supercapacitors, nanogenerators, and fuel cells; in food packaging for enhancing shelf life and storage; and in electronics for flexible sensors. In the agriculture field, biopolymer nanocomposites are widely used for the development of slow-release fertilisers, biodegradable films, and different types of sensors, specifically moisture sensors, to improve crop yields. These composites are also used for the fabrication of biodegradable nets and coatings in the marine field. The comparable properties of biopolymer composites, including their thermal stability, chemical resistance, and easy processability, make them a superior candidate in different industrial applications, like use in paints, adhesives, filtration membranes, and insulation foams, which ultimately decreases the dependence on petroleum-derived materials (Xiong et al. 2018; Díez-Pascual 2019).

Researchers are also focusing on the development of biopolymer-based nanocomposites for advanced materials, which include shape-memory polymers that can change their structure with external stimuli and self-healing polymers that autonomously heal the damage caused to them. The unique characteristics of biopolymer composites, such as adaptability, long durability, and high reliability, make them a superior choice in different industries like aerospace, defence technologies, and robotics. Also, the sensitivity of those composites to external stimuli like pH, light, humidity, and temperature makes them appropriate for the development of various sensors to detect environmental parameters and in wearable electronics. The combination of various functionalities in a single material system widens the range of applications for next-generation intelligent devices (Thomas et al. 2017; Lendlein et al. 2019; Khalid et al. 2024).

A variety of processing methods, including solution casting, melt blending, freeze-drying, electrospinning, and 3D printing, enable the transformation of these materials into films, foams, fibres, or hydrogels, depending on the requirements of the application. These methods are scalable and compatible with modern industrial processes. In addition, such processes are likely to consume less energy and less application of hazardous chemicals than traditional polymer processing, thus enabling greener production processes (Sampath et al. 2016; Stanisz et al. 2020; Tang et al. 2025; Akhila et al. 2025).

This chapter discusses the growing relevance of biopolymer-based macro, micro, and nanocomposites in various applications. Packaging, agricultural, automotive, aerospace, biomedical, pharmaceutical, environmental, energy harvesting and storage, marine, and other advanced applications like self-healing and shape-memory materials are just a few of them. Since the world is moving towards a circular economy, these composites can potentially occupy a central role in the development of high-performance materials that are aligned with the United Nations Sustainable Development Goals (UN SDGs).

1.2 Classification: Macro, Micro, and Nanocomposites from Biopolymers

Depending on the size and scale of reinforcing fillers, the biopolymer-based composites are typically classified as macro, micro, and nanocomposites. Such reinforcement materials can include natural fibres (e.g., jute, sisal, flax, and hemp) in a dimension range of millimeters that are incorporated into biopolymer matrix materials such as polylactic acid (PLA), starch, and polyhydroxyalkanoates (PHAs) to form macro composites. These composites are also widely used in structural, car, and packaging aspects worldwide because of their biodegradability and good mechanical properties (Satyanarayana et al. 2009; Kumar et al. 2023).

Micro composites, with fillers of micrometer sizes, have superior dispersion and interfacial bonding compared with macro composites, resulting in augmented mechanical strength, dimensional stability, and thermal resistance. The applications of these products include a large number of moderate-performance products, like agricultural films, biodegradable containers, and consumer products (Averous and Boquillon 2004).

The most modern group is nanocomposites, which are reinforced with reinforcing resources with a nano-dimension, e.g., nanocellulose, layered silicates, carbon nanotubes, graphene oxide, and nanoparticles of metal and metal oxide. With a high surface area-to-volume ratio and surface reactivity, these fillers make the barrier properties, mechanical, thermal stability, and functional activity (e.g., antimicrobial, UV shielding) of the biopolymer matrix substantially better, even at low loading levels (Norozi et al. 2025). Nanocomposites based on biopolymers are gaining momentum in more developed applications in food packaging, controlled release of drugs, tissue engineering, and environmental clean-up. This may not only indicate the magnitude of reinforcement in such systems but also in highlighting the vast area of application available and performance improvement that can be realized by use of biopolymer-related composite technologies (Wani et al. 2023; Bharathi and Jayas 2024; Zarrintaj et al. 2023).

1.3 Applications of Biopolymers

Recently, biopolymers have been employed for a wide range of industrial and technological applications due to their versatile characteristics. Advanced and innovative research in bio-based nanocomposites has been developed to satisfy the functional requirements in the field of packaging, construction, agriculture, medicine, energy, and environmental remediation. Some of the different applications of biopolymer-based nanocomposites are discussed below.

1.3.1 Packaging Applications

The fast growth of the world's population, combined with the strain on natural resources that is resulting from climate change, has increased the need for sustainable food systems. One of the strategic efforts towards increasing food security and reducing food loss is the application of sophisticated packaging technology to increase shelf life and active properties of food products (Fig. 1.1).

Traditional packaging materials, which are typically petroleum-derived synthetic materials, cause enormous environmental threats, particularly through their inputs into marine pollution and long-term ecological effects (Bhaskar et al. 2023; Juikar and Warkar 2023; Liu et al. 2021). To counter such challenges, researchers have developed a variety of packaging materials for food, such as glass, metal, plastic, paperboard, and flexible laminates, whose main purpose is to preserve food integrity and reduce spoilage. However, the ecological and economic disadvantages of traditional materials, particularly non-renewable petroleum-based resources and economic disadvantages, have initiated the quest for alternative sustainable solutions (Juikar and Warkar 2023; Abdullah et al. 2022; Rhim and Ng 2007). Materials derived from biopolymers like chitosan (Cazón et al. 2019; Priyadarshi and Rhim 2020; Liu et al. 2022; Wang et al. 2018; Haghighi et al. 2020), gelatin (Said et al. 2023; Lu et al. 2022; Ashrafi et al. 2023; Luo et al. 2022; Ramos et al. 2016; Riahi et al. 2021; Roy and Rhim 2020), starch (García-Guzmán et al. 2022; Lauer and Smith 2020; Cheng et al. 2021; Khan et al. 2017), cellulose (Liu et al. 2021; Ghaderi et al. 2014; Yekta et al.

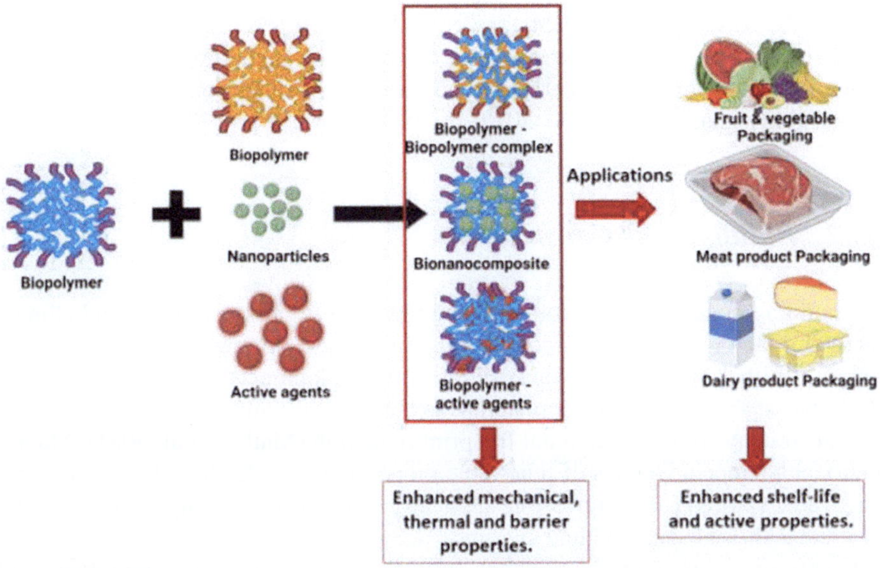

Fig. 1.1 Biopolymers for food packaging applications (Perera et al. 2023)

2023; Saedi et al. 2021; Asim et al. 2022), and bacterial nanocellulose (Cazón and Vázquez 2021; Azeredo et al. 2019; Ludwicka et al. 2020; Efthymiou et al. 2022) are increasingly viewed as practically feasible and eco-friendly substitutes. Their superior mechanical, barrier, and antimicrobial properties make them suitable for the preservation of food items. The functionality of such composites can be further increased by the integration of various substances like active agents, nanoparticles, etc. (Liu et al. 2021; Abdullah et al. 2022; Khezerlou et al. 2021). Polylactic acid (PLA) (Friné et al. 2019; Guinault et al. 2010; Busolo et al. 2010; Malek et al. 1892; Shao et al. 2022; Marano et al. 2022), and polyhydroxyalkanoates (PHA) (Mirpoor et al. 2023; Fabra et al. 2016; Yeo et al. 2024; Patiño Vidal et al. 2024; Bhardwaj et al. 2014), having superior biodegradability, along with the ability to seal food items from external factors like moisture, oxygen, etc. makes them a better substitute for bioplastics. Polysaccharide films consisting of starch, xanthan gum, carrageenan, pectin, and alginates are found to possess strong properties, including selective gas permeability and low toxicity, rendering them extremely relevant to food packaging (Rhim and Ng 2007).

New packaging technology now incorporates functional components into bio-based materials. Protein-derived coatings like zein, gluten, whey, casein, and gelatin can be designed to release active ingredients in response to biochemical alterations, thus imparting food product freshness and safety. Similarly, lipid-based films from plant and animal waxes, resins, and essential oils possess improved properties in the realms of water resistance and antioxidant functionality as well as beauty, attractiveness, and affordability (Bhaskar et al. 2023). Despite these advances, there are still issues to be solved for the stability, safety, and effectiveness of food and biodegradable packaging throughout their entire life cycle. However, these advances are a huge leap forward towards next-generation food packaging that is compatible with global sustainability goals (Yadav et al. 2018; Hu et al. 2014).

1.3.2 Building and Construction

The building construction sector is among the biggest carbon emitters and climate degrading factors, and also among the biggest energy and raw material consumers globally. With the transition towards sustainable development targets, the application of bio-based and biodegradable materials in building systems is in the limelight as a critical strategy for environmental footprint reduction (Mathur et al. 2021). Macro, micro, and nanocomposites of renewable biopolymers are leading the innovations, which are explored for their possible applications in structural components, mortars, concrete additives, and cladding systems.

Composites based on biopolymer sources, from agriculture, microbial, and marine, have superior properties like renewability, biodegradability, non-toxicity, and high mechanical and functional characteristics, bringing numerous benefits to building constructions (Phiri et al. 2023). They are widely used in load-bearing and non-load-bearing applications, which include their usage as additives in concretes and mortar and as strengthening agents for building blocks, coatings, and cladding products (Firoozi et al. 2024; Bailly et al. 2024).

1.3.2.1 Sustainable Admixtures for Concrete and Mortars

The use of biomaterials in the construction sector dates back to ancient civilizations. For example, Romans used various kinds of protein-rich materials to suppress the gypsum setting, and also used some animal-based materials for making ventilation. Recently, microbial polysaccharides, modified plant extracts, etc. have been used for improving the performance of cement-based systems (Li et al. 2020; Lubczak and Lubczak 2022). Different kinds of materials, including lignosulfonates as dispersive agents (Bhardwaj et al. 2014), xanthan gum, guar gum, and welan gum as thickeners and viscosity modifiers (Yadav et al. 2018), casein as binders and water-retention agents (Hu et al. 2014), are used as modifiers in cements. Additionally, starch ethers are utilized for workability control and slump retention (Crépy et al. 2014). These biopolymers frequently exhibit performance comparable to, or superior to, synthetic alternatives, particularly in terms of controlled hydration, reduction of shrinkage, and enhanced adhesion to substrates. Methyl hydroxypropyl cellulose and hydroxypropyl starch exhibit significant water retention in dry-mix mortars, which mitigates cracking and enhances durability (Chen et al. 2020; Zhou et al. 2019).

The recent environmental laws and regulations implemented by governments across the world and the customer preferences towards sustainable materials increase the market shares of bio-admixtures. The advancements in the production methods, like fermentation, biorefining, etc., in the manufacture of bio-admixtures, especially in the case of microbial polysaccharides like xanthan and curdlan, boost their industrial applications due to their reliable quality and customized rheological properties (Philip and Reshmy 2024; Kim et al. 2024).

1.3.2.2 Biopolymers in Earth-Based Building Systems

Traditional earthen construction is common; the utilization of biopolymer composites is being researched to enhance the mechanical properties and durability of soil-based building units. Tourtelot et al. reported that adding biopolymers like cellulose fibres and wheat starch to earthen blends provided enhanced compressive strength, from 3.5 to 5.5 MPa, without compromising the initial material properties. Enhanced adhesion of biopolymer molecules to clay particles resulted in enhanced particle cohesion and increased resistance to weathering (Navier Lab and Organisations 2021). Bio-stabilized earth blocks provide a more sustainable option than cement-stabilized blocks for low-income and climate-exposed conditions. They reduce embodied energy in buildings and facilitate the utilization of nearby waste farm material, thus promoting circular construction processes (Sinha and Sudarsan 2025; Bredenoord and Kulshreshtha 2023; Valenzuela et al. 2024).

1.3.2.3 Sustainable Cladding Materials

Cladding systems protect the buildings from sunlight, heat, and water and provide thermal efficiency and appearance. Previously, the materials employed for cladding were metal panels, synthetic polymers, fibre cement, and composites, which are all resource-intensive and difficult to recycle. As a result, researchers have started to investigate biopolymers as an environmentally friendly substitute for cladding panels. Nazrun et al. (2024) recently reported that biopolymers such as polylactic acid (PLA) (Rebelo et al. 2019), polyhydroxyalkanoates (PHAs) (Nazrun et al. 2024), polybutylene succinate (PBS), cellulose (Tetlow et al. 2017; Hafez and Tajvidi 2020; Musitelli et al. 2013), and starch (Pilla 2011)-based materials are better options as bio-cladding materials. They have low embodied energy, biodegradability, mechanical strength, and fire resistance, and are hence appropriate for use in frontages of green buildings (Rebelo et al. 2019). PLA is defined by its good mechanical properties and thermoplastic nature. PLA panels, made from fermented vegetable sugars, can be extruded or moulded and have considerably lower carbon emissions than comparable products derived from petroleum (Rebelo et al. 2019; Nazrun et al. 2024).

Similarly, PHAs derived from bacterial fermentation are fully biodegradable and also resistant to a broad spectrum of environmental conditions. Starch thermoplastics, when combined with other biopolymers or plasticizers, have the potential to be cheap, mouldable claddings. Their water sensitivity remains an issue, with further modification or laminating treatment needed to provide long-term strength. Although biopolymer-based materials are widely used in cladding systems, certain drawbacks, like lower durability, higher cost, and lower fire retardancy, limit their wide usage in the construction industry (Nazrun et al. 2024). The addition of biopolymer-based composites to building materials leads to the advancement towards sustainability. The significant potential of biopolymers, including low environmental impact, reduced energy usage, and following circular construction practices, makes them suitable for cladding systems. Interdisciplinary researchers are now focusing on overcoming various constraints of biopolymer-based construction materials, including less durability, low fire resistance, and issues with scalable production. The construction industry is increasingly adopting greener solutions, with biopolymer composites set to significantly influence the development of resilient, low-carbon, and sustainable infrastructure in the future.

1.3.3 Agriculture Applications

The advancements made by recent research on biopolymer-based composites have the potential to revolutionize the modern agriculture system with eco-friendly and multi-purpose applications. Biopolymers like starch, cellulose, chitosan, alginate, and polylactic acid (PLA) are derived from renewable sources and are distinguished by biodegradability, non-toxicity, and eco-friendliness (Patel et al. 2022). The application of biopolymers in agriculture is widely seen in mulch films (Menossi et al. 2021; Li et al. 2024), seed coatings (Basumatary et al. 2022; Riseh et al. 2024; Sohail et al. 2022), soil conditioners (Fatehi et al. 2021; Chang et al. 2016; Akbar et al. 2024) carriers for biopesticides, and controlled-release fertilisers (Priya et al. 2024; Firmanda et al. 2023). Mulch films made of starch and PLA-based materials have the ability to suppress weeds and soil temperature, and their higher biodegradability results in their removal from the soil after the harvest and thereby minimizes the accumulation of plastic waste within the fields.

In addition to mulch films, biopolymer composites are used in other applications, like the development of seed coatings. Seed coatings made of chitosan, alginate, and other protein-based blends are used to improve the germination rate and seedling vigour and to protect the crops against pathogens. The controlled nutrient release provided by seed coatings reduces the dependence on chemical fertilisers and improves the yields in a cost-effective manner (Liyakat 2024; Gamage et al. 2022).

Biopolymer-based composites are also used as soil conditioners and water retention agents in their hydrogel forms. A variety of biopolymers, like chitosan, guar gum, gelatin, and alginates, are used in their hydrogel form as controlled nutrient delivery agents, to improve the water retention capacity of soils, and also to minimize the evaporation rates and leaching. These specific applications benefit farmers in drought-prone and semi-arid regions by improving the water stress tolerance, resulting in high productivity and conservation of water resources. However, when it comes to commercialization, problems exist in the scalability of these composites, which require optimization of formulation and crosslinking processes (Tariq et al. 2023).

Conversion of agricultural waste to a variety of products like bio-based fertilisers, bio stimulants, and biopolymer feedstocks can induce a circular economy. Agricultural wastes like crop husks, lignocellulosic biomass, and food waste are increasingly used to create biofertilisers and biopolymer precursors. It is a waste-free process that avoids waste disposal problems and facilitates nutrient cycle closure in agroecosystems. Composted agricultural residues, when mixed with microbial consortia, act as a conditioner of the soil and as a source of carbon-rich biopolymers for the manufacture of biodegradable films and coatings (Patel et al. 2022; Gamage et al. 2022). Starch polymers are being viewed as cost-saving and scalable solutions for numerous agricultural applications. Thermoplastic nature can be tailored for application in fertiliser encapsulation, biodegradable packaging, and irrigation. Starch composites can be engineered with different mechanical strength, oxygen barrier properties, and hydrophilic properties critical for the development of active and intelligent films that can react to external stimuli (Puglia et al. 2021).

Biopolymer-based sustainable composites have great potential to improve resilience, productivity, and environmental sustainability of agro-systems. These composites decrease the use of petroleum-based products in agriculture and thereby improve the soil health, water retention, and valorization of agricultural wastes. The agriculture industry is confronted with issues like climate change, loss of resources, and increased food demand. Integration of sustainable composite technologies could prove to be pivotal towards ensuring long-term sustainability.

1.3.4 Automotive Industry

The automobile industry is witnessing a dramatic change based on environmental friendliness and efficiency in the use of resources. With rising emissions standards and growing customer pressure for environmental-friendly technologies, industries are increasingly embracing biopolymer-based composites in automobile vehicles. These green materials allow for the substantial reduction of the environmental footprint while satisfying relevant technical performance requirements for structural and semi-structural components in the automobile industry (Prasanth et al. 2021; Roy et al. 2019). Natural fibres like flax, hemp, jute, kenaf, and sisal are widely used as reinforcements in bio-based polymeric matrices like PLA, PHS, PBS, and bio-based epoxies to impart a high strength-to-weight ratio, high impact strength, improved thermal resistance, etc., for applications in the automotive field. Weight savings of 20% to 30% result in enhanced fuel efficiency and reduced carbon emissions, in accordance with global targets outlined in the EURO 6 standards and End-of-Life Vehicle (ELV) Directive, requiring 95% recyclability of automotive parts (Prakash et al. 2022; Islam et al. 2022).

Another significant development in this field is the development of bioprepregs, which are defined by pre-impregnated fabrics with bio-based reinforcement and matrix systems. These materials enable accurate alignment of fibres, fibre volume fraction control, and enhanced fatigue strength and vibration-damping properties. Bioprepregs are produced by technologies such as hot-melt impregnation, film stacking, and powder-based technologies, and their use is becoming a common practice in the manufacture of lightweight interior panels, dashboards, trims, and structural reinforcement. Their high processing efficiency and better compatibility with automated manufacturing methods result in lower energy consumption and processing time compared to traditional composite manufacturing methods (Carvalho et al. 2024).

In vehicles, natural fibres, including kenaf, flax, and hemp, are used as reinforcements with bioplastics in making structural parts like door panels and trunk liners, which gained wide commercial acceptance in vehicles manufactured by large original equipment manufacturers (OEMs). The fibres are renowned for their biodegradability, acoustic insulation, accessibility of renewable raw materials, and cost-effectiveness. The heterogeneity of the properties of these fibres, based on factors such as agricultural origin, water absorption, and matters of interfacial bonding with the matrix, is still a major research area. Chemical and physical treatments such as silane, alkali hydrolysis plasma, and heat are used to increase fibre-matrix compatibility (Dhakal et al. 2024; Agarwal et al. 2020; Gupta et al. 2022).

At the same time, bio-based thermosets such as bio-epoxy, bio-polyurethanes, and bio-polyesters are being developed as alternatives to petroleum-based resins. These resins, based on vegetable oils, sugars, and crop residues, have lower VOC emissions, are less harmful, and offer enhanced mechanical strength. The advancements in nanotechnology enable the development of biopolymer-based nanocomposites containing nano-clays, nanocellulose, or graphene that results in the enhancement of thermal, mechanical, and flame-resistance properties of bio-based car composites (Mori 2023).

Nowadays, bio-based fuels like bioethanol, bio diesel, biobutanol, etc. are widely used in hybrid vehicles for sustainability. These substitute fuels are derived from microbial fermentation and thermochemical processing of lignocellulosic biomass or crop waste to reduce fossil fuel consumption and ensure cleaner combustion. Issues of compatibility with current engine technologies, reduced energy density, and scalable production requirements require policy incentives and technological innovation solutions. End-of-life approaches and circular economy of bio-composites in automotive uses are a dynamic research area. In contrast to traditional composites that are extremely challenging to recycle, bio-composites are more amenable to mechanical, chemical, and biological degradation processes. Certain pilot studies are exploring the recycling of fibres and biopolymer resins by composting and pyrolysis processes, which could lead to enhanced sustainability of the automotive life cycle (Rajesh Kumar and Saravanan 2016; Yusoff et al. 2015; Demirbas 2007).

Currently, biofuels and bio-based composites are widely used in automotive industries as an eco-friendly alternative to conventional materials and fuels. Advances made in the sector of bioprepreg, natural fibre-reinforced composites, and bio-based matrices yield materials that are superior to their conventional equivalents in performance while minimizing environmental problems. While long-standing issues in the form of moisture sensitivity, lifespan, and economics remain challenges, advances in surface treatment, hybridization, and recycling methods are increasingly contributing to their commercial attractiveness. With regulation tightening and shifting consumer attitudes moving in the direction of greener options, biopolymer-based composites are set to be the stars of the green mobility of the future.

1.3.5 Energy-Harvesting Systems

The evolution of materials to harvest ambient mechanical energy has progressed with the trend towards sustainable and renewable energy resources. Biopolymer-based piezoelectric and triboelectric nanogenerator systems are emerging as new energy-harvesting technologies with flexible, biocompatible, and eco-friendly solutions in the field of electronics (Maiti et al. 2019).

1.3.5.1 Biopolymer-Based Piezoelectric Nanogenerators (BP-PENG)

The biopolymers showing the property of transforming mechanical energy into electrical signals are of significant interest. This property is beneficial for use in self-powered sensors, wearable electronics, and implantable medical devices. The biopolymers gain electrical charge when they are mechanically deformed, which is called the piezoelectric property. This has been explored well in synthetic polymers like polyvinylidene fluoride (PVDF) and poly (L-lactic acid) (PLLA), and natural biopolymers like cellulose, chitin, collagen, and silk (Ali et al. 2023).

The incorporation of piezo-active nanofillers like ZnO, MoS_2, CuO, and TiO_2 into natural polymers enables the creation of biodegradable materials that show highly promising power densities and voltage outputs, and have the potential to be used in powering low-energy wearable devices and sensors. In addition, protein-based materials such as silk fibroin and collagen inherently possess piezoelectric properties because of their orderly structured fibrillar morphologies. The crystalline β-sheet conformation in silk and the triple-helix conformation in collagen are accountable for the electromechanical coupling. Interestingly, mechanically drawn silk films have been shown to exhibit higher piezoelectric constants and have been considered as promising bio-integrated energy harvesting systems (Mahanty et al. 2024; Xu et al. 2021).

1.3.5.2 Biopolymer-Based Triboelectric Nanogenerators (BP-TENG)

Triboelectric nanogenerators (TENGs) were first developed in 2012 by Wang et al. who worked on the principles of triboelectric effect and electrostatic induction. They were applied to generate electrical energy from various mechanical motions, including vibration and touch. Biopolymer-based triboelectric nanogenerators (BP-TENGs) utilize bio-based materials like cellulose, lignin, chitin, silk fibroin, alginate, and gelatin since they are more sustainable in comparison to petroleum-based polymers like PTFE, PVDF, Nylon, etc. These materials facilitate the design of TENGs for specific applications, such as wearable electronics, smart textiles, and implantable biosensors (Zhu et al. 2024).

In addition to TENGs, piezoelectric nanogenerators (PENGs) are also used to harvest electrical energy from mechanical vibrations. Normally, PENGs are made of inorganic and organic materials, including polymers like PVDF, nylon, etc. The combination of piezoelectric and triboelectric NGs is employed to improve the efficiency and responsiveness of biopolymer-based energy harvesters. Hybrid devices are utilized for application in smart wearables, healthcare diagnostics, environmental sensors, and energy-autonomous devices. While these advances are being made, there are still challenges to be overcome, such as enhancing mechanical durability, optimizing charge generation, and achieving long-term material stability. Molecular engineering, nanostructuring, and functional additive addition are all viable methods to improve piezoelectric and triboelectric performance (Liu et al. 2025).

1.3.6 Defence and Military Applications

The sustainable, biodegradable, and multifunctional characteristics of biopolymers make them an interesting material for use in defence and military applications. The versatile functional properties of biopolymer-based materials have the potential to be used in the manufacture of structural components in the defence field. The field of global defence is embracing green technology and multifunctional lightweight systems at a very rapid pace, and biopolymers are emerging as a key region of innovation interest (Olosho 2024).

Biopolymers are increasingly undergoing severe examination for their potential uses in protective equipment and structural applications. Nanocellulose, a tough biopolymer derived from plant fibres, has better mechanical properties than many metals, with tensile strengths up to eight times higher than stainless steel. The addition of nanocellulose as a reinforcement material into polymer matrices greatly improves impact strength and tensile properties, and is ideal for the manufacture of lightweight body armour, combat helmets, and structural elements of military vehicles. The superior thermal insulation properties and flame retardancy of bio-composites made by reinforcing high-performance fibres like Kevlar can ensure enhanced protection under combat conditions (Hemanth Kumar et al. 2022; Faiz Norrahim et al. 2021).

Biopolymers are gaining importance in the field of military textiles. Clothing made of nanocellulose and chitosan-coated fabrics possesses a variety of functional properties, such as resistance to antimicrobial attack, flame resistance, and electroconductivity. Such biopolymer-based textiles can be integrated with a variety of sensors to identify chemical or bio threats, thus making them suitable for application in advanced uniforms and protective clothing. In addition, the lightness and breathability of biopolymer-based materials enhance the comfort and usability of such fabrics in high-stress environments without compromising the safety and durability of the products (Kammeta et al. 2023; Zecheru 2010; Ferreira et al. 2019).

Biodegradable polymers, such as acid-hydrolysed cellulose microfibrils, find promise as more environmentally friendly substitutes for synthetic polyacrylamides for stabilizing dry fine soils prevalent in military environments. They function to minimize sediment runoff and suppress dust, thus realizing key advantages in the construction of temporary helipads and landing zones in desert environments. Biopolymer use is found to improve visibility and reduce equipment wear from airborne particulates during helicopter operations (Orts et al. 2007).

Biopolymers are revolutionizing treatment protocols in battlefield medicine and combat casualty care through their application in 3D printing. Biopolymer-derived naturally occurring hydrogels like alginate, hyaluronic acid, and collagen are used as bioinks in additive manufacturing for the development of complex tissue scaffolds. Bio-printed constructs are designed for use in a range of applications, such as on-demand skin grafts, bone tissue engineering, and vascularized tissues for wound healing. The potential to print functional tissues at or close to the site of injury has the immense potential to maximize recovery outcomes while simultaneously minimizing logistical complexities of medical evacuations (Betz et al. 2020; Rushdha Hiba et al. 2025).

In response to the growing demand for the application of sustainable defence logistics, these materials offer a closed-loop process that can be utilized in the military environment, including packaging and equipment design. Through this process, the environmental footprint of military operations is minimized, and the security of energy and material resources is increased. While biopolymers possess many benefits, they continue to experience enormous challenges in their large-scale use in defence applications. Military materials of the future will witness increased use of biopolymers in multifunctional systems, autonomous platforms, and regenerative medical technology, and they will form a fundamental component of future defence technologies (Kammeta et al. 2023; Zecheru 2010; Oladele et al. 2020).

1.3.7 Energy Storage Applications

The use of biopolymers for energy storage devices is a significant step towards the creation of green and sustainable technologies. The growing global demand for clean energy, fuelled by the explosion of portable electronics, electric vehicles, and renewable energy devices, has positioned biopolymer-based materials at the forefront of next-generation energy storage device development. The graphical abstract for the energy storage applications is given in Fig. 1.2. Biopolymers offer green alternatives to petroleum-derived, synthetic polymers in conventional energy storage technologies due to their biodegradable, renewable, chemically diverse, and green chemistry principle compatibility (Gopinath et al. 2023).

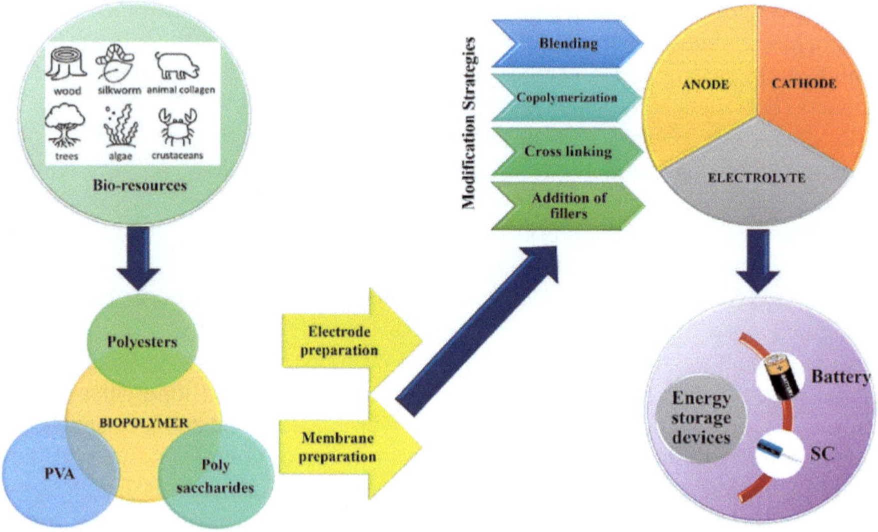

Fig. 1.2 Energy storage applications of biopolymers (Gopinath et al. 2023)

Biopolymers such as cellulose, chitosan, starch, silk fibroin, gelatin, pectin, and their derivatives are widely investigated as promising candidates for energy storage devices such as lithium-ion batteries (LIBs), zinc-ion batteries, supercapacitors, and hybrid capacitive systems. The multifunctional properties of biopolymers make them a superior candidate for different components like electrolytes, electrode binders, separators, and encapsulating matrices, each serving a specific purpose to support the performance, safety, and sustainability of the devices. They play a key role in the fabrication of solid polymer electrolytes (SPEs) and hydrogel-based electrolytes. They provide a variety of advantages over conventional liquid electrolytes, such as improved thermal and electrochemical stability, leakage prevention, better mechanical strength, and reduced flammability. Cellulose and chitosan have been employed to design hydrogel electrolytes that are characterized by high ionic conductivity and high flexibility, which are highly suitable for flexible and wearable energy devices, which are increasingly used in military, aerospace, and medical electronic devices. The hydrophilic functional groups in biopolymers imparts high ion transport ability to the materials which results in high energy and high-power densities for the energy storage systems (Beg et al. 2025; Rasheed et al. 2022; Xu et al. 2022; Mahajan and Sharma 2024). Their tuneable properties prevent the growth of dendrites and internal short circuits in lithium-based batteries. The tuneable porosity and mechanical properties of biopolymer composites control the ion transport and minimize the risk of degradation caused to anodes. In addition, the incorporation of different inorganic nanoparticles, conductive polymers, etc., to the composites improves the electrochemical performances of the storage systems by enhancing their capacitance, charge–discharge efficiency, and cycling stability (Lizundia and Kundu 2021; Prasanna et al. 2015). Biopolymers reinforced with different fillers are used in the production of electrode materials and binders for super capacitors which involves rapid charging and discharging mechanisms. Recently, modified cellulose, lignin composites, and silk-based carbons having high surface area and pseudo-capacitance are used in energy storage devices. These materials can be used to prepare lightweight and compact energy storage devices that are in high demand now. Separators made from biopolymers have been discovered to be of immense value in increasing not just the thermal but also the mechanical safety of battery cells. These separators are physical barriers present between the electrodes and permit selective ion passage. Natural polymers such as gelatin and silk have excellent film-forming qualities and thermal stability, and hence can be excellent alternatives to synthetic polyethylene or polypropylene separators, which are not biodegradable and are environmentally toxic (Rahman et al. 2024; Appiah et al. 2022; Lee et al. 2025).

In short, biopolymers hold vast potential in creating energy storage applications more efficient, sustainable, and secure. Sensible material design and composite addition enable the enhanced performance and other properties of these natural materials as compared to their synthetic analogue, thus providing effective solutions for a wide variety of energy storage applications, especially where ecologically sustainable and biodegradable conditions are the top priority.

1.3.8 Self-Healing and Shape-Memory Applications

1.3.8.1 Self-Healing Biopolymers

Biopolymer-based self-healing systems have attracted a lot of attention as advanced materials become more and more utilized in healthcare and engineering domains. Inspired by biological repair systems, self-healing biopolymers present the means to independently heal from physical damage, thereby improving material lifetime, performance, and dependability. Applications include wound healing, drug administration, and other structural materials for advanced applications. Because of their biocompatibility, biodegradability, and chemical modifiability, natural polymers, including chitosan, alginate, guar gum, methylcellulose, and gelatin, have become crucial components in self-healing hydrogel matrices. Dynamic Schiff base bonds created by the use of aldehyde-modified oxidized guar gum (OGG) and carboxymethyl chitosan (CMCS) enable injectable hydrogels with outstanding self-healing and sticky characteristics. As seen in vivo employing systems loaded with vascular endothelial growth factor (VEGF), these hydrogels not only restore mechanical integrity after damage but also encourage angiogenesis and collagen deposition (Biobased polyurethane: an overview. 2008; Kong et al. 2023).

Likewise, exosome-loaded methylcellulose-chitosan (MC-CS) hydrogels have been produced for synergistic wound healing. Particularly in diabetic wound models, physical integrity, 3D porous structure, and biological signal transmission all promote faster skin regeneration. This degrades safely after healing, enhances tissue adhesion, and facilitates vascular reconstruction, thereby addressing the three primary phases of wound healing: inflammation, proliferation, and remodelling. Beyond biomedical purposes, self-healing capabilities are being integrated into biopolymer composites for structural applications, where external stimuli such as moisture, pH, and mechanical stress trigger healing. Reactive compatibilization of polysaccharide-based biopolymers with biodegradable matrices has demonstrated efficacy in modifying mechanical characteristics and providing self-repair capability, claims Kumar et al. These systems allow materials to recover microcracks and fractures over several cycles by mostly depending on dynamic covalent bonds or reversible supramolecular interactions (Biobased polyurethane: an overview 2008; Wang et al. 2019).

Self-healing biopolymer systems provide multifunctional solutions, sustainability, and circular material design. To further adjust healing speed, mechanical strength, and environmental responsiveness, future directions consider the incorporation of nanofillers, smart stimuli-responsive domains, and bioactive chemicals. In next-generation medical and industrial applications, these developments open the path for strong, biodegradable, and application-specific materials.

1.3.8.2 Shape-Memory Biopolymers

Shape-memory polymers (SMPs) are a class of smart materials able to "remember" a particular shape and restore it from a distorted condition by external stimuli in the form of thermal energy, humidity, light, pH, or electric and magnetic fields. From health care, aerospace, and robotics to textiles and military, the ability has opened a great spectrum of uses in almost all domains of activity. Programming the material into a temporary form, stabilized by crosslinking or crystallization, and then subsequent recovery of the original shape upon stimulation is the fundamental mechanism. More attention has lately been paid to bio-based shape-memory polymers (SMPs) and green substitutes for conventional petroleum-based SMPs. Bio-based shape-memory polymers are made from renewable fuels such as natural rubber, plant oils, starch, lignin, and proteins. They provide the double advantages of sustainability and functional performance. To increase their mechanical performance and thermal stability, however, bio-based SMPs are reinforced with natural fibres or nanoparticles, producing bio-based shape-memory polymer composites (bio-SMPCs) (Uyan and Celiktas 2023; Jamaludin et al. 2025).

Research has shown that formulations of both thermoplastic and thermoset bio-based SMPs may be tailored against various stimuli. Among the examples are poly (lactic acid), polyhydroxyalkanoates, and poly (butylene succinate) as thermoplastics; vanillin, epoxidized plant oils, and lignin derivatives as thermosets. These have shown good shape recovery ratios and variable transition temperatures, so offering possible temperature-sensitive and multi-responsive uses. Fascinatingly, nature herself offers cues for shape-memory behaviour. Processes including swelling, hydration, and viscoelastic recovery give indications for biomimetic SMP design to operate in wet or aquatic environments by means of this natural shape-memory behaviour (Uyan and Celiktas 2023; Liu et al. 2015).

SMPs are used in the biomedical field to create minimally invasive tools like drug-delivery systems, sutures, and self-expanding stents. Once implanted in the folded form, SMPs extend or unfold in response to body temperature or humidity stimuli. Their high prospects for temporary implants and tissue scaffolds are derived from biocompatibility, biodegradability, and stimulus responsiveness. The discovery of shape-memory hydrogels (SMHs) is a soft, water-swollen material with minimal volume change, able to change form. Especially in watery settings when conventional SMPs operate poorly, SMHs can be utilized for biomedical applications such as cell manipulation, drug delivery, and soft actuators with the addition of reversible crosslinks and molecular switches (Lendlein et al. 2019; Löwenberg et al. 2017; Delaey et al. 2020).

Ultimately, shape-memory elastomers (SMEs) have acquired momentum because of their elasticity, softness, and recoverability of form under mild conditions. In wearable biomedical sensors, flexible electronics, and soft robotics, these materials are utilized since they give comfort and durability. SMEs respond faster and are more responsive than traditional SMPs (Suethao et al. 2022). Basically, shape-memory materials based on biopolymers present an interesting mix of stimuli-responsiveness, environmental friendliness, and adaptability. Thus, these next-generation smart materials have applications in the field of the environment, medicine, industry, and defence.

1.3.8.3 Biomedical Applications of Biopolymers

Among the most promising and sustainable materials available in the field of biomedicine are nanocomposites based on biopolymers. Natural biopolymers like chitosan, alginate, collagen, cellulose, and gelatin find application when combined with functional nanomaterials, including graphene, metal nanoparticles like silver, copper oxide, zinc oxide, and nano-hydroxyapatite. Biodegradable, biocompatible, functionally flexible materials with good mechanical properties, having a synergic effect, are suitable for both therapeutic and diagnostic applications (Shaikh et al. 2024).

In the field of drug delivery, bio-nanocomposites show lower side effects and improved treatment efficacy by means of regulated, targeted, continuous release. For example, chitosan-based nanocomposites can target tumour tissues by means of appropriate loading with antibacterial agents and anticancer drugs, further improving delivery efficiency with graphene-based bio-nanocomposites with their large surface area and interaction with biomolecules. Especially, composites containing graphene oxide and silver nanoparticles in conjunction with biopolymers have been shown to be quite successful in delivering anticancer medications, such as paclitaxel and doxorubicin, with higher cell absorption and less cytotoxicity to normal cells.

In tissue engineering and regenerative medicine, biopolymer nanocomposites replicate the structure and purpose of the native extracellular matrix (ECM), therefore enabling cell adhesion and proliferation (Shaikh et al. 2024; Pandey et al. 2025).

Introducing nanofillers in hydrogels based on alginate, cellulose, or gelatin can be used in the regeneration of a variety of tissues, including skin and neural tissues (Fig. 1.3). Hydrogels significantly affect the controlled release of medicinal agents. These systems respond to a variety of biological stimuli, including pH, temperature, and enzyme activity, by means of temporally limited and targeted drug delivery. Also, hydrogels reinforced with nanocellulose have been employed in the repair of bone defects, and collagen-based nanocomposites reinforced with hydroxyapatite mimic the native bone tissue composition. Moreover, by means of their adaptability in a spectrum of therapeutic uses, from minimally invasive surgery and wound healing, biopolymers can be generated in a variety of forms, including sponges, films, membranes, and injectable gels. Hydrogel-based nanocomposites are also used in insulin delivery, cancer treatment, and growth factor release, because of their stimulus-responsive therapeutic efficiency (Ulery et al. 2011; Mahmood et al. 2022).

These systems with antibiotics and growth factors can accelerate healing and lower the infection rate, and are very helpful in chronic wound care. Biopolymer

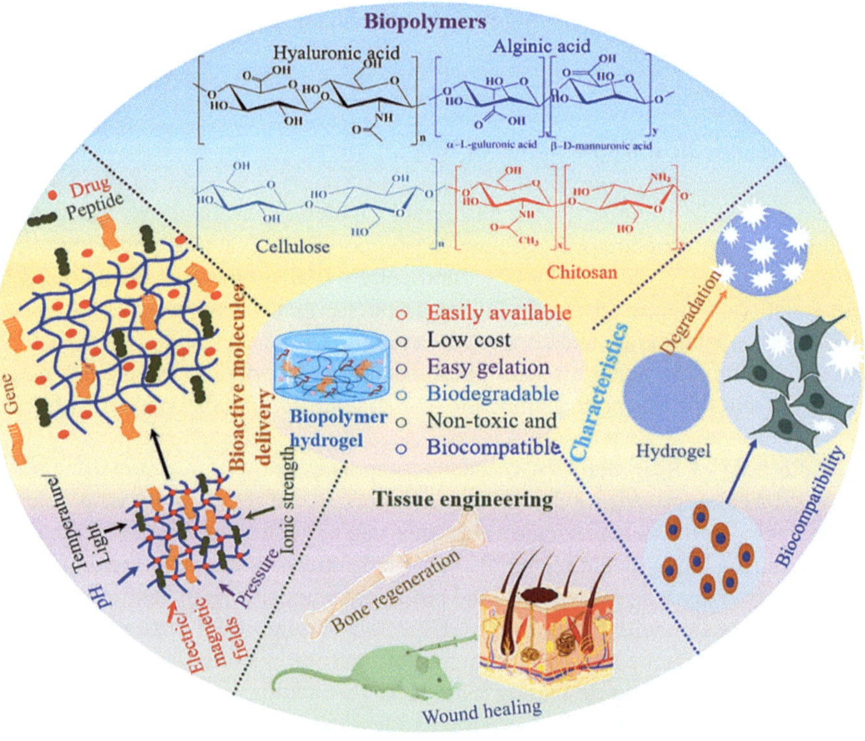

Fig. 1.3 Biopolymer hydrogels for various biomedical applications (Patel et al. 2024)

coatings and films help to further extend the biomedical applications for nanocomposites. Applied on implants or medical instruments, these coatings can considerably enhance surface properties like biocompatibility, corrosion resistance, and antibacterial action. For example, including nanoparticles in polymer coatings helps limit bacterial colonization on orthopaedic implants or catheters. Some coatings also release bioactive compounds, thereby promoting tissue integration and reducing inflammation at the implantation location. Another field is opened by the synthesis of sulphated polysaccharide-based nanocomposites with antiviral, anticoagulant, anti-inflammatory, and immunomodulating properties. Because their surface chemistry and charge features enable good interaction with biological systems, these compounds have considerable potential for therapeutic purposes, biosensing, and regenerative applications. Their great bioresorbability, modifiability, and biological activity enable their design for several therapeutic purposes (Pandey et al. 2025; Mahmood et al. 2022; Nathanael and Oh 2020; Alfinaikh et al. 2025).

Recent advances in graphene-enhanced biopolymer nanocomposites have opened doors in biosensing, bioimaging, and precision medicine. It can be used for the development of implanted sensors, diagnostic tools, and live monitoring systems with good surface reactivity and electrical conductivity. These technologies detect disease development and real-time monitoring of drug release, therefore bringing biomedical technology closer to tailored health care. Biopolymer-based nanocomposites hold great promise in the field of drug delivery, tissue engineering, wound healing, diagnostics, and implant technology. Their practical effectiveness, environmental friendliness, and adaptability help to shape the trend of safe and individualized medical treatments (Sionkowska 2016).

1.3.9 Aerospace Applications

Biopolymer-based nanocomposites are progressing as significant materials in the aerospace industry because of their sustainable and lightweight characteristics. Conventionally, these materials are used in the aerospace industry as an alternative to synthetic composite materials produced from petroleum-based resins like epoxy and polyester. The approach towards sustainability has pushed the growth of bio-based alternatives for non-structural and interior applications. Research on bio-based epoxy resins derived from soybean oil, linseed oil, castor oil, and canola oil has shown promising results in recent years. These bio-resins can be used as a substitute for conventional epoxy matrices in composite materials used in the interior, side walls, floor panels, and secondary structural parts of an aircraft (Ramon et al. 2018). These materials offer satisfactory mechanical strength, thermal stability, and flexibility, even though they are not suitable for the high-performance demands of materials for the manufacture of primary aircraft structures.

Thermosetting matrices for fibre-reinforced composites are produced from epoxidized soybean oil (ESO) and epoxidized linseed oil (ELO). These bio-synthesized epoxy resins exhibit good impact strength and thermal resistance, which can be further enhanced by adjusting the curing agent or blending with conventional resins. A comparable tensile strength and flexural properties to petroleum-based epoxies are shown by modified bio-epoxies that makes them suitable for the resin transfer moulding (RTM) process in the aerospace industry (Ramon et al. 2018). Biopolymer matrices reinforced with magnetic nanoparticles can be used for electromagnetic interference shielding (EMI), structural health monitoring (SHM), and thermal protection systems. The properties, such as lightweight and corrosion resistance, reduce the fuel consumption by reducing the overall mass of the aircraft, thereby supporting the development of safer and more efficient aircraft systems (Hariz et al. 2021; Romero-Fierro et al. 2022).

The limitations of biopolymers, in mechanical, thermal, and flame-retardant properties, can be improved by the incorporation of chemical modifiers and nanofillers to make bio-based composites more suitable for auxiliary systems, aerospace interior structures, and eventually for exterior or semi-structural applications. Thus, these sustainable aerospace composites are likely to become more competitive, offering new innovations for green technologies in aviation.

1.3.10 Marine Applications

There is a growing need for sustainability in the marine sector, in order to support a sustainable marine environment. Marine structures traditionally use the glass fibre-reinforced thermoset polymer, but the recent and rising environmental regulations and demands of extended life service have driven the industry to look elsewhere.

Plant-based composites have become a durable alternative to man-made fillers in the maritime industry. They have attracted so much attention since they possess certain advantages over ordinary synthetic fillers. Some of these include reduced cost of flexibility, environmental sustainability, biodegradability, renewability, and reduced density. Over the past few years, composite materials made of plant biomass have become an excellent solution to massive problems in most industries, particularly the maritime industry. These are biomass-based renewable composites and can provide a huge stride in fashioning composites that will be much more environment friendly. Individuals are concerned about the environmental impact created by the conventional products, such as steel, fibreglass, and artificial fibres, because they require excessive energy in the fabrication process and are difficult to discard. Conversely, composites based on plant biomass employ biomass of plants and byproducts of agriculture, which can be reused over and over and are environmental friendly.

The environmental responsibility and sustainability that it has been going all out about have given way to the plant biomass-based composites that are a more friendly alternative. They contribute greatly to carbon optimization because, compared to traditional materials, especially steel, a great deal of energy is involved and emission of greenhouse gases is involved. The mentioned composites are special due to the fact that they are light, strong, and corrosion-resistant. They question the antiquated

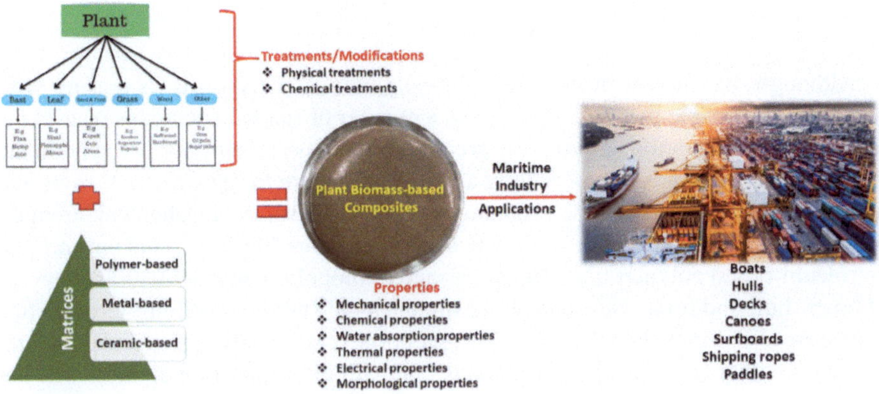

Fig. 1.4 Plant biomass base composites for maritime industry applications (Iwuozor et al. 2024)

maritime approaches and bring the industry a step closer to the materials that are more environmentally friendly and cost-effective. Plant-based composites (Fig. 1.4) demonstrate that they can be applied in many maritime environments. They enhance structural strength, reduce maintenance expenses, and leave a lower imprint on the environment. The release of these products is at the turning point in the transition of the industry to sustainable operations, so they are important in the resolution of worldwide issues such as climate change and resource preservation (Iwuozor et al. 2024).

El Mahamdi et al. (2024) aimed at revealing the synthesis of new coatings based on a biopolymer epichlorohydrin-modified alginate, and alginate-epichlorohydrin-$SrTiO_3$ nanocomposites, in which nanoparticles $SrTiO_3$ (STO) are embedded in the alginate (Alg) layer (Alg-Ep-STO). The results have shown that adding STO nanoparticles into the alginate grid with epichlorohydrin contributes significantly to the resistance of the metal to corrosion. Computational analyses showed that the Alg-Ep-STO nanocomposite possesses significant attachment on the copper surface with a flat arrangement, and with an alignment that is profoundly dependent on the existence of STO nanoparticles (Mahamdi et al. 2024).

Zhang et al. (2024) fabricated a thin-film composite (terrein-PA) membrane by covalently coupling marine fungi-derived (+)-terrein to reverse osmosis polyamide membrane through a glutaraldehyde treatment. This membrane exhibits increased water permeability, antifouling properties, and antibacterial properties. Compared to the conventional reverse osmosis membranes, the Terrein-PA membrane appeared with a 51.0% increase in permeability (7.4 L/m^2 h bar) because of its beneficial hydroxyl groups. It also markedly inhibits the growth of gram-positive and gram-negative bacteria since it has an antibacterial activity of 98.0 and 94.9%, respectively. This reduces the attachment of microbes extending the life of the membrane. It would be more useful to exploit natural (+)-terrein to treat human health than employ chlorine. Further, under the bovine serum albumin solution, the Terrein-PA membrane reaches a 90.9% recovery rate of fluxes and total contamination of 12.3% (Zhang et al. 2024).

Although the interest in the use of biopolymers as a reason is increasing with regard to sustainable marine applications, a number of fundamental gaps restrict the feasible implementation. Among the greatest dilemmas is their poor life span in the rugged marine conditions where long and high exposure to seawater, UV rays, and microbial activities may cause premature wear out. Also, the mechanical strength, impact resistance, and stiffness of many biopolymers are lower than in conventional petroleum-based polymers, limiting the usage of biopolymers in structural elements of boats: hull and foils. The changeable dimensional stability and long-term performance are also diminished due to being susceptible to water absorption and moisture. There is another serious gap in processing compatibility; most biopolymers will not readily be modified to the current methods of making marine composites, such as resin infusion or autoclave curing. Cost and scale are also obstacles, with biopolymers commonly both costlier and not possible to obtain in commercial quantities. Moreover, the commercialized biodegradability of biopolymers is variable, and their degradation is often uneven in the marine environment. In addition, the degradation pathways depend on several factors that do not correlate with a standardized test scheme, and end-of-life management guidelines have not been established. Lastly, the majority of marine regulatory systems are based on traditional materials, with no direct way of certifying biopolymers. The solution to these gaps involves effective work on the innovation of materials, life-cycle assessment, and coordination with stakeholders of marine industries to achieve all the benefits of biopolymers on the technology of marine sustainability (Samalens et al. 2022; Samalens et al. 2022).

1.3.10.1 Environmental Applications

Pollution of natural resources is getting worse, and we need to deal with it right away if we want the world to stay a place where people can live for years to come. One part of this problem that needs our urgent attention is the damage to freshwater resources caused by the release of industrial waste that hasn't been handled or treated well enough. In wastewater treatment, people typically use coagulation-flocculation, adsorption, ion-exchange, membrane separation, evaporation, flotation, and electro-precipitation. Traditional techniques either have no potential to attain the purification level required to meet the strict discharge standards or might prove too costly. Various processes that are carried out can be done using new resources, thereby enhancing acceptability. The removal of such pollutants (Fig. 1.5) as toxic dyes and metal ions via wastewater adsorption is widely used. The costly, energy-intensive commercial activated carbon could be replaced with the alternative low-cost adsorbent based on biopolymers. As natural coagulants and flocculants, biopolymers can be used to clarify wastewater (Pinaeva and Noskov 2024).

A research group described the green synthesis and characterization of bimetallic zinc oxide-gold nanoparticle (ZnO@AuNPs) and incorporation into carboxymethylcellulose (CMC) to produce a biopolymer-based nanocomposite. The objective of

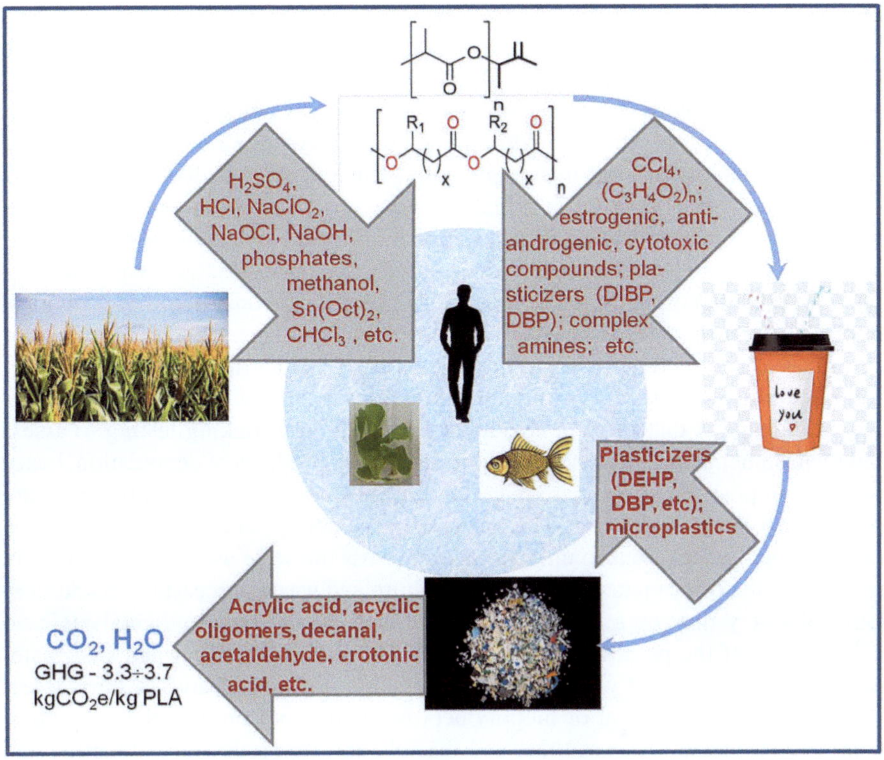

Fig. 1.5 Graphical abstract of various environmental pollutants (Pinaeva and Noskov 2024)

the formulation was to improve stability, bioactivity, and functional properties of the nanoparticles by embedding them with biopolymers. The nanoscale morphology of both ZnO@AuNPs (R1 15 nm) and the nanocomposite (R2 25 nm) was established using the characterization methods, and characterization using dynamic light scattering (DLS) revealed constant nanoparticle sizes of 27 nm and 93 nm respectively. Life tests of the CMC-based nanocomposite showed their higher antimicrobial effect on various bacterial and fungal strains than ZnO@AuNPs. Also, the nanocomposite demonstrated an increased antioxidant activity (IC 50 = 32.4 microg/mL) and an impressive anti-diabetic potential (IC 50 = 7.4 microg/mL) compared with the bimetallic nanoparticles as well as acarbose that is used as a standard drug. Photocatalytic experiments showed that the nanocomposite could effectively degrade both RR195 and RB dyes with more than 91 percent degradation in 100 min, which may be applicable to the environmental clean-up due to its high level of degradation (Al Abboud et al. 2024).

In the context of depleting oil supplies and the global warming caused by the use of synthetic polymers, biopolymers produced either on the basis of plant, animal, or microbial biomass become very important. They can be polysaccharides, lipopolysaccharides, glycolipids, proteins, or polyhydroxyalkanoates and are extremely fit to be used in the environment. Among the biopolymers, cellulose and chitosan, the most widespread biopolymers on the planet, are preferred (Ravindran et al. 2022).

Although there is an emerging spread of biopolymers being used as green alternatives to traditional plastics in environmental uses, there are still some gaps that are essential in both research and their application. The primary drawback is that they do not always behave in terms of degradation under various environmental conditions, including soil, fresh, and marine ecosystems. A large percentage of the biopolymers advertised as biodegradable in fact do not fully degrade in natural environments due to the potential of microplastics becoming possible and unanticipated ecological effects. Also, mechanical and thermal properties of biopolymers tend to be inferior to those of a synthetic counterpart, limiting them to applications in load-bearing or long-term environmental use, e.g., structural geotextiles or sturdy packaging. The other major challenge is that there is no standard mode of undertaking testing to assess their environmental performance, toxicity, and the by-products of degradation, hence making it impossible to compare materials across studies or ascertain that they are possibly regulatory-compliant. The inadequate scalability and the high production costs of biopolymers are also unfavourable towards the wide use of biopolymers in areas where the price matters, such as agriculture and waste processing. In addition, chemical safety may be a factor in that additives, plasticizers, and catalysts used in the course of the processing may present an issue, especially when said materials seep into the soil or water during degradation. Future research to fully assess the environmental potential of biopolymers needs to work on material properties, ecological sustainability over time, and feasible and cost-reducing production and recycling processes.

1.3.11 Computational and Theoretical Modelling

Besides explaining multiscale dependency features, computational and theoretical models offer a convenient tool to design, predict, understand, and optimize the structure and features of systems based on biopolymers. Length scales typically included in biopolymer materials and in biological processes include a single chemical bond at about the angstrom scale, to the monomer and individual chain of the polymer, and to the formation of domains, as well as the higher order, larger scale three-dimensional lattices of these systems (Burla et al. 2019). In classical terms, structural order at the 10–8 and 10–5-m scales is traced upwards in crystallization zones and mixing behaviour of polymer composites and mixtures. The formation of structure in biopolymers is moreover affected by the higher order structure assembly via the hierarchical build-up of higher order structures, like proteins and polysaccharides. The continuous scale also applies in the case of biopolymer systems. The dynamics of interest are in the femtosecond regime relating to chemical bond formation and breakage, to the hour's regime relating to phenomena like large-scale phase separation and protein aging, which involves modification of secondary and tertiary structures in crowded conditions. Relevant phenomena in polymeric systems would require a wide range of time and space scales, implying the need for various theoretical and computational formulations spanning the entire range of phenomena of interest (Schmid 2022).

The inability to combine various computing techniques at length and time scales (Fig. 1.6) is a difficulty when it comes to comprehending them. At the lowest level, the electronic structure and reactivity of molecules can be described with quantum mechanics (QM) techniques [Density Functional Theory (DFT) and Coupled Cluster methods [CCSD(T)]] to high levels of accuracy. These methods work at the angstrom/femtosecond level, and, hence, they are good for the examination of chemical bonding and reaction dynamics, while QM techniques are mostly restricted to small systems and short time scales because of their high computational cost. When one wishes to model larger systems over long periods, molecular dynamics (MD) techniques are involved. MD can perform detailed classical physics-based simulation of atomic movement and provides nanoseconds to microseconds simulation at a nanometer scale. Improved exploration of the conformational space or increased speed in sampling the representation of systems such as proteins, membranes, and polymers is achieved via variants, including Steered Molecular Dynamics (SMD) and Replica Exchange Molecular Dynamics (REMD).

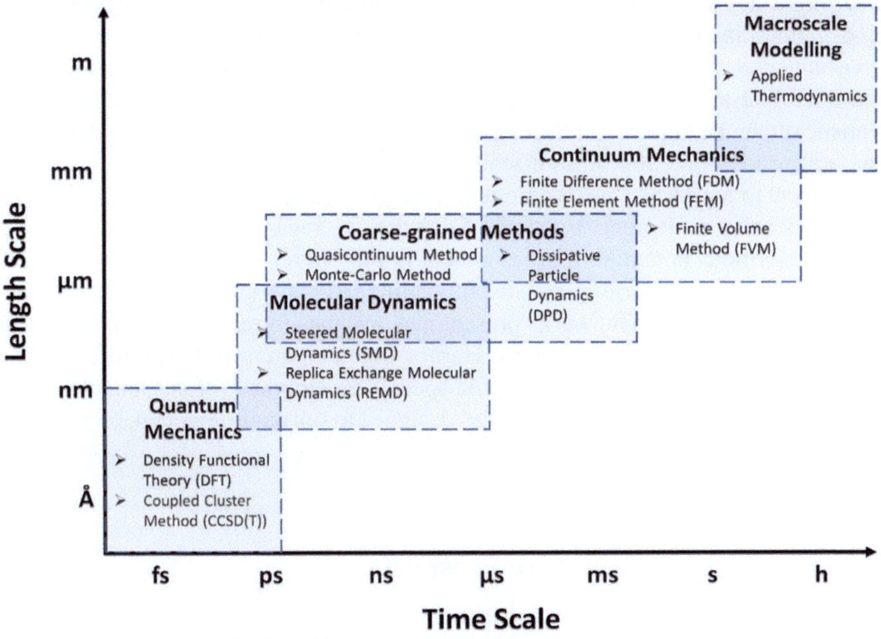

Fig. 1.6 Typical time and length/size scales of different modelling methods for biopolymers (Laycock et al. 2024)

Looking at larger scales, coarse-grained (CG) methods reduce computer computations by incorporating molecules into bulkier, or even primary, structures formed by the combination of atoms into larger structures, or beads. Such algorithms as the Quasi-continuum Method and the Monte Carlo simulations, as well as Dissipative Particle Dynamics (DPD), are applicable to this mesoscale range, nanometers to micrometers, and milliseconds of time. They are particularly applicable in modelling soft matter, complex liquids, and large biomolecular structures. Continuum mechanics beyond the molecular scale matter is described using continuum mechanics, which models matter as a continuous medium and not a collection of discrete particles. In fluid dynamics, structural analysis, and thermal simulation, common methods include the Finite Difference Method (FDM), Finite Element Method (FEM), and Finite Volume Method (FVM). They work on micrometer to millimeter levels and run from microseconds to seconds in time scale and are used extensively in engineering and applied physics (Laycock et al. 2024).

Despite its busy development, computational and theoretical modelling of biopolymers is currently characterized by the important drawbacks hindering its reasonable use. The largest such gap is the absence of accurate, biopolymer-specific force fields that would allow capturing the richness of molecular architecture and conformational flexibility and chemistry of natural polymers. The current models have a tendency to be modified based on the conventional synthetic polymers; thus, they might not accurately predict the significant behaviours of the polymers, like biodegradability, water uptake, and mechanical behaviour, when they are exposed to environmental conditions. Moreover, pH, temperature variations, enzymatic action, and microbial-microbe interaction are some of the factors that should be simulated realistically in the majority of the ongoing simulations, but are usually not done. The incomplete incorporation of multiscale modelling, on the one hand, from quantum mechanical to mesoscale, also complicates the possibility of predicting macroscopic properties using molecular structures. Nevertheless, regardless of such drawbacks, there is enormous potential of development in the future. The new modelling tools that include machine learning-assisted modelling, AI material design, and improved multiscale modelling frameworks present new avenues to realize more precise and predictive modelling of biopolymers. Harnessing computers and experimental data with life-cycle studies could help in the speedy development of sustainable, high-performance green biopolymer resources designed to suit environmental, biomedical, and industrial applications.

1.4 Challenges and Future Prospects

There are various challenges associated with the development of sustainable macro, micro, and nanocomposites, most of which come during issues of compatibility, processing, scalability, and end-of-life. Among the most significant ones is the high interfacial bond strength between the matrix and the reinforcement material, particularly when utilizing natural or biodegradable materials, since most of them have poor matching properties. Also, sustainable composites can be inferior in terms of mechanical strength, thermal stability, or moisture resistance in comparison with conventional composites. The processing techniques will also need to be changed to suit the green materials without compromising the performance levels, and mass production is yet to deal with affordability, replicability, and standardization. In addition, the absence of uniform experimental procedures to test their performance properties and the difficulty in completing the recycling process of bio-based composites will not promote them in the industries.

Nevertheless, the future appears bright for sustainable composites because of ever-growing environmental regulations, consumer pressure, and the growth of new technologies. The desired end in the research is to functionalize natural fibres, bio-based polymers with improved properties, and new methods of fabrication like 3D printing techniques and green chemistry. New opportunities are opening up regarding the development of sustainable materials with high performance due to hybrid composites and nanotechnology. Besides, the design of materials is being informed by life-cycle assessments and circular economy approaches, rendering them effective both during the use phase and after the retirement (which can be via biodegradation or recycling). Through the ongoing interdisciplinary research, the sustainable macro, micro, and nanocomposites have the potential to bring change in various industries like the packaging, construction, and electronics, as well as the biomedical applications.

1.5 Conclusions

This study reveals the emergence of an increased importance of sustainable macro, micro, and nanocomposites as environmental-friendly substitutes of traditional composite materials. They are the renewable-source-derived or biodegradable-containing composites with a hopeful avenue in lowering environmental degradation, along with preserving favourable mechanical, thermal, and functional attributes. Natural fibre polymer/bio-based polymer and nanomaterials have demonstrated significant improvement in the formulation of composites in terms of enhanced performance of composites in many applications, such as automotive, packaging, bio-medical, and construction industries. Nonetheless, there still exist some barriers in the form of interfacial compatibility, limitations during processing, and end-of-life recyclability.

Through these obstacles, the sustainable composites become increasingly viable and more efficient through the development of new material science, surface modification processes, and green fabrication methods. Future studies aim at streamlining the processing procedure, normalization of the evaluation protocol, and designing cost-effective, scalable applications. The transition to sustainable composite technologies will be an urgent issue demanding a multidisciplinary approach entailing materials engineering, environmental science, and industrial design to speed up the process. On the whole, sustainable macro, micro, and nanocomposites could be considered an important step in achieving the sustainability objectives globally through innovative materials.

References

Abdullah, Cai J, Hafeez MA, Wang Q, Farooq S, Huang Q, Tian W, Xiao J (2022) Biopolymer-based functional films for packaging applications: a review. Front Nutr 9:1000116. https://doi.org/10.3389/fnut.2022.1000116

Agarwal J, Sahoo S, Mohanty S, Nayak SK (2020) Progress of novel techniques for lightweight automobile applications through innovative eco-friendly composite materials: a review. https://doi.org/10.1177/0892705718815530

Akbar WA, Rahim HU, Rutigliano FA (2024) Microbial and seaweed-based biopolymers: Sources, extractions and implications for soil quality improvement and environmental sustainability—a review. J Environ Manage 359:120964. https://doi.org/10.1016/j.jenvman.2024.120964

Akhila B et al (2025) Innovations in core–shell electrospinning: a comprehensive review in recent advances of core–shell electrospun polylactic acid nanocomposite fibers for potential biomedical applications. ACS Biomater Sci Eng. https://doi.org/10.1021/acsbiomaterials.5c00194

Alfinaikh RS, Alamry KA, Hussein MA (2025) Sustainable and biocompatible hybrid materials-based sulfated polysaccharides for biomedical applications: a review. RSC Adv 15:4708–4767. https://doi.org/10.1039/D4RA07277D

Al Abboud MA, Mashraqi A, Qanash H, Gattan HS, Felemban HR, Alkorbi F, Alawlaqi MM, Abdelghany TM, Moawad H (2024) Green biosynthesis of bimetallic ZnO@AuNPs with its formulation into cellulose derivative: biological and environmental applications. Bioresourc Bioprocess. https://doi.org/10.1186/s40643-024-00759-3

Ali M et al (2023) Biodegradable piezoelectric polymers: recent advancements in materials and applications. Adv Healthc Mater. Wiley Online Library. https://doi.org/10.1002/adhm.202300318

Appiah ES, Dzikunu P, Mahadeen N, Ampong DN, Mensah-Darkwa K, Kumar A, Gupta RK, Adom-Asamoah M (2022) Biopolymers-derived materials for supercapacitors: recent trends, challenges, and future prospects. Molecules 27:6556. https://doi.org/10.3390/molecules27196556

Ashrafi A, Babapour H, Johari S, Alimohammadi F, Teymori F, Nafchi AM, Shahrai NN, Huda N, Abedinia A (2023) Application of poultry gelatin to enhance the physicochemical, mechanical, and rheological properties of fish gelatin as alternative mammalian gelatin films for food packaging. Foods 12:670. https://doi.org/10.3390/foods12030670

Asim N, Badiei M, Mohammad M (2022) Recent advances in cellulose-based hydrophobic food packaging. Emergent Mater 5:703–718. https://doi.org/10.1007/s42247-021-00314-2

Averous L, Boquillon N (2004) Biocomposites based on plasticized starch: thermal and mechanical behaviours. Carbohyd Polym 56:111–122. https://doi.org/10.1016/j.carbpol.2003.11.015

Azeredo HMC, Barud H, Farinas CS, Vasconcellos VM, Claro AM (2019) Bacterial cellulose as a raw material for food and food packaging applications. Front Sustain Food Syst 3. https://doi.org/10.3389/fsufs.2019.00007

Bailly GC et al (2024) Advancing earth-based construction: a comprehensive review of stabilization and reinforcement techniques for adobe and compressed earth blocks. https://www.mdpi.com/2673-4117/5/2/41

Basumatary IB, Mukherjee A, Katiyar V, Kumar S (2022) Biopolymer-based nanocomposite films and coatings: recent advances in shelf-life improvement of fruits and vegetables. Crit Rev Food Sci Nutr 62:1912–1935. https://doi.org/10.1080/10408398.2020.1848789

Beg M, Saju J, Alcock KM, Mavelil AT, Markapudi PR, Yu H, Manjakkal L (2025) Biodegradable biopolymers for electrochemical energy storage devices in a circular economy. RSC Sustain 3:37–63. https://doi.org/10.1039/D4SU00468J

Betz JF, Ho VB, Gaston JD (2020) 3D bioprinting and its application to military medicine. Mil Med 185:e1510–e1519. https://doi.org/10.1093/milmed/usaa121

Bharathi VSK, Jayas DS (2024) Evolution of bionanocomposites: innovations and applications in food packaging. Foods 13:3787. https://doi.org/10.3390/foods13233787

Bhardwaj U, Dhar P, Kumar A, Katiyar V (2014) Polyhydroxyalkanoates (PHA)-cellulose based nanobiocomposites for food packaging applications. In: Food additives and packaging. American Chemical Society, pp 275–314

Bhaskar R, Zo SM, Narayanan KB, Purohit SD, Gupta MK, Han SS (2023) Recent development of protein-based biopolymers in food packaging applications: a review. Polym Test 124:108097. https://doi.org/10.1016/j.polymertesting.2023.108097

Biobased polyurethane: an overview. Polyesters and Polyamides (2008). 1-s2.0-S2589234723001173-main

Bredenoord J, Kulshreshtha Y (2023) Compressed stabilized earthen blocks and their use in low-cost social housing. Sustainability 15:5295. https://doi.org/10.3390/su15065295

Busolo MA, Fernandez P, Ocio MJ, Lagaron JM (2010) Novel silver-based nanoclay as an antimicrobial in polylactic acid food packaging coatings. Food Addit Contam Part A 27:1617–1626. https://doi.org/10.1080/19440049.2010.506601

Burla F, Mulla Y, Vos BE, Aufderhorst-Roberts A, Koenderink GH (2019) From mechanical resilience to active material properties in biopolymer networks. Nat Rev Phys. https://www.nature.com/articles/s42254-019-0036-4

Carvalho D, Ferreira N, França B, Marques R, Silva M, Silva S, Silva E, Macário D, Barroso L, Silva CJ, Oliveira C (2024) Advancing sustainability in the automotive industry: bioprepregs and fully bio-based composites. Compos Part C Open Access 14:100459. https://doi.org/10.1016/j.jcomc.2024.100459

Cazón P, Vázquez M (2019) Applications of chitosan as food packaging materials. In: Crini G, Lichtfouse E (eds) Sustainable agriculture reviews 36: chitin and chitosan: applications in food, agriculture, pharmacy, medicine and wastewater treatment. Springer International Publishing, Cham, pp 81–123

Cazón P, Vázquez M (2021) Bacterial cellulose as a biodegradable food packaging material: a review. Food Hydrocolloids 113:106530. https://doi.org/10.1016/j.foodhyd.2020.106530

Chang I, Im J, Cho G-C (2016) Introduction of microbial biopolymers in soil treatment for future environmentally-friendly and sustainable geotechnical engineering. Sustainability 8:251. https://doi.org/10.3390/su8030251

Chen N, Wang P, Zhao L, Zhang G (2020) Water retention mechanism of HPMC in cement mortar. Materials 13:2918. https://doi.org/10.3390/ma13132918

Cheng H, Chen L, McClements DJ, Yang T, Zhang Z, Ren F, Miao M, Tian Y, Jin Z (2021) Starch-based biodegradable packaging materials: a review of their preparation, characterization and diverse applications in the food industry. Trends Food Sci Technol 114:70–82. https://doi.org/10.1016/j.tifs.2021.05.017

Cherian RM, Antony T, Varghese RT, Akhila B, Abhijith V, Vipina Vinod TN, Kargarzadeh H, Radhakrishnan EK, Chirayil CJ, Thomas S (2025) A sustainable one-step approach for the functionalized cellulose nanocrystal production using recyclable organic acids: exploring structure-property dynamics. Int J Biol Macromol 305:141192. https://doi.org/10.1016/j.ijbiomac.2025.141192

Crépy L, Petit J-Y, Wirquin E, Martin P, Joly N (2014) Synthesis and evaluation of starch-based polymers as potential dispersants in cement pastes and self leveling compounds. Cem Concr Compos 45:29–38. https://doi.org/10.1016/j.cemconcomp.2013.09.004

Delaey J, Dubruel P, Van Vlierberghe S (2020) Shape-memory polymers for biomedical applications. Adv Funct Mater 30:1909047. https://doi.org/10.1002/adfm.201909047

Demirbas A (2007) Progress and recent trends in biofuels. Prog Energy Combust Sci 33:1–18. https://doi.org/10.1016/j.pecs.2006.06.001

Díez-Pascual AM (2019) Synthesis and applications of biopolymer composites. IJMS 20:2321. https://doi.org/10.3390/ijms20092321

Dhakal HN et al (2024) Potential of date palm fibers (DPFs) as a sustainable reinforcement for bio-composites and its property enhancement for key applications: a review. Macromol Mater Eng. Wiley Online Library. https://doi.org/10.1002/mame.202400081

Efthymiou M-N, Tsouko E, Papagiannopoulos A, Athanasoulia I-G, Georgiadou M, Pispas S, Briassoulis D, Tsironi T, Koutinas A (2022) Development of biodegradable films using sunflower protein isolates and bacterial nanocellulose as innovative food packaging materials for fresh fruit preservation. Sci Rep 12:6935. https://doi.org/10.1038/s41598-022-10913-6

El Mahamdi M, Daoudi W, Dagdag O, Kim H, Eddaoudy F, Verma DK, Gupta S, Berisha A, Loutou M, Noureddine B, El Aatiaoui A (2024) Integrating experimental and theoretical studies in the development of a novel alginate-based bio-composite for copper anticorrosion in 3.5 % NaCl environments. International Journal of Biological Macromolecules. 257, 128600 (2024). https://doi.org/10.1016/j.ijbiomac.2023.128600

Fabra MJ, López-Rubio A, Ambrosio-Martín J, Lagaron JM (2016) Improving the barrier properties of thermoplastic corn starch-based films containing bacterial cellulose nanowhiskers by means of PHA electrospun coatings of interest in food packaging. Food Hydrocolloids 61:261–268. https://doi.org/10.1016/j.foodhyd.2016.05.025

Fatehi H, Ong DEL, Yu J, Chang I (2021) Biopolymers as green binders for soil improvement in geotechnical applications: a review. Geosciences 11:291. https://doi.org/10.3390/geosciences11070291

Ferreira DP, Costa SM, Felgueiras H, Fangueiro R (2019) Smart and sustainable materials for military applications based on natural fibres and silver nanoparticles. Key Eng Mater 812:66–74. https://doi.org/10.4028/www.scientific.net/KEM.812.66

Friné VC et al (2019) Development and characterization of a biodegradable PLA food packaging hold monoterpene–cyclodextrin complexes against *Alternaria alternate*. https://www.mdpi.com/2073-4360/11/10/1720

Firmanda A, Fahma F, Syamsu K, Sari YW, Suryanegara L, Wood K, Saito Y (2023) Factors influencing the biodegradability of agro-biopolymer based slow or controlled release fertilizer. J Polym Environ 31:1706–1724. https://doi.org/10.1007/s10924-022-02718-5

Firoozi AA, Firoozi AA, Oyejobi DO, Avudaiappan S, Flores ES (2024) Emerging trends in sustainable building materials: technological innovations, enhanced performance, and future directions. Result Eng 24:103521. https://doi.org/10.1016/j.rineng.2024.103521

Gamage A, Liyanapathiranage A, Manamperi A, Gunathilake C, Mani S, Merah O, Madhujith T (2022) Applications of starch biopolymers for a sustainable modern agriculture. Sustainability 14:6085. https://doi.org/10.3390/su14106085

García-Guzmán L, Cabrera-Barjas G, Soria-Hernández CG, Castaño J, Guadarrama-Lezama AY, Rodríguez Llamazares S (2022) Progress in starch-based materials for food packaging applications. Polysaccharides 3:136–177. https://doi.org/10.3390/polysaccharides3010007

Ghaderi M, Mousavi M, Yousefi H, Labbafi M (2014) All-cellulose nanocomposite film made from bagasse cellulose nanofibers for food packaging application. Carbohyd Polym 104:59–65. https://doi.org/10.1016/j.carbpol.2014.01.013

Gopinath G, Ayyasamy S, Shanmugaraj P, Swaminathan R, Subbiah K, Kandasamy S (2023) Effects of biopolymers in energy storage applications: a state-of-the-art review. J Energy Storage 70:108065. https://doi.org/10.1016/j.est.2023.108065

Gupta P et al (2022) Recent developments and research avenues for polymers in electric vehicles. The Chemical Record. Wiley Online Library. https://doi.org/10.1002/tcr.202200186

Islam MZ, Sarker ME, Rahman MM, Islam MR, Ahmed ATMF, Mahmud MS, Syduzzaman M (2022) Green composites from natural fibers and biopolymers: A review on processing, properties, and applications. https://doi.org/10.1177/07316844211058708

Hafez I, Tajvidi M (2020) Laminated wallboard panels made with cellulose nanofibrils as a binder: production and properties. Materials 13:1303. https://doi.org/10.3390/ma13061303

Haghighi H, Licciardello F, Fava P, Siesler HW, Pulvirenti A (2020) Recent advances on chitosan-based films for sustainable food packaging applications. Food Packag Shelf Life 26:100551. https://doi.org/10.1016/j.fpsl.2020.100551

Hariz HM, Sapuan SM, Ilyas RA (2021) Advanced composite in aerospace application: a review on future aspect of fiber-reinforced polymer (FRP) in aerospace industry

Hu B (2014) Biopolymer-based lightweight materials for packaging applications. In: Yang Y, Xu H, Yu X (eds) ACS symposium series. American Chemical Society, Washington, DC, pp 239–255

Guinault A et al (2010) Influence of crystallinity on gas barrier and mechanical properties of PLA food packaging films. Int J Mater. https://doi.org/10.1007/s12289-010-0842-9

Iwuozor KO, Micheal TT, Micheal KT, Emmanuel SS, Emenike EC, Adeniyi AG (2024) Plant biomass-based composites in the maritime industry: a review. https://www.sciencedirect.com/science/article/pii/S0951833924000376?via%3Dihub

Jamaludin FH, Nordin NS, Li X, Wong J-W, Lim Z, Wang L, Ma C, Zhao Q, Wong T-W, Neffe AT, Li T (2025) Recent advances in the field of shape-memory polymer from bio-based precursors: review and perspective. Polym Rev 65:199–249. https://doi.org/10.1080/15583724.2024.2427827

Juikar SK, Warkar SG (2023) Biopolymers for packaging applications: an overview. Packag Technol Sci 36:229–251. https://doi.org/10.1002/pts.2707

Kammeta M, Mukhopadhyay A, Sikka M (2023) A review on challenges in developing the military uniforms. https://doi.org/10.1007/978-981-97-5902-6_2

Khalid MY, Arif ZU, Al Rashid A, Bukhari SMZS, Hossain M, Koç M (2024) Shape-memory and self-healing properties of sustainable cellulosic nanofibers-based hybrid materials for novel applications. Giant 19:100299. https://doi.org/10.1016/j.giant.2024.100299

Khan B, Niazi MBK, Samin G, Jahan Z (2017) Thermoplastic starch: a possible biodegradable food packaging material—a review. J Food Process Eng 40:e12447. https://doi.org/10.1111/jfpe.12447

Khezerlou A, Tavassoli M, Alizadeh Sani M, Mohammadi K, Ehsani A, McClements DJ (2021) Application of nanotechnology to improve the performance of biodegradable biopolymer-based packaging materials. Polymers 13:4399. https://doi.org/10.3390/polym13244399

Kim J, Jeong J, Kim Y, Jung S (2024) Physicochemical and rheological properties of succinoglycan overproduced by *Sinorhizobium meliloti* 1021 mutant. Polymers 16:244. https://doi.org/10.3390/polym16020244

Kong B et al (2023) Natural biopolymers derived hydrogels with injectable, self-healing, and tissue adhesive abilities for wound healing. Nano Res 2798–2807. https://doi.org/10.1007/s12274-022-4936-8

Kumar MH, Babu NBK, Sharma A, Bal S, Suresh NS (2022) Applications of Kevlar fabric composites. Presented at the 10 June 2022

Kumar S, Gupta D, Sharma V, Chaudhary AK, Meena ML (2023) Recent development in natural fiber composites, testing and fabrication methods: a review. Mater Today Proc. https://doi.org/10.1016/j.matpr.2023.02.073

Liyakat KKS (2024) Review of biopolymers in agriculture application: an eco-friendly alternative. IJCCM

Lauer MK, Smith RC (2020) Recent advances in starch-based films toward food packaging applications: physicochemical, mechanical, and functional properties. Compr Rev Food Sci Food Saf 19:3031–3083. https://doi.org/10.1111/1541-4337.12627

Laycock BG, Chan CM, Halley PJ (2024) A review of computational approaches used in the modelling, design, and manufacturing of biodegradable and biobased polymers. Prog Polym Sci 157:101874. https://doi.org/10.1016/j.progpolymsci.2024.101874

Lee S, Park JY, Yoon H, Park J, Lee J, Hwang B, Padil VVT, Cheong JY, Yun TG (2025) Long-lasting supercapacitor with stable electrode-electrolyte interface enabled by a biopolymer conjugate electrolyte additive. Energy Storage Mater 77:104195. https://doi.org/10.1016/j.ensm.2025.104195

Lendlein A, Balk M, Tarazona NA, Gould OEC (2019) Bioperspectives for shape-memory polymers as shape programmable active materials. Biomacromol 20:3627–3640. https://doi.org/10.1021/acs.biomac.9b01074

Li X, Bond P, O'Moore L, Wilkie S, Hanžič L, Johnson I, Mueller K, Yuan Z, Jiang G (2020) Increased resistance of nitrite-admixed concrete to microbially induced corrosion in real sewers. Environ Sci Technol. https://doi.org/10.1021/acs.est.9b06680

Li X, Zheng G, Li Z, Fu P (2024) Formulation, performance and environmental/agricultural benefit analysis of biomass-based biodegradable mulch films: a review. Eur Polym J 203:112663. https://doi.org/10.1016/j.eurpolymj.2023.112663

Liu ZQ, Jiao D, Zhang ZF (2015) Remarkable shape memory effect of a natural biopolymer in aqueous environment. Biomaterials 65:13–21. https://doi.org/10.1016/j.biomaterials.2015.06.032

Liu Y, Ahmed S, Sameen DE, Wang Y, Lu R, Dai J, Li S, Qin W (2021) A review of cellulose and its derivatives in biopolymer-based for food packaging application. Trends Food Sci Technol 112:532–546. https://doi.org/10.1016/j.tifs.2021.04.016

Liu T, Li J, Tang Q, Qiu P, Gou D, Zhao J (2022) Chitosan-based materials: an overview of potential applications in food packaging. Foods 11:1490. https://doi.org/10.3390/foods11101490

Liu Z, Chen X, Wang ZL (2025) Biopolymer and biomimetic techniques for triboelectric nanogenerators (TENGs). Adv Mater 37:2409440. https://doi.org/10.1002/adma.202409440

Lizundia E, Kundu D (2021) Advances in natural biopolymer-based electrolytes and separators for battery applications. Adv Func Mater 31:2005646. https://doi.org/10.1002/adfm.202005646

Löwenberg C, Balk M, Wischke C, Behl M, Lendlein A (2017) Shape-memory hydrogels: evolution of structural principles to enable shape switching of hydrophilic polymer networks. Acc Chem Res 50:723–732. https://doi.org/10.1021/acs.accounts.6b00584

Lu Y, Luo Q, Chu Y, Tao N, Deng S, Wang L, Li L (2022) Application of gelatin in food packaging: a review. Polymers 14:436. https://doi.org/10.3390/polym14030436

Lubczak J, Lubczak R (2022) Increased thermal stability and reduced flammability of polyurethane foams with an application of polyetherols. Presented at the 1 June 2022

Ludwicka K, Kaczmarek M, Białkowska A (2020) Bacterial nanocellulose—a biobased polymer for active and intelligent food packaging applications: recent advances and developments. Polymers 12:2209. https://doi.org/10.3390/polym12102209

Luo Q, Hossen MA, Zeng Y, Dai J, Li S, Qin W, Liu Y (2022) Gelatin-based composite films and their application in food packaging: a review. J Food Eng 313:110762. https://doi.org/10.1016/j.jfoodeng.2021.110762

Mahajan P, Sharma M (2024) A comprehensive review on developments and future perspectives of biopolymer-based materials for energy storage. Energy Storage 6:e634. https://doi.org/10.1002/est2.634

Mahanty B, Kumar Ghosh S, Lee D-W (2024) Advancements in polymer nanofiber-based piezoelectric nanogenerators: revolutionizing self-powered wearable electronics and biomedical applications. Chem Eng J 495:153481. https://doi.org/10.1016/j.cej.2024.153481

Mahmood A, Patel D, Hickson B, DesRochers J, Hu X (2022) Recent progress in biopolymer-based hydrogel materials for biomedical applications. IJMS 23:1415. https://doi.org/10.3390/ijms23031415

Malek NSA, Faizuwan M, Khusaimi Z, Bonnia NN, Rusop M, Asli NA (2021) Preparation and characterization of biodegradable polylactic acid (PLA) film for food packaging application: a review. J Phys Conf Ser 1892:012037. https://doi.org/10.1088/1742-6596/1892/1/012037

Marano S, Laudadio E, Minnelli C, Stipa P (2022) Tailoring the barrier properties of PLA: a state-of-the-art review for food packaging applications. Polymers 14:1626. https://doi.org/10.3390/polym14081626

Mathur V, Farouq M, Labaran YH (2021) The carbon footprint of construction industry: A review of direct and indirect emission. J Sustain Constr Mater Technol 6:101–115. https://doi.org/10.29187/jscmt.2021.66

Menossi M et al (2021) Current and emerging biodegradable mulch films based on polysaccharide bio-composites: a review. Agron Sustain Dev. https://doi.org/10.1007/s13593-021-00685-0

Mirpoor SF, Patanè GT, Corrado I, Giosafatto CVL, Ginestra G, Nostro A, Foti A, Gucciardi PG, Mandalari G, Barreca D, Gervasi T, Pezzella C (2023) Functionalization of polyhydroxyalkanoates (PHA)-based bioplastic with phloretin for active food packaging: characterization of its mechanical, antioxidant, and antimicrobial activities. Int J Mol Sci 24:11628. https://doi.org/10.3390/ijms241411628

Mori R (2023) Replacing all petroleum-based chemical products with natural biomass-based chemical products: a tutorial review. RSC Sustain 1:179–212. https://doi.org/10.1039/D2SU00014H

Musitelli M, Garbagnoli P, Diamanti MV, Curto BD, Pedeferri M, Proserpio C (2013) Improvement of cladding insulation panels for sustainable buildings

Nathanael AJ, Oh TH (2020) Biopolymer coatings for biomedical applications. Polymers 12:3061. https://doi.org/10.3390/polym12123061

Maiti S et al (2019) Nature driven bio-piezoelectric/triboelectric nanogenerator as next-generation green energy harvester for smart and pollution free society. Adv Energy Mater. Wiley Online Library. https://doi.org/10.1002/aenm.201803027

Navier Lab, University Gustave Eiffel, ENPC, CNRS, Champs-sur-Marne, France, Tourtelot J, Laboratoire de Recherche des Monuments Historiques, Champs-sur-Marne, France, Bourgès A, C2RMF, Centre de Recherche et de Restauration des Musées de France, France-Institut de Recherche Chimie Paris, PSL Research University, Chimie ParisTech-CNRS, Paris, France, Keita E, Navier Lab, University Gustave Eiffel, ENPC, CNRS, Champs-sur-Marne, France (2021) Influence of biopolymers on the mechanical behavior of earth-based building materials. RPM 3:1–1. https://doi.org/10.21926/rpm.2103031

Nazrun T, Hassan MK, Hossain MD, Ahmed B, Hasnat MR, Saha S (2024) Application of biopolymers as sustainable cladding materials: a review. Sustainability 16:27. https://doi.org/10.3390/su16010027

Norozi M, Peighambardoust SH, Khorram S, Hamishehkar H (2025) Cold plasma-enhanced polylactic acid/chitosan films incorporating zein nanoparticles: a promising strategy for developing food packaging materials. Carbohydr Polym Technol Appl 10:100852. https://doi.org/10.1016/j.carpta.2025.100852

Norrrahim MNF, Kasim NAM, Knight VF, Ujang FA, Janudin N, Razak MAIA, Shah NAA, Noor SAM, Jamal SH, Ong KK, Yunus WMZW (2021) Nanocellulose: the next super versatile material for the military. Mater Adv 2:1485–1506. https://doi.org/10.1039/D0MA01011A

Oladele IO, Omotosho TF, Adediran AA (2020) Polymer-based composites: an indispensable material for present and future applications. Int J Polym Sci 2020:1–12. https://doi.org/10.1155/2020/8834518

Olosho AI (2024) Sustainable biopolymer materials for defense and surveillance applications. In: AUN international conference, vol 2, pp 390–396

Optica Publishing Group. https://opg.optica.org/optica/viewmedia.cfm?uri=optica-4-2-209&html=true

Orts WJ, Roa-Espinosa A, Sojka RE, Glenn GM, Imam SH, Erlacher K, Pedersen JS (2007) Use of synthetic polymers and biopolymers for soil stabilization in agricultural, construction, and military applications. J Mater Civ Eng 19:58–66. https://doi.org/10.1061/(ASCE)0899-1561(2007)19:1(58)

Pandey M, Nazar R, Elella MHA, Praharaj S, Makhado E, Rout D, Gilani EH (2025) A comprehensive review of recent developments in biomedical materials based on graphene-modified bio-nanocomposites. BioNanoSci 15:125. https://doi.org/10.1007/s12668-024-01757-7

Patel N, Feofilovs M, Blumberga D (2022) Agro biopolymer: a sustainable future of agriculture – state of art review. Environ Clim Technol 26:499–511. https://doi.org/10.2478/rtuect-2022-0038

Patel DK, Jung E, Priya S, Won S-Y, Han SS (2024) Recent advances in biopolymer-based hydrogels and their potential biomedical applications. Carbohyd Polym 323:121408. https://doi.org/10.1016/j.carbpol.2023.121408

Patiño Vidal C, Muñoz-Shugulí C, Guivier M, Puglia D, Luzi F, Rojas A, Velásquez E, Galotto MJ, López-de-Dicastillo C (2024) PLA- and PHA-biopolyester-based electrospun materials: development, legislation, and food packaging applications. Molecules 29:5452. https://doi.org/10.3390/molecules29225452

Perera KY, Jaiswal AK, Jaiswal S (2023) Biopolymer-based sustainable food packaging materials: challenges, solutions, and applications. Foods 12:2422. https://doi.org/10.3390/foods12122422

Philip E, Reshmy R (2024) Applications of biopolymers in construction and civil engineering. In: Applications of biopolymers in science. Biotechnology, and engineering. John Wiley & Sons, Ltd, pp 265–289

Phiri R, Mavinkere Rangappa S, Siengchin S, Oladijo OP, Dhakal HN (2023) Development of sustainable biopolymer-based composites for lightweight applications from agricultural waste biomass: a review. Adv Ind Eng Polym Res 6:436–450. https://doi.org/10.1016/j.aiepr.2023.04.004

Pilla S (2011) Handbook of bioplastics and biocomposites engineering applications. John Wiley & Sons

Pinaeva LG, Noskov AS (2024) Biodegradable biopolymers: real impact to environment pollution. Sci Total Environ 947:174445. https://doi.org/10.1016/j.scitotenv.2024.174445

Prasanna K, Subburaj T, Jo YN, Lee WJ, Lee CW (2015) Environment-friendly cathodes using biopolymer chitosan with enhanced electrochemical behavior for use in lithium ion batteries. ACS Appl Mater Interfaces 7:7884–7890. https://doi.org/10.1021/am5084094

Prasanth SM, Kumar PS, Harish S, Rishikesh M, Nanda S, Vo D-VN (2021) Application of biomass derived products in mid-size automotive industries: a review. Chemosphere 280:130723. https://doi.org/10.1016/j.chemosphere.2021.130723

Priya E, Sarkar S, Maji PK (2024) A review on slow-release fertilizer: Nutrient release mechanism and agricultural sustainability. J Environ Chem Eng 12:113211. https://doi.org/10.1016/j.jece.2024.113211

Priyadarshi R, Rhim J-W (2020) Chitosan-based biodegradable functional films for food packaging applications. Innov Food Sci Emerg Technol 62:102346. https://doi.org/10.1016/j.ifset.2020.102346

Prakash SO et al (2022) A review on natural fibre-reinforced biopolymer composites: properties and applications. Int J Polym Sci. Wiley Online Library. https://doi.org/10.1155/2022/7820731

Puglia D, Pezzolla D, Gigliotti G, Torre L, Bartucca ML, Del Buono D (2021) The opportunity of valorizing agricultural waste, through its conversion into biostimulants, biofertilizers, and biopolymers. Sustainability 13:2710. https://doi.org/10.3390/su13052710

Rahman A, Hasan K, Imran AB (2024) Biopolymers for supercapacitors. In: Bio-based polymers: farm to industry, vol 3. Emerging Trends and Applications. American Chemical Society, pp 57–80

Rajesh Kumar B, Saravanan S (2016) Use of higher alcohol biofuels in diesel engines: a review. Renew Sustain Energy Rev 60:84–115. https://doi.org/10.1016/j.rser.2016.01.085

Ramon E, Sguazzo C, Moreira PMGP (2018) A Review of recent research on bio-based epoxy systems for engineering applications and potentialities in the aviation sector. Aerospace 5:110. https://doi.org/10.3390/aerospace5040110

Ramos M, Valdés A, Beltrán A, Garrigós MC (2016) Gelatin-based films and coatings for food packaging applications. Coatings 6:41. https://doi.org/10.3390/coatings6040041

Rasheed T, Anwar MT, Naveed A, Ali A (2022) Biopolymer based materials as alternative greener binders for sustainable electrochemical energy storage applications. ChemistrySelect 7:e202203202. https://doi.org/10.1002/slct.202203202

Rebelo HB, Lecompte D, Cismasiu C, Jonet A, Belkassem B, Maazoun A (2019) Experimental and numerical investigation on 3D printed PLA sacrificial honeycomb cladding. Int J Impact Eng 131:162–173. https://doi.org/10.1016/j.ijimpeng.2019.05.013

Ravindran L, Jesitha K, Megha PU, Anilkumar S, Sreekala MS, Harikumar PS (2022) Nanosilica entrapped alginate beads for the purification of groundwater contaminated with bacteria. Silicon. https://doi.org/10.1007/s12633-021-01544-z

Rhim J-W, Ng PKW (2007) Natural biopolymer-based nanocomposite films for packaging applications. Crit Rev Food Sci Nutr 47:411–433. https://doi.org/10.1080/10408390600846366

Riahi Z, Priyadarshi R, Rhim J-W, Bagheri R (2021) Gelatin-based functional films integrated with grapefruit seed extract and TiO_2 for active food packaging applications. Food Hydrocolloids 112:106314. https://doi.org/10.1016/j.foodhyd.2020.106314

Riseh RS, Vazvani MG, Vatankhah M, Kennedy JF (2024) Chitosan coating of seeds improves the germination and growth performance of plants: a review. Int J Biol Macromol 278:134750. https://doi.org/10.1016/j.ijbiomac.2024.134750

Romero-Fierro D, Bustamante-Torres M, Bravo-Plascencia F, Esquivel-Lozano A, Ruiz J-C, Bucio E (2022) Recent trends in magnetic polymer nanocomposites for aerospace applications: a review. Polymers 14:4084. https://doi.org/10.3390/polym14194084

Roy S, Rhim J-W (2020) Preparation of antimicrobial and antioxidant gelatin/curcumin composite films for active food packaging application. Colloids Surf, B 188:110761. https://doi.org/10.1016/j.colsurfb.2019.110761

Roy P, Tadele D, Defersha F, Misra M, Mohanty AK (2019) Environmental and economic prospects of biomaterials in the automotive industry. Clean Techn Environ Policy 21:1535–1548. https://doi.org/10.1007/s10098-019-01735-8

Rushdha Hiba T, Abhijith V, Akhila B, Thomas S, Sreekala MS (2025) Recent advances in polymer based energy harvesting systems for efficient wound healing applications. Mater Today Commun 47:113152. https://doi.org/10.1016/j.mtcomm.2025.113152

Saedi S, Garcia CV, Kim JT, Shin GH (2021) Physical and chemical modifications of cellulose fibers for food packaging applications. Cellulose 28:8877–8897. https://doi.org/10.1007/s10570-021-04086-0

Said NS, Howell NK, Sarbon NM (2023) A review on potential use of gelatin-based film as active and smart biodegradable films for food packaging application. Food Rev Int 39:1063–1085. https://doi.org/10.1080/87559129.2021.1929298

Samalens F, George Thomas M, Claverie M, Castejón N, Zhang Y, Thierry P, Blanc S, Fernandes SCM (2022) Progresses and future prospects in biodegradation of marine biopolymers and emerging biopolymer-based materials for sustainable marine ecosystems. Green Chem 24. https://doi.org/10.1039/D1GC04327G

Sampath UGTM, Ching YC, Chuah CH, Sabariah JJ, Lin P-C (2016) Fabrication of porous materials from natural/synthetic biopolymers and their composites. Materials 9:991. https://doi.org/10.3390/ma9120991

Satyanarayana KG, Arizaga GGC, Wypych F (2009) Biodegradable composites based on lignocellulosic fibers—AN overview. Prog Polym Sci 34:982–1021. https://doi.org/10.1016/j.progpolymsci.2008.12.002

Schmid F (2022) Understanding and modeling polymers: the challenge of multiple scales. ACS Polyme Au. https://doi.org/10.1021/acspolymersau.2c00049

Shaikh AA, Datta P, Dastidar P, Majumder A, Das MD, Manna P, Roy S (2024) Biopolymer-based nanocomposites for application in biomedicine: a review. J Polym Eng 44:83–116. https://doi.org/10.1515/polyeng-2023-0166

Shao L, Xi Y, Weng Y (2022) Recent advances in PLA-based antibacterial food packaging and its applications. Molecules 27:5953. https://doi.org/10.3390/molecules27185953

Sinha S, Sudarsan JS (2025) Building a greener future: how earth blocks are reshaping sustainability and circular economy in construction. Architecture 5:25. https://doi.org/10.3390/architecture5020025

Sionkowska A (2016) Biopolymeric nanocomposites for potential biomedical applications. Polym Int 65:1123–1131. https://doi.org/10.1002/pi.5149

Sohail M, Pirzada T, Opperman CH, Khan SA (2022) Recent advances in seed coating technologies: transitioning toward sustainable agriculture. Green Chem 24:6052–6085. https://doi.org/10.1039/D2GC02389J

Stanisz M, Klapiszewski Ł, Jesionowski T (2020) Recent advances in the fabrication and application of biopolymer-based micro- and nanostructures: a comprehensive review. Chem Eng J 397:125409. https://doi.org/10.1016/j.cej.2020.125409

Suethao S, Prasopdee T, Buaksuntear K, Shah D, Smitthipong W (2022) Recent developments in shape memory elastomers for biotechnology applications. Polymers 14:3276. https://doi.org/10.3390/polym14163276

Tang Y, Zhang Y, Zou L, Sun C, Tang W, Zou Y, Zhou A, Fu W, Wang F, Li K, Zhang Q, Zhang X (2025) Review of 3D-printed bioceramic/biopolymer composites for bone regeneration: fabrication methods, technologies and functionalized applications. Biofabrication 17:032002. https://doi.org/10.1088/1758-5090/adcbd7

Tariq Z, Iqbal DN, Rizwan M, Ahmad M, Faheem M, Ahmed M (2023) Significance of biopolymer-based hydrogels and their applications in agriculture: a review in perspective of synthesis and their degree of swelling for water holding. RSC Adv 13:24731–24754. https://doi.org/10.1039/D3RA03472K

Tetlow D, De Simon L, Liew SY, Hewakandamby B, Mack D, Thielemans W, Riffat S (2017) Cellulosic-crystals as a fumed-silica substitute in vacuum insulated panel technology used in building construction and retrofit applications. Energy Build 156:187–196. https://doi.org/10.1016/j.enbuild.2017.08.058

Thomas D, Cabibihan J-J, Kumar S, Khadheer Pasha SK, Mandal D, Laad M, Yadav BC, Patil SI, Ghule A, Mazumdar P, Rattan S, Sadasivuni KK (2017) Biodegradable nanocomposites for energy harvesting, self-healing, and shape memory. In: Ponnamma D, Sadasivuni KK, Cabibihan J-J, Al-Maadeed MA-A (eds) Smart polymer nanocomposites: energy harvesting, self-healing and shape memory applications. Springer International Publishing, Cham, pp 377–397

Ulery BD, Nair LS, Laurencin CT (2011) Biomedical applications of biodegradable polymers. J Polym Sci B Polym Phys 49:832–864. https://doi.org/10.1002/polb.22259

Uyan M, Celiktas MS (2023) Evaluation of the bio-based materials utilization in shape memory polymer composites production. Eur Polym J 195:112196. https://doi.org/10.1016/j.eurpolymj.2023.112196

Valenzuela M, Ciudad G, Cárdenas JP, Medina C, Salas A, Oñate A, Pincheira G, Attia S, Tuninetti V (2024) Towards the development of performance-efficient compressed earth blocks from industrial and agro-industrial by-products. Renew Sustain Energy Rev 194:114323. https://doi.org/10.1016/j.rser.2024.114323

Wang H, Qian J, Ding F (2018) Emerging chitosan-based films for food packaging applications. J Agric Food Chem 66:395–413. https://doi.org/10.1021/acs.jafc.7b04528

Wang C et al (2019) Fabrication of highly efficient self-healing hydrogel from natural biopolymers loaded with exosomes for the synergistic promotion of severe wound healing. Biomater Sci 8(1). https://doi.org/10.1039/C9BM01207A

Wani NR, Dar AH, Dash KK, Pandey VK, Srivastava S, Jan SY, Deka P, Sabahi N (2023) Recent advances in the production of bionanomaterials for development of sustainable food packaging: a comprehensive review. Environ Res 237:116948. https://doi.org/10.1016/j.envres.2023.116948

Xiong R, Grant AM, Ma R, Zhang S, Tsukruk VV (2018) Naturally-derived biopolymer nanocomposites: interfacial design, properties and emerging applications. Mater Sci Eng R Rep 125:1–41. https://doi.org/10.1016/j.mser.2018.01.002

Xu Q, Wen J, Qin Y (2021) Development and outlook of high output piezoelectric nanogenerators. Nano Energy 86:106080. https://doi.org/10.1016/j.nanoen.2021.106080

Xu T, Liu K, Sheng N, Zhang M, Liu W, Liu H, Dai L, Zhang X, Si C, Du H, Zhang K (2022) Biopolymer-based hydrogel electrolytes for advanced energy storage/conversion devices: properties, applications, and perspectives. Energy Storage Mater 48:244–262. https://doi.org/10.1016/j.ensm.2022.03.013

Yadav A, Mangaraj S, Singh R, Das SK, Arora S (2018) Biopolymers as packaging material in food and allied industry. Int J Chem Stud

Yekta R, Abedi-Firoozjah R, Azimi Salim S, Khezerlou A, Abdolmaleki K (2023) Application of cellulose and cellulose derivatives in smart/intelligent bio-based food packaging. Cellulose 30:9925–9953. https://doi.org/10.1007/s10570-023-05520-1

Yeo JCC, Muiruri JK, Fei X, Wang T, Zhang X, Xiao Y, Thitsartarn W, Tanoto H, He C, Li Z (2024) Innovative biomaterials for food packaging: unlocking the potential of polyhydroxyalkanoate (PHA) biopolymers. Biomater Adv 163:213929. https://doi.org/10.1016/j.bioadv.2024.213929

Yusoff MNAM, Zulkifli NWM, Masum BM, Masjuki HH (2015) Feasibility of bioethanol and biobutanol as transportation fuel in spark-ignition engine: a review. RSC Adv 5:100184–100211. https://doi.org/10.1039/C5RA12735A

Zarrintaj P, Seidi F, Youssefi Azarfam M, Khodadadi Yazdi M, Erfani A, Barani M, Chauhan NPS, Rabiee N, Kuang T, Kucinska-Lipka J, Saeb MR, Mozafari M (2023) Biopolymer-based composites for tissue engineering applications: a basis for future opportunities. Compos Part B Eng 258:110701. https://doi.org/10.1016/j.compositesb.2023.110701

Zecheru T (2010) Biopolymers for military use: opportunities and environment implications—a review. In: Elnashar M (ed) Biopolymers. Sciyo

Zhang L, Yang Q, Jiang Y, Yu L, Song N, Zhao D, Sui L, Dong L (2024) Antifouling and antibacterial bioactive metabolites of marine fungus (terrein)/polyamide thin-film composite reverse osmosis membranes for desalination applications. Desalination 572:117140. https://doi.org/10.1016/j.desal.2023.117140

Zhou SC et al (2019) Effects of chemical admixtures on the working and mechanical properties of ordinary dry-mixed mortar. Adv Mater Sci Eng. Wiley Online Library. https://doi.org/10.1155/2019/5978089

Zhu Q, Sun E, Zhao Z, Wu T, Meng S, Ma Z, Shoaib M, Ur Rehman H, Cao X, Wang N (2024) Biopolymer materials in triboelectric nanogenerators: a review. Polymers 16:1304. https://doi.org/10.3390/polym16101304

Chapter 2
Processing and Structure–Property Relationships of Green Composites

Samia Nouira, Abir Abdessalem, Joseph Fitoussi, and Mohammadali Shirinbayan

Abstract This chapter presents a comprehensive analysis of green composites, emphasizing the use of natural fibers and biopolymer matrices as environmentally conscious alternatives to conventional, synthetically reinforced composites. It systematically reviews the mechanical properties, durability, and environmental impact of various green composites, highlighting their suitability for applications such as automotive, civil engineering, packaging, and consumer goods. The chapter examines how fiber-matrix interactions, surface treatments, and processing conditions affect the overall performance of these materials, focusing on factors such as fiber morphology, interfacial bonding, and matrix crystallinity to identify optimal material combinations. It also addresses the challenges associated with green composites, in particular reduced mechanical performance, and discusses strategies for improvement, including hybridization with synthetic fibers and advanced treatments. Current findings indicate that while green composites typically match the properties of traditional composites in many applications, strategic hybridization and ongoing research developments provide pathways to potentially meet or exceed these traditional benchmarks in specific, targeted applications. Such advances support wider adoption in both general and specialized sectors, contributing meaningfully to sustainable practices within the composites industry.

S. Nouira · J. Fitoussi · M. Shirinbayan (✉)
Arts et Metiers Institute of Technology, CNAM, CNRS, PIMM, Paris, France
e-mail: mohammadali.shirinbayan@ensam.eu

S. Nouira
e-mail: samia.nouira@univ-lorraine.fr

J. Fitoussi
e-mail: joseph.fitoussi@ensam.eu

S. Nouira
Université de Lorraine, Laboratoire LEM3 UMR 7239, Metz, France

A. Abdessalem
Ecole Nationale d'Ingénieurs de Sousse (ENISo), Laboratoire LMS-Université de Sousse, Pôle Technologique de Sousse, Sousse, Tunisie
e-mail: abir.abdessalem@eniso.u-sousse.tn

L. Ravindran et al. (eds.), *Macro, Micro and Nanocomposites from Sustainable Sources*, Advanced Structured Materials 244, https://doi.org/10.1007/978-981-95-2469-3_2

Keywords Green composites · Sustainable materials · Composite processing · Durability of composites

2.1 Introduction

Composite materials are engineered by combining distinct components with complementary properties, resulting in enhanced characteristics unattainable by individual constituents alone. The reinforcement, often composed of fibers, is crucial for bearing mechanical loads, minimizing thermal stresses, and providing the necessary rigidity and structural strength. The matrix encapsulates these fibers, ensuring cohesion, protecting against environmental factors, and facilitating efficient load transfer (Yadav and Rathore 2024; Akroum-Amrouche et al. 2024).

Typically, the matrix comprises two main categories of polymers: thermosetting (TS) polymers and thermoplastic (TP) polymers (Han et al. 2023; Sonnenfeld et al. 2017; Islam et al. 2025). Thermosetting polymers, including epoxy and phenolic resins, undergo a curing process that results in the formation of a rigid three-dimensional network. Once cured, these materials are unable to be remelted or reshaped, which consequently restricts their recyclability. In contrast, thermoplastic polymers, such as polyethylene (PE) and polyamide (PA), feature flexible linear chains that can be softened and reformed multiple times, making them more amenable to recycling and aligning with sustainability objectives.

Reinforcing materials are typically synthetic fibers, such as glass, carbon, and aramid. Despite their advantages, synthetic composites are confronted with considerable challenges. Synthetic composites are primarily derived from petroleum-based products, contributing to natural resource depletion and environmental degradation (Mori 2023; Yang et al. 2023; Raj et al. 2023). Furthermore, the non-biodegradable nature of synthetic polymers gives rise to significant concerns regarding their disposal at the end of their lifecycle. Additionally, the energy-intensive manufacturing processes entail health risks and environmental impacts, underscoring the necessity for sustainable alternatives (Singh et al. 2023; Angellier-Coussy et al. 2024).

In light of these challenges, green composites represent a compelling solution to the aforementioned issues, as they incorporate natural fibers, such as jute, hemp, flax, and kenaf in combination with bio-based or recycled matrices. These composites offer several advantages over traditional synthetic materials, including a reduced environmental impact, a reduction in energy consumption during production, and an improvement in biodegradability (Ammar 2023; Park et al. 2024; Cakmak 2024). Furthermore, they facilitate a more environmentally conscious methodology within the field of materials science.

As we delve deeper, it is important to note that the selection of reinforcing fibers and their orientation, along with the interactions at the fiber-matrix interface, represent critical factors influencing the mechanical properties and overall performance of green composites. Therefore, optimizing the fiber-matrix interface is essential to enhancing the mechanical performance, as this region is primarily responsible for the transfer of loads.

Historically, green composites have gained traction over several decades due to their advantages over conventional materials, especially in terms of performance and weight reduction. These advantages, coupled with their sustainable nature, have attracted numerous industries, including aerospace, automotive, infrastructure, and sports (Ramesh et al. 2022; Arpitha et al. 2024; Sabet 2024). The evolution of green composite technology has been characterized by both incremental improvements and transformative innovations, enabling improved mechanical properties and broader applications. As the demand for green solutions continues to grow, the development of green composites offers a viable path forward that balances performance with sustainability goals.

This chapter focuses on the influence of processing techniques on the microstructure of green composites and their relationship to the resulting mechanical, thermal, and durability properties. By understanding these relationships, we can highlight the benefits of green composites in promoting sustainable engineering practices.

2.2 Green Composites: Fundamental Constituents and Material Types

"Green composites" is a broad term that refers to materials made from both bio-based and biodegradable components. Bio-based materials are derived from renewable resources, while biodegradable materials are designed to decompose naturally over time. Within the green composites category, "biocomposites" specifically refer to composites that incorporate bio-based materials, either in the matrix or reinforcement. In addition, "biofiber-reinforced composites" are a subset of biocomposites that use natural fibers, such as plant fibers, as reinforcements paired with a matrix that may be bio-based or synthetic.

2.3 Matrix Material

In the production of biofiber-reinforced composites, traditional matrix materials have typically been fossil-based thermoplastics, including polypropylene (PP) and polyethylene (PE). However, with growing awareness of environmental issues, climate change, and the depletion of fossil resources, there is an increasing drive to identify more eco-friendly alternatives. This pursuit has intensified interest in biodegradable matrices, which offer benefits such as recyclability and a reduced environmental footprint. The use of green matrices in biofiber-reinforced composites has become a key focus for both the scientific community and industry, paving the way for more sustainable solutions.

2.4 Classification of Biopolymers

Biodegradable polymers are generally divided into two categories according to their origin: bio-based and synthetic polymers (Sharma and Bhende 2024; Sinha 2024). The classification is illustrated in Fig. 2.1.

Bio-based polymers can be grouped as biomass and microbial synthesized products, while synthetic polymers can be grouped as renewable resource-based and petro-products (Zhou et al. 2024; Talukder et al. 2024; Ajitesh et al. 2024).

Biomass polymers used as matrices in green composites are typically derived from renewable resources such as plant materials or agricultural waste. Common biomass polymer matrices include starch, cellulose, and proteins, which are valued for their biodegradability and environmental benefits. They are used to produce composites with enhanced mechanical properties suitable for various applications such as packaging, construction, and automotive parts. For example, starch-based matrices are widely used because of their abundance and ability to form films, while cellulose is preferred for its high mechanical strength and flexibility in composites.

Biomass polymers offer several advantages that make them attractive for use in green composites. They are derived from renewable resources, making them a sustainable alternative to petroleum-based polymers. Their biodegradability helps reduce environmental pollution and long-term waste accumulation. Additionally, they often have a lower carbon footprint, contributing to efforts in combating climate change. Biomass polymers can also be engineered to exhibit specific mechanical and physical properties, making them adaptable for various industrial applications, including packaging, construction, and automotive components.

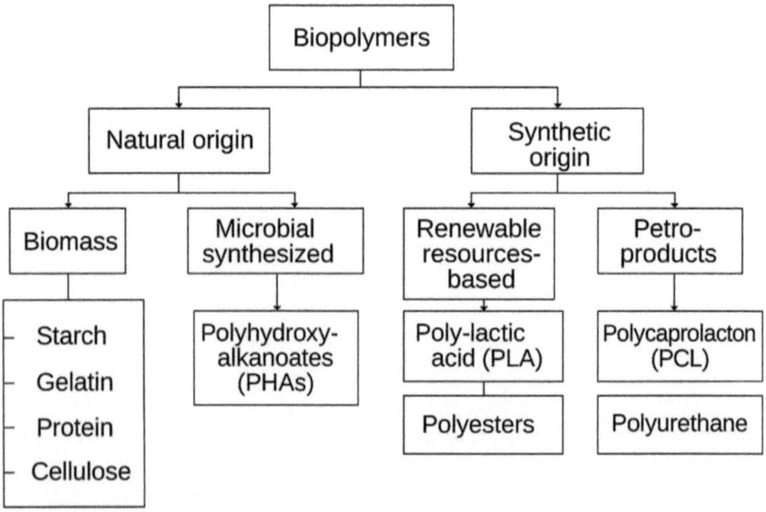

Fig. 2.1 Classification of biopolymers (Talukder et al. 2024; Ajitesh et al. 2024; Wu et al. 2024)

However, biomass polymers also present some disadvantages. They can have limited durability, with reduced resistance to moisture, heat, and UV exposure, which may restrict their use in demanding environments. Their mechanical properties, while sufficient for many applications, may not always match those of conventional synthetic polymers. Processing and manufacturing biomass-based composites can be more complex and costly due to the need for specialized techniques. Furthermore, the use of agricultural crops for polymer production can raise ethical and environmental concerns, particularly when it competes with food production and affects land use.

Microbial-based polymers, such as polyhydroxyalkanoates (PHAs) and polyhydroxybutyrate (PHB), are produced through bacterial fermentation processes and offer excellent biocompatibility, making them ideal for biomedical applications and agricultural films (Zhang et al. 2024; Bravo-Porras et al. 2024).

Microbial-based polymers offer significant advantages as sustainable materials in green composite applications. They are produced by microorganisms through fermentation processes, often using renewable feedstocks, which reduces reliance on fossil resources. These polymers, such as polyhydroxyalkanoates (PHAs) and polylactic acid (PLA), are biodegradable and compostable, making them environmentally friendly alternatives to conventional plastics. Their production can be tailored by adjusting microbial strains and fermentation conditions, allowing for the customization of polymer properties to suit specific applications in packaging, agriculture, and biomedical fields. Additionally, microbial-based polymers can contribute to a circular economy by enabling the recycling of organic waste into valuable materials.

Despite their benefits, microbial-based polymers also face certain disadvantages. Their production costs are generally higher than those of traditional synthetic polymers, primarily due to the complex fermentation and downstream processing steps involved. This can limit their competitiveness in large-scale industrial applications. Moreover, while they are biodegradable, their degradation may require specific environmental conditions, such as industrial composting facilities, which are not always widely available. In terms of performance, some microbial-based polymers may have lower mechanical strength or thermal resistance compared to petrochemical-based alternatives, which can restrict their use in high-performance or long-term applications.

Bioengineered synthetic polymers, also known as renewable resource-based polymers, are often produced using advanced biotechnological methods such as genetic engineering, recombinant DNA technology, and enzyme-catalyzed polymerization. These techniques allow the synthesis of polymers that mimic natural biopolymers but with enhanced properties tailored for specific applications. For example, polylactic acid (PLA) is one of the most common synthetic biopolymers, produced by fermenting starch or sugar using Lactobacillus species. PLA is widely used in industries such as automotive, textiles, tissue engineering, and food packaging due to its favorable properties. However, it has limitations, including brittleness, poor energy absorption, and slow degradation. Researchers have explored various modifications and blends to improve the durability and mechanical properties of PLA. Additionally, life cycle assessments have shown that PLA production uses 25–55% less energy than traditional petrochemical-based polymers, making it a more sustainable option.

Despite its benefits, ongoing research is focused on optimizing PLA's performance and expanding its commercial use.

Petrochemical-based biopolymers, such as polycaprolactone (PCL) and polybutylene adipate terephthalate (PBAT), are synthesized from fossil resources but are designed to biodegrade under specific environmental conditions. Polycaprolactone (PCL) is a versatile polymer often combined with natural materials like starch to reduce costs and improve its environmental profile. It is known for its stability and controlled degradation, making it an ideal choice for drug delivery applications. PBAT, on the other hand, is a flexible copolymer with properties similar to polyethylene, making it a popular option for compostable packaging. Despite their biodegradability, these polymers still rely on non-renewable petrochemical sources, creating a trade-off between their environmental benefits and resource sustainability.

2.5 Mechanical Properties of Biopolymers

The mechanical properties of biopolymers exhibit considerable variation contingent on their chemical structure and the conditions under which they are processed. Typically, biopolymers exhibit reduced tensile strength and modulus compared to conventional polymers. However, these properties can be enhanced through the processes of blending, cross-linking, or reinforcing with other materials. Table 2.1 and Fig. 2.2 summarize the mechanical properties of several biopolymers.

2.5.1 Mechanical Properties of Natural Origin Polymers

Gelatin hydrogels, which are commonly employed in biomedical applications, are renowned for their flexibility and high elongation at break. However, their stiffness and tensile strength are relatively low. The incorporation of reinforcing agents, such

Table 2.1 Properties of biopolymers (Taib et al. 2023; Omar et al. 2021; Rech et al. 2021; Elsayad et al. 2020; Nuvoli et al. 2021)

Biopolymer	Young's Modulus (GPa)	Tensile Strength (MPa)	Density (g/cm^3)	Elongation at Break (%)
PLA	2.7	60	1.25	4.0
PHB	3.5	40	1.2	5.0
Cellulose	15.0	100	1.5	2.0
Starch	0.01	10	1.5	2.5
Gelatin	0.05	5	1.35	3.0
PCL	0.4	16	1.15	300

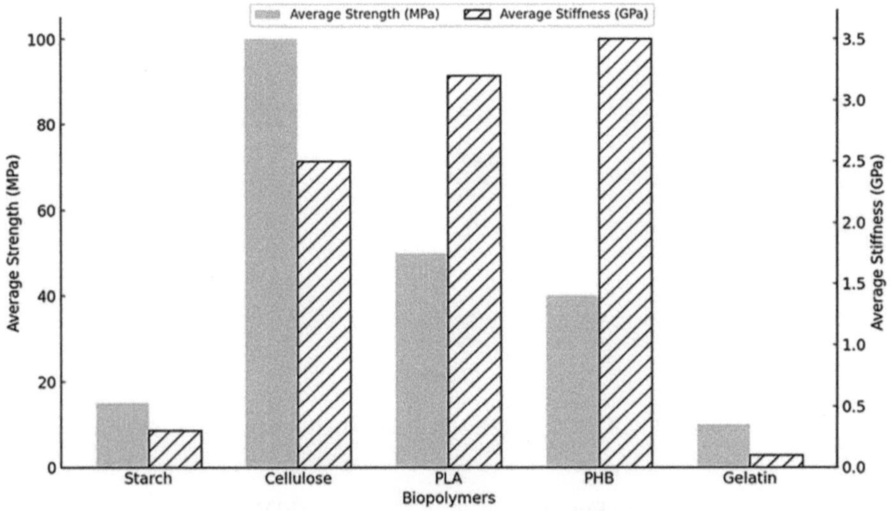

Fig. 2.2 Comparison of strength and stiffness of variable biopolymers

as chitin whiskers, has been demonstrated to markedly enhance these properties, rendering them more suitable for structural applications.

Beyond gelatin, a plethora of alternative biopolymers offer a diverse array of mechanical behaviors that can be customized for specific applications. To illustrate this point, cellulose and starch-based materials are intrinsically stiff (E = 10–15 GPa for cellulose) but brittle. However, they can be rendered more compliant by plasticization (e.g., glycerol) or copolymer blending (e.g., starch-PVA blends) to enhance elongation and toughness. Copolymers of PHAs (i.e., PHB-co-HHx) can be produced, or, alternatively, the material can be plasticized with citrate esters, thus allowing a compromise to be made between strength and ductility. In the case of PLA, toughening is often achieved through melt-blending with PCL or the addition of nanoclay and graphene oxide to raise its fracture resistance. Finally, gelatin hydrogels, which are characterized by their high elongation, exhibit significant increases in stiffness and tensile strength when reinforced with chitin whiskers, nanocellulose, or silicate nanosheets, or when their cross-link density is elevated through enzymatic or UV-mediated chemistry. Through the judicious selection of fillers, plasticizers, copolymers, and cross-linking strategies, the mechanical profile of a wide range of biopolymers can be tailored, ranging from soft and flexible to rigid and load-bearing.

Poly(lactic acid) (PLA) is a widely utilized biopolymer, renowned for its elevated tensile strength (approximately 60 MPa) and Young's modulus (ranging from 2.5 to 3.5 GPa), which render it a rigid yet fragile material. This brittleness represents a significant limitation in terms of its suitability for applications that require both flexibility and toughness. To overcome these limitations, modifications such as copolymerization or blending with plasticizers such as polycaprolactone (PCL) are commonly employed. The study conducted by Kouassi et al. (2024a, b) on

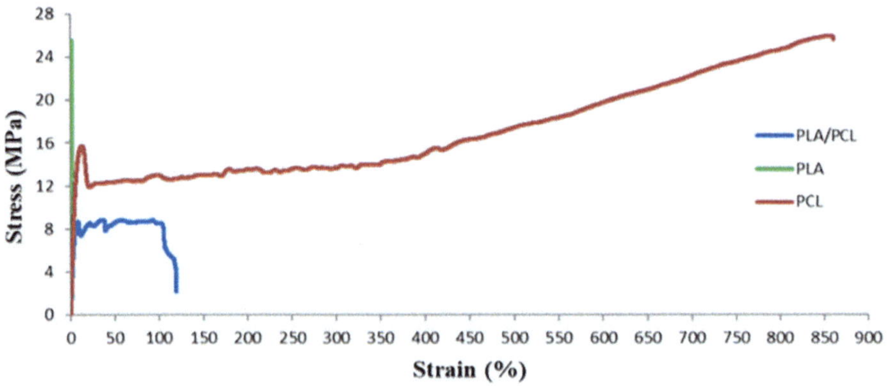

Fig. 2.3 Tensile curves of PLA, PCL, and PLA/PCL (70/30) (Kouassi et al. 2024b)

microneedle printing using Fused Deposition Modeling (FDM) demonstrated that the incorporation of 30% PCL into PLA resulted in a notable enhancement in the material's mechanical properties, as shown in Fig. 2.3. This was evidenced by an increase in elongation at break from 4% for pure PLA to 18% for the blend, thereby improving its ductility and rendering it more suitable for biomedical applications such as transdermal drug delivery. In comparison, pure PCL, with its lower tensile strength of around 20 MPa but much higher elongation at break (up to 600%), provides superior flexibility but lacks the structural integrity of PLA. However, blending PCL with PLA results in a composite material with a balanced combination of strength and flexibility, offering enhanced toughness and a reduced susceptibility to fracture under mechanical stress.

Moreover, when compared to other biopolymers such as polyhydroxybutyrate (PHB), which exhibits high stiffness but poor ductility (with an elongation at break of around 5%), the PLA/PCL blend demonstrates a superior mechanical profile for applications requiring both rigidity and flexibility.

The incorporation of PCL into the PLA matrix not only augmented the elongation at break but also enhanced the toughness, rendering the PLA/PCL composite less susceptible to fracture under mechanical stress. The study demonstrated that the optimized blend was capable of withstanding the mechanical forces encountered during microneedle penetration without cracking, a common limitation observed in pure PLA applications. Furthermore, the incorporation of PCL was observed to influence the morphology of the microneedles, as evidenced by scanning electron microscopy (SEM) images presented in Fig. 2.4. These images demonstrated that the incorporation of 30% PCL resulted in the formation of smoother needle surfaces and an improvement in the dimensional accuracy of the microneedles, which is of paramount importance for effective skin penetration and the reduction of pain sensation. To illustrate, the diameter of the microneedle tip was observed to decrease from 50 μm in pure PLA to approximately 30 μm in the PLA/PCL blend. This

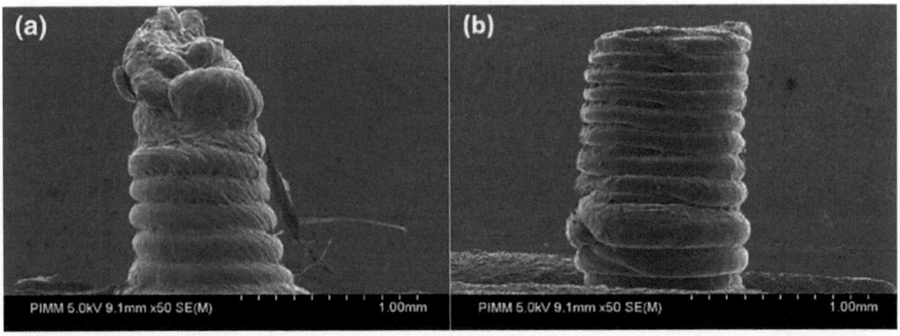

Fig. 2.4 Scanning electron micrographs of **a** PLA/PCL-based microneedles and **b** PLA-based microneedles (Kouassi et al. 2024b)

enhancement in precision and effectiveness is beneficial for transdermal delivery applications.

This example illustrates how incorporating softer polymers like PCL can transform the mechanical profile of PLA, making it more suitable for dynamic and load-bearing applications. By fine-tuning the blend ratio and processing parameters, as demonstrated in the study, researchers achieved a biocompatible and biodegradable microneedle material with superior mechanical performance and optimized geometry compared to pure PLA. Furthermore, the PCL blend reduced the brittleness index by up to 50%, indicating a substantial improvement in toughness and flexibility, making it ideal for applications requiring both strength and resilience.

2.5.2 *Mechanical Properties of Synthetic Biopolymer*

Polyhydroxyalkanoates (PHAs) and polyhydroxybutyrate (PHB), although microbially produced, are here classified as synthetic-origin biopolymers. They exhibit mechanical properties—high tensile strength and modulus—that rival those of polypropylene, though their inherent brittleness and limited processability often necessitate blending with more ductile polymers or direct chemical modification (e.g., copolymerization, grafting) to broaden their application range.

In fact, polypropylene (PP) and polyhydroxybutyrate (PHB) are two semi-crystalline polymers widely used in various applications, but their mechanical properties and crystallinity differ significantly. Polypropylene is a fossil-based polymer known for its good balance between strength, flexibility, and processability. It typically exhibits a crystallinity of 40–60%, which results in moderate stiffness (Young's Modulus of 1.5–2 GPa) and good tensile strength ranging from 30 to 40 MPa. The crystallization behavior of isotactic PP is highly influenced by processing conditions such as cooling rate and crystallization temperature, and according to Nouira et al. (2024, 2022) variations in spherulite size and structure can lead to different

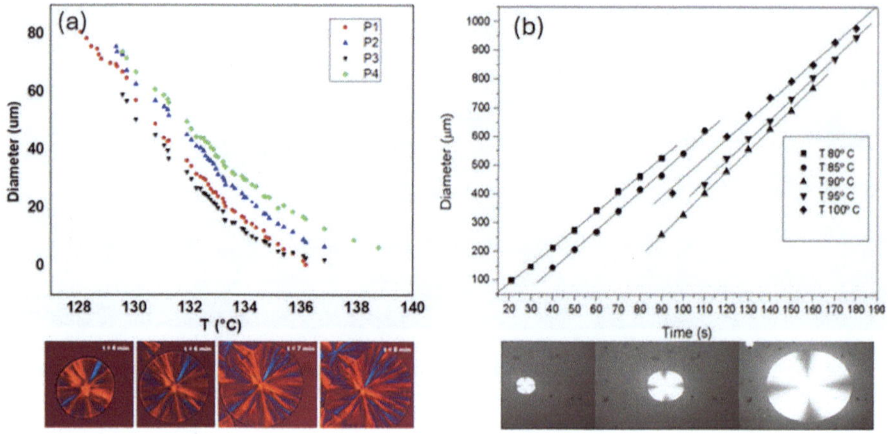

Fig. 2.5 Microstructure development of **a** PP and **b** PHB

mechanical responses, where higher crystallinity results in increased rigidity and brittleness as shown in Fig. 2.5a. The diameter of the spherulites in PP can reach up to 100 μm, depending on the conditions process. Smaller spherulites facilitate the uniform stress distribution, rendering the polymer less susceptible to catastrophic failure under tensile stress. Consequently, the principal fracture mechanism observed in PP is characterized by plastic deformation and a gradual increase in crack length, resulting in enhanced elongation at break and superior impact resistance.

In contrast, PHB displays higher crystallinity (50–80%) compared to PP, which translates into higher stiffness (Young's Modulus of 3–4 GPa) and similar tensile strength (~40 MPa). However, its increased crystallinity contributes to its inherent brittleness, resulting in a low elongation at break of 5–10%, compared to the ductility of PP, which can achieve elongations of up to 200–500%. PHB can form much larger spherulites, reaching sizes up to 900 μm, as shown in Fig. 2.5b (D'Amico et al. 2012). This larger spherulitic structure promotes more brittle fracture behavior as crack propagation through the large crystalline regions occurs more readily, reducing the material's ability to absorb energy during deformation, as shown in Fig. 2.6. Brittle fracture in PHB is further exacerbated by its higher crystallinity, which creates a dense network of crystalline lamellae, limiting chain mobility and increasing the likelihood of rapid crack propagation.

2.6 Types of Reinforcements in Bio-based Composites

Reinforcing agents—ranging from nanoparticles and whiskers to continuous fibers—are added to polymer matrices to boost stiffness, strength, and toughness without substantially increasing weight. Among these, fibers stand out for their high aspect ratio and load-bearing capacity. The classification of natural fibers is typically based

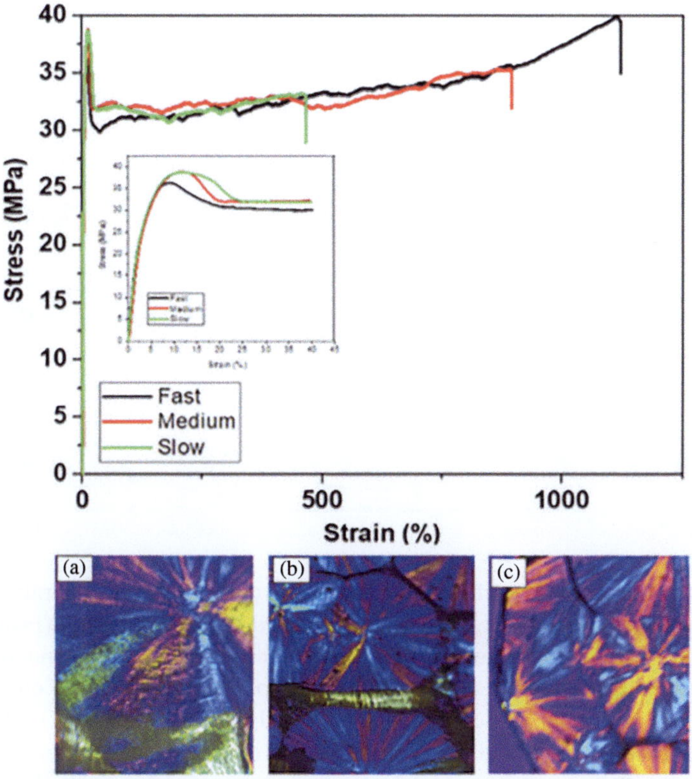

Fig. 2.6 Effect of processing parameter (cooling rate) on the rupture of PP: **a** sample after slow cooling, **b** medium, and **c** fast cooling

on their origin and can be divided into three main categories: animal, plant, and mineral fibers, as presented in Fig. 2.7. Plant-based fibers, which are primarily composed of cellulose, typically exhibit superior mechanical properties, including higher tensile strength and rigidity, compared to animal-derived fibers, which are predominantly protein-based. Animal fibers, such as wool and silk, are widely employed in the textile industry due to their elasticity, softness, and thermal insulation properties.

Plant fibers are further categorized based on their source into subfamilies, which include seed fibers (e.g., cotton), fruit fibers (e.g., coir), bark fibers (e.g., kenaf), leaf fibers (e.g., sisal), and stem fibers (e.g., flax, hemp, and jute). Cotton, a seed fiber, is the most widely produced natural fiber globally, with an annual yield of approximately 18,450 thousand tons according to the FAO (2022b). Bamboo, a stem fiber, has a production volume of around 10,000 thousand tons (FAO 2022a). Other stem fibers, including jute, kenaf, and flax, contribute approximately 2,300, 970, and 830 thousand tons, respectively (International Jute Study Group 2021; CELC, 2021).

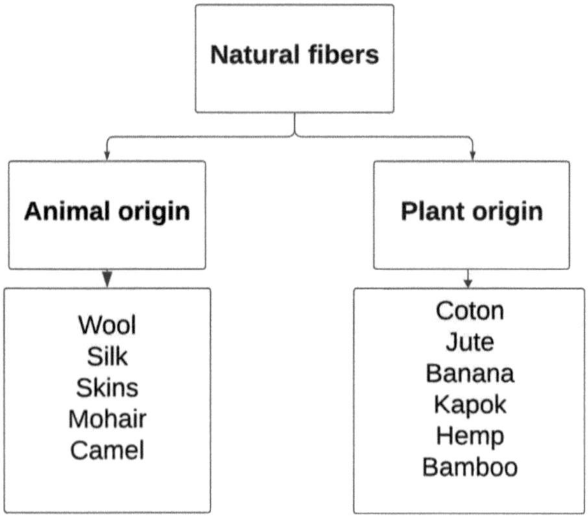

Fig. 2.7 Biofiber classification

In contrast, leaf fibers such as sisal and abaca demonstrate comparatively lower production levels, with about 378 and 70 thousand tons, respectively (FAO 2022b). Similarly, hemp, a versatile stem fiber, accounts for roughly 214 thousand tons (European Industrial Hemp Association 2022), while coir, derived from the outer husk of coconuts, has an annual production of around 100 thousand tons (FAO 2022a).

The degradation of natural fibers is influenced by several factors, including their chemical composition, environmental exposure, and processing methods. Cellulose-rich fibers like flax, hemp, and jute typically degrade more rapidly under composting or soil conditions compared to lignin-rich fibers such as coir.

2.7 Classification of Biofibers

2.7.1 Plant Origin Fibers

The primary components of plant-based biofibers are cellulose, hemicellulose, and lignin, which contribute to their mechanical properties. Cellulose, the main structural component, provides high tensile strength, while hemicellulose and lignin influence the flexibility and thermal stability of the fibers. However, the hydrophilic nature of these fibers can sometimes reduce their compatibility with hydrophobic polymer matrices, resulting in poor interfacial bonding in composites. To overcome this, chemical treatments such as silanization or acetylation are often applied to improve the surface properties of the fibers and increase their adhesion to polymer matrices. Table 2.2 presents the main applications of some common plant fibers.

Table 2.2 Applications and properties of several biofibers

Natural fiber	Primary applications	Key properties
Cotton	Textiles, apparel, medical bandages, home furnishings	Soft, high absorbency, breathable
Jute	Packaging, sacks, carpets	High strength, low cost, biodegradable
Flax	Linen fabrics, paper industry	High tensile strength, stiffness, lightweight
Hemp	Textiles, construction materials, automotive composites	Durable, high specific strength, low density
Sisal	Ropes, twine, agriculture products	High rigidity, resistant to wear and moisture

The growing interest in jute fiber-reinforced composites as an environmentally friendly alternative to synthetic fiber composites is due to their unique properties and sustainable benefits. Jute is a bast fiber, mainly grown in tropical regions, and is considered one of the strongest natural fibers available. Its high strength-to-weight ratio, biodegradability, and low density make it popular for reinforcing various polymer matrices. These composites are frequently utilized in the automotive, construction, and packaging industries, where they are employed instead of conventional materials such as glass fiber in non-structural components. Jute composites exhibit excellent mechanical properties, including high tensile strength and stiffness, which are comparable to those of glass fiber composites. Additionally, they offer the advantages of reduced weight and lower cost.

Cotton, a cellulose-based fiber, offers good tensile strength and flexibility, rendering it an appropriate reinforcement for the creation of robust yet lightweight biocomposites. These composites frequently utilize polymer matrices, including epoxy, PLA, and geopolymer (Uniyal et al. 2025). The integration of cotton fibers can markedly enhance the mechanical strength and impact resistance of the material, particularly when combined with other natural fibers, such as flax or sisal.

2.7.2 *Animal Origin Fibers*

The classification of animal fibers is typically organized into two principal categories, distinguished by their provenance: hair fibers and secretion fibers. Hair fibers encompass wool, mohair, and cashmere, which are derived from the fur of animals such as sheep, goats, and alpacas. These fibers are employed primarily in the textile industry due to their insulating properties, elasticity, and softness. Wool is a material that is widely used in the production of clothing, upholstery, and carpets due to its ability to retain heat and resist creasing. In addition to their use in textiles, some hair fibers are employed in composite materials to enhance thermal and acoustic insulation (Prabhu et al. 2025).

Secretory fibers, including silk, are produced by insects such as the silkworm and certain species of spider. Silk, composed of the proteins sericin and fibroin, is highly esteemed for its luster, smooth texture, and exceptional tensile strength. The structure of silk fibers can vary depending on the species, resulting in different mechanical properties. For example, spider silk, particularly Nephila silk, has been demonstrated to possess a tensile strength of up to 1.1 GPa, which positions it among the strongest natural fibers; mulberry silk, a material frequently utilized in the textile and biomedical industries, exhibits a tensile strength of approximately 600 MPa and thermal stability up to 250 °C, rendering it an optimal choice for applications such as sutures and medical scaffolding (Chen et al. 2023).

Chicken feathers, representing a further example of animal fiber, are primarily composed of keratin and exhibit distinctive absorbent characteristics. Due to their lightweight composition and capacity to retain air, they are employed in the manufacture of sponges for the absorption of oil and other environmental cleaning applications (Chau et al. 2025). Furthermore, chicken feather composites exhibit favorable thermal and acoustic damping properties, rendering them optimal insulation materials.

2.8 Extraction Techniques for Biofibers

The extraction of biofibers is conducted through a variety of traditional and modern techniques, each tailored to the specific fiber type and its source within the plant or animal. The three principal methods employed for the extraction of fibers are mechanical, chemical, and biological (Bavaliya et al. 2024).

- **Mechanical Extraction** is a process whereby fibers are extracted from a material using mechanical means. This method is a common means of extracting fibers, including those derived from jute, flax, and hemp. The process typically entails physical methods such as decortication, which involves subjecting the stems of the plant to crushing and beating to separate the fibers from the woody core. The utilization of mechanical equipment for decortication has the potential to markedly enhance production efficiency in comparison to manual techniques, with yields increasing by a factor of 20 to 25. Nevertheless, this approach may occasionally result in the residual accumulation of compounds such as lignin and pectin, necessitating further processing to eliminate them.
- **The Process of Chemical Extraction** involves the use of chemical solutions, such as alkalis (e.g., sodium hydroxide) or acids, to break down the binding materials that hold the fiber bundles together. These materials include pectin, hemicellulose, and lignin. In this method, chemical solutions, including alkalis (e.g., sodium hydroxide) and acids, are employed to disrupt the binding materials, such as pectin, hemicellulose, and lignin, which hold the fiber bundles together. While chemical extraction is an effective method for producing high-purity fibers,

it can also give rise to environmental concerns due to the utilization of harmful reagents and the generation of waste by-products.

- **The Process of Biological Extraction** involves the use of microorganisms and enzymes to break down non-cellulosic components. This environmentally friendly method relies on the use of microorganisms and enzymes to facilitate the breakdown of non-cellulosic components. To illustrate, enzymes such as pectinase and xylanase can efficiently degrade pectin within fiber bundles, facilitating the separation of fibers while preserving the integrity of the cellulose structure. The process of organic retting is frequently employed to extract bast fibers, including those derived from flax and hemp. Although this method is more environmentally friendly, it is slower than mechanical and chemical processes.

In general, each extraction technique has both advantages and limitations. Mechanical methods are rapid, yet they may result in the residual presence of impurities. Chemical methods yield fibers of superior quality, but they give rise to environmental concerns. Biological methods are environmentally benign, but they are often time-consuming. The selection of an appropriate technique is dependent upon the quality of fiber that is required and the intended application of the fiber.

2.9 Morphology of Natural Fibers

The primary constituents of natural fibers are cellulose, hemicellulose, lignin, and pectin, each of which plays a distinctive role in the formation and mechanical characteristics of the fiber.

Cellulose is the principal structural component of natural fibers, accounting for between 40 and 80% of their composition, as presented in Fig. 2.8. This is the primary structural component, representing between 40 and 80% of the composition of natural fibers, as presented in Fig. 2.8. Cellulose is constituted of linear chains of β-glucose molecules, which confer rigidity, tensile strength, and structural stability to the fibers (Kurien et al. 2023). Due to its crystalline structure, cellulose endows the fiber with considerable strength and stiffness, thereby constituting a pivotal determinant of the mechanical performance of biofibers.

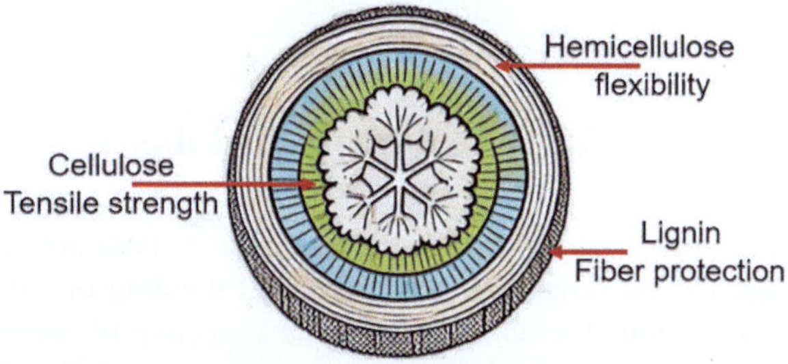

Fig. 2.8 Section cut for a natural fiber

Hemicellulose is a polysaccharide that is found in the secondary cell wall of plants. It is a heteropolysaccharide comprising a backbone of β-1,4-linked D-glucose units with various side chains. Hemicellulose is a branched polymer that forms a matrix around cellulose fibers and is found in smaller proportions (typically 15–30%). The hydrophilic nature of hemicellulose renders the fibers more susceptible to moisture absorption than cellulose. Hemicellulose plays a role in determining the flexibility and thermal properties of the fiber, which in turn affects its thermal stability and degradation characteristics.

Lignin is a natural polymer that is found in the cell walls of plants. It represents between 5 and 30% of the fiber's composition and acts as a natural binder, filling the spaces between cellulose and hemicellulose. Lignin constitutes between 5 and 30% of the fiber's composition and serves as a natural binder, filling the spaces between cellulose and hemicellulose. Lignin endows the fiber with rigidity and protection, while also contributing to its resistance to microbial degradation and environmental stress. Due to its amorphous structure, lignin enhances the impact resistance and rigidity of natural fibers.

The characteristics of natural fibers vary depending on their classification, including bast fibers (such as flax and jute), leaf fibers (such as sisal), seed fibers (such as cotton), and animal fibers (such as wool and silk). The primary morphological characteristics of natural fibers are length, diameter, surface texture, and microfibril orientation, all of which influence their mechanical properties and suitability for diverse applications. Natural fibers can be classified according to their length, with short fibers including those derived from cotton and wool, and long fibers from jute and linen. The length of the fiber is a critical factor influencing the strength and quality of fiber-reinforced composites. The use of longer fibers typically results in the production of stronger and more durable composites. Natural fibers generally vary in diameter from 10 to 50 microns. Finer fibers, such as silk, tend to be softer and more flexible, while thicker fibers, such as jute and sisal, are stiffer and more suitable for heavy-duty applications. Uniformity of fiber diameter is also important to ensure consistent mechanical performance in composites.

Natural fibers have a rough surface texture with voids and pits along their length. This roughness increases the surface area available for bonding with matrix materials in composites, but can also result in poor interfacial adhesion if not properly treated. Chemical treatments such as alkalization are often used to modify the surface texture and improve fiber-matrix bonding.

2.10 Surface Treatments for Fiber Enhancement

The utilization of surface treatments is of paramount importance in enhancing the mechanical properties and interfacial bonding of natural fibers when employed as reinforcement in composites. The objective of these treatments is to modify the surface of the fiber, thereby improving wettability, increasing surface roughness, and reducing moisture absorption. This contributes to enhanced compatibility with hydrophobic polymer matrices. Physical treatments, such as plasma and corona treatments, modify the surface morphology of the fiber without altering its chemical structure, thereby rendering the surface more reactive and improving mechanical interlock. For instance, plasma treatment introduces polar functional groups to the fiber surface, which enhances fiber-matrix adhesion and results in composites that are more resistant to traction and impact (Samant et al. 2023; Wang et al. 2023).

- **Chemical treatments** are pervasive due to their direct impact on fiber composition. The most prevalent method is alkaline treatment, which employs sodium hydroxide (NaOH) to eliminate surface impurities and non-cellulosic components, including lignin and hemicellulose. This process enhances the binding surface and reveals a greater number of reactive sites on the cellulose. The aforementioned treatment has the potential to enhance tensile strength and flexibility while simultaneously reducing moisture absorption. Other chemical methods, such as silanization and acetylation treatments, modify the hydrophilic nature of the fiber, creating a more hydrophobic surface that is compatible with polymer matrices such as polypropylene or epoxy. Silane, in particular, forms covalent bonds with both the fiber and the matrix, thereby greatly improving interfacial adhesion (Fig. 2.9).
- **Physico-chemical treatments**, which entail the combination of physical and chemical processes, encompass techniques such as heat treatment and ultraviolet (UV) radiation. These methods simultaneously modify the surface morphology and chemical composition, thereby improving thermal stability, reducing degradation, and enhancing the mechanical properties of the fibers. Such treatments are especially beneficial for applications that necessitate enhanced durability and long-term functionality in challenging environmental circumstances.
- **Biological surface treatments**, which employ natural agents such as enzymes and fungi, present an environmentally benign alternative to the conventional chemical and physical methodologies employed in the modification of natural fibers. These

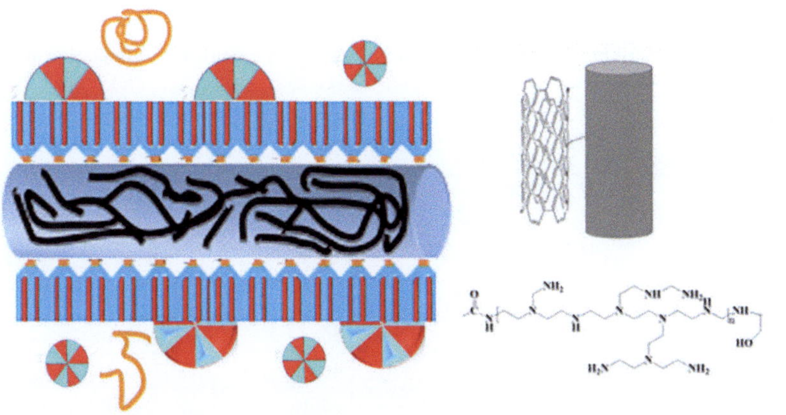

Fig. 2.9 Presentation of surface treatment

treatments selectively remove non-cellulosic components, such as hemicellulose and pectin, resulting in the production of fibers with a cleaner, finer composition and improved mechanical and interfacial properties. It has been demonstrated that enzymes such as esterase and lipase can enhance the surface texture and fineness of fibers. Microbial treatments utilizing fungi such as Aspergillus niger have also been shown to refine fiber surfaces without damaging the cellulose core. The utilization of biological methods not only enhances the adhesion and crystallinity of fibers but also necessitates the consumption of less energy and the production of fewer environmental pollutants than conventional techniques. Consequently, these methods are optimal for the sustainable processing of biofibers.

The selection of suitable surface treatments enables natural fiber composites to attain enhanced mechanical characteristics, resistance to moisture, and interfacial adhesion, thereby rendering them competitive with synthetic fiber composites in a multitude of industrial applications.

2.11 Mechanical Properties of Biofibers

The mechanical properties of natural fibers vary considerably depending on their composition, structure, and the environmental conditions under which they grow. The key characteristics of natural fibers, including their density, tensile strength, Young's modulus, and elongation at break, are of great importance in determining their suitability for a range of applications. Fibers with high tensile strength and Young's modulus, such as flax and hemp, are frequently selected for structural applications due to their stiffness, whereas softer fibers, such as cotton and wool, are preferred in textiles due to their flexibility. Table 2.3 provides a comparative overview of these properties for ten common natural fibers.

Table 2.3 Comparison of key properties for ten commonly used natural fibers

Fiber	Density (g/cm^3)	Tensile strength (MPa)	Young's modulus (GPa)	Elongation at break (%)
Cotton	1.54 ± 0.02	400	5.5 ± 0.5	8.0
Jute	1.46 ± 0.03	393 ± 10	26.5	1.8
Flax	1.50	500	60 ± 2	2.7
Hemp	1.48	550	70	1.6
Sisal	1.45 ± 0.05	600 ± 15	34 ± 1.5	2.0
Coir	1.20 ± 0.03	175 ± 5	4–6	30.0
Kenaf	1.29	350	35	2.4
Bamboo	1.10 ± 0.05	500	40	3.0
Ramie	1.50	938	44 ± 3	3.0
Wool	1.30 ± 0.04	180 ± 8	3.0	35.0

2.12 Green Composite Manufacturing Processes

Biocomposite processing involves the transformation of raw materials, such as natural fibers and resins, into composites with specific shapes, sizes, and mechanical properties. The choice of an appropriate manufacturing process plays a crucial role in determining the final performance of the biocomposite. Many of the techniques traditionally used for synthetic fiber-reinforced composites, such as extrusion, injection, compression molding, and filament winding, are also applicable to biofiber-reinforced composites, particularly those based on thermoplastics. For thermoset-based biocomposites, processes such as resin transfer molding (RTM) and sheet molding (SMC) are commonly used. Each processing method must be carefully selected according to several factors, including the desired mechanical properties, dimensions, production rate, ease of handling, and overall cost. Optimization of these parameters is essential to ensuring efficient and sustainable production of high-performance biocomposites.

2.12.1 Conventional Techniques

The conventional processing techniques for biocomposites are analogous to those employed for synthetic fiber-reinforced composites, with methodologies adapted to the distinctive attributes of natural fibers. The principal techniques encompass compression molding, injection molding, extrusion, filament winding, resin transfer molding (RTM), and manual lamination. Each technique possesses distinctive advantages and applies to diverse types of matrix and fiber configurations.

The wet/manual lamination process is a straightforward and prevalent method of producing biocomposites, largely due to its simplicity and relatively low cost. The open mold manufacturing technique involves the manual deposition of fibers and resin into a mold until the desired shape and thickness are achieved. Reinforcing materials, such as layers of woven, knitted, or sewn fabric, are impregnated with liquid resin using a brush or roller in order to ensure complete wetting and bonding. The process is well-suited to the production of large, complex shapes with low to medium production volumes, rendering it an optimal choice for applications in sectors such as wind turbine blades, automotive parts, and marine systems.

Filament winding is a manufacturing process utilized for the production of cylindrical or round composite structures, including tanks, pipes, and pressure vessels, which are required to possess a high degree of structural integrity. The process entails the winding of resin-impregnated fibers, such as glass or carbon, around a rotating mandrel in a predetermined pattern to optimize the strength of the resulting structure in the desired directions. This technique results in a high fiber load and excellent mechanical properties, rendering it suitable for the production of components such as yacht masts, large-diameter tanks, and natural gas cylinders.

2.12.2 *Innovative Techniques for Manufacturing*

The majority of conventional manufacturing techniques necessitate the utilization of molds, which can impede design flexibility and augment manufacturing costs. Conversely, additive manufacturing (AM), also designated as 3D printing, facilitates the generation of composite structures in a layer-by-layer fashion through the employment of a computer-aided design (CAD) process. This approach obviates the necessity for molds, enables the production of more intricate geometries, and curtails material wastage and production time, thereby rendering it a cost-effective and versatile alternative for the manufacture of biocomposite materials.

Among these methods, FFF is widely used for biocomposites as it offers an efficient and low-cost means of producing parts with minimal waste. The versatility of AM has facilitated the incorporation of a range of biological fillers and fibers, resulting in the development of biocomposites with tailored properties, including enhanced mechanical strength, thermal resistance and biodegradability. Innovations such as multi-material printing, the incorporation of nanofillers, and surface modifications (using silane or coupling agents) have further enhanced the performance of biocomposites, making them a competitive alternative to synthetic composites.

Furthermore, the development of novel raw materials for AM, including biopolymer filaments for FFF and photocurable resins for SLA, has facilitated the fabrication of lightweight, site-specific biocomposites. Such flexibility is essential for applications in the automotive, biomedical, and aerospace sectors, which require the production of complex shapes, multifunctional properties, and a reduced environmental impact. These developments in the field of additive manufacturing (AM) for biocomposites not only enhance the durability of the resulting products but also facilitate the utilization of bio-based materials in high-performance sectors.

2.13 Impact of Processing Parameters on the Quality and Properties of the Final Composite

The intrinsic variability and low thermal stability of bio-based raw materials present significant challenges in manufacturing biocomposites, as most conventional processing techniques have been optimized for synthetic fibers. It is therefore essential to pre-treat or modify these materials in order to improve their processability and the quality of the final composite. It is crucial to carefully controlling key factors such as thermal stability, hydrophilicity, fiber type, distribution, and content to achieve the desired mechanical and physical properties in the final product and ensure optimum performance during conventional manufacturing processes.

2.13.1 Thermal Stability

Thermal stability represents a pivotal attribute of biocomposites, exhibiting a pronounced dependency on the crystallinity and microstructure of the matrix. In highly crystalline matrices, the ordered arrangement of polymer chains enhances thermal resistance, thereby rendering the material more stable at elevated temperatures. Conversely, matrices with low crystallinity or an amorphous structure tend to exhibit reduced thermal stability, as the irregular structure is more susceptible to thermal degradation. The microstructure, including the interaction and distribution of fibers within the matrix, also exerts an influence on stability. Well-bonded interfaces offer greater resistance to thermal stress and prevent premature degradation.

2.13.2 Fiber Type

The specific type of fiber and the characteristics of the matrix material collectively influence the microstructure and performance of biocomposites. The incorporation of fibers with high crystallinity and well-organized microfibrillar structures, such as flax and ramie, contributes to the overall stiffness, tensile strength, and thermal stability of the composite by promoting the formation of a more ordered crystalline network within the matrix. Similarly, the crystallinity and molecular arrangement of the polymer matrix are of significant importance in defining the mechanical properties of the composite. A highly crystalline matrix can enhance the rigidity and thermal resistance of the composite as a whole, rendering it more suitable for high-temperature processes such as compression molding. Conversely, the utilization of low crystallinity fibers, such as coir, in conjunction with a more amorphous matrix structure, gives rise to the formation of a composite that is comparatively softer and less thermally stable. Such combinations are more compatible with low-temperature processing techniques, such as manual lamination or resin transfer molding, to avoid thermal degradation and preserve fiber integrity. It is therefore essential to optimizing the combination of fiber and matrix types and to select appropriate processing methods in order to achieve the desired mechanical and thermal performance in biocomposites. The appropriate alignment of fiber and matrix characteristics ensures enhanced compatibility and structural integrity under diverse processing conditions.

To assess the effect of fiber type in a composite, a comparison between the studies of Nouira et al. and Bouzouita (2011), as seen in Fig. 2.10, reveals interesting differences in the crystallization and nucleation behavior around synthetic and natural fibers in the same polypropylene (PP) matrix. Nouira et al. used synthetic glass fibers as reinforcement, which act as effective nucleating agents for the polypropylene matrix. The presence of these fibers promotes the formation of small, uniformly distributed spherulites, which leads to a higher rate of crystallization around the surface of the fibers. The result is a well-ordered crystalline structure that improves

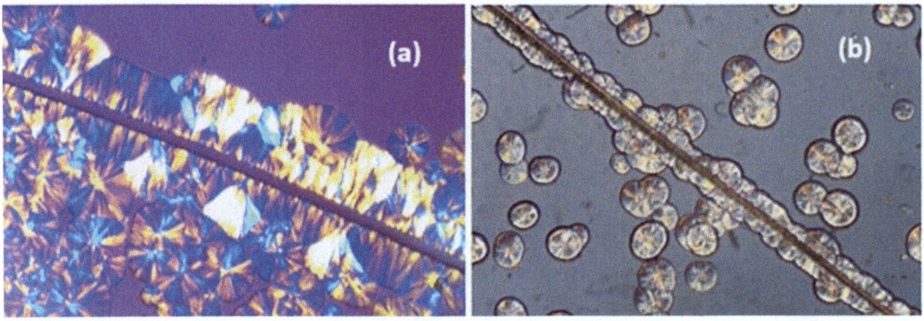

Fig. 2.10 Matrix-fiber interface: **a** PP reinforce with glass fiber; **b** PP reinforced with plant fiber

the stiffness and strength of the composite, thanks to the strong interfacial interaction between the glass fibers and the PP matrix.

In contrast, Bouzouita et al. employed the use of natural fibers, such as flax or hemp, within the same PP matrix. The nucleation efficiency of natural fibers is inferior to that of glass fibers. This results in the formation of larger and less uniform spherulites around the fibers. This effect is attributable, at least in part, to the presence of hemicellulose and lignin, which disrupt the regular crystalline structure of the polymer chains. The formation of larger spherulites consequently diminishes the composite's mechanical properties, including tensile strength and stiffness. Conversely, these larger spherulites enhance the composite's flexibility and impact resistance.

However, Bouzouita's study indicates that surface treatments, such as alkali treatment or silane coupling agents, can be employed to modify the surface properties of natural fibers, thereby enhancing fiber-matrix interfacial adhesion. This enhanced interfacial adhesion facilitates a more uniform nucleation and growth pattern, akin to that observed in synthetic fiber-reinforced composites. Consequently, the mechanical properties attained are analogous to those exhibited by synthetic fibers. Consequently, through the implementation of suitable surface modifications, natural fiber-reinforced PP composites are capable of attaining comparable levels of performance to their synthetic counterparts, thereby rendering them a more resilient alternative for a multitude of applications.

2.13.3 Fiber Orientation

The orientation and distribution of fibers in a biocomposite have a significant impact on the composite's overall mechanical properties, including tensile strength, stiffness, and impact resistance. When fibers are aligned in a direction parallel to that of the applied load, the composite exhibits significantly greater strength and stiffness than when fibers are oriented randomly or perpendicularly. This is due to the fact that aligned fibers are better able to transfer and distribute the load within the matrix, thereby improving the composite's ability to withstand stress. Conversely, random fiber orientations tend to absorb impact energy more effectively, making the composite more resilient but less rigid.

As an illustration, in composites reinforced with natural fibers, such as lyocell or cellulose, the longitudinal alignment of the fibers enhances the tensile strength properties in comparison to composites where the fibers are oriented at $\pm 45°$ or $90°$. It has been demonstrated that variations of up to 50% in ultimate tensile strength may occur as a function of the angle of fiber orientation in relation to the direction of loading (Shirinbayan et al. 2024).

The selection of fiber orientation is inextricably linked to the manufacturing process. Techniques such as filament winding and automated fiber placement facilitate precise fiber alignment in high-performance applications, whereas methods such as injection molding often result in less controlled fiber distribution. To illustrate, composites manufactured via thermoforming or hot-pressing with hybrid yarn structures demonstrate a more uniform wetting and distribution of fibers within the matrix, thereby enhancing the transmission of load between the fibers and the polymer. As demonstrated in the study by Nouira et al., the application of a tensile load perpendicular to the fiber orientation results in the rapid failure and delamination of the fibers and matrix. This is due to inadequate load transfer between the matrix and fibers, which results in premature separation at the interface and ultimately leads to delamination. Conversely, when a tensile load is applied in alignment with the fibers, the fibers themselves assume the primary load, which can result in the progressive fragmentation of the fibers while maintaining the overall cohesion of the composite. This alignment enables the load to be distributed efficiently, thereby enhancing the material's capacity to withstand elevated mechanical stresses and postponing the advent of complete failure. The distinction in behavior as a function of load orientation underscores the pivotal significance of regulating fiber alignment and distribution to optimize the mechanical performance of biocomposites.

2.13.4 Fiber Diameter

The diameter of fibers plays a crucial role in determining the mechanical properties of biocomposites, particularly in terms of interfacial bonding between the matrix and reinforcement. Finer fibers, typically in the range of 10–20 μm, tend to exhibit higher surface area-to-volume ratios, which can significantly enhance the interfacial bonding between the fibers and the polymer matrix. This improved bonding is often reflected in the increased mechanical strength, stiffness, and overall performance of the composite material. For instance, as seen in Table 2.4, biocomposites with fiber diameters of 10–20 μm show a marked improvement in tensile strength and flexural modulus compared to composites with larger fiber diameters (e.g., 30–50 μm). Specifically, biocomposites reinforced with finer fibers demonstrated a 15% increase in tensile strength and a 20% improvement in flexural modulus, compared to those with coarser fibers.

Table 2.4 Mechanical properties of biocomposites (Vilaseca et al. 2007; Wang et al. 2017; Mukherjee and Kao 2011; Sharma et al. 2019)

Type of composite	Processing technique	Tensile strength (MPa)	Young's modulus (GPa)	Elongation at break (%)
Starch + 30% jute	Injection molding	26.3 ± 0.55	2.5 ± 0.23	2 ± 0.2
PLA + 30% ramie	Hot-pressing sheet molding	66.8 ± 1.7	–	4.8 ± 0.2
PLA + 30% jute	Injection molding	81.9 ± 2.9	9.6 ± 0.36	1.8
PHBV + 30% jute	Injection molding	35.2 ± 1.3	7 ± 0.26	0.8
PLLA + 30% flax	Compression molding	98 ± 12	9.5 ± 0.5	2.3 ± 0.2
PHB + 30% flax	Compression molding	40 ± 2.5	4.7 ± 0.3	7 ± 1.5
PLA + 30% flax	Twin-screw extruder + compression molding	53 ± 3.1	8.3 ± 0.6	1 ± 0.2
PLA + 30% abaca	Film stacking	74	5.84	–
PLA + 40% jute	Film stacking	100.5	9.4	1.6
PLA + 30% man-made cellulose	Injection molding	92	8.032	–
PLA + 30 vol.% denim fabric	Hand layup	75.76	4.65	–
PLA + lyocell	–	81.8	6.78	4.09
PLA + 5% silk	Injection molding	62	4.2	–
PLA + 10% sugar beet pulp	Compression molding	37.5 ± 0.5	1.043 ± 0.0395	6.7 ± 0.5

The finer fibers enhance the matrix-fiber interaction, allowing the load to be more efficiently transferred from the matrix to the fibers under stress, which results in better mechanical performance. Additionally, the smaller diameter fibers facilitate more uniform dispersion within the matrix, reducing the occurrence of voids or weak spots that could otherwise compromise the composite's structural integrity. This explains the improved durability and impact resistance observed in biocomposites with finer fiber diameters, as shown by the data in Table 2.4. Overall, optimizing fiber diameter is a key factor in tailoring biocomposite properties for specific applications, especially those that require superior mechanical performance and durability.

2.13.5 Mechanical Properties of Biocomposite

The mechanical properties of biocomposites are of critical importance and are largely dependent on the characteristics of reinforcing fibers, the polymer matrix, and the fiber/matrix interfacial adhesion. An understanding of these properties, in conjunction with the utilization of appropriate characterization techniques, is imperative to fully actualize the potential of biocomposites across a multitude of applications. Table 2.4 provides a summary of the mechanical properties of several biocomposites, offering insight into their performance and suitability for specific applications. By evaluating these properties, researchers and manufacturers can enhance the strength, stiffness, and durability of biocomposites following their intended applications.

The mechanical properties of biocomposites are critically important and largely determined by the characteristics of the reinforcing fibers, the polymer matrix, and the fiber/matrix interfacial adhesion. A comprehensive understanding of these properties, combined with the use of appropriate characterization techniques, is essential for maximizing the performance of biocomposites in various applications. The table provided offers a summary of the mechanical properties of several biocomposites, offering valuable insights into their performance and suitability for specific uses. By thoroughly analyzing these properties, researchers and manufacturers can optimize the strength, stiffness, and durability of biocomposites to meet the demands of their intended applications.

Tensile properties are a primary consideration for structural applications of biocomposites. Tensile strength, Young's modulus, and elongation at break are influenced by several factors, including fiber type, matrix material, fiber/matrix adhesion, and processing conditions. The inclusion of natural fibers such as sisal, flax, and hemp in a polypropylene (PP) matrix has been shown to significantly enhance the tensile strength and stiffness of the resulting composite. However, tensile performance is highly sensitive to variables such as fiber alignment, fiber volume fraction, and chemical treatments. Optimal chemical modifications, such as alkalization and esterification, can improve fiber-matrix adhesion, thereby increasing tensile strength and reducing water absorption. By controlling these parameters, the mechanical properties of biocomposites can be fine-tuned to rival or even surpass those of conventional synthetic fiber composites.

Flexural properties are typically assessed through three-point bending tests, which measure the material's stiffness and resistance to deformation. The flexural response is significantly affected by factors such as fiber orientation, stacking sequence, fiber waviness, and the presence of voids. Flexural properties vary widely across thermoplastic-based biocomposites depending on the polymer type and fiber loading. For example, composites made from kenaf/PLA, jute/PP, and hemp/epoxy show excellent flexural strength and stiffness due to strong interfacial bonding and good fiber-matrix compatibility. Thermoset-based biocomposites, such as flax/epoxy and banana/polyester, generally offer superior flexural performance compared to thermoplastic matrices, which is attributed to better load distribution and reduced fiber pull-out under flexural stress.

The research on **fatigue properties** of green composites is relatively limited compared to the extensive studies on their static mechanical properties. Natural fibers like hemp, sisal, and coir have shown comparable fatigue resistance to synthetic fiber composites when treated with coupling agents such as silane. Fatigue behavior is primarily influenced by fiber orientation, interfacial adhesion, and fiber volume fraction. In some cases, the fatigue strength of treated natural fiber composites can even exceed that of glass fiber composites, demonstrating their potential for applications requiring resistance to dynamic loading.

By understanding and optimizing these mechanical properties, biocomposites can be tailored for a wide range of applications, from structural components to impact-resistant materials. Though the performance of biocomposites may not always match that of synthetic fiber-reinforced composites, they present unique advantages in terms of sustainability, weight reduction, and environmental impact. With ongoing research and the development of advanced chemical treatments and hybridization techniques, green composites continue to close the gap in mechanical performance, making them an increasingly viable alternative to traditional composite materials.

2.13.6 Degradation and Durability of Green Composites

The durability of green composites is a crucial factor that determines their long-term performance and suitability for various applications. Durability refers to a material's ability to resist degradation when exposed to environmental stressors such as weathering, moisture, UV radiation, chemical attack, and mechanical wear. The performance of green composites over time is heavily influenced by the nature of the fibers, matrix properties, fiber-matrix interfacial adhesion, and processing conditions.

A study conducted by Bouzouita et al. investigated the durability of PP/hemp composites under various environmental conditions, including different temperatures, humid atmospheres, and immersion in distilled water. Water and moisture absorption kinetics were monitored over three months. Both virgin and aged composites were characterized using static three-point bending tests and dynamic mechanical analysis (DMA). The results showed that for virgin materials, enhancements at the fiber-matrix interface led to improved overall mechanical and hygroscopic properties. Composites impregnated using the Fibroline technique without prior drying exhibited the highest mechanical performance, with a tensile strength of 68.7 MPa and a modulus of 5.4 GPa, as well as superior hygroscopic behavior, characterized by lower mass gain and delayed onset of degradation. In comparison, composites processed by compression molding showed slightly lower mechanical performance, with a tensile strength of 62.3 MPa and a modulus of 4.8 GPa, indicating that the Fibroline method provided a more effective fiber-matrix integration and better moisture resistance.

Moisture absorption kinetics confirmed the high sensitivity of natural fiber-based composites to water. Temperature exacerbated degradation phenomena, accelerating the aging process. The study found that immersion conditions were more severe compared to those at relative humidity (80% RH), resulting in higher mass gains. Flexural moduli and rupture stresses decreased significantly, depending on the amount of water absorbed and the extent of degradation. Overall, the findings highlight the importance of controlling environmental factors to ensure the long-term durability of natural fiber-reinforced composites.

2.14 Applications of Green Composites

Biocomposites are fast emerging as a viable alternative to metal, conventional fiber-reinforced composites, and ceramic-based materials in a wide range of industries, including biomedical engineering, aerospace, automotive, marine, sports, packaging, and electronics. In addition, biocomposites are used in specialist applications such as coatings, packaging, electrical components, medical devices, magnetic and optical systems, and electromagnetic shielding. This section gives an overview of the various applications of biocomposites in different sectors.

- **Automotive industry** has been a principal catalyst for the uptake of green composites. Natural fiber-reinforced biocomposites are employed extensively in automotive applications, including interior and exterior car components such as door panels, dashboards, seat backs, headliners, and storage shelves. The principal advantages of utilizing green composites in automotive applications are the reduction of weight, the enhancement of fuel efficiency, and the optimization of acoustic properties in comparison to traditional materials. For instance, hemp/PP and flax/epoxy composites are employed in the manufacture of lightweight interior components, thereby contributing to a reduction in vehicle emissions and an improvement in durability. Furthermore, green composites exhibit excellent vibration-damping properties, rendering them suitable for the reduction of noise in vehicle interiors (Ullah et al. 2025).
- **Construction Industry**: Green composites are being employed as sustainable alternatives to synthetic composites for applications including wall panels, roofing sheets, insulation boards, and decking materials. The high strength-to-weight ratio of natural fiber composites, coupled with their biodegradability and aesthetic appeal, renders them optimal for structural and decorative applications. For example, jute/epoxy and kenaf/PLA composites have been successfully employed in load-bearing structures and as reinforcing elements in concrete. Additionally, green composites are employed in the production of environmentally friendly cementitious materials, which serve to mitigate the environmental impact of conventional construction materials. Also, incorporating natural fibers like hemp into biocomposites has been shown to decrease the carbon footprint of construction materials by 25 to 40% relative to traditional options. Furthermore, a polyamide-starch biocomposite revealed that increasing the starch content to 70 wt% led to a 42–43% reduction in total carbon footprint compared to the pure polyamide matrix (Äkräs et al. 2024).

- **Packaging industry** is witnessing the emergence of green composites as an available alternative to conventional plastic materials. These materials are employed in the fabrication of biodegradable films, containers, and trays. Composites comprising starch, PLA, or PHB reinforced with cellulose or natural fibers represent a sustainable option for food packaging, offering biodegradability, a low environmental impact, and sufficient barrier properties. The utilization of these materials serves to diminish reliance on petroleum-based plastics, thereby contributing to the reduction of plastic waste. Recent developments have concentrated on enhancing the barrier properties of green composites to prevent the permeation of moisture and gases, thereby rendering them more competitive in high-performance packaging applications.
- **Sporting Goods and Leisure Equipment**: Green composites are also used in the production of sports and leisure equipment, including bicycles, surfboards, tennis rackets, and skis. Their excellent mechanical properties, low weight, and environmental benefits make them ideal for products where performance and durability are key considerations. For example, flax/epoxy and hemp/PLA composites are used in the manufacture of lightweight bicycles and surfboards, providing the necessary strength and flexibility. In addition, green composites are used in leisure products such as furniture and toys, offering an environmentally friendly alternative to conventional materials.
- **Aerospace and Marine Sectors**: The potential of green composites for non-structural applications is being investigated due to their favorable weight-to-strength ratio and environmentally beneficial properties. Components such as interior panels, seat frames, and non-load-bearing structures are manufactured using natural fiber-reinforced composites. The primary challenge in these sectors is to achieve the requisite mechanical performance and durability. This is why hybrid composites (for example, combining natural fibers with carbon or glass fibers) are frequently employed as a means of enhancing mechanical properties while simultaneously reducing the environmental impact (Ullah et al. 2025).
- **Biomedical Applications**: The exploration of green composites for medical and biomedical applications is driven by their biocompatibility, biodegradability, and non-toxic nature. For instance, PLA- and PHB-based composites reinforced with natural fibers are employed in the fabrication of surgical implants, bone fixation devices, and drug delivery systems. These materials can be designed to degrade harmlessly in the human body over time, obviating the necessity for surgical removal. Furthermore, natural fiber composites are utilized in the production of prostheses and orthoses due to their low weight and ease of customization.

2.15 Conclusions

This chapter provided a comprehensive analysis of green composites, covering their mechanical properties, durability, environmental impact, processing techniques, and diverse industrial applications. The main objective was to explore how natural fibers, when combined with biodegradable and non-biodegradable matrices, can serve as sustainable alternatives to conventional composites reinforced with synthetic fibers such as glass or carbon. Due to their renewable origin, biodegradability, low environmental footprint, and cost-effectiveness, green composites present significant advantages, making them ideal candidates for various engineering applications. In recent years, research efforts have been focused on enhancing the performance of biocomposites based on both thermoplastic and thermoset matrices, leading to substantial progress in sectors with high demand, such as civil engineering and automotive manufacturing. This advancement is attributed to a deeper understanding of material extraction methods, fiber modification techniques, and interfacial treatments, which have collectively improved the mechanical properties, durability, and overall functionality of these materials. For example, in the automotive industry, biocomposites are increasingly used to manufacture lightweight, impact-resistant interior components, thereby reducing vehicle weight and improving fuel efficiency. Similarly, in civil engineering, high-strength biocomposites are being developed for load-bearing structures, offering a sustainable alternative to conventional construction materials. These developments have been further supported by innovations in surface treatment techniques, such as salinization and alkalization, which improve fiber-matrix bonding, resulting in enhanced tensile strength, moisture resistance, and long-term stability. Consequently, biocomposites are becoming more competitive with synthetic composites, paving the way for broader adoption in industrial applications that demand high performance and sustainability.

The mechanical performance of biocomposites is significantly influenced by several factors, including fiber architecture, stacking sequence, interfacial adhesion, fiber hybridization, and processing conditions. A deeper understanding of fiber morphology is essential for optimizing fiber dispersion and enhancing interfacial bonding. Therefore, comprehensive studies on the mechanical behavior and surface modification of individual fibers are crucial for fully leveraging the potential of biocomposites.

A comparative analysis of commonly used biofibers such as jute, flax, sisal, and hemp has revealed substantial potential for incorporating novel biofibers into biopolymer matrices, thereby expanding the range of biocomposite applications. The integration of new biofibers not only optimizes the manufacturing process but also diversifies the scope of applications in various consumer products, contributing to greater industrial utilization.

Future advancements in biocomposites should focus on developing new, cost-effective biopolymers with superior mechanical properties, enhanced thermal stability, improved recyclability, and increased stability during transport and storage. Addressing these aspects is essential for enabling biocomposites to compete more effectively with synthetic materials. Currently, conventional biopolymers such as PHBV and PHA face limitations in terms of low mechanical performance and high production costs, which restrict their widespread implementation. However, some underutilized biopolymers have shown promising properties, with tensile moduli surpassing those of synthetic counterparts, suggesting their potential to bridge the gap between synthetic and biodegradable matrices.

As research continues to evolve, the development of biopolymer-based products with improved dimensional stability, moisture resistance, durability, and fire resistance will enable their integration into a broader range of sectors. A key breakthrough in this field has been the incorporation of nanotechnology to enhance the properties of biocomposites. For instance, nanocrystalline cellulose, a material stronger than steel and stiffer than aluminum, can be incorporated into polymer matrices to produce composites with exceptional strength, rigidity, and environmental sustainability. This innovative approach opens new possibilities for overcoming the current limitations of biopolymers, paving the way for high-performance biocomposites that rival traditional materials in both structural and functional applications.

However, despite these advances, key challenges remain in terms of scalability and cost. Large-scale production of biocomposites is often hampered by inconsistent raw material quality, limited infrastructure, and the need for process optimization, all of which contribute to higher manufacturing costs compared to conventional composites. Recent techno-economic analyses have shown that while natural fibers and biopolymers are renewable and potentially lower in cost per unit mass, overall production costs can be 20–40% higher due to labor-intensive processing, variable feedstock supply, and less mature production technologies. For example, a techno-economic study based on a 2023 LCA (Akbarian-Saravi et al. 2025) found that while hemp-based composites significantly reduce CO_2 emissions, their cost per functional unit is still higher than glass fiber composites in automotive panel production. Addressing these economic and scalability barriers through automation, supply chain development, and standardized fiber treatment methods is critical to enabling widespread industrial adoption and long-term competitiveness of green composites.

Overall, while significant progress has been made in the development of green composites, further research and innovation are necessary to fully optimize their performance and broaden their application scope. By addressing existing challenges such as interfacial adhesion, processing efficiency, and durability, green composites have the potential to become a cornerstone of sustainable materials engineering, contributing to a greener and more eco-friendly future for a wide range of industries.

References

Ajitesh VA, Preet Singh JI, Singh S, Dhawan V, Sharma V (2024) A review on natural fibres: its applications, chemical and mechanical properties. AIP Conf Proc 2962(1):020025

Akbarian-Saravi N, Sowlati T, Ahmad H, Hewage K, Sadiq R, Milani AS (2025) 12—Life cycle assessment of hemp-based biocomposites production for agricultural emission mitigation strategies: a case study. In: Milani AS, Takaffoli M (eds) Biocomposites and the circular economy [Internet]. Woodhead Publishing Series in Composites Science and Engineering. Woodhead Publishing, pp 261–85 [cited 2025 Apr 21]. https://www.sciencedirect.com/science/article/pii/B9780443237188000120

Äkräs L, Silvenius F, Baniasadi H, Vahvaselkä M, Ilvesniemi H, Seppälä J (2024) A cradle-to-gate life cycle assessment of polyamide-starch biocomposites: carbon footprint as an indicator of sustainability. Clean Technol Environ Policy 26(10):3297–3312

Akroum-Amrouche D, Akroum H, Lounici H (2024) Challenges and opportunities in green hydrogen production materials for biological hydrogen production. In: Singh P, Agarwal AK, Thakur A, Sinha RK (eds) Challenges and opportunities in green hydrogen production [Internet]. Springer Nature, Singapore, pp 329–51 [cited 2024 Oct 10]. https://doi.org/10.1007/978-981-97-1339-4_14

Ammar EE (2023) Environmental impact of biodegradation. In: Ali GAM, Makhlouf ASH (eds) Handbook of biodegradable materials [Internet]. Springer International Publishing, Cham, pp 721–60 [cited 2024 Oct 10]. https://doi.org/10.1007/978-3-031-09710-2_27

Angellier-Coussy H, Gastaldi E, Gontard N, Guillaume C, Guillard V, Peyron S (2024) Converting agro-industrial by-products into biodegradable composite materials for food packaging: presentation of an eco-reasoned approach. In: Baumberger S (ed) Green chemistry and agro-food industry: towards a sustainable bioeconomy [Internet]. Springer Nature Switzerland, Cham, pp 237–267 [cited 2024 Oct 10]. https://doi.org/10.1007/978-3-031-54188-9_10

Arpitha GR, Jain N, Verma A, Madhusudhan M (2024) Corncob bio-waste and boron nitride particles reinforced epoxy-based composites for lightweight applications: fabrication and characterization. Biomass Convers Biorefin 14(11):12531–12538

Bavaliya KJ, Vala NS, Raj M, Raj L (2024) A review on biodegradable composites based on poly (lactic acid) with various bio fibers. Chem Pap [Internet] [cited 2024 Oct 10]. https://doi.org/10.1007/s11696-023-03298-x

Bouzouita S (2011) Optimisation des interfaces fibre-matrice de composites à renfort naturel [Internet] [These de doctorat]. Ecully, Ecole centrale de Lyon [cited 2024 Oct 10]. https://theses.fr/2011ECDL0052

Bravo-Porras G, Fernández-Güelfo LA, Álvarez-Gallego CJ, Carbú M, Sales D, Romero-García LI (2024) Influence of the total concentration and the profile of volatile fatty acids on polyhydroxyalkanoates (PHA) production by mixed microbial cultures. Biomass Convers Biorefin 14(1):239–253

Cakmak OK (2024) Biodegradable polymers—a review on properties, processing, and degradation mechanism. Circular Econ Sustain [Internet] [cited 2024 Oct 10]. https://doi.org/10.1007/s43615-023-00277-y

Chau MQ, Truong TT, Hoang AT, Le TH (2025) Oil spill cleanup by raw cellulose-based absorbents: a green and sustainable approach. Energy Sources Part Recov Util Environ Eff 47(1):8269–8282

Chen D, Jones OG, Campanella OH (2023) Plant protein-based fibers: fabrication, characterization, and potential food applications. Crit Rev Food Sci Nutr 63(20):4554–4578

D'Amico DA, Manfredi LB, Cyras VP (2012) Relationship between thermal properties, morphology, and crystallinity of nanocomposites based on polyhydroxybutyrate. J Appl Polym Sci 123(1):200–208

Elsayad K, Urstöger G, Czibula C, Teichert C, Gumulec J, Balvan J et al (2020) Mechanical properties of cellulose fibers measured by brillouin spectroscopy. Cellulose 27(8):4209–4220

European Industrial Hemp Association (2022) Hemp—European Commission [Internet] [cited 2025 Apr 21] https://agriculture.ec.europa.eu/farming/crop-productions-and-plant-based-products/hemp_en

FAO (2022a) Non wood forest products Bamboo|Statistics|Non-wood forest products|Food and Agriculture Organization of the United Nations [Internet] [cited 2025 Apr 21]. https://www.fao.org/forestry/nwfp/statistics/bamboo/en

FAO (2022b) World food and agriculture—statistical yearbook 2022 [Internet] [cited 2025 Apr 21]. https://openknowledge.fao.org/items/ffc06bd1-b7e9-4b94-a6a7-9ba45352976b

Han L, Song Y, Qi H, Yang JS, Li S, Liu PA (2023) A comparative study on the mechanical properties of open-hole carbon fiber-reinforced thermoplastic and thermosetting composite materials [Internet]. Polymers [cited 2024 Oct 10]. https://www.mdpi.com/2073-4360/15/22/4468

International Jute Study Group (2021) International Jute Study Group|UIA yearbook profile|Union of International Associations [Internet] [cited 2025 Apr 21]. https://uia.org/s/or/en/1100035644

Islam S, Karim FE, Islam MR, Saeed MA, Alam KA, Khatun MM (2025) Thermoset and thermoplastic polymer composite with date palm fiber and its behavior: a review. Islam. SPE Polym Wiley Online Library [Internet] [cited 2024 Oct 10]. https://doi.org/10.1002/pls2.10157

Kouassi M, Kallel A, Abdallah AB, Nouira S, Ballut S, Fitoussi J et al (2024a) Assessment of fused deposition modeling (FDM) parameters for fabrication of solid and hollow microneedles using polylactic acid (PLA). Polym Adv Technol 35(8):e6548

Kouassi M, Abdallah AB, Nouira S, Ballut S, Fitoussi J, Shirinbayan M et al (2024b) FDM printing of microneedles based on biocompatible and biodegradable thermoplastic polymers. J Appl Polym Sci 141(41):e56067

Kurien RA, Maria Anil M, Sharan Mohan SL, Anna Thomas J (2023) Natural fiber composites as sustainable resources for emerging applications—a review. Mater Today Proc [Internet] [cited 2024 Oct 10]. https://www.sciencedirect.com/science/article/pii/S2214785323022642

Mori R (2023) Replacing all petroleum-based chemical products with natural biomass-based chemical products: a tutorial review. RSC Sustain (RSC Publishing) https://doi.org/10.1039/D2SU00014H [Internet] [cited 2024 Oct 10]. https://pubs.rsc.org/en/content/articlehtml/2023/su/d2su00014h

Mukherjee T, Kao N (2011) PLA based biopolymer reinforced with natural fibre: a review. J Polym Environ 19(3):714–725

Nouira S, Hassine T, Fitoussi J, Shirinbayan M, Gamaoun F, Tcharkhtchi A (2022) Non-isothermal crystallization kinetics and its effect on the mechanical properties of homopolymer isotactic polypropylene. J Polym Res 29(1):26

Nouira S, Shirinbayan M, Peixinho J, Benfriha K, Hassine T, Fitoussi J (2024) Effect of processing conditions on morphology and mechanical damage in glass-reinforced polypropylene composite. Polym Compos 45(9):8251–8263

Nuvoli L, Conte P, Fadda C, Reglero Ruiz JA, García JM, Baldino S et al (2021) Structural, thermal, and mechanical properties of gelatin-based films integrated with tara gum. Polymer 1(214):123244

Omar AA, Hanafi MHM, Razak NH, Ibrahim A, Razak NAA (2021) A best-evidence review of bio-based plasticizer and the effects on the mechanical properties of PLA. Chem Eng Trans 15(89):241–246

Park D, Lee H, Won W (2024) Unveiling the environmental gains of biodegradable plastics in the waste treatment phase: a cradle-to-crave life cycle assessment. Chem Eng J ScienceDirect [Internet] [cited 2024 Oct 10]. https://www.sciencedirect.com/science/article/abs/pii/S1385894724020278

Prabhu P, Chokkalingam V, Arunkumar S, Perumal P (2025) The investigation on mechanical, thermal conductivity and water absorption characteristics of coir fiber-reinforced vinyl ester biocomposites. Biomass Convers Biorefinery. 15(3):4199–4209

Raj H, Tripathi S, Bauri S, Choudhury AM, Mandal SS, Maiti P (2023) Green composites using naturally occurring fibers: a comprehensive review. Sustain Polym Energy SCIEPublish [Internet] [cited 2024 Oct 10]. https://www.sciepublish.com/article/pii/60

Ramesh D, Karthikeyan S, Subramanian P, Ammayappan L, Anand KV (2022) Green composites from renewable feedstocks. In: Baskar C, Ramakrishna S, Daniela La Rosa A (eds) Encyclopedia of green materials [Internet]. Springer Nature, Singapore, pp 1–9 [cited 2024 Oct 10]. https://doi.org/10.1007/978-981-16-4921-9_157-1

Rech CR, Brabes KC, Silva BE, Martines MA, Silveira TF, Alberton J, Amadeu CA, Caon T, Arruda EJ, Martelli SM (2021) Antimicrobial and physical–mechanical properties of polyhydroxybutyrate edible films containing essential oil mixtures. J Polym Environ [Internet] [cited 2024 Oct 10]. https://doi.org/10.1007/s10924-020-01943-0

Sabet M (2024) Exploring biodegradable polymer composites for sustainable packaging: a review on properties, manufacturing techniques, and environmental impacts. Iran Polym J [Internet] [cited 2024 Oct 10]. https://doi.org/10.1007/s13726-024-01365-y

Samant L, Goel A, Mathew J, Jose S, Thomas S (2023) Effect of surface treatment on flax fiber reinforced natural rubber green composite. J Appl Polym Sci 140(12):e53651

Sharma S, Bhende M (2024) An overview: non-toxic and eco-friendly polysaccharides—its classification, properties, and diverse applications. Polym Bull 81(14):12383–12429

Sharma H, Komal UK, Singh I, Misra JP, Rakesh PK (2019) Introduction to green composites. In: Rakesh PK, Singh I (eds) Processing of green composites [Internet]. Springer, Singapore, pp 1–13 [cited 2024 Oct 10]. https://doi.org/10.1007/978-981-13-6019-0_1

Shirinbayan M, Nouira S, Imaddahen MA, Fitoussi J (2024) Microstructure-sensitive investigation on the plastic deformation and damage initiation of fiber-reinforced polypropylene composite. Compos Part B Eng 1(286):111790

Singh AA, Bharathi SD, Jacob BS (2023) The impacts of plastics on environmental sustainability and ways to degrade microplastics. In: Samuel Jacob B, Ramani K, Vinoth Kumar V (eds) Applied biotechnology for emerging pollutants remediation and energy conversion [Internet]. Springer Nature, Singapore, pp 17–35 [cited 2024 Oct 10]. https://doi.org/10.1007/978-981-99-1179-0_2

Sinha S (2024) An overview of biopolymer-derived packaging material. Polym Renew Resour 15(2):193–209

Sonnenfeld C, Mendil-Jakani H, Agogué R, Nunez P, Beauchêne P (2017) Thermoplastic/thermoset multilayer composites: a way to improve the impact damage tolerance of thermosetting resin matrix composites. Compos Struct ScienceDirect [Internet] [cited 2024 Oct 10]. https://www.sciencedirect.com/science/article/abs/pii/S0263822316320384

Taib NAAB, Rahman MR, Huda D, Kuok KK, Hamdan S, Bakri MKB et al (2023) A review on poly lactic acid (PLA) as a biodegradable polymer. Polym Bull 80(2):1179–1213

Talukder MdS, Gokul G, Kaur H, Dutta S, Jaiswal KS, Ramaswamy AP et al (2024) Biomass resources and potential for its applications in energy and platform chemicals with 3G biorefinery approaches. In: Kumar S, Sundaramurthy S, Kumar D, Chandel AK (eds) Clean energy transition-via-biomass resource utilization: a way to mitigate climate change [Internet]. Springer Nature, Singapore, pp 29–58 [cited 2024 Oct 10]. https://doi.org/10.1007/978-981-97-6321-4_2

Ullah S, Akhter Z, Palevicius A, Janusas G (2025) Review: natural fiber-based biocomposites for potential advanced automotive applications. J Eng Fibers Fabr 1(20):15589250241311468

Uniyal P, Gaur P, Yadav J, Bhalla NA, Khan T, Junaedi H, Sebaey TA (2025) A comprehensive review on the role of nanosilica as a toughening agent for enhanced epoxy composites for aerospace applications. ACS Omega [Internet] [cited 2025 Apr 21]. https://doi.org/10.1021/acsomega.4c10073

Vilaseca F, Mendez JA, Pèlach A, Llop M, Cañigueral N, Gironès J et al (2007) Composite materials derived from biodegradable starch polymer and jute strands. Process Biochem 42(3):329–334

Wang P, Chen F, Zhang H, Meng W, Sun Y, Liu C (2017) Large-scale preparation of jute-fiber-reinforced starch-based composites with high mechanical strength and optimized biodegradability. Starch Stärke 69(11–12):1700052

Wang F, Wang J, Fang D, Zhou S, Huang J, Zhao G et al (2023) Surface sizing introducing carbon nanotubes for interfacial bond strengthening of basalt fiber–reinforced polymer composites. Adv Compos Hybrid Mater 6(3):117

Wu D, Gao W, Zhao Y, Wei Z, Song C, Qu F, Wang F (2024) Elaborating the microbial mechanism of humic substance formation during lignocellulosic biomass composting by inoculation with different functional microbes. Ind Crops Products ScienceDirect [Internet] [cited 2024 Oct 10] https://www.sciencedirect.com/science/article/abs/pii/S0926669023016035

Yadav RS, Rathore VK (2024) Advancements and innovations in green hydrogen technologies. In: Singh P, Agarwal AK, Thakur A, Sinha RK (eds) Challenges and opportunities in green hydrogen production [Internet]. Springer Nature, Singapore, pp 211–38 [cited 2024 Oct 10]. https://doi.org/10.1007/978-981-97-1339-4_9

Yang Y, Zhang L, Zhang J, Ren Y, Huo H, Zhang X, Huang K, Zhang Z (2023) Reengineering waste boxwood powder into light and high-strength biodegradable composites to replace petroleum-based synthetic materials. ACS Appl Mater Interfaces [Internet] [cited 2024 Oct 10]. https://doi.org/10.1021/acsami.2c19844

Zhang G, Zheng W, Bai X, Xu L, Li K, Zhang M et al (2024) Polyhydroxyalkanoates (PHAs) biological recovery approaches and protein-mediated secretion model hypothesis. J Clean Prod 10(440):140851

Zhou SP, Ke X, Jin LQ, Xue YP, Zheng YG (2024) Sustainable management and valorization of biomass wastes using synthetic microbial consortia. Bioresour Technol 1(395):130391

Chapter 3
Revolutionizing Packaging with Sustainable Polymer Composites

Erika Blanzeanu, Ionut-Cristian Radu, Derniza-Elena Cozorici, and Catalin Zaharia

Abstract Packaging, widely used across industries such as food, cosmetics, and health care, contributes significantly to pollution due to waste accumulation in landfills and energy-intensive production processes. Achieving sustainability in packaging requires implementing clean production technologies, utilizing recyclable materials, and sourcing renewable, natural resources. Composites, which combine fibers and powders to enhance material properties, present additional challenges in terms of production and recycling compared to common materials like polyethylene and polyethylene terephthalate (PET). This chapter highlights the importance of sustainability in polymeric composites for packaging, exploring both current and emerging materials and technologies. Topics covered include biopolymers from natural sources, such as chitosan, starch, and cellulose, as well as polyhydroxyalkanoates and biodegradable polymers like polylactic acid (PLA), poly(ε-caprolactone) (PCL), and polybutylene succinate (PBS). The chapter also examines recyclability, process efficiency strategies, and the role of bioplastics such as bio-PE and bio-PP in promoting a circular economy in packaging.

Keywords Sustainability · Circular economy · Biodegradable polymers · Biocomposites · Active packaging · Intelligent packaging

E. Blanzeanu · I.-C. Radu · D.-E. Cozorici · C. Zaharia (✉)
Faculty of Chemical Engineering and Biotechnology, Department of Bioresources and Polymer Science, Advanced Polymer Materials Group, National University of Science and Technology POLITEHNICA Bucharest, Bucharest, Romania
e-mail: catalin.zaharia@upb.ro

E. Blanzeanu
e-mail: erika.blanzeanu@upb.ro

I.-C. Radu
e-mail: ionut_cristian.radu@upb.ro

D.-E. Cozorici
e-mail: derniza.cozorici@upb.ro

L. Ravindran et al. (eds.), *Macro, Micro and Nanocomposites from Sustainable Sources*, Advanced Structured Materials 244, https://doi.org/10.1007/978-981-95-2469-3_3

3.1 Introduction

Evolution of packaging: A brief history

The primary function of packaging is to preserve the quality of an item from production to consumption, shielding it from environmental factors like dust, humidity, temperature changes, and microorganisms during transportation and storage (Khandeparkar et al. 2024; Zhu et al. 2022; Tajeddin et al. 2020).

Although packaging is widely believed to have originated during the industrial era, its history dates back much earlier. During the Roman and Byzantine empires, valuable goods were preserved in woven rattan baskets or carved ebony boxes to maintain their quality over time. The origins of food packaging can be traced back to antiquity or prehistoric times, when materials such as leaves, birch bark, human-made, rudimentary containers, and even animal pelts and hides were used for protection. In the modern age, paper and cardboard have gained widespread popularity as packaging materials. A major turning point occurred in 1810 with the patenting of the first tin-based metal can, which revolutionized the food packaging industry. The era of plastic packaging began with the introduction of polyvinylidene chloride, also known as Saran, which was initially used as a wrap due to its effective barrier against humidity. This advancement was followed by further innovations in the packaging industry, including the invention of polyethylene terephthalate (PET) bottles by DuPont (Historical Perspective and Evolution 2020; Meyers and Gerstman 2005). As highlighted by this brief look at packaging history, advancements in this field have led to the development of new materials with improved properties, replacing outdated packaging solutions such as leaves and hides.

Packaging materials: An overview

Currently, four main classes of packaging materials are widely used: polymers (commonly known as plastics), glass, metal, and paper, as illustrated in Fig. 3.1. The following section provides a brief overview of these materials and the industries they are used in, highlighting their advantages and drawbacks (Tajeddin et al. 2020).

Glass is a traditional, transparent packaging material with several advantages, including chemical inertness and excellent barrier properties against gases and moisture. However, it also has its drawbacks, such as allowing light to interact with its contents, potentially leading to degradation, and being relatively fragile and dense, which makes transportation more challenging and expensive compared to other packaging options (Khandeparkar et al. 2024; Furszyfer Del Rio et al. 2022). In the beverage industry, glass is predominantly utilized for bottles that store wine, beer, spirits, and water. It is also extensively employed in the food sector, pharmaceutical industry, cosmetics, personal care products, and various chemical and laboratory applications (Debeaufort 2021a).

Paper, derived from wood-based sources, offers the advantage of being biodegradable. However, its weak barrier properties often require coating with wax or polymers to enhance functionality (Khandeparkar et al. 2024). It is widely used in the food and beverage industry for products like cereal boxes and juice cartons, as well as in retail

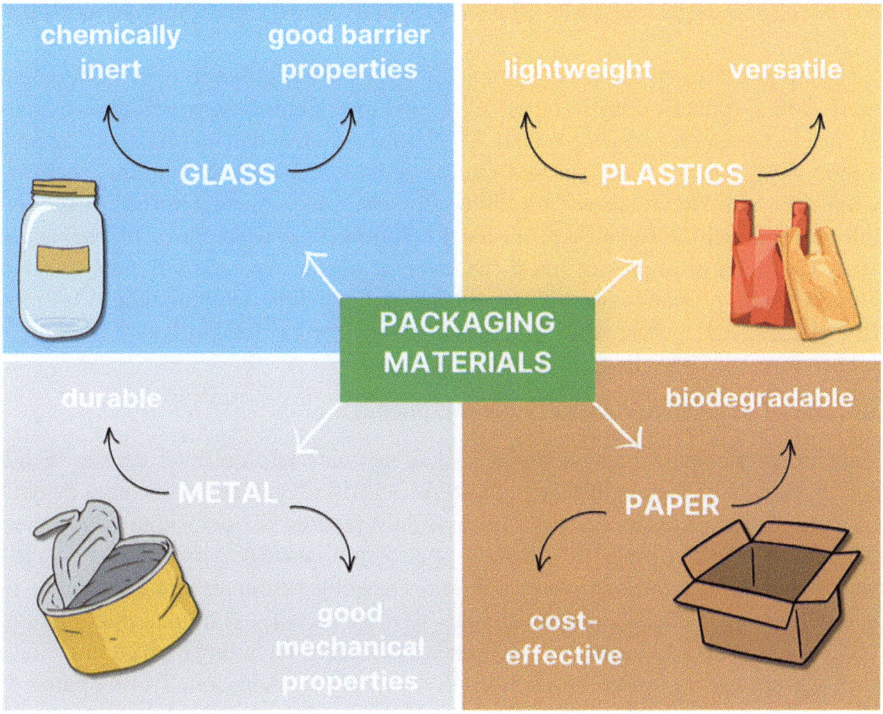

Fig. 3.1 Classes of materials used in packaging

for shopping bags and gift wrapping, in the pharmaceutical sector, and for shipping purposes (Deshwal et al. 2019).

Metal is generally used in packaging due to its durability and ability to protect against light, gases, or humidity, making it ideal for long-distance transportation. To prevent corrosion, the interior of metal cans is typically coated with a polymer layer; however, this coating can sometimes lead to the migration of substances like bisphenol A, particularly when epoxy linings are used (Khandeparkar et al. 2024). A major health risk associated with metal packaging, especially in food applications, is the migration of dangerous metals such as lead, cadmium, and mercury (Deshwal and Panjagari 2020). Metal packaging is widely used across various industries, including food and beverages, as well as in aerosol products like deodorants and insect repellents, and the pharmaceutical sector, where metal tubes are often used for creams (Debeaufort 2021b).

Aluminum foil is widely used in households due to its heat resistance and conductivity, but it presents potential health risks, particularly as high levels of aluminum in the brain have been linked to Alzheimer's disease. When exposed to distilled water or acidic and alkaline environments, aluminum can begin to leach from the foil (Khandeparkar et al. 2024). Despite these concerns, it remains prevalent in packaging for products such as chocolate, ready-to-eat meals, dairy items, and in the pharmaceutical sector for blister packs (Kerry et al. 2012).

Polymers are often the preferred choice for packaging due to their lightweight nature, which reduces transportation costs, and their durability, which ensures reliable product protection during transit. They offer effective barrier properties against various gases, such as oxygen and carbon dioxide, as well as humidity, helping to maintain product freshness. Additionally, their low manufacturing costs and high customizability make them an ideal solution for a wide range of packaging needs. Common examples of fossil fuel-derived plastics include polyethylene and polypropylene, primarily used for wrapping; polyethylene terephthalate (PET) for drink bottles; and polystyrene or expanded polystyrene for cups and containers (Sangroniz et al. 2019; Babaremu et al. 2023).

Limitations of conventional polymers in packaging

Considering the main categories of packaging materials outlined earlier in the chapter, along with their respective advantages and disadvantages, it becomes evident that polymer-based materials have been widely favored for packaging since their discovery. However, despite their benefits, petroleum-based polymers present significant challenges: they are derived from non-renewable resources and accumulate in the environment, taking hundreds of years to degrade without human intervention. Their recycling also has notable environmental impacts, including high carbon emissions and energy consumption (Upadhyay et al. 2024). Additionally, compared to other packaging materials such as glass or metal, plastics exhibit significantly lower recycling rates, often resulting in lower quality products with diminished properties (Khandeparkar et al. 2024).

Another important drawback of polymeric materials lies in the health threats to which workers are subjected during production and processing, as well as those affecting consumers, since unreacted monomers or additives can sometimes migrate from the packaging. This is why packaging that comes into contact with food is subjected to rigorous toxicity assessments. However, despite these evaluations, under certain conditions, such as exposure to heat or UV radiation, small molecules may still be released from the packaging and enter the body. The broad applicability of plastic packaging across nearly all industries has contributed to the accumulation of microplastics, tiny particles ranging from 1 μm to 5 mm in the environment, including seawater, soil, and eventually in the human body. Packaging microplastics enter the body through contaminated water, either during production or through packaging used in storage and transportation, as well as through food, especially in the case of pre-packaged meals in supermarkets (Aloysius et al. 2015; Jadhav et al. 2021; Aslam et al. 2023). There are several mechanisms through which microplastics can harm the human body:

1. Direct hazards caused by exposure through inhalation during packaging processing, ingestion, or dermal contact.
2. Release of chemicals by microplastics, such as phthalates, bisphenols, and halogenated products during long-term storage or improper storage conditions.

3. Interaction with other harmful substances—once accumulated in the human body through packaging, microplastics can intensify the hazards produced by other dangerous compounds (Aloysius et al. 2015; Verla et al. 2019).

The presence of microplastics in the human body can potentially lead to health risks, including inflammation, oxidative stress, cell damage, DNA alterations, and an increased likelihood of respiratory and neurological diseases, potentially even cancer (Aslam et al. 2023). Strategies to mitigate the risks of microplastic accumulation in the human body, as well as in the environment, should be developed with the potential risks to the population in mind. The prevalent reasons include inadequate safety measures during packaging production, improper storage conditions that lead to leakage of harmful substances from packaging (Sun and Wang 2023), and poor waste management strategies that lead to landfill disposal.

During the pandemic, the demand for single-use protective equipment such as masks, gloves, and coats led to a corresponding increase in petroleum-based polymeric waste. With no comprehensive strategy in place to manage this waste, most of it, as well as its packaging materials, ended up in landfills (Zhang et al. 2022). This highlights how packaging, widely used in industries such as food, cosmetics, chemicals, health care, and manufacturing, has become a persistent source of environmental pollution. While packaging's primary role is to protect products from environmental factors, it often ends up harming the environment because of waste materials disposed of in landfills, the harsh production processes, and the significant energy consumption involved.

Transition to biopolymers: A sustainable shift in packaging

In the past decades, and particularly after the pandemic, these concerns, combined with new regulations surrounding petroleum-based products and their recyclability, have shifted the focus toward sustainable packaging solutions. As defined by the European Commission in 2019, sustainability means meeting the needs of the current generation without compromising the ability of future generations to meet their needs.

The EU Commission's approach to achieving sustainability encourages transitioning to a cost-effective, climate-neutral, resource-efficient, and biodiversity-friendly economy, with a growing reliance on renewable resources. Between 2012 and 2015, for example, the EU's income from clean energy technology exports exceeded its imports, demonstrating that sustainable practices can stimulate economic growth. Businesses are increasingly focusing on creating cutting-edge materials and technologies for recycling, repair, and reuse to reduce the need for new resource extraction at both significant financial and environmental costs (https://commission.europa.eu/publications/sustainable-europe-2030_en).

To sum up, sustainability in the context of packaging refers to the use of bio-based materials from natural, renewable sources, employing clean production technologies that minimize energy consumption, and ensuring these materials can be reused or recycled for the same or different purposes, as illustrated in Fig. 3.2 (Dörnyei et al. 2023).

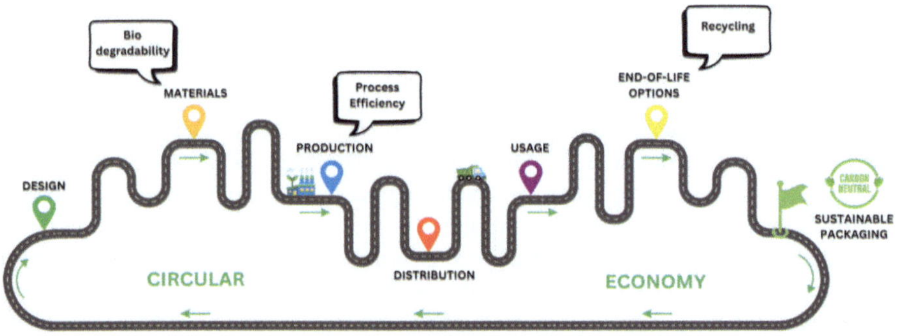

Fig. 3.2 Sustainability process flow in circular packaging

Considering all the discussed aspects and drawbacks of polymer utilization in packaging, a shift to biodegradable polymers derived from natural sources would reduce health and environmental harm caused by packaging, while simultaneously leading to better preservation of the current planet resources.

Biocomposites in packaging: Significance and emerging classes

The advantage of using bio-based materials is clear, but these materials typically do not exhibit exceptional properties; some of them may present poor mechanical properties, while others lack barrier properties, requiring the incorporation of additional materials or active ingredients to achieve the desired performance and properties (Wang et al. 2022). These multiphasic materials, composed of at least two different ingredients with distinct properties, mixed to obtain better properties than the separate materials, are known as **composite materials** (Sumithra et al. 2023). In particular, there is a growing interest in biopolymer-based composites in food packaging, with various studies on polyvinyl alcohol (PVA), polylactic acid (PLA), chitosan, poly(3-hydroxybutyrate) (PHB), poly(3-hydroxybutyrate-co-3-hydroxyvalerate) (PHBV), poly(ε-caprolactone) (PCL), and starch mixed with different fillers, leading to extraordinary results in the field (Wang et al. 2022; Khalid and Arif 2022).

Various packaging materials developed by researchers can be classified as composites and include improved, active, and intelligent packaging.

Improved packaging aims to achieve superior material performance in terms of mechanical properties or gas/UV-barrier properties by incorporating special nanomaterials and optimizing their arrangement within the composite. Enhanced mechanical properties can be achieved through the addition of nanofillers like clays, fibers, or other polymers, or by crosslinking the polymer matrix. To achieve improved barriers against gases or humidity, the focus is shifted toward the incorporation and dispersion of the nanofiller or using multiple layers of different biopolymers. UV-barrier properties are generally improved by the addition of metal oxides (Zhang et al. 2022).

Active packaging is a class of packaging based on a technology that either releases compounds into, or absorbs compounds from, the environment to extend the product's shelf life (Upadhyay et al. 2024). Active packaging technologies that scavenge oxygen and hydrogen usually include iron or platinum and palladium compounds capable of catalyzing the conversion of hydrogen and oxygen into water, creating a favorable medium for maintaining the quality of the contents. In food packaging sector, ethylene is chemically neutralized with titanium oxide, which transforms it into carbon dioxide and water under UV radiation. Gases can also be physically eliminated in the presence of very porous structures. Moisture content must be controlled, as it can promote microbial growth. Therefore, systems including multilayer materials with a layer of NaCl or hygroscopic substances are being utilized. Another approach for active packaging, as mentioned before, is to use compounds that release in the environment, such as silver or other metal nanoparticles, as well as graphene and other carbon-based materials for antimicrobial effects, or polyphenols and plant extracts for their antioxidant effects (Zhang et al. 2022; Kumari et al. 2024; Thirupathi Vasuki et al. 2023).

Intelligent packaging is a technology that is based on the continuous monitoring of product quality without interfering with the product itself (Upadhyay et al. 2024), and refers to the development of indicators that show the current quality of the product through color or intensity modification. Additionally, sensors that detect the presence of certain substances related to the freshness of the product are included in this intelligent packaging category. For example, it can show, through real-time coloration, if the oxygen or carbon dioxide level, which are indicators for bacterial growth, are suitable for consuming the product. They can also indicate if the content was subjected to a temperature higher or lower than a chosen threshold or can even indicate the product's temperature history based on irreversible changes upon exposure to temperature (Zhang et al. 2022; Thirupathi Vasuki et al. 2023; Azeredo and Correa 2021).

These classes of packaging were intended to serve as a sustainable alternative to conventional packaging because they help reduce waste caused by either improper storage, handling, or shipping, or by internal modifications caused by enzymes, chemical reactions, or microorganisms. Additionally, most of the materials used are derived from agriculture waste, thus contributing to further waste reduction. Examples such as banana fibers, rice husk, calcium oxide from eggshell, and feather proteins are increasingly being valorized in current studies (Yin and Woo 2024). These naturally sourced materials, used as biopolymer-based composites in food packaging applications, help reduce the environmental impact associated with the energy required for recycling and the carbon footprint, thanks to their biodegradability and non-toxic properties (Upadhyay et al. 2024).

Nowadays, the goal of any manufacturing process is to create sustainable products. This modern goal highlights the crucial need for manufacturing practices that not only satisfy current demands but also consider the long-term impact on the environment and resources.

In terms of packaging, sustainability refers to implementing clean, non-harmful production technologies, creating and using recyclable packaging, utilizing natural raw materials from renewable sources, reducing the quantities of raw materials required, and optimizing the use of energy and materials. In essence, it emphasizes the wise use of resources to ensure they remain available for future generations. Achieving sustainability is more challenging with composites, compared to common packaging materials like polyethylene and polyethylene terephthalate (PET), because of the more complex production, processing, and recycling techniques that make sustainable practices harder to implement.

3.2 Sustainability Directions

3.2.1 Types of Materials in Sustainable Packaging

As emphasized in the previous chapter, the urgent demand for sustainable packaging materials has spurred extensive research on various bio-based polymers. Before classifying biopolymers, it is important to clarify the distinction between **bio-based polymers** and **biodegradable polymers**. Bio-based polymers, as their name suggests, are obtained from natural sources, while biodegradable polymers are compounds that can be degraded under environmental conditions and transformed into non-harmful compounds. A polymer can be bio-based but not biodegradable; for example, bio-polyethylene, bio-polypropylene, and bio-polyethylene terephthalate are derived from renewable resources but do not break down naturally. On the other hand, biodegradable polymers, such as polycaprolactone (PCL), may degrade over time but are not necessarily bio-based, as they are typically synthesized from petrochemical sources.

The widely accepted classification of biodegradable polymers includes three categories: 1. polymers synthesized chemically; 2. polymers extracted from biomass; 3. polymers produced through microbial fermentation, as illustrated in Fig. 3.3 (Zhang et al. 2022). These classes of biodegradable polymers will be discussed in greater detail in the upcoming subchapters.

3.2.1.1 Synthetically Derived Biodegradable Polymers

This section will primarily focus on polylactic acid (PLA), polybutylene succinate (PBS), and poly(ε-caprolactone) (PCL), highlighting the properties that make them suitable for packaging applications, along with current research trends aimed at developing sustainable composites.

Polylactic Acid (PLA)

Due to the chirality of lactic acid, which exists in both d-type and l-type isomers, three forms of polylactic acid can be produced: poly-L-lactic acid (PLLA), poly-D-lactic acid (PDLA), and poly-D, L-lactic acid (PDLLA) (Ramezani Dana and

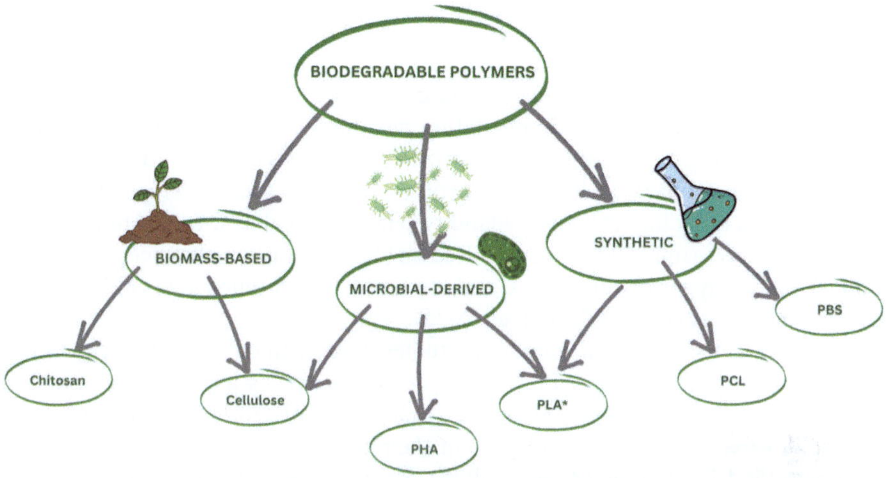

Fig. 3.3 Classification of polymers based on their source

Ebrahimi 2023; Swetha et al. 2023). While chemical methods can achieve large-scale continuous production of lactic acid (LA) and the resulting products are U.S. Food and Drug Administration (FDA)-approved, the raw materials used in these methods are typically toxic and do not align with the principles of green chemistry. Additionally, chemical routes yield a racemic mixture containing both lactic acid optical isomers. Lactic acid is preferably produced through bacterial fermentation of carbohydrates such as corn, sugar beet, sweet potato, straw cellulose, rice, and other starch containing natural sources (Zhang et al. 2022; Yu et al. 2023).

Three methods for synthesizing PLA with a molecular weight greater than 10,000 g/mol have been reported: (a) direct condensation polymerization, (b) azeotropic dehydration condensation, and (c) lactide ring-opening polymerization (Ramezani Dana and Ebrahimi 2023). Among these, lactide ring-opening polymerization is the most commonly employed in industrial applications due to its ability to produce high purity lactide and, subsequently, high-purity PLA.

PLA has several properties that make it suitable for food packaging applications: it is transparent, biodegradable, and offers good mechanical strength (Silva Pens et al. 2024), can be thermally processed, and is relatively low cost (Donkor et al. 2023). Its monomer, lactide, is approved by the FDA as a safe food preservative, since it is derived from lactic acid, a substance naturally occurring also in the human body, making PLA safe for use in applications where it comes in direct contact with food (Silva Pens et al. 2024). As such, PLA is currently found in markets in the form of cutlery, cups for cold drinks, or bottles (Khouri et al. 2024).

To serve as a viable replacement for synthetic polymers in packaging, PLA must overcome its drawbacks, including its poor gas and moisture permeability, high UV transmission, rigidity (brittleness), low elongation, and low temperature resistance (Ganjeh et al. 2024; Han Lyn et al. 2024). PLA's barrier properties are strongly influenced by crystallinity: the higher the percentage of crystallinity, the better the barrier properties. PLA thermal degradation usually happens when the material is processed without being dried, as moisture leads to molecular weight reduction. PLA's biodegradability can be improved by blending it with other bio-based polymers, thus reducing its crystallinity and favoring hydrolysis reactions. There is a very fine line between enhancing PLA's properties by incorporating different substances/nanofillers and reducing its biodegradability (Han Lyn et al. 2024). To sum up, PLA weaknesses lead to multiple studies focused on blending PLA with other polymers or adding different types of substances, making it part of a composite to achieve better food packaging properties (Zhao et al. 2025).

As shown in the literature, numerous studies have been conducted on enhancing the antibacterial properties of PLA films. These properties are typically achieved either by preventing or minimizing bacterial adhesion to the films or by incorporating bactericides that kill bacteria and prevent their proliferation. Both strategies offer advantages and limitations (Zhao et al. 2025). As a result, there is a high interest in this area, with researchers focusing on using natural compounds from renewable sources, even different types of waste, to inhibit microbial growth.

For example, a biopolymeric composite film containing PLA, curcumin, and fenugreek essential oil for packaging applications was developed by Subbuvel and Kavan (2022) through solvent-casting method. Curcumin and fenugreek oil were added for their antibacterial, antifungal, and antioxidant activity. Curcumin's antibacterial activity had been demonstrated previously, but the novelty of this study lies in the addition of the essential oil, which had not been employed before. This oil aimed to supplement the composite properties by providing synergistic antimicrobial effects, complemented by antioxidant and antifungal features. Their study revealed that the composite presented antioxidant properties, antimicrobial activity against *E. coli* and *S. aureus*, while the tensile strength of the composite was improved when compared to PLA/curcumin composite, pristine PLA, and commercial packaging material made from polyethylene (Subbuvel and Kavan 2022).

Another study followed a green route for modifying cellulose nanofibers (CNF) with rosin, for a better incorporation into a PLA matrix. The modified cellulose nanofibers were added for their antibacterial properties, as well as their role as a reinforcement agent. A film containing PLA and modified CNF was obtained by solvent casting over a chitosan film, ultimately resulting in a multilayer composite film based on chitosan and PLA reinforced with modified CNF. Antimicrobial properties against *E. coli* and *B. subtilis* were shown to be enhanced for the developed composite compared with PLA film; these properties were attributed to both the presence of rosin and chitosan (Niu et al. 2018).

Dejene et al. (2024) worked on a study aimed at reducing food waste caused by spoiling of Injera, a traditional food in East Africa. In this case, the composite was designed by solvent-casting, followed by melting. Zinc oxide nanoparticles were employed for their antimicrobial activity, while Enset fibers were used as reinforcing agent, all incorporated in the PLA matrix by melting, followed by compression to obtain a film. The final composite exhibited improved mechanical properties due to

Enset fibers and antifungal properties, extending the shelf life of Injera with 4–5 days (Dejene et al. 2024).

Another study focused on improvement of PLA mechanical and antibacterial properties by reinforcing with natural halloysite nanotubes (HNT) modified with copper (Cu^{2+}). The composite was processed through electrospinning, eventually helping to prolong the shelf life of fruits (Zhao et al. 2024).

Other materials used in research articles in combination with PLA for achieving antibacterial properties in active food packaging include titanium oxide (Goñi-Ciaurriz and Vélaz 2022), silver oxide (Liao et al. 2023a), cerium lactate (Wu et al. 2022), graphene oxide-zinc oxide (Liew et al. 2023), tea polyphenols (Ren et al. 2022), polyhexamethylene guanidine hydrochloride (PHMG) (Zhao et al. 2025), cinnamaldehyde (Khanjari et al. 2023; Gonon et al. 2023; Wang et al. 2023a), and cinnamon essential oil (Sharma et al. 2023).

Improving PLA's barrier properties is another area of focus for PLA composites, with several articles published on the topic. For example, avocado seed waste could be valorized for starch production, in this way replacing other food-based sources and contributing to a sustainable lifecycle. *Zuniga* et al. (Godoy Zúniga et al. 2024) worked on producing a PLA-based composite filled with starch extracted from avocado waste, which is a good alternative for reducing the cost of producing this type of composite. The starch was further modified with acetoacetate to improve its compatibility with PLA, then mixed within the PLA matrix by melt blending. The composite was proven to possess higher moisture, oxygen, and UV-barrier properties (Godoy Zúniga et al. 2024).

Addressing the problem of excessive moisture into packaged fruits or vegetables, three-layered PLA composites made from zinc oxide nanoparticles, cellulose acetate, and PLA were designed to ensure a contrasting behavior: hydrophilic and hydrophobic in the composite film. Electrospinning was employed to produce a relatively hydrophobic mesh made of PLA and zinc oxide and a hydrophilic mesh made of PLA, zinc oxide, and cellulose acetate on the two sides of a non-woven fabric. The composite film addressed two problems usually encountered in food packaging: bacterial growth and high moisture inside packaging. Its efficiency in maintaining fruits fresh was proven to be relatively high: from 8 days to up to 12 days, depending on the type of fruit (Guo et al. 2024).

Another approach to improve the gas barrier properties of PLA was proposed by Patel et al. (2022). In addition to incorporating chitin nanocrystals as a nucleating agent to enhance PLA crystallinity, a PLA-based composite material was developed by adding another biodegradable polymer with higher crystallinity—poly(hydroxybutyrate) (PHB). As expected, the improvement in the crystallization degree of the composite resulted in better barrier characteristics, with the best results obtained when using a PLA:PHB = 3:1 ratio (Patel et al. 2022).

Less research has been conducted on improving the UV-barrier properties of PLA compared to the other two areas previously discussed: enhancement of antimicrobial and gas barrier properties. Eugenol, sourced from cinnamon essential oil, was found to enhance the UV-barrier properties of a PLA-PBAT composite film designed to preserve the quality of cheese. In this study, cinnamon oil was used to modify titanium oxide, resulting in a synergistic effect on the composite's UV-blocking properties. Compared to the PLA-PBAT blend, the cinnamon-modified TiO_2 PLA-PBAT composite showed a 99.63% improvement in UV-barrier performance. Additionally, the composite exhibited antibacterial properties against *E. coli* and *S. aureus*. However, the addition of cinnamon oil must be carefully controlled, as the study indicates that its concentration is inversely proportional to the mechanical and thermal properties of the composite, while also affecting its transparency (Sharma et al. 2023).

As noted in the previously mentioned studies, PLA composites for food packaging applications are typically processed through melt blending, where the filler or reinforcing agent is directly added into a mixer. This is followed by further processing using methods such as compression molding, extrusion, injection molding, and blown film molding. Another widely used technique for fabricating PLA composites is solvent casting. In this method, PLA is dissolved in a solvent, and the filler can either be dispersed in a solvent or directly added to the PLA solution. The mixture is then poured into a Petri dish or mold, allowing the solvent to evaporate and form the composite film (Jacob et al. 2020). The focus on using PLA in so many applications instead of other biopolymers is also due to its processability through existing processing technologies, without any further intervention (Kumari et al. 2022).

Polybutylene succinate (PBS)

Polybutylene succinate (PBS) is an aliphatic polyester typically obtained through the polycondensation of 1,4-butanediol with succinic acid. Both monomers can be sourced either from non-renewable sources, such as fossil fuels, or renewable sources, like bacterial fermentation of biomass derived feedstock (Aliotta et al. 2022). Both types of PBS are available on the market, with the one originating from bacteria being the more expensive option. PBS is widely used due to its biodegradability, which is independent of the monomers' origin. Another factor that makes PBS a suitable candidate for packaging applications is its easy thermo-processability, which requires no modifications to conventional polymer processing equipment. It can be processed through injection molding, thermoforming, casting, and blown extrusion (Barletta et al. 2022). However, PBS is a semi-crystalline polymer that exhibits low mechanical properties and thermostability, with a melting point ranging from 88 to 120 °C (Ghasemlou et al. 2024). As a result, PBS is often utilized in blends and composites, where fillers are incorporated to enhance thermal conductivity, mechanical properties, gas barrier performance, or flame resistance (Fredi and Dorigato 2021).

Although there are several reviews and book chapters on the use of PBS as a packaging material, the experimental research in this area has not been as extensively explored as it has for other polymers, such as PLA or cellulose.

A significant advancement in the field of food packaging was made with the development of PBS-based vacuum packaging for raw poultry meat. Composite films made from PBS, PBSA, and talc were compared to commercially available films used for meat packaging, such as those based on PA/LDPE. The study concluded that PBS-based films could serve as an alternative to the commercially available packaging films, with the added benefit of being biodegradable and derived from a renewable source (Vytejčková et al. 2017).

Another study focused on prolonging the shelf life of bread by producing a film made of lignin nanoparticles and cinnamaldehyde in a PBS matrix. The lignin nanoparticles, derived from waste, improves the mechanical properties of PBS, while cinnamaldehyde was used for its antifungal effect (Moe et al. 2023). A similar approach was employed to develop packaging for mangoes (Nuamduang et al. 2024).

In addition to blending with other polymers, PBS can be used in smart packaging, where different active ingredients are added. Essential oils, which have antibacterial and antifungal properties, are commonly used as active ingredients in food packaging. One study explored the effects of three essential oils on the properties of PBS film, finding that not only antibacterial properties but also the mechanical and thermal properties were influenced by the addition of the oils (Mohamad et al. 2022). Another study proposed a composite made of PBS/PBAT and cinnamon or citrus essential oil for trimmed coconut packaging. The optimal percentage of PBAT was determined by considering the mechanical and thermal properties of the mixtures. The mechanical properties of the blend film decreased with the addition of PBAT. The addition of essential oils affected the water and oxygen transmission rates, with higher percentages of essential oil resulting in increased water and oxygen transmission rates (Pattaraudomchok et al. 2024).

In the field of smart and active packaging, other active ingredients used with PBS include gallic acid (Pothinuch et al. 2024) and Lesser galangal essential oil (Wongphan et al. 2023) for their antibacterial effect, and halloysite nanotubes for their ability to scavenge ethylene (Ucpinar Durmaz et al. 2024).

Poly(ε-caprolactone) (PCL)

Poly(ε-caprolactone) (PCL) is an aliphatic polyester, mainly produced by ionic and metal catalyzed ring-opening polymerization of ε-caprolactone. PCL is a biodegradable and semi-crystalline polymer that is widely used due to its thermal properties, with its low melting temperature making it easy to process. Additionally, it is miscible with several polymers and is frequently found in blends to overcome its stress crack resistance and adhesion limitations. Its biocompatibility, combined with slow degradation rates, makes it a suitable material for biomedical applications such as implants and scaffolds for tissue engineering and repair. The degradation of PCL is faster in low molecular weight and low crystallinity PCL, and follows two mechanisms: enzymatic or hydrolytic, which often occur at the same time. Enzymatic degradation takes place at the surface of PCL and is generated by esterase and lipase, while hydrolytic degradation takes place in bulk and starts when water reaches the amorphous regions and breaks the ester bonds, followed by infiltration into crystalline areas (Thakur et al. 2021; Guarino et al. 2002).

Considering all the above-mentioned properties, it becomes obvious why the interest in developing packaging materials made of PCL has increased, despite its high cost. The literature provides several examples of composites designed for food packaging applications, but usually, PCL is mixed with other polymers to increase its elasticity and delay its degradation (Sady et al. 2021). For example, fibers of PCL reinforced with carboxymethyl chitosan and antibacterial components like curcumin, thyme, and natamycin were produced directly on cherry tomatoes in order to delay their ripening (Shen et al. 2024). Numerous studies have utilized electrospinning to produce PCL-based active packaging, owing to the high porosity and surface area-to-volume ratio of the fibers, which enhance the release of active ingredients and, consequently, provide better protection against bacteria, fungi, and other foodborne pathogens (Hemmati et al. 2021).

Other natural polymers used in packaging applications in combination with PCL include gelatin (Du et al. 2024), zein (Ullah et al. 2024), cellulose or cellulosic derivatives (Ahmadi et al. 2024a, b), pullulan (Rodríguez-Sánchez et al. 2023), and starch (Khalid et al. 2018).

In terms of ingredients used in combination with PCL to obtain active food packaging, there are studies that reported using nisin (Opálková Šišková et al. 2022), palladium nanoparticles (Fabra et al. 2016a), silver nanoparticles (Ahmadi et al. 2024a), carbon dots (Ahmadi et al. 2024b), oregano essential oil (Shi et al. 2022), β-carotene (Fabra et al. 2016b), and different plant and fruit extracts (Ullah et al. 2024; Khalid et al. 2018) (Liu et al. 2021).

3.2.1.2 Biodegradable Polymers from Biomass

This section will focus on chitosan and cellulose, highlighting their natural sources, the extraction processes involved, the properties that make them suitable for packaging, and the current strategies involving sustainable packaging.

Chitosan

Chitin is a polysaccharide placed on the second place in terms of availability, after cellulose. It is usually biomass-derived, being found in the exoskeleton of marine sources such as shrimp, crabs, and cuttlefish shells (Hajji et al. 2014), and also in mussels (Varma and Vasudevan 2020) and fungi (Huq et al. 2022). Chitosan is a biopolymer obtained from chitin, usually by following a three-step chemical treatment using strong acidic or alkali media, starting with demineralization to eliminate calcium carbonate and calcium chloride, then deproteinization to remove proteins, and finally deacetylation to remove acetyl groups. A less environmentally harmful extraction method is the biological route that uses bacteria from proteases for the first two steps, while deacetylation is performed using chitin deacetylases (Knidri et al. 2018).

The properties of chitosan that recommend it for use in sustainable food packaging are its biodegradability, low cost, and antibacterial and antioxidant properties as well as relatively good film-forming properties (Jin et al. 2024). Chitosan's antibacterial properties are attributed to its positively charged nature, which binds to bacteria. These properties are strongly influenced by several factors, including the source of the chitosan, its molecular weight, degree of deacetylation, pH, and the type of bacteria (Sarfraz et al. 2024).

Despite its advantages and bio-based nature, chitosan has some drawbacks, such as poor mechanical and barrier properties, that are not compatible with packaging materials. Thus, researchers have sought other materials to produce chitosan composites with enhanced properties (Jin et al. 2024). Another serious drawback that affects chitosan utilization in food packaging is its solubility only in acidic solutions. Thus, a more recurring alternative to improve water solubility of chitosan is its chemical modification at the functional groups—either amino or hydroxyl. For example, carboxymethyl chitosan is employed more and more in studies in the field of food packaging, usually combined with other polymers or active ingredients to achieve better barrier and mechanical properties (Su et al. 2025).

Various active packaging materials with antibacterial properties containing chitosan have been developed and are continuously being improved. Although chitosan itself possesses antibacterial properties, researchers often incorporate additional substances and extracts to further enhance its effectiveness.

Chitosan bioactive films were developed in a study by reinforcing chitosan with kaolin clay, to improve mechanical and barrier properties. Bio-vanillin, a polyphenol known for its antibacterial activity, was further added. The composite film demonstrated high antioxidant properties, inhibited *E. coli* and *S. aureus* bacteria development, and presented lower moisture permeability. However, the addition of bio-vanillin caused a yellow coloration, which may influence consumer preference when compared to conventional colorless packaging options (Subramani and Manian 2024).

Another approach was followed by Stefanowska et al. (2024) who designed a chitosan composite film with varied percentages of nanocellulose in the form of cellulose nanofibrils and propolis for applications in food packaging. For composite fabrication, a solution-casting preparation route was employed. Upon propolis addition, the composite film exhibited improved antibacterial activity, water vapor transmission, and adhesion, but also improved oxygen permeability. Nanocellulose was found to improve the oxygen barrier properties of the composite, while the variations in its concentration had no impact on the film tensile strength, probably because of the small percentage used—ranging from 0.25 to 1%. An important outcome of the study was that water vapor transmission of chitosan composite films could be adjusted by varying the percentage of added propolis (Stefanowska et al. 2024).

Reinforcing agents, such as fibers or particles, are commonly used in chitosan composites to enhance their mechanical properties. Other approaches include the use of plasticizers and crosslinking agents. For example, a recent study showed that glycerol can play the role of a plasticizer, increasing the flexibility of the film, while gallic acid can be used as a crosslinking agent, enhancing the composite's mechanical properties (Bhowmik et al. 2024).

The intelligent packaging field is also being explored in the case of chitosan composite films. A pH-dependent multilayer film made from chitosan and sodium alginate was developed to track changes related to food quality. The pH-dependent color mechanism is based on structural modification of anthocyanins, which undergo color changes from red to yellow when exposed to alkaline conditions. Additionally, it presents antioxidant, antibacterial, and anti-UV properties. The study presents a comparison between the same composite using the casting method and the layer-by-layer method, with the latter presenting higher mechanical strength and lower light transmittance properties (He et al. 2024).

Advances in the production of edible films were made, with chitosan composites comprising polymers like guar gum (Dehankar et al. 2023), gelatin (Wang et al. 2021; Eranda et al. 2024), and active ingredients such as ZnO nanoparticles (Dehankar et al. 2023), different sources of polyphenols (Dehankar et al. 2023), curcumin (Jiang et al. 2024), rosemary and ginger essential oil (Pires et al. 2018), and fruits extracts (Loukri et al. 2024; Sady et al. 2021).

Chitosan is a versatile polysaccharide that can be processed through various techniques, as observed in the literature. Probably the most frequently used method in research articles is casting due to its ease of operation and low cost (no special equipment needed), followed by extrusion, a method that limits the use of solvents, which are typically environmentally harmful. Other preparation methods include electrospinning, thermoforming, spraying, and dipping (Barik et al. 2024).

Cellulose

Cellulose is a biodegradable and biocompatible polysaccharide (Asim et al. 2022) that occupies the first place in terms of abundance in nature (Liu et al. 2023a). Cellulose is usually extracted from wood, forestry wastes, vegetable fibers, agriculture crops, algae, or tunicates through either mechanical methods, such as grinding and ball milling or chemical methods such as hydrolysis (Avila et al. 2021). Unlike plant-derived cellulose, which contains lignin, hemicellulose, or other components, cellulose can also be biosynthesized by aerobic bacteria in a highly pure form (Cazón and Vázquez 2021).

Cellulose consists of β-(1,4)-linked D-glucose units, which form hydrogen bonds that help organize the molecular chains into microfibrils. This arrangement gives the relatively good mechanical properties exhibited by cellulose (Liu et al. 2023a). Its properties are closely linked to the extraction source, which should be taken into consideration when designing sustainable food packaging composites. For example, cellulose produced by bacteria presents higher purity, degree of polymerization, and crystallinity, therefore better mechanical and barrier properties when compared to cellulose extracted from agricultural waste, which contains impurities, has lower crystallinity, and implicitly poorer mechanical properties (Nath et al. 2024). Additionally, given its abundance in nature and in different types of waste, cellulose has the advantage of low cost (Liu et al. 2021).

Despite its advantages and widespread availability, cellulose lacks antibacterial properties and is insoluble in conventional solvents due to its extensive hydrogen bonding (Liu et al. 2023a). Various solvents and solvent systems have been developed to dissolve cellulose, but issues such as toxicity, high volatility, hydrophobicity, environmental pollution, and high cost limit their use at the industrial level. Examples of promising solvents for cellulose include ionic liquids, alkali/urea, N, N-dimethylacetamide/lithium chloride (DMAc/LiCl), N-methyl morpholine-N-oxide (NMMO), molten inorganic salt solutions, metal complex solutions, and tetrabutylammonium fluoride/dimethyl sulfoxide (DMSO) (Liu et al. 2023a).

Cellulose can be processed in several ways, but the traditional and most used methods, particularly in industry, are wet molding and thermocompression of dried cellulose powder or fibers, as well as dried microfibrillated cellulose. Novel technologies include dry molding, deep-drawing, development of cellulose foams, injection-molding, and even 3D printing (Freville et al. 2024). Traditional methods are continuously studied and improved with the final goal of developing performing materials, in a smarter, greener, and less expensive way. For example, cellulose functionalization, specifically phosphorylation can be performed during the thermoforming process, leading to an increase in material crystallinity and thus, improved water barrier properties (Semlali et al. 2023).

Different types of cellulose derivatives are currently employed in the food packaging industry depending on the cellulose source and processing methods (Avila et al. 2021). Examples include cellulose esters, such as cellulose acetate, cellulose nitrate, and cellulose sulfate, as well as cellulose ethers like carboxymethylcellulose and nanocellulose (Avila et al. 2021; Nath et al. 2024).

Cellulose acetate is a biodegradable polymer approved for food contact and classified as Generally Recognized as Safe (GRAS) by the FDA. It is mainly used in food packaging as a wrapping material, but is often blended with other polymers to enhance its gas and water barrier properties (Sid et al. 2021). Cellulose nitrate, also known as nitrocellulose, exhibits good mechanical properties and solubility, but it is sensitive to heat and light. Cellulose sulfate has the advantage of being soluble in water even at high concentrations, along with exhibiting antibacterial properties. It was studied more for its applications in drug delivery systems, but its properties recommend it for use also in packaging (Romão et al. 2022). Carboxymethylcellulose is a water-soluble derivative of cellulose that addresses its hydrophobic limitations. It has been extensively studied for packaging applications but is often mixed with other polymers or fillers to reduce its brittleness (Ramakrishnan et al. 2024). Nanocellulose is found in the form of cellulose nanocrystals and cellulose nanofibrils, which are used as reinforcing agents, as well as in the form of bacterial nanocellulose. It was proven to improve gas and water barrier properties, as well as mechanical properties when added in different compositions as a filler (Ahankari et al. 2021).

Cellulose or cellulose derivatives are usually employed in studies for their availability, low cost, and biodegradability. Even if cellulose derivatives were designed to exhibit better solubility, and mechanical and barrier properties, they have their own weaknesses and thus need to be blended or reinforced with fillers and different types of agents for specific applications in packaging, especially in the food sector.

Several studies utilize cellulose solely for its film-forming properties as a polymeric matrix. For instance, a cellulose composite containing curcumin was developed to monitor food freshness through pH-responsive color changes. Curcumin is responsible for the pH-dependent coloration mechanism, improving films' thermal stability and mechanical strength, as well as providing good antioxidant and anti-biofilm activity, while cellulose has no impact on these properties (Miao et al. 2024).

The literature presents many studies in the field of packaging on cellulose and cellulose derivatives blended with other bio-sourced or biodegradable polymers like sodium alginate (Tan et al. 2024), polyvinyl alcohol (Wang et al. 2025), gelatin (Khan et al. 2024), polylactic acid (Chen et al. 2024), natural rubber latex (Viswanathan et al. 2024), polydiacetylene (Yue et al. 2024), or chitosan (Zhao et al. 2023).

Cellulose derivatives may also undergo chemical modification to enhance their compatibility with other composite components. For example, oxidized cellulose nanocrystals bearing carboxyl groups were found to physically interact with gelatin's amino groups, improving dispersion and yielding a flexible and more uniform film. This highlights the importance of polymer-filler interactions in influencing composite performance (Leite et al. 2024). One major disadvantage of cellulose acetate is the small difference between its glass transition temperature and degradation temperature, making it difficult to process. Additives, such as plasticizers, can be incorporated to allow for higher processing temperatures without risking damage to the polymer (Oliviero et al. 2023).

Active ingredients, like sorbates, are often incorporated into composites for their antimicrobial properties; however, direct blending with the polymer can lead to an uncontrollable release of these ingredients. Approaches such as LDH modification with sorbates lead to a controlled release of these active ingredients but also to a better dispersion and interaction with the cellulosic matrix (Oliviero et al. 2023). Other types of clays, such as sodium montmorillonite, have also been added in cellulosic matrices for improved mechanical and barrier properties (Baniasadi et al. 2025).

Ongoing efforts in cellulose extraction focus on utilizing food waste via biorefining approaches, which convert biomass into value-added products. For example, waste from kimchi production has been explored as a cellulose source for making edible packaging films, reflecting the push toward waste valorization and sustainability (Lee et al. 2024).

Although extensive research has been devoted to cellulose in food packaging, recent developments have expanded its application into electronic packaging. One innovative study aimed at producing "brick–mortar" structures with good heat dissipation properties, by combining cellulose nanocrystals and MoS_2 as "brick," with polypropylene carbonate serving as the, plays the role of "mortar." Here, cellulose nanocrystals acted as spacers for MoS_2 layer, blocking them in the exfoliated structure (Wu et al. 2024).

3.2.1.3 Biodegradable Polymers from Microbial Fermentation

This chapter explores the various types of polyhydroxyalkanoates (PHAs), highlighting their synthesis, structural diversity, and applications in the packaging industry.

Polyhydroxyalkanoates (PHAs) are a class of naturally occurring polymers, synthesized by hundreds of bacterial species as a survival mechanism when exposed to stress conditions or nutrient deficiencies, such as a lack of oxygen, nitrogen, and phosphorus. PHAs are stored as granules in the cytoplasm of cells when there is an excess of carbon in the environment (Thamarai et al. 2024; Bhatia et al. 2024; Abate et al. 2024; Mai et al. 2024).

When it comes to industrial production, there are some limitations linked with PHA production that make its cost considerably higher when compared to conventional polymers like polyethylene or PET. On one side, the operation of the bioreactors needs to be performed in sterile condition, the carbon source is generally carefully chosen and corresponds to approxiately 40% of the production expenses, and the purification process involves a high amount of environmentally harmful solvents (Donno et al. 2021).

There are some strategies for making the production of PHA more sustainable, thereby contributing to waste valorization and sustainable materials development (Bhatia et al. 2024). For example, current approaches involve the valorization of renewable carbon sources derived from agriculture or the food industries for PHA production, such as cheese whey, orange, grape or pineapple peels, apple pulp, coffee ground, waste cooking oil, waste fish fat, and other animal or plant-derived waste (Bhatia et al. 2024). Additionally, numerous green alternatives are available for solvents used in PHA purification, helping to reduce its environmental impact.

Regardless of the production method chosen, PHAs generally possess several desirable properties, including biocompatibility, non-toxicity, and mechanical flexibility, making them suitable for a wide range of applications, from packaging materials to biomedical devices (Thamarai et al. 2024).

The chemical structure and properties of PHAs are strongly influenced by the bacterial species that synthesize them, as well as the growth conditions and carbon source used (Masood et al. 2017). Based on the number of carbon atoms in the monomeric unit, PHA can be divided into short-chain length (with 3–5 carbon atoms), medium-chain length (with 6–14 carbon atoms), and long-chain length (with more than 14 carbon atoms), as well as copolymers that include combinations of the already mentioned groups. Short-chain length PHAs present a high percentage of crystallinity, between 60 and 80%, which makes them brittle, while medium-chain length and their copolymers present a lower crystallinity, making them more elastic, with lower tensile strength and processing temperatures (Ragaert et al. 2019). Examples of short-chain length PHAs include poly(3-hydroxybutyrate) (PHB), poly(3-hydroxybutyrate-co-3-hydroxyvalerate) (PHBV), and poly(4-hydroxybutyrate) (P4HB), while poly(3-hydroxybutyrate-co-3-hydroxyhexanoate) (PHBH) is part of the long-chain length PHAs (Ragaert et al. 2019).

Advancements in improving PHA performance, particularly for food packaging applications, often focus on enhancing their barrier properties, thermal processability, stability, and mechanical strength by blending them with other natural polymers or additives (Yeo et al. 2024). Examples of polymers used for blending with PHAs include PLA, PVA, PEG, and cellulose (Kumar et al. 2021). This process typically results in less crystalline structures with increased biodegradability (Adak et al. 2024).

For example, a strategy for improving PHA UV-barrier, and mechanical and antibacterial properties was offered by Wang et al. They developed a nanofiller in the form of custom-made quantum dots, which functioned as nucleating agents for PHAs. The quantum dots were developed using nature-derived raw materials and, due to their amino and carboxy groups, exhibited very good dispersion in the polymeric matrix. The film was produced through solvent casting and is presented as a possible green packaging and intelligent sensing packaging due to its pH-responsiveness (Wang et al. 2023b).

Remarkable improvements in polyhydroxybutyrate (PHB) properties were achieved by incorporating graphene nanoplatelets. By adding very low concentrations of graphene, following a solution-casting procedure, the resulting nanocomposite exhibited better thermal properties, tensile strength, and improved barrier properties against oxygen and water (Manikandan et al. 2020).

Studies on the effects of various essential oils—such as grapeseed, bergamot, and ginger—on PHB films have shown that these oils not only provide antibacterial properties but also enhance flexibility, water vapor barrier, and UV-blocking capabilities (Kumari et al. 2023). MoS_2 was also studied as an active ingredient for use in active packaging (Maureira et al. 2023).

PHAs are sometimes used to enhance the properties of other polymers. For instance, PHA was incorporated into PLA to improve its elasticity and reduce brittleness in an active packaging film with antibacterial properties. To prevent the rapid release of the active ingredient—oregano oil, montmorillonite filler was also included (Zheng et al. 2022). Another example demonstrates how PHBV, combined with fillers such as talc and starch, improves the poor water vapor and oxygen barrier properties of bio-PBS, resulting in the development of composites with tunable mechanical and thermal properties (Rodriguez-Uribe et al. 2021). In another study, PHBV was grafted onto cellulose nanofibers (CNF) to enhance their compatibility with PBAT. This modification addresses PBAT's drawbacks, such as poor gas barrier properties and low crystallinity, by improving the interaction between CNF and the polymer (Dai et al. 2024).

In the case of PHBV, studies report the production of electrospun nanocomposite fibers incorporating TiO_2 and humic nanoparticles, with potential applications in food packaging due to their enhanced properties (Venezia et al. 2023). Other sustainable nanocomposites made of PHBV, PBAT, and nanoclays were produced by compression molding and cast film extrusion. The two fabrication methods were compared in terms of properties of the resulting materials, indicating that cast extrusion leads to chain orientation and thus improved barrier properties (Pal et al. 2020).

The interest in discovering and employing different biopolymers in various domains, from biotechnology to packaging and electronics, has increased with the awareness of the current global situation regarding non-regenerable resources, recyclability, and sustainability. More and more articles on the use of bio-sourced polymers are released every year, but in many situations, the manufacturing cannot be scaled up to industry because of the complicated technology and, implicitly, the high cost when compared to fossil-fuel-derived plastics. Additionally, bio-sourced polymers are in most of the cases mixed with other polymer, particles, or fibers to either reduce their cost or to improve some properties. Close attention should be paid to how the biodegradability of the polymer is affected when combined with other compounds in a composite. Even if composites exhibit better properties, at the end of their lifecycle they become more difficult to recycle (Sangroniz et al. 2019).

3.3 Recyclability of Biocomposite Packaging: Key Considerations for End Products

Numerous examples of biodegradable polymers, both natural and synthetic, are being explored in current research with the aim of developing revolutionary sustainable materials. Sustainability does not only refer to the biodegradability of the raw materials or final product, but also to product valorization after its utilization through different recycling approaches. In this context, closed-loop recycling plays a crucial role by reducing the amount of new plastic needed and preserving existing resources for future generations.

At this moment, bioplastics derived from renewable sources represent only a small part of the total market of plastics because of their high processing cost and reduced performance when compared to conventional fossil-derived polymers. The growing demand for sustainability, along with strict regulations and high taxes on conventional plastics (Rasheed et al. 2024), will likely drive industries to focus more on materials derived from renewable sources in the near future. Recycling methods for these types of materials are being explored and implemented now to prepare for the transition to a more sustainable way of life (Tamizhdurai et al. 2024). Similar to conventional plastics, polymers derived from renewable sources can be recycled according to their characteristics: mechanical recycling for high molecular weight polymers, composting for biodegradable materials, advanced recycling methods for lower molecular weight polymers, and incineration for polymers that cannot be recycled (Parveen et al. 2024).

Recycling composites is more challenging due to their multiple components, which often require costly and complex separation processes, if separation is even possible. Researchers have developed and proposed several recycling approaches to address these challenges. For example, a study focused on separating the materials used in Tetra Pak® packaging by means of a mechano-chemical recycling process. Cellulose, polyethylene, and aluminum were successfully separated, and characterization techniques indicated that the recovered materials could be further used in other applications (Borges et al. 2024).

Probably the most effective way of dealing with composite materials recycling would be to consider the recyclability of composite materials right from the design stage, even before they are manufactured. An example in this case is an antibacterial film made of carbon nitride, chitosan, and PVA with applications in fruit packaging. The authors proved that the film could be recycled by dissolution in water followed by solvent-casting, ultimately obtaining properties similar to the original material (Liu et al. 2023b). A similar recycling method was applied to an antimicrobial packaging film with a polymeric matrix made of ethyl cellulose, containing rosin-based quaternary ammonium salt as the active ingredient. After use, the waste films could be recycled twice through solvent casting with chloroform, while retaining their antibacterial and mechanical properties at levels comparable to the original film (Liao et al. 2023b).

A sustainable way of producing composite polyurethane foams, filled with blast slag from a furnace, from bio-based polyols was developed by Recupido et al. (2024). At the same time, they proposed a chemical recycling route for polyurethane depolymerization using bio-based amines, such as putrescine. The depolymerized product was then used to produce second-generation foam with potential applications in packaging (Recupido et al. 2024). Another study that develops packaging films and suggests a recycling route refers to a gelatin composite film incorporating lignosulfonate and fish oil to enhance mechanical strength and water repellency. This composite film can be recycled up to three times through solvent-casting with water, while maintaining its mechanical properties and water-repellent qualities (Chen et al. 2023).

Literature studies on recycling composites produced from renewable sources with applications in packaging use easy recycling techniques such as dissolution, followed by solvent-casting. In practice, not all composite materials can be processed using this technique, as they often require more complex technologies and large amounts of solvents, which may result in materials with lower performance that are unsuitable for high-value applications. Although the processes can be costly, the energy required for recycling is often significantly lower than that needed to produce new materials, making it a more sustainable option. In the case of recycling complex composites, priority should be given to applying the cascade principle, utilizing the material for its next most valuable use before allowing it to biodegrade (Khalid et al. 2022; Hubbe 2023).

3.4 Process Efficiency in Biocomposite Packaging Production

In the rush to achieve sustainability, it is crucial that people, as well as companies, remain mindful of the importance of efficiently using all types of resources, including human resources, energy, time, or materials (Schilirò et al. 2019). Figure 3.4 highlights the key factors that should be considered to enhance process efficiency in sustainable packaging production. Circular economy models are closely aligned with these efficiency principles, offering a more energy and cost-effective approach to manufacturing compared to traditional linear systems (Nikolakis et al. 2024). The principles of circular economy refer to:

1. Waste elimination, either by waste valorization or by implementing more efficient production processes.
2. Circulating products and maintaining their high quality by employing effective recycling approaches.
3. Regenerate nature by restoring natural ecosystems and developing materials and technologies inspired by nature (Nikolakis et al. 2024).

Strategies have been implemented to ensure the efficient use of limited raw materials such as petroleum, while also promoting waste valorization. Natural products

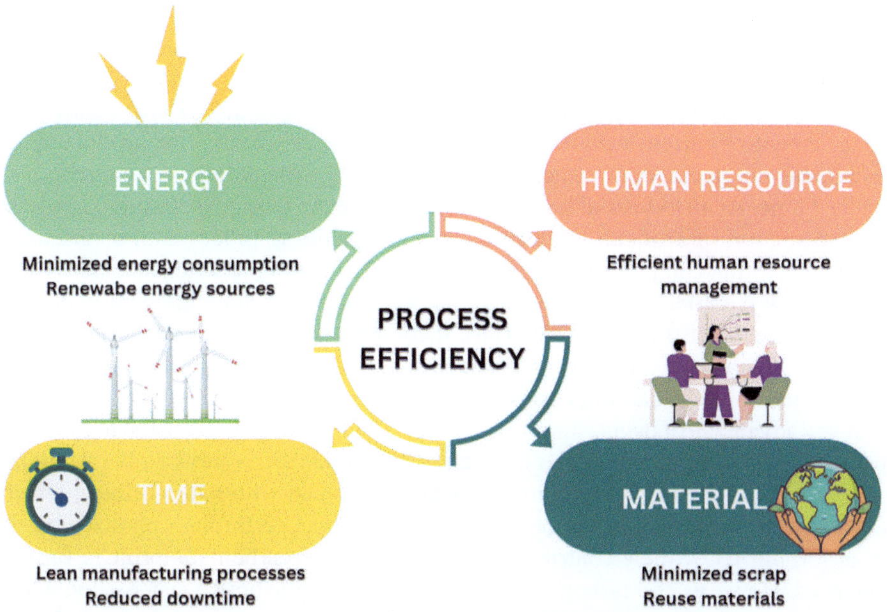

Fig. 3.4 Key factors for process efficiency in sustainable packaging production

derived from feedstocks are now used in the manufacturing of conventional polymers like PE, PP, and PET, leading to products called bio-PE, bio-PP, and bio-PET. Glucose serves as the starting raw material for all the polymers mentioned earlier. The production of monomers for bio-PET is more complex than for bio-PP and bio-PE, as it involves two monomers instead of one. However, once the monomers are obtained, the polymerization process follows the same pathway as that for petroleum-based polymers, and the resulting products are reported to have properties identical to those derived from petroleum sources (Siracusa and Blanco 2020).

In 2007, Braskem became the first company to produce bio-based ethylene from sugarcane ethanol, expanding its portfolio with the development of bio-based ethylene vinyl acetate copolymer (EVA) in 2018 (https://www.coca-colacompany.com/media-center/100-percent-plant-based-plastic-bottle). Meanwhile, LyondellBasell offers several bio-based alternatives under the brand name CirculenRenew, including bio-PE and bio-PP made from waste and residue oils supplied by Neste (https://www.coca-colacompany.com/media-center/100-percent-plant-based-plastic-bottle). Coca-Cola was the first to produce PET bottles containing 30% plant-based material with their PlantBottle™. More recently, they introduced the world's first beverage bottle made entirely from 100% plant-based plastic, excluding the cap and label (https://www.coca-colacompany.com/media-center/100-percent-plant-based-plastic-bottle). Another notable polyester produced at an industrial scale from monomers derived from natural sources is bio-based polybutylene succinate (bio-PBS), which is also biodegradable (https://www.coca-colacompany.com/media-center/100-percent-plant-based-plastic-bottle; Kato et al. 2024).

Another important factor in improving process efficiency is the continuous enhancement of current manufacturing techniques to ensure that the production of sustainable polymeric composites requires less energy, fewer human resources, and reduced time. At an industrial level, this can be achieved through automation and the implementation of artificial intelligence in manufacturing processes. For instance, machine learning can reduce unplanned stops by predicting maintenance needs and forecast raw material requirements, leading to a faster and more efficient manufacturing process (Zatsu et al. 2024). Thus, there are several factors affecting process efficiency when speaking about manufacturing sustainable composite materials and all of them should be carefully considered before production.

A key advantage of using composites is their reduced weight compared to products made entirely from pure polymers, which helps lower their overall global warming potential. Additionally, when composites are reinforced with natural fibers, which are considered carbon–neutral because they are produced through photosynthesis by absorbing CO_2, their contribution to reducing the carbon footprint is further enhanced. However, the carbon footprint should not be the only consideration when designing sustainable composites. Other important factors to address include the depletion of natural resources, ozone layer degradation, eutrophication of natural waters, air and water acidification, photochemical oxidation, and the toxic effects on humans, soil, water, and animals. Moreover, attention should be given to creating long-lasting products, as these can offer a net reduction in global warming potential compared to similar items that need to be replaced more frequently (Jadhav et al. 2021).

3.5 Conclusions and Future Perspectives

The field of packaging is continuously expanding, with the focus on developing biodegradable or compostable materials that take advantage of different types of natural polymers or substances, as well as food and agriculture-derived waste. Examples of natural polymers extensively studied by researchers for their potential in packaging applications include, but are not limited to, PLA, PBS, PCL, chitosan, cellulose, and polyhydroxyalkanoates. Sustainable packaging alternatives for food industry are already being produced by different companies such as Braskem, Coca-Cola, DOW, and Lyondell Basell, proving that these environmentally friendly packaging alternatives can be economically feasible and exhibit good properties at the same time.

The challenges related to sustainable composite materials are diverse and involve many factors, such as their higher cost compared to conventional materials, particularly when scaling up for industrial production. Additionally, improvements are often needed in their performance and durability, specifically regarding mechanical and barrier properties. Finally, consumer education is crucial, as misconceptions about natural polymers and biodegradability could result in inappropriate disposal in landfills (Sabet 2024).

The opportunities and positive outlooks for sustainable composite materials are evident in several key areas: the industry's acknowledgment of their environmental benefits in reducing plastic pollution and fostering a circular economy; ongoing innovation and research that are enhancing the performance and cost-effectiveness of these materials, making them more appealing to businesses; and increasing regulatory support through government initiatives and policies that promote sustainable packaging solutions (Sabet 2024). However, while these advances are promising, large-scale implementation will require overcoming significant logistical and technological challenges, including the establishment of efficient recycling infrastructures and the reduction of production costs. One of the emerging challenges is the integration of novel processing technologies, such as 3D printing and bio-manufacturing, which may offer potential solutions but also introduce additional complexities in terms of cost and scalability.

When it comes to polymeric composite materials used in the packaging industry, their recycling becomes more challenging, requiring complex technologies and significant investment. The recycling costs of these materials should be assessed against the expenses associated with producing the same material from scratch, considering factors such as energy consumption and carbon footprint. Innovative recycling technologies, such as chemical recycling and enzymatic degradation, may hold promise, but their feasibility at large scales needs further exploration. If recycling is considered insufficiently sustainable, a more effective solution for these materials could be to design them to be fully biodegradable or compostable. An example in this case is TIPA, a company that makes their flexible and multifilm packaging solutions compostable, thus advancing efforts toward achieving a truly circular economy.

In conclusion, the reviewed literature shows significant progress in the development of sustainable composites for packaging. Most studies focus on food packaging, especially active or intelligent packaging, with the primary goal of reducing waste. However, substantial research is still required to address the pressing environmental challenges posed by technological advancements. Future research should prioritize solutions that not only reduce carbon footprints and address other life-cycle impacts but also ensure durability, enable full reuse or recycling of materials, preserve their value across multiple uses, and demonstrate strong potential for large-scale implementation (Hubbe 2023). Emerging technologies, coupled with increased consumer education and the scaling of sustainable production techniques, will be key drivers in overcoming these challenges and achieving a truly sustainable packaging industry.

Acknowledgements This work was funded by the EU's NextGenerationEU instrument through the National Recovery and Resilience Plan of Romania—Pillar III-C9-I8, managed by the Ministry of Research, Innovation and Digitalization, within the project entitled "Advanced 3D Printing for Designer Catalysts Applied to Biomass Conversion," contract no.760086/23.05.2023, code CF 43/14.11.2023.

References

A Sustainable Europe by 2030 (2019) https://commission.europa.eu/publications/sustainable-europe-2030_en

Abate T, Amabile C, Muñoz R, Chianese S, Musmarra D (2024) Polyhydroxyalkanoate recovery overview: properties, characterizations, and extraction strategies. Chemosphere 356:141950. https://doi.org/10.1016/j.chemosphere.2024.141950

Adak S, Kayalvizhi R, Bishai M, Jacob S, Kundu D (2024) Advancements in microbial production of polyhydroxyalkanoates (PHA) from wastes for sustainable active food packaging: an eclectic review. Biocatal Agric Biotechnol 60:103288. https://doi.org/10.1016/j.bcab.2024.103288

Ahankari SS, Subhedar AR, Bhadauria SS, Dufresne A (2021) Nanocellulose in food packaging: a review. Carbohyd Polym 255:117479. https://doi.org/10.1016/j.carbpol.2020.117479

Ahmadi N, Ahari H, Anvar A, Khosravi-Darani K, Gharachorloo M (2024a) Polycaprolactone (PCL)-based films integrated with hairy cellulose nanocrystals and silver nanoparticles for active Tilapia packaging applications. Food Chem: X 22:101490. https://doi.org/10.1016/j.fochx.2024.101490

Ahmadi S, Khoshkalampour A, Ghorbani M, Ramezani S, Ghasempour Z, Ghareaghajlou N (2024b) Development of active packaging material based on polycaprolactone/hydroxypropyl methylcellulose nanofibers containing carbon dot nanoparticles for meat preservation. LWT 197:115913. https://doi.org/10.1016/j.lwt.2024.115913

Aliotta L, Seggiani M, Lazzeri A, Gigante V, Cinelli P (2022) A brief review of poly (butylene succinate) (PBS) and its main copolymers: synthesis, blends, composites, biodegradability, and applications. Polymers 14(4):844

Aloysius H, Desaulniers D, Bandiera S (2015) Evaluating toxic endpoints of exposure to a commercial PCB mixture: an in vivo laboratory study. AIMS Environ Sci 2:314–332. https://doi.org/10.3934/environsci.2015.2.314

Asim N, Badiei M, Mohammad M (2022) Recent advances in cellulose-based hydrophobic food packaging. Emerg Mater 5(3):703–718. https://doi.org/10.1007/s42247-021-00314-2

Aslam S, Khurram A, Hussain R, Qadir A, Ahmad SR (2023) Sources, distribution, and incipient threats of polymeric microplastic released from food storage plastic materials. Environ Monit Assess 195(6):638. https://doi.org/10.1007/s10661-023-11242-5

Azeredo HMC, Correa DS (2021) Smart choices: mechanisms of intelligent food packaging. Curr Res Food Sci 4:932–936. https://doi.org/10.1016/j.crfs.2021.11.016

Babaremu K, Oladijo OP, Akinlabi E (2023) Biopolymers: a suitable replacement for plastics in product packaging. Adv Ind Eng Polym Res 6(4):333–340. https://doi.org/10.1016/j.aiepr.2023.01.001

Baniasadi H, Fathi Z, Lizundia E, Cruz CD, Abidnejad R, Fazeli M et al (2025) Development and characterization of pomegranate peel extract-infused carboxymethyl cellulose composite films for functional, sustainable food packaging. Food Hydrocolloids 158:110525. https://doi.org/10.1016/j.foodhyd.2024.110525

Barik M, BhagyaRaj GVS, Dash KK, Shams R (2024) A thorough evaluation of chitosan-based packaging film and coating for food product shelf-life extension. J Agric Food Res 16:101164. https://doi.org/10.1016/j.jafr.2024.101164

Barletta M, Aversa C, Ayyoob M, Gisario A, Hamad K, Mehrpouya M et al (2022) Poly(butylene succinate) (PBS): materials, processing, and industrial applications. Prog Polym Sci 132:101579. https://doi.org/10.1016/j.progpolymsci.2022.101579

Bhatia SK, Patel AK, Yang Y-H (2024) The green revolution of food waste upcycling to produce polyhydroxyalkanoates. Trends Biotechnol 42(10):1273–1287. https://doi.org/10.1016/j.tibtech.2024.03.002

Bhowmik S, Agyei D, Ali A (2024) Enhancement of mechanical, barrier, and functional properties of chitosan film reinforced with glycerol, COS, and gallic acid for active food packaging. Sustain Mater Technol 41:e01092. https://doi.org/10.1016/j.susmat.2024.e01092

Borges R, Brondi M, Elias AM, Farinas CS, Ribeiro C (2024) Mechanochemical recycling of cellulose multilayer carton packages to produce micro and nanocellulose from the perspective of techno-economic and environmental analysis. J Environ Manage 363:121254. https://doi.org/10.1016/j.jenvman.2024.121254

Cazón P, Vázquez M (2021) Bacterial cellulose as a biodegradable food packaging material: a review. Food Hydrocolloids 113:106530. https://doi.org/10.1016/j.foodhyd.2020.106530

Chen L, Qiang T, Ren W, Tian Q, Zhang X, Zhang HJ (2023) Strong, water-repellent, and recyclable gelatin-based bioplastic film as sustainable express packaging film. J Clean Prod 385:135705. https://doi.org/10.1016/j.jclepro.2022.135705

Chen C, Wang L, Shams Es-haghi S, Tajvidi M, Wang J, Gardner DJ (2024) Biodegradable and recyclable bio-based laminated films of poly (lactic acid) and cellulose nanocrystals for food barrier packaging. Food Packag Shelf Life 42:101244. https://doi.org/10.1016/j.fpsl.2024.101244

da Silva Pens CJ, Klug TV, Stoll L, Izidoro F, Flores SH, de Oliveira RA (2024) Poly (lactic acid) and its improved properties by some modifications for food packaging applications: a review. Food Packag Shelf Life 41:101230. https://doi.org/10.1016/j.fpsl.2023.101230

Dai Z, Li B, He M, Li J, Zou X, Wang Y et al (2024) Biodegradable and high-performance composites of poly (butylene adipate-co-terephthalate) with PHBV chain-functionalized nanocellulose. Prog Org Coat 187:108136. https://doi.org/10.1016/j.porgcoat.2023.108136

De Donno NL, Moreno Sayavedra S, Rene ER (2021) Polyhydroxyalkanoate (PHA) production via resource recovery from industrial waste streams: a review of techniques and perspectives. Biores Technol 331:124985. https://doi.org/10.1016/j.biortech.2021.124985

de Avila Delucis R, de Cademartori PHG, Fajardo AR, Amico SC (2021) Cellulose and its derivatives: properties and applications. Polysaccharides:221–252

Debeaufort F (2021a) Glass packaging. In: Packaging materials and processing for food, pharmaceuticals and cosmetics, pp 49–73

Debeaufort F (2021b) Metal packaging. In: Packaging materials and processing for food, pharmaceuticals and cosmetics, pp 75–104

Dehankar HB, Mali PS, Kumar P (2023) Edible composite films based on chitosan/guar gum with ZnONPs and roselle calyx extract for active food packaging. Appl Food Res 3(1):100276. https://doi.org/10.1016/j.afres.2023.100276

Dejene BK, Gudayu AD, Abtew MA (2024) Development and optimization of sustainable and functional food packaging using false banana (Enset) fiber and zinc-oxide (ZnO) nanoparticle-reinforced polylactic acid (PLA) biocomposites: a case of Injera preservation. Int J Biol Macromol 279:135092. https://doi.org/10.1016/j.ijbiomac.2024.135092

Deshwal GK, Panjagari NR (2020) Review on metal packaging: materials, forms, food applications, safety and recyclability. J Food Sci Technol 57(7):2377–2392. https://doi.org/10.1007/s13197-019-04172-z

Deshwal GK, Panjagari NR, Alam T (2019) An overview of paper and paper based food packaging materials: health safety and environmental concerns. J Food Sci Technol 56(10):4391–4403. https://doi.org/10.1007/s13197-019-03950-z

Donkor L, Kontoh G, Yaya A, Bediako JK, Apalangya V (2023) Bio-based and sustainable food packaging systems: relevance, challenges, and prospects. Appl Food Res 3(2):100356. https://doi.org/10.1016/j.afres.2023.100356

Dörnyei KR, Uysal-Unalan I, Krauter V, Weinrich R, Incarnato L, Karlovits I et al (2023) Sustainable food packaging: an updated definition following a holistic approach. Front Sustain Food Syst:7. https://doi.org/10.3389/fsufs.2023.1119052

Du Y, Zhang Y, Pang J (2024) Electrospun konjac glucomannan-gelatin/polycaprolactone bilayer nanofibrous films with improved hydrophobicity and mechanical properties for food packaging. Food Biosci 61:104544. https://doi.org/10.1016/j.fbio.2024.104544

El Knidri H, Belaabed R, Addaou A, Laajeb A, Lahsini A (2018) Extraction, chemical modification and characterization of chitin and chitosan. Int J Biol Macromol 120:1181–1189. https://doi.org/10.1016/j.ijbiomac.2018.08.139

Eranda DHU, Chaijan M, Panpipat W, Karnjanapratum S, Cerqueira MA, Castro-Muñoz R (2024) Gelatin-chitosan interactions in edible films and coatings doped with plant extracts for biopreservation of fresh tuna fish products: a review. Int J Biol Macromol 280:135661. https://doi.org/10.1016/j.ijbiomac.2024.135661

Fabra MJ, López-Rubio A, Cabedo L, Lagaron JM (2016a) Tailoring barrier properties of thermoplastic corn starch-based films (TPCS) by means of a multilayer design. J Colloid Interface Sci 483:84–92. https://doi.org/10.1016/j.jcis.2016.08.021

Fabra MJ, López-Rubio A, Sentandreu E, Lagaron JM (2016b) Development of multilayer corn starch-based food packaging structures containing β-carotene by means of the electrohydrodynamic processing. Starch Stärke 68(7–8):603–610. https://doi.org/10.1002/star.201500154

Fredi G, Dorigato A (2021) Recycling of bioplastic waste: a review. Adv Ind Eng Polym Res 4(3):159–177. https://doi.org/10.1016/j.aiepr.2021.06.006

Freville E, Sergienko JP, Mujica R, Rey C, Bras J (2024) Novel technologies for producing tridimensional cellulosic materials for packaging: a review. Carbohyd Polym 342:122413. https://doi.org/10.1016/j.carbpol.2024.122413

Furszyfer Del Rio DD, Sovacool BK, Foley AM, Griffiths S, Bazilian M, Kim J et al (2022) Decarbonizing the glass industry: a critical and systematic review of developments, sociotechnical systems and policy options. Renew Sustain Energy Rev 155:111885. https://doi.org/10.1016/j.rser.2021.111885

Ganjeh AM, Saraiva JA, Pinto CA, Casal S, Gonçalves I (2024) Recent advances in nano-reinforced food packaging based on biodegradable polymers using layer-by-layer assembly: a review. Carbohydr Polym Technol Appl 7:100395. https://doi.org/10.1016/j.carpta.2023.100395

Ghasemlou M, Barrow CJ, Adhikari B (2024) The future of bioplastics in food packaging: an industrial perspective. Food Packag Shelf Life 43:101279. https://doi.org/10.1016/j.fpsl.2024.101279

Godoy Zúniga MM, Ding R, Oh E, Nguyen TB, Tran TT, Nam J-D et al (2024) Avocado seed starch utilized in eco-friendly, UV-blocking, and high-barrier polylactic acid (PLA) biocomposites for active food packaging applications. Int J Biol Macromol 265:130837. https://doi.org/10.1016/j.ijbiomac.2024.130837

Goñi-Ciaurriz L, Vélaz I (2022) Antibacterial and degradable properties of β-cyclodextrin-TiO_2 cellulose acetate and polylactic acid bionanocomposites for food packaging. Int J Biol Macromol 216:347–360. https://doi.org/10.1016/j.ijbiomac.2022.06.202

Gonon H, Srisa A, Promhuad K, Chonhenchob V, Bumbudsanpharoke N, Jarupan L et al (2023) PLA thermoformed trays incorporated with cinnamaldehyde and carvacrol as active biodegradable bakery packaging. Food Packag Shelf Life 38:101123. https://doi.org/10.1016/j.fpsl.2023.101123

Guarino V, Gentile G, Sorrentino L, Ambrosio L (2002) Polycaprolactone: synthesis, properties, and applications. In: Encyclopedia of polymer science and technology, pp 1–36

Guo X, Li L, Qi Y, Su J, Ou X, Lv M et al (2024) Green-synthesized antibacterial and unidirectional water-permeable polylactic acid/ZnO composite film for enhanced preservation of perishable fruits. Mater Today Chem 40:102284. https://doi.org/10.1016/j.mtchem.2024.102284

Hajji S, Younes I, Ghorbel-Bellaaj O, Hajji R, Rinaudo M, Nasri M et al (2014) Structural differences between chitin and chitosan extracted from three different marine sources. Int J Biol Macromol 65:298–306. https://doi.org/10.1016/j.ijbiomac.2014.01.045

Han Lyn F, Ismail-Fitry MR, Noranizan MA, Tan TB, Nur Hanani ZA (2024) Recent advances in extruded polylactic acid-based composites for food packaging: a review. Int J Biol Macromol 266:131340. https://doi.org/10.1016/j.ijbiomac.2024.131340

He M, Zheng Y, Shen J, Shi J, Zhang Y, Xiao Y et al (2024) Chitosan/sodium alginate multilayer pH-sensitive films based on layer-by-layer self-assembly for intelligent packaging. J Renew Mater 12(2):215–233

Hemmati F, Bahrami A, Esfanjani AF, Hosseini H, McClements DJ, Williams L (2021) Electrospun antimicrobial materials: Advanced packaging materials for food applications. Trends Food Sci Technol 111:520–533. https://doi.org/10.1016/j.tifs.2021.03.014
Historical perspective and evolution. Packaging technology and engineering 2020, pp 1–25
https://www.coca-colacompany.com/media-center/100-percent-plant-based-plastic-bottle
Hubbe MA (2023) Sustainable composites: a review with critical questions to guide future initiatives. Sustainability 15(14):11088
Huq T, Khan A, Brown D, Dhayagude N, He Z, Ni Y (2022) Sources, production and commercial applications of fungal chitosan: a review. J Bioresour Bioprod 7(2):85–98. https://doi.org/10.1016/j.jobab.2022.01.002
Jacob J, Lawal U, Thomas S, Valapa RB (2020) Chapter 4—biobased polymer composite from poly(lactic acid): processing, fabrication, and characterization for food packaging. In: Zhang Y (ed) Processing and development of polysaccharide-based biopolymers for packaging applications. Elsevier, pp 97–115
Jadhav EB, Sankhla MS, Bhat RA, Bhagat DS (2021) Microplastics from food packaging: an overview of human consumption, health threats, and alternative solutions. Environ Nanotechnol Monit Manag 16:100608. https://doi.org/10.1016/j.enmm.2021.100608
Jiang Y, Sun Y, Wei C, Li X, Deng W, Wu S et al (2024) Development and characterization of curcumin-loaded chitosan/egg yolk freshness-keeping edible films for chilled fresh pork packaging application. Int J Biol Macromol 276:133907. https://doi.org/10.1016/j.ijbiomac.2024.133907
Jin J, Luo B, Xuan S, Shen P, Jin P, Wu Z et al (2024) Degradable chitosan-based bioplastic packaging: design, preparation and applications. Int J Biol Macromol 266:131253. https://doi.org/10.1016/j.ijbiomac.2024.131253
Kato S, Ueda T, Aoshima T, Kosaka N, Nitta S (2024) BioPBS™ (Polybutylene Succinate). In: Künkel A, Battagliarin G, Winnacker M, Rieger B, Coates G (eds) Synthetic biodegradable and biobased polymers: industrial aspects and technical products. Springer International Publishing, Cham, pp 269–304
Kerry J (2012) Aluminium foil packaging. In: Emblem A, Emblem H (eds) Packaging technology. Woodhead Publishing, pp 163–177
Khalid MY, Arif ZU (2022) Novel biopolymer-based sustainable composites for food packaging applications: a narrative review. Food Packag Shelf Life 33:100892. https://doi.org/10.1016/j.fpsl.2022.100892
Khalid S, Yu L, Feng M, Meng L, Bai Y, Ali A et al (2018) Development and characterization of biodegradable antimicrobial packaging films based on polycaprolactone, starch and pomegranate rind hybrids. Food Packag Shelf Life 18:71–79. https://doi.org/10.1016/j.fpsl.2018.08.008
Khalid MY, Arif ZU, Ahmed W, Arshad H (2022) Recent trends in recycling and reusing techniques of different plastic polymers and their composite materials. Sustain Mater Technol 31:e00382. https://doi.org/10.1016/j.susmat.2021.e00382
Khan A, Riahi Z, Kim JT, Rhim J-W (2024) Carboxymethyl cellulose/gelatin film incorporated with eggplant peel waste-derived carbon dots for active fruit packaging applications. Int J Biol Macromol 271:132715. https://doi.org/10.1016/j.ijbiomac.2024.132715
Khandeparkar AS, Paul R, Sridhar A, Lakshmaiah VV, Nagella P (2024) Eco-friendly innovations in food packaging: a sustainable revolution. Sustain Chem Pharm 39:101579. https://doi.org/10.1016/j.scp.2024.101579
Khanjari A, Esmaeili H, Hamedi M (2023) Shelf life extension of minced squab using poly-lactic acid films containing Cinnamomum verum essential oil. Int J Food Microbiol 385:109982. https://doi.org/10.1016/j.ijfoodmicro.2022.109982
Khouri NG, Bahú JO, Blanco-Llamero C, Severino P, Concha VOC, Souto EB (2024) Polylactic acid (PLA): properties, synthesis, and biomedical applications—a review of the literature. J Mol Struct 1309:138243. https://doi.org/10.1016/j.molstruc.2024.138243
Kumar V, Sehgal R, Gupta R (2021) Blends and composites of polyhydroxyalkanoates (PHAs) and their applications. Eur Polym J 161:110824. https://doi.org/10.1016/j.eurpolymj.2021.110824

Kumari SVG, Pakshirajan K, Pugazhenthi G (2022) Recent advances and future prospects of cellulose, starch, chitosan, polylactic acid and polyhydroxyalkanoates for sustainable food packaging applications. Int J Biol Macromol 221:163–182. https://doi.org/10.1016/j.ijbiomac.2022.08.203

Kumari SVG, Pakshirajan K, Pugazhenthi G (2023) Facile fabrication and characterization of novel antimicrobial and antioxidant poly (3-hydroxybutyrate)/essential oil composites for potential use in active food packaging applications. Int J Biol Macromol 252:126566. https://doi.org/10.1016/j.ijbiomac.2023.126566

Kumari S, Debbarma R, Nasrin N, Khan T, Taj S, Bhuyan T (2024) Recent advances in packaging materials for food products. Food Bioeng 3(2):236–249. https://doi.org/10.1002/fbe2.12096

Lee H-G, Lee SY, Yoo S (2024) Innovative food-upcycling solutions: Comparative analysis of edible films from kimchi-extracted cellulose, sorbitol, and citric acid for food packaging applications. Food Chem 450:139267. https://doi.org/10.1016/j.foodchem.2024.139267

Leite LSF, Le Gars M, Azeredo HMC, Moreira FKV, Mattoso LHC, Bras J (2024) Insights into the effect of carboxylated cellulose nanocrystals on mechanical and barrier properties of gelatin films for flexible packaging applications. Int J Biol Macromol 280:135726. https://doi.org/10.1016/j.ijbiomac.2024.135726

Liao M, Pan Y, Fu X, Wu S, Gan S, Wu Z et al (2023a) Electrospun polylactic acid nanofiber film modified by silver oxide deposited on hemp fibers for antibacterial fruit packaging. Int J Biol Macromol 253:126569. https://doi.org/10.1016/j.ijbiomac.2023.126569

Liao L, Ke Z, Wang X, Li S, Wang S, Rao X (2023b) Recyclable ethyl cellulose film modified by rosin based quaternary ammonium salt for antimicrobial packaging. Ind Crops Prod 198:116692. https://doi.org/10.1016/j.indcrop.2023.116692

Liew WC, Muhamad II, Chew JW, Karim KJA (2023) Synergistic effect of graphene oxide/zinc oxide nanocomposites on polylactic acid-based active packaging film: properties, release kinetics and antimicrobial efficiency. Int J Biol Macromol 253:127288. https://doi.org/10.1016/j.ijbiomac.2023.127288

Liu Y, Ahmed S, Sameen DE, Wang Y, Lu R, Dai J et al (2021) A review of cellulose and its derivatives in biopolymer-based for food packaging application. Trends Food Sci Technol 112:532–546. https://doi.org/10.1016/j.tifs.2021.04.016

Liu X, Qin Z, Ma Y, Liu H, Wang X (2023a) Cellulose-based films for food packaging applications: review of preparation, properties, and prospects. J Renew Mater 11(8):3203–3225

Liu S, Gao X, Fan H, Zhang M, Waterhouse GIN, Zhu S (2023b) Green and recyclable graphitic carbon nitride/chitosan/polyvinyl alcohol photocatalytic films with efficient antibacterial activity for fruit packaging. Int J Biol Macromol 236:123974. https://doi.org/10.1016/j.ijbiomac.2023.123974

Loukri A, Kyriakoudi A, Oliinychenko Y, Stratakos AC, Lazaridou A, Mourtzinos I (2024) Preparation and characterization of chitosan-citric acid edible films loaded with Cornelian cherry pomace extract as active packaging materials. Food Hydrocolloids 150:109687. https://doi.org/10.1016/j.foodhyd.2023.109687

Mai J, Kockler K, Parisi E, Chan CM, Pratt S, Laycock B (2024) Synthesis and physical properties of polyhydroxyalkanoate (PHA)-based block copolymers: a review. Int J Biol Macromol 263:130204. https://doi.org/10.1016/j.ijbiomac.2024.130204

Manikandan NA, Pakshirajan K, Pugazhenthi G (2020) Preparation and characterization of environmentally safe and highly biodegradable microbial polyhydroxybutyrate (PHB) based graphene nanocomposites for potential food packaging applications. Int J Biol Macromol 154:866–877. https://doi.org/10.1016/j.ijbiomac.2020.03.084

Masood F (2017) Chapter 8—polyhydroxyalkanoates in the food packaging industry. In: Oprea AE, Grumezescu AM (eds) Nanotechnology applications in food. Academic Press, pp 153–177

Maureira J, Olate-Moya F, Bastías R, Farias S, Alvarez R, Rosales-Cuello N et al (2023) Multifunctional poly(3-hydroxybutyrate) composites with MoS2 for food packaging applications. Eur Polym J 188:111914. https://doi.org/10.1016/j.eurpolymj.2023.111914

Meyers H, Gerstman R (2005) The evolution of packaging. In: The visionary package: using packaging to build effective brands. Palgrave Macmillan UK, London, pp 8–24

Miao Z, Yang M, Abdalkarim SYH, Yu H-Y (2024) In situ growth of curcumin-loaded cellulose composite film for real-time monitoring of food freshness in smart packaging. Int J Biol Macromol 279:135090. https://doi.org/10.1016/j.ijbiomac.2024.135090

Moe NC, Basbasan A Jr, Winotapun C, Hararak B, Wanmolee W, Suwanamornlert P et al (2023) Application of lignin nanoparticles in polybutylene succinate based antifungal packaging for extending the shelf life of bread. Food Packag Shelf Life 39:101127. https://doi.org/10.1016/j.fpsl.2023.101127

Mohamad N, Mazlan MM, Tawakkal ISMA, Talib RA, Kian LK, Jawaid M (2022) Characterization of active polybutylene succinate films filled essential oils for food packaging application. J Polym Environ 30(2):585–596. https://doi.org/10.1007/s10924-021-02198-z

Nath PC, Sharma R, Mahapatra U, Mohanta YK, Rustagi S, Sharma M et al (2024) Sustainable production of cellulosic biopolymers for enhanced smart food packaging: an up-to-date review. Int J Biol Macromol 273:133090. https://doi.org/10.1016/j.ijbiomac.2024.133090

Nikolakis N, Catti P, Chaloulos A, van de Kamp W, Coy MP, Alexopoulos K (2024) A methodology to assess circular economy strategies for sustainable manufacturing using process eco-efficiency. J Clean Prod 445:141289. https://doi.org/10.1016/j.jclepro.2024.141289

Niu X, Liu Y, Song Y, Han J, Pan H (2018) Rosin modified cellulose nanofiber as a reinforcing and co-antimicrobial agents in polylactic acid/chitosan composite film for food packaging. Carbohyd Polym 183:102–109. https://doi.org/10.1016/j.carbpol.2017.11.079

Nuamduang P, Auras R, Winotapun C, Hararak B, Wanmolee W, Leelaphiwat P (2024) Enhanced antifungal properties of poly(butylene succinate) film with lignin nanoparticles and trans-cinnamaldehyde for mango packaging. Int J Biol Macromol 267:131185. https://doi.org/10.1016/j.ijbiomac.2024.131185

Oliviero M, Lamberti E, Cafiero L, Pace B, Cefola M, Gorrasi G et al (2023) Biodegradable cellulose acetate/layered double-hydroxide composite film for active packaging of fresh food. Mater Chem Phys 310:128469. https://doi.org/10.1016/j.matchemphys.2023.128469

Opálková Šišková A, Mosnáčková K, Musioł M, Opálek A, Bučková M, Rychter P et al (2022) Electrospun nisin-loaded poly(ε-caprolactone)-based active food packaging. Materials 15(13):4540

Pal AK, Wu F, Misra M, Mohanty AK (2020) Reactive extrusion of sustainable PHBV/PBAT-based nanocomposite films with organically modified nanoclay for packaging applications: compression moulding versus cast film extrusion. Compos Part B: Eng 198:108141. https://doi.org/10.1016/j.compositesb.2020.108141

Parveen N, Naik SVCS, Vanapalli KR, Sharma HB (2024) Bioplastic packaging in circular economy: a systems-based policy approach for multi-sectoral challenges. Sci Total Environ 945:173893. https://doi.org/10.1016/j.scitotenv.2024.173893

Patel MK, Hansson F, Pitkänen O, Geng S, Oksman K (2022) Biopolymer blends of poly(lactic acid) and poly(hydroxybutyrate) and their functionalization with glycerol triacetate and chitin nanocrystals for food packaging applications. ACS Appl Polym Mater 4(9):6592–6601. https://doi.org/10.1021/acsapm.2c00967

Pattaraudomchok P, Winotapun C, Tameesrisuk M, Hararak B (2024) Essential oils loaded biodegradable PBAT/PBS films as young coconut packaging after harvest. Food Packag Shelf Life 44:101322. https://doi.org/10.1016/j.fpsl.2024.101322

Pires JRA, de Souza VGL, Fernando AL (2018) Chitosan/montmorillonite bionanocomposites incorporated with rosemary and ginger essential oil as packaging for fresh poultry meat. Food Packag Shelf Life 17:142–149. https://doi.org/10.1016/j.fpsl.2018.06.011

Pothinuch P, Promsorn J, Sablani SS, Harnkarnsujarit N (2024) Antioxidant release, morphology and packaging properties of gallic acid incorporated biodegradable PBAT blended PBS active packaging. Food Packag Shelf Life 43:101304. https://doi.org/10.1016/j.fpsl.2024.101304

Ragaert P, Buntinx M, Maes C, Vanheusden C, Peeters R, Wang S et al (2019) Polyhydroxyalkanoates for food packaging applications. In: Reference module in food science. Elsevier

Ramakrishnan R, Kim JT, Roy S, Jayakumar A (2024) Recent advances in carboxymethyl cellulose-based active and intelligent packaging materials: a comprehensive review. Int J Biol Macromol 259:129194. https://doi.org/10.1016/j.ijbiomac.2023.129194

Ramezani Dana H, Ebrahimi F (2023) Synthesis, properties, and applications of polylactic acid-based polymers. Polym Eng Sci 63(1):22–43. https://doi.org/10.1002/pen.26193

Rasheed T, Vattathurvalappil SH, Shaukat MM, Theravalappil R, Ali U, Ummer AC et al (2024) Recent updates on biodegradability and recyclability of bioplastics—towards a new era in sustainability. Sustain Mater Technol 41:e01051. https://doi.org/10.1016/j.susmat.2024.e01051

Recupido F, Lama GC, Steffen S, Dreyer C, Seidlitz H, Russo V et al (2024) Efficient recycling pathway of bio-based composite polyurethane foams via sustainable diamine. Ecotoxicol Environ Saf 269:115758. https://doi.org/10.1016/j.ecoenv.2023.115758

Ren J, Li Y, Lin Q, Li Z, Zhang G (2022) Development of biomaterials based on plasticized polylactic acid and tea polyphenols for active-packaging application. Int J Biol Macromol 217:814–823. https://doi.org/10.1016/j.ijbiomac.2022.07.154

Rodríguez-Sánchez IJ, Rivera-Monroy ZJ, García-Castañeda JE, Clavijo-Grimaldo D, Fuenmayor CA, Zuluaga-Domínguez CM (2023) Multilayer polycaprolactone—pullulan nanofiber mats incorporated with the antimicrobial palindromic peptide LfcinB (21–25)Pal as a potential application in active packaging. Food Packag Shelf Life 38:101110. https://doi.org/10.1016/j.fpsl.2023.101110

Rodriguez-Uribe A, Wang T, Pal AK, Wu F, Mohanty AK, Misra M (2021) Injection moldable hybrid sustainable composites of BioPBS and PHBV reinforced with talc and starch as potential alternatives to single-use plastic packaging. Compos Part c: Open Access 6:100201. https://doi.org/10.1016/j.jcomc.2021.100201

Romão S, Bettencourt A, Ribeiro IAC (2022) Novel features of cellulose-based films as sustainable alternatives for food packaging. Polymers 14(22):4968

Sabet M (2024) Exploring biodegradable polymer composites for sustainable packaging: a review on properties, manufacturing techniques, and environmental impacts. Iran Polym J. https://doi.org/10.1007/s13726-024-01365-y

Sady S, Błaszczyk A, Kozak W, Boryło P, Szindler M (2021) Quality assessment of innovative chitosan-based biopolymers for edible food packaging applications. Food Packag Shelf Life 30:100756. https://doi.org/10.1016/j.fpsl.2021.100756

Sangroniz A, Zhu J-B, Tang X, Etxeberria A, Chen EYX, Sardon H (2019) Packaging materials with desired mechanical and barrier properties and full chemical recyclability. Nat Commun 10(1):3559. https://doi.org/10.1038/s41467-019-11525-x

Sarfraz MH, Hayat S, Siddique MH, Aslam B, Ashraf A, Saqalein M et al (2024) Chitosan based coatings and films: a perspective on antimicrobial, antioxidant, and intelligent food packaging. Prog Org Coat 188:108235. https://doi.org/10.1016/j.porgcoat.2024.108235

Schilirò D (2019) Sustainability, innovation, and efficiency: a key relationship. In: Ziolo M, Sergi BS (eds) Financing sustainable development: key challenges and prospects. Springer International Publishing, Cham, pp 83–102

Semlali F-Z, Ait Benhamou A, El Bourakadi K, Qaiss AEK, Bouhfid R, Jacquemin J et al (2023) Thermo-compression process-mediated in-situ cellulose microfibers phosphorylation enables high performant cellulosic paper packaging. Chem Eng J 473:145268. https://doi.org/10.1016/j.cej.2023.145268

Sharma S, Byrne M, Perera KY, Duffy B, Jaiswal AK, Jaiswal S (2023) Active film packaging based on bio-nanocomposite TiO_2 and cinnamon essential oil for enhanced preservation of cheese quality. Food Chem 405:134798. https://doi.org/10.1016/j.foodchem.2022.134798

Shen C, Yang X, Wang D, Li J, Zhu C, Wu D et al (2024) Carboxymethyl chitosan and polycaprolactone-based rapid in-situ packaging for fruit preservation by solution blow spinning. Carbohyd Polym 326:121636. https://doi.org/10.1016/j.carbpol.2023.121636

Shi C, Zhou A, Fang D, Lu T, Wang J, Song Y et al (2022) Oregano essential oil/β-cyclodextrin inclusion compound polylactic acid/polycaprolactone electrospun nanofibers for active food packaging. Chem Eng J 445:136746. https://doi.org/10.1016/j.cej.2022.136746

Sid S, Mor RS, Kishore A, Sharanagat VS (2021) Bio-sourced polymers as alternatives to conventional food packaging materials: a review. Trends Food Sci Technol 115:87–104. https://doi.org/10.1016/j.tifs.2021.06.026

Siracusa V, Blanco I (2020) Bio-polyethylene (Bio-PE), bio-polypropylene (Bio-PP) and bio-poly(ethylene terephthalate) (Bio-PET): recent developments in bio-based polymers analogous to petroleum-derived ones for packaging and engineering applications. Polymers 12(8):1641

Stefanowska K, Bucher M, Reichert CL, Sip A, Woźniak M, Schmid M et al (2024) Chitosan-based films with nanocellulose and propolis as active packaging materials. Ind Crops Prod 219:119112. https://doi.org/10.1016/j.indcrop.2024.119112

Su J, Zhang W, Moradi Z, Rouhi M, Parandi E, Garavand F (2025) Recent functionality developments of carboxymethyl chitosan as an active food packaging film material. Food Chem 463:141356. https://doi.org/10.1016/j.foodchem.2024.141356

Subbuvel M, Kavan P (2022) Preparation and characterization of polylactic acid/fenugreek essential oil/curcumin composite films for food packaging applications. Int J Biol Macromol 194:470–483. https://doi.org/10.1016/j.ijbiomac.2021.11.090

Subramani G, Manian R (2024) Bioactive chitosan films: integrating antibacterial, antioxidant, and antifungal properties in food packaging. Int J Biol Macromol 278:134596. https://doi.org/10.1016/j.ijbiomac.2024.134596

Sumithra G, Reddy RN, Dheeraj Kumar G, Ojha S, Jayachandra G, Raghavendra G (2023) Review on composite classification, manufacturing, and applications. Mater Today Proc. https://doi.org/10.1016/j.matpr.2023.04.637

Sun A, Wang W-X (2023) Human exposure to microplastics and its associated health risks. Environ Health 1(3):139–149. https://doi.org/10.1021/envhealth.3c00053

Swetha TA, Bora A, Mohanrasu K, Balaji P, Raja R, Ponnuchamy K et al (2023) A comprehensive review on polylactic acid (PLA)—synthesis, processing and application in food packaging. Int J Biol Macromol 234:123715. https://doi.org/10.1016/j.ijbiomac.2023.123715

Tajeddin B, Arabkhedri M (2020) Chapter 16—polymers and food packaging. In: AlMaadeed MAA, Ponnamma D, Carignano MA (eds) Polymer science and innovative applications. Elsevier, pp 525–543

Tamizhdurai P, Mangesh VL, Santhosh S, Vedavalli R, Kavitha C, Bhutto JK et al (2024) A state-of-the-art review of multilayer packaging recycling: challenges, alternatives, and outlook. J Clean Prod 447:141403. https://doi.org/10.1016/j.jclepro.2024.141403

Tan W, McClements DJ, Chen J, Ma D (2024) Novel biopolymer-based active packaging material for food applications: cinnamaldehyde-loaded calcium nanoparticles incorporated into alginate-carboxymethyl cellulose films. Food Packag Shelf Life 45:101351. https://doi.org/10.1016/j.fpsl.2024.101351

Thakur M, Majid I, Hussain S, Nanda V (2021) Poly(ε-caprolactone): A potential polymer for biodegradable food packaging applications. Packag Technol Sci 34(8):449–461. https://doi.org/10.1002/pts.2572

Thamarai P, Vickram AS, Saravanan A, Deivayanai VC, Evangeline S (2024) Recent advancements in biosynthesis, industrial production, and environmental applications of polyhydroxyalkanoates (PHAs): a review. Bioresour Technol Rep 27:101957. https://doi.org/10.1016/j.biteb.2024.101957

Thirupathi Vasuki M, Kadirvel V, Pejavara NG (2023) Smart packaging—an overview of concepts and applications in various food industries. Food Bioeng 2(1):25–41. https://doi.org/10.1002/fbe2.12038

Ucpinar Durmaz B, Ugur Nigiz F, Aytac A (2024) Active packaging films based on poly(butylene succinate) films reinforced with alkaline halloysite nanotubes: production, properties, and fruit packaging applications. Appl Clay Sci 259:107517. https://doi.org/10.1016/j.clay.2024.107517

Ullah A, Yang H, Takemae K, Wang F, Lee S, Kim IS (2024) Sustainable bioactive food packaging based on electrospun zein-polycaprolactone nanofibers integrated with aster yomena extract loaded halloysite nanotubes. Int J Biol Macromol 267:131375. https://doi.org/10.1016/j.ijbiomac.2024.131375

Upadhyay A, Agbesi P, Arafat KMY, Urdaneta F, Dey M, Basak M et al (2024) Bio-based smart packaging: fundamentals and functions in sustainable food systems. Trends Food Sci Technol 145:104369. https://doi.org/10.1016/j.tifs.2024.104369

Varma R, Vasudevan S (2020) Extraction, characterization, and antimicrobial activity of chitosan from horse mussel modiolus modiolus. ACS Omega 5(32):20224–20230. https://doi.org/10.1021/acsomega.0c01903

Venezia V, Prieto C, Evtoski Z, Marcoaldi C, Silvestri B, Vitiello G et al (2023) Electrospun hybrid TiO_2/humic substance PHBV films for active food packaging applications. J Ind Eng Chem 124:510–522. https://doi.org/10.1016/j.jiec.2023.05.006

Verla AW, Enyoh CE, Verla EN, Nwarnorh KO (2019) Microplastic–toxic chemical interaction: a review study on quantified levels, mechanism and implication. SN Appl Sci 1(11):1400. https://doi.org/10.1007/s42452-019-1352-0

Viswanathan VP, Kulandhaivelu SV, Manivasakan K, Ramakrishnan R (2024) Development of biodegradable packaging films from carboxymethyl cellulose and oxidised natural rubber latex. Int J Biol Macromol 262:129980. https://doi.org/10.1016/j.ijbiomac.2024.129980

Vytejčková S, Vápenka L, Hradecký J, Dobiáš J, Hajšlová J, Loriot C et al (2017) Testing of polybutylene succinate based films for poultry meat packaging. Polym Test 60:357–364. https://doi.org/10.1016/j.polymertesting.2017.04.018

Wang H, Ding F, Ma L, Zhang Y (2021) Edible films from chitosan-gelatin: physical properties and food packaging application. Food Biosci 40:100871. https://doi.org/10.1016/j.fbio.2020.100871

Wang W, Niu B, Liu R, Chen H, Fang X, Wu W et al (2023a) Development of bio-based PLA/cellulose antibacterial packaging and its application for the storage of shiitake mushroom. Food Chem 429:136905. https://doi.org/10.1016/j.foodchem.2023.136905

Wang Q, Xu P, Lv D, Xu Y, Zhang J, Liu T et al (2023b) A novel strategy for the preparation of multifunctional and robust polyhydroxyalkanoates/GQDs nanocomposites with UV-resistance, pH-responsive and antibacterial properties. Chem Eng J 477:147106. https://doi.org/10.1016/j.cej.2023.147106

Wang S, Ma Y, Wang F, Lu C, Liu Y, Zhang S et al (2025) Development of cellulose-based self-healing hydrogel smart packaging for fish preservation and freshness indication. Carbohyd Polym 348:122806. https://doi.org/10.1016/j.carbpol.2024.122806

Wang Q, Chen W, Zhu W, McClements DJ, Liu X, Liu F (2022) A review of multilayer and composite films and coatings for active biodegradable packaging. npj Sci Food 6(1):18. https://doi.org/10.1038/s41538-022-00132-8

Wongphan P, Nampanya P, Chakpha W, Promhuad K, Laorenza Y, Leelaphiwat P et al (2023) Lesser galangal (Alpinia officinarum Hance) essential oil incorporated biodegradable PLA/PBS films as shelf-life extension packaging of cooked rice. Food Packag Shelf Life 37:101077. https://doi.org/10.1016/j.fpsl.2023.101077

Wu Y, Hao X, Lin F, Wang S, Chen L, Lin X et al (2022) Developing a cerium lactate antibacterial nucleating agent for multifunctional polylactic acid packaging film. Int J Biol Macromol 220:56–66. https://doi.org/10.1016/j.ijbiomac.2022.08.082

Wu S, Abdalkarim SYH, Zhu J, Worku Alebachew A, Yu H-Y, Asad RAM et al (2024) Functionalized cellulose nanocrystals-MoS2 nanosheets to tailor brick–mortar architecture of green composite for next-generation electronic packaging. Chem Eng J 499:156328. https://doi.org/10.1016/j.cej.2024.156328

Yeo JCC, Muiruri JK, Fei X, Wang T, Zhang X, Xiao Y et al (2024) Innovative biomaterials for food packaging: unlocking the potential of polyhydroxyalkanoate (PHA) biopolymers. Biomater Adv 163:213929. https://doi.org/10.1016/j.bioadv.2024.213929

Yin Y, Woo MW (2024) Transitioning of petroleum-based plastic food packaging to sustainable bio-based alternatives. Sustain Food Technol 2(3):548–566. https://doi.org/10.1039/D4FB00028E

Yu J, Xu S, Liu B, Wang H, Qiao F, Ren X et al (2023) PLA bioplastic production: from monomer to the polymer. Eur Polymer J 193:112076. https://doi.org/10.1016/j.eurpolymj.2023.112076

Yue C, Wang M, Zhou Z, You Y, Wang G, Wu D (2024) Cellulose-based intelligent packaging films with antibacterial, UV-blocking, and biodegradable properties for shrimp freshness monitoring. Chem Eng J 488:150975. https://doi.org/10.1016/j.cej.2024.150975

Zatsu V, Shine AE, Tharakan JM, Peter D, Ranganathan TV, Alotaibi SS et al (2024) Revolutionizing the food industry: the transformative power of artificial intelligence-a review. Food Chem: X 24:101867. https://doi.org/10.1016/j.fochx.2024.101867

Zhang M, Biesold GM, Choi W, Yu J, Deng Y, Silvestre C et al (2022) Recent advances in polymers and polymer composites for food packaging. Mater Today 53:134–161. https://doi.org/10.1016/j.mattod.2022.01.022

Zhao Y, Li C, Xia X, Tan M, Wang H, Lv Y et al (2023) Eco-friendly and intelligent cellulosic fibers-based packaging system for real-time visual detection of food freshness. Chem Eng J 474:146013. https://doi.org/10.1016/j.cej.2023.146013

Zhao X, Zou D, Liu Y, Xia Y, Tao J, Zeng Q et al (2024) Electrospun polylactic acid nanofibers membrane with copper ion-loaded clay nanotubes for fresh-keeping packaging. Int J Biol Macromol 267:131651. https://doi.org/10.1016/j.ijbiomac.2024.131651

Zhao X, Chen T, Liu J, Wang X, Weng Y (2025) Development of antifouling antibacterial polylactic acid (PLA)—based packaging and application for chicken meat preservation. Food Chem 463:141116. https://doi.org/10.1016/j.foodchem.2024.141116

Zheng H, Tang H, Yang C, Chen J, Wang L, Dong Q et al (2022) Evaluation of the slow-release polylactic acid/polyhydroxyalkanoates active film containing oregano essential oil on the quality and flavor of chilled pufferfish (Takifugu obscurus) fillets. Food Chem 385:132693. https://doi.org/10.1016/j.foodchem.2022.132693

Zhu Z, Liu W, Ye S, Batista L (2022) Packaging design for the circular economy: a systematic review. Sustain Prod Consump 32:817–832. https://doi.org/10.1016/j.spc.2022.06.005

Chapter 4
Pioneering Sustainable Building Solutions with Green Polymer Composites

Md Arifuzzaman, Abdulrahman Alfuhaid, and Md Enamul Hoque

Abstract This chapter explores the increasing need for eco-friendly materials in the construction industry to reduce environmental impact, highlighting how polymer composites, engineered materials with superior properties, offer a key solution. The chapter discusses the history of composite materials in construction and introduces modern green polymer composites, which utilize renewable or recycled resources. It defines green polymer composites, examines their types, advantages, challenges, and opportunities, and looks at future trends and research directions in this evolving field.

Keywords Sustainable · Eco-friendly material · Green polymer composite · Lightweight · Construction

4.1 Introduction

The construction industry is confronted with the pressing challenge of reducing its environmental impact and transitioning towards sustainable practices. This transition necessitates the adoption of eco-friendly materials that provide both high performance and a diminished ecological footprint. Polymer composites, which are materials engineered by combining two or more distinct constituent phases to achieve excellent properties associated to their distinctive components, are increasingly being

Md Arifuzzaman · A. Alfuhaid
Department of Civil and Environmental Engineering, College of Engineering, King Faisal University, Al Ahsa 31982, Saudi Arabia
e-mail: marifuzzaman@kfu.edu.sa

A. Alfuhaid
e-mail: aalfuhaid@kfu.edu.sa

Md E. Hoque (✉)
Department of Biomedical Engineering, Military Institute of Science and Technology, Dhaka, Bangladesh
e-mail: enamul1973@gmail.com

L. Ravindran et al. (eds.), *Macro, Micro and Nanocomposites from Sustainable Sources*, Advanced Structured Materials 244, https://doi.org/10.1007/978-981-95-2469-3_4

investigated as a key solution. The concept of composite materials is not novel in construction. Ancient civilizations over 2,000 years ago utilized straw to reinforce bricks, a practice that persists today. Evidence also indicates the use of metallic substances to reinforce concrete beam structures in Greece nearly 1,000 years ago, demonstrating the long-standing history of composite materials in construction. Modern polymer composites, a significant class of these materials, are created by combining polymer resins (such as epoxy, polyester, or vinylester) with fillers and reinforcing fibers. This combination results in enhanced material properties. Fillers can reduce cost and weight and improve aesthetics, whereas fibers provide reinforcement, increasing strength and stiffness. Carbon, aramid and glass fibers are commonly used as principal load-bearing components. The polymer resin acts like binder, protecting the fibers and creating a cohesive structural unit. Fiber composite types of materials, traditionally consumed in non-structural purposes for example baths, cladding, and decoration, are now increasingly employed in structural roles within the construction sector. Their potential as competitive alternatives to steel in reinforced concrete and for the rehabilitation and retrofitting of existing infrastructure underscores their transformative potential for modern construction (Abera 2024; Cabrera 2021; dos Santos et al. 2022; Maiti et al. 2022; Skosana et al. 2025).

The construction industry faces the urgent challenge of minimizing its environmental impact and transitioning towards sustainable practices. This necessitates a shift towards eco-friendly materials that offer both high performance and a reduced ecological footprint. Polymer composites, which are materials engineered by combining two or more distinct constituent phases to achieve grander properties associated to their discrete components, are increasingly being explored as a key solution. The concept of composite materials is not new in construction. Ancient civilizations over 2,000 years ago used straw to reinforce bricks, a practice that continues today. Evidence also suggests the use of metallic substances to reinforce concrete beam structures in Greece nearly 1,000 years ago, demonstrating the long-standing history of the application of composites in construction. Modern polymer composites materials, a significant category of these materials, are created by combining polymer resins (such as epoxy, polyester, or vinylester) with fillers and reinforcing fibers. This combination results in enhanced material properties. Fillers can reduce cost and weight and improve aesthetics, whereas fibers provide reinforcement, increasing strength and stiffness. High-strength possessing fibers, such as glass, aramid, and carbon, mostly serve as principle load-bearing components. The polymer resin acted as a binder, protecting the fibers and creating a cohesive structural unit. Fiber composite materials, which are traditionally used in non-structural applications such as baths, cladding, and decoration, are now being increasingly used in structural roles within the construction sector. Their potential as competitive alternatives to steel in reinforced concrete and for the rehabilitation and retrofitting of existing infrastructure highlights their transformative potential for modern construction (Rupal et al. 2022; Salami et al. 2024; Singh et al. 2022).

This chapter explored the use of green polymer composites for sustainable construction. It defines green polymer composites, discusses their various types, benefits, challenges, and opportunities, and examines future trends and research directions in this evolving field.

4.2 Understanding Green Polymer Composites (GPC)

A type of composite material known as "green polymer composites" is made from recycled or renewable resources. Such materials are invented to diminish the environmental effect while offering comparable or superior performance to conventional composites. A schematic diagram of GPC is shown in Fig. 4.1.

Table 4.1 relates the mechanical properties and costs of three commonly used reinforcement fibers: glass fiber, carbon fiber, and aramid fiber. Carbon fiber exhibits the topmost tensile strength (3000–7000 MPa) and modulus (230–700 GPa), making it supreme for high-performance functions, though its cost is high. Glass fiber offers an impartial performance with medium strength (2000–3500 MPa) and a lower cost, making it widely used in construction and automotive industries. Aramid fiber, known for its lightweight nature and good impact resistance, provides a tensile strength of 3000–4000 MPa but possesses relatively lower modulus value of 60–130 GPa.

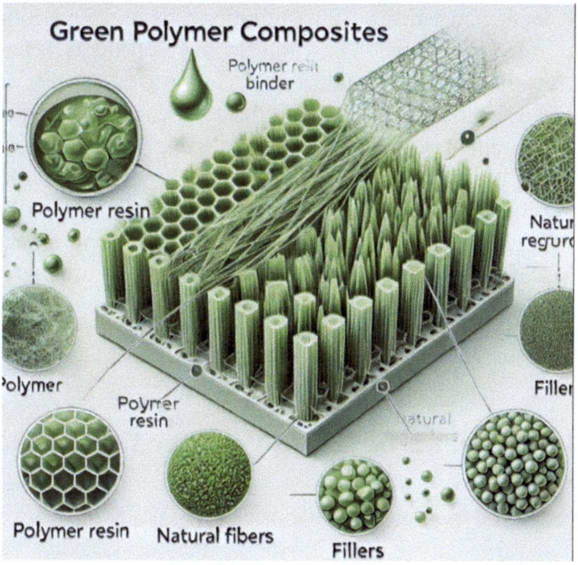

Fig. 4.1 Material components in green polymer composites

Table 4.1 Comparison of properties of common reinforcing fibers (Bledzki and Gassan 1999a; Mallick 2007)

Property	Glass fiber	Carbon fiber	Aramid fiber
Tensile strength (MPa)	2000–3500	3000–7000	3000–4000
Tensile modulus (GPa)	70–85	230–700	60–130
Elongation at break (%)	2–4	1–2	2–4
Density (g/cm^3)	2.5–2.6	1.7–2.1	1.4–1.5
Cost (relative)	Low	High	Medium

4.3 Types of Green Polymer Composites

- **Bio-based Composites**: The aforementioned materials science domain focuses on the application of natural fibers and biopolymers, originating from the renewable supplies such as flora, fauna, and microbes, in composite materials fabrication. Illustrative examples include epoxy resin matrices reinforced with flax fibers and composite systems based on soy protein. This approach underscores a commitment to sustainable material development (Bairwan et al. 2025; Ghalme et al. 2025; Islam et al. 2025)
- **Recycled Composites**: In the realm of composite material development, a significant focus lies on the integration of recycled constituents, encompassing post-consumer plastics and industrial byproducts. This strategic utilization aims to mitigate the consumption of virgin resources and curtail the generation of waste streams. Illustrative instances include composite matrices formulated with reclaimed polyethylene terephthalate (PET) from discarded bottles and the incorporation of recycled tire rubber as a reinforcing phase. Such approaches underscore a commitment to circular economy principles within materials science (Liu et al. 2024; Radu et al. 2025; Rani et al. 2024; Seier et al.2024; zu Castell-Rüdenhausen 2025). Figure 4.2 illustrates the process of Composite Manufacturing focused on sustainable and green materials.

4.4 Benefits of Green Polymer Composites

Green polymer composites offer several advantages over conventional ones.

- **Sustainability**: The fabrication of these composite materials prioritizes the incorporation of renewable resources and/or recycled constituents. This strategic approach serves to diminish the reliance on finite fossil fuel reserves and consequently minimizes their ecological burden, aligning with principles of sustainable engineering and materials science.

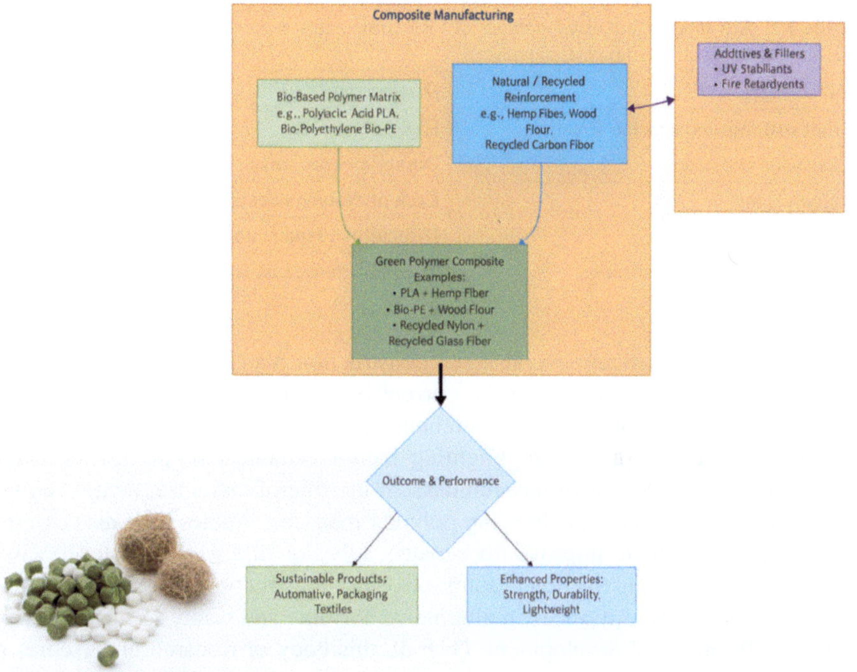

Fig. 4.2 Structural integration of green polymer composites in sustainable construction

- **Biodegradability**: A notable characteristic of numerous green composite materials is their biodegradability (Bhat et al. 2021). This implies that these materials are susceptible to decomposition by microorganisms, transforming them into naturally occurring compounds (Jha et al. 2019). Such a process effectively mitigates waste accumulation and diminishes environmental pollution associated with conventional, non-degradable materials (Rydz et al. 2024).
- **Cost-effectiveness**: An important consideration regarding green composite materials is their potential for cost-effectiveness when juxtaposed with conventional counterparts. This economic viability becomes particularly apparent upon the inclusion of comprehensive life cycle assessments that account for long-term environmental externalities and broader societal costs associated with traditional materials.
- **Performance**: Notably, green composite materials demonstrate performance characteristics that are either comparable to or, in some instances, superior to those exhibited by conventional composites. This includes critical mechanical properties such as flexural stiffness, tensile strength and long-term durability. Such attributes render green composites viable alternatives across a diverse range of engineering applications, potentially replacing traditional materials in demanding structural roles (Table 4.2).

Table 4.2 Advantages and disadvantages of FRP composites in construction (Sathishkumar et al. 2014a; Zaman et al. 2013a)

Advantages	Disadvantages
High stiffness and specific strength	High initial cost
Corrosion resistance	Durability concerns in harsh environments
Lightweight	Lack of design codes and standards
Tailorable properties	Difficulty in repairs and modifications
Fatigue and impact resistance	Fire performance issues

Table 4.3 summarizes several green construction materials that can be used to get a very quick idea regarding the current book chapter discussions. The listed articles collectively provide a comprehensive overview of bio-based and natural fiber-reinforced composites, highlighting their mechanical properties, durability, and sustainability. Research has emphasized the role of cellulose, hemp, bamboo, and other natural fibers in enhancing polymer matrices. Studies have also explored chemical treatments to improve fiber-matrix adhesion and water resistance. Green composites are gaining traction in automotive and construction sectors, aligning with sustainability goals. Advances in alternative solvents and catalysis further support eco-friendly material development. Overall, this body of research underscores the growing consequence of bio-based composites materials in reducing environmental impact while conserving structural integrity.

4.5 Challenges and Opportunities

Despite their promise, the development of green composites faces several challenges:

- **Hydrophilic Nature of Biofibers**: The absorption of moisture by several biofibers can result in poor interfacial adhesion with hydrophobic phenomenon in the matrices of polymer. This can result in reduced mechanical properties and compromised durability in the composite materials.
- **Limited Processing Temperatures**: The processing temperatures for many biofibers are limited, which poses challenges for composite fabrication processes that require high temperatures.

These issues are being aggressively addressed by ongoing research and development initiatives. Several surface modification practices such as chemical treatments and physical modifications have been employed to improve the compatibility between biofibers and protein-based matrices. Additionally, novel processing methods such as compression molding, extrusion and injection molding are being optimized to accommodate the specific requirements of bio-based materials.

Table 4.3 Summary of key research on green construction materials from several sources

Title of article	Authors	Year	Citation number
Biocomposites based on plasticized starch: thermal and mechanical behaviours	Avérous, L., and Boquillon, N.	2004	Averous and Boquillon (2004)
Composites reinforced with cellulose based fibres	Bledzki, A. K., and Gassan, J.	1999	Bledzki and Gassan (1999b)
Green composites: a review of adequate materials for automotive applications	Koronis, G., Silva, A., and Fontul, M.	2013	Koronis et al. (2013)
Effect of water absorption on the mechanical properties of hemp fibre reinforced unsaturated polyester[2] composites	Dhakal, H. N., Zhang, Z. Y., and Richardson, M. O. W.	2007	Dhakal et al. (2007)
Biocomposites reinforced with natural fibers: 2000–2010	Faruk, O., Bledzki, A. K., Fink, H. P., and Sain, M.	2012	Faruk et al. (2012)
Cellulosic/synthetic fibre reinforced polymer hybrid composites: a review	Jawaid, M., and Khalil, H. P. S. A.	2011	Jawaid and Khalil (2011)
Chemical treatments on plant-based natural fibre reinforced polymer composites: an overview	Kabir, M. M., Wang, H., Lau, K. T., and Cardona, F.	2012	Kabir et al. (2012))
Sustainable bio-composites from renewable resources: opportunities and challenges in the green materials world	Mohanty, A. K., Misra, M., and Drzal, L. T.	2002	Mohanty et al. (2002)
Natural fibres as reinforcement in polylactic acid (PLA) composites	Oksman, K., Skrifvars, M., and Selin, J. F.	2003	Oksman et al. (2003)
A review of recent developments in natural fibre composites and their mechanical performance	Pickering, K. L., Efendy, M. G. A., and Le, T. M.	2016	Pickering et al. (2016)
Natural fiber polymer composites: a review	Saheb, D. N., and Jog, J. P.	1999	Saheb and Jog (1999)
A review on interface modification and characterization of natural fiber reinforced plastic composites	George, J., Sreekala, M. S., and Thomas, S.	2001	George et al. (2001)
A study of the mechanical properties of short natural fiber reinforced composites	Herrera-Franco, P. J., and Valadez-González, A.	2005	Herrera-Franco and Valadez-Gonzalez (2005)
Mechanical and water absorption behavior of composites made from a biodegradable matrix and alkaline treated sisal fibers	Alvarez, V. A., Ruscekaite, R. A., and Vazquez, A.	2003	Alvarez et al. (2003)

(continued)

Table 4.3 (continued)

Title of article	Authors	Year	Citation number
Bio-based polymers and composites	Wool, R. P., and Sun, X. S.	2005	Wool and Sun (2005)
Green chemistry: theory and practice	Anastas, P. T., and Warner, J. C.	1998	Anastas and Warner (1998)
2, 2, 5, 5-Tetramethyltetrahydrofuran (TMTHF): a non-polar, non-peroxide forming ether replacement for hazardous hydrocarbon solvents	Byrne, F., Forier, B., Bossaert, G., Hoebers, C., Farmer, T. J., Clark, J. H., and Hunt, A. J.	2017	Byrne et al. (2017)
Alternative solvents: shades of green	Clark, J. H., and Tavener, S. J.	2007	Clark and Tavener (2007)
Alternative solvents for green chemistry	Kerton, F. M., and Marriott, R.	2013	Kerton and Marriott (2013)
Atom efficiency and catalysis in organic synthesis	Sheldon, R. A.	2000	Sheldon (2000)
Natural-fiber-reinforced polymer composites in automotive applications	Holbery, J., and Houston, D.	2006	Holbery and Houston (2006)
Glass fiber-reinforced polymer composites–a review	Sathishkumar, T. P., Satheeshkumar, S., and Naveen, J.	2014	Sathishkumar et al. (2014b)
A review on FRP composites applications and durability concerns in the construction sector	Zaman, A., Gutub, S. A., and Wafa, M. A.	2013	Zaman et al. (2013b)

These challenges present opportunities for innovation. For example, the development of new surface modification techniques and processing methods can head to noteworthy improvements in the enactment and versatility of green composites.

4.6 Future Trends and Research Directions

The field of green composites derived from protein-based polymers is constantly evolving, with new research and innovations emerging. Current trends include:

- **Exploration of Novel Protein Sources**: Ongoing research endeavors are focused on investigating unconventional protein sources, such as algae and insect proteins, to broaden the spectrum of bio-based materials suitable for green composite fabrication. This exploration aims to diversify the feedstock options beyond traditional plant-based sources, potentially enhancing material properties and sustainability (Jagadeesh et al. 2017, Umesh et al. 2023, Gürbüz et al. 2025).

- **Advanced Composite Architectures**: The evolution of sophisticated composite architectures, including the integration of nanoscale reinforcements in nanocomposites and the synergistic combination of different material phases in hybrid composites, has significantly augmented both the performance features and the span of potential applications in green composite materials. These advancements enable the creation of materials with customized properties for particular engineering demands (Simões 2024).
- **3D Printing with Green Composites**: 3D printing technology is increasingly being used to fabricate complex shapes and structures from green composites, opening up new possibilities for design and application. The integration of additive manufacturing technologies, particularly three-dimensional printing, is gaining prominence in the fabrication of intricate geometries and complex structural elements from green composite materials. This synergy facilitates unprecedented design freedom and inflates the possible concentrations of these sustainable materials across diverse sectors, enabling the creation of customized and optimized components (Baskar et al. 2024; Khosravani and Reinicke 2020; Scaffaro et al. 2023; Tabassum and Mir 2023; Vazquez-Martel et al. 2024).
- **AI in Material Design**: The application of artificial intelligence methodologies is increasingly pivotal in accelerating the research and development of advanced green composite materials. AI algorithms facilitate the rapid screening of potential material combinations, predict their resulting properties and performance characteristics, and optimize formulations, thereby significantly reducing the time and resources required for the discovery of novel sustainable materials with tailored functionalities (Athanasiou et al. 2024; Ahmad et al. 2021; Nachtane et al. 2023; Pande et al. 2025; Kibrete et al. 2023).

4.7 Conclusion

Green composites derived from protein-based polymers are promising for the global pursuit of a sustainable future. Their possible to revolutionize industries reaching from packaging and construction to automotive and aerospace is immense. These materials offer a compelling pathway towards reducing our reliance on finite resources, minimizing environmental burdens, and fostering a circular economy. Their inherent sustainability and biodegradability address critical global challenges related to resource depletion and waste management, whereas their potential for cost-effectiveness makes them an increasingly viable alternative to conventional composites across diverse economic contexts.

The construction sector, a major consumer of resources and significant contributor to global emissions, plays a pivotal role in this transformation. Green composites offer the opportunity to construct high-performance, environmentally responsible buildings, and infrastructure that not only meet the demands of a growing global population, but also contribute to a healthier planet. As research and innovation continue to break new ground, we anticipate the widespread adoption of green composites to accelerate, flooring the way for a beyond resilient, resource-efficient, as well as sustainable built environment on a global scale. Through collaborative efforts and continued advancements, we can fully unlock the transformative power of these materials and build a truly sustainable world for future generations.

Funding This work was supported by the Deanship of Scientific Research, Vice Presidency for Graduate Studies and Scientific Research, King Faisal University, Saudi Arabia (Grant No. 251291).

References

Abera YA (2024) Sustainable building materials: a comprehensive study on eco-friendly alternatives for construction. Compos Adv Mater 33:26349833241255957

Ahmad MR, Chen B, Dai JG, Kazmi SMS, Munir MJ (2021) Evolutionary artificial intelligence approach for performance prediction of bio-composites. Constr Build Mater 290:123254

Alvarez VA, Ruscekaite RA, Vazquez A (2003) Mechanical properties and water absorption behavior of composites made from a biodegradable matrix and alkaline-treated sisal fibers. J Compos Mater 37(17):1575–1588

Anastas PT, Warner JC (2000) Green chemistry: theory and practice. Oxford University Press

Athanasiou C, Deng B, Hassen AA (2024) Integrating experiments, simulations, and artificial intelligence to accelerate the discovery of high-performance green composites. In: AIAA SCITECH 2024 Forum p 0041

Averous L, Boquillon N (2004) Biocomposites based on plasticized starch: thermal and mechanical behaviours. Carbohydr Polym 56(2):111–122

Bairwan RD, Khalil HPS A, Nuryawan A, Ahmad M I, Kassim MHM, Ahmad A (2025) Enhancing biopolymer materials with coffee waste-derived reinforcements. Polym Eng Sci 65(2):455–477

Baskar C, Ramakrishna S, Rosa ADL (eds) (2024) Encyclopedia of green materials. Springer Nature Singapore, Singapore

Bhat KM, Rajagopalan J, Mallikarjunaiah R, Rao NN, Sharma A (2021) Eco-friendly and biodegradable green composites. IntechOpen, London, UK

Bledzki AK, Gassan J (1999) Composites reinforced with cellulose based fibres. Prog Polym Sci 24(2):221–274

Byrne F, Forier B, Bossaert G, Hoebers C, Farmer TJ, Clark JH, Hunt AJ (2017) 2, 2, 5, 5-Tetramethyltetrahydrofuran (TMTHF): a non-polar, non-peroxide forming ether replacement for hazardous hydrocarbon solvents. Green Chem 19(15):3671–3678

Cabrera FC (2021) Eco-friendly polymer composites: a review of suitable methods for waste management. Polym Compos 42(6):2653–2677

Clark JH, Tavener SJ (2007) Alternative solvents: shades of green. Org Process Res Dev 11(1):149–155

Dhakal HN, Zhang ZA, Richardson MO (2007) Effect of water absorption on the mechanical properties of hemp fibre reinforced unsaturated polyester composites. Compos Sci Technol 67(7–8):1674–1683

dos Santos GZB, Caldas LR, de Almeida Melo Filho J, Monteiro NBR, Rafael SIM, da Silva NM (2022) Circular alternatives in the construction industry: an environmental performance assessment of sisal fiber-reinforced composites. J Build Eng 54:104603

Faruk O, Bledzki AK, Fink HP, Sain M (2012) Biocomposites reinforced with natural fibers: 2000–2010. Prog Polym Sci 37(11):1552–1596

George J, Sreekala MS, Thomas S (2001) a review on interface modification and characterization of natural fiber reinforced plastic composites. Polym Eng Sci 41(9):1471–1485

Ghalme S, Hayat M, Harne M (2025) A comprehensive review of natural fibers: bio-based constituents for advancing sustainable materials technology. J Renew Mater 13(2)
Gürbüz BN, Pastrana LM, Pereira RN, Cerqueira MA (2025) Alternative protein-based meat and fish analogs by conventional and novel processing technologies: a systematic review and bibliometric analysis. Foods 14(3):498
Herrera-Franco P, Valadez-Gonzalez A (2005) a study of the mechanical properties of short natural-fiber reinforced composites. Compos Part B Eng 36(8):597–608
Holbery J, Houston D (2006) Natural-fiber-reinforced polymer composites in automotive applications. Jom 58(11):80–86
Islam T, Hossain MM, Covington SM (2025) Natural fibers towards fashion sustainability: a review of raw materials, production, application, and perspective. J Nat Fibers 22(1):2462218
Jagadeesh D, Kanny K, Prashantha K (2017) A review on research and development of green composites from plant protein-based polymers. Polym Compos 38(8):1504–1518
Jawaid MHPS, Khalil HA (2011) Cellulosic/synthetic fibre reinforced polymer hybrid composites: a review. Carbohydr Polym 86(1):1–18
Jha K, Kataria R, Verma J, Pradhan S (2019) Potential biodegradable matrices and fiber treatment for green composites: a review. AIMS Mater Sci 6(1):119–138
Kabir MM, Wang H, Lau KT, Cardona F (2012) Chemical treatments on plant-based natural fibre reinforced polymer composites: an overview. Compos Part B Eng 43(7):2883–2892
Kerton FM, Marriott R (2013) Alternative solvents for green chemistry, vol 20. Royal Society of Chemistry
Khosravani MR, Reinicke T (2020) On the environmental impacts of 3D printing technology. Appl Mater Today 20:100689
Kibrete F, Trzepieciński T, Gebremedhen HS, Woldemichael DE (2023) Artificial intelligence in predicting mechanical properties of composite materials. J Compos Sci 7(9):364
Koronis G, Silva A, Fontul M (2013) Green composites: a review of adequate materials for automotive applications. Compos Part B Eng 44(1):120–127
Liu Q, Martinez-Villarreal S, Wang S, Tien NNT, Kammoun M, De Roover Q, Richel A (2024) The role of plastic chemical recycling processes in a circular economy context. Chem Eng J 155227
Maiti S, Islam MR, Uddin MA, Afroj S, Eichhorn SJ, Karim N (2022) Sustainable fiber-reinforced composites: a review. Adv Sustain Syst 6(11):2200258
Mallick PK (2007) Fiber-reinforced composites: materials, manufacturing, and design. CRC press
Mohanty AK, Misra M, Drzal LT (2002) Sustainable bio-composites from renewable resources: opportunities and challenges in the green materials world. J Polym Environ 10:19–26
Nachtane M, Tarfaoui M, Abichou MA, Vetcher A, Rouway M, Aâmir A, Naanani H (2023) An overview of the recent advances in composite materials and artificial intelligence for hydrogen storage vessels design. J Compos Sci 7(3):119
Oksman K, Skrifvars M, Selin JF (2003) Natural fibres as reinforcement in polylactic acid (PLA) composites. Compos Sci Technol 63(9):1317–1324
Pande PB, Dhengare SW, Raut JM, Bhagat RM, Bahoria BV, Shelke N, Vairagade VS (2025) Optimizing sustainability and resilience of composite construction materials using life cycle assessment and advanced artificial intelligence techniques. Asian J Civ Eng 26(2):471–489
Pickering KL, Efendy MA, Le TM (2016) A review of recent developments in natural fibre composites and their mechanical performance. Compos Part A Appl Sci Manuf 83:98–112
Radu IC, Vadureanu AM, Cozorici DE, Blanzeanu E, Zaharia C (2025) Advancing sustainability in modern polymer processing: strategies for waste resource recovery and circular economy integration. Polymers 17(4):522
Rani A, Negi S, Fan C, Lam SS, Kim H, Pan SY (2024) Revitalizing plastic wastes employing bio-circular-green economy principles for carbon neutrality. J Hazard Mater 134394
Rupal A, Meda SR, Gupta A, Tank I, Kapoor A, Sharma SK, Murugan P (2022) Utilization of polymer composite for development of sustainable construction material. Adv Mater Sci Eng 2022(1):1240738

Rydz J, Sikorska W, Musioł M (2024) Biosynthesis and biodegradation—eco-concept for polymer materials. Int J Mol Sci 25(5):2674

Saheb DN, Jog JP (1999) Natural fiber polymer composites: a review. Adv Polym Technol J Polym Process Inst 18(4):351–363

Salami BA, Bahraq AA, ul Haq MM, Ojelade OA, Taiwo R, Wahab S, Ibrahim M (2024) Polymer-enhanced concrete: a comprehensive review of innovations and pathways for resilient and sustainable materials. Next Mater 4:100225

Sathishkumar TP, Satheeshkumar S, Naveen J (2014) Glass fiber-reinforced polymer composites–a review. J Reinf Plast Compos 33(13):1258–1275

Scaffaro R, Citarrella MC, Morreale M (2023) Green composites based on mater-Bi® and solanum lycopersicum plant waste for 3D printing applications. Polymers 15(2):325

Seier M, Archodoulaki VM, Koch T (2024) The morphology and properties of recycled plastics made from multi-layered packages and the consequences for the circular economy. Resour Conserv Recycl 202:107388

Sheldon RA (2000) Atom efficiency and catalysis in organic synthesis. Pure Appl Chem 72(7):1233–1246

Simões S (2024) High-performance advanced composites in multifunctional material design: state of the art, challenges, and future directions. Materials 17(23):5997

Singh A, Charak A, Biligiri KP, Pandurangan V (2022) Glass and carbon fiber reinforced polymer composite wastes in pervious concrete: material characterization and lifecycle assessment. Resour, Conserv Recycl 182:106304

Skosana SJ, Khoathane C, Malwela T (2025) Driving towards sustainability: a review of natural fiber reinforced polymer composites for eco-friendly automotive light-weighting. J Thermoplast Compos Mater 38(2):754–780

Tabassum T, Mir AA (2023) A review of 3D printing technology-the future of sustainable construction. Mater Today Proc 93:408–414

Umesh M, Sarojini S, Basheer T, Suresh S, Santhosh AS, Stanly LM, James N (2023) Sustainable biodegradable and bio-based materials. In: In handbook of sustainable materials: modelling, characterization, and optimization. CRC Press, pp 71–92

Vazquez-Martel C, Florido Martins L, Genthner E, Almeida C, Martel Quintana A, Bastmeyer M, Blasco E (2024) Printing green: microalgae-based materials for 3D printing with light. Adv Mater 36(33):2402786

Wool R, Sun XS (2011) Bio-based polymers and composites. Elsevier

Zaman A, Gutub SA, Wafa MA (2013) A review on FRP composites applications and durability concerns in the construction sector. J Reinf Plast Compos 32(24):1966–1988

zu Castell-Rüdenhausen M (2025) Material flow analysis and modelling of the feedstock potential for recycling polystyrene. Circ Econ 4(1):100127

Chapter 5
Advancements and Opportunities of Green Polymer Composites in Agriculture

Pedro Minasi Brandão, Micheli Legemann Monte, Janaína Oliveira Gonçalves, Renata Rodrigues de Moura, Bruna Silva de Farias, Tito Roberto Sant'Anna Cadaval Junior, Luiz Antônio de Almeida Pinto, Débora Pez Jaeschke, and Anelise Christ-Ribeiro

Abstract In recent years, green polymer composites have emerged as a sustainable and promising agricultural alternative, replacing conventional plastics associated with adverse environmental impacts. This chapter comprehensively analyzes the development and various applications of these materials derived from renewable and biodegradable sources. Some examples of green polymer composites are polylactic acid, polyhydroxyalkanoate, thermoplastic starch, polycaprolactone *and green polymeric composites.* The intrinsic relationship between the molecular structure of these polymers and their specific properties has allowed for a wide range of applications, such as biodegradable soil covers, greenhouse films, and more efficient irrigation systems. In addition to the environmental benefits, green polymer composites have significant economic advantages, such as reducing the costs of managing agricultural waste. However, challenges still need to be addressed, such as the considerable production costs and adequate infrastructure for large-scale implementation of these materials. Prospects are also discussed, including the development of promising new materials and their inclusion in emerging agricultural technologies such as precision agriculture. However, it is concluded that, with continued investment in research and development, green polymer composites have promising potential in global farming practices, promoting advances in operational efficiency and economic viability.

P. M. Brandão · M. L. Monte · R. R. de Moura · B. S. de Farias · T. R. S. C. Junior · L. A. de Almeida Pinto · D. P. Jaeschke · A. Christ-Ribeiro (✉)
Industrial Technology Laboratory, School of Chemistry and Food, Federal University of Rio Grande, Rio Grande, Brazil
e-mail: anelise.christ@furg.br

J. O. Gonçalves
Department of Civil and Environmental, Universidad de La Costa, Barranquilla, Atlántico, Colombia

L. Ravindran et al. (eds.), *Macro, Micro and Nanocomposites from Sustainable Sources*, Advanced Structured Materials 244, https://doi.org/10.1007/978-981-95-2469-3_5

5.1 Introduction

Global food security faces heightened challenges due to climate change, negatively impacting food production. The global population is estimated to reach around 10.3 billion by 2100, further increasing the need for sustainable solutions. Additionally, the excessive use of agrochemicals and their accumulation in the environment are emerging and critical concerns for the future (Bouteska et al. 2024; Mitra et al. 2023; Intergovernmental Panel on Climate Change (IPCC) 2023). Finding sustainable and innovative agricultural practices can be an alternative to increasing productivity, preserving natural resources, mitigating the effects of climate change, and ensuring food supply for future generations.

Currently, new agricultural applications are being developed to minimize the use of conventional methods and reduce the dependence on synthetic chemicals. Among the most promising innovations is the utilization of polymers, such as mulching films, and nanotechnology, which improve water retention (Mubaraka et al. 2022), control (Iqbal et al. 2020), optimize nutrient delivery (Aziz et al. 2022), and reduce soil erosion (Tariq and Akhtar 2022). However, these practices are generally carried out using polymers, such as in soil coverage through mulching or as matrices for encapsulating and releasing fertilizers, pesticides, and herbicides in the form of nanoparticles. While these polymers are highly efficient in reducing chemical inputs, minimizing leaching, and reducing the accumulation of harmful substances in the environment, the excessive use of non-biodegradable synthetic polymers can present long-term challenges (Neme et al. 2021). The persistence of these polymers in the environment poses risks, contributing to plastic pollution and affecting soil health. The incomplete degradation of these materials releases microplastics, which accumulate in ecosystems, negatively impacting biodiversity and potentially entering the food chain (Chouchene et al. 2022). As a result, green polymeric materials, including some synthetic biodegradable polymers, have garnered considerable attention in recent years. These materials are recognized for being environmentally safe, high-quality, and efficient alternatives.

Green polymers, or biopolymers, are derived from natural sources such as plants, animals, and bacteria. Some are generated as co-products or by-products during food processing and can be utilized to produce renewable biomaterials, offering a sustainable alternative to non-biodegradable synthetic materials (Borba et al. 2023). Among the most widely used green polymers in agriculture are polylactic acid (Zimmermann et al. 2020), polyhydroxyalkanoate (Adeleye et al. 2020), thermoplastic starch (Nakaramontri et al. 2022), polycaprolactone (Hejna et al. 2015), and green polymeric composites such as chitosan and cellulose. These polymers present important characteristics for agricultural applications, such as biodegradability, renewable origin, lower toxicity, and sustainable cycles (Harada et al. 2019a).

5.1.1 *Origin of the Application of Synthetic Polymers in Agriculture*

Farmers and agricultural producers widely adopt agrochemicals to control pests, weeds, and insects in agriculture. It is estimated that more than 2 million tons of agrochemicals globally to increase agricultural productivity. Of this total, approximately 50% are herbicides, 30% are insecticides, and 20% are fungicides. In addition to agrochemicals, farmers use fertilizers to provide essential nutrients for plant growth and development, including macronutrients such as nitrogen, phosphorus, and potassium and micronutrients such as zinc and manganese. However, the indiscriminate use of these often-toxic products accumulates emerging contaminants in soil and water, negatively affecting the metabolism of plants, animal species, aquatic organisms, and even humans (Ore et al. 2023; Akash et al. 2022; Hui et al. 2022).

The excessive use of agrochemicals and fertilizers also contaminates air, with volatile compounds and particulates spreading to adjacent areas, affecting ecosystems far from the application sites. This is exacerbated by the lack of strict regulation and the inefficient use of products, leading to the chronic accumulation of chemical substances that interfere with soil health and affect groundwater and surface water quality. Additionally, the leaching of these compounds into water bodies can contribute to eutrophication, a phenomenon that promotes excessive algae growth and reduces oxygen availability, negatively impacting aquatic biodiversity. Recent studies indicate the urgency of controlling agrochemical applications, as these compounds' long-term effects and chemical interactions with the environment are largely unknown. The potential toxicity and persistence of these pollutants in ecosystems represent significant risks, both for biodiversity and human health (Ore et al. 2023; Mali et al. 2023; Peluso et al. 2023).

In this context, introducing of non-biodegradable synthetic in agriculture has emerged as an alternative to minimize and replace agrochemicals. These materials, defined as agro-composites, aim to offer more sustainable plant protection and nutrition solutions. The overuse of non-biodegradable polymers raises serious concerns. The fragmentation of these materials generates micro and nanoplastics that persist in the environment, accumulating in soil, water, and living organisms. These microplastics can be ingested by animals and eventually enter the food chain, causing negative health impacts such as toxicity, hormonal interference, and developmental problems. The slow degradation of these polymers means they can remain in the environment for decades, exacerbating pollution and posing long-term risks to biodiversity and ecosystems. Currently, combined with nanotechnology, other promising technologies include the use of controlled-release polymers, the development of polymer networks, and super-adsorbent hydrogels, which may be options for minimizing ecological damage and implementing more sustainable and safe agriculture (Cherwoo et al. 2024).

However, the extensive use of synthetic polymers in agriculture is not without environmental concerns. Studies have shown that these materials can fragment into micro- and nanoplastics, which persist in the environment and accumulate in living organisms. These microparticles can cause adverse effects in aquatic and terrestrial organisms, such as toxicity, hormonal disruption, and impacts on reproduction and development. The slow degradation of synthetic polymers means they can remain in soil and water for decades, exacerbating environmental pollution. Furthermore, many of these materials are derived from non-renewable fossil resources, raising questions about the sustainability of their long-term use (Rani and Sharma 2022; Govil et al. 2024; Rajagopalan et al. 2022).

5.2 Development of Green Polymer Composites for Agriculture

5.2.1 Types of Green Polymers Used in Agriculture

The use of green polymers in agriculture has expanded significantly, driven by the need to reduce the environmental impacts of using petroleum-derived polymers. Conventional polymers such as polypropylene (PP), polyethylene (PE), and polystyrene (PS), although highly durable and cost-effective, present significant environmental challenges due to being persistent and accumulative compounds. However, the growing demand for sustainable solutions has driven the development of alternative materials from natural and renewable sources that offer equivalent or superior performance to conventional polymers. Green polymers minimize pollution and promote the conservation of natural resources (Dou et al. 2024; Wang et al. 2024; Kumar and Gupta 2024; Abdel-Raouf et al. 2018).

Green polymers can include synthetic ones, such as polylactic acid (PLA), polyhydroxyalkanoate (PHA), and thermoplastic starch (TPS), which have gained prominence due to their biodegradable properties and applications in agriculture. PLA, produced from sources such as cornstarch, has properties similar to conventional plastics and is used in mulching and biodegradable packaging (Al-Khairy et al. 2022a). However, the decomposition rate of PLA can vary depending on environmental conditions. According to Vaverkova et al. (2014a), although PLA can decompose in composting environments, its degradation under natural conditions, such as burial in soil, tends to be slower, which may limit some of its applications. PHAs, derived from bacterial action on organic waste, have high biodegradability (less than 1 year) and can be molded by various processes, such as extrusion and 3D printing. In addition, this biopolymer is versatile, and in its applications, it is ideal for agricultural films, slow-release fertilizers, and seed coatings (Puppi and Chiellini 2020; Mukherjee and Koller 2023a; Shahzad et al. 2017a). TPS, derived from starch sources like corn, potato, or cassava, is another widely used polymer due to its low cost, easy processing, and biodegradability within a relatively short time frame of 3 months to 1 year, which makes it suitable for applications such as soil coverings and containers (Table 5.1). TPS is a biodegradable polymer from starch sources such as potato, corn, and cassava. It has high availability, low cost, and ease of processing. The biopolymer can be used to develop films for soil cover, biodegradable containers, and protection against soil erosion. In addition, it can be enhanced by incorporating antioxidant agents, such as alpha-tocopherol, to create coatings that extend the shelf life of agricultural products. Ferreira et al. (2021a) indicated the potential of TPS combined with chitosan and bentonite for protecting green coffee beans, evidencing its effectiveness in preserving quality during storage.

On the other hand, natural polymers like chitin, chitosan, and cellulose stand out for their biodegradability and abundance in nature (Table 5.1) (Bagheri and Ghaedi 2020; Das et al. 2022). Chitin and chitosan are extracted primarily from the exoskeletons of crustaceans, such as shrimp and crabs, while cellulose is obtained from plant-based sources like wood and agricultural residues. It is estimated that the number of biopolymers accumulated annually can reach up to 2 million tons, derived from agro-industrial waste (Varghese et al. 2019). Guar gum is another example of a natural polymer obtained from plant seeds. It is known for its thickening properties and water retention capacity, which makes it widely used in various applications (Sharahi and Bahrami 2024).

Developing new materials through the modification or functionalization of biodegradable polymers for agricultural applications has been gaining prominence (Abdel-Raouf et al. 2018; Ostovan et al. 2018; Farooq et al. 2022; Ansari and Masoum 2021; Sole et al. 2021). Applying polymeric films for soil cover and polymeric hydrogels for water retention are examples of sustainable alternatives that help significantly reduce the accumulation of waste in the environment. Song et al. (2020) demonstrate that using polymeric hydrogels, such as those derived from lignin and humic acid, can enhance soil water retention, reduce water loss, and aid in the recovery of soil quality. Kumar et al. (2023) highlighted that lignin-based hydrogels, in addition to being used for seed coating and the absorption of trace metal ions, can improve water use efficiency in plants and release agrochemicals in a controlled manner. Therefore, besides being biodegradable, these three-dimensional polymeric networks improve the soil's physical and chemical properties, support healthy plant development, and help mitigate erosion.

Renewable biopolymer composites, such as starch, lignin, and cellulose derivatives, can significantly improve soil quality, promote microbial diversity, and reduce greenhouse gas emissions using slow/controlled-release fertilizers. These fertilizers are designed to encapsulate the necessary nutrients within the polymeric matrix, allowing gradual and efficient release into the soil. They can also help maintain soil moisture levels, improving water retention, especially in arid and semiarid regions (Firmanda et al. 2024). A promising example of this composite type is the chitosan-–kaolinite clay biocomposite developed by Eddarai et al. (2023a) as a coating material for slow-release NPK fertilizers, the authors reported that this coating enhanced the mechanical strength of the NPK granules and increased nutrient availability to the tomato fruits, resulting in higher chlorophyll content, biomass, and improved nutritional quality. Additionally, the composite allowed for a 50% reduction in the amount of NPK fertilizer needed, which led to a significant increase in tomato metabolism, with a 44% increase in total phenols, 56% in flavonoids, and 94% in vitamin C.

Table 5.1 Green polymers and sustainable applications in agriculture

Green polymer	Source	Characteristics	Agricultural applications	Biodegradation time	References
PLA	Corn starch	Biodegradable, similar mechanical properties to conventional plastics; slower decomposition rate in compost; not suitable for natural burial conditions	Mulching films, biodegradable packaging, disposable cutlery	6 months to 2 years	Al-Khairy et al. (2022b) Vaverkova et al. (2014b)
PHA	Animal residues, bacteria	Biocompatible, moldable through various methods (e.g., injection, extrusion, 3D printing); maintains chemical structure during processing; environmentally friendly when produced with renewable energy	Biodegradable films, slow-release fertilizers, seed coatings	Less than 1 year	Shahzad et al. (2017b); Puppi and Chiellini (2020); Mukherjee and Koller (2023b)
Chitin/ chitosan	Exoskeletons of crustaceans	Antimicrobial, biostimulant, and biodegradable can form biocomposites with clays	Coating for slow-release fertilizers, water retention, pest control	3 to 6 months	Eddarai et al. (2023b); Ferreira et al. (2021b)
Xanthan gum	Bacterial fermentation of sugars	Water-soluble, thickening agent, biodegradable, can be modified for controlled release	Edible coatings, spray coatings for fresh-cut produce, controlled-release hydrogels	Less than 6 months	Hafezi Moghaddam et al. (2024); Lara et al. (2023)

(continued)

Table 5.1 (continued)

Green polymer	Source	Characteristics	Agricultural applications	Biodegradation time	References
Starch, lignin, and cellulose derivatives	Plant-based (corn, wood, agricultural residues)	Renewable, biodegradable, improve soil quality, promote microbial diversity, reduce greenhouse gas emissions	Slow/controlled-release fertilizers (SRF/CRF) for gradual nutrient release and improved soil moisture retention	3 months to 2 years	Firmanda et al. (2024)
TPS	Corn, potato	Low-cost, biodegradable, enhanced with additives like alpha-tocopherol for specific uses	Active coatings for green coffee beans to improve shelf life and protection	3 months to 1 year	Ferreira et al. (2021b)

Another notable green polymer is xanthan gum, produced by bacterial fermentation. In agricultural cultivation, xanthan gum in hydrogel form has been shown to improve the growth of forage corn using electron beam irradiation instead of toxic chemicals, making this material more biocompatible with soil and plants. The hydrogel has been noted for increasing soil moisture, corroborating the availability of water and fertilizer to the roots due to its gradual and slow release (Moghaddam et al. 2024). Lara et al. (2020a) investigated using spray-applied edible xanthan gum coatings to extend the shelf life of freshly cut lotus roots. Xanthan gum was combined with citric acid and glycerol and applied via a spray system. The results showed that the coating significantly reduced enzymatic browning and inhibited the growth of Bacillus subtilis bacteria, improving product quality and stability during storage. This approach highlights the effectiveness of green polymers, such as xanthan gum, in increasing the preservation of fresh agricultural products.

5.2.2 Synthesis and Processing of Biopolymers

5.2.2.1 Obtaining Biopolymers

Biopolymers can be obtained directly from natural macromolecules and their derivatives, or they can be produced by the polymerization of natural resources that have been transformed into monomers (Getahun et al. 2024). Examples of natural biopolymers are polysaccharides, such as cellulose, chitosan, starch, alginate, and agarose, among others. The production process of these natural polymers involves their extraction and purification. In addition, bacteria can synthesize many biopolymers, converting different carbon sources into several different polymers. In this case, the production process involves biopolymer synthesis by bacteria fermentation, extraction from bacteria cell or extraction medium, purification, and polymerization (Fasciotti 2017).

Polylactic Acid

PLA is produced by the polymerization of the monomer lactic acid (LA), which has two active configurations, L(+)- or (S)- and D(-)- or (R)-. Therefore, depending on the production process, it is possible to produce poly-L-lactic acid (PLLA), poly-D-lactic acid (PDLA), and poly-D, L-lactic acid (PDLLA) (Albuquerque et al. 2021). LA production can be performed by chemical synthesis from acetaldehyde or by fermentation processes of carbohydrates. Chemical synthesis involving petrochemical materials has high production costs, resulting in low-quality lactic acid due to the inability to process only the L(+) isomer (Khouri et al. 2024). The fermentation process accounts for 90% of the total PLA production globally (Swetha et al. 2023). This method has low raw material cost, milder operation temperatures, and lower energy consumption (Albuquerque et al. 2021).

The fermentation is performed by LA bacteria using sugar as raw material, such as hexose and pentose, to produce lactic acid, acetic acid, and ethanol. Lactobacillus (used primarily for industrial production) and Leuconostoc are examples of lactic acid bacteria. The first step in fermentation is the carbohydrate extraction from food sources, such as sugarcane, corn, or beet. Depending on the bacteria strain, the fermentation can be homofermentative or heterofermentative. The homofermentative process is the most used because it results in lower co-product production. In this method, 100 g of glucose leads to 90 g of LA (Khouri et al. 2024). PLA's physical, mechanical, and thermal properties depend on the optical purity of LA and the molecular distribution and mass. In the solid state, PLA can be amorphous or semi-crystalline. High crystalline PLA is only possible to synthesize by using optical pure lactic acid isomers, which are only produced by fermentation and depend on the bacteria strain and the composition of the fermentation medium (Swetha et al. 2023). PLA with 90% PLLA is crystalline, and lower amounts of L-LA result in an amorphous polymer. Other properties, such as melting temperature and glass transition temperature, decrease with the decrease of PLLA.

After fermentation, LA must be separated from other by-products and the fermentation medium. Different techniques have been used for this purpose, such as molecular distillation, adsorption, precipitation, solvent extraction, and membrane separation. After getting a purified LA, a polymerization step is required to produce PLA. Various polymerization techniques have been developed, and ring-opening polymerization, direct polycondensation of lactic acid, and, more recently, enzymatic ring-opening polymerization are the most used techniques. The polymerization process also influences PLA properties since high molecular weight PLA has more excellent commercial value. In the ring-opening polymerization, the lactide (lactic acid cylinder dimer) opens its ring during the reaction when a catalyst is added. PLA catalysts like aluminum, tin, iron, and zinc have generated high molecular weight. In direct polycondensation, each polymerization step produces a water molecule that breaks the polymer molecule. Therefore, it is necessary to remove the water to make a high molecular weight PLA. These reactions are performed at high temperatures for an extended period, resulting in low molecular weight PLA. Hydrolases and lipases do enzymatic ring-opening polymerization and are an alternative to conventional techniques since the enzymes act as catalysts. This polymerization method has the advantage of high selectivity and low toxicity. However, although the enzymes are considered safe, organic solvents are also needed in this process. Moreover, PLA obtained by enzymatic polymerization has a lower molecular weight than PLA obtained from other methods (Khouri et al. 2024; Swetha et al. 2023; Yu et al. 2023).

Polyhydroxyalkanoate

Chemical and biological methods can produce PHA. The chemical methods are polymerization of β-lactones, carbonylation of epoxides, condensation reactions, and enzymatic polymerization (Adak et al. 2024). However, although PHA produced by chemical methods is biodegradable, it is not obtained from natural sources. Bacteria also synthesize this polymer during fermentation under nutrient-limiting conditions (such as nitrogen and phosphorus deprivation) and surplus carbon conditions. PHA is stored in the cytoplasm of different bacteria strains, mainly Bacillus megaterium, Alcaligenes latus, Pseudomonas, Comamonas, and Cupriavidus. In the cell cytoplasm, PHA is an amorphous polymer surrounded by a phospholipid monolayer with phases and is used by the bacteria as an energy reserve. Different sources of carbohydrates have been used as fermentation substrates, such as glucose, sucrose, glycerol, xylose, and agricultural wastes (Thamarai et al. 2024; Rajgadia and Debnath 2023). PHA production yield depends on the bacteria strain, media composition, and physical and chemical fermentation conditions, varying from 20 to 86% of the microorganism's dry cell weight. The carbon source selection is one of the main factors affecting PHA productivity and cost (Yadav et al. 2021). The molecular weight of PHA is also affected by fermentation conditions, such as the carbon source, nutritional limitation, pH, and osmolarity. Nitrogen limitation and high temperature favor the production of low molecular weight PHA, and an increase in the fermentation period may favor high molecular weight PHA (Thamarai et al. 2024; Raza et al. 2024).

Polycaprolactone

PCL is a synthetic polymer derived from petrochemical sources. However, it is considered a biopolymer since it is biodegradable, biocompatible, and bioabsorbable by microorganisms (Raza et al. 2024). PCL is produced from ε-caprolactone and 6-hydroxyhexanoic acid. These molecules are intermediates in the oxidation reaction of cyclohexanol into adipic acid by microorganisms. Industrially, ε-caprolactone is produced from oxidation of cyclohexanone by peracetic acid. PCL is synthesized mainly by ring-opening polymerization from ε-caprolactone and lactide in the presence of catalysts such as tin (II) 2-ethyl hexanoate or stannous octoate. The polymerization occurs at 100–200 °C and atmospheric pressure. This method allows the production of high molecular weight and low polydispersity. Another method to synthesize the polymer is by polycondensation of 6-hydroxy hexanoic acid. Polycondensation of 6-hydroxyhexanoic acid may be performed by removing water and displacing the equilibrium towards the polymer formation. The reaction is performed at 80–150 °C and may be assisted by enzymes, such as lipases (Sukwijit et al. 2024a).

Natural Polysaccharides: Cellulose and Chitosan

Cellulose is chemically and physically linked with lignin and hemicellulose in vegetable sources. Cellulose extraction is performed through pre-hydrolysis, pulping, and bleaching steps. The most used method is the chemical method, in which cellulose extraction is performed by sodium hydroxide at high temperatures (80 °C) to remove lignin and hemicellulose. Afterward, the blanching is performed with sodium hypochlorite solution (10%) at 90 °C. The material is then dried to obtain white cellulose powder (Liang et al. 2024). Cellulose has a high degree of crystallinity and is not soluble in water or other common solvents. Therefore, researchers have studied methods to improve its solubility. These methods involve the disruption of the natural hydrogen bond network. Some examples of solvents used for this purpose are inorganic molten salts, sodium hydroxide-based solutions, and ionic liquids. Despite their efficiency, using these solvents has the drawbacks of high cost, limited solubility, and environmental pollution (Liang et al. 2024).

The most used method of chitosan extraction involves demineralization, deproteinization, and deacetylation. The demineralization step is an acid treatment to remove the main inorganic compounds, such as calcium carbonate and chloride. The deproteinization step is performed with an alkali solution. Usually, sodium hydroxide (0.1–5 M) hydrolyzes the amide bonds in proteins and generates soluble amino acids, and pigments, mainly carotenoids, are removed using sodium hypochlorite and hydrogen peroxide, among other solvents. The deacetylation is done to remove acetyl groups from chitin and hydrolyze amides. The degree of acetylation is determined by the percentage of free amino groups (Liang et al. 2024; Hisham et al. 2024). Due to various drawbacks of the chemical methods, such as toxicity and high energy consumption, other techniques have been explored to obtain chitosan. In this context, enzymatic deproteinization and fermentation processes using different microorganisms and microwave treatment have been studied (Hisham et al. 2024).

5.2.2.2 Production Process of Mulching Films from Biopolymers

Mulch films and packing materials are obtained by extrusion, casting, spraying, or dipping. Extrusion is used chiefly for PLA-, PHA-, and mixtures of these polymers with natural polysaccharides-based films. Poly (butylene adipate-co-terephthalate) (PBAT)/PLA films at different blend concentrations (100/0, 70/30, 80/20, 90/10, 0/100, weight %) were produced using a screw extruder at 140 °C (PBAT) and 170 °C (PLA and blends) (Lyu and Han 2023a). The films used a chain extender at 0.5 and 1 phr concentrations. The polymer pellets were dried (60 °C) before extrusion to avoid degradation. After extrusion, the material was transformed into films using a heat press machine (30 MPa, 90 s, 140–170 °C). The average thickness of the films was 0.19 mm. The authors concluded that the strength of the 80/20% PBAT/PLA film with 1 phr of chain extender was superior to commercial biodegradable films and comparable to polyethylene-based mulch films.

PLA/PBAT films with different blend compositions (100/0, 90/10, 75/25, 50/50, 0/100) with and without the addition of epoxidized soybean oil (ESO), a reactive compatibilizer, and epoxy-functionalized styrene acrylic copolymer (ESA), a chain extender, were studied. PLA and PBAT granules were dried at 60 °C for 4 h to remove the moisture content. The mixture was melted and extruded using a co-rotation twin screw extruder at 155–170 °C with a screw speed of 120 rpm. The blends were dried at the same conditions described earlier before film preparation. Thin films were prepared in a blown film extruder at 155–170 °C, 10–100 rpm, by adding the blends. The PLA/PBAT 75:25 blend film with 5phr of ESO and 0.5 phr of ESA was considered the most suitable for mulching application.

Othman et al. (2022) prepared PHA/polycaprolactone (PCL) dual-layer films consisting of 100% PHA in the first layer and with different polymer formulations in the second. The films were prepared by melt-compounding and hot-press. Before melting, the polymer granules were pre-dried at 40 °C for 24 h in a vacuum oven. PHA and PCL were blended by mixing, and the films were prepared by hot press at 185 °C and pre-heated for 10 min before pressing. The applied pressure was 15 MPa for 5 min. Afterward, films were submitted to radiation for crosslinking (30 kGy of irradiation dose, 2 MeV, 10 mA current).

Natural polysaccharides, such as cellulose, chitosan, starch, alginate, and agarose, have been used for aqueous base film formulation, casting, spraying, or dipping. These methods are performed by mixing water and polymers, degassing, and film preparation. All steps are done at low-temperature conditions to avoid polymer degradation. Casting consists of pouring the film solution into a mold and letting the solvent evaporate to produce the film. The spray technique is particularly interesting for agriculture applications because the film solution can be sprayed directly on the soil or fruits. The dipping method is performed by dipping fruits or legumes into the film solution, removing the vegetables, and allowing the solvent to evaporate (Gonçalves et al. 2024).

Chemical modifications on natural polymer structure have improved natural polysaccharide film characteristics, such as rigidity, brittleness, high water adsorption capacity, and poor wet stability. Chemical modification methods include substitution, oxidation, graft copolymerization, and cross-linking. The substitution method involves substituting polysaccharides functional groups, such as hydroxyl, amino, and carboxyl, with other groups, such as carboxymethyl, phosphate, sulfate, alkyl, and acetyl groups. Grafting is performed by adding hydrophobic polymers into polysaccharide chemical structure. This process is done by ring-opening polymerization and free radical process. Crosslinking is done by adding a chemical modifier, such as citric acid, genipin, sucrose, vanillin, tannin acid, ferric, and calcium ions. Physical modifications of natural polymer films have also been performed to improve their properties. These have been done by adding plasticizers and blending them with other polymers. Plasticizers, such as glycerol, sorbitol, and ethylene glycol, have decreased brittleness and improved flexibility for mulch films (Saberi Riseh 2024).

Sukwijit et al. (2024a) prepared films with different blends of cellulose and PLA. Cellulose was extracted from pineapple leaves using sodium hydroxide (4%) at 80 °C for 2 h. The extracted material was dried at 60 °C for 24 h, bleached with sodium hypochlorite (10%) at 90 °C for 1 h, and dried. Afterward, sulfuric acid (64%) was added to perform the hydrolysis, which was conducted at 40 °C for 1 h. The material was further neutralized using a dialysis membrane and freeze-dried to obtain cellulose nanocrystals (PNCs). PNCs were modified with a solution of triethoxyvinylsilane (4%), ethanol (80%), and acetic acid (pH 3.5). The treatment was performed at 50 °C for 120 min. The material was centrifuged, washed with water and ethanol, and freeze-dried. The PLA/cellulose films were produced by casting. For that, PLA was solubilized in dichloromethane at 45 °C, and, to this solution, a PNC-dichloromethane solution at different PNC concentrations (1 – 8%) was added. The suspension was stirred, sonicated, and poured into Petri dishes. The films were let at ambient temperature for 24 h for dichloromethane evaporation.

5.2.2.3 Production Process of Biopolymer-Coated Fertilizers

Coating fertilizers have been produced by dipping, phase inversion, and spray methods. The most common methods are immersion and spraying the coating solution directly into the substrate. Dipping is an immersion method in which the fertilizers are dipped into the polymer solution to allow the polymer solution to adhere to the fertilized granule. The granules are then separated from the liquid by centrifugation or filtration and dried to allow the fixation of the coating layer on the fertilizer surface. The spray method is performed using a rotatory drum or coating pan. This method is based on shaking the fertilizer granules and spraying the coating solution using a spray gun, usually atomized through a nozzle. The fertilizer granules are normally pre-heated before the spraying, and after spraying, hot air is also applied in the chamber to allow solvent evaporation. A similar process is also performed using a fluidized bed, in which air is introduced through a distribution plate. The fertilizer granules are pre-heated, and then the coating solution is sprayed through a nozzle placed at the bottom or top of the reactor (Fertahi et al. 2021; Kassem et al. 2024).

Boutriouia et al. (2024) prepared a coating material of poly(ε-caprolactone) (PCL) and chitosan (CS) and incorporated granular diammonium phosphate fertilizer using a rotatory drum. The graft copolymer CS-g-PCL was prepared by ring-opening polymerization of ε-caprolactone with chitosan, using tetra(phenylethynyl)tin as a catalyst. The product was separated by ultrasonication and centrifugation, and films were prepared by casting. The fertilizer particles were added to the rotatory pan with the copolymer dissolved in tetrahydrofuran, which was slowly added with a syringe. Solvent evaporation was done with a flow tube inside the pan at 30 °C.

Messa et al. (2020) produce chitosan/cellulose microparticles for NPK fertilizer coating. Chitosan (1.0 g) was dissolved in 0.5% acetic acid solution and stirred for 24 h. NPK fertilizer was prepared by mixing ammonium sulfate, sodium phosphate, and potassium chloride in a 1:1:1 ratio. Then, 1.0 g of this mixture was added to the chitosan solution and stirred for 1 h. Separately, formulations with individual salts were also prepared. A 10% cellulose nanofiber suspension was sonicated and gradually added to the chitosan/NPK mixtures. The formulations were spray-dried using specific parameters: 180 °C inlet air temperature, 95–110 °C outlet temperature, 7 mL/min feed rate, and 473 L/h of drying airflow. The different materials were transformed into tables with a hydraulic press.

5.2.3 Structure–Property of Green Polymeric Composites

5.2.3.1 Chitosan

Chitosan is obtained through the deacetylation process of chitin from the exoskeleton of arthropods or cell walls of fungi and yeast, resulting in a polycationic polymer composed of N-acetyl-d-glucosamine and d-glucosamine linked to β-(1 → 4) (Alves and Mano 2008; Kumari and Kishor 2020; Moura et al. 2015). The chitosan functionality depends on the high degree of chemically reactive amino ($-NH_2$) and hydroxyl (–OH) groups, giving the molecule hydrophilicity (Kumari and Kishor 2020; Rinaudo 2006). Both groups interact electrostatically in acidic environments, allowing chitosan molecules to interact with negatively charged molecules, such as nucleic acids and proteins, forming complexes and drug delivery (Kumari and Kishor 2020; Huang et al. 2022). Another essential characteristic of $-NH_2$ and –OH groups is that the hydrogens in these groups form hydrogen bonds with other molecules that contain oxygen and nitrogen in their structure, increasing the molecule's structural stability (Kumari and Kishor 2020).

Within this context, chitosan can confer stability to gels and hydrogels (Gonçalves et al. 2017) for example, the study of adsorbent based on chitosan hydrogel scaffold modified with carbon nanotubes to food dye removal in single and binary aqueous systems (Gonçalves et al. 2020); survey of fibroin nanofiber containing chitosan in the hydrogel as wound dressing, biomaterials, and drug delivery (Gonçalves et al. 2020; Tran et al. 2025); films for use as adsorbent (Moura et al. 2015; Carvalho et al. 2019); and emulsions that can protect antioxidant-rich foods (Yin et al. 2024).

Therefore, the $-NH_2$ and –OH groups present in the chitosan chemical structure confer reactivity, antioxidant and antimicrobial activity, and high adsorbent capacity to chitosan, enabling the development of new materials for use in the food and pharmaceutical industries and effluent treatment.

5.2.3.2 Cellulose

Cellulose is a biopolymer composed of combining several 1000 anhydroglucose units. Among the main functional groups in the cellulose molecule are the hydroxyl –OH and ether (R-O-R') groups, which provide structural stability and chemical reactivity (Pasquini et al. 2010; Yu 2015).

Although cellulose is insoluble in water, the hydroxyls present in each glucose unit give cellulose solubility in polar solvents. They form hydrogen bonds that, together with the ether groups, contribute to the stability of the three-dimensional structure. The reaction of –OH of cellulose with acids or anhydrides results in the formation of cellulose esters, which is a derivative that is more soluble in organic solvents and has more excellent resistance to heat and humidity, making it more versatile than cellulose in the development of products in the pharmaceutical, food, textile, and packaging industries (Yu et al. 2022; Wei et al. 2024; Kassie et al. 2024).

The glycosidic bonds that connect the β-D-glucose units form the ether group in cellulose. It gives the plant cell wall structural linearity, resistance to enzymatic degradation, and the formation of microfibrils that, together with hydrogen bonds, give cellulose its characteristic strength and rigidity (Cabiac et al. 2011).

5.2.4 Biodegradability and Sustainability

The growing concern about the environmental impacts of conventional plastics has led to an intense search for sustainable alternatives, especially in the agricultural sector, which represents about 4% of the global plastic market. The use of plastics in agriculture is estimated to reach 6.6 million tons per year, with a 64% increase projected by 2030, driven by the need to meet the growing food demand resulting from population growth. Plastic mulching films are crucial in agricultural applications, covering approximately 6,500 km^2 of arable land worldwide. These films are widely used due to their multifunctional benefits, such as weed control, insect protection, soil moisture maintenance, and soil temperature control (Somanathan et al. 2022; Hayes et al. 2019; Moine and Ferry 2019).

In this context, the biodegradability and sustainability of materials become fundamental. According to EN 13,432:2002, biodegradable plastics are defined as materials that can be rapidly decomposed by microorganisms into natural elements such as water, CO_2, methane, and biomass under natural conditions (aerobic and anaerobic) (Goel et al. 2021). The European Organization for Bioplastics defines bioplastics as materials that are bio-based, biodegradable, or both, as is the case with polylactic acid (PLA) (Zimmermann et al. 2020).

The transition to using biodegradable polymers in agriculture is not limited only to the issue of biodegradability but also offers wider positive environmental impacts. By reducing reliance on conventional plastics derived from fossil fuels, these materials help to reduce the carbon footprint throughout their life cycle, from production to degradation. This is particularly relevant in a global scenario where greenhouse gas emissions associated with conventional plastics continue to rise. Thus, biodegradable polymers emerge as an effective alternative to mitigate the environmental impacts caused by plastics in agriculture (Chen et al. 2024).

5.2.4.1 Advantages of Green Polymers Over Conventional Polymers in Agriculture

The transition to green polymers in agriculture has numerous advantages over conventional plastics, especially regarding biodegradability and environmental impact. One of the main benefits of bioplastics is their ability to decompose rapidly in natural environments, which contrasts with the slow degradation of conventional plastics, which can remain in the environment for hundreds of years (Webber and Tibbitt 2022). This slow degradation results in microplastics forming, which, as observed in studies, have a much greater potential for environmental contamination due to their persistence and the risks associated with their impact on marine life and ecosystems (Degli-Innocenti et al. 2022).

Green polymers, such as bioplastics, not only prevent the accumulation of plastic waste but also offer a more sustainable alternative in terms of production. PLA, for example, has a significant reduction in greenhouse gas emissions during its production and degradation. Studies indicate that PLA production consumes about two-thirds of the energy needed for conventional plastics manufacturing and does not result in a net increase in CO_2 during biodegradation (Atiwesh et al. 2021). In addition, replacing traditional plastics with PLA from corn can reduce greenhouse gas emissions by up to 25% (Porta 2017).

Another relevant aspect is that the biodegradability of plastics depends on environmental factors, such as the physical and chemical characteristics of the soil and the presence of microorganisms, which influence the degradation rate of plastic films (Francioni et al. 2021). Green polymers are designed to degrade in agricultural environments, where moisture and microbial activity are critical for their efficiency. The biodegradable mulching residues positively influenced soil field capacity and water droplet penetration time, providing better conditions for plant growth (Qi et al. 2020).

Conversely, conventional plastics have significant disadvantages. Often, these materials accumulate dirt and soil debris, making recycling difficult and more complex. In addition, the fragmentation of plastic films after harvesting results in particles that become micro and nano plastic over time (Kasirajan and Ngouajio 2012). This fragmentation aggravates pollution and exposes soil and plants to toxic waste.

Developing bioplastics that mix starch with biodegradable aliphatic polyesters has been a promising approach to overcoming bioplastics' mechanical limitations, ensuring they retain their properties while being environmentally sound (Mansoor et al. 2022). Thus, combining mechanical properties and biodegradability is crucial to increasing the acceptance and use of green polymers in agriculture.

Therefore, green polymers are a significant shift towards more sustainable agricultural practices. These practices mitigate the environmental impacts associated with conventional plastics and promote soil health and the farm ecosystem. The transition to these materials benefits the environment and ensures a more sustainable future for agriculture by aligning agricultural production with the planet's and society's needs.

5.2.4.2 Positive Environmental Impact: Reduction of the Use of Conventional Plastics

Replacing conventional plastics with biodegradable polymers in agriculture offers several significant environmental benefits, especially in reducing pollution caused by plastic waste. One of the main problems with traditional plastics is their low degradation rate, which results in the accumulation of waste in the soil and the formation of micro- and nanoplastics over time. Studies indicate that conventional plastics used as agricultural films are often contaminated with soil and debris, making it difficult to recycle them and aggravating environmental pollution (Kasirajan and Ngouajio 2012). Fragmentation of these films during and after use releases tiny particles that can remain in the soil for long periods, affecting soil quality and plant health.

On the other hand, biodegradable polymers such as PLA have a much faster and more efficient decomposition rate. The production of PLA, for example, can reduce greenhouse gas emissions by up to 70% compared to conventional plastics such as polyethylene (PE). This is because PLA is derived from renewable biomass, such as corn, which, during its growth, absorbs CO_2 from the atmosphere, contributing to the reduction of carbon footprint in the life cycle of this polymer (Chen et al. 2024). In addition, biodegradable plastics decompose faster and can integrate back into the soil, turning into beneficial nutrients and compounds rather than persistent pollutants.

Using biodegradable plastics also helps mitigate other environmental impacts associated with the production and disposal of conventional plastics. Global plastic production is expected to continue growing, with the expectation that by 2030, plastic-related greenhouse gas emissions will exceed coal sector emissions (Kane et al. 2022). By reducing dependence on petroleum-based plastics and adopting biodegradable polymers, the agricultural sector can be crucial in reducing these emissions and promoting more sustainable farming practices.

The transition to using biodegradable polymers in agriculture offers significant ecological advantages, such as reducing microplastic pollution. In addition, it contributes to climate change mitigation by reducing greenhouse gas emissions associated with the production and degradation of plastics (Kane et al. 2022). These benefits position biodegradable plastics as an indispensable alternative in the global effort for more sustainable and environmentally responsible agriculture.

5.2.5 Applications in Multiple Agricultural Sectors

Agricultural production faces water scarcity, declining soil quality, and excessive chemical fertilizers and pesticide use. With rapid population growth and demand for food, there is an urgent need to find sustainable solutions (Qin et al. 2024). Plastic materials play a fundamental role in modern agriculture. However, most techniques still rely on fossil-based petrochemical plastics due to their low cost, lightweight, ease of processing, and durability. PE and PP are the most used materials in agricultural activities, mainly in mulch, low tunnels, greenhouse covers, solarization film, fumigation film, and packaging (Gamage et al. 2022).

Biodegradable films have been developed for agricultural purposes over the last decade, particularly for mulching applications (Briassoulis and Giannoulis 2018; Mola et al. 2021; Harada et al. 2019b; Serrano-Ruiz et al. 2021; Sukwijit et al. 2024b; Thakur and Kumar 2021). Mulch films have revolutionized modern agrarian practices by converting arid lands, including deserts, into fertile areas and more efficient irrigation systems, representing a substantial increase in food production (Gamage et al. 2022; Serrano-Ruiz et al. 2021). Furthermore, nanocomposites from natural sources with advanced agrochemical activities are another example that has transformed conventional agriculture, elevating food quality and quantity without causing the long-term detrimental effect of synthetic agrochemicals on soil and ecosystem health (Kumar et al. 2021; Lang et al. 2021).

5.2.5.1 Biodegradable Mulching Films

Mulching is a critical agronomic practice that involves covering the soil with a plastic film for a variable period, even up to 6 weeks or more (Mola et al. 2021). Materials with characteristics opposite to the soil profile are applied to cover the surface and change the microclimate. It is a water conservation technique that enhances the water infiltration capacity of the soil, retards soil erosion, and hence reduces surface runoff. It increases the top soil temperature and fertility, changes microbial biomass, and improves soil quality that enhances seed germination, root production, and plant development and thus increases crop yields in low water input of the world (Thakur and Kumar 2021).

Agronomic practices that promote the water economy of crops are essential since water used for irrigation in agriculture accounts for 88% of the total global freshwater consumption. Furthermore, a significant problem of increasing water scarcity is present, which can substantially impact crop production (Qin et al. 2024). Therefore, mulching is a simple and promising method to be applied in agricultural production.

However, most plastic mulches are made of low-density polyethylene (LDPE), contributing to worldwide plastic pollution (Serrano-Ruiz et al. 2021). These plastic films take about 100 years to decompose fully, and recycling is challenging. The cleaning steps required to eliminate soil, plants, and agrochemicals adsorbed by the films increase complexity and are costly, and the agrochemicals they may release are harmful to the environment (Serrano-Ruiz et al. 2021; Li et al. 2024). Furthermore, removing polyethylene plastics from fields may only sometimes be complete due to environmental factors that cause the plastics to become brittle and fragmented. The accumulation of microplastics (<5 mm in diameter) and nanoplastics (<1 μm) resulting from mulch decomposition emerged as a hidden problem with potential risks to soil quality, ecosystem multifunctionality, and human health (Huang et al. 2023).

Biodegradable mulch films (BDM) have been introduced as a sustainable alternative to traditional polyethylene films as part of a series of climate-conscious innovations because, after their use, they can be degraded progressively in the soil without releasing toxic residues into the environment (Mola et al. 2021). Biodegradable films have already been tested as soil mulching films on several crops, such as zucchini (Mola et al. 2019) tomato (Moreno and Moreno 2008), strawberry (Costa et al. 2014) lettuce (Cozzolino et al. 2020), and melon (Sellami et al. 2024). The global BDM market is expected to grow at a compound annual growth rate of 8.4% between 2022 and 2027 (Huang et al. 2023).

The main biodegradable polymers used in the production of BDM are polysaccharides (cellulose and starch), aliphatic polyesters (PLA, PHB, PHV), and aromatic polyesters (PBAT). However, biodegradable polymers have limitations such as shorter processing range, greater fragility, high moisture absorption, low gas barrier, and low thermal and mechanical resistance (Harada et al. 2019b). Poly (butylene adipate *co*-terephthalate) (PBAT) is the most used copolymer for producing biodegradable mulch. It is a ductile and flexible polymer that exhibits outstanding elongational properties. It has attracted much interest because of its processability and mechanical characteristics, which are comparable to traditional polyethylene films. Aromatic groups improve mechanical properties but are not prone to biological degradation and must be limited to allow complete biodegradation, while aliphatic chains contribute to biodegradability. They are often mixed with starch or PLA to reduce stiffness and gain hydrophilicity (Serrano-Ruiz et al. 2021; Jian et al. 2020).

Poly (lactic acid) (PLA) is a relatively rigid polymer that can be applied in mixtures with PBAT and polycaprolactone (PCL), which is flexible, promoting a biodegradable, stable, tenacious mixture, and applicable to the cultivation of mulch. The blends of PBAT and PLA are considered complementary because of the stiffness of PLA and the elastomeric characteristics of PBAT (Lyu and Han 2023b).

Plastic mulch additives can be added to improve limitations to facilitate mechanical installation of the material (e.g., elasticity), to achieve an efficient functional performance during the crop cycle (e.g., mulch stability), to provide specific mulch characteristics (e.g., color), and to facilitate the in situ (bio)degradation, into the soil, following the mulch use. Bio-based additives can be added to improve limitations in biodegradable mulch production. Sliding agents such as fatty acid amides, glycerol oleates, and stearates can reduce polymer surface friction and facilitate processing. Benzophenone can act as a stabilizing agent against UV radiation and atmospheric conditions. Glycerol and sorbitol can be used as plasticizers and improve the flexibility of plastic films (Serrano-Ruiz et al. 2021).

Additionally, dyes are used, such as carbon black (black), TiO_2 (white), Fe_2O_3 (red), and $CaCO_3$ (white). Dark colors inhibit weed growth under the plastic mulch. White or clear colors provide more significant soil warming than dark films but poor inhibition of weed growth. Other colors or color combinations are used for specific purposes (e.g., green colors for aesthetic purposes) (Serrano-Ruiz et al. 2021).

Furthermore, minor natural fillers can be incorporated into the processing of these films, reducing the use of agrochemicals in organic agriculture (Briassoulis and Giannoulis 2018). In recent studies, Harada et al. (Harada et al. 2019b) evaluated the incorporation of natural fillers (carbon black, organic fertilizer, and rice husk ash silica) into biodegradable mulching films composed of PBAT/PLA prepared by twin-screw extrusion. The results showed that adding these natural fillers resulted in gains in sorption and biodegradation, which were superior to pure mixtures, making PBAT/PLA mulching films more beneficial to the growth of terrestrial crops.

Sukwijit et al. (2024b) produced mulch films from polylactic acid (PLA) with cellulose nanocrystals (PNC) extracted from pineapple leaves. PNC was modified by incorporating 4 wt% triethoxyvinylsilane (TEVS), designated 4PNC, to increase its interaction with PLA. The results indicated that higher concentrations of PNC increased the water vapor permeability (WVP) and biodegradability of the composite films while reducing light transmission. Additionally, these films showed reduced light transmission, correlating with reduced weed growth, reduced WVP, and improved barrier properties, indicative of improved soil moisture retention. PLA films with 4PNC demonstrated more excellent thermal degradation stability than those with unmodified PNC, suggesting more excellent heat resistance. This study confirmed that TEVS-modified cellulose significantly improved the properties of bio-composite films, making them more suitable for mulch film applications.

Some commercial examples of biodegradable mulch films are Mater-Bi M15 (mulch grade, 15 μm thick) produced by Novamont and Ecovio M12 (mulch grade, 12 μm thick) produced by BASF. Mater-Bi is based on biodegradable aromatic/aliphatic polyesters and starch (minimum C14 content: 40%; non-food derivative), and Ecovio is composed of the biodegradable copolyester ecoflex ® and polylactic acid. Both comply with the European biodegradability standard (Briassoulis and Giannoulis 2018). Several polymers accomplishing the properties BDM requires have been identified, together with their main characteristics, and are already used for producing biodegradable plastic mulches.

The soil solarization technique is another sustainable approach that uses heat-based biodegradable mulches to eradicate fungi, weeds, and nematodes. The solarization process involves covering the ground with clear plastic, allowing shortwave solar radiation to penetrate the plastic. In a covered moist soil, the high temperature (45–55 °C) at a 5 cm soil depth causes the death of several soil-borne plant pathogens, such as *Verticillium dahliae*, *Fusarium* spp., *Rhizoctonia solani* and *Sclerotinia minor* (Mola et al. 2021).

Di Mola et al. (2021) evaluated the effect of soil solarization using biodegradable mulch or traditional polyethylene plastic film in a greenhouse. The biodegradable film avoided a high ammonia concentration in the soil, thanks to lower soil water content and slightly lower temperature than the polyethylene film. These conditions were optimal for the growth of nitrifying bacteria, confirming the film's biodegradable potential.

5.2.5.2 Encapsulated Fertilizers and Pesticides

Conventional applications of synthetic agrochemicals have drawbacks such as rapid degradation, poor solubility, and increased pesticide runoff that pollute the environment. Biopolymeric substances such as polylactic acid (PLA), poly(3-hydroxybutyrate) (PHB), polycaprolactone (PCL), polyethylene glycol (PEG), polyglycolic acid (PGA), chitosan (CS), alginate (ALG), and starch have been widely investigated for the development of different nanoforms of agrochemicals such as nano pesticides, nano-fabricated fertilizers, and nanocapsules (Lang et al. 2021; Victoria et al. 2023).

PLA has previously been used to encapsulate the herbicide metazachlor to control grass and weeds in crops such as soybeans and potatoes. Smaller nanoparticles appeared to release the herbicide in an initial burst, while a diffusion-controlled process occurred for the larger microparticles (Stloukal et al. 2012). Meyer et al. (2015) demonstrated functional polylactic acid (PLA) use for imidacloprid encapsulation at lower concentrations (about 200 times) than commercial anti-vector plant pathogen formulations. PHB has also been used to engineer particles for the slow, controlled release of herbicides to minimize the toxic effects. The diffusion process release of ametrine, an s-triazine herbicide encapsulated in ~5–6 μm microparticles, was slowed by 75% (Grillo et al. 2011) and was likewise investigated as a coating for urea granules to improve the retention of macro and micronutrients, and to develop a slow controlled release system for fertilizers (Lang et al. 2021; Costa et al. 2013).

Numerous studies have been conducted on using nanomaterials in the agricultural sector (Table 5.2). The encapsulation of agrochemicals has provided good biocompatibility over conventional applications by inhibiting the early degradation of active ingredients, improving delivery, and reducing environmental impact and human exposure. Therefore, it is an essential part of agricultural operations to increase productivity and achieve sustainable ecological agriculture (Costa et al. 2013).

Table 5.2 Summary of the biopolymers and applications in the agricultural sector

Biopolymers	Active ingredient	Effects	Ref
PLA	Metazachlor	The herbicide release rate decreased with increasing particle size of 92, 56, and 34% for 0.6, 0.8, and 8 μm sized nanoparticles, respectively	Stloukal et al. (2012)
PHB/PHBV	Ametryn	The release profile of ametrine in microparticles was reduced to 75% and 87%, respectively, compared to the release of pure ametrine	Grillo et al. (2011)
PCL	Metribuzin	During 7 days of exposure to water, 81 to 96% of the initially loaded metribuzin was released; the highest release was detected for 40% of charged forms	Boyandin and Kazantseva (2021)
CS/ALG	NPK	Percentage of cumulative release of N, P, and K ions of 33.2, 47.8, and 68.3%, alternatively in 30 days	Hamed et al. (2024)
CS	Silicon	Applying 0.08% CS-Si nano fertilizer resulted in a 43.4% higher yield, and 0.04% resulted in a 45% higher test weight than SiO_2	Kumaraswamy et al. (2021)
Guar gum	Boron	The half-life period of boron in guar gum (96.25 days) in the soil was almost triple the half-life of commercial fertilizer (33.32 days), which indicates a slow and controlled release	Thombare et al. (2021)

5.3 Conclusion

The application of green and sustainable polymer composites in agriculture has shown great potential to reduce the environmental impact of agricultural practices. The use of mulching and the incorporation of nanocomposites in planting systems offer a promising alternative to increasing resource efficiency, such as water and fertilizers, especially in semi-arid areas. Additionally, the development of biopolymers as nanopesticides and nanofertilizers opens new possibilities for pest control and increased agricultural productivity in a more sustainable way.

Although technological advances in this area are promising, there are still challenges to overcome, such as adapting techniques to field conditions and ensuring economic viability on a large scale. Integrating new technologies with traditional approaches also requires a deep understanding of the interactions between polymeric materials and the agricultural environment.

In the coming years, further research is expected to explore the properties of green polymers and natural composites, focusing on improving their mechanical characteristics, biodegradability, and performance in various agricultural scenarios. The search for sustainable solutions in the agricultural sector is essential to reducing synthetic inputs, mitigating the impacts of climate change, and ensuring global food security.

References

Abdel-Raouf ME, El-Saeed SM, Zaki EG, Al-Sabagh AM (2018) Green chemistry approach for preparation of hydrogels for agriculture applications through modification of natural polymers and investigating their swelling properties. Egypt J Pet 27:1345–1355

Adak S, Kayalvizhi R, Bishai M, Jacob S, Kundu D (2024) Advancements in microbial production of polyhydroxyalkanoates (PHA) from wastes for sustainable active food packaging: an eclectic review. Biocatal Agric Biotechnol 60:103288. https://doi.org/10.1016/j.bcab.2024.103288

Adeleye AT, Odoh CK, Enudi OC, Banjoko OO, Osiboye OO, Toluwalope Odediran E, Louis H (2020) Sustainable synthesis and applications of polyhydroxyalkanoates (PHAs) from biomass. Process Biochem 96:174–193. https://doi.org/10.1016/j.procbio.2020.05.032

Akash S, Sivaprakash B, Rajamohan N, Pandiyan CM, Vo D-VN Pesticide pollutants in the environment—a critical review on remediation techniques, mechanism and toxicological impact. Chemosphere 301:134754. https://doi.org/10.1016/j.chemosphere.2022.134754

de Albuquerque TL, Marques Júnior JE, de Queiroz LP, Ricardo ADS, Rocha MVP (2021) Polylactic acid production from biotechnological routes: a review. Int J Biol Macromol 186:933–951. https://doi.org/10.1016/j.ijbiomac.2021.07.074

Al-Khairy D, Fu W, Alzahmi AS, Twizere J-C, Amin SA, Salehi-Ashtiani K, Mystikou A (2022a) Closing the gap between bio-based and petroleum-based plastic through bioengineering. Microorganisms 10:2320

Al-Khairy D, Fu W, Alzahmi AS, Twizere J-C, Amin SA, Salehi-Ashtiani K, Mystikou A (2022b) Closing the gap between bio-based and petroleum-based plastic through bioengineering. Microorganisms 10:2320. https://doi.org/10.3390/microorganisms10122320

Alves NM, Mano JF (2008) Chitosan derivatives obtained by chemical modifications for biomedical and environmental applications. Int J Biol Macromol 43:401–414. https://doi.org/10.1016/j.ijbiomac.2008.09.007

Ansari S, Masoum S (2021) Recent advances and future trends on molecularly imprinted polymer-based fluorescence sensors with luminescent carbon dots. Talanta 223:121411

Atiwesh G, Mikhael A, Parrish CC, Banoub J, Le T-AT (2021) Environmental impact of bioplastic use: a review. Heliyon 7:e07918. https://doi.org/10.1016/j.heliyon.2021.e07918

Aziz A, Tahir MA, Ahmad A, Hamza A, Ramzan H, Khan BA (2022) Effect of mulches on mineral fertilizer (N, P & K) management and fertilizer use efficiency. In: Mulching in agroecosystems: plants, soil and environment. Springer, pp 3–13

Bagheri AR, Ghaedi M (2020) Green preparation of dual-template chitosan-based magnetic water-compatible molecularly imprinted biopolymer. Carbohydr Polym 236:116102

Bouteska A, Sharif T, Bhuiyan F, Abedin MZ (2024) Impacts of the changing climate on agricultural productivity and food security: evidence from Ethiopia. J Clean Prod 449:141793. https://doi.org/10.1016/j.jclepro.2024.141793

Boutriouia EH, El Assimi T, Qayouh H, Raihane M, El Meziane A, Baouab MHV, Youcef HB, El Kadib A, Lahcini M (2024) Poly(ε-Caprolactone)-grafted-chitosan copolymers: synthesis and use as tunable and biodegradable coating for water soluble fertilizers. React Funct Polym 198:105887. https://doi.org/10.1016/j.reactfunctpolym.2024.105887

Boyandin AN, Kazantseva EA (2021) Constructing slow-release formulations of herbicide metribuzin using its co-extrusion with biodegradable polyester Poly-ε-Caprolactone. J Environ Sci Health B 56:467–476

Briassoulis D, Giannoulis A (2018) Evaluation of the functionality of bio-based plastic mulching films. Polym Test 67:99–109

Cabiac A, Guillon E, Chambon F, Pinel C, Rataboul F, Essayem N (2011) Cellulose reactivity and glycosidic bond cleavage in aqueous phase by catalytic and non catalytic transformations. Appl Catal A Gen 402:1–10. https://doi.org/10.1016/j.apcata.2011.05.029

Carvalho VVL, Gonçalves JO, Silva A, Cadaval TR, Pinto LAA, Lopes TJ (2019) Separation of anthocyanins extracted from red cabbage by adsorption onto chitosan films. Int J Biol Macromol 131:905–911. https://doi.org/10.1016/j.ijbiomac.2019.03.145

Chen G, Li J, Sun Y, Wang Z, Leeke GA, Moretti C, Cheng Z, Wang Y, Li N, Mu L et al (2024) Replacing traditional plastics with biodegradable plastics: impact on carbon emissions. Engineering 32:152–162. https://doi.org/10.1016/j.eng.2023.10.002

Cherwoo L, Gupta I, Bhatia R, Setia H (2024) Improving agricultural practices: application of polymers in agriculture. Energy Ecol Environ 9:25–41. https://doi.org/10.1007/s40974-023-00295-4

Chouchene K, Nacci T, Modugno F, Castelvetro V, Ksibi M (2022) Soil contamination by microplastics in relation to local agricultural development as revealed by FTIR, ICP-MS and pyrolysis-GC/MS. Environ Pollut 303:119016

Costa MME, Cabral-Albuquerque ECM, Alves TLM, Pinto JC, Fialho RL (2013) Use of polyhydroxybutyrate and ethyl cellulose for coating of urea granules. J Agric Food Chem 61:9984–9991

Costa R, Saraiva A, Carvalho L, Duarte E (2014) The use of biodegradable mulch films on strawberry crop in Portugal. Sci Hortic 173:65–70

Cozzolino E, Giordano M, Fiorentino N, El-Nakhel C, Pannico A, Di Mola I, Mori M, Kyriacou MC, Colla G, Rouphael Y (2020) Appraisal of biodegradable mulching films and vegetal-derived biostimulant application as eco-sustainable practices for enhancing lettuce crop performance and nutritive value. Agronomy 10:427

Das O, Babu K, Shanmugam V, Sykam K, Tebyetekerwa M, Neisiany RE, Försth M, Sas G, Gonzalez-Libreros J, Capezza AJ (2022) Natural and industrial wastes for sustainable and renewable polymer composites. Renew Sustain Energy Rev 158:112054 de Borba VS, Gropelli VM, Silveira CO, Alves JB, Badiale-Furlong E, Ribeiro AC (2023) Starch as a promising replacement for synthetic polymers. In: Handbook of natural polymers, Vol 1, Elsevier, pp 61–76

Degli-Innocenti F, Barbale M, Chinaglia S, Esposito E, Pecchiari M, Razza F, Tosin M (2022) Analysis of the microplastic emission potential of a starch-based biodegradable plastic material. Polym Degrad Stab 199:109934. https://doi.org/10.1016/j.polymdegradstab.2022.109934

Del Sole R, Mele G, Bloise E, Mergola L (2021) Green aspects in molecularly imprinted polymers by biomass waste utilization. Polymers (Basel) 13:2430

Di Mola I, Cozzolino E, Ottaiano L, Duri LG, Riccardi R, Spigno P, Leone V, Mori M (2019) The effect of novel biodegradable films on agronomic performance of zucchini squash grown under open-field and greenhouse conditions. Aust J Crop Sci 13:1810–1818

Di Mola I, Ventorino V, Cozzolino E, Ottaiano L, Romano I, Duri LG, Pepe O, Mori M (2021) Biodegradable mulching vs traditional polyethylene film for sustainable solarization: chemical properties and microbial community response to soil management. Appl Soil Ecol 163:103921

Dou Y, Liu C, Chen X, Yang X, Hao L, Wang Q, Wang Z, Wu Q, Wang C (2024) Green synthesis of azo-linked porous organic polymer for enrichment of nitroimidazoles from water, shrimp and basa fish. Food Chem 446:138875

Eddarai EM, El Mouzahim M, Ragaoui B, El Addaoui S, Boussen R, Warad I, Bellaouchou A, Zarrouk A (2023a) Chitosan/kaolinite clay biocomposite as a sustainable and environmentally eco-friendly coating material for slow release NPK fertilizers: effect on soil nutrients and tomato growth. Int J Biol Macromol 242:125019

Eddarai EM, El Mouzahim M, Ragaoui B, El Addaoui S, Boussen R, Warad I, Bellaouchou A, Zarrouk A (2023b) Chitosan/kaolinite clay biocomposite as a sustainable and environmentally eco-friendly coating material for slow release NPK fertilizers: effect on soil nutrients and tomato growth. Int J Biol Macromol 242:125019. https://doi.org/10.1016/j.ijbiomac.2023.125019

Farooq S, Wu H, Nie J, Ahmad S, Muhammad I, Zeeshan M, Khan R, Asim M (2022) Application, advancement and green aspects of magnetic molecularly imprinted polymers in pesticide residue detection. Sci Total Environ 804:150293

Fasciotti M (2017) Perspectives for the use of biotechnology in green chemistry applied to biopolymers, fuels and organic synthesis: from concepts to a critical point of view. Sustain Chem Pharm 6:82–89. https://doi.org/10.1016/j.scp.2017.09.002

Ferreira LF, Figueiredo LP, Martins MA, Luvizaro LB, Baldone de bLara BR, de Oliveira CR, Júnior MG, Tonoli GHD, Dias MV (2021) Active coatings of thermoplastic starch and chitosan with alpha-tocopherol/bentonite for special green coffee beans. Int J Biol Macromol 170:810–819

Ferreira LF, Figueiredo LP, Martins MA, Luvizaro LB, bLara de BRB, Oliveira de CR, Júnior MG, Tonoli GHD, Dias MV (2021) Active coatings of thermoplastic starch and chitosan with alpha-tocopherol/bentonite for special green coffee beans. Int J Biol Macromol 170:810–819. https://doi.org/10.1016/j.ijbiomac.2020.12.199.

Fertahi S, Ilsouk M, Zeroual Y, Oukarroum A, Barakat A (2021) Recent trends in organic coating based on biopolymers and biomass for controlled and slow release fertilizers. J Control Release 330:341–361. https://doi.org/10.1016/j.jconrel.2020.12.026

Firmanda A, Fahma F, Syamsu K, Mahardika M, Suryanegara L, Munif A, Gozan M, Wood K, Hidayat R, Yulia D (2024) Biopolymer-based slow/controlled-release fertilizer (SRF/CRF): nutrient release mechanism and agricultural sustainability. J Environ Chem Eng 112177

Francioni M, Kishimoto-Mo AW, Tsuboi S, Takada Hoshino Y (2021) Evaluation of the mulch films biodegradation in soil: a methodological review. Ital J Agron. https://doi.org/10.4081/ija.2021.1936

Gamage A, Liyanapathiranage A, Manamperi A, Gunathilake C, Mani S, Merah O, Madhujith T (2022) Applications of starch biopolymers for a sustainable modern agriculture. Sustainability 14:6085

Getahun MJ, Kassie BB, Alemu TS (2024) Recent advances in biopolymer synthesis, properties, and commercial applications: a review. Process Biochem 145:261–287. https://doi.org/10.1016/j.procbio.2024.06.034

Goel V, Luthra P, Kapur GS, Ramakumar SSV (2021) Biodegradable/bio-plastics: myths and realities. J Polym Environ 29:3079–3104. https://doi.org/10.1007/s10924-021-02099-1

Gonçalves JO, Santos JP, Rios EC, Crispim MM, Dotto GL, Pinto LAA (2017) Development of chitosan based hybrid hydrogels for dyes removal from aqueous binary system. J Mol Liq 225:265–270. https://doi.org/10.1016/j.molliq.2016.11.067

Gonçalves JO, da Silva KA, Rios EC, Crispim MM, Dotto GL, de Almeida Pinto LA (2020) Chitosan hydrogel scaffold modified with carbon nanotubes and its application for food dyes removal in single and binary aqueous systems. Int J Biol Macromol 142:85–93. https://doi.org/10.1016/j.ijbiomac.2019.09.074

Gonçalves JO, de Farias BS, Rios EC, Jaeschke DP, Ribeiro AC, da Silva MD, Vieira MLG, Carvalho de VVL, Cadaval TRS, Pinto de LAA (2024) Advances in chitosan-based materials for application in catalysis and adsorption of emerging contaminants. Sustainability 16:8321. https://doi.org/10.3390/su16198321

Govil S, Van Duc Long N, Escribà-Gelonch M, Hessel V (2024) Controlled-release fertiliser: recent developments and perspectives. Ind Crops Prod 219:119160. https://doi.org/10.1016/j.indcrop.2024.119160

Grillo R, Santo Pereira do AE, de Melo NFS, Porto RM, Feitosa LO, Tonello PS, Dias Filho NL, Rosa AH, Lima R, Fraceto LF (2011) Controlled release system for Ametryn using polymer microspheres: preparation, characterization and release kinetics in water. J Hazard Mater 186:1645–1651

Hafezi Moghaddam R, Dadfarnia S, Shabani AMH, Shirmardi SP (2021) Design and manufacture of new hybrid hydrogel and superabsorbent polymer for controlled release of fulvic acid based on grafted xanthan gum/gelatin using electron irradiation and its use in fodder corn cultivation. Int J Biol Macromol 266:131360. https://doi.org/10.1016/j.ijbiomac.2024.131360

Hamed R, Jodeh S, Alkowni R (2024) Nano bio fertilizer capsules for sustainable agriculture. Sci Rep 14:13646

Harada J, de Souza AG, de Macedo JRN, Rosa DS (2019a) Soil culture: influence of different natural fillers incorporated in biodegradable mulching film. J Mol Liq 273:33–36. https://doi.org/10.1016/j.molliq.2018.09.109

Harada J, de Souza AG, de Macedo JRN, Rosa DS (2019b) Soil culture: influence of different natural fillers incorporated in biodegradable mulching film. J Mol Liq 273:33–36

Hayes DG, Anunciado MB, DeBruyn JM, Bandopadhyay S, Schaeffer S, English M, Ghimire S, Miles C, Flury M, Sintim HY (2019) Biodegradable plastic mulch films for sustainable specialty crop production. In Polymers for agri-food applications. Springer International Publishing, Cham, pp 183–213

Hejna A, Formela K, Saeb MR (2015) Processing, mechanical and thermal behavior assessments of polycaprolactone/agricultural wastes biocomposites. Ind Crops Prod 76:725–733. https://doi.org/10.1016/j.indcrop.2015.07.049

Hisham F, Maziati Akmal MH, Ahmad F, Ahmad K, Samat N (2024) Biopolymer chitosan: potential sources, extraction methods, and emerging applications. Ain Shams Eng J 15:102424. https://doi.org/10.1016/j.asej.2023.102424

Huang M, Xu Y, Xu L, Bai Y, Xu X (2022) Interactions of water-soluble myofibrillar protein with chitosan: phase behavior, microstructure and rheological properties. Innov Food Sci Emerg Technol 78:103013. https://doi.org/10.1016/j.ifset.2022.103013

Huang F, Zhang Q, Wang L, Zhang C, Zhang Y (2023) Are biodegradable mulch films a sustainable solution to microplastic mulch film pollution? A biogeochemical perspective. J Hazard Mater 132024

Hui K, Xi B, Tan W, Song Q (2022) Long-term application of nitrogen fertilizer alters the properties of dissolved soil organic matter and increases the accumulation of polycyclic aromatic hydrocarbons. Environ Res 215:114267. https://doi.org/10.1016/j.envres.2022.114267

Intergovernmental Panel on Climate Change (IPCC) (2023) Climate change 2021—the physical science basis. Cambridge University Press. ISBN 9781009157896

Iqbal R, Raza MAS, Valipour M, Saleem MF, Zaheer MS, Ahmad S, Toleikiene M, Haider I, Aslam MU, Nazar MA (2020) Potential agricultural and environmental benefits of mulches—a review. Bull Natl Res Cent 44:1–16

Jian J, Xiangbin Z, Xianbo H (2020) An overview on synthesis, properties and applications of poly (butylene-adipate-co-terephthalate)–PBAT. Adv Ind Eng Polym Res 3:19–26

Kane S, Van Roijen E, Ryan C, Miller S (2022) Reducing the environmental impacts of plastics while increasing strength: biochar fillers in biodegradable, recycled, and fossil-fuel derived plastics. Compos Part C: Open Access 8:100253. https://doi.org/10.1016/j.jcomc.2022.100253

Kasirajan S, Ngouajio M (2012) Polyethylene and biodegradable mulches for agricultural applications: a review. Agron Sustain Dev 32:501–529. https://doi.org/10.1007/s13593-011-0068-3

Kassem I, Ablouh E-H, El Bouchtaoui F-Z, Jaouahar M, El Achaby M (2024) Polymer coated slow/ controlled release granular fertilizers: fundamentals and research trends. Prog Mater Sci 144:101269. https://doi.org/10.1016/j.pmatsci.2024.101269

Kassie BB, Daget TM, Tassew DF (2024) Synthesis, functionalization, and commercial application of cellulose-based nanomaterials. Int J Biol Macromol 278:134990. https://doi.org/10.1016/j.ijbiomac.2024.134990

Khouri NG, Bahú JO, Blanco-Llamero C, Severino P, Concha VOC, Souto EB (2024) Polylactic acid (PLA): properties, synthesis, and biomedical applications—a review of the literature. J Mol Struct 1309:138243. https://doi.org/10.1016/j.molstruc.2024.138243

Koller M (ed) (2020) The handbook of polyhydroxyalkanoates. CRC Press. ISBN 9781003087663

Kumar A, Choudhary A, Kaur H, Mehta S, Husen A (2021) Smart nanomaterial and nanocomposite with advanced agrochemical activities. Nanoscale Res Lett 16:1–26

Kumar D, Gupta RK (2024) Natural polymers-humic acid and lignin based hydrogels: in agriculture, environment and energy storage. Ind Crops Prod 219:119029

Kumaraswamy RV, Saharan V, Kumari S, Choudhary RC, Pal A, Sharma SS, Rakshit S, Raliya R, Biswas P (2021) Chitosan-Silicon nanofertilizer to enhance plant growth and yield in maize (Zea Mays L.). Plant Physiol Biochem159:53–66

Kumari S, Kishor R (2020) Chitin and chitosan: origin, properties, and applications. Elsevier, In Handbook of chitin and chitosan, pp 1–33

Lang C, Mission EG, Fuaad AA-HA, Shaalan M (2021) Nanoparticle tools to improve and advance precision practices in the agrifoods sector towards sustainability-a review. J Clean Prod 293:126063

Lara G, Yakoubi S, Villacorta CM, Uemura K, Kobayashi I, Takahashi C, Nakajima M, Neves MA (2020a) Spray technology applications of xanthan gum-based edible coatings for fresh-cut lotus root (nelumbo nucifera). Food Res Int 137:109723

Lara G, Yakoubi S, Villacorta CM, Uemura K, Kobayashi I, Takahashi C, Nakajima M, Neves MA (2020b) Spray technology applications of xanthan gum-based edible coatings for fresh-cut lotus root (nelumbo nucifera). Food Res Int 137:109723. https://doi.org/10.1016/j.foodres.2020.109723

Li J, Liu X, Zhang S, Chen L, Jiang Y, Li R, Yin X, Long Y, Ma Y, Chen Z (2024) Liquid mulch films based on polysaccharide/furfural schiff base adducts with ph-controlled degradability. Ind Crops Prod 216:118756

Liang X, Guo S, Kuang X, Wan X, Liu L, Zhang F, Jiang G, Cong H, He H, Tan SC (2024) Recent advancements and perspectives on processable natural biopolymers: cellulose, chitosan, eggshell membrane, and silk fibroin. Sci Bull (Beijing). https://doi.org/10.1016/j.scib.2024.08.023

Lyu JS, Han J (2023) Scale-up fabrication of a biodegradable PBAT/PLA composite film compatibilized with a chain extender for industrial agricultural mulch film application. Compos Part C: Open Access 12:100397

Lyu JS, Han J (2023a) Scale-up fabrication of a biodegradable pbat/pla composite film compatibilized with a chain extender for industrial agricultural mulch film application. Compos Part C: Open Access 12:100397. https://doi.org/10.1016/j.jcomc.2023.100397

Mali H, Shah C, Raghunandan BH, Prajapati AS, Patel DH, Trivedi U, Subramanian RB (2023) Organophosphate pesticides an emerging environmental contaminant: pollution, toxicity, bioremediation progress, and remaining challenges. J Environ Sci 127:234–250. https://doi.org/10.1016/j.jes.2022.04.023

Mansoor Z, Tchuenbou-Magaia F, Kowalczuk M, Adamus G, Manning G, Parati M, Radecka I, Khan H (2022) Polymers use as mulch films in agriculture—a review of history, problems and current trends. Polymers (Basel) 14:5062. https://doi.org/10.3390/polym14235062

Messa LL, Faez R (2020) Spray-dried chitosan/nanocellulose microparticles: synergistic effects for the sustained release of NPK fertilizer. Cellulose 27:10077–10093. https://doi.org/10.1007/s10570-020-03482-2

Meyer WL, Gurman P, Stelinski LL, Elman NM (2015) Functional nano-dispensers (FNDs) for delivery of insecticides against phytopathogen vectors. Green Chem 17:4173–4177

Mitra D, Adhikari P, Djebaili R, Thathola P, Joshi K, Pellegrini M, Adeyemi NO, Khoshru B, Kaur K, Priyadarshini A et al (2023) Biosynthesis and characterization of nanoparticles, its advantages, various aspects and risk assessment to maintain the sustainable agriculture: emerging technology in modern era science. Plant Physiol Biochem 196:103–120. https://doi.org/10.1016/j.plaphy.2023.01.017

Moghaddam RH, Dadfarnia S, Shabani AMH, Shirmardi SP (2024) Design and manufacture of new hybrid hydrogel and superabsorbent polymer for controlled release of fulvic acid based on grafted xanthan gum/gelatin using electron irradiation and its use in fodder corn cultivation. Int J Biol Macromol 266:131360

Le Moine B, Ferry X (2019) Plasticulture: economy of resources. Acta Hortic 121–130. https://doi.org/10.17660/ActaHortic.2019.1252.17

Moreno MM, Moreno A (2008) Effect of different biodegradable and polyethylene mulches on soil properties and production in a tomato crop. Sci Hortic 116:256–263

Moura JM, Farias BS, Rodrigues DAS, Moura CM, Dotto GL, Pinto LAA (2015) Preparation of chitosan with different characteristics and its application for biofilms production. J Polym Environ 23:470–477. https://doi.org/10.1007/s10924-015-0730-y

Mubaraka R, Hussain MB, Tariq H, Qayyum M, Tariq I (2022) Mulches effects on soil physical properties ie porosity, aggregate stability, infiltration rates, bulk density, compaction. In: Mulching in Agroecosystems: Plants, Soil & Environment. Springer, pp 41–58

Mukherjee A, Koller M (2023a) Microbial polyhydroxyalkanoate (PHA) biopolymers—intrinsically natural. Bioengineering 10:855

Mukherjee A, Koller M (2023b) Microbial polyhydroxyalkanoate (PHA) biopolymers—intrinsically natural. Bioengineering 10:855. https://doi.org/10.3390/bioengineering10070855

Nakaramontri Y, Boonluksiri Y, Sornsri P, Duangkhaw S, Udompongpaiboon P, Johns J, Klinnawee L (2022) Composites of thermoplastic starch/natural rubber blends for fertilizer-releasing in agriculture. Ind Crops Prod 187:115522. https://doi.org/10.1016/j.indcrop.2022.115522

Query ID="Q4" Text="" Neme K, Nafady A, Uddin S, Tola YB (2021) Application of nanotechnology in agriculture, postharvest loss reduction and food processing: food security implication and challenges. Heliyon 7

Ore OT, Adeola AO, Bayode AA, Adedipe DT, Nomngongo PN (2023) Organophosphate pesticide residues in environmental and biological matrices: occurrence, distribution and potential remedial approaches. Environ Chem Ecotoxicol 5:9–23. https://doi.org/10.1016/j.enceco.2022.10.004

Ostovan A, Ghaedi M, Arabi M (2018) Fabrication of water-compatible superparamagnetic molecularly imprinted biopolymer for clean separation of baclofen from bio-fluid samples: a mild and green approach. Talanta 179:760–768

Othman NAF, Selambakkannu S, Seko N (2022) Biodegradable dual-layer polyhydroxyalkanoate (Pha)/polycaprolactone (Pcl) mulch film for agriculture: preparation and characterization. Energy Nexus 8:100137. https://doi.org/10.1016/j.nexus.2022.100137

Pasquini D, Teixeira de EM, Curvelo da AAS, Belgacem MN, Dufresne A (2010) Extraction of cellulose whiskers from cassava bagasse and their applications as reinforcing agent in natural rubber. Ind Crops Prod 32:486–490. https://doi.org/10.1016/j.indcrop.2010.06.022

Peluso J, Chehda AM, Olivelli MS, Ivanic FM, Pérez Coll CS, Gonzalez F, Valenzuela L, Rojas D, Cristos D, Butler M et al (2023) Metals, pesticides, and emerging contaminants on water bodies from agricultural areas and the effects on a native amphibian. Environ Res 226:115692. https://doi.org/10.1016/j.envres.2023.115692

Porta R (2017) Plastic pollution and the challenge of bioplastics. J Appl Biotechnol Bioeng 2. https://doi.org/10.15406/jabb.2017.02.00033

Puppi D, Chiellini F (2020) Additive manufacturing of PHA. CRC Press, In the handbook of polyhydroxyalkanoates, pp 119–136

Qi Y, Beriot N, Gort G, Huerta Lwanga E, Gooren H, Yang X, Geissen V (2020) Impact of plastic mulch film debris on soil physicochemical and hydrological properties. Environ Pollut 266:115097. https://doi.org/10.1016/j.envpol.2020.115097

Qin C, Wang H, Zhao Y, Qi Y, Wu N, Zhang S, Xu W (2024) Recent advances of hydrogel in agriculture: synthesis, mechanism, properties and applications. Eur Polym J 113376

Rajagopalan K, Christyraj JRSS, Karthikeyan SC, Jeevanandam M, Ganesan H, Mathews MGR, Selvan Christyraj JD (2022) Biodegradation of microplastics and synthetic polymers in agricultural soils. In: Microbes and microbial biotechnology for green remediation. Elsevier, pp 563–573

Rajgadia N, Debnath M (2023) Biodegradable mulch utilizing bioplastic biopolymer polyhydroxyalkanoates. Mater Today Proc 79:411–419. https://doi.org/10.1016/j.matpr.2022.12.167

Rani GU, Sharma S (2022) Biopolymers, bioplastics and biodegradability: an introduction. In: Encyclopedia of materials: plastics and polymers. Elsevier, pp 474–486

Raza MA, Sharma MK, Nagori K, Jain P, Ghosh V, Gupta U (2024) Ajazuddin recent trends on polycaprolactone as sustainable polymer-based drug delivery system in the treatment of cancer: biomedical applications and nanomedicine. Int J Pharm 666:124734. https://doi.org/10.1016/j.ijpharm.2024.124734

Rinaudo M (2006) Chitin and chitosan: properties and applications. Prog Polym Sci 31:603–632. https://doi.org/10.1016/j.progpolymsci.2006.06.001

Saberi Riseh R (2024) Advancing agriculture through bioresource technology: the role of cellulose-based biodegradable mulches. Int J Biol Macromol 255:128006. https://doi.org/10.1016/j.ijbiomac.2023.128006

Sellami MH, Di Mola I, Ottaiano L, Cozzolino E, del Piano L, Mori M (2024) Evaluation of biodegradable mulch films on melon production and quality under mediterranean field conditions. Agronomy 14:2075

Serrano-Ruiz H, Martin-Closas L, Pelacho AM (2021) Biodegradable plastic mulches: impact on the agricultural biotic environment. Sci Total Environ 750:141228

Shahzad K, Narodoslawsky M, Sagir M, Ali N, Ali S, Rashid MI, Ismail IMI, Koller M (2017a) Techno-economic feasibility of waste biorefinery: using slaughtering waste streams as starting material for biopolyester production. Waste Manag 67:73–85

Shahzad K, Narodoslawsky M, Sagir M, Ali N, Ali S, Rashid MI, Ismail IMI, Koller M (2017b) Techno-economic feasibility of waste biorefinery: using slaughtering waste streams as starting material for biopolyester production. Waste Manag 67:73–85. https://doi.org/10.1016/j.wasman.2017.05.047

Sharahi M, Bahrami SH, Karimi A (2024) A comprehensive review on guar gum and its modified biopolymers: their potential applications in tissue engineering. Carbohydr Polym 122739

Somanathan H, Sathasivam R, Sivaram S, Mariappan Kumaresan S, Muthuraman MS, Park SU (2022) An update on polyethylene and biodegradable plastic mulch films and their impact on the environment. Chemosphere 307:135839. https://doi.org/10.1016/j.chemosphere.2022.135839

Song B, Liang H, Sun R, Peng P, Jiang Y, She D (2020) Hydrogel synthesis based on lignin/sodium alginate and application in agriculture. Int J Biol Macromol 144:219–230

Stloukal P, Kucharczyk P, Sedlarik V, Bazant P, Koutny M (2012) Low molecular weight poly (Lactic Acid) microparticles for controlled release of the herbicide metazachlor: preparation, morphology, and release kinetics. J Agric Food Chem 60:4111–4119

Sukwijit C, Seubsai A, Charoenchaitrakool M, Sudsakorn K, Niamnuy C, Roddecha S, Prapainainar P (2024a) Production of PLA/cellulose derived from pineapple leaves as bio-degradable mulch film. Int J Biol Macromol 270:132299. https://doi.org/10.1016/j.ijbiomac.2024.132299

Sukwijit C, Seubsai A, Charoenchaitrakool M, Sudsakorn K, Niamnuy C, Roddecha S, Prapainainar P (2024) Production of PLA/cellulose derived from pineapple leaves as bio-degradable mulch film. Int J Biol Macromol 270:132299

Swetha TA, Ananthi V, Bora A, Sengottuvelan N, Ponnuchamy K, Muthusamy G, Arun A (2023) A review on biodegradable polylactic acid (PLA) production from fermentative food waste—its applications and degradation. Int J Biol Macromol 234:123703. https://doi.org/10.1016/j.ijbiomac.2023.123703

Tariq M, Akhtar K (2022) Mulching is an approach for a significant decrease in soil erosion. In: Mulching in agroecosystems: plants, soil and environment. Springer, pp 59–70

Thakur M, Kumar R (2021) Mulching: boosting crop productivity and improving soil environment in herbal plants. J Appl Res Med Aromat Plants 20:100287

Thamarai P, Vickram AS, Saravanan A, Deivayanai VC, Evangeline S (2024) Recent advancements in biosynthesis, industrial production, and environmental applications of polyhydroxyalkanoates (PHAs): a review. Bioresour Technol Rep 27:101957. https://doi.org/10.1016/j.biteb.2024.101957

Thombare N, Mishra S, Shinde R, Siddiqui MZ, Jha U (2021) Guar gum based hydrogel as controlled micronutrient delivery system: mechanism and kinetics of boron release for agricultural applications. Biopolymers 112:e23418

Tran NP, Okahisa Y, Okubayashi S (2025) Chitosan hydrogel containing silk fibroin nanofibrils for controllable properties and its application to drug delivery system. Next Mater 6:100288. https://doi.org/10.1016/j.nxmate.2024.100288

Varghese AG, Paul SA, Latha MS (2019) Remediation of heavy metals and dyes from wastewater using cellulose-based adsorbents. Environ Chem Lett 17:867–877

Vaverková M, Adamcová D, Kotovicová J, Toman F (2014a) Evaluation of biodegradability of plastics bags in composting conditions. Ecol Chem Eng S 21(1):45–57

Vaverková M, Adamcová D, Kotovicová J, Toman F (2014b) Evaluation of biodegradability of plastics bags in composting conditions. Ecol Chem Eng S 21:45–57. https://doi.org/10.2478/eces-2014-0004

Victoria J, Tripathi S, Prakash V, Tiwari K, Mahra S, Sharma A, Rana S, Kandhol N, Sahi S, Tripathi DK (2023) Encapsulated nanopesticides application in plant protection: quo Vadis?Plant Physiol Biochem108225

Wang C, Zhao B, Wang Q, Zhang S, Wu Q, Shi X (2024) Green construction of magnetic azo porous organic polymer for highly efficient enrichment and detection of phenolic endocrine disruptors. J Hazard Mater 465:133050

Webber MJ, Tibbitt MW (2022) Dynamic and reconfigurable materials from reversible network interactions. Nat Rev Mater 7:541–556. https://doi.org/10.1038/s41578-021-00412-x

Wei J, Long Y, Li T, Gao H, Nie Y (2024) Exploring hydrogen-bond structures in cellulose during regeneration with anti-solvent through two-dimensional correlation infrared spectroscopy. Int J Biol Macromol 267:131204. https://doi.org/10.1016/j.ijbiomac.2024.131204

Yadav B, Talan A, Tyagi RD, Drogui P (2021) Concomitant production of value-added products with polyhydroxyalkanoate (PHA) synthesis: a review. Bioresour Technol 337:125419. https://doi.org/10.1016/j.biortech.2021.125419

Yin X, Lu J, Du W, Wu Q, Han L, Su S (2024) Encapsulation of β-carotene in pickering emulsions stabilized by self-aggregated chitosan nanoparticles: factors affecting β-carotene stability. Int J Biol Macromol 277:133696. https://doi.org/10.1016/j.ijbiomac.2024.133696

Yu C (2015) Natural textile fibres. Elsevier, In textiles and fashion, pp 29–56

Yu T-Y, Tseng Y-K, Lin T-H, Wang T-C, Tseng Y-H, Chang Y-H, Wu M-C, Su W-F (2022) Effect of cellulose compositions and fabrication methods on mechanical properties of polyurethane-cellulose composites. Carbohydr Polym 291:119549. https://doi.org/10.1016/j.carbpol.2022.119549

Yu J, Xu S, Liu B, Wang H, Qiao F, Ren X, Wei Q (2023) PLA bioplastic production: from monomer to the polymer. Eur Polym J 193:112076. https://doi.org/10.1016/j.eurpolymj.2023.112076

Zimmermann L, Dombrowski A, Völker C, Wagner M (2020) Are bioplastics and plant-based materials safer than conventional plastics? In vitro toxicity and chemical composition. Environ Int 145:106066. https://doi.org/10.1016/j.envint.2020.106066

Chapter 6
Sustainable Polymer Composites for Automotive Industry

Merve Yurdakul, Mutahire Tok, Gizem Suzan Döner, and Çisem Kırbıyık Kurukavak

Abstract Polymers are widely used in the automotive industry due to their versatility, lightweight properties, cost-effectiveness, and ability to be molded into complex shapes. As technology advances, the use of polymers is expected to grow, driven by innovations in materials science and a focus on sustainability. Sustainable polymer composites are materials composed of a polymer matrix reinforced with fibers or particles that are derived from renewable resources. In the automotive industry, especially with the growing adoption of electric vehicles, these composites are gaining attraction due to their potential to reduce vehicle weight, improve fuel efficiency, and minimize environmental impact by complying with new regulations such as the European Union Green Deal. This chapter aims to provide a comprehensive overview of the potential applications of sustainable polymer composites in the automobile industry. It summarizes the knowledge of sustainable polymer composites, determines the research needs based on current limitations and regulations, and gives insights into future perspectives.

Keywords Sustainable polymer composites · Automotive industry · Sustainability

6.1 Introduction

Polymers, which are formed by small molecules called monomers, are a crucial component of modern life. They are formed into long chains with covalent bonds and are used extensively in our daily lives. Although the use of polymer derivatives

M. Yurdakul (✉) · G. S. Döner
Ozer Group Company, Organized Industrial Zone, Eskisehir, Turkey
e-mail: merveyur@gmail.com

M. Tok
Pharmacy Services, Vocational School, Yuksek Ihtisas University, Ankara, Turkey

M. Yurdakul · M. Tok · Ç. Kırbıyık Kurukavak
Department of Chemical Engineering, Konya Technical University, Konya, Turkey

L. Ravindran et al. (eds.), *Macro, Micro and Nanocomposites from Sustainable Sources*, Advanced Structured Materials 244, https://doi.org/10.1007/978-981-95-2469-3_6

has a long historical precedent, it was not until after World War II that their significance was fully appreciated. Until that time, the construction industry employed a restricted range of materials, including steel, glass, wood and brick, while the textile sector utilized cotton and jute (Namazi 2017). The introduction of polymers has accelerated in parallel with the growth of industrial production and in response to the demands of consumers on a large scale. They are now widely used in a variety of areas, including the production of synthetic fibers and kitchenware. At the time of their initial discovery, polymers were created through a process of chemical modification of natural polymers, conducted in an experimental manner. However, by the nineteenth century, they were produced entirely through a synthetic approach (Rasmussen 2018).

The life cycle of polymeric materials comprises seven stages, as follows: production, polymer utilization, post-consumer waste generation, waste reuse, waste recovery, waste recycling, and waste storage. Although polymeric materials are ubiquitous in modern life, their production and use have the potential to create significant environmental impacts. These include the depletion of natural resources, the non-renewability of these resources, and the generation of solid waste throughout the product's life cycle. It is, therefore, crucial to implement effective strategies for the management of polymer consumption and the widespread use of sustainable polymers in the market. This section will focus on the utilization of sustainable polymers in the automotive industry.

6.1.1 General Information on Composite Structures

Composite materials can be defined as new and artificial materials formed by combining two or more of the same or different types of materials for a specific purpose. This is done to combine the properties of two or more of the same or different types of materials in a new and single material at the macro-level. These composite materials are formed by combining different materials, such as ceramics, metals, and polymers, in a way that protects their own boundaries. The final product exhibits a range of beneficial properties, including high rigidity and durability, low weight and ease of production, excellent wear and corrosion resistance, high temperature capacity, and good thermal and heat conductivity (Sönmez 2009; Şahin 2000). Composite materials are composed of two distinct phases: the relatively less rigid and weaker matrix (continuous phase) and the relatively stiffer and stronger reinforcement (discontinuous phase) (Hancox 1996). In composite materials, the load-bearing mechanism is based on the interaction between the matrix and reinforcing elements, whereby the former provides the structural integrity, and the latter provides the necessary reinforcement. The reinforcement of materials is typically achieved through the utilization of high-strength and high-stiffness materials, which serve to enhance the mechanical properties of the matrix. In the automotive industry, composite materials can be employed in both interior design and the production of structural components, thereby reducing costs and imparting advanced mechanical properties to the product.

6.1.2 Properties of Polymer Composites

A composite material is defined as a substance consisting of a combination of two or more materials, the individual components of which possess distinct physical and chemical properties (Strong 2008). These materials are composed of two principal components: the matrix and the reinforcing elements. The matrix facilitates the material's shape acquisition and enhances the load-carrying capacity of the reinforcing elements. The reinforcing elements serve to enhance the mechanical properties of the material, including strength and stiffness. The combination of these two elements results in the emergence of a new material with a multitude of superior properties, effectively overcoming the limitations of traditional materials. The mechanical, thermal, electrical and chemical properties of composite materials render them optimal for advanced engineering applications. Because of advances in technology and materials science, these materials are now employed extensively across a diverse range of industries, particularly in the aerospace, automotive, construction, biomedical and defense sectors. A more detailed discussion of these properties can be found below (Doruk 2023).

6.1.2.1 Lightness

One of the most readily apparent advantages of composite materials is their lightness. Composites have a significantly lower density than metals and ceramics yet exhibit similar strength. To illustrate, carbon fiber-reinforced polymer (CFRP) composites exhibit comparable levels of strength despite being approximately five times lighter than steel. The advantage of light weight is particularly pertinent in the aerospace, automotive and space industries. The use of lightweight materials in aircraft and vehicles results in reduced fuel consumption, which in turn leads to a reduction in operating costs and a minimization of environmental impact.

6.1.2.2 High Strength

The utilization of composite materials enables the attainment of elevated strength levels in relation to their respective densities. The incorporation of fibers, such as glass, carbon or aramid fibers, serves to markedly enhance the load-bearing capacity of the material. Carbon fiber-reinforced polymers exhibit a superior strength-to-weight ratio in comparison to conventional steel. Composite materials that are capable of withstanding diverse stresses can function without deformation even when subjected to substantial loads. This attribute has contributed to the proliferation of the utilization of composite materials in structural and civil engineering, particularly in bridge, building and infrastructure projects. Furthermore, these materials are preferred for use in critical applications such as aircraft structural elements, rockets, and satellite components.

6.1.2.3 High Hardness and Flexibility

The term "stiffness" is defined as the ability of a material to resist deformation under external forces. Composite materials can exhibit high stiffness due to the reinforcing elements they contain. This increases the material's resistance to external influences by minimizing its tendency to bend and deform. However, composites are not only rigid; they can also have a flexible structure. In this case, composite materials exhibit a stretchable property against applied forces, thus minimizing the risk of cracking or fracture. The orientation of the fibers and the modulus of elasticity of the matrix allow the material to offer both high strength and sufficient flexibility. This balance allows composites to be used in a wide range of applications.

6.1.2.4 High Wear Resistance

Wear is defined as the loss or deformation of the surfaces of materials over time by exposure to abrasive effects. Composites exhibit resistance to harsh environmental conditions and chemicals, enabling them to provide long-lasting performance even in corrosive environments. Polymer matrix composites, for instance, are highly resistant to friction and abrasion, a property that causes composites to be preferred in tribological applications (e.g. bearings, sliding surfaces and rotors). Furthermore, in shipbuilding and the marine industry, composite materials are widely used to provide resistance to the corrosive effects of salt water.

6.1.2.5 High Corrosion Resistance

One of the most significant shortcomings of metals is their proclivity towards corrosion. The processes of rusting and oxidation can lead to the deterioration of metal structures over time. However, composite materials exhibit enhanced chemical stability and offer superior corrosion resistance. Polymer matrix composites provide long-term durability, particularly in the presence of moisture, saltwater, chemicals, and acidic environments. This makes them well suited for use in aggressive environments such as the chemical industry, water treatment plants, and submarine structures. Furthermore, corrosion resistance reduces the necessity for maintenance and lowers long-term costs.

6.1.2.6 High Electrical and Thermal Conductivity

The electrical and thermal conductivity properties of composite materials can be influenced by the composition of their components. The incorporation of conductive reinforcements, such as carbon fibers, can enhance the electrical conductivity of composite materials. This property is of particular significance in the context of electronic devices and antistatic applications. Furthermore, composites that exhibit high thermal conductivity can be employed as effective heat dissipation components. Such composites are widely preferred in computer processors, light-emitting diodes (LEDs) and cooling systems.

6.1.2.7 High Sound Insulation

The utilization of composite materials in construction also proves highly effective in the context of sound insulation, due to the diverse range of components that they comprise. Polymer matrix composites offer insulation by impeding the propagation of sound waves. This characteristic enables the utilization of composites in the construction industry, particularly in enhancing the acoustic characteristics of edifices. Moreover, sound-insulating composites are commonly employed in the automotive and aviation industries to mitigate engine noise.

6.1.2.8 High Fire Resistance

Some composite materials are designed to demonstrate high resistance to fire. It has been demonstrated that ceramic-based composites and polymer composites reinforced with fire retardant additives can maintain their structural integrity even when exposed to fire. The ability to withstand fire is a crucial safety consideration, particularly in the construction, transportation and defense sectors. Furthermore, fire-resistant composite structures contribute to enhanced safety by limiting the emission of smoke and toxic gases during combustion.

6.1.2.9 High Biocompatibility

The biocompatibility of composite materials represents a significant aspect of consideration within the field of biomedical profession. The term "biocompatibility" is used to describe a material's capacity to function safely within the human body, effectively eliminating any possibility of an allergic or harmful reaction. Several medical devices, including heart valves, dental prostheses and orthopedic implants, rely heavily on composite materials. The use of composites in medical devices allows for extended periods of implantation within the human body due to their biocompatibility.

6.1.2.10 Dimensional Stability

The dimensions of materials may be subject to alteration because of environmental factors, including temperature and humidity. Composite materials demonstrate high dimensional stability in the face of these factors, exhibiting minimal expansion and contraction and thus maintaining their shape. This property is of particular significance in the context of precision engineering applications. The capacity of composites to withstand temperature fluctuations enables the maintenance of dimensional stability over extended periods of use.

6.1.3 Sustainable Polymer Composites

Polymers are materials that form interconnected structures with strong carbon–carbon bonds and are widely used in various industries. The robust nature of these bonds enhances the mechanical durability of polymers; however, they also give rise to environmental concerns due to their prolonged degradation time in the natural environment, which ranges from 20 to 500 years. The increasing population and industrialization are responsible for the generation of 400 million tons of polymer waste on an annual basis. When considered alongside the recycling rates, which currently stand at only 9%, this presents a significant environmental threat (Advances in sustainable polymer composites-A volume in Woodhead Publishing Series in Composites Science and Engineering 2020). The biodegradation time of polymers has become a significant concern in relation to their ease of use and durability, as well as their impact on waste management and the environment. The residence time of traditional polymers in the natural environment has a detrimental impact on ecosystems, resulting in water and soil contamination. Furthermore, the degradation of these materials into microplastics over time represents a significant threat to marine and terrestrial life. The relatively low rate of polymer recycling underscores the necessity for the development of sustainable solutions for the management of these materials. The current recycling systems are frequently only capable of processing specific types of polymers, which constrains the achievable recycling rates. It is therefore imperative that innovative methods and technologies be developed to transform the polymer industry into a sustainable model. In the current era, the mounting environmental challenges have prompted scientists to undertake a multitude of investigations into biodegradable polymers. In the context of sustainable development, the issue of fossil fuel depletion has emerged as a significant topic of concern. It has been demonstrated by scientific evidence that the reserves of fossil fuels are finite and are undergoing depletion. Furthermore, the release of CO_2 into the atmosphere because of the combustion of fossil fuels has been identified as a significant contributor to global warming and the acceleration of climate change (Dodds and Gross 2007; Abhilash and Thomas 2017). This situation is regarded as a threat to environmental and social sustainability, and it has become imperative to review fossil fuel consumption and transition to sustainable energy sources. In this context, the fact that synthetic polymers are largely produced from fossil fuels represents another significant factor contributing to increased environmental pressures. The production of polymeric materials necessitates the consumption of both energy and resources and is frequently disposed of in landfills without the implementation of recycling processes. The extensive utilization of these materials, particularly in single-use items, has a considerable impact on the generation of solid waste, with long-lasting adverse effects on the environment. The depletion of resources throughout the life cycle of polymers, the limitations of recycling processes and problems in waste management serve to accentuate the environmental impact of these materials. It is therefore imperative to identify more sustainable methods for the production and utilization of polymers (Mohanty et al. 2022).

6.1.4 The Production of Sustainable Polymer Composites

Composite materials are currently employed in a multitude of sectors within the field of modern manufacturing technologies. Consequently, a variety of techniques must be employed to manufacture composites with the requisite properties and shape. A variety of techniques are employed in the production of polymer-based composite materials with the requisite properties and shapes. This section will provide a brief overview of the methods (Wibowo et al. 2004; Gao and Mäder 2006).

6.1.4.1 Production with Open Mold Methods

In this method, a permanent mold or guide is employed to produce a substantial number of composite structures that can be utilized on multiple occasions. Molds can be classified into two main categories: perishable and permanent. The following section provides a comprehensive overview of the production techniques employed in open mold methods.

Hand Lay-Up Technique

The reinforcement fabrics, which have been prepared with either woven or chopped fibers, are then laid on the mold by hand. Following this, the liquid resin is applied, and the resin is impregnated between the fiber layers. Prior to the deposition of the fibers, the mold is subjected to a cleaning process and a gelcoat is applied. Once the gelcoat has undergone a sufficient degree of polymerization, the fiber layers are then laid. The resin is applied at the final stage, after the deposition and preparation of the composite reinforcement material. It is important to ensure that the resin penetrates the fiber fabric effectively throughout the process. In addition to polyester and epoxy, vinyl ester and phenolic resins are also preferred as reinforcement materials. Although hand lay-up necessitates a considerable investment of labor, it is an optimal technique for low-volume production. This is the most straightforward technique, and it is employed for the fabrication of large-scale structural components, including boat hulls, tanks, building panels, and crates (Hoa 2009).

Spraying Technique

The spraying method may be regarded as an instrumental variant of the manual lay-up technique. The chopped fiber is applied to the mold surface with resin and hardener via a specialized spraying device. The trimming of the fiber is conducted via a trimming mechanism integrated into the spray gun, which operates independently.

Following the spraying process, the final product is prepared by means of surface smoothing with a roller. The solidification of the product is typically achieved through room temperature or heat-based methods. Prior to spraying the main material, silicone is applied into the mold, resulting in enhanced surface quality. Polyesters are commonly utilized as resin. This method is more advantageous than the hand lay-up technique when the part complexity is high. Other benefits of this technique include straightforward, cost-effective, portable equipment and no part size limitations.

Fiber Winding Technique

The fiber winding method represents a significant advancement in the production of products with superior surface quality. This method is applicable to the mass production of products with bespoke shapes. The fiber winding method can be defined as the process of winding continuous fibers on a rotating mold by pulling them from a spool after they have been soaked in resin. The winding of continuous fibers into a mold at varying angles allows the production of products with differing mechanical properties. Once enough fiber plies have been wound, the product undergoes a hardening process. Subsequently, the final product is detached from the rotary mold. The products resulting from this method of production are typically cylindrical pipes, car shafts, aircraft water tanks, yacht masts, and circular pressure tanks. Although this approach is more costly, it offers greater control of the resin and allows for flexibility in fiber placement in complex-shaped products. While the method offers several advantages, including comprehensive control over product properties, the ability to produce a range of strength values, minimal labor costs, an extensive selection of reinforcement materials and resin options, high levels of process adaptability, and the use of thermosets and thermoplastics, it also presents certain disadvantages. These include the limitation of the method to axial products, the necessity for significant plant investment, and the requirement for mechanical controls due to the emission of volatile organics (Hoa 2009).

Layered Compounding (Bag Molding) Technique

In the bag molding process, fibers and thermoset resin are piled on the mold, and the mold is covered with a flexible diaphragm (bag). Subsequently, pressure and temperature are employed to harden the system. There are three methods for molding bags: autoclave bag molding, vacuum bag molding, and pressure bag molding.

- *Vacuum Bag Molding*: This method represents an advanced form of hand molding. It entails the application of thermosetting resin to the fiber layers, which are then subjected to a 15-min heating process. Once the desired thickness has been attained, the upper surface is covered with a bag made of polyvinyl alcohol, and the edges are closed and connected to the vacuum system. Once the molded components have been connected to the primary vacuum system, the temperature of the oven is adjusted to the curing temperature of the resin and maintained for a designated period. The application of a vacuum creates a pressure effect on the layers, facilitating the escape of air and excess resin.
- *Bag Molding in Autoclave*: The temperature of the autoclave can be increased up to 175 °C, according to the requirements of the process. The pressure is provided by air and/or CO_2 gas. In autoclave molding, a vacuum is applied with the objective

of removing air and other volatiles that may be trapped between the layers. The primary production variables are autoclave pressure, temperature, and curing time, which are contingent upon the resin system and the thickness of the molded part.

- *Pressure Bag Molding*: It is the most cost-effective method among the techniques employed in the production of composite materials. The layering process is analogous to the hand molding technique. The layering process is continued until the desired thickness is achieved. Subsequently, the surface of the top layer is covered with cellophane, and the mold cover is closed. The rubber bag attached to the cover is inflated with air or steam pressure, thereby applying pressure to the layers. The application of pressure facilitates the expulsion of air and excess resin entrapped between the layers, establishing direct contact and enabling resin hardening.

6.1.4.2 Production with Closed Molds Methods

Closed mold production methods employ a mold that completely encircles the product. It is possible to obtain superior products than those produced using open mold techniques. The following section provides a comprehensive overview of production techniques utilizing closed mold methods (Hoa 2009).

The Compression Transfer Molding Technique

The method is suitable to produce complex and high-strength glass fiber-reinforced plastics under high-volume, high-pressure conditions. The fibers utilized in this process are typically derived from glass, graphite, and asbestos. The resins employed are polyester, epoxy, and phenolics. The molds are composed of two distinct parts, the male (positive) and the female (negative), and are typically electrically heated. The plate or casting is placed in the open mold in the press with the amount of molding compound or resin added. This process is used to produce small parts with low pressure, room temperature, and inexpensive molds. Glass fibers and thermosetting resin are deposited into the molds and subjected to pressure between 130 and 340 kPa. Molds are made of metal, plaster, and glass fiber-reinforced plastics.

Pultrusion Technique

The pultrusion method bears resemblance to extrusion. In contrast to extrusion, in which the raw material is pushed into the mold, pultrusion involves the pulling of the material into the mold. Subsequently, the continuous reinforcing fibers are conveyed through a resin tank, whereupon the fiber bundles are hardened in a heated mold. The method allows to produce a variety of profiles, rods, flat sheets, and corrugated sheets with fixed cross-sections. The production speed is considerable. A single-step process is available to produce layered polymer matrix composite materials. Two distinct pultrusion techniques exist: wet and dry. In the case of wet pultrusion, the fibers that have been passed through the resin tank are immediately molded and hardened. In contrast, in the case of dry pultrusion, the polymer matrix composite material is shaped in a state of annealing and softening. The wet method is employed when cost is a primary consideration, whereas the dry method is utilized in applications where strength is of paramount importance. The pultrusion technique is employed with the use of a variety of resins, including polyesters, epoxy, polyamide, and polysulfide, which are reinforced with glass and graphite (Hoa 2009).

Injection Technique

This method is preferred to produce fiber-reinforced thermoplastics. The molding compounds that are fed to the injection molding machines are premixes in the form of granules, with a length ranging from 3 to 6 mm. In the context of injection molding, the utilization of pallet-shaped premixes is a prevalent practice (Bulut and Erdoğan 2011; Gay 2022).

Centrifuge (Centrifugal) Method

This method is an appropriate means of producing cylindrical components. The composite material is produced by filling a cylindrical container with a mixture of chopped fibers, resin, and hardener. The container is then rotated, causing centrifugal forces to disperse the material on the container walls. Finally, the material is hardened with hot air.

Pressure Molding Techniques

The method is rapid and employs a serial molding process, with molding times ranging from three to six minutes, contingent on the dimensions of the product in question. The mixture of glass fiber, polyester, and filler, in addition to additives, is shaped in steel molds at a temperature of 150–170 °C and under a pressure of 50–120 kgf/cm^2. This mixture is pre-prepared in the form of pulp or dough. The method offers several advantages, including the ability to produce complex shapes, embed metal components within the structure, and achieve varying wall thicknesses. Moreover, the product is shaped on both sides with molds. It is possible to produce complex shapes, such as holes, which are not feasible using alternative composite material manufacturing techniques. The initial financial outlay is considerable. The process yields a multitude of products with high production speed. The system is also referred to as the co-molding method. The polymerization (hardening) process is completed under conditions of elevated temperature and pressure. In the production of sheet-shaped parts, the sheet molding compound (SMC) method is employed when the objective is to produce three-dimensional (monolithic) parts. The bulk molding compound (BMC) method is employed (Hoa 2009).

6.2 Application of Sustainable Polymer Composites in the Automotive Industry

The increasing need for sustainable and lightweight polymer composites due to the environmental concerns and global regulations in world effects the automotive industry, in which polymers have a very large amount of use in the total mass. Polymers are used in different components of a vehicle including interior (door panels, dashboards, seat frames, and other), exterior (bumpers, fenders, and other) and/or structural (under-the-hood applications, such as engine covers). As an alternative to metallic counterparts such as iron, steel, and aluminum, polymer composites create a great economic difference thanks to their lightness advantages. However, to their general acceptance in the automotive sector, their production costs need to be reduced. In material science, several investigations have been performed to overcome this issue. This section will cover the exploring of sustainable polymers as alternatives toward the reduction in weight of vehicles to decrease the fuel consumption and the emission of carbon dioxide.

Door panels, dashboards, steering, glove compartments, seat frames, seat belts or headliners can be made from sustainable polymer composites, providing a balance of aesthetics, strength, and lightweight properties. Bumpers, fenders, doors, sunroofs, mirrors, and other exterior components benefit from the durability and lightweight characteristics of these composites. Under-the-hood applications, such as engine covers, radiators, chassis, brake systems, and other structural parts, can also utilize sustainable polymer composites to reduce weight and improve performance. Even though there are some limitations on the use of sustainable polymer composites in the automotive industry, many deficiencies have been characterized thanks to new research. Very shortly, it can be said that human and environmentally friendly biodegradable polymer composites will be used in vehicle interior and other components.

The idea of using bio-based materials in cars dates back to the 1930s when Henry Ford sought to support farmers by using agricultural resources for vehicle parts and fuels. Ford's research led to the 1941 T**concept car**, made with natural fiber composites and powered by hemp oil, but further development stalled due to the abundance of oil and World War II. In 1957, East Germany's **Trabant** used biocomposites in its construction, and in 1982, FIAT introduced wood-filled door panels in the **FIAT Punto**. The practical use of natural fibers in vehicles gained momentum in the late twentieth century, driven by concerns over fossil fuel depletion and environmental impact, with companies like Daimler-Benz leading efforts to replace traditional materials with bio-based alternatives (Akampumuza et al. 2017).

6.2.1 Types of Sustainable Polymer Composites Used in Vehicle Components

In recent years, the sustainable reinforced polymer composites have been intensively studied to use them as a valid lightweight and high mechanical strength alternative. For different purpose, multiple types of reinforcement materials are used in various components of vehicles. Carbon and glass fibers are the most popular reinforcement materials for polymer composites. In addition, studies conducted in textile, construction, automotive and other industrial fields using natural fibers-based polymer composites have shown that natural fibers provide good acoustic/thermal insulation and mechanical properties, high recycling properties, less health hazards, and reduced tool wear in comparison with traditional reinforcement materials properties. Natural fibers are obtained from three main types of sources: cellulose/

lignocellulose fibers, mineral fibers, and animal fibers (given in Fig. 6.1) (Samanth and Subrahmanya Bhat 2023).

Plant based natural fibers are constituted mainly cellulose, hemicellulose, glucose, lignin, pectin, and small number of impurities. Cellulose/lignocellulose fibers are bast (e.g., flax, jute, kenaf, hemp, ramie, roselle), seed (cotton, kapok, loofah, milk weed), leaf (e.g., sisal, manila, banana, palm), stalk (e.g., wheat, rice, oat), wood, grass (e.g., reed, bamboo, bagasse). Mineral fibers are mainly asbestos and fibrous brucite. Lastly, animal fibers are hair/wool and silk. The average weights of natural fibers used in various components of a vehicle are listed in Table 6.1. One of the biggest reasons for incorporating natural fibers into production is that natural fibers can contribute up to a 32% weight reduction.

After bio-composites, as a combination of natural fiber and bio-resin, were first used to produce automotive door panels in 1980, natural fiber-reinforced polymer composites are now used in more components (Olhan et al. 2021). The type of natural fibers used is dependent on the manufacturers. For example, panels from South America contain mostly ramie or sisal fibers, while panels from Europe contain hemp or flax fibers. Natural fiber types have a key role because of different requirements in different components. For instance, plant-based fibers such as hemp, coir, or sisal are most popular reinforcement materials since it has been proven that the production costs can be reduced up to 21%, while replacing glass fibers (Dunne et al. 2016). Table 6.2 shows an example list of interior, exterior, and structural components of a vehicle produced from natural fibers.

As seen in Table 6.2, natural fiber-reinforced polymer composites are primarily utilized for interior component, as natural fibers are hydrophilic by nature, which makes them restricted for the application in exterior components. After all, some car manufacturers such as, Volkswagen, Mercedes, Ford, BMW, or Audi are looking for a way to make more components of their vehicles more recyclable or biodegradable. Exemplary, Mercedes A-class model applied abaca fiber-reinforced composites to underfloor paneling and Travego model used flax fibers for the application in exterior parts. The comparison of flax fiber/PP composites and glass fiber/PP composites was also investigated from the point of life-cycle assessment (LCA) and micromechanical analysis by Le Duigou and Baley (2014). It was found that the flax fiber/PP composite (30% w/w) is lighter than glass/PP composites, which can reduce the environmental impact by 10–20% by reduced global warming, acidification, etc. Park and Park also investigated the preparation of flax/vinyl ester composite applied resin transfer molding fabrication technique for the automobile bonnet structure designing (Park and Park 2018). The flax/vinyl ester panel showed a reduction in weight by 31.7%, when the researchers compared it with traditional steel panel. Although, the mechanical properties of samples were acceptable, they noted that the repair and maintenance could be needed, when impact exceeded 10 J. In a similar study, a flax/vinyl ester composite based automobile hood was subjected to the structural analyses (Kong et al. 2016). The experimental outcomes showed that the designed hood provided a structural integrity and safety, and acceptable stability.

Even though flax fibers are the most preferred reinforcement material in European industry, other natural fibers provide good properties. In a report published, coir

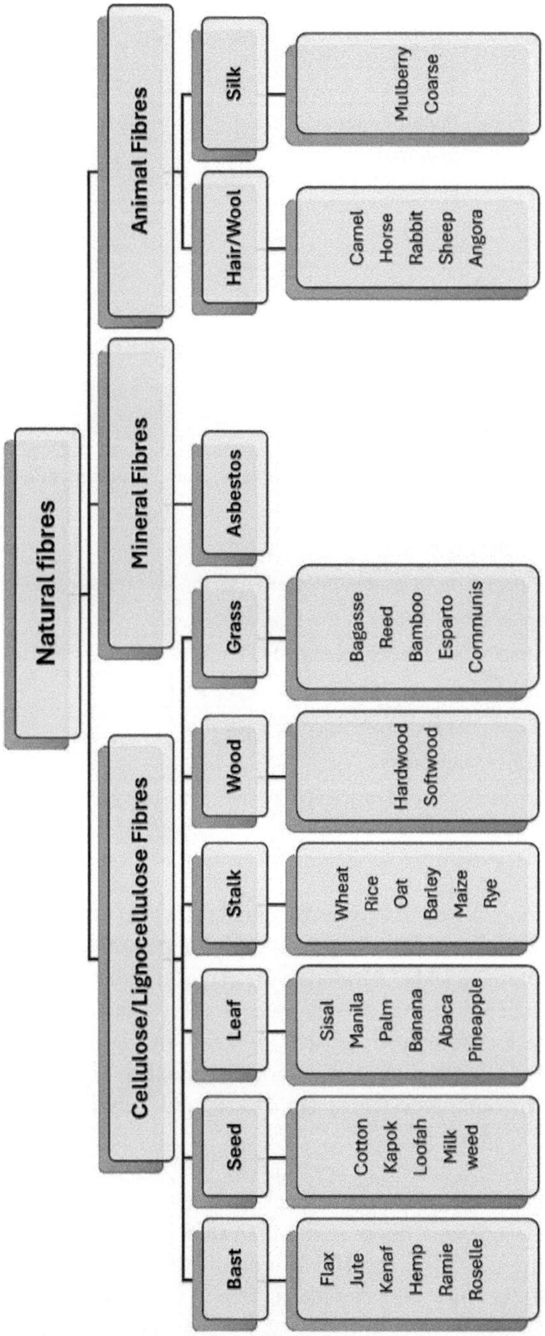

Fig. 6.1 Natural fibers classification (Jabu et al. 2024)

Table 6.1 The average weights of natural fibers used for the production of components (Society 2003)

Vehicle component	Weight (Kg)
Door liner	1.2–1.8 (Front door)
	0.8–1.5 (Rear door)
Boot liner	1.5–2.5
Sunroof interior shields	0.6
Seat back	1.6–2.0
Headrest	2.5
Parcel shelve	2.0

Table 6.2 Examples of vehicle components—interior, exterior, and structural—manufactured using natural fiber materials. (Naik and Kumar 2021; Holbery and Houston 2006; Witayakran et al. 2017; Boland et al. 2016)

Natural fibers	Components of vehicle
Cotton	Trunk panel, trunk floor, insulation material, sound proofing, glove box
Kenaf	Door bolsters, insulation
Jute	Dashboard, door panels
Flax	Seat back panels, floor panels, door panel cover, arm rests
Flax/Sisal	Interior panels, door panels, glove box, door linings,
Coir	Seat cushions, seat covers, outer hood
Hemp	Dashboard, door panels, engine cover, seat back panels
Abaca	Body and floor panels, spare-wheel compartment
Wool	Seat coverings, upholstery
Wood	Seat back panels, foamed instrument panels, rear shelves, trims for decklid, glove box

fiber and date palm fibers based micro-porous materials was stated that it presented good acoustics properties at a low frequency, which makes them consumable to the conventional alternatives (Abd Alrahman 2013). Mercedes Benz used coir fiber obtained from the mesocarp of mature coconuts in the production of seats in Class A automobile and trucks (Salazar et al. 2011). Yazli and Ismail reported that the preparation of natural fiber composites using coir fiber and epoxy resin with different weight ratios to use in outer hood of a car (Yazli and Ismail 2024). The results revealed that 20 wt% coir fiber addition could meet the mechanical performance and special requirements for the car hood application. Two or more natural fibers can be mixed for the preparation of natural fiber composite. Okafor et al. studied the combination of banana and coir fibers particulate in polyester matrix for designing and application of vehicle components (Okafor et al. 2021). After the optimal design of composite material, they used it as an alternative to car window regulator handle. It was revealed that the prototype resulted in 12.9% reduced weight in comparison to traditional Nylon window regulator models, while deformation and stress was within linear range without material failure according to the stress analysis.

Kenaf is also one of the most favorable natural fibers, especially for the application in interior components of vehicles. Toyota has preferred it in over 25 car models (Fatchurrohman et al. 2021). Typically, kenaf reinforced polylactic acid (PLA) or polypropylene (PP) are used for this purpose. PLA/kenaf fibers composite is a bio-composite obtaining from plants. It is a sustainable, low density, and high-strength polymer composite. Recent years, various investigations have been conducted on the field of the using kenaf fibers to replace the glass fibers with thermoplastic polymers to apply in vehicle components. Davoodi et al. have reported that the hybrid kenaf/glass fiber-reinforced epoxy composite presented competitive mechanical properties, such as Young's modulus, flexural modulus, tensile strength. However, it was noted that the impact property should be still improved to be preferred (Davoodi et al. 2010). In a different study, it was reported that the untreated kenaf/PP composite showed excellent tensile strength (692 MPa), Young's modulus (44 GPa), and low density (1.027 g/cm^3) (Mohd Radzuan et al. 2020). The researchers concluded that the kenaf/PP composite can be used as an alternative for automotive in engine cover and other components. In addition, there have been several reports to use metal oxide nanoparticles to improve the mechanical performance of natural fiber-reinforced polymer composites to replace the glass-fiber sheet molding compound in cars. Wu et al. prepared natural fiber-reinforced biocomposite consisting of kenaf fiber with ZnO nanoparticles for use instead of Meridian's SLI 323IF (Wu et al. 2020). It was revealed that the fabrication of natural fiber-reinforced biocomposite with ZnO reduced the energy consumption by 9% and life-cycle environmental impacts by 33%. Akubueze et al. investigated the preparation of thermosetting hybrid kenaf/banana fibers composites for using in car bumper (Emmanuel et al. 2019). It was found that the hybrid kenaf/banana reinforced unsaturated polyester composites prepared with the addition of 50% fiber value provided better mechanical properties than those of composite of kenaf and banana fibers reinforcement. They concluded that the synergetic effect of each component leads to performance improvement.

6.2.2 *Driving Innovation: The Collaborative Efforts of Automakers, Suppliers, and Academia in Advancing Biocomposites for Automotive Use*

6.2.2.1 Contribution of Original Equipment Manufacturers (OEMs), Tier1, and Tier2 Suppliers

Faurecia, a leading automotive equipment supplier based in Nanterre, France, specializes in manufacturing seats, exhaust systems, and interior components such as dashboards, center consoles, and door panels. The company supplies major automakers including Volkswagen, Ford, Renault-Nissan, Toyota, and Hyundai, with products that account for 15–20% of a vehicle's weight. Faurecia is committed to enhancing vehicle life-cycle assessment by focusing on weight reduction, lower fuel consumption, and reduced environmental impact. Through its **BioAttitude** program, Faurecia introduced technologies like **NAFILean**, which uses hemp to create lightweight interior components, and **LignoLite**, a wood fiber-based material used in Mercedes-Benz S-Class door panels, reducing their weight by up to 45%. The company also developed **Flaxpreg**, a long flax fiber-reinforced composite used for load-bearing components, which offers significant weight savings and high load-bearing capabilities.

Goodyear, a global tire manufacturer, has partnered with **DuPont** and **Genencor** to develop renewable alternatives to petroleum-based chemicals. In 2009, they introduced **Bio-Isoprene**, a biomass-derived material used in Ford's European market Fiestas, reducing tire weight and improving fuel efficiency. **BMW** is also considering using these tires in its vehicles. Another bio-based tire, **GT-3 BioTred**, developed with Italian firm **Novamont**, reduces fuel consumption by 5%, cuts rolling noise by 50%, and improves wet braking performance by 10%. This innovation won Goodyear the "Environmental Achievement of the Year Award" at the 2010 Geneva Motor Show. Goodyear is also exploring biomass sources like switchgrass, sugarcane, and woody biomass to produce **Bio-Isoprene**, providing a sustainable alternative to natural rubber, which is limited in supply and vulnerable to political and environmental challenges.

Rieter Automotive, a global parts supplier, began producing spare wheel pan covers for the Mercedes A-Class in 2004, replacing most glass fibers with **abaca** (banana) natural fiber reinforcements. This initiative, which took 18 months to implement, successfully reduced both cost and weight while maintaining product quality, achieving a 10% decrease in primary energy use. Rieter also created a load floor for the Maserati Quattroporte using wood fiber-reinforced polypropylene and has developed eco-friendly sound packages under its "bioacoustics" project. These efforts aim to enhance recycling, reduce environmental impact, and lower energy consumption in component manufacturing. Rieter was recognized with the **JEC Composites Award** in 2005 for its design of natural fiber-reinforced thermoplastic composites used in under-floor modules with integrated functions (Society 2003).

6.2.2.2 Contribution of Automotive Makers

Ford: The company has leveraged its historical partnership with agriculture to develop automotive foams and composites from renewable resources, focusing on reducing weight and carbon footprints. Collaborating with farmers, researchers, and various industry stakeholders, Ford's R&D centers in the U.S., Germany, and Brazil have shown that soy-based foams reduce carbon dioxide emissions by 5.5 pounds per pound of soy oil used compared to traditional petroleum-based foams. Currently, over 8 million vehicles feature soy foam seats, resulting in an annual reduction of over 5 million pounds of petroleum oil usage.

In 2008, Ford partnered with **Lear Corporation** to create soy-based seat foam for the Ford Mustang, and by 2010, bio-based cushions were incorporated into more than a million Lincoln and Mercury vehicles. Other Ford models, including the **Expedition**, **F-150**, and **Focus**, also adopted similar materials. Additionally, Ford has utilized wheat straw in the **2010 Ford Flex** for injection-molded storage bins, hemp-propylene composites in the front grille of the **Montana**, and 100% castor bean oil-derived nylon-11 in fuel tank tubes, collectively achieving nearly 1 million pounds of carbon dioxide emissions reduction annually (Akampumuza et al. 2017).

Schmidt and Beyer worked on life-cycle analysis of two designs of insulation components for a Ford car. The reference component ethylene propylene diene copolymer (EPDM), polypropylene (PP) and reinforced with glass fibers. The alternative component is also made from EPDM and PP but reinforced with hemp fibers (30%). In summary, their results demonstrated that net benefits of switching to the hemp fiber component from glass fiber reinforced component. They reported that the hemp fiber component showed net benefits of 88.9 MJ in cumulative energy demand, 8.18 kg of CO_2 emissions in basic scenario (Schmidt and Beyer 1998).

Volkswagen: Volkswagen, headquartered in Wolfsburg, Germany, is the largest automaker in Germany and the third largest in the world. The company has utilized natural fiber-reinforced composites in various components, such as door inserts, seat backs, rear flap lining, and parcel trays in the Golf A4 and its 1998 variant. In 2000, Audi launched the A2, which featured door trim panels made from polyurethane reinforced with flax and sisal hybrid mats (Akampumuza et al. 2017).

Wötzel, Wirth and Flake presented life cycle assessments of a side panel for Audi A3 car made from akrilonitril bütadien stiren copolymer and an alternative material hemp fiber 66% (v/v)- epoxy resin composite. The cumulative energy use and selected emissions from production of one component, reported in this research are given in Table 6.3 (Wötzel et al. 1999; Joshi 2004).

Mercedes Benz: This company began incorporating biocomposites into its vehicles in 1994 with jute-reinforced plastics for the interior door panels of the **E-Class**, achieving a 20% weight reduction. By 2000, the company expanded its use of natural fibers, integrating polyester/flax-reinforced composites in engine encapsulations for

Table 6.3 Environmental impacts from production of an auto side panel

Component material environmental indicator	ABS copolymer	Hemp-epoxy
Total energy (MJ)	132	73
CO_2 emissions (kg)	4.97	4.19
Methane (g)	17.43	16.96
SO2 (g)	17.54	10.70
NOx (g)	14.14	18.64
CO (g)	4.44	2.14
Phosphate emissions to water (g)	0	0.09
Nitrate emissions to water (g)	0.08	12.05

several bus models, resulting in a 5% weight reduction. In the 2006 **Class**, Mercedes increased the use of natural fibers by 73%, utilizing wood and flax fibers in various interior components, contributing nearly 43 kg to the car's weight. The **C-Class** models manufactured in South Africa featured rear panel shelves made from sisal-reinforced composites, leading to a 10% weight reduction and an 80% decrease in production energy (Akampumuza et al. 2017).

6.3 Challenges and Future Prospects of Sustainable Polymers in Automotive Industry

The automotive industry is under increasing pressure to focus not only on reducing tailpipe emissions but also on embracing both the green and digital transitions. These shifts are driven by the broader goals of the European Green Deal and the Circular Economy Action Plan, which call for reducing the environmental impact of products across their entire lifecycle—from production to end-of-life.

Green Transition: This involves moving towards greater sustainability by reducing carbon emissions, conserving resources, and minimizing environmental harm. It's not just about lowering vehicle emissions anymore, but also about cutting the overall environmental footprint throughout the manufacturing process, supply chains, and during recycling or disposal. The circular economy is central to this transition, promoting waste reduction, material reuse, and closing the loop on resources.

Digital Transition: This focuses on using technologies like IoT, AI, machine learning, big data, and blockchain to improve automotive processes. This includes digitalizing manufacturing, enhancing vehicle connectivity, and applying data-driven management to track resource use, energy consumption, and emissions throughout the value chain. Digital tools help improve efficiency, predict maintenance, and optimize product lifecycles, supporting both sustainability and competitiveness.

6.3.1 Important EU Policies Influencing the Materials Used in Automotive Industry

EURO 7: The European Parliament and the council of the European Union (EU) reached agreement on new emission standards for light and heavy-duty vehicles on December, 2023. The Euro 7 regulation include requirements fort h type approval of vehicles, brake systems, and tires intended for sale in EU regarding their emissions behavior. It is a regulatory framework for controlling emission limits from vehicles, durability requirements to ensures that vehicles remain compliant over their entire operational life, and defines requirements for on (like on board diagnostics systems and off board (such as periodic testing and inspections) compliance verification methods.

EU Green Deal and European Climate Pact: Launched by the European Commission as part of the European Green Deal, the pact is a movement of people united by a shared mission to build a more sustainable Europe and help the EU to become climate neutral by 2050.

The Climate pact aims to learn about climate change and develop and implement solution, to connect with others and maximize the impact of these solutions.

EU Green Deal: Climate change and environmental degradation are an existential threat to Europe and the World. To overcome these challenges, the European Green Deal will transform the EU into a modern resource efficient and competitive economy, ensuring:

- No net emissions of greenhouse gases by 2050
- Economic growth decoupled from resource use
- No person and no place left behind

One third of the €1.8 trillion investments from the NextGenerationEU Recovery Plan, and the EU's seven-year budget will finance the European Green Deal. The European Commission has adopted a set of proposals to make the EU's climate, energy, transport and taxation policies fit for the reducing net greenhouse gas emissions by at least 55% by 2030, compared to 1990 levels.

With these frameworks, the EU now has legally binding climate targets covering all key sectors of the economy including automotive industry. The overall package includes to make transport sustainable for all. With the new CO_2 Standards, all new cars and vans registered in Europe will be zero-emission by 2035. As an intermediary step towards zero emissions, average emissions of new cars will have to come down by 55% by 2030, and new vans by 50% by 2030. This will put road transport on a firm path to zero-emission mobility in 2050 (Dornoff 2024).

6.3.2 Challenges Faced by the Automotive Industry with Respect to Sustainable Materials

The automotive industry faces significant challenges in aligning with circularity and sustainability throughout the entire product life cycle, from raw material sourcing to end-of-life disposal. Due to the long production and usage periods, along with the need for legacy spare parts (lasting 10 to 15 years), it is critical to rethink solutions for the sector (Ildefonsoa 2023). Here is some challenges and suggestions for these problems.

6.3.2.1 Balancing Material Durability and Recyclability

- Design for Disassembly: This involves designing parts and vehicles in such a way that they can be easily disassembled at the end of life, thus facilitating the recycling process. Modular design is one approach that can make both durability and recyclability more achievable.
- Material Innovation: Developing alternative materials that are both durable and recyclable without containing restricted substances is challenging but essential. Collaborating with material scientists and regulatory bodies can lead to new composites that meet these needs.

6.3.2.2 Avoiding Legacy Hazardous Substances

- Improved Tracking Systems: Implementing advanced tracking systems to monitor the presence of substances of very high concern can help recyclers and manufacturers stay updated. Digital systems such as blockchain could be used to trace material composition from production to end-of-life, ensuring better transparency.
- Proactive Compliance Audits: Regular audits of materials and processes, with a focus on upcoming regulations, can prevent the use of non-compliant substances. Companies could establish internal guidelines to identify and eliminate high-risk materials early in the product development cycle.

6.3.2.3 Addressing the Time Lag in Regulations

- Anticipating Future Regulations: Developing future-proof materials by staying ahead of anticipated changes in chemical regulations is one approach. This can be done by studying trends in regulatory developments and shifting toward safer alternatives even before legal mandates require it.
- Pre-regulatory Collaboration: Collaborating with regulatory agencies and stakeholders within the industry may allow manufacturers and recyclers to share insights and develop materials that are compliant with future requirements. Early engagement with policy makers can result in more achievable and flexible regulatory frameworks.

6.4 Conclusion

The main conclusions of this chapter can be summarized as:

- As the circular economy movement gains attention especially in Europe, addressing the carbon footprint of the vehicles' upstream supply chain and overall environmental impact is quickly becoming a top priority for suppliers and manufacturers. Regulations such as the EU Green Deal, EPA rules in the U.S., and carbon neutrality goals in Asia are shaping industry behavior. OEMs like Stellantis, Ford, BMW, Toyota, BYD will demand carbon footprint reporting and material traceability throughout the supply chain, posing both challenges and opportunities for sustainable polymer suppliers.
- From a technical point of view, focusing on bio-based composites as sustainable materials will help to increase the mechanical strength, reduce material weight and fuel consumption, decrease the production cost, and improve biodegradability for the interior parts of the vehicles'.
- Identifying sustainable alternative materials for the automotive industry is crucial to enhancing the resilience of the value chain and improving competitiveness. While the EU has established a framework to address general industry needs, the automotive sector faces specific challenges that require tailored solutions.
- Sustainability is no longer just about being eco-friendly, but also about delivering functional performance. It is predicted that smart polymers that are self-healing, temperature-responsive, or reactive to environmental stimuli will find applications in both interior and exterior automotive parts.
- Global supply chain disruptions are pushing for more localized polymer production. All around the world investments in domestically sourced bio-materials will grow, reducing emissions while enhancing supply security.
- Decision-makers are increasingly relying on LCA (life-cycle assessment) reports to evaluate materials. The future of sustainable polymers will rely heavily on digitization and material analytics, enabling stakeholders to quantify and compare the environmental impacts of different materials.

All in all, it can be said that companies invest in these areas will gain a significant competitive edge.

References

Abd Alrahman RRIL, Abdul Rahman R (2013) Experimental study on natural fibers for green acoustic absorption materials. Am J Appl Sci 10(10):1307–1314

Abhilash M, Thomas D (2017) Biopolymers for biocomposites and chemical sensor applications, Elsevier

Advances in sustainable polymer composites-A volume in Woodhead Publishing Series in Composites Science and Engineering, Elsevier (2020)

Akampumuza O, Wambua PM, Ahmed A, Li W, Qin X-H (2017) Review of the applications of biocomposites in the automotive industry. Polym Compos 38(11):2553–2569

Boland CS, De Kleine R, Keoleian GA, Lee EC, Kim HC, Wallington TJ (2016) Life cycle impacts of natural fiber composites for automotive applications: effects of renewable energy content and lightweighting. J Ind Ecol 20(1):179–189

Bulut Y, Erdoğan ÜH (2011) Selüloz Esasli Doğal Liflerin Kompozit Üretiminde Takviye Materyali Olarak Kullanimi. Tekstil Ve Mühendis 18(82):26–35

Davoodi MM, Sapuan SM, Ahmad D, Ali A, Khalina A, Jonoobi M (2010) Mechanical properties of hybrid kenaf/glass reinforced epoxy composite for passenger car bumper beam. Mater des 31(10):4927–4932

Dodds DR, Gross RA (2007) Chemicals from biomass. Science 318(5854):1250–1251

Dornoff RFJ (2024) Euro 7: The new emission standard for light- and heavy-duty vehicles in the European Union

Doruk SB (2023) Elastomer takviyeli epoksi kompozitlerinin üretimi ve karakterizasyonu, Trakya Üniversitesi

Dunne R, Desai D, Sadiku R, Jayaramudu J (2016) A review of natural fibres, their sustainability and automotive applications. J Reinf Plast Compos 35(13):1041–1050

Emmanuel A, Nwaeche C, Muniru O, Ezeanyanaso C, Nwaze IO, Nwauzor G, Mafe O, Okeke C, Afolabi A, Igwe C (2019) Hybrid thermosetting composite of kenaf/banana fibres for utilisation in a passenger car bumper. Int J Mater Sci Innov 4:1–6

Fatchurrohman N, Jun HX, Muhida R, Adelino MI (2021) Performance simulation of bio-reinforced composite car door panel using finite element analysis. In: International conference on computer science and engineering (IC2SE), IEEE, pp 1–5

Gao S-L, Mäder E (2006) Jute/polypropylene composites I. Eff Matrix Modif Compos Sci Technol 66(7–8):952–963

Gay D (2022) Composite materials: design and applications, CRC press

Hancox N (1996) Engineering mechanics of composite materials. Mater des 2(17):114

Hoa SV (2009) Principles of the manufacturing of composite materials. DEStech Publications, Inc

Holbery J, Houston D (2006) Natural-fiber-reinforced polymer composites in automotive applications. JOM 58(11):80–86

Ildefonsoa SDB (2023) State-of-the-art analysis on alternative materials for the automotive industry. Transp Res Procedia 72:1560–1567

Jabu MA, Alugongo A, Nkomo N (2024) Application of natural fibre composites in interior panels in the automotive industry: a review. Int J Eng Trends Technol 72(3):91–98

Joshi DLTSV, Mohanty AK, Arora S (2004) Are natural fiber composites environmentally superior to glass fiber reinforced composites? Compos Part A: Appl Sci Manuf 35(3):371–376

Kong C, Lee H, Park H (2016) Design and manufacturing of automobile hood using natural composite structure. Compos B Eng 91:18–26

Le Duigou A, Baley C (2014) Coupled micromechanical analysis and life cycle assessment as an integrated tool for natural fibre composites development. J Clean Prod 83:61–69

Mohanty AK, Wu F, Mincheva R, Hakkarainen M, Raquez J-M, Mielewski DF, Narayan R, Netravali AN, Misra M (2022) Sustainable polymers. Nat Rev Methods Prim 2(1):46

Mohd Radzuan NA, Tholibon D, Sulong AB, Muhamad N, Haron CHC (2020) New processing technique for biodegradable kenaf composites: a simple alternative to commercial automotive parts. Compos Part B: Eng 184:107644

Naik V, Kumar M (2021) A review on natural fiber composite material in automotive applications. Eng Sci 18:1–10

Namazi H (2017) Polymers in our daily life. BioImpacts: BI 7(2):73

Okafor CE, Onovo AC, Imoisili PE, Kulkarni KM, Ihueze CC (2021) Optimal route to robust hybridization of banana-coir fibre particulate in polymer matrix for automotive applications. Materialia 16:101098

Olhan S, Khatkar V, Behera BK (2021) Review: textile-based natural fibre-reinforced polymeric composites in automotive lightweighting. J Mater Sci 56(34):18867–18910

Park G, Park H (2018) Structural design and test of automobile bonnet with natural flax composite through impact damage analysis. Compos Struct 184:800–806

Rasmussen SC (2018) Revisiting the early history of synthetic polymers: critiques and new insights. Ambix 65(4):356–372

Şahin Y (2000) Kompozit malzemelere giriş, Gazi Kitabevi

Salazar VLP, Leão AL, Rosa DS, Gomez JGC, Alli RCP (2011) Biodegradation of coir and sisal applied in the automotive industry. J Polym Environ 19(3):677–688

Samanth M, Bhat KS (2023) Conventional and unconventional chemical treatment methods of natural fibres for sustainable biocomposites. Sustain Chem Clim Action 3

Schmidt W-P, Beyer H-M (1998) Life cycle study on a natural fibre reinforced component. SAE Trans 107:2095–2102

Society FP (2004) Seventh international conference on woodfiber-plastic composites (and Other Natural Fibers): May 19–20, 2003, Monona terrace community and convention center, Madison, Wisconsin, Forest products society

Sönmez M (2009) Polimer Matrisli Kompozitlerin Endüstri Ürünleri Tasarımında Önemi ve Geleceği: Türkiye'den Dört Örnek Firma Üzerine Bir İnceleme, İstanbul Teknik Üniversitesi

Strong AB (2008) Fundamentals of composites manufacturing: materials, methods and applications, Society of manufacturing engineers

Wibowo AC, Mohanty AK, Misra M, Drzal LT (2004) Chopped industrial hemp fiber reinforced cellulosic plastic biocomposites: thermomechanical and morphological properties. Ind Eng Chem Res 43(16):4883–4888

Witayakran S, Smitthipong W, Wangpradid R, Chollakup R, Clouston P (2017) Natural fiber composites: review of recent automotive trends

Wötzel K, Wirth R, Flake M (1999) Life cycle studies on hemp fibre reinforced components and ABS for automotive parts. Die Angewandte Makromolekulare Chemie 272(1):121–127

Wu Y, Cai L, Mei C, Lam SS, Sonne C, Shi SQ, Xia C (2020) Development and evaluation of zinc oxide-blended kenaf fiber biocomposite for automotive applications. Mater Today Commun 24:101008

Yazli DIM, Ismail N (2024) Study on the characteristics of natural fibre composite using coir fibre composite for car outer hood. Prog Eng Appl Technol 5(1):296–302

Chapter 7
Green Polymer Composites for Energy Harvesting Applications

V. Abhijith, B. Akhila, Sabu Thomas, and M. S. Sreekala

Abstract The fabrication of nanogenerators requires more innovative materials for their optimal performance and superior efficiency. Over the past few years, biopolymers have become an interesting material for the energy field due to their inherent properties like biocompatibility, mechanical strength, and dielectric properties. This book chapter summarizes the advancements in the development of nanogenerators based on biopolymers, with special focus on cellulose, chitosan, silk fibroin, and alginate. The chapter begins by introducing the different types of nanogenerators, and in the subsequent sections, it gives more importance to piezoelectric and triboelectric nanogenerators for energy harvesting applications. The mechanism of piezoelectric and triboelectric nanogenerators is briefly explained in the chapter. The source, structure, and mechanism of charge generation in different biopolymers are also discussed. The chapter also includes the main fabrication techniques used for nanogenerators. Finally, the major applications of biopolymer-based nanogenerators in different fields like energy harvesting, healthcare monitoring, gas sensing, and self-powered devices are illustrated.

Keywords Biopolymers · Piezoelectric nanogenerators · Triboelectric nanogenerators · Healthcare monitoring · Gas sensing

V. Abhijith · B. Akhila · M. S. Sreekala (✉)
School of Chemical Sciences, Mahatma Gandhi University, Kerala, India
e-mail: sreekalams@mgu.ac.in

S. Thomas · M. S. Sreekala
International and Inter University Centre for Nanoscience and Nanotechnology, Mahatma Gandhi University, Kerala, India

M. S. Sreekala
School of Polymer Science and Technology, Mahatma Gandhi University, Kerala, India

L. Ravindran et al. (eds.), *Macro, Micro and Nanocomposites from Sustainable Sources*, Advanced Structured Materials 244, https://doi.org/10.1007/978-981-95-2469-3_7

7.1 Introduction

Over the past few years, there has been a rapid increase in the production of macro- and microelectronics for a variety of applications, including environmental sensing and health monitoring. This is due to the explosive growth of the fields of artificial intelligence, machine learning, and the Internet of Things (IoTs). Each of these devices is dependent on external power sources, such as batteries and fossil fuels, which leads to significant environmental issues. Researchers are currently concentrating on the advancement of sustainable energy sources within this context. Despite the existence of renewable energy sources such as solar energy, the pollution generated by them after their lifespan continues to pose a threat to the environment (Jeong et al. 2025; Sandhya et al. 2023). Sustainable solutions to these issues are achievable through the development of autonomous systems that operate independently. Nanogenerators capable of producing electrical energy from heat and mechanical vibrations made a significant impact in the field of nanogenerators and self-powered electronics. These nanogenerators can be integrated with different IoTs to make them self-powered devices without the need for any external power sources like batteries or capacitors to overcome their limitations and challenges (Shen et al. n.d.).

Nanogenerators, especially piezoelectric nanogenerators (PENG), were first reported by Wang et al. (2006) in 2006 and triboelectric nanogenerators (TENG) in 2012 by the same group (Fan et al. 2012). Numerous research projects have been concentrated on enhancing the output performance, durability and stability of these devices, from the time of their invention to the present. Nanogenerators are classified into various categories based on the energy source. Piezoelectric and triboelectric nanogenerators produce electricity through mechanical vibrations, while pyroelectric and thermoelectric nanogenerators (PyNG and ThNG) generate electricity through heat change and temperature gradient (Rushdha Hiba et al. 2025). Additionally, research is being conducted to integrate all of these effects in order to create hybrid nanogenerators that operate at their highest efficiency. The structural modifications and material selection are the two critical factors that contribute to the enhancement of a nanogenerator's performance (Wu et al. 2025). The primary objective of nanogenerator fabrication is to achieve high output performance. Various materials, including polyvinylidene fluoride (PVDF), barium titanate (BTO), lead zirconate titanate (PZT), and zinc oxide (ZnO), have been selected for PENG and PyNG due to their inherent piezoelectric and pyroelectric properties. Polymers such as polytetrafluoroethylene (PTFE), polydimethylsiloxane (PDMS), and fluorinated ethylene propylene (FEP) were employed in the fabrication of TENGs (Qian and Yang n.d.; Abhijith et al. 2025). The high production cost, toxicity, and non-biodegradability continue to be a challenge for them. A variety of sustainable materials have been investigated in order to address these issues. Biopolymers were identified as an exceptional solution for their biodegradability, cost-effectiveness, ease of processing, and reduced toxicity. Nanogenerators have been extensively manufactured using a variety of biopolymers, including cellulose, chitosan, alginate, silk fibroin, starch, and protein-based polymers. All these materials are produced from natural sources like plants, animals and microbes (Zhu et al. 2024). The Fig. 7.1 illustrates overall picture, including the origin of raw materials to high-end applications of nanogenerators.

This book chapter concentrates on the development of green composite-based piezoelectric and triboelectric nanogenerators for energy harvesting applications. The first section provides a concise explanation of the structure and energy harvesting mechanism of both PENG and TENG. Consequently, the synthesis and properties of the four major biopolymers—cellulose, chitosan, alginate, and silk fibroin—that are frequently employed in the production of nanogenerators are explained. The subsequent section provides a list of the various fabrication methods that are employed to fabricate nanogenerators. Lastly, the applications of biopolymer-based nanogenerators in various fields, including energy harvesting, healthcare monitoring, gas sensing,

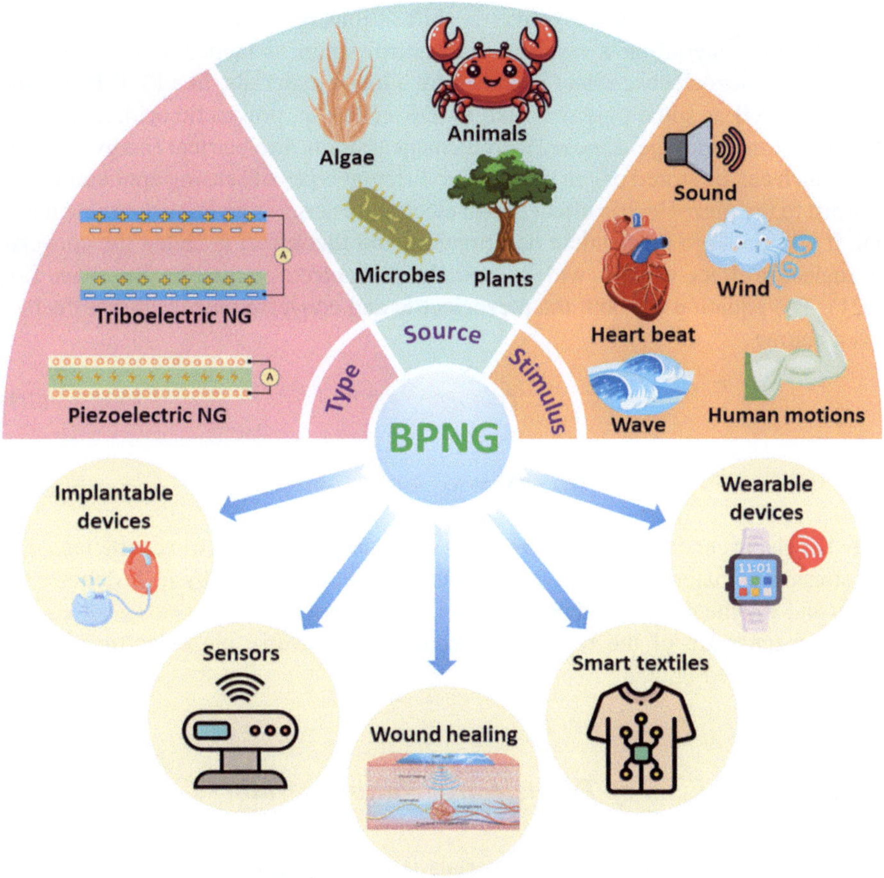

Fig. 7.1 Source, types, triggering force and applications of biopolymer-based nanogenerators (BPNGs)

and electronic devices, are discussed. This chapter aims to explain the role of biopolymers as an intriguing material for the development of sustainable nanogenerators for a foreseeable tomorrow.

7.2 Structure and Mechanism of Nanogenerators

7.2.1 Mechanism of Piezoelectric Nanogenerator

Piezoelectric nanogenerators work on the principle of the piezoelectric effect, which was first discovered by Pierre and Jacques Curie in 1880 on quartz crystals. In piezoelectric nanogenerators, the asymmetric shift of ions or charges developed inside a piezoelectric material, as a result of mechanical strain, is tapped with an external circuit. The piezoelectric effect is shown by a lot of materials, like PZT, BTO, and quartz, as well as some polymers like PVDF, cellulose, polylactic acid, and so on. When these materials get exposed to a mechanical strain, an electrical energy is generated which can be stored or can be used for different types of sensing applications. In contrast to the piezoelectric effect, when electrical energy is given to the same material, it will undergo mechanical deformation, and that effect is called the converse piezoelectric effect, which is widely applied in piezoelectric actuators (Sezer and Koç 2021). The equation for both the piezoelectric and converse piezoelectric effects is given below.

$$D = dT + \varepsilon E \tag{7.1}$$

$$X = sT + dE \tag{7.2}$$

Here the electrical displacement is given by D. d is the piezoelectric coefficient, T is the stress, ε is material permittivity, the electrical field is given by E, X denotes the strain and s is the mechanical compliance.

The mechanism of the both piezoelectric and converse piezoelectric effect is illustrated in the Fig. 7.2.

7.2.2 Mechanism of Triboelectric Nanogenerator

Triboelectrification between two materials is the main principle of a triboelectric nanogenerator. The triboelectrification property is well explained with triboelectric series where the materials are arranged on the basis of the electron-donating and electron-accepting nature. When two materials come in contact with each other or undergo frictional sliding, opposite charges are developed on the surface of each material, which results in a potential difference. This potential difference can be

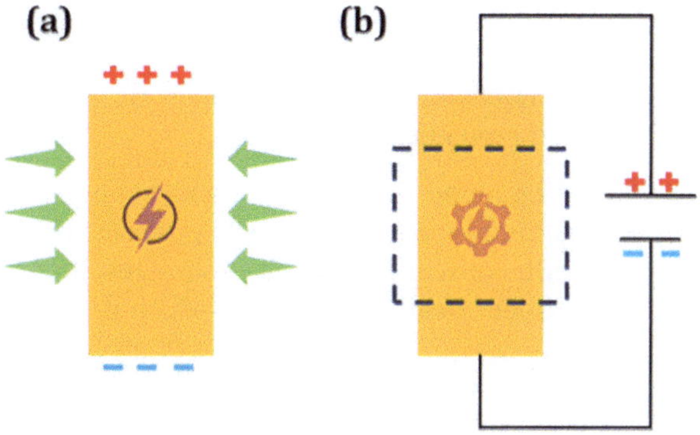

Fig. 7.2 Mechanism of piezoelectric and converse piezoelectric effect

easily tapped with two electrodes to generate electrical voltage for various applications. Depends on the structure, TENGs are classified into four types, which are contact separation mode, lateral sliding mode, single-electrode mode and finally free-standing mode, as illustrated in the Fig. 7.3.

Contact separation mode: In contact separation mode, electron transfer between two materials takes place when the two triboelectric layers come in contact with each other. The electrodes attached to each layer assist in the movement of electrons through the external circuit and thereby balance the electrical charge. The electrons flow back to their initial position when the materials get separated, and this process is repeated to generate an alternating current.

Lateral sliding mode: In this mode, an electrical voltage is generated as a result of the potential difference caused by the sliding between two triboelectric materials. Here the lateral sliding between two materials results in a periodic change in contact area between the two materials.

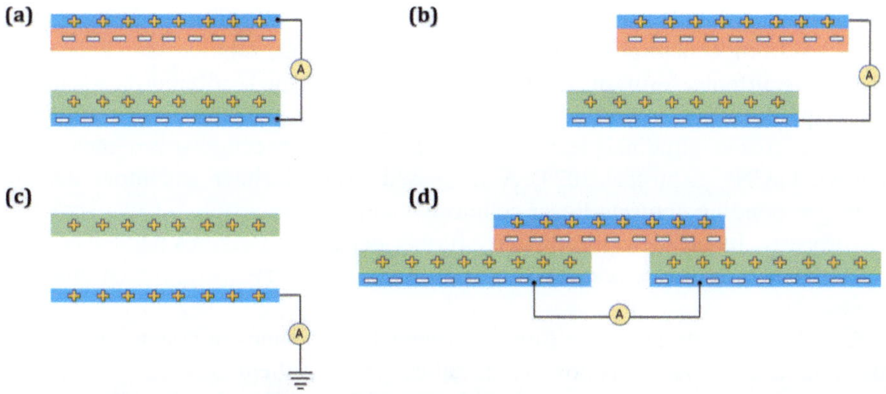

Fig. 7.3 Different modes of triboelectric nanogenerator **a** contact separation mode, **b** lateral sliding mode, **c** single-electrode mode, **d** free-standing mode

Single-electrode mode: Different from contact separation mode and lateral sliding mode, single-electrode mode has only one electrode. When the moving triboelectric layer comes in contact with the grounded electrode, transfer of charge between the electrode and ground takes place.

Free-standing mode: Free-standing mode relies on electrostatic induction and triboelectric charging between two fixed electrodes. Here, the triboelectric layer moves back and forth between the two electrodes, resulting in an asymmetric charge distribution due to electrostatic induction, which drives the charge through the external circuit.

7.3 Biopolymer Used for Nanogenerator Fabrication

7.3.1 Cellulose

Being one of the major constituents of lignocellulosic biomass, cellulose is made of anhydroglucose (AGU) monomeric units, which are linked by 1,4-β-glycosidic linkages (Freitas et al. 2024). Among different biopolymers produced on planet Earth, cellulose found its first place with an annual production of around 50 billion tons. The higher abundance, economical manufacturing, and renewability of cellulose make it a promising material for the development of a variety of products for different kinds of applications (Zhou et al. 2022). The flexible nature of cellulose toward various chemical modifications and its tunable mechanical, electrical, and physical properties make it an excellent material for a variety of applications ranging from packaging to high-end military-grade products (Annamalai et al. 2021). The inter- and intramolecular hydrogen bonding connects the cellulose chains and makes them a linear polymer. In the nanoform, cellulose is of two types: the rod-like, highly crystalline cellulose nanocrystals (CNC) and the filament-like, moderately crystalline cellulose nanofibers (CNF) (Zhang 2024; Cherian et al. 2025). The synthesis of CNF and CNC from its natural raw materials is depicted in Fig. 7.4.

Among different forms of cellulose, like I_α, I_β, and I_γ, the I_β polymorph is found to exhibit piezoelectric properties. However, when compared to piezoelectric polymers like PVDF, the longitudinal piezoelectric coefficient of pure cellulose is found to be as low as 0.4 pCN^{-1} (Zhu et al. 2024). As discussed earlier, the inter- and intramolecular hydrogen bonding between linear cellulose chains is responsible for its piezoelectric property. Its piezoelectric property further depends on various parameters like crystallite nature, elastic property, orientation of elementary fibers, etc. (Annamalai et al. 2021).

Cellulose has the ability to lose electrons when it comes in contact with other materials and become tribopositive in nature, making them an excellent replacement for petroleum-based synthetic polymers (Wang 2024). Its flexibility toward chemical modification gives the opportunity to tailor the surface charges, which will be beneficial for TENGs. Moreover, the morphological properties of the active

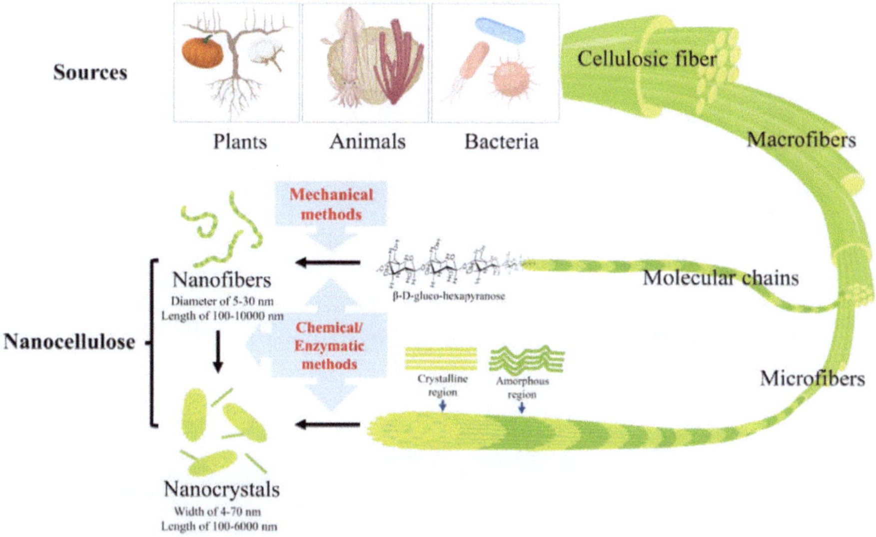

Fig. 7.4 Sources, synthesis, and structure of CNF and CNC (Zhang et al. 2023)

layer composed of cellulose materials can be easily tailored to enhance the electric output characteristics of the TENG (Fang et al. 2024). Different derivatives of cellulose, like cellulose acetate (CA), methyl cellulose (MC), carboxymethyl cellulose (CMC), bacterial cellulose, etc., are explored for piezoelectric and triboelectric energy harvesting applications (Zhou et al. 2022).

7.3.2 *Chitosan*

Chitosan is produced by the deacetylation of chitin, which is found in the exoskeletons of marine organisms like shrimp, crabs, etc., and in the cell walls of certain fungi. It is a novel linear polysaccharide composed of repeating monomeric units, glucosamine and N-acetyl glucosamine, which are linked together by β-(1 → 4)-glycosidic bonds (Torres et al. 2023). The inherent antibacterial and antifungal properties, along with the superior mechanical characteristics, non-toxicity, and biocompatible nature of chitosan, made it an excellent material for various industrial and biomedical applications (Sharkawy et al. 2020). The abundant hydroxyl (OH) and secondary amine groups (NH_2) present in chitosan allow a variety of chemical modifications, which helps to fine-tune its physical, chemical, and electrical properties (Thadathil and Velappan 2014). The extraction of chitin from natural raw materials, the synthesis of chitosan, and its structure are depicted in the Fig. 7.5.

The piezoelectric property of chitosan is due to the generation of electric dipoles resulting from unequal charge distribution created by hydroxyl and amine functional

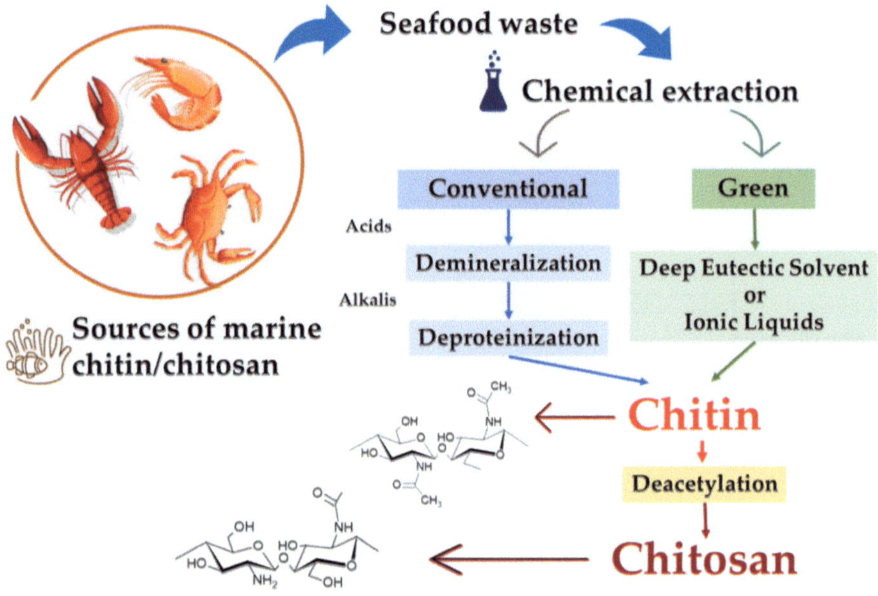

Fig. 7.5 Sources, synthesis and structure of chitosan (Teixeira-Costa and Andrade 2021)

groups. The application of a mechanical stress results in the alignment of these electric dipoles, and hence a net polarization is generated in the system. The ability of chitosan to form crystalline domains further improves its piezoelectric properties. To be specific, the crystalline domain's orientation can influence the magnitude and direction of electric charge produced by the system in response to a mechanical stress (Chakraborty et al. 2025).

The abundant primary hydroxyl and secondary amine groups present in chitosan give them the property of losing electrons when they come in contact with another material, and hence they are placed in the positive part of the triboelectric series (Gao et al. 2023). However, the low triboelectric characteristics of the chitosan are due to its low-surface charge density. This limitation can be easily covered up by the chemical modification of chitosan, which not only improves the surface charge density but also makes its surface rougher and thus improves its surface area and finally the triboelectric output characteristics (Shyju 2024).

7.3.3 Silk Fibroin

In the electronic field, silk fibroin is an emerging material due to its exceptional mechanical, thermal, and electric properties. Silk fibroin is a protein derived from the cocoons of Bombyx mori worms, also known as silk worms. The structure of silk fibroin consists of a heavy chain of 12 domains, which are made of three amino acids, namely glycine and alanine or serine, which are arranged in the sequence G-X-G-X-G-X, where G = glycine and X = alanine or serine (Silk Fibroin—an overview. ScienceDirect Topics n.d.). Being biocompatible and biodegradable in nature, these materials are widely used for the fabrication of flexible, wearable devices, mainly bioelectronics, due to their tunable mechanical and electrical properties. According to the specific applications, fibroin can be processed into different forms like hydrogels, aerogels, nanofibrous mats, composite membranes, etc. Figure 7.6 shows the extraction and solubilization of silk fibroin (Yadav et al. 2024).

The semicrystalline structure of silk fibroin is responsible for its inherent piezoelectricity. Silk materials having a piezoelectric constant of 1 pC/N can be further increased by elongating them. The piezoelectricity of these materials also depends on various environmental conditions like temperature, humidity, etc., and any change in those parameters will adversely affect its properties (Niu et al. 2022). A study on the effect of humidity on the piezoelectric coefficient of silk fibroin found that the presence of water molecules improves the internal rotation of dipoles around the peptide bonds, which results in less relaxation time and easy piezoelectric polarization (Fukada 1983). Another factor that affects the piezoelectric properties of silk fibroin is the fabrication methods. Silk-based PENG's active layer can be fabricated by a number of methods, including solvent casting, spin coating, electrospinning, etc.

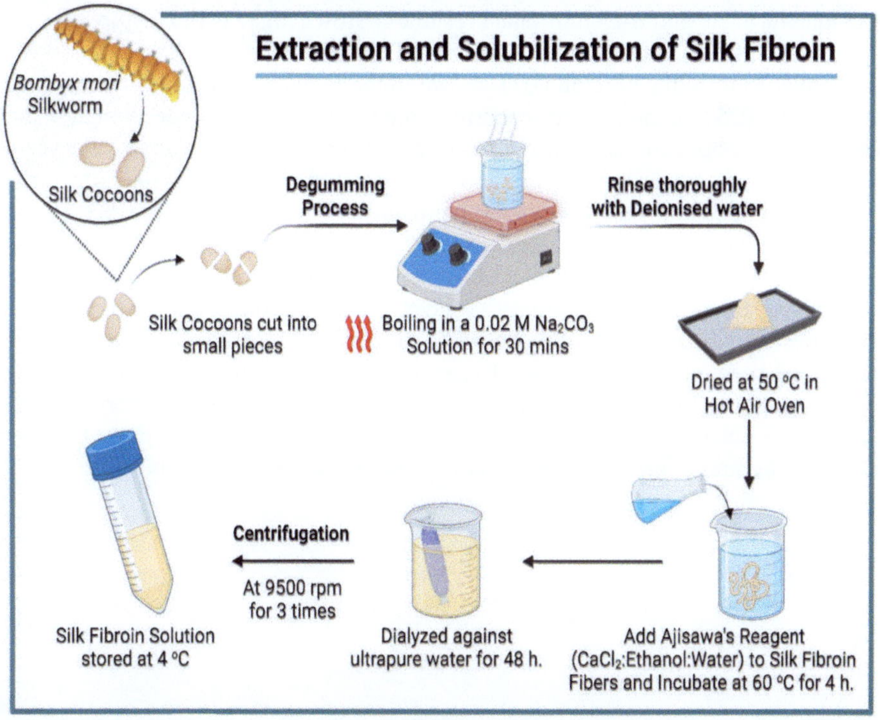

Fig. 7.6 Extraction and solubilization of silk fibroin (Yadav et al. 2024)

Among them, the electrospinning technique is found to be the superior one because it produces aligned fibers, and also the presence of a high electric field during fabrication results in the orientation of dipoles in the fibers (Sencadas et al. 2019). The introduction of organic and inorganic fillers to the fibroin matrix is another strategy to improve its piezoelectricity.

The electric output performance of silk fibroin-based TENGs depends on the mesoscale structure, design, and polarity between the tribo layers. In the triboelectric series, silk fibroin is placed in the positive part due to its easy electron-losing nature (Xu et al. 2012). The tribopositive character of fibroin is due to the presence of carboxy and amino groups present in their structure (Candido et al. 2023). The triboelectric property of fibroin-based TENG can be further enhanced by choosing the right fabricating techniques. For example, an electrospun fiber mat-based TENG made of fibroin is found to deliver higher voltage and current. Additionally, incorporation of various fillers into the fibroin matrix is also found to improve its triboelectric properties (Sencadas et al. 2019).

7.3.4 Alginate

Alginates are a naturally occurring anionic linear block copolymer linked by (1-4) β-D-mannuronate and α-L-guluronate residues. Brown algae is the major source of alginate, which is extracted with an alkali treatment followed by the addition of a crosslinker to form the precipitate as shown in the Fig. 7.7 (Lee and Mooney 2012; Khan et al. 2025). In the current scenario, more researchers are focused on alginates for electronic applications, mainly in the field of bioelectronics, due to their versatile properties like biocompatibility, low cost, less toxicity and easy manufacturing process. Alginate is widely used in the fabrication of nanogenerators, especially for triboelectric nanogenerators (Pang et al. 2025). The presence of abundant hydroxyl groups (–OH) in alginate assists in the easy donation of electrons, which makes them a promising material as a tribopositive layer (Li et al. 2023). The morphological properties of the active layer can be further modified by choosing the appropriate fabrication method. The presence of reactive functional groups in alginate makes them susceptible to chemical modifications, which improves the properties of the material. Its electrical properties can be further modified by the addition of various fillers like ceramics, carbon-based fillers, or inorganic, organic and bio fillers.

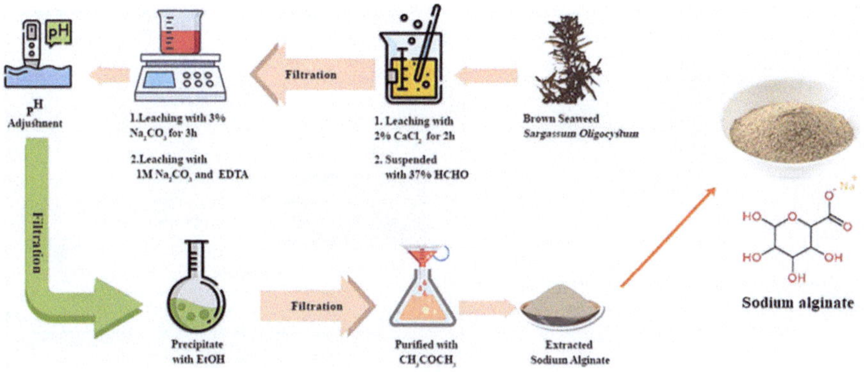

Fig. 7.7 Extraction of sodium alginate from sea weed (Khan et al. 2025)

7.4 Fabrication Methods for the Production of Nanogenerators

7.4.1 Inkjet Printing

Inkjet printing (Fig. 7.8a) is a versatile technology developed recently for the fabrication of electronic devices, mainly sensors and solar cells, with absolute controllability and accuracy. Piezoelectric print, utilizing piezoelectric actuators, and thermal print, utilizing thermal film resistors, are the two types of inkjet printing (Lorber et al. 2013). Among them, piezoelectric print is the efficient method because of its non-destructive nature toward bio samples and easy controllability on spray direction and droplet shape, whereas the higher temperature condition in thermal print destroys the sample (Kim et al. 2010). The printing of layers in piezoelectric print follows two steps: the preparation of ink and its storage in the ink tank, followed by the dispensing of the ink with the aid of an electrically driven piezoelectric actuator. In inkjet printing, the two major factors that need to be taken care of are the printhead and the ink formulation (Lee et al. 2012). In nanogenerator technology, the inkjet printing technology can be utilized for the fabrication of both the active layer and the electrode. Printing parameters like print distance and number of layers printed should be optimized for the best system. Seol et al. (2018) used polymethyl methacrylate (PMMA) and silver nanoparticles to fabricate an all-printed TENG device by inkjet printing technology. The fabricated TENG exhibited a very good energy harvesting performance when subjected to mechanical stress (Seol et al. 2018). Another research team led by Zhou et al. (2018) printed the hole-patterned Ag electrode over the PDMS layer to fabricate a transparent and flexible TENG device for realizing an autonomous raindrop-sensing technology for smart windows (Zhou et al. 2018).

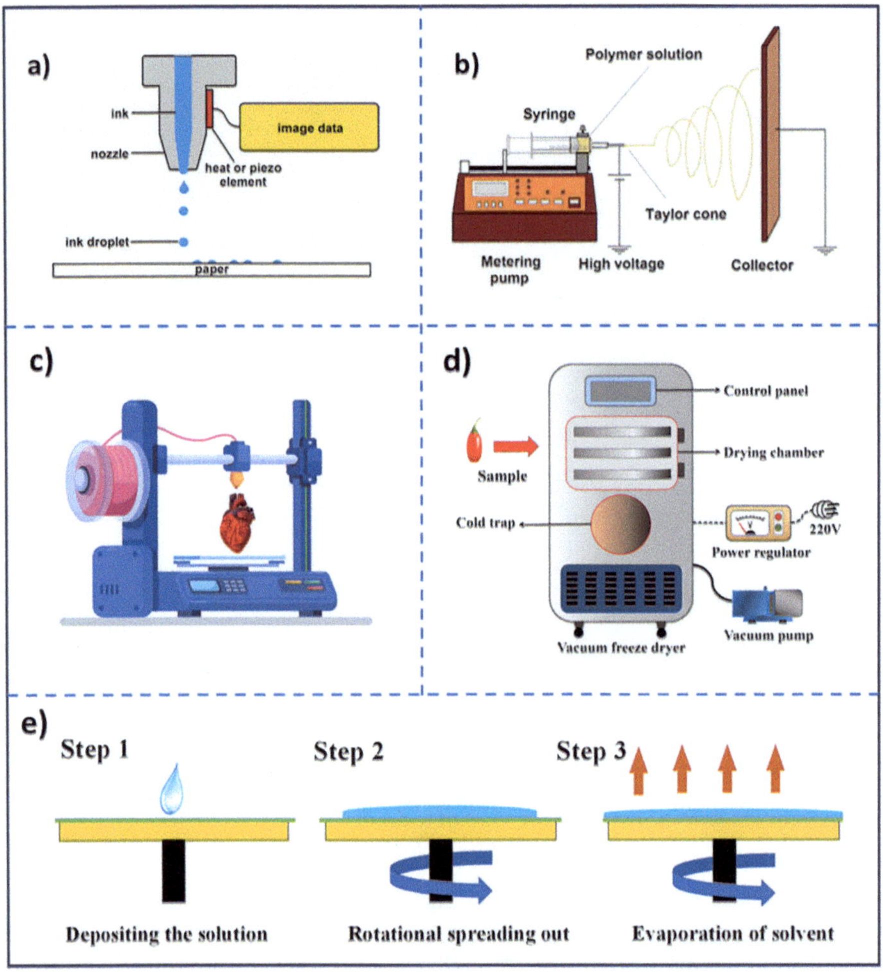

Fig. 7.8 Illustrating different fabrication methods for nanogenerator **a** inkjet printing (Samiei 2020), **b** electrospinning (Islam et al. 2019), **c** 3D printing, **d** freeze-drying (Zeng 2023), and **e** spin coating (Butt 2022)

7.4.2 Electrospinning

Electrospinning (Fig. 7.8b) is a highly efficient technology for the fabrication of nanofibrous biopolymer-based nanogenerators. A high-voltage power supply unit, syringe pump, needle and collector are the four major components of an electrospinning unit. The biopolymers that can be dissolved in a solvent or a mixture of solvents can be utilized for the fabrication of nanofibrous membranes by electrospinning technology (Bai et al. 2021). By controlling the various parameters like spinning voltage, distance between the spinneret and collector and flow rate, the width of the nanofibers can be manipulated, and continuous, uniform nanofibers can be produced (Moon et al. 2025). In the electrospinning method, the choosing of an appropriate solvent system is a crucial part because it should be able to dissolve the biopolymers, which are crystalline in nature. Moreover, the solvent should be easily vaporizing as soon as it ejects from the spinneret (Shi et al. 2025). For different biopolymers, different solvents are used, which include N-methyl morpholine N-oxide (NMMO)/H_2O (Han et al. 2008), ionic liquids (Ahn et al. 2012), lithium chloride/N,N-dimethylacetamide (DMAc) (Ao et al. 2017), sodium hydroxide (NaOH)/urea for cellulose, trifluoroacetic acid (TFA) (Shiffman and Schauer n.d.) or aqueous acetic acid (Schiffman and Schauer 2007) solvent for chitosan, formic acid (Wang et al. 2016) and 1,1,1,3,3,3-hexafluoro-2-propanol (Jin et al. 2002) for silk fibroin, etc. While preparing the electrospinning solution, the concentration of the biopolymers should be optimized because the viscosity of the solution plays an important role in forming continuous and uniform nanofibers (Mahanty et al. 2024). The molecular structure as well as the arrangement of biopolymers within the nanofibers affects their properties. In order to further improve the properties of the nanofibers, synthetic as well as biobased active components are introduced during the solution preparation (Li et al. 2025). This technique allowed the fabrication of nanofibrous membranes with different structures, including aligned fibers, hollow fibers, hierarchical structures, core-sheath structures and nanofibers coated by nanoparticles (Akhila et al. n.d.).

7.4.3 3D Printing

Recently, 3D printing (Fig. 7.8c), also called additive manufacturing, has gained significant attention in the biomedical field, mainly for the fabrication of flexible and wearable bioelectronic devices, including micro sensors, electronic skin patches, etc. 3D printing technology makes use of a computer-aided (CAD) model to fabricate a three-dimensional prototype (Dutta et al. n.d.). In the nanogenerator field, 3D printing is widely used for the fabrication of complex origami-like active layers and electrodes with enormous surface areas for improving the output characteristics. Unlike traditional fabrication methods such as spray coating, screen printing, patterning, or drop casting, 3D printing allows for the detailed patterning of polymer ink, resulting in enhanced properties (Yang et al. 2020). Biopolymers like cellulose, chitosan, and silk fibroin are 3D printed to make diverse materials. However, materials fabricated using biomaterials by conventional 3D printing technology have high porosity; they lack surface smoothness and need post-fabrication treatments like heat treatment to improve their mechanical properties. In order to overcome these constraints, the filaments are reinforced with nanomaterials, or a filament made of blends of more than one material is used (Dutta et al. n.d.). Currently, researchers are focusing on printing biomaterials for tissue engineering, drug delivery, and implantable devices, as well as nanogenerators. Among different biopolymers, nanocellulose is widely explored for the fabrication of nanogenerators due to its scalability and efficiency compared to conventional techniques. 3D-printed nanocellulose films have a very high mechanical strength of 139.94 MPa compared to normal solvent-casted nanocellulose film (66.29 MPa). Also, the nanogenerator made out of this film exhibits the ability to produce high voltage toward various human motions like walking and running (Cho et al. 2025). Quian et al. (2019) printed an aerogel-like structure with cellulose nanofibers using 3D printing technology to sense mechanical vibrations as well as to detect the real-time change in humidity. The surface roughness and contact area of the fabricated TENG were sufficient to harvest enough electrical energy for lighting 88 light-emitting diodes (LEDs) and to detect the humidity with a response ratio of 5:1 (Qian et al. 2019).

7.4.4 Freeze-Drying

The freeze-drying (Fig. 7.8d) technique is commonly used for the fabrication of aerogel structures with very high porosity and surface area, which are beneficial for nanogenerator applications. Freeze-drying, also known as lyophilization, consists of two steps. The first step is the freezing of the sample solution, followed by sublimation of the material under high vacuum. As a result of this, the solvent, specifically water, which converted to ice in the initial step, gets converted directly to vapor phase, leaving a porous structure to the material (Mahmud and Rahman 2024). Maintaining a very low temperature during the freeze-drying technique is the crucial part of freeze-drying because a small change in the temperature will result in spoilage of the sample. The freeze-drying technique is considered one of the best methods for the fabrication of nanogenerators because of the porous nature of the samples, aiding improvement in surface area and thereby the output characteristics. Among different polymers, cellulose is the most common polymer used for the fabrication of nanogenerators. Song et al. (2022) fabricated a filler-free triboelectric nanogenerator utilizing a freeze-drying technique to make TEMPO-oxidized cellulose nanofibrils as the tribopositive layer. The technique imparts a very high surface area of about 134.1 m^2/g, which results in a superior power density of 156 Mw/m^2 (Song et al. 2022). The freeze-drying technique is also used for the fabrication of piezoelectric nanogenerators. A combination of a freeze-dried cellulose nanofibrous aerogel layer with a layer of PDMS coating is found to produce a very high piezoelectric output voltage of 60.1 V with a power density of 6.3 mW/cm^3. This nanogenerator maintained superior stability and durability and was able to light up 19 LEDs and charge a capacitor to 3.7 V (Zheng et al. 2016).

7.4.5 Spin Coating

The spin-coating technique (Fig. 7.8.e) is used for the fabrication of thin-film membranes by applying a high-speed spinning action to the polymer solution dropped on the surface of a substrate. In this technique the thickness of the membrane can be controlled by adjusting the spinning speed of the substrate placed on a vacuum chuck connected to a rotor. The uniformity of the material coated on the substrate makes it a promising method for the active layer fabrication as well as the deposition of electrodes (Ahmed et al. 2018). A study by Peng et al. (2017) reveals that the high centrifugal force during spinning imparts a shear force to the polymer solution during spin coating, which results in parallel orientation of cellulose. They had spin-coated the PDMS/cellulose nanocrystal flake composite and found out that the output performance was enhanced by 10 times compared to the pristine PDMS composite (Peng et al. 2017). Chen et al. (2022) used this versatile technique to fabricate a cellulose-based TENG, where a uniform dispersion of cellulose nanofiber and MXene were spin-coated on PDMS matrix as the tribopositive layer. The TENG was able to produce an output voltage of 24.9 V with an output power density of 1.2 mWm^{-2} (Yang et al. 2022). Another work by Alghamdi et al. (2025) used this technique to deposit cellulose nanocrystals over graphene-coated PET substrate to formulate a single-electrode TENG for harvesting electrical energy from piano playing motion. Spin coating allows them to precisely control the layer thickness; were they studied the effect of layer thickness on the output performance of the TENG (Alghamdi et al. 2025).

7.5 Applications of Biopolymer-Based Nanogenerators

7.5.1 Healthcare Monitoring

Conventional healthcare monitoring systems experience a lot of limitations, including their limited mobility and lack of comprehensive, real-time data analysis. A large number of research projects are happening around the world, focusing on the development of healthcare monitoring devices. Nanogenerators are a promising candidate for real-time health monitoring because they can produce electrical signals from the body's feeble mechanical vibrations. Integrating nanogenerators into different body parts like the wrist, knee, abdomen, chest, throat and face allows the monitoring of joint movements, respiration rate and heartbeat, which will be helpful for diagnosing various conditions like arthritis, pulmonary diseases and cardiovascular diseases. TENGs consist of supramolecular nanofibers of sodium alginate in combination with polyacrylamide and polyaniline, which produced a very high voltage of 166 V. The excellent sensitivity of this TENG toward stimuli made it an excellent device for monitoring different human motions like finger bending, elbow bending, forehead frowning and knee bending (Zhang et al. 2024). Placing the nanogenerators in knee joints can collect various knee movements which will be helpful for the diagnosis of diseases like arthritis. Techniques like Tai Chi involving different body movements that are performed to alleviate various diseases like cancer, diabetes, and depression have a high risk for knee injuries due to incorrect postures. A self-powered wireless monitoring system that can distinguish correct and incorrect postures during Tai Chi has been realized with sodium alginate-based TENG. The TENG produces different output voltages for correct and incorrect postures due to the different angles, and these output voltages are collected and transmitted to a supercomputer equipped with special software to analyze the knee movement, which reduces the risk of knee injuries as shown in the (Fig. 7.9a–d). The TENG was also able to produce an output voltage of 49 V with 108.8 μW peak power (Feng et al. 2025). Han et al. reported a novel hydrogel-type TENG capable of detecting human movements like walking, running, and finger movements, as well as facial expressions and recognition of different voices. The TENG was made of a composite consisting of sodium alginate in combination with polyvinyl alcohol and polyacrylamide and KCl as a conductive filler. The impressive sensitivity and robust durability of this TENG with a gauge factor of about 6.78 and prompt response time of 268 ms make it superior for wearable electronics (Han et al. 2024). In another work sodium alginate was used to fabricate an antibacterial TENG to act as a self-powered tactile sensor for various physiological actions like pressure mapping, human–machine interfacing and human motion monitoring. The addition of glycerol to the system improves its flexibility and makes the composite transparent. Beyond tactile sensing, the TENG produces an output voltage of 53 V with a power of 4 μW, which is sufficient for small wearable electronic devices (Li et al. 2023).

Wang et al. designed a humidity-resistant TENG for respiratory monitoring applications using amino-oxidized sodium alginate as depicted in the Fig. 7.9e. The output performance and hydrophilicity of the alginate matrix were enhanced by its modification with amino groups. The TENG maintains its output performance when the relative humidity is increased from 15 % to 85 %. Normal breathing, deep breathing and rapid breathing were easily distinguished with this novel TENG by observing its output signals, where normal breath produces 4 output peaks in 10 s, whereas rapid breath gives 6 peaks in 10 s (Fig. 7.9f). The property of this TENG to distinguish these three types of breathing made it a diagnostic tool for various respiratory diseases. For example, the highest breath rate (38.2 breaths/min) recorded by this TENG implies the patient has pneumonia (Wang et al. 2025).

Different types of bioindicators, like heartbeat and pulse wave, exist to understand various conditions associated with cardiovascular diseases. However, the current diagnostic tools involving sophisticated machines remain as a hurdle for common people for regular checkups. The ability of nanogenerators to produce electrical signals from minute physiological movements has been exploited to develop a real-time autonomous cardiac monitoring system (Bhaduri et al. n.d.). Maity et al. (2023) developed a piezoelectric nanogenerator capable of precisely detecting heartbeat as well as respiratory movements without using batteries or any kind of energy storage devices. The PENG consists of a cellulose nanocrystal active layer fabricated by a well-known 3D printing technique. When attached to a face mask, it produces different voltage signals corresponding to inspiration and expiration, which is helpful for the detection of various respiratory diseases (Maity et al. 2023). The performance of the biopolymer-based nanogenerators can be further improved by various techniques, like the incorporation of different organic and inorganic fillers and by its

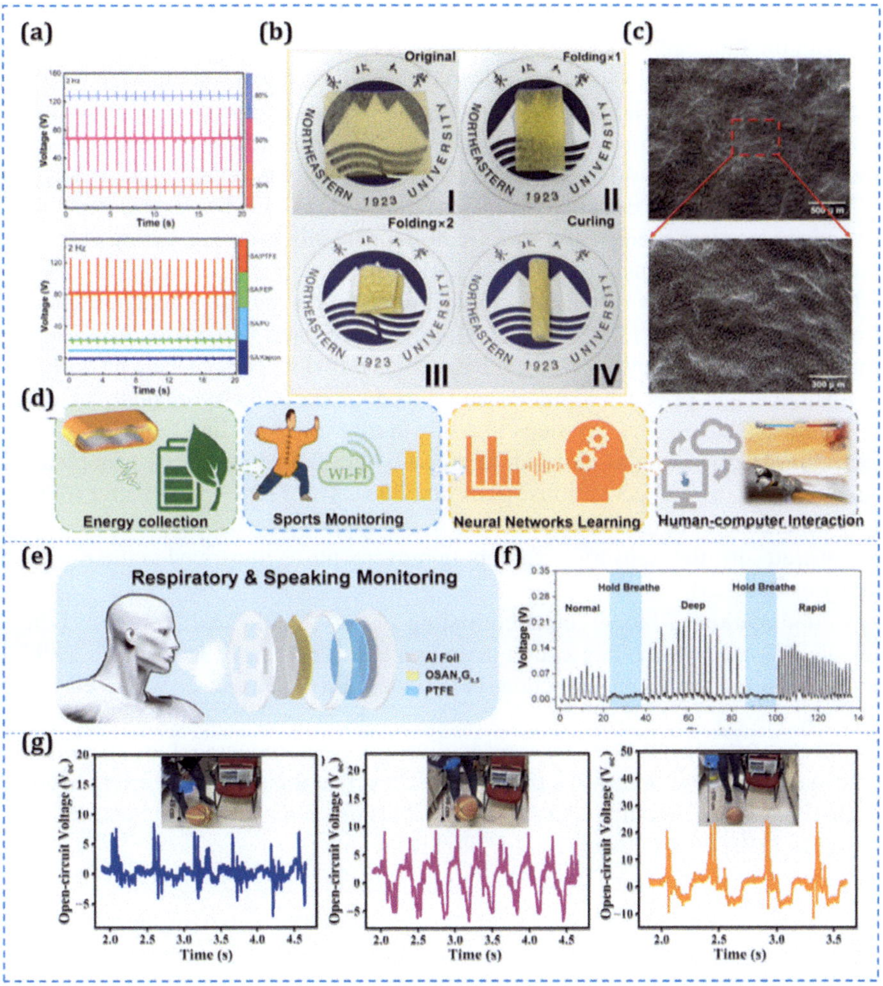

Fig. 7.9 **a** Voltage versus time curve for alginate with different negative materials, **b** optical images of the fabricated alginate film, **c** SEM image of alginate film, **d** application of NG in different fields like energy harvesting, sports monitoring and human machine interaction (Feng et al. 2025), **e** diagram of the respirators monitoring system, **f** voltage time curve for different respiratory types (Wang et al. 2025), **g** voltage versus time curve for basketball dripping from low, medium and high height (Maity et al. 2025)

modification by different physical and chemical methods. Modification of cellulose paper with polyethylene oxide imparts excellent triboelectric properties to the nanogenerator. The TENG generated out of it produces a very high output voltage of 222.1 V with a power density of 217.3 mW m^{-2}. The high performance of this TENG toward feeble movements like the movement of eye muscles, heartbeat, respiration and pulse gives it the high potential to be used in healthcare devices (Lin et al. 2023).

The complex synthesis procedures and toxicity are the two main problems associated with the fabrication of energy harvesting devices. These unavoidable problems can be easily sorted out by using the materials in their natural form. PENG, consisting of untreated onion peel, is an example of this. Onion peel has abundant cellulose fibers in it which are aligned in a proper direction and hence exhibits a piezoelectric coefficient of 2.8 pC/N. The PENG fabricated out of it produces an output voltage of 18 V along with a power density of 1.7 $\mu W/cm^2$. Also, the device was capable of producing electrical energy from various mechanical vibrations, including heartbeat, voice and throat movement (Maiti et al. 2017).

Monitoring of sports activities is another application of nanogenerators. Integration of both PENG and TENG with different body parts will allow the real-time monitoring of various physical activities associated with sports. This data will help the team manager or coach to understand the condition of the sportsman, including his stamina, to take necessary steps like substitution in sports like football. A triboelectric layer consisting of chitosan reinforced with 2 wt% graphene nanoplates is found to accurately detect various sports activities. The TENG constructed with the layer generated an output voltage of 166 V and a power density of 44 mW/m^2 that demonstrates its high sensitivity for taping movements as illustrated in the Fig. 7.9g (Maity et al. 2025).

A combination of biopolymers consisting of poly (lactic acid)/chitosan/sodium alginate was used for the fabrication of an implantable pressure sensor with 99.99% antibacterial properties and excellent biocompatibility. The sensor was used to tap electrical energy from biomechanical movements of animals and humans by the triboelectric effect, and as a real-time application, it was integrated in rats to detect abnormal respiration in rats. An interesting fact is that this sensor is easily absorbed within 5 days and becomes a part of the body, causing no harm to the animal body as well as to the environment (Li et al. 2023).

7.5.2 Gas Sensing

As a result of development, problems like air pollution are increasing day by day. Air pollution is caused by a number of pollutants like dust particles, toxic gases and microbes. Ammonia gas emitted from nature as well as from industries, agricultural sites and so on are causing serious health issues for mankind, including respiratory disease and blindness. The presence of ammonia in very low concentrations can itself cause serious health issues. In this scenario, the development of ammonia sensors capable of detecting ammonia at very low concentrations is necessary. A variety of materials like metal oxides, conducting polymers, etc., are widely used for the fabrication of ammonia sensors. But the dependency of these sensors on external power sources like batteries limits their applications. Current research is focusing on utilizing nanogenerators as self-powered ammonia sensors. For that purpose, both synthetic and bio-based materials are used for the nanogenerator production. Among different biopolymers used for nanogenerator fabrication, cellulose is widely exploited for gas sensing applications due to its easy production and porous nature. Conducting polymers like PANI are coated on a cellulosic paper to develop a TENG that is highly sensitive toward ammonia gas. The system has a hierarchical structure that helps excellent permeability toward ammonia gas, and the availability of a large number of adsorption sites made the system highly sensitive. In the TENG, PANI acts as the tribomaterial as well as the electrode, and protonation and deprotonation of PANI imparts an impedance to the system that enhances the sensing characteristics. The nanogenerator produces an output voltage of 23 V and a current of 0.17 μA and has a minimum detection limit of 100 ppb and a maximum limit of 500 ppm toward ammonia gas (Yang et al. 2023). Zhang et al. (2022) used a modified cellulose nanofibril (triethoxy-1H,1H,2H,2H-tridecafluoro-n-octylsilane cellulose nanofiber) to build a layered TENG for autonomous ammonia gas sensing. The modification improves the adsorption sites on cellulose, and the reinforcement of this matrix with Ti_3C_2Tx MXene results in a high ammonia sensitivity with a 12:14 s response: recovery ratio and a 37.6% selectivity response. The TENG was capable of accurately detecting concentration changes in ammonia gas within the 10-120 ppm range and transferring the data to software, thereby enabling real-time monitoring of ammonia gas (Zhang et al. 2022). In another work by Sardana et al. (2022), the electrospinning technique was utilized to fabricate a nanofibrous TENG consisting of cellulose acetate and MXene. The TENG was capable of generating sufficient power of 1361 mW/m^2. The incorporation of TiO_2 into the above system enhances its activity and exhibits superior sensitivity toward ammonia gas as shown in the Fig. 7.10a. With a response: recovery ratio of 76:62 s, the gas sensor provides excellent sensitivity and reproducibility toward ammonia sensing in a range of 1–100 ppm. The integration of this sensor with an LED using an equivalent circuit aids the realization of a self-powered real-time detection of ammonia leakage in different scenarios (Sardana et al. 2022).

7.5.3 Electronic Devices

Human–machine interface is another hot area where the nanogenerator technology can be implemented. Different types of interfaces like mouse, touch screens, keyboards, etc., exist to exchange data between humans and machines. These devices can be made independent of an external power source by integrating nanogenerators that can generate electrical energy from mechanical movements. For instance, a TENG developed with a bacterial cellulose/Ag NWs/BTO composite film has the ability to generate an electrical energy of 25 V when integrated into a mouse and keyboard as illustrated in the Fig. 7.10b. Moreover, the output voltage of the TENG was found to increase with click speed and give a maximum of 55 V (Yu et al. 2021). Silk fibroin is another biopolymer used for the fabrication of nanogenerator-based portable electronic devices. Liu et al. (2022) made a silk fibroin-based single-electrode TENG that acts as a keyboard less input system. The TENG integrated into 5 fingers senses the motion of fingers and generates different electrical signals for

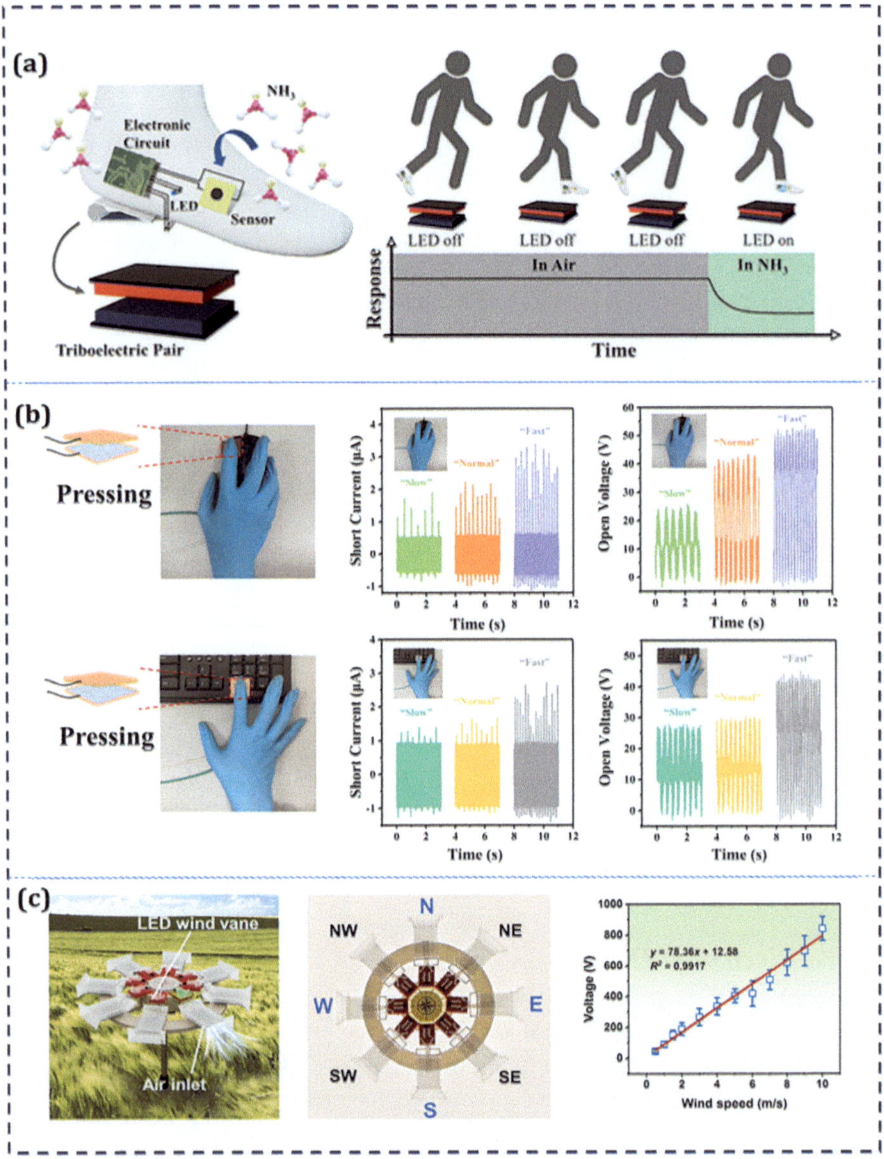

Fig. 7.10 **a** Self-powered ammonia sensor based on cellulose/Mxene TENG (Sardana et al. 2022), **b** short-circuit current and open-circuit voltage generated by BC/AgNWs/BTO PENG on mouse and keyboard operation (Yu et al. 2021), **c** Wind energy harvester TENG and its output characteristics (Dai et al. 2022)

different finger positions. With the help of a printed circuit board, the signals were processed and decoded to English alphabets with a superior accuracy of 92% (Liu et al. 2022).

7.5.4 Energy Harvesting

From our day-to-day activities, like walking, running, jumping and so on, a lot of mechanical energy gets wasted. The integration of nanogenerators in different things like shoe soles, fabrics, smartwatches and so on helps to generate electrical energy from these mechanical vibrations. For instance, Yao et al. (2016) developed a cellulose nanofiber-based TENG to make fiber optic plates. The film reinforced with cardboard fibers produced a high output voltage of 30 V with a short-circuit current of 90 μA under walking. It was capable of lighting 35 LEDs, and the fiber optic plate exhibited a high charge transfer efficiency of 98% (Yao et al. 2016). In another work, rabbit hair was used to integrate with cellulose nanofiber to fabricate an aerogel-type TENG. The TENG, when placed in a shoe sole, generates an output voltage of 58 V during normal walking speed (Mi et al. 2018).

Different types of natural energy sources, like human movement, water flow, wind energy and so on, are utilized to harvest electrical energy using nanogenerators. For example, natural wind or wind-blown using the mouth was used to generate an output voltage of 130 V with the aid of a TENG composed of cellulose nanocrystals. The TENG was used to color print various substrates without relying on external power sources (Chen et al. 2018). Wind-powered TENGs are also used for the real-time monitoring of various environmental parameters like wind speed, wind direction, etc., which are crucial for sectors like agriculture, fishing, etc. A slit-effect-based TENG developed by Dai et al. (2022) is an example of this, where they used hydroxyethyl ethyl cellulose to fabricate the active layer for the TENG. The output characteristics of the wind energy harvester are given in Fig. 7.10c (Dai et al. 2022) (Table 7.1).

7.6 Conclusion

The recent developments achieved by biopolymer-based nanogenerators are critically analyzed in this chapter. This chapter mainly focused on biopolymer-based piezoelectric and triboelectric nanogenerators for sustainable energy harvesting applications. BPNGs are emerged as an excellent candidate for the green and flexible energy harvesting technologies due to their biocompatibility, biodegradability and environment friendly nature. From the literature it is clear that in the past few years more efforts are given on biopolymers for the fabrication of nanogenerators. In that, more focus is given to triboelectric nanogenerators due to their high electric output characteristics. However, it is observable that the biopolymer-based nanogenerators are still in their developing stage. Future research should solve the various problems faced by

Table 7.1 Overview of different BPNGs including the materials and techniques used for fabrication, output characteristics and real-world applications

Type of NG	Biopolymer	Technique	Voltage (V)	Current	Application	Refs.
TENG	Sodium alginate	Casting	166	–	Human motion detection	Zhang et al. (2024)
TENG	Sodium alginate	Casting	49	–	Sports injury monitoring	Feng et al. (2025)
TENG	PVA/alginate	Casting	50	–	Strain sensing	Han et al. (2024)
TENG	Sodium alginate	Casting	53	0.37 μA	Tactile sensing	Li et al. (2023)
TENG	Sodium alginate	Casting	20.19	0.54 μA	Respiratory monitoring	Wang et al. (2025)
PENG	Cellulose nanocrystals	3D printing	15	0.2 μA	Cardiorespiratory monitoring	Maity et al. (2023)
TENG	Cellulose	Freeze-drying	222.1	4.3 μA	Human health monitoring	Lin et al. (2023)
PENG	Onion skin	–	18	166 nA	Health care monitoring	Maiti et al. (2017)
TENG	Chitosan	Casting	166.25	13.56 μA	Sports monitoring	Maity et al. (2025)
TENG	PLA/Sodium alginate/ Chitosan	Casting	9.2	–	Respiratory monitoring	Li et al. (2023)
TENG	Cellulose	Casting	23	0.17 μA	Ammonia sensing	Yang et al. (2023)
TENG	Cellulose	Vacuum filtration	84	–	Ammonia sensing	Zhang et al. (2022)
TENG	Cellulose acetate	Electrospinning	140	92 μA	Ammonia sensing	Sardana et al. (2022)
TENG	Silk fibroin	Coating	4	18 nA	Human computer interaction	Liu et al. (2022)
TENG	Hydroxy ethyl cellulose	Casting	600	–	Environmental sensing	Dai et al. (2022)

this field for their practical applications and large-scale production. These includes the processibility of materials, stability of TENG etc. In future researches, focus should be given to optimize the processing techniques and parameters for achieving scalable, cost-effective and compatible BPNGs. It is also important to give focus on developing BPNGs with long-term durability and environmental resilience. When comes to applications, BPNGs hold vast potential in different areas like biomedical sensors, wearable and flexible devices, hybrid energy harvesting systems and environmental sensing modules. More researches should be focused on coupling the BPNGs with other energy harvesting technologies like thermoelectric NGs, solar-cell, etc. to realize hybrid energy harvesting systems with improved output characteristics. Additionally combining the NGs with energy storage devices will helps to achieve self-powered systems for versatile applications. Further modifications in the design and integration of biofillers with multiple properties are essential for the multifaceted applications of nanogenerators.

References

Abhijith V, Akhila B, Pramod K, Mohanty TR, Sivakumar C, Ramakrishnan S, Thomas S, Sreekala MS (2025) Recent advances in metal oxide reinforced polymer composites for efficient piezoelectric energy harvesting applications. Applied Energy 397:126335. https://doi.org/10.1016/j.apenergy.2025.126335

Ahmed U, Alizadeh M, Rahim NA, Shahabuddin S, Ahmed MS, Pandey AK (2018) A comprehensive review on counter electrodes for dye sensitized solar cells: A special focus on Pt-TCO free counter electrodes. Solar Energy 174:1097–1125. https://doi.org/10.1016/j.solener.2018.10.010

Ahn Y, Hu D-H, Hong JH, Lee SH, Kim HJ, Kim H (2012) Effect of co-solvent on the spinnability and properties of electrospun cellulose nanofiber. Carbohydr Polym 89:340–345. https://doi.org/10.1016/j.carbpol.2012.03.006

Akhila B, Abhijith V, Sreedharan M, Ravindran L, Sathian A, Thomas S, Sadasivan SM (n.d.) Innovations in core–shell electrospinning: A comprehensive review in recent advances of core–shell electrospun polylactic acid nanocomposite fibers for potential biomedical applications. ACS Biomater Sci & Eng. https://doi.org/10.1021/acsbiomaterials.5c00194

Alghamdi MS, Morgan JJ, Walsh K, Shin DW, Nigmatullin R, Saadi Z, Routledge J, Neves AIS, Russo S, Eichhorn SJ, Craciun MF (2025) Triboelectric nanogenerator based on cellulose nanocrystals and graphene for energy harvesting from piano playing motion. Nano Energy 138:110816. https://doi.org/10.1016/j.nanoen.2025.110816

Annamalai PK, Nanjundan AK, Dubal DP, Baek J-B (2021) An overview of cellulose-based nanogenerators. Adv Mater Technol 6:2001164. https://doi.org/10.1002/admt.202001164

Ao C, Yuan W, Zhao J, He X, Zhang X, Li Q, Xia T, Zhang W, Lu C (2017) Superhydrophilic graphene oxide@electrospun cellulose nanofiber hybrid membrane for high-efficiency oil/water separation. Carbohydr Polym 175:216–222. https://doi.org/10.1016/j.carbpol.2017.07.085

Bai L, Li Q, Yang Y, Ling S, Yu H, Liu S, Li J, Chen W (2021) Biopolymer nanofibers for nanogenerator development. Research 2021. https://doi.org/10.34133/2021/1843061

Bhaduri A, Ha T-J (n.d.) Biowaste-derived triboelectric nanogenerators for emerging bioelectronics. https://doi.org/10.1002/advs.202405666

Butt MA (2022) Thin-film coating methods: A successful marriage of high-quality and cost-effectiveness—a brief exploration. Coatings 12:1115. https://doi.org/10.3390/coatings12081115

Candido ICM, da Oliveira GS, Ribeiro SJL, Cavicchioli M, Barud HS, Silva LG, de Oliveira HP (2023) PVA-silk fibroin bio-based triboelectric nanogenerator. Nano Energy 105:108035. https://doi.org/10.1016/j.nanoen.2022.108035

Chakraborty S, Debnath S, Mahipal Malappuram K, Parasuram S, Chang H-T, Chatterjee K, Nain A (2025) Flexible and robust piezoelectric chitosan films with enhanced bioactivity. Biomacromolecules 26:1128–1140. https://doi.org/10.1021/acs.biomac.4c01464

Chen B, Yang N, Jiang Q, Chen W, Yang Y (2018) Transparent triboelectric nanogenerator-induced high voltage pulsed electric field for a self-powered handheld printer. Nano Energy 44:468–475. https://doi.org/10.1016/j.nanoen.2017.12.026

Cherian RM, Antony T, Varghese RT, Akhila B, Abhijith V, Vipina Vinod TN, Kargarzadeh H, Radhakrishnan EK, Chirayil CJ, Thomas S (2025) A sustainable one-step approach for the functionalized cellulose nanocrystal production using recyclable organic acids: Exploring structure-property dynamics. Int J Biol Macromol 305:141192. https://doi.org/10.1016/j.ijbiomac.2025.141192

Cho HC, Latif M, Jiang Y, Kim J (2025) Environment-friendly triboelectric nanogenerator based on 3d-printed nanocellulose films for human motion monitoring. ACS Appl Energy Mater 8:267–275. https://doi.org/10.1021/acsaem.4c02319

Dai S, Li X, Jiang C, Zhang Q, Peng B, Ping J, Ying Y (2022) Omnidirectional wind energy harvester for self-powered agro-environmental information sensing. Nano Energy 91:106686. https://doi.org/10.1016/j.nanoen.2021.106686

Dutta SD, Ganguly K, Randhawa A, Patil TV, Kim H, Acharya R, Lim K-T (n.d.) 3D Printable hydrogel bioelectronic interfaces for healthcare monitoring and disease diagnosis: Materials, design strategies, and applications. https://doi.org/10.1002/admt.202301874

Fan F-R, Tian Z-Q, Lin Wang Z (2012) Flexible triboelectric generator. Nano Energy 1:328–334. https://doi.org/10.1016/j.nanoen.2012.01.004

Fang S, Ji X, Wang H, Jiang H, Gao M, Liu H, Liu Y, Cheng B (2024) Cellulose-based green triboelectric nanogenerators: materials, form designs, and applications. J Mater Chem A 12:9322–9344. https://doi.org/10.1039/D3TA07294K

Feng Q, Xie Z, Wen Y, Cheng Z, Zhang M, Wang Y, Liu D, Cao Y, Mao Y, Zhao C (2025) An eco-friendly, sodium alginate degradable conformal triboelectric nanogenerator for self-powered sensing and real-time injury monitoring. Sustain Mater Technol 43:e01262. https://doi.org/10.1016/j.susmat.2025.e01262

de Freitas FA, de Sá Barros S, Saron C, Lobo WV, dos Santos RIL, Las-Casas B, Yupanqui-Mendoza SL, de Souza LKC (2024) Lignocellulose characterization and exploitation. In: Rahimpour MR (ed) Encyclopedia of renewable energy, sustainability and the environment, 1st edn. Elsevier, Oxford, pp 565–576. https://doi.org/10.1016/B978-0-323-93940-9.00022-0

Fukada E (1983) Piezoelectric properties of biological polymers. Q Rev Biophys 16:59–87. https://doi.org/10.1017/S0033583500004923

Gao C, Tong W, Liu S, Wang X, Zhang Y (2023) Fully degradable chitosan–based triboelectric nanogenerators applying in disposable medical products for information transfer. Nano Energy 117:108876. https://doi.org/10.1016/j.nanoen.2023.108876

Han D, Wang G, Xu X, Chen J, Lu M, Liu X, Zhang L, Lai L (2024) Ultra-stretchable, self-healing, bonding, and skin-inspired conductive triple network hydrogel for wearable strain sensors and friction nanogenerators. Polymer 305:127169. https://doi.org/10.1016/j.polymer.2024.127169

Han SO, Son WK, Youk JH, Park WH (2008) Electrospinning of ultrafine cellulose fibers and fabrication of poly(butylene succinate) biocomposites reinforced by them. J Appl Polym Sci 107:1954–1959. https://doi.org/10.1002/app.26643

Islam MS, Ang BC, Andriyana A, Afifi AM (2019) A review on fabrication of nanofibers via electrospinning and their applications. Discover Applied Sciences. https://link.springer.com/article/10.1007/s42452-019-1288-4#rightslink (accessed 17 June 2025)

Jeong J, Jo E, Choi JA, Kang Y, Pyo (2025) Structural design strategies of triboelectric nanogenerators for omnidirectional wind energy harvesting. Micro Nano Syst Lett. https://link.springer.com/article/10.1186/s40486-025-00224-6 (accessed 18 April 2025)

Jin H-J, Fridrikh SV, Rutledge GC, Kaplan DL (2002) Electrospinning Bombyx mori Silk with Poly(ethylene oxide). Biomacromolecules 3:1233–1239. https://doi.org/10.1021/bm025581u

Khan MAA, Hasan MM, Hossain MK, Adhikery D, Hakim M, Mohanta LC, Sharif ASM, Sarker AK (2025) Extraction and characteristic properties analyses of sodium alginate derived from the *Sargassum oligocystum* brown seaweed alga of the Bay of Bengal. Next Materials 6:100417. https://doi.org/10.1016/j.nxmate.2024.100417

Kim JD, Choi JS, Kim BS, Chan Choi Y, Cho YW (2010) Piezoelectric inkjet printing of polymers: Stem cell patterning on polymer substrates. Polymer 51:2147–2154. https://doi.org/10.1016/j.polymer.2010.03.038

Lee BK, Yun YH, Choi JS, Choi YC, Kim JD, Cho YW (2012) Fabrication of drug-loaded polymer microparticles with arbitrary geometries using a piezoelectric inkjet printing system. Int J Pharm 427:305–310. https://doi.org/10.1016/j.ijpharm.2012.02.011

Lee KY, Mooney DJ (2012) Alginate: Properties and biomedical applications. Prog Polym Sci 37:106–126. https://doi.org/10.1016/j.progpolymsci.2011.06.003

Li H, Xu Z, Zhang J, Li S, Li S, Chen X, Lu L, Pan Z, Mao Y (2025) Recent progress of flexible electrospun nanofibers based triboelectric nanogenerators for self-powered electronics. J Phys.: Condens Matter 37:183001. https://doi.org/10.1088/1361-648X/adbfed

Li Y, Chen S, Yan H, Jiang H, Luo J, Zhang C, Pang Y, Tan Y (2023) Biodegradable, transparent, and antibacterial alginate-based triboelectric nanogenerator for energy harvesting and tactile sensing. Chem Eng J 468:143572. https://doi.org/10.1016/j.cej.2023.143572

Li Z, Li C, Deng Y (2023) Bioresorbable pressure sensor and its applications in abnormal respiratory event identification. ACS Appl Electron Mater 5:1761–1769. https://doi.org/10.1021/acsaelm.2c01782

Lin C, Zhao H, Huang H, Ma X, Cao S (2023) PEO/cellulose composite paper based triboelectric nanogenerator and its application in human-health detection. Int J Biol Macromol 228:251–260. https://doi.org/10.1016/j.ijbiomac.2022.12.237

Liu J, Chen J, Dai F, Zhao J, Li S, Shi Y, Li W, Geng L, Ye M, Chen X, Liu Y, Guo W (2022) Wearable five-finger keyboardless input system based on silk fibroin electronic skin. Nano Energy 103:107764. https://doi.org/10.1016/j.nanoen.2022.107764

Lorber B, Hsiao W-K, Hutchings IM, Martin KR (2013) Adult rat retinal ganglion cells and glia can be printed by piezoelectric inkjet printing. Biofabrication 6:015001. https://doi.org/10.1088/1758-5082/6/1/015001

Mahanty B, Kumar Ghosh S, Lee D-W (2024) Advancements in polymer nanofiber-based piezoelectric nanogenerators: Revolutionizing self-powered wearable electronics and biomedical applications. Chem Eng J 495:153481. https://doi.org/10.1016/j.cej.2024.153481

Mahmud RU, Rahman MZ (2024) 13.13—Synthesis and characterization of nanocomposites for tissue engineering. In: Hashmi S (ed) Comprehensive materials processing, 2nd edn. Elsevier, Oxford, pp 241–269. https://doi.org/10.1016/B978-0-323-96020-5.00183-7

Maiti S, Kumar Karan S, Lee J, Kumar Mishra A, Bhusan Khatua B, Kon Kim J (2017) Bio-waste onion skin as an innovative nature-driven piezoelectric material with high energy conversion efficiency. Nano Energy 42:282–293. https://doi.org/10.1016/j.nanoen.2017.10.041

Maity K, Mondal A, Saha MC (2023) Cellulose nanocrystal-based All-3D-printed pyro-piezoelectric nanogenerator for hybrid energy harvesting and self-powered cardiorespiratory monitoring toward the human-machine interface. ACS Appl Mater Interfaces 15:13956–13970. https://doi.org/10.1021/acsami.2c21680

Maity SK, Tyagi U, Sharma AK, Bisht P, Sirohi S, Kumar K, Sheoran N, Singh S, Kumar G (2025) Enhancing the electrical performance of chitosan-based triboelectric nanogenerator using graphene nanoplatelets for real-time sports application. Cellulose 32:1787–1804. https://doi.org/10.1007/s10570-025-06373-6

Mi H-Y, Jing X, Cai Z, Liu Y, Turng L-S, Gong S (2018) Highly porous composite aerogel based triboelectric nanogenerators for high performance energy generation and versatile self-powered sensing. Nanoscale 10:23131–23140. https://doi.org/10.1039/C8NR05872E

Moon CJ, Choi GJ, Sohn SH, Kim SJ, Park IK (2025) Enhanced performance of triboelectric nanogenerator based on TiO_2 Nanoparticle/Nylon 66 composite nanofibers. Int J Precis Eng Manuf-Green Technol. https://doi.org/10.1007/s40684-025-00736-x (accessed 14 April 2025)

Niu Q, Wei H, Hsiao BS, Zhang Y (2022) Biodegradable silk fibroin-based bio-piezoelectric/triboelectric nanogenerators as self-powered electronic devices. Nano Energy 96:107101. https://doi.org/10.1016/j.nanoen.2022.107101

Pang Y, Xi F, Luo J, Liu G, Guo T, Zhang C (2018) An alginate film-based degradable triboelectric nanogenerator. RSC Advances (RSC Publishing). https://doi.org/10.1039/C7RA13294H. https://pubs.rsc.org/en/content/articlehtml/2017/sc/c7ra13294h (accessed 14 April 2025)

Peng J, Zhang H, Zheng Q, Clemons CM, Sabo RC, Gong S, Ma Z, Turng L-S (2017) A composite generator film impregnated with cellulose nanocrystals for enhanced triboelectric performance. Nanoscale 9:1428–1433. https://doi.org/10.1039/C6NR07602E

Qian W, Yang Y (n.d.) Cellulose-templated nanomaterials for nanogenerators and self-powered sensors. Adv Mater n/a:2412858. https://doi.org/10.1002/adma.202412858

Qian C, Li L, Gao M, Yang H, Cai Z, Chen B, Xiang Z, Zhang Z, Song Y (2019) All-printed 3D hierarchically structured cellulose aerogel based triboelectric nanogenerator for multi-functional sensors. Nano Energy 63:103885. https://doi.org/10.1016/j.nanoen.2019.103885

Rushdha Hiba T, Abhijith V, Akhila B, Thomas S, Sreekala MS (2025) Recent advances in polymer based energy harvesting systems for efficient wound healing applications. Mater Today Commun 47:113152. https://doi.org/10.1016/j.mtcomm.2025.113152

Samiei (2020) Recent trends on applications of 3D printing technology on the design and manufacture of pharmaceutical oral formulation: a mini review. Beni-Suef Univ J Basic Appl Sciences. https://bjbas.springeropen.com/articles/10.1186/s43088-020-00040-4 (accessed 17 June 2025)

Sandhya PK, Abhijith V, Sreekala MS, Thomas S (2023) 9—Graphene-based polymer blend nanocomposites for energy storage applications. In: Thomas S, Ajitha AR, Jaroszewski M (eds) Polymer blend nanocomposites for energy storage applications. Elsevier, pp 271–291. https://doi.org/10.1016/B978-0-323-99549-8.00014-5

Sardana S, Kaur H, Arora B, Aswal DK, Mahajan A (2022) Self-powered monitoring of Ammonia Using an MXene/TiO2/cellulose nanofiber heterojunction-based sensor driven by an electrospun triboelectric nanogenerator. ACS Sens 7:312–321. https://doi.org/10.1021/acssensors.1c02388

Schiffman CL, Schauer (n.d.) One-step electrospinning of cross-linked chitosan fibers. Biomacromolecules. https://doi.org/10.1021/bm7006983 (accessed April 14, 2025)

Schiffman JD, Schauer CL (2007) Cross-linking chitosan nanofibers. Biomacromolecules 8:594–601. https://doi.org/10.1021/bm060804s

Sencadas V, Garvey C, Mudie S, Kirkensgaard JJK, Gouadec G, Hauser S (2019) Electroactive properties of electrospun silk fibroin for energy harvesting applications. Nano Energy 66:104106. https://doi.org/10.1016/j.nanoen.2019.104106

Seol M-L, Han J-W, Moon D-I, Yoon KJ, Hwang CS, Meyyappan M (2018) All-printed triboelectric nanogenerator. Nano Energy 44:82–88. https://doi.org/10.1016/j.nanoen.2017.11.067

Sezer N, Koç M (2021) A comprehensive review on the state-of-the-art of piezoelectric energy harvesting. Nano Energy 80:105567. https://doi.org/10.1016/j.nanoen.2020.105567

Sharkawy A, Barreiro MF, Rodrigues AE (2020) Chitosan-based Pickering emulsions and their applications: A review. Carbohydr Polym 250:116885. https://doi.org/10.1016/j.carbpol.2020.116885

Shen Y, Jiang Z, Huang H, Wang S, Wu S, Wang J, Sun X, Liu Y, Wen Z (n.d.) Advances in textile-based triboelectric sensors for physiological signal monitoring. Adv Funct Mater n/a:2426081. https://doi.org/10.1002/adfm.202426081

Shi X, Peng B, Cai X, Zhang S, Chen Y (2025) Ultrasound-enhanced piezoelectric nanogenerators for wireless electrostimulation therapy in the wound healing. Nano Energy 139:110940. https://doi.org/10.1016/j.nanoen.2025.110940

Shyju S, Chandran AM, Varun S, Kumar MVP, Mural PKS (2024) Chitosan nanocomposite-based triboelectric nanogenerators with enhanced electrical performance: An opportunity for bioelectronics. ACS Appl Electron Mater 6:887–900. https://doi.org/10.1021/acsaelm.3c01428

Silk Fibroin—an overview. ScienceDirect Topics. https://www.sciencedirect.com/topics/nursing-and-health-professions/silk-fibroin (accessed 12 April 2025)

Song Y, Bao J, Hu Y, Xu M, Yang Z, Liu Y, Yang Q, Xiong C, Shi Z (2022) Ultra-porous cellulose nanofibril aerogel films as excellent triboelectric positive materials via direct freeze-drying of dispersion. Nano Energy 103:107832. https://doi.org/10.1016/j.nanoen.2022.107832

Teixeira-Costa BE, Andrade CT (2021) Chitosan as a valuable biomolecule from seafood industry waste in the design of green food packaging. Biomolecules 11:1599. https://doi.org/10.3390/biom11111599

Thadathil N, Velappan SP (2014) Recent developments in chitosanase research and its biotechnological applications: A review. Food Chemistry 150:392–399. https://doi.org/10.1016/j.foodchem.2013.10.083

Torres FG, Gonzales KN, Troncoso OP, Corman-Hijar JI, Cornejo G (2023) A review on the development of biopolymer nanocomposite-Based Triboelectric Nanogenerators (Bio-TENGs). ACS Appl Electron Mater 5:3546–3559. https://doi.org/10.1021/acsaelm.3c00621

Wang (2024) Improved electrical output performance of cellulose-based triboelectric nanogenerators enabled by negative triboelectric materials. Small—Wiley Online Library. https://onlinelibrary.wiley.com/doi/full/10.1002/smll.202308195 (accessed 10 April 2025)

Wang C, Wu S, Jian M, Xie J, Xu L, Yang X, Zheng Q, Zhang Y (2016) Silk nanofibers as high efficient and lightweight air filter. Nano Res 9:2590–2597. https://doi.org/10.1007/s12274-016-1145-3

Wang X, Li N, Yang A, Zhang N, Qiu W, Yang D (2025) Amino oxidized sodium alginate-based humidity-resistant triboelectric nanogenerator for human motion and respiration monitoring. Chem Eng J 506:159842. https://doi.org/10.1016/j.cej.2025.159842

Wang ZL, Song J (2006) Piezoelectric nanogenerators based on zinc oxide nanowire arrays. Science 312:242–246. https://doi.org/10.1126/science.1124005

Wu Z, Ding X, Chen X, Chen J, Chang X, Liu Z, Song L, Huang J, Zhu Y (2025) Recent progress of polymer-based piezoelectric nanogenerators. Adv Compos Hybrid Mater. https://link.springer.com/article/10.1007/s42114-025-01225-0 (accessed 18 April 2025)

Xu J, Su H, Han J, Chen Y, Song W, Gu Y, Moon W-J, Zhang D (2012) In situ deposition of flower-like ZnO on silk fibroin fibers. Appl Phys A 108:235–238. https://doi.org/10.1007/s00339-012-6882-x

Yadav RH, Kenchegowda M, Angolkar M, TS M, Osmani RA, Palaksha S, Gangadharappa HV (2024) A review of silk fibroin-based drug delivery systems and their applications. Eur Polym J 216:113286. https://doi.org/10.1016/j.eurpolymj.2024.113286

Yang H, Fan FR, Xi Y, Wu W (2020) Bio-derived natural materials based triboelectric devices for self-powered ubiquitous wearable and implantable intelligent devices. Adv Sustain Syst 4:2000108. https://doi.org/10.1002/adsu.202000108

Yang M, Liu J, Hu C, Zhang W, Jiao J, Cui N, Gu L (2023) Highly sensitive self-powered ammonia gas detection enabled by a rationally designed PANI/commercial cellulosic paper based triboelectric nanogenerator. J Mater Chem A 11:21937–21947. https://doi.org/10.1039/D3TA04974D

Yang W, Chen H, Wu M, Sun Z, Gao M, Li W, Li C, Yu H, Zhang C, Xu Y, Wang J (2022) A flexible triboelectric nanogenerator based on cellulose-reinforced MXene composite film. Adv Mater Interfaces 9:2102124. https://doi.org/10.1002/admi.202102124

Yao C, Hernandez A, Yu Y, Cai Z, Wang X (2016) Triboelectric nanogenerators and power-boards from cellulose nanofibrils and recycled materials. Nano Energy 30:103–108. https://doi.org/10.1016/j.nanoen.2016.09.036

Yu H, Shao Y, Luo C, Li Y, Ma H, Zhang Y, Yin B, Shen J, Yang M (2021) Bacterial cellulose nanofiber triboelectric nanogenerator based on dielectric particles hybridized system. Compos Part A: Appl Sci Manuf 151:106646. https://doi.org/10.1016/j.compositesa.2021.106646

Zeng (2023) Effects of different drying methods on quality and water distribution of Lycium barbarum. J Food Process Eng. Wiley Online Library. https://onlinelibrary.wiley.com/doi/full/10.1111/jfpe.14482 (accessed 17 June 2025)

Zhang F, Shen R, Li N, Yang X, Lin D (2023) Nanocellulose: An amazing nanomaterial with diverse applications in food science. Carbohydr Polym 304:120497. https://doi.org/10.1016/j.carbpol.2022.120497

Zhang R (2024) The opportunities of cellulose for triboelectric nanogenerators: A critical review. Nanoenergy Adv 4:209–220. https://doi.org/10.3390/nanoenergyadv4030013

Zhang W, Zhao J, Cai C, Qin Y, Meng X, Liu Y, Nie S (2022) Gas-sensitive cellulosic triboelectric materials for self-powered ammonia sensing. Adv Sci 9:2203428. https://doi.org/10.1002/advs.202203428

Zhang Y, Chen Z, Zou J, Feng P, Jing X (2024) Sodium alginate supramolecular nanofibers in synergy with surface crack engineering to prepare tough and highly sensitive hydrogels. Int J Biol Macromol 279:135507. https://doi.org/10.1016/j.ijbiomac.2024.135507

Zheng Q, Zhang H, Mi H, Cai Z, Ma Z, Gong S (2016) High-performance flexible piezoelectric nanogenerators consisting of porous cellulose nanofibril (CNF)/poly(dimethylsiloxane) (PDMS) aerogel films. Nano Energy 26:504–512. https://doi.org/10.1016/j.nanoen.2016.06.009

Zhou J, Wang H, Du C, Zhang D, Lin H, Chen Y, Xiong J (2022) Cellulose for sustainable triboelectric nanogenerators. Adv Energy Sustain Res 3:2100161. https://doi.org/10.1002/aesr.202100161

Zhou Q, Park JG, Kim KN, Thokchom AK, Bae J, Baik JM, Kim T (2018) Transparent-flexible-multimodal triboelectric nanogenerators for mechanical energy harvesting and self-powered sensor applications. Nano Energy 48:471–480. https://doi.org/10.1016/j.nanoen.2018.03.074

Zhu Q, Sun E, ZhaoZ, Wu T, Meng S, Ma Z, Shoaib M, Ur Rehman H, Cao X, Wang N (2024) Biopolymer materials in triboelectric nanogenerators: A review. Polymers 16:1304. https://doi.org/10.3390/polym16101304

Zhu Q, Song X, Chen X, Li D, Tang X, Chen J, Yuan Q (2024) A high performance nanocellulose-PVDF based piezoelectric nanogenerator based on the highly active CNF@ZnO via electrospinning technology. Nano Energy 127:109741. https://doi.org/10.1016/j.nanoen.2024.109741

Chapter 8
Biopolymer-Based Materials for Advanced Military Applications

Hassan Tawsif Tazwar, Maisha Farzana Antora, Adib Bin Rashid, and Md Enamul Hoque

Abstract Synthetic polymer-based materials contribute significantly to plastic pollution, highlighting the need for sustainable alternatives. This chapter explores the use of biopolymer-based materials in advanced military applications, emphasizing their classification, structure, and advantages such as biodegradability, lightweight properties, and mechanical strength. Key biopolymers, including cellulose, chitin, and polylactic acid (PLA), are examined for their potential in protective gear, packaging, medical applications, and environmental remediation. While biopolymers offer environmental and functional benefits, challenges such as scalability, cost, and performance limitations remain. Case studies of successful implementations are discussed, along with future trends in nanotechnology and sustainability-driven innovations. The chapter concludes by highlighting the need for continued research and investment to enhance their feasibility and effectiveness.

Keywords Biopolymer · Cellulose · Chitin · PLA · Military applications

8.1 Introduction

The widespread issue of persistent plastic waste has ignited global concern, prompting a search for sustainable alternatives. Biopolymers could be an effective solution to this exacerbating situation. Biopolymers can be both protein-based or

H. T. Tazwar · M. F. Antora
Department of Aeronautical Engineering, Military Institute of Science and Technology, Dhaka, Bangladesh

A. B. Rashid (✉)
Department of Industrial and Production Engineering, Military Institute of Science and Technology, Dhaka, Bangladesh
e-mail: adib8809@gmail.com

M. E. Hoque (✉)
Department of Biomedical Engineering, Military Institute of Science and Technology, Dhaka, Bangladesh
e-mail: enamul1973@gmail.com

L. Ravindran et al. (eds.), *Macro, Micro and Nanocomposites from Sustainable Sources*, Advanced Structured Materials 244, https://doi.org/10.1007/978-981-95-2469-3_8

carbohydrate-based polymers that can be sourced from animals or plants. They have diverse applications across industries; however, their commercialization is often hindered by costly production processes and inefficient extraction methods. Their emerging applications in agriculture, food, cosmetics, and therapeutics are driving renewed research efforts to address these challenges and unlock the full potential of biopolymers as a sustainable and versatile resource (Rao et al. 2014).

Among the biopolymers, cellulose and starch are particularly valuable for materials applications. However, there is a growing interest in more complex hydrocarbon polymers synthesized by bacteria and fungi, such as polysaccharides like xanthan, curdlan, pullulan, chitin, chitosan, and hyaluronic acid (Janssen and Leszek 2009). The rising demand for biodegradable polymers has fueled research into the development of novel materials. These polymers, which can either be synthesized or naturally occur in organisms, offer promising environmental benefits. In addition, microorganisms and enzymes capable of degrading these materials have been identified, enhancing their sustainability. As a result, biodegradable polymers are often classified based on factors such as their chemical structure, origin, and biodegradability, which help guide their development and application (Burande et al. n.d.).

The annual global production of synthetic polymers has reached an astounding 140 million tons, yet their exceptional stability poses a significant challenge for biosphere degradation. The accumulation of synthetic polymers, including plastic waste and water-soluble polymers, in the environment has become a pressing environmental concern. While polymers are an integral part of modern life, their resistance to degradation leads to their accumulation in landfills at alarming rates, accounting for approximately 8% by weight and 20% by volume (Doble 2005). Non-biodegradable petroleum-based polymers, such as polypropylene (PP), polyethylene (PE), polyvinyl chloride (PVC), and polyethylene terephthalate (PET), are major contributors to plastic pollution. However, some petroleum-derived polymers are biodegradable, such as polybutylene adipate-co-terephthalate (PBAT), which decomposes in soil, and polyvinyl alcohol (PVA), which dissolves in water and can be broken down by bacteria. Similarly, biomass-derived polymers can be either biodegradable or non-biodegradable. Biodegradable variants include polylactic acid (PLA), which degrades under composting conditions, and polyhydroxyalkanoate (PHA), a bacteria-produced polymer with high biodegradability. A subtype, poly (3-hydroxybutyric acid), even degrades in marine environments. In contrast, non-biodegradable biomass polymers include biomass polyethylene (PE), made from bioethanol, and biomass polyurethane (PU), widely used in biomedical applications (Mori 2023).

Biodegradable polymers have found widespread applications in various fields, leading to the development of numerous commercial products. Compared to other industries, the application of biodegradable polymers in the medical industry has grown quickly because of their high level of specialization and value. Homopolymers of β-glucans, known for their low toxicity, are a significant group of biopolymers with promising applications in treating various diseases. Advancements in medical research have elucidated the mechanisms of glucan effects and the relationship between their structure and activity. While traditional drug research has focused on synthesizing or discovering new drugs, biopolymer materials offer a promising approach for drug delivery, minimizing tissue reactions and enabling non-conventional administration methods. Although biopolymer formulations have been limited to a narrow range of applications, current efforts are focused on registering drug formulations incorporating polymer drug delivery systems, which are expected to make a significant impact on the market soon. These novel sustained-release systems are particularly valuable in treating cancers and diseases affecting the elderly, as well as hormone disorders and geriatric diseases, allowing for less frequent drug administration (Rao et al. 2014).

Military power significantly influences international affairs, with nations heavily investing in research and development to maintain an edge in defense technology. A key driver of modern military advancements is materials innovation, allowing the fabrication of lighter, stronger, and more resilient equipment. Advanced materials enhance combat effectiveness through applications such as rugged packaging, camouflage, self-healing structures, and EMI shielding materials. These innovations support upgrades to existing land, sea, and air forces while also advancing emerging technologies like UAVs and first-person view (FPV) drones. To fully leverage these materials, collaboration between defense sectors, industry, and academia is essential, along with substantial financial investment to address operational and environmental challenges (Lopez Vicente n.d.; Siengchin 2023). Biopolymer composites reinforced with fibers or particles offer a promising solution to meet evolving military needs, combining strength, lightweight properties, and biodegradability for enhanced performance and sustainability.

In this chapter, we begin with a brief overview of classification, followed by an introduction to commonly used biopolymers, highlighting their structures and properties. The significance and performance of various biopolymer-based materials, providing a foundation for evaluating their potential in military applications has also been discussed in this chapter. We then explore their practical uses before concluding by highlighting their limitations and potential future research directions.

8.2 Overview of Biopolymers

Biopolymers are biological macromolecules formed by linking together smaller units called monomers. These monomers can be proteins, nucleic acids, or polysaccharides (Pattanashetti et al. 2017). Biopolymers can be primarily categorized into three types: natural biopolymers, synthetic biopolymers, and hybrid biopolymers. An overview of different types of biopolymers has been presented in Fig. 8.1.

8.2.1 Natural Biopolymers

Natural biopolymers, such as starch, cellulose, chitosan, carrageenan, gelatin, alginates, and their derivatives, offer a sustainable alternative to non-biodegradable

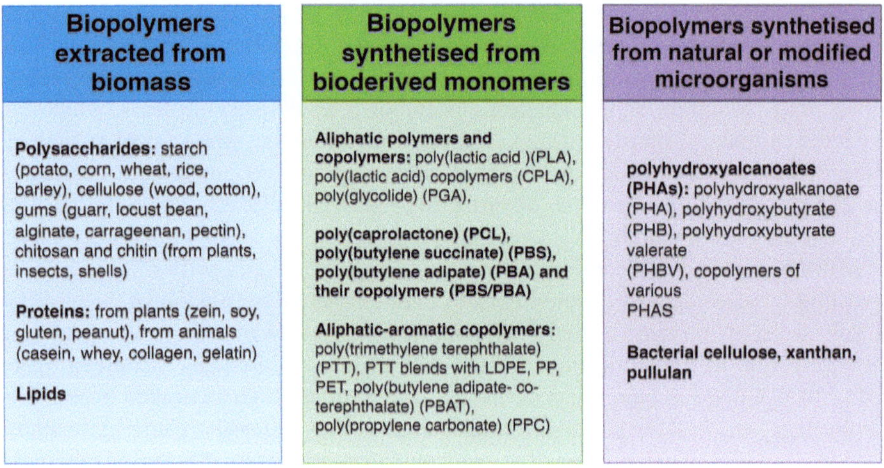

Fig. 8.1 Biobased polymer materials

petroleum-based plastics. These biopolymers are well-suited for various applications due to their abundance, renewability, affordability, biocompatibility, and biodegradable nature. Researchers are actively exploring unconventional sources, including bacterial or fungal components, soy protein, and fish skin gelatin, to isolate biopolymers and create advanced materials (Pattanashetti et al. 2017). Some common natural biopolymers are discussed below.

8.2.1.1 Cellulose

The polysaccharide cellulose, which is regarded as the most significant skeletal material in plants, is a polymeric raw material which has a unique structure and properties and almost never runs out. It is composed of the linkage of D-glucose building blocks repetitively forming a–broadly functionalized and linear stiff chain homopolymer which is hydrophilic, chiral, biodegradable, highly chemically modifiable and render different semicrystalline fiber shapes. The growing interest in renewable resources has spurred a renewed focus on cellulose research worldwide in the last decade (Klemm et al. 2005). Anselme Payen pioneered the discovery of cellulose, determining its molecular formula and that of starch. Due to their biocompatibility and environmental sustainability, cellulose and its derivatives have garnered significant attention and are considered to hold substantial potential. Cellulose is an affordable material with adequate mechanical properties and a high global demand (Liu et al. 2021a). Cellulose has emerged as a leading candidate for producing biobased polymeric materials. In recent years, significant advancements have been made in developing biopolymers from various forms of cellulose (Shaghaleh et al. 2018). Figure 8.2 shows the cellular structure of cellulose. From the figure, it can be observed that, cellulose contains numerous hydroxyl (–OH) groups, which can be chemically

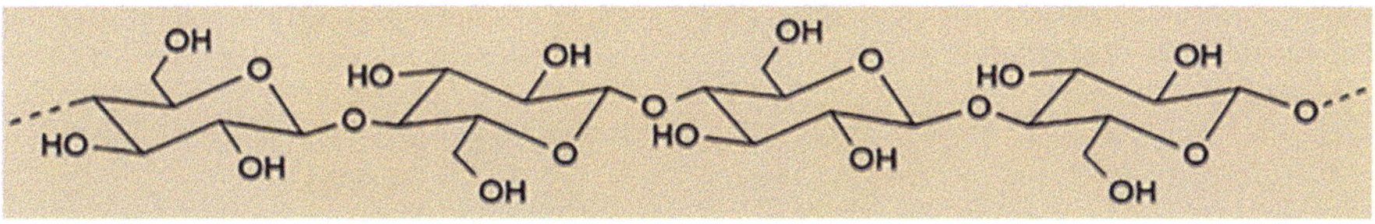

Fig. 8.2 Cellular structure of cellulose (Klemm et al. 2005)

modified with functional groups like phosphate, carboxyl, and amino to enhance its pollutant removal capabilities from wastewater. Cellulose can also be processed into various forms, including chains, fibers, nanocrystals, nanosheets, films, and sponges. These properties make cellulose an ideal starting material for developing adsorbents (Qi et al. 2021).

8.2.1.2 Chitin and Chitosan

Chitin and its deacetylated derivative, chitosan, are among the most abundant structural biopolymers in nature, finding applications across various human endeavors, including materials science, nutrition, health, and energy. These polymers offer unique properties with significant economic potential, yet their full potential remains largely untapped. Chitosan is primarily produced from crustacean shells, accounting for 85% of the market, while fungal and insect sources are less exploited. Efforts to replace crustacean-derived chitosan with fungal or insect-based alternatives have encountered challenges due to limitations in conventional chitin biosynthesis methods (Philibert et al. 2017). After cellulose, chitin is the second most abundant naturally occurring polysaccharide in the world. Figure 8.3 shows the cellular structure of chitin and chitosan. N-acetylglucosamine monomers are β-(1, 4) glycosidically linked to this high-molecular-weight linear homopolymer. Chitosan, its main derivative is a copolymer of 2-amino-2-deoxy-D-glucose and N-acetyl-2-amino-2-deoxy-D-glucoside linked by 1–4 glucosidic linkages. Chitin is the primary component of crustacean's shells, exoskeletons of insects and fungal cell walls for durability and rigidity. On the other hand, naturally occurring chitosan can only be found in specific kinds of fungi, those in the *Mucoraceae* family (Roberts and Roberts 1992). Chitin and Chitosan are natural biopolymers, which serve as excellent substrates and show a unique range of useful features like biodegradability, non-toxicity, biocompatibility and various biomedical applications such as hemostatic and antimicrobial activities. Several studies have reported a wide range of biomedical applications for chitin and chitin-based materials, which include biomedical and military applications (Singh et al. 2017).

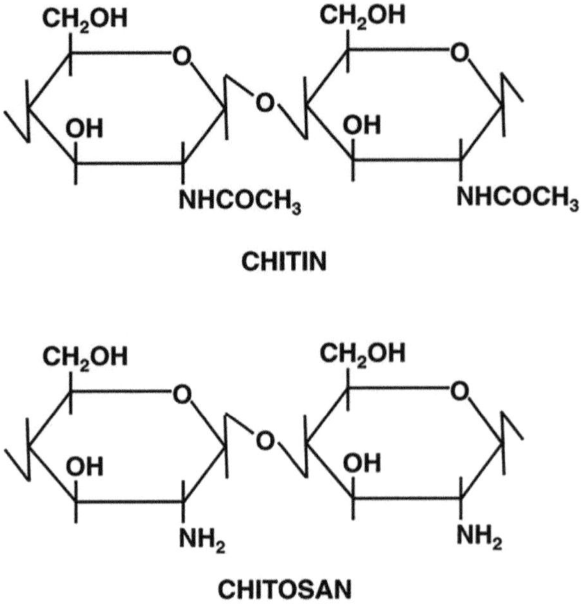

Fig. 8.3 Cellular structure of chitin and chitosan (Singh et al. 2017)

8.2.1.3 Starch

Starch-based biopolymers offer a promising alternative to petroleum-derived plastics due to their non-toxicity, biocompatibility, and comparable mechanical properties. Derived from agricultural waste, these biopolymers can be chemically modified to create bio composites suitable for numerous applications. While biodegradable, starch-based biopolymers face challenges such as brittleness, moisture sensitivity, and poor thermal and mechanical properties. Researchers are actively exploring strategies to address these limitations and improve the material's performance to compete with conventional plastics. Although a few companies have begun using starch-based biopolymers commercially, significant scientific research is needed to enhance their properties, reduce production costs, and boost the overall economic viability of the biopolymer sector (Agarwal et al. 2023). Starch is easily available in nature as it is stored in plants and algae. Starch synthase forms and plastidial glucan phosphorylases create α-1,4 linkages by transferring glucosyl residues to the non-reducing ends of oligo- or polyglucans. This process results in the unidirectional growth of glucan chains. Over 90% of interglucosyl bonds are α-1,4 linkages, forming chains of varying lengths. Branching points, known as α-1,6 linkages, are created by various branching isozymes. Consequently, the fundamental biochemistry of assimilatory and reserve starch granules is similar (Compart et al. 2021). Figure 8.4 presents the cellular structure of starch.

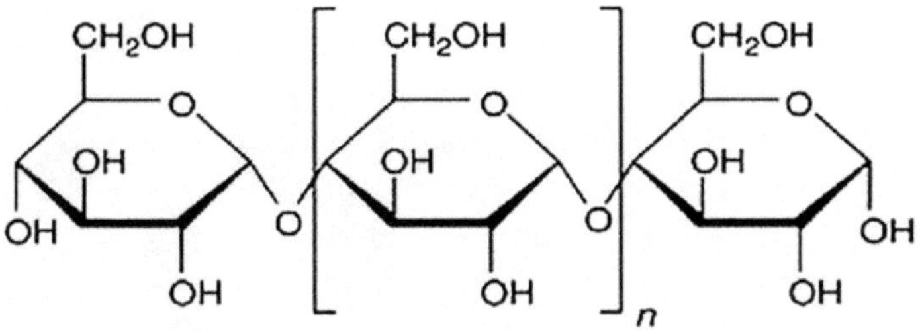

Fig. 8.4 Cellular structure of starch (Samir et al. 2022)

8.2.2 *Synthetic Biopolymers*

Synthetic biodegradable polymers are engineered substances, which decompose into less complex substances after some time with the help of microorganisms or enzymes. These polymers have been produced to counter the harmful effects of plastics which do not decompose after use. Unlike adrenaline starch masking tape or cellulose, these substances are not found in nature but rather created by chemical means. Their applications are noticed in different areas such as packaging, agriculture, military and in medicine. Synthetic biodegradable polymers undergo enzymatic degradation in aqueous environments. Polyesters, with their hydrolysable linkages, are the most prominent group of these polymers. PLA, PCL, PBS, PBSA, and their copolymers, particularly PLA-PCL, are among the most commercially available biodegradable polymers (Siracusa et al. 2017; Huang 2005). An overview of different types of synthetic biopolymers has been presented here.

8.2.2.1 Polylactic Acid (PLA)

Polylactic acid (PLA) is a synthetic plastic made from lactic acid with a chemical formula of $(C_3H_4O_2)_n$ or $[–C(CH_3)HC(=O)O–]_n$. It can be produced by either combining lactic acid molecules or opening the rings of lactide molecules ("Polylactic acid", Wikipedia n.d.). Figure 8.5 shows the chemical composition of PLA. Biodegradable plastics, including polylactic acid (PLA), offer a sustainable alternative to conventional petroleum-based plastics, which contribute to environmental problems like pollution and global warming. PLA is widely available, safe to decompose, and possesses properties comparable to traditional plastics like PP and PET, making it suitable for various industrial applications. The growing demand for PLA, driven by its desirable characteristics, has led to a steady expansion of the global PLA market. Overall, PLA-based bioplastics present a promising solution for replacing conventional plastics, offering environmental benefits and economic advantages (Taib et al. 2023). Polylactic acid (PLA) is a thermoplastic polymer made from lactic acid, which is typically produced through fermentation of agricultural sources like corn, sugarcane, potato, or sugar beet. Additionally, lactic acid can

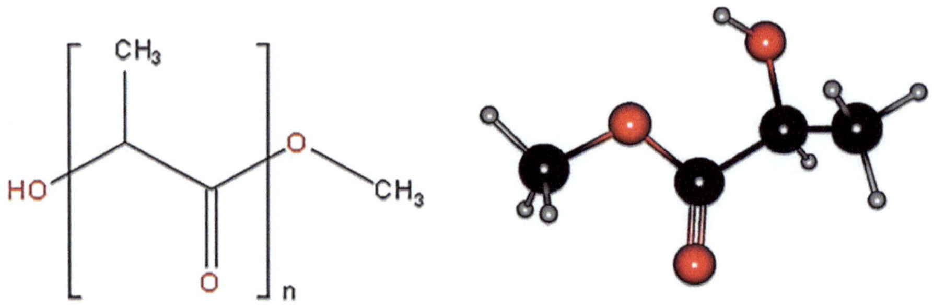

Fig. 8.5 Cellular structure of PLA (Martinez et al. 2022)

be derived from second-generation livestock residues such as corn stover, bagasse, and straw (Beltrán et al. 2018; Aryan et al. 2021). While PLA is biodegradable under composting conditions, this process can have environmental consequences, including emissions, land use, and potential genetic toxicity of residues (Cosate de Andrade et al. 2016). In 2022, PLA accounted for 20.7% of global bioplastic production and 40.2% of biodegradable bioplastics (Bioplastics Market Data 2022). The global PLA market was valued at an estimated USD 786 million in 2022, with projections for growth to USD 1,756 million by 2025, representing a compound annual growth rate of 17.4% (Lactic Acid Market 2020).

8.2.2.2 Polyhydroxyalkanoates (PHAs)

Polyhydroxyalkanoates (PHAs) are naturally occurring biological polymers produced by microorganisms under specific conditions of excess carbon and nutrient deficiency (Zhang et al. 1997). Microbial polyhydroxyalkanoates (PHAs) are biodegradable, non-toxic biopolyesters produced by microorganisms as intracellular granules in response to nutrient stress. These bio polyesters offer a sustainable alternative to traditional plastics, thanks to their excellent properties like melting temperature, glass transition temperature, crystallinity, and mechanical strength. PHAs can be used to produce bioplastics for packaging and disposable products, as well as biofuels. These versatile biopolymers have applications across various industries, including agriculture, pharmaceuticals, and healthcare. Numerous bacterial, microalgal, and fungal species can produce PHAs, with notable examples including *Bacillus megaterium*, *Pseudomonas aeruginosa*, *P. oleovorans*, *P. stutzeri*, and *Cupriavidus necator* (Muneer et al. 2020). These polymers accumulate as granules within the cells of various microorganisms, including archaea and both Gram-positive and Gram-negative bacteria. Over 150 different types of PHAs have been identified, and they can be produced both aerobically and anaerobically (Mousavioun et al. 2012). Figure 8.6 presents the chemical composition of PHA. Polyhydroxyalkanoates (PHAs) have recently gained significant interest due to their desirable properties such

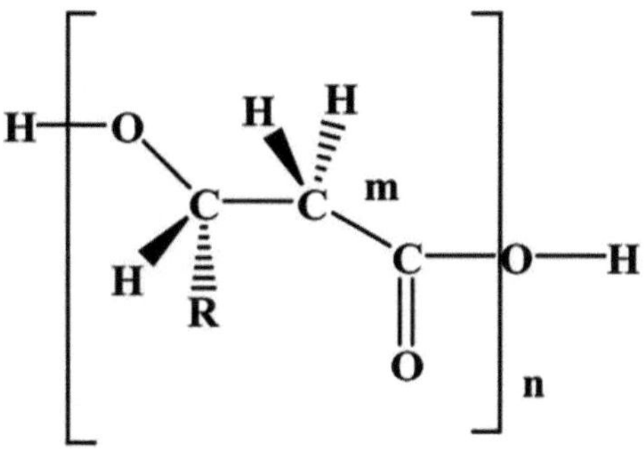

Fig. 8.6 Cellular structure of PHA (Tariq et al. 2019)

as biodegradability, biocompatibility, and hydrophobicity. However, these biopolymers also have limitations that hinder their competitiveness with synthetic polymers. To address these shortcomings, researchers are modifying PHAs to enhance their properties for various applications (Sharma et al. 2021).

8.2.2.3 Polycaprolactone (PCL)

Polycaprolactone (PCL) is a biodegradable, biocompatible thermoplastic with rising demand in biomedical, packaging, and sustainable applications. It has a chemical formula of (1,7)-Polyoxepan-2-one. The chemical composition of PCL has been presented in Fig. 8.7. Its applications include controlled drug release, tissue engineering, bone scaffolds, packaging, and compost bags (Mohamed and Yusoh 2016). Polycaprolactone (PCL) offers long-term stability due to its slow degradation rate, making it suitable for applications requiring structural support over extended periods, such as tissue regeneration. PCL's biocompatibility can be enhanced through surface modifications, such as adding bioactive molecules or coatings, to promote cell interaction and integration with surrounding tissues. PCL can also be combined with other biomaterials or additives to improve its properties, like adding ceramic particles for enhanced mechanical strength or bioactivity. However, PCL has limitations, including slow degradation, limited mechanical strength, poor cell adhesion, hydrophobicity, and potential toxicity (Ding et al. 2021).

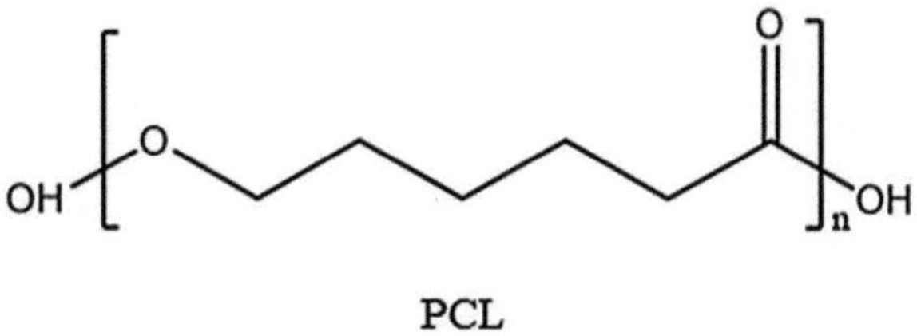

PCL

Fig. 8.7 Cellular structure of PCL (Kim et al. 2023)

8.2.3 Hybridization

Nanomaterials and nanofabrication are emerging as promising alternatives to traditional approaches. Hybrid biopolymers are those materials comprising both natural and artificial components together so that the properties of both are exploited. The result of this is generally a material with better performance, easier acceptance in living organisms and more environmentally friendly. Vilma Conceição Costa et al. (2007) fabricated a hybrid biopolymer with PVA and bioactive glass hybrid foams. This hybrid composite showed better performance, and the pore inside the composite was more homogeneous. There are different types of hybrid biopolymers such as: Starch-based blends, Chitosan-based blends, Gelatin-based blends, Cellulose–nanocrystal composites, Chitin–nanoparticle composites, Protein–mineral composites, Cellulose–graphene composites, Chitosan–graphene composites, Protein–graphene composites, Starch–nanoparticle composites, Chitosan–nanoparticle composites, Protein–nanoparticle composites, Starch–cellulose blends, Chitosan–gelatin blends, Protein–polysaccharide blends, etc. A variety of such hybrid biopolymers exhibit numerous properties and applications such as packaging, biomedical device manufacturing, electronics and environmental protection. The materials and processing techniques employed may greatly affect the attributes and effectiveness of the resultant product. Thus, hybrid bio composites are the improved versions of traditional biopolymers. Table 8.1 summarizes different hybrid biopolymer combinations.

8.3 Properties of Biopolymers

Biopolymer-based composites offer a sustainable alternative to traditional synthetic materials. Their unique properties, including biodegradability, mechanical strength, lightweight nature, chemical resistance, biocompatibility, thermal properties, and fire resistance, make them attractive for various applications, from biomedical and automotive industries. This section will provide an overview of these topics.

8.3.1 Biodegradability

Biopolymers are eco-friendly and biodegradable, though degradation varies by material and conditions. Challenges include balancing biodegradability, strength, and cost. Research advances bio composites as sustainable plastic alternatives, with PLA emerging as a leading biopolymer (Deeraj et al. 2022). In military applications, biodegradability is crucial due to the uncertainty of equipment disposal. For example, food packaging based on bio composites can be buried on-site, eliminating the need to carry waste and improving efficiency. Moreover, in active combat, the fate

Table 8.1 Hybridization of biopolymers

Sl	Material	Fabrication technique	Notable properties	Refs.
1	PLA + Chitosan and ZnO/TiO_2	Mixing + molding/casting	Effective against bacteria and fungi Medium tensile strength, modulus, and elongation Two-stage decomposition for chitosan-ZnO and chitosan-TiO_2, single-stage decomposition for PLA-based bioplastics Low barrier ability to water vapor	Suryanegara et al. (2021)
2	PVA + Chitosan + Selenium nanoparticles (SeNPs)	Laser ablation + solution casting	Increased antimicrobial properties due to SeNPs	Menazea et al. (2020)
3	Purple cabbage anthocyanins (PCA) + PVA + cellulose nanocrystals (CNC)	–	Enhanced UV–vis barrier properties Improved mechanical properties Increased moisture resistance pH-sensing capabilities Volatile ammonia sensing Visual indication of food freshness	He et al. (2022)
4	Cellulose + Polyhydroxybutyrate (PHB)	Melt blending + extrusion	Good filler dispersion Poor interfacial adhesion No impact on melting behavior and crystallinity Improved thermal stability Minimized negative impact of extrusion Enhanced mechanical performance compared to processed PHB	D'Arienzo et al. (2024)
5	Starch + clay nanoparticles	Casting	Increased tensile strength Higher thermal degradation temperature Enhanced resistance to microbial growth	Pinto et al. (2021)
6	PVA/Bioactive glass	Sol–gel	Hierarchical macroporous structure Good mechanical properties Biocompatibility	Mansur and Costa (2008)
7	PVA/Chitosan/ Bioactive glass			

(continued)

Table 8.1 (continued)

Sl	Material	Fabrication technique	Notable properties	Refs.
8	PVA/46S6 Bioglass composite scaffold	Lyophilization	Porous structure, bioactivity Controlled drug release Enhanced mechanical properties (with increasing glass content)	Mabrouk et al. (2014)
9	Chitosan/PVA/ Bioactive glass	Sol–gel + Lyophilization + Crosslinking	Porous structure (96% porosity) Interconnected pores Biocompatibility Controlled degradation	Silva et al. (2016)

of ammunition boxes and other gear is unpredictable. Biopolymer-based materials allow quick, hassle-free disposal, improving battlefield operations. However, certain biopolymers require definite soil conditions for proper degradation which may or may not be present during disposal (Mori 2023).

In a study conducted by Ejaz et al. (2020), the biodegradability tests confirmed that the PLA-based flax, jute, and hybrid composites exhibit enhanced biodegradability, making them suitable for sustainable applications. Tanjung et al. (2020) studied the biodegradation of PLA/coconut shell (CS) bio composites, finding that chemically treated composites degraded more slowly than untreated ones due to improved interfacial adhesion, which reduced enzyme hydrolysis exposure. In another interesting study, Said et al. (2024) found that adding 3% cellulose nano whiskers to chitosan-glycerol biodegradable plastic films improved tensile strength, water resistance, and biodegradation rate, though excessive addition led to aggregation and reduced performance.

8.3.2 Mechanical Strength

Biopolymers exhibit diverse mechanical properties but are generally weaker than synthetic polymers. Their strength depends on molecular structure, weight, processing, and additives. To enhance performance, techniques like chemical modification, blending, compositing, and nanotechnology are used. Despite their lower strength, biopolymers' sustainability and biocompatibility justify these efforts. Water absorption is a key concern, especially for hydrophilic biopolymers like PLA, which requires drying below 250 ppm for optimal processing. In composites, water uptake weakens fiber/polymer interfaces, particularly in flax reinforcement, causing swelling and strength loss. Modifying flax to reduce water absorption may be necessary. Additionally, the interaction between biopolymers and nanofillers plays a crucial role in composite strength (Velde and Kiekens 2002; Sadasivuni et al. 2020).

In the military, material strength is a critical factor, especially impact and compressive strength, as equipment often undergoes rough handling during transportation. Ensuring adequate durability is important to withstand harsh conditions. Biopolymeric composites have the potential to revolutionize military applications by offering a decent balance between strength, lightweight properties, and sustainability. These advanced composites could enhance the resilience of military gear, vehicles, and protective equipment while reducing environmental impact. Lei et al. (2021) investigated catechol-functionalized chitosan (C-CS)/PVA composite films for active food packaging, finding that they exhibited a 46.3% increase in tensile strength and a 25.4% increase in elongation at break compared to pure PVA, while maintaining transparency and improving antibacterial and UV barrier properties. Srivastava et al. (2021) investigated PVA–banana pseudo stem fiber composite films reinforced with nanocellulose, achieving a 14.3% increase in tensile strength and a 29.7% improvement in water vapor permeability, while reducing elongation at break by 9.1%, making them a promising alternative to plastic packaging films.

8.3.3 Lightweight Nature

Biopolymers have lower densities than synthetic polymers, making them ideal for lightweight applications. Their reduced weight stems from the low-density biopolymer matrix and reinforcing natural fibers, some of which are hollow, further enhancing weight savings. These bio composites offer benefits such as improved fuel efficiency, lower pollution, and easier handling. However, they may have inferior mechanical properties and higher production costs. Density influences transport costs and energy consumption although these factors do not really affect military operations due to their high budget.

Flax fiber (1.45 g/cm^3) is preferred over glass fiber (2.54 g/cm^3) for weight reduction in composites (Velde and Kiekens 2002). Rath et al. (2023) investigated the mechanical properties of chitosan–flax composites, finding that increasing molecular weight led to higher porosity, which reduced tensile and flexural strength but contributed to a lower density, making the material competitive for lightweight applications. Sometimes, a hybrid approach could be very effective as well. Pragadish et al. (2025) studied the effect of infill percentage on the lightweight properties of 3D printed PLA wood-glass fiber composites. Honeycomb structures with 60% infill showed a high strength-to-weight ratio, ideal for lightweight, robust components. Triangular structures with 40% infill offered competitive mechanical properties and efficient load distribution, making them suitable for lightweight applications like protective gear.

8.3.4 Biocompatibility

Biopolymers are known for their excellent biocompatibility, allowing safe interaction with biological systems, which is essential in both biomedical and military applications. Factors like origin, chemical structure, molecular weight, and fiber processing methods influence biocompatibility, ensuring no adverse reactions and promoting cell adhesion and tissue growth. Rigorous testing of bio composites is crucial, especially in sensitive applications such as implants and medical devices. Biopolymers, derived from plants, animals, bacteria, and fungi, offer natural biocompatibility and sustainability, making them ideal for biomedical coatings. Their properties make them increasingly relevant for military uses, including biodegradable medical devices, wound care, and biocompatible protective coatings (Song et al. 2020; Zhao et al. 2023a; John et al. 2023).

8.3.5 Thermal Property

Thermal stability is a critical factor in military applications, as operations and exercises often take place in extreme heat, whether in tropical or desert environments. However, biopolymers inherently lack thermal stability, posing a challenge for their widespread use in such conditions. Norrrahim et al. (2021), in their review paper discussed that, incorporating cellulose can significantly enhance the thermal stability of biopolymers, making them more viable for military applications.

Montero et al. (2017) studied the thermal stability of thermoplastic starch (TPS) films made from starches of different vegetal sources, plasticized with glycerol, and reinforced with cellulose nanocrystals. The results showed that high amylopectin starches exhibited good thermal stability, while the addition of cellulose nanoparticles further improved the thermal stability and moisture resistance of the TPS films. This work aimed to develop biobased TPS films as an alternative to petroleum-derived materials in packaging applications. In another work, a new eco-friendly method was proposed to enhance the thermal stability of PHB for food packaging applications. PHB was modified with bacterial cellulose nanofibers (BC) using melt compounding, followed by plasma treatment or zinc oxide (ZnO) nanoparticle plasma coating. Plasma treatment preserved the thermal stability, crystallinity, and melting behavior of PHB-BC nanocomposites, even with varying BC content, while significantly improving stiffness and strength. This approach offers a sustainable solution for food packaging with enhanced thermal properties (Panaitescu et al. 2018).

Li et al. (2023) developed a high-performance PLA/PHB composite using a weak shear elongational rheology technique and a melt quenching process. This approach minimized molecular weight degradation, enhancing thermal stability. The in-situ fibrillated PHB improved interfacial bonding, while melt quenching preserved fibrillation and accelerated PLA crystallization. The resulting composite exhibited superior thermal stability, making it suitable for applications in biomedical, agricultural, and food packaging fields.

8.3.6 Fire Retardant

Biocomposites, made from biodegradable polymers and natural or manmade fibers, generally have lower fire resistance than conventional plastics. However, recent advancements have improved their fire resistance through factors like material choice, additives, and processing methods. Plant-based biopolymers, often vulnerable to fire, require enhanced flame resistance, leading to the incorporation of flame retardants (FRs). Traditional halogenated and inorganic FRs are toxic and non-biodegradable, prompting interest in eco-friendly alternatives like nanocellulose and lignin, which are gaining attention in scientific research for their safer and more sustainable properties (Taib et al. 2022). In the military, fabrication of fire-resistant materials that can handle explosions is a top priority, even more so than for commercial use. There is also a need for these materials to be strong and durable as they will be used in protective gears and packaging.

In their study, Zhang et al. (2021) investigated the use of urea- and phosphate-grafted bamboo charcoal as a flame-retardant filler in PLA composites, focusing on its effect on thermal stability. The addition of 10–30 wt % bamboo charcoal significantly enhanced the limiting oxygen index and reduced both the peak heat release rate and total heat release by more than 50% compared to pure PLA. Bamboo charcoal contributed to the formation of a dense, honeycomb-like char layer, which acted as a protective barrier in the condensed phase, improving the flame retardancy and thermal stability of the PLA composites.

Biopolymers can enhance flame retardancy of other materials as well. Zhu et al. (2022) developed a novel aerogel composite with excellent flame retardancy, achieving a high limiting oxygen index of 50.8 and a low peak heat release rate of 24.12 kW m^{-2}, earning a UL-94 V-0 rating. The enhanced flame resistance is attributed to the synergistic effects of Magnesium Hydroxide and chitosan. However, synthetic materials may sometimes supersede their performance. Yang et al. (2021) developed organic–inorganic composite aerogels with enhanced flame retardancy, achieving increased limiting oxygen index values of 35% to 41.1%, reduced heat and smoke release, and improved fire resistance due to the formation of alumina-rich hybrid chars. The aerogels showed superior fire performance compared to chitosan/PVA aerogels.

8.4 Applications in Advanced Military Systems

Biopolymers have a wide spectrum of potential uses in military applications that go from protection to environmental concerns (Fig. 8.8). Regarding protective equipment, novel lightweight armor materials derived from biopolymers can offer satisfactory ballistic capability with minimal load on the soldier. They can also be used in packaging and storage to provide biodegradable packaging for munitions and food products. In medical applications, biopolymers can be used in wound healing, tissue engineering, and drug delivery systems meeting the basic medical needs in harsh conditions (Rahman et al. 2023). In environmental applications, biopolymers can be employed in the removal of oil slicks and in reinforcing the ground to reduce the effects of military operations on the environment. Furthermore, biopolymers have prospects in electromagnetic shielding and wearable sensors for providing enhanced technologies for military use. Biopolymers are proving to be an asset in the military, offering innovative and sustainable solutions to a variety of challenges. They provide a greener alternative to traditional materials, demonstrating their potential to revolutionize military technology (Gowthaman et al. 2021).

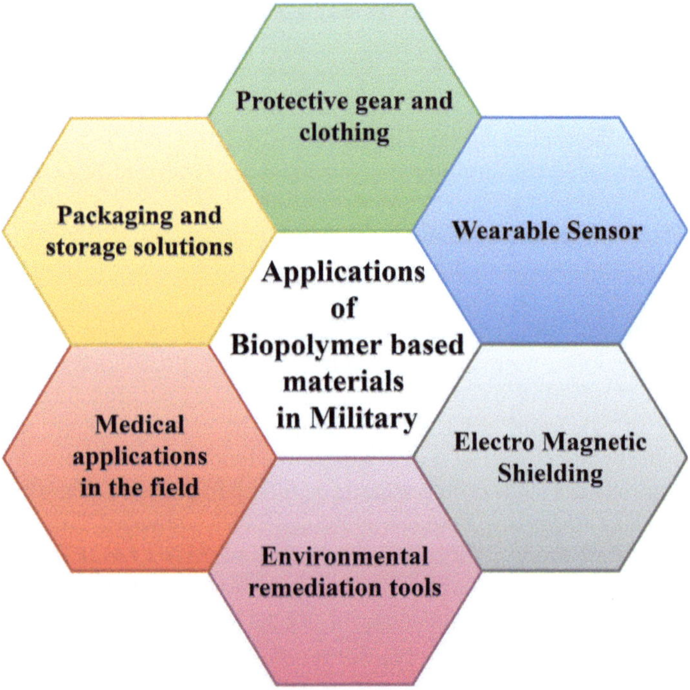

Fig. 8.8 Applications of biopolymer-based materials in military

8.4.1 Protective Gear and Clothing

The advancement in lightweight and efficient armor materials has been one of the most important areas of focus in the military and police force. Metals and ceramics are conventional armor materials with great ballistic performance, but they have one drawback; weight and large dimensions that hamper movements and comfort for the end user. This has created a need to look for other materials such as biopolymers and their composites. Natural polymers such as plants, animal and microbial origin have properties that make them suitable for armor uses. They are usually small, environmentally friendly and non-toxic to human skin, thus causing little or no skin reactions. Furthermore, biopolymers can be filled with reinforcing agents such as carbon nanotubes or ceramic nanoparticles where these composites exhibit improved mechanical properties and ballistic resistance (Norrahim et al. 2021).

A biopolymer of armor interest is polylactic acid (PLA), which is a biodegradable thermoplastic that is produced from renewable resources such as cornstarch or sugarcane. The tensile strength and modulus of PLA are relatively high hence its suitability to be used in armor structures (Ma et al. 2024; Moyo et al. 2020). Incorporation of PLA with reinforcing agents can provide ballistic protection that is as efficient as conventional materials with a much lighter weight. Sebastian Quintero et al. (2018) in their work discovered that the PLA composite reinforced with natural fabric from *Manicaria saccifera* may be utilized to protect against secondary debris from secondary explosions or debris from other sources (Fig. 8.9). Another biopolymer which has been identified is polyhydroxybutyrate (PHB), a natural polyester synthesized by some bacteria. PHB is an environmentally friendly material that is biodegradable and biocompatible and has good mechanical characteristics. In its use in composite structures, it has been demonstrated to offer good ballistic resistance. The property of PHB can be reinforced with different reinforcements like carbon fiber or glass fiber with the armor application in mind (Yan et al. 2024).

Besides PLA and PHB, other biopolymers are also being considered for use in armor systems. For example, chitosan, a derivative of potential chitin polymer from crustaceans' chitin exoskeleton, has been reported to have magnificent potentiality as a ballistic when reinforced by agents. Chitosan is also biodegradable and antimicrobial and therefore it is one of the candidate materials for self-healing armor. Moreover, investigations are being carried out into the application of biopolymers in conjunction with other materials to develop composite armor systems (Souza et al. 2022). For example, biopolymer-based foams can be employed as energy dissipating layers in combination with stiffer ballistic mats. It can offer better security and at the same time offer a slim and comfortable look. However, it is necessary to mention that the creation of biopolymer-based armor materials is still in the process of investigation. There are still some issues regarding the improvement of ballistic resistance, the mechanical properties and the durability of these materials. But there are certain advantages of using biopolymers in armor applications and as such the future research in this area is expected to bring further development in this field (Odesanya et al. 2021).

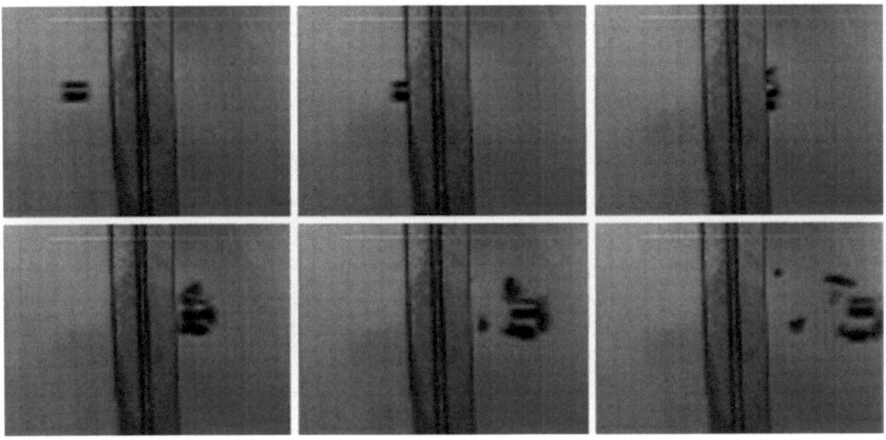

Fig. 8.9 *Manicaria saccifera* natural fabric reinforced PLA composite undergoing ballistic projectile test (Quintero et al. 2018)

Therefore, biopolymers can be considered as coherent candidates for substitution of the traditional armor materials because biopolymers have many advantages such as low density, biodegradability and biocompatibility in the condition of military and policemen's usage. Incorporating reinforcing agents into biopolymers and utilizing hybrid material systems enables researchers to develop advanced armor that enhances protection while improving wearer comfort and mobility. With further development of research in this area, biopolymer-based armor materials are expected to have a significant role in protecting personnel in high-risk areas.

8.4.2 Packaging and Storage Solutions

Military packaging materials serve distinct purposes compared to commercial packaging. Certain military equipment requires long-term storage. Military preservation involves safeguarding objects during transportation, storage, and distribution. Military packaging prioritizes materials with high mechanical strength, waterproofing, and barrier qualities over commercial options. Nanocellulose, a biopolymer with outstanding characteristics that can greatly improve the efficiency and eco-friendliness of military materials (Taherimehr et al. 2021). Because of its high strength and low density, it is used in applications where weight reduction and strength are paramount. Military goals of environmental sustainability are met by nanocellulose as it is biodegradable and derived from renewable resources. Further, the enhancement of barrier properties can prevent moisture, oxygen or any other contaminant to affect sensitive equipment (Jafarzadeh et al. 2024). These properties make nanocellulose suitable for use in combination with other polymers like biopolymers like polylactic acid (PLA), starch, and polyhydroxyalkanoate (PHA) to develop

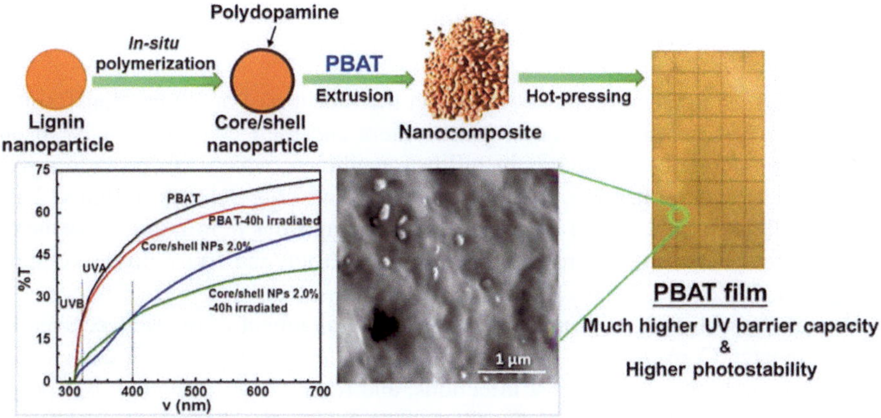

Fig. 8.10 Development of Lignin nanoparticle and PBAT-based food packaging material (Xing et al. 2019)

smart packaging that is functional and has enhanced properties. For instance, the use of nanocellulose in PLA can give high strength, lightweight, and biodegradable material for packaging electronics. In the same way, starch modified with nanocellulose can be a more sustainable and moisture barrier material for food packaging applications (Liu et al. 2021a). Biopolymers, such as thermoplastic starch, PLA, cellulose, and Polybutylene adipate terephthalate (PBAT), offer promising alternatives to synthetic polymers in food packaging. Xing et al. (2019) used core–shell lignin–melanin nanoparticles to make inexpensive, biodegradable nanofillers that improve the strength, UV protection, and durability of biodegradable PBAT films, making them more useful for packaging and agriculture (Fig. 8.10). Their biodegradability, eco-friendliness, and non-toxicity make them attractive options. However, their limitations in mechanical, barrier, and thermal properties can be addressed by incorporating reinforcement agents like nanofillers and active agents.

Biopolymers are quite brittle and will absorb water which makes it less suitable for direct use in some packaging uses (Abdullah et al. 2022). However, there is current research to solve these problems, for example, by applying multilayer structures or by adding substances enhancing the mechanical characteristics and water resistance of military packaging. Nevertheless, these challenges must be overcome and further applications developed to make biopolymers the future of military packaging. It provides a long-lasting, efficient, and multifunctional solution which can improve the coverage, sturdiness, and eco-friendliness of military equipment.

8.4.3 Medical Applications in the Field

Biomaterials as a discipline have evolved in the recent past and have transformed many areas of medicine. Of these, biopolymers and their composites have been identified as potential candidates in military medicine, which can provide novel approaches to wound healing, tissue engineering and drug delivery systems. This section will discuss the use of biopolymers and composites in military medicine in detail with respect to wound dressings and tissue scaffolds and drug delivery systems (Peng et al. 2022).

Wound dressings are of great importance in the treatment of injuries received in the war theater. Conventional dressings are sometimes confined to their capacity to heal the wound, protect it against infections, and reduce scar formation. Bio polymers are then natural polymers with origins from plant, animal or microbial sources and have some benefits over synthetic polymers. They are generally biocompatible, biodegradable, and have inherent properties which are suitable for wound healing. Among biopolymers used in wound dressings, chitosan is perhaps the most well-known. Derived from chitin, a natural component of crustacean exoskeletons, it is widely recognized for its beneficial properties (Zhao et al. 2023b; Ji et al. 2022).

Chitosan has antibacterial activity, stimulates cell division and can also take up excess exudate, thus providing a good environment for tissue repair. In Fig. 8.11a, it is observed that Chitosan stimulates macrophages to have tumoricidal effects and provides a non-protein matrix for 3D tissue development. It promotes the organization of histoarchitectural tissue and cell proliferation. As a hemostat, chitosan reduces pain by blocking nerve terminals and promoting spontaneous blood coagulation. N-acetyl-b-D-glucosamine, which is released as chitosan progressively depolymerizes, promotes fibroblast proliferation, aids in the orderly deposition of collagen, and raises the amount of natural hyaluronic acid production at the wound site. It promotes quicker wound healing and the avoidance of scarring (Paul and Sharma 2004). There was no discernible difference in the rate of wound healing between the PLA and PLA/PVA/SA groups. Figure 8.11b demonstrated that wounds treated with dressings generally healed more quickly than the control group. According to the test results, PLA's biocompatibility and porous nature promoted wound healing while preserving the fiber's desirable morphology as a dressing. Because of its high hygroscopicity and biocompatibility, SA can absorb wound exudate and keep the surrounding environment moist, which reduces the inflammatory phase of wound recovery and speeds up skin wound healing. However, SA damages the dressing's microstructure by absorbing interstitial fluid (Bi et al. 2020). Figure 8.11c illustrates the degree of wound healing seen while applying various dressings on various days which makes it evident that using PVA/chitosan nanomembrane wound dressings reduces the area of the wound (Chellamani et al. 2012).

Moreover, the incorporation of other biopolymers or inorganic nanoparticles with chitosan is possible to improve the mechanical characteristic and usefulness of the material. Another biopolymer that has potential for use in wound dressing is alginate derived from brown seaweed. Alginate is water-swellable, which forms hydrogel,

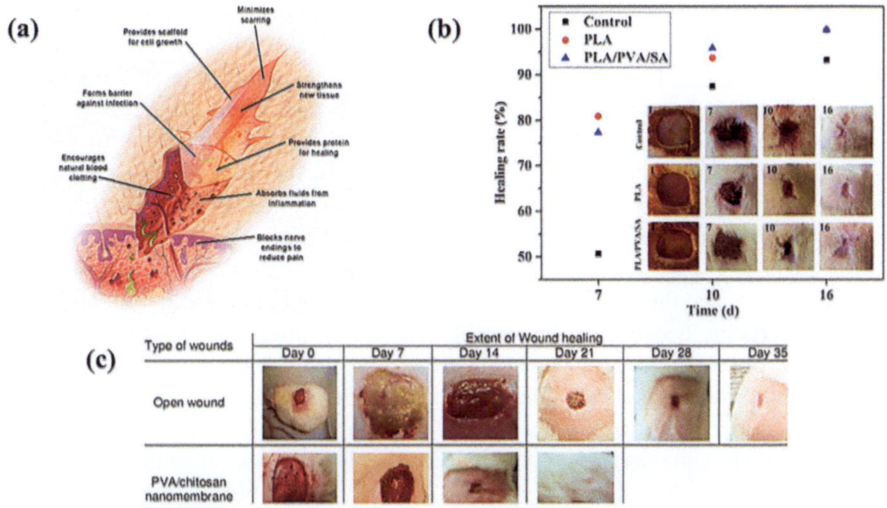

Fig. 8.11 **a** Advantages of using chitosan as a wound dressing material (Paul and Sharma 2004). **b** Healing rate and wound closure images in PLA fiber-treated and PLA/PVA/SA fiber-treated membranes (Bi et al. 2020) and **c** PVA/chitosan nano membrane for wound healing (Chellamani et al. 2012)

a suitable environment for wound healing once immersed in water. It is also very hydrophilic and can be incorporated into dressings that stick to the wound but do not cause pain when they are removed. Also, alginate can be further chemically altered to include antibiotics or other growth promoting factors to enhance tissue healing (Moeini et al. 2020).

Tissue scaffolds act as 'smart neutral' structures which facilitate the growth and remodeling of damaged tissues. In the military, tissue scaffolds are useful in the management of complicated injuries including burn and soft tissue injuries. Biopolymers have been widely used for constructing tissue scaffolds because of the opportunity to create structures with necessary characteristics. For example, the extra cellular matrix component called collagen can be employed for fabricating scaffolds that will provide an environment like that of tissues (Pita-Vilar et al. 2023; Zhong et al. 2021). Collagen-based scaffolds offer a three-dimensional template for cell adhesion and mobility and can be bio activated with growth factors for tissue regeneration. Like other proteins that are derived from collagen, gelatin can also be employed to produce scaffolds with variable mechanical characteristics and biodegradability (Barbu et al. 2021; Varaprasad et al. 2020).

The management of wounds and control of infection in military situations requires efficient drug delivery. Some of the drawbacks of conventional systems include short circulation half-life, off-target side effects, and poor drug release in the targeted area. Biopolymers and their related composites present new ways for dealing with such issues. Microspheres and nanoparticles produced from biopolymers are useful in encapsulation and delivery of different forms of drugs. These carriers can be designed to deliver the drugs in a controlled way, so that there are always therapeutic concentrations of the drug at the wound site. Furthermore, biopolymer-based carriers can be modified with targeting ligands to improve the drug delivery to certain cells or tissue. Another potential drug delivery system is the use of biopolymer hydrogels. Hydrogels can be incorporated with drug and can be applied directly to the wound site which will act as a controlled release system. Furthermore, hydrogels can be prepared to have stimuli sensitive properties including pH and temperature, which enables the controlled release of drugs (Dreiss 2020; Pushpamalar et al. 2021; Thang et al. 2023).

Biopolymers and their composites have been identified as important materials for Military Medicine because of their ability to provide novel solutions to problems related to wound healing, tissue engineering, and drug delivery. These materials have inherent properties of biocompatibility, biodegradability and functionality that mirror the properties of the tissues needed at the apt site of infection and healing. Biopolymers are suitable for varying medical applications because they are tacky, disposable, renewable, and biodegradable making it possible for researchers and clinicians to introduce advanced medical technologies to treat injured soldiers and enhance their quality of life.

8.4.4 Electro Magnetic Shielding

Electromagnetic (EM) radiation has proven to be a very useful asset in modern warfare, with the help of different EM energy such as radio waves, microwaves, infrared radiation among others in different military systems. However, unwanted exposure to EM radiation is a threat to humans as well as functioning electronic devices which are classified as sensitive. This has prompted a need for materials that are able to easily shield or block EM radiation, especially in military applications. Copper, aluminum and steel are some of the metals that are typically known to be used as shielding materials against EM. It is true that this offer shielding performance is efficient, but they don't come cheap and are generally bulky which makes handling a problem. This is likely to pose problems in a military setup whereby lightweight materials are needed (Sun et al. 2023).

An interesting alternative material for this application would be polymers, biopolymers, in this case that are obtained from plants, animal and microbe sources. For the most part, biopolymers are lightweight, biodegradable and with favorable interactions with human bodies hence ideal shielding materials for military use. In addition, biopolymers can be doped with skins such as carbon nanotube jelly and silver nano jelly to improve their EM shielding capabilities. Chitin derived from crustaceans' exoskeletons, which can further be processed into chitosan forms, is being put forward as a hopeful candidate for polymers in EM shielding. Tan et al. (2022) their work presented a series of CS/MXene alternating layered films that are inspired by the electrostatic interaction between oppositely charged CS solution and

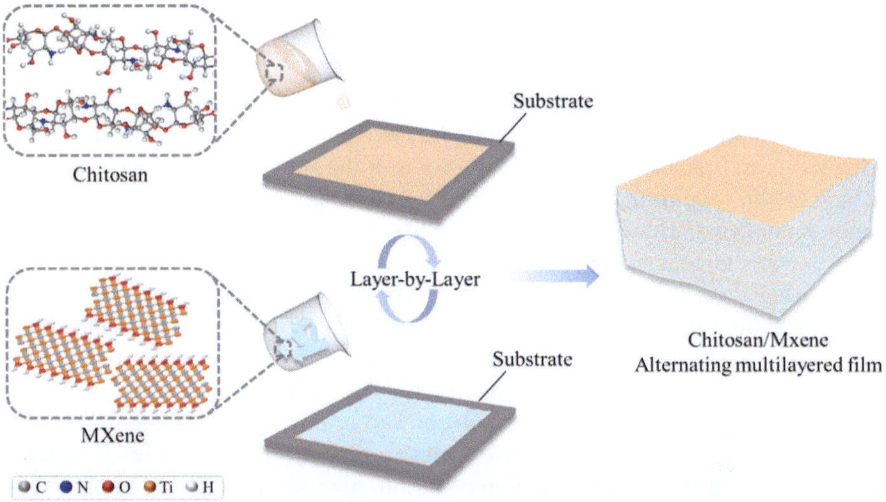

Fig. 8.12 Chitosan/MXene multilayered film and EMI shielding (Tan et al. 2022)

transition metal carbide (MXene) solution, using a layer-by-layer assembly technique which has been shown in Fig. 8.12. The resulting 35-µm thick CS/MXene multilayer film exhibits good Joule heating capacity, superior thermal conductivity (6.3 W m^{-1} K^{-1}), and a maximum EMI SE of 40.8 dB with a high specific EMI SE (SSEt) of 10 650 dB cm^2 g^{-1}. Chitosan exhibits natural antibacterial activity, and its structure can easily be changed to add conductive fillers. Researchers have confirmed that chitosan matrix composites can effectively shield electric fields of wide range of frequencies including radar, communication systems, and electronic warfare (Li et al. 2022).

Cellulose, the most common organic substance, is another biopolymer worthy of attention. Cellulose is obtained from a wide range of sources, including trees, cotton and bamboo. When combined with conductive fillers, cellulose-based composites demonstrate outstanding EM shielding efficiency. Cellulose is also biodegradable and renewable, which can contribute to making military practices more eco-friendly. Biopolymer composites demonstrate better performance than many other available materials for military applications as EM shielding materials also. They are lightweight and flexible, hence improving comfort and, therefore, the effectiveness of soldiers using protective equipment. Biopolymers resist corrosion and degrading; hence, the EM shielding material can be relied upon to last in extreme military conditions (Gebrekrstos et al. 2022).

8.4.5 *Wearable Sensor*

Chitosan, gelatin, silk, PLA, and PHAs are well-suited for military wearable sensors because they are compatible with the body, flexible, can break down naturally, are sustainable, and can incorporate sensors. Some of the applications of these materials include health monitoring, environmental monitoring, communication and performance monitoring. However, some issues like the small size of the sensor, sensitivity, power source, and data processing remain to be solved to make biopolymer-based military wearable sensors effective. These challenges can be overcome to provide solutions that improve soldier performance and safety, as well as the ability to effectively understand their environment (Liu et al. 2018; Gan et al. 2020).

To find a solution to the problem, MWCNTs/Ag/PLA nanofibrous membrane (MAPNM) sensor was developed which showed promising parameters. The resistance change signals for 50 consecutive cycles are displayed in Fig. 8.13a–d as a function of the folding and unfolding actions of various body sections. The MAPNM could react quickly, precisely, and reliably to all performed body actions (Gan et al. 2020). Liu et al. (2018) designed and fabricated a new biosensor system utilizing indium oxide nanoribbons. Figure 8.13e shows the chitosan-SWCNT sensor performance for eyeball movement, artificial hand and glucose concentration. The high-performance In_2O_3 nanoribbon sensing platform showed great potential to serve as essential components for wearable healthcare electronics since the glucose sensors exhibited a very wide detection range spanning at least 5 orders of magnitude and a detection limit down to 10 nM with electrodes functionalized with glucose oxidase, chitosan, and single-walled carbon nanotubes. All the shadow-mask-fabricated devices had good electrical performance including high mobility, on–off ratio, and mechanical stability. The biosensors designed from chitosan, glucose oxidase, and SWCNT had a broad linear range and low detection limit. This new technology has great potential in wearable healthcare electronics as a useful device that can measure health status and facilitate individualized medicine.

8.5 Research and Development Challenges

Biopolymers, which are synthesized from renewable biological resources, can be used as an alternative to conventional petroleum-based plastics, which are a cause of environmental degradation and depletion of natural resources. But their use is limited by several research and development challenges (Fig. 8.14). One of the major challenges is the difference in performance between biopolymers and traditional plastics. Biopolymers possess some unique characteristics such as biodegradability and renewability, but they lack some properties such as mechanical strength, thermal stability and chemical resistance. This limitation limits their use in important areas of use especially in the aerospace, automotive and electronics industries. To address

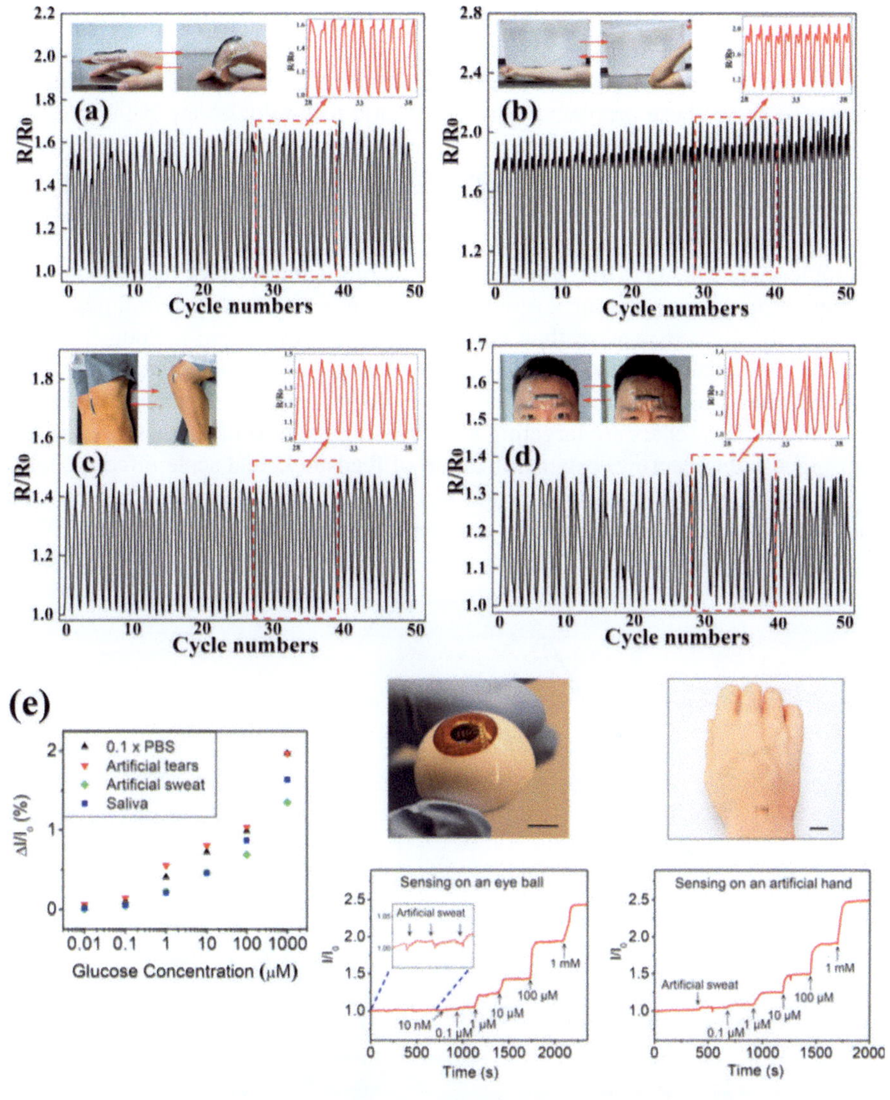

Fig. 8.13 Real-time electrical resistance response of MWCNTs/Ag/PLA nanofibrous membrane for **a** finger bending, **b** inner elbow bending, **c** knee bending, and **d** forehead movement for 50 cycles (Gan et al. 2020), **e** chitosan-SWCNT sensor performance (Liu et al. 2018)

this issue, researchers are currently trying to develop ways on how the mechanical properties of biopolymers can be improved. This includes polymer blending, formation of composites and chemical modification. When biopolymers are blended with synthetic polymers or reinforced with other materials, it is easy to achieve the desired performance characteristics (Ding et al. 2021). Another challenge is the challenge of scaling up the synthesis of biopolymers. The current manufacturing processes are generally energy-demanding and need specific structures to be put in place. However, the feedstock that can be used in the production of these polymers is renewable but may be scarce depending on the geographical location of the country. To solve these problems, new production technologies are being created. For example, biorefineries are being developed to transform a vast array of biomass into biopolymers and other desirable products. Further, biopolymers are produced through fermentation processes due to the development in biotechnology, which can be more effective and sustainable (Ding et al. 2021).

Another major factor is the cost-effectiveness of the biopolymer-based materials. Although biopolymers present several advantages in terms of the environment, they are generally more expensive than conventional plastics. These include the need to source specialized feedstocks, the processes involved in manufacturing as well as the relatively low volumes of production. To increase the competitiveness of biopolymers, researchers are targeting at decreasing the cost of production through better process engineering, better utilization of feedstock, and scale effects. Moreover, looking for new opportunities and niches in which biopolymers can provide more benefits than their competitors can explain the higher costs (Ji et al. 2022).

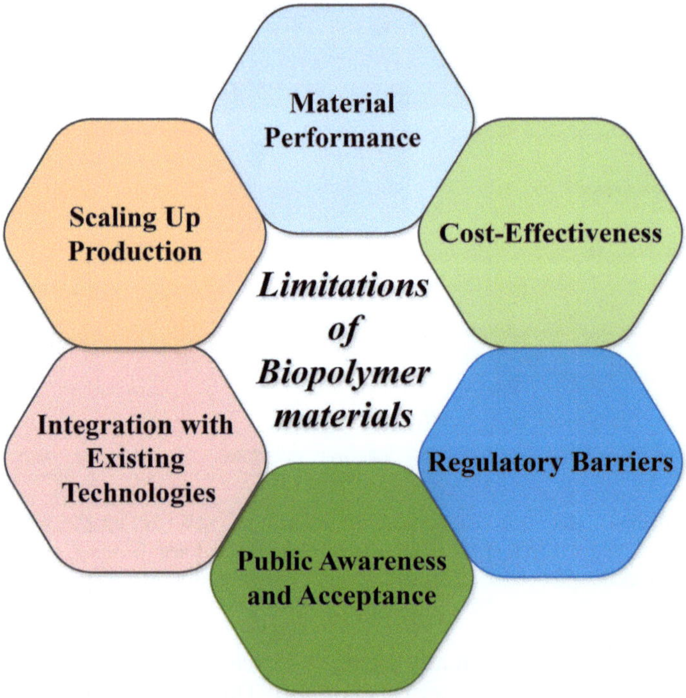

Fig. 8.14 Research and development challenges associated with biopolymer-based materials

The application of biopolymer-based materials into the existing military technologies has its own advantages and disadvantages. Military applications require materials with specific performance requirements including high durability, light weight and ability to withstand adverse conditions. Biopolymers may not be effective for all military uses but they can be effective in packaging, insulation and protective clothing. To integrate biopolymers into military applications, a collaborative relationship between military technologists and material scientists is essential. This partnership will help identify suitable applications, develop specialized materials, and address performance and reliability challenges (Isogai 2020).

Therefore, the following research and development challenges affect the development and adoption of biopolymer-based materials. Addressing these challenges requires collaboration between academia, industry, and government. Efforts should focus on improving biopolymer performance, increasing production, reducing costs, and integrating them into existing technologies. If these problems are solved, biopolymers can contribute greatly to the shift toward a circular economy.

8.6 Future Trends in Biopolymer Development

The growth of biopolymers is expected to increase in the future due to technological development, changes in consumer trends and consciousness of environmental impacts. There is great potential for improvement in the properties of biopolymers using nanotechnology. Thus, the use of nanomaterials including carbon nanotubes, graphene, and nanoparticles can enhance mechanical properties, thermal stability, electrical conductivity, and barrier characteristics. This creates new opportunities for the use of biopolymers in electronics, energy storage, and medicine. Another area of current research interest is biopolymer composites where biopolymers are reinforced with other materials such as cellulose, hemp, or flax. These composites provide a good combination between the environmental aspect and mechanical properties; thus, they can be used in automotive parts, packaging materials, and construction products (Liu et al. 2021b). With increasing awareness of environmental issues, more attention will be paid to the sustainability of biopolymer material. This entails evaluating their whole life cycle right from feedstock production to their disposal. Evaluations of the life cycle will assist in determining possible environmental effects and guide choices in materials and their processing. The future of biopolymers will be sustained by the development of new biorefining technologies. The transformation of biomass into numerous valuable products at biorefineries can help minimize waste and enhance the use of resources. Further, attempts will be made to produce biopolymers that are biodegradable and biodegradable and compostable to reduce their impact on the environment (Taib et al. 2023).

Biopolymers can be used in many civilian fields apart from packaging and agriculture, which have been the major uses of the material. For example, in the automotive industry, biopolymer-based components can help to cut the weight of the car and emissions. In the textile industry, biopolymer fibers can be used as a replacement to synthetic fibers which are environmentally unfriendly. However, biopolymers can also be used in other fields of medical technology, for instance, in tissue engineering, drug delivery and in medical devices. For such purposes, biocompatibility and biodegradability features of such polymers are rather compelling. Scientists are developing biopolymers that offer performance characteristics that are at least equivalent to those of conventional plastics. This will increase their use in challenging sectors such as aerospace and electronics (Zhao et al. 2022).

3D printing can incorporate biopolymers and produce complex structures tailor-made products. This technology has the potential to change the way manufacturing processes are done, and the number of wastes produced. With focus on biopolymers in the development of new forms of advanced biotechnology, it is possible to mention the creation of biopolymers for specialized uses such as in medical implants and in wearables. This could result in us finding more suitable and appropriate solutions to be provided to the customers. Biopolymers can be of great importance in a circular economy since they can be recycled and used again. This is because when designing biopolymer products, it is possible to consider the end of life of the product and therefore reduce wastage of resources. Thus, the further prospects for biopolymer development are quite promising, and there are many opportunities for further research and development. We can contribute to the development of a more circular economy if we can overcome the obstacles and seize the potential these materials provide.

8.7 Conclusion

Biopolymer-based materials offer a promising alternative to traditional plastics, helping to reduce environmental pollution while providing useful properties like biodegradability and renewability. However, their widespread use, especially in military applications, faces challenges such as lower mechanical strength, thermal stability, and chemical resistance compared to synthetic plastics. Researchers are working on ways to improve these properties by blending biopolymers with other materials, creating composites, and modifying their chemistry. Additionally, large-scale production is still expensive and energy intensive, making it necessary to develop more efficient and cost-effective manufacturing processes. Collaboration between scientists, industries, and government agencies will be key to overcoming these challenges and finding the best ways to use biopolymers in defense applications.

Looking ahead, new technologies like nanotechnology and composite materials will help improve the performance of biopolymers, making them more competitive with traditional plastics. Advances in areas like 3D printing and biorefineries will also create new opportunities for using these materials in military gear, medical devices, and electronics. As awareness of environmental issues grows and regulations push for greener solutions, biopolymers have the potential to play a bigger role in creating more sustainable products. However, continued research, investment, and collaboration will be essential to fully unlock their potential and expand their use in both military and civilian applications.

References

Abdullah et al (2022) Biopolymer-based functional films for packaging applications: A review. Front Nutr 9:1000116

Agarwal S, Singhal S, Godiya CB, Kumar S (2023) Prospects and applications of starch based biopolymers. Int J Environ Anal Chem 103(18):6907–6926. https://doi.org/10.1080/03067319.2021.1963717

Aryan V, Maga D, Majgaonkar P, Hanich R (2021) Valorisation of polylactic acid (PLA) waste: A comparative life cycle assessment of various solvent-based chemical recycling technologies. Resour Conserv Recycl 172:105670

Barbu A, Neamtu B, Zăhan M, Iancu GM, Bacila C, Mireşan V (2021) Current trends in advanced alginate-based wound dressings for chronic wounds. J Pers Med 11(9):890

Beltrán FR, Lorenzo V, Acosta J, De La Orden MU, Urreaga JM (2018) Effect of simulated mechanical recycling processes on the structure and properties of poly (lactic acid). J Environ Manage 216:25–31

Bi H, Feng T, Li B, Han Y (2020) In Vitro and In Vivo comparison study of electrospun PLA and PLA/PVA/SA fiber membranes for wound healing. Polymers (Basel) 12(4). https://doi.org/10.3390/polym12040839

Burande BC, Dhakite PA, Rawat SG National conference on recent trends in synthesis and characterization of futuristic material in science for the development of society (NCRDAMDS-2018) in association with." [Online]. www.ijsrst.com

Bioplastics Market Data (2022) European bioplastics

Chellamani K, Palanisamy S, Suresham T (2012) Wound dressing made out of Poly Vinyl Alcohol/Chitosan nanomembranes. J Acad Ind Res 1:342–347

Compart J, Li X, Fettke J (2021) Starch-A complex and undeciphered biopolymer. J Plant Physiol 258–259:153389. https://doi.org/10.1016/j.jplph.2021.153389

Conceição Costa V et al (2007) Preparation of hybrid biomaterials for bone tissue engineering

Cosate de Andrade MF, Souza PMS, Cavalett O, Morales AR (2016) Life cycle assessment of poly (lactic acid) (PLA): Comparison between chemical recycling, mechanical recycling and composting. J Polym Environ 24:372–384

D'Arienzo L, Acierno S, Patti A, Di Maio L (2024) Cellulose/Polyhydroxybutyrate (PHB) composites as a sustainable bio-based feedstock to 3d-printing applications. Materials 17(4). https://doi.org/10.3390/ma17040916

Deeraj BDS, Jayan JS, Saritha A, Joseph K (2022) 8—PLA-based blends and composites. In: Mavinkere Rangappa S, Parameswaranpillai J, Siengchin S, Ramesh M (eds) Biodegradable polymers, blends and composites. Woodhead Publishing, pp 237–281. https://doi.org/10.1016/B978-0-12-823791-5.00014-4

Ding S, Wang Y, Li J, Chen S (2021) Progress and prospects in chitosan derivatives: Modification strategies and medical applications. J Mater Sci Technol 89:209–224

Doble M (2005) Biodegradation of polymers

Dreiss CA (2020) Hydrogel design strategies for drug delivery. Curr Opin Colloid Interface Sci 48:1–17

Ejaz M, Azad MM, Shah AUR, Afaq SK, Song J (2020) Mechanical and biodegradable properties of jute/flax reinforced PLA composites. Fibers and Polymers 21:2635–2641

Gan L et al (2020) Antibacterial, flexible, and conductive membrane based on MWCNTs/Ag Coated Electro-Spun PLA nanofibrous scaffolds as wearable fabric for body motion sensing. Polymers (Basel) 12(1). https://doi.org/10.3390/polym12010120

Gebrekrstos A, Orasugh JT, Muzata TS, Ray SS (2022) Cellulose-based sustainable composites: A review of systems for applications in EMI shielding and sensors. Macromol Mater Eng 307(9):2200185

Gowthaman NSK, Lim HN, Sreeraj TR, Amalraj A, Gopi S (2021) Advantages of biopolymers over synthetic polymers: social, economic, and environmental aspects. In: Biopolymers and their industrial applications. Elsevier, pp 351–372

He Y et al (2022) Intelligent pH-sensing film based on polyvinyl alcohol/cellulose nanocrystal with purple cabbage anthocyanins for visually monitoring shrimp freshness. Int J Biol Macromol 218:900–908. https://doi.org/10.1016/j.ijbiomac.2022.07.194

Huang SJ (2005) Poly (lactic acid) and copolyesters. Handb Biodegrad Polyms: 287–294

Isogai A (2020) Cellulose nanofibers: Recent progress and future prospects. J Fiber Sci Technol 76(10):310–326

Jafarzadeh S et al (2024) Green synthesis of nanomaterials for smart biopolymer packaging: challenges and outlooks. J Nanostructure Chem 14(2):113–136

Janssen LPBM, Moscicki L (2009) Thermoplastic starch : a green material for various industries. Wiley-VCH

Ji M et al (2022) Advances in chitosan-based wound dressings: Modifications, fabrications, applications and prospects. Carbohydr Polym 297:120058

John A, Črešnar KP, Bikiaris DN, Zemljič LF (2023) Colloidal solutions as advanced coatings for active packaging development: focus on PLA systems. Polymers (Basel) 15(2):273

Kim J, Park S, Park K, Kim B-K (2023) Non-toxic natural additives to improve the electrical conductivity and viscosity of polycaprolactone for melt electrospinning. Appl Sci 13:1844. https://doi.org/10.3390/app13031844

Klemm D, Heublein B, Fink H-P, Bohn A (2005) Cellulose: Fascinating biopolymer and sustainable raw material. Angew Chem Int Ed 44(22):3358–3393. https://doi.org/10.1002/anie.200460587

Lei Y, Mao L, Yao J, Zhu H (2021) Improved mechanical, antibacterial and UV barrier properties of catechol-functionalized chitosan/polyvinyl alcohol biodegradable composites for active food packaging. Carbohydr Polym 264:117997. https://doi.org/10.1016/j.carbpol.2021.117997

Li M et al (2022) Recent progress in the application of cellulose in electromagnetic interference shielding materials. Macromol Mater Eng 307(7):2100899

Li X et al (2023) Efficient fabrication of PLA/PHB composites with enhanced mechanical properties, excellent thermal stability, fast crystallization ability, and degradation rate via the synergistic of weak shear field and melt quenching technique. Ind Crops Prod 196:116516. https://doi.org/10.1016/j.indcrop.2023.116516

Liu K et al (2021b) Recent advances in cellulose and its derivatives for oilfield applications. Carbohydr Polym 259:117740

Liu Q et al (2018) Highly sensitive and wearable In2O3 nanoribbon transistor biosensors with integrated on-chip gate for glucose monitoring in body fluids. ACS Nano 12(2):1170–1178

Liu Y et al (2021a) A review of cellulose and its derivatives in biopolymer-based for food packaging application. Trends Food Sci Technol 112:532–546. https://doi.org/10.1016/j.tifs.2021.04.016

Lactic Acid Market (2020) Markets and markets

Ma Q et al (2024) Effect of infill pattern of polylactide acid (PLA) 3D-printed integral sandwich panels under ballistic impact loading. Mater Today Commun 38:107626

Mabrouk M, Mostafa AA, Oudadesse H, Mahmoud AA, El-Gohary MI (2014) Effect of ciprofloxacin incorporation in PVA and PVA bioactive glass composite scaffolds. Ceram Int 40(3):4833–4845. https://doi.org/10.1016/j.ceramint.2013.09.033

Mansur HS, Costa HS (2008) Nanostructured poly(vinyl alcohol)/bioactive glass and poly(vinyl alcohol)/chitosan/bioactive glass hybrid scaffolds for biomedical applications. Chem Eng J 137(1):72–83. https://doi.org/10.1016/j.cej.2007.09.036

Martinez K, Arias-Tapia M, Useche J, Escobar Macías D (2022) "Thermoplastic Starch (TPS)/Polylactic Acid (PLA) blending methodologies: a review. J Polym Environ 30. https://doi.org/10.1007/s10924-021-02207-1

Menazea AA, Ismail AM, Awwad NS, Ibrahium HA (2020) Physical characterization and antibacterial activity of PVA/Chitosan matrix doped by selenium nanoparticles prepared via one-pot laser ablation route. J Market Res 9(5):9598–9606. https://doi.org/10.1016/j.jmrt.2020.06.077
Moeini A, Pedram P, Makvandi P, Malinconico M, d'Ayala GG (2020) Wound healing and antimicrobial effect of active secondary metabolites in chitosan-based wound dressings: A review. Carbohydr Polym 233:115839
Mohamed RM, Yusoh K (2016) A review on the recent research of Polycaprolactone (PCL). Adv Mat Res 1134:249–255. https://doi.org/10.4028/www.scientific.net/AMR.1134.249
Montero B, Rico M, Rodríguez-Llamazares S, Barral L, Bouza R (2017) Effect of nanocellulose as a filler on biodegradable thermoplastic starch films from tuber, cereal and legume. Carbohydr Polym 157:1094–1104. https://doi.org/10.1016/j.carbpol.2016.10.073
Mori R (2023) Replacing all petroleum-based chemical products with natural biomass-based chemical products: a tutorial review. RSC Sustain. 1(2):179–212. https://doi.org/10.1039/D2SU00014H
Mousavioun P, George GA, Doherty WOS (2012) Environmental degradation of lignin/poly(hydroxybutyrate) blends. Polym Degrad Stab 97(7):1114–1122. https://doi.org/10.1016/j.polymdegradstab.2012.04.004
Moyo M, Kanny K, Velmurugan R (2020) Performance of kenaf non-woven mat/PLA biocomposites under medium velocity impact. Fibers and Polymers 21:2642–2651
Muneer F, Rasul I, Azeem F, Siddique MH, Zubair M, Nadeem H (2020) Microbial Polyhydroxyalkanoates (PHAs): Efficient replacement of synthetic polymers. J Polym Environ 28(9):2301–2323. https://doi.org/10.1007/s10924-020-01772-1
Norrrahim MNF et al (2021) Nanocellulose: The next super versatile material for the military. Mater Adv 2(5):1485–1506
Odesanya KO, Ahmad R, Jawaid M, Bingol S, Adebayo GO, Wong YH (2021) Natural fibre-reinforced composite for ballistic applications: A review. J Polym Environ 29(12):3795–3812
Panaitescu DM et al (2018) Poly(3-hydroxybutyrate) Modified by nanocellulose and plasma treatment for packaging applications. Polymers (Basel) 10(11). https://doi.org/10.3390/polym10111249
"Polylactic acid," Wikipedia
Lopez Vicente P Future advanced materials for defence applications. European Defence Matters
Pattanashetti NA, Heggannavar GB, Kariduraganavar MY (2017) Smart biopolymers and their biomedical applications. Procedia Manuf 12:263–279. https://doi.org/10.1016/j.promfg.2017.08.030
Paul W, Sharma CP (2004) Chitosan and Alginate wound dressings: A short review. [Online]. http://www.sbaoi.org
Peng W et al (2022) Recent progress of collagen, chitosan, alginate and other hydrogels in skin repair and wound dressing applications. Int J Biol Macromol 208:400–408
Philibert T, Lee BH, Fabien N (2017) Current status and new perspectives on chitin and chitosan as functional biopolymers. Appl Biochem Biotechnol 181(4):1314–1337. https://doi.org/10.1007/s12010-016-2286-2
Pinto L et al (2021) Biopolymer hybrid materials: Development, characterization, and food packaging applications. Food Packag Shelf Life 28:100676. https://doi.org/10.1016/j.fpsl.2021.100676
Pita-Vilar M, Concheiro A, Alvarez-Lorenzo C, Diaz-Gomez L (2023) Recent advances in 3D printed cellulose-based wound dressings: A review on in vitro and in vivo achievements. Carbohydr Polym:121298
Pragadish N, Bakkiyaraj M, Rajakrishnamoorthy P, Ganapathy Srinivasan R, Selvam M, Thamarai Selvan A (2025) Infill structure analysis of FDM-printed PLA wood glass fiber composite for commercial product applications. Smart Mater Struct 34(3):035016
Pushpamalar J et al (2021) Development of a polysaccharide-based hydrogel drug delivery system (DDS): An update. Gels 7(4):153

Qi X, Tong X, Pan W, Zeng Q, You S, Shen J (2021) Recent advances in polysaccharide-based adsorbents for wastewater treatment. J Clean Prod 315:128221. https://doi.org/10.1016/j.jclepro.2021.128221

Quintero S, Porras A, Hernandez C, Maranon A (2018) The Response of Manicaria saccifera natural fabric reinforced pla composites to impact by fragment simulating projectiles. In: Fangueiro R, Rana S (eds) Advances in natural fibre composites. Springer International Publishing, Cham, pp 89–98

Rahman MZ, Rahman M, Mahbub T, Ashiquzzaman M, Sagadevan S, Hoque ME (2023) Advanced biopolymers for automobile and aviation engineering applications. J Polym Res 30(3):106

Rao MG, Bharathi P, Akila RM (2014) A comprehensive review on biopolymers. Sci Revs Chem Commun 4(2):61–68. [Online]. www.sadgurupublications.com

Rath A et al (2023) Fabrication of chitosan-flax composites with differing molecular weights and its effect on mechanical properties. Compos Sci Technol 235:109952

Roberts GAF, Roberts GAF (1992) Chitin chemistry. Springer

Sadasivuni KK, Saha P, Adhikari J, Deshmukh K, Ahamed MB, Cabibihan J-J (2020) Recent advances in mechanical properties of biopolymer composites: a review. Polym Compos 41(1):32–59. https://doi.org/10.1002/pc.25356

Said L, Lisdiana H, Saefurahman G (2024) Synthesis of biodegradable plastic film from chitosan-glycerol with cellulose nanowhiskers of coconut coir as reinforcing agent. In: IOP conference series: Earth and environmental science. IOP Publishing, p 012028

Samir A, Ashour F, Hakim A, Bassyouni M (2022) Recent advances in biodegradable polymers for sustainable applications. Npj Mater Degrad 6. https://doi.org/10.1038/s41529-022-00277-7

Shaghaleh H, Xu X, Wang S (2018) Current progress in production of biopolymeric materials based on cellulose, cellulose nanofibers, and cellulose derivatives. RSC Adv 8(2):825–842

Sharma V, Sehgal R, Gupta R (2021) Polyhydroxyalkanoate (PHA): Properties and modifications. Polymer (Guildf) 212:123161. https://doi.org/10.1016/j.polymer.2020.123161

Siengchin S (2023) A review on lightweight materials for defence applications: Present and future developments. Defence Technology 24:1–17. https://doi.org/10.1016/j.dt.2023.02.025

Da Silva ARP, Macedo TL, Coletta DJ, Feldman S, de Pereira MM (2016) Synthesis, characterization and cytotoxicity of Chitosan/Polyvinyl alcohol/bioactive glass hybrid scaffolds obtained by Lyophilization. Revista Materia 21(4):964–973. https://doi.org/10.1590/S1517-707620160004.0089

Singh R, Shitiz K, Singh A (2017) Chitin and chitosan: biopolymers for wound management. Int Wound J 14(6):1276–1289. https://doi.org/10.1111/iwj.12797

Siracusa V, Dalla Rosa M, Iordanskii AL (2017) Performance of poly (lactic acid) surface modified films for food packaging application. Materials 10(8):850

Song J, Winkeljann B, Lieleg O (2020) Biopolymer-based coatings: promising strategies to improve the biocompatibility and functionality of materials used in biomedical engineering. Adv Mater Interfaces 7(17):2000850. https://doi.org/10.1002/admi.202000850

Souza AT et al (2022) Ballistic properties and Izod impact resistance of novel epoxy composites reinforced with caranan fiber (Mauritiella armata). Polymers (Basel) 14(16):3348

Srivastava KR et al (2021) Effect of nanocellulose on mechanical and barrier properties of PVA–banana pseudostem fiber composite films. Environ Technol Innov 21:101312. https://doi.org/10.1016/j.eti.2020.101312

Sun Y-X et al (2023) Eco-friendly silver nanoparticles/chitosan/poly (vinyl alcohol) composites exhibit remarkable EMI shielding capabilities and outstanding thermal conductivities. ACS Appl Mater Interfaces 15(29):35631–35638

Suryanegara L et al (2021) Novel antimicrobial bioplastic based on PLA-chitosan by addition of TiO_2 and ZnO. J Environ Health Sci Eng 19(1):415–425. https://doi.org/10.1007/s40201-021-00614-z

Taherimehr M, YousefniaPasha H, Tabatabaeekoloor R, Pesaranhajiabbas E (2021) Trends and challenges of biopolymer-based nanocomposites in food packaging. Compr Rev Food Sci Food Saf 20(6):5321–5344

Taib MNAM et al (2022) Current progress of biopolymer-based flame retardant. Polym Degrad Stab 205:110153. https://doi.org/10.1016/j.polymdegradstab.2022.110153
Taib N-AAB et al (2023) A review on poly lactic acid (PLA) as a biodegradable polymer. Polym Bull 80(2):1179–1213. https://doi.org/10.1007/s00289-022-04160-y
Tan Z et al (2022) Fabrication of Chitosan/MXene multilayered film based on layer-by-layer assembly: Toward enhanced electromagnetic interference shielding and thermal management capacity. Compos Part A Appl Sci Manuf 155:106809. https://doi.org/10.1016/j.compositesa.2022.106809
Tanjung FA, Arifin Y, Husseinsyah S (2020) Enzymatic degradation of coconut shell powder–reinforced polylactic acid biocomposites. J Thermoplast Compos Mater 33(6):800–816
Tariq M, Majeed M, Banat I (2019) Recent developments in bioreactor scale production of bacterial polyhydroxyalkanoates. Bioprocess Biosyst Eng 42. https://doi.org/10.1007/s00449-019-02093-x
Thang NH, Chien TB, Cuong DX (2023) Polymer-based hydrogels applied in drug delivery: An overview. Gels 9(7):523
Van de Velde K, Kiekens P (2002) Biopolymers: overview of several properties and consequences on their applications. Polym Test 21(4):433–442. https://doi.org/10.1016/S0142-9418(01)00107-6
Varaprasad K, Jayaramudu T, Kanikireddy V, Toro C, Sadiku ER (2020) Alginate-based composite materials for wound dressing application: A mini review. Carbohydr Polym 236:116025
Xing Q, Buono P, Ruch D, Dubois P, Wu L, Wang W-J (2019) Biodegradable UV-blocking films through core-shell lignin–melanin nanoparticles in poly(butylene adipate-co-terephthalate). ACS Sustain Chem Eng 7(4):4147–4157. https://doi.org/10.1021/acssuschemeng.8b05755
Yan K et al (2024) Study on the mechanical properties and ballistic resistance of interface-modified fabrics with environmentally friendly shear thickening liquid. Compos Struct 337:118036
Yang Z, Li H, Niu G, Wang J, Zhu D (2021) Poly(vinylalcohol)/chitosan-based high-strength, fire-retardant and smoke-suppressant composite aerogels incorporating aluminum species via freeze drying. Compos B Eng 219:108919. https://doi.org/10.1016/j.compositesb.2021.108919
Zhang L et al (2021) Intumescent-grafted bamboo charcoal: A natural nontoxic fire-retardant filler for polylactic acid (PLA) composites. ACS Omega 6(41):26990–27006
Zhang L, Deng X, Huang Z (1997) Miscibility, thermal behaviour and morphological structure of poly(3-hydroxybutyrate) and ethyl cellulose binary blends. Polymer (Guildf) 38(21):5379–5387. https://doi.org/10.1016/S0032-3861(97)84642-3
Zhao J et al (2023) Chitosan-based hydrogel wound dressing: From mechanism to applications, a review. Int J Biol Macromol 244:125250
Zhao J et al (2023) Chitosan-based hydrogel wound dressing: From mechanism to applications, a review. Int J Biol Macromol 244:125250. https://doi.org/10.1016/j.ijbiomac
Zhao X, Liu J, Li J, Liang X, Zhou W, Peng S (2022) Strategies and techniques for improving heat resistance and mechanical performances of poly (lactic acid) (PLA) biodegradable materials. Int J Biol Macromol 218:115–134
Zhong Y et al (2021) Antimicrobial/biocompatible hydrogels dual-reinforced by cellulose as ultrastretchable and rapid self-healing wound dressing. Biomacromol 22(4):1654–1663
Zhu Z et al (2022) Magnesium hydroxide coated hollow glass microspheres/chitosan composite aerogels with excellent thermal insulation and flame retardancy. J Colloid Interface Sci 612:35–42. https://doi.org/10.1016/j.jcis.2021.12.138

Chapter 9
Biopolymer-Based Materials for Sustainable Energy Storage Applications

A. F. Fuzlin, N. M. Ghazali, S. Aynharan, M. Diantoro, and A. S. Samsudin

Abstract In pursuing sustainable and environmentally acceptable energy solutions, biopolymer-based materials have emerged as attractive options for energy storage applications. Eco-friendly biomass-derived biopolymers and other natural materials have attracted a lot of interest as a replacement for conventional petroleum-based materials, given the constantly increasing environmental issues caused by synthetic substances. Biopolymers are an emerging class of innovative materials with a wide range of applications and features, including great sustainability and tunability. Biopolymer applications are defined in terms of energy storage devices, specifically supercapacitors, rechargeable batteries, and fuel cells. This chapter focuses on the development, characterization, and application of biopolymer-based macro, micro, and nanocomposites, with a focus on their ability to reduce environmental footprints. In the last part of the chapter, the focus will be on recent challenges and future research directions, emphasizing the potential of biopolymer-based materials to contribute to more sustainable and efficient energy storage solutions.

A. F. Fuzlin · N. M. Ghazali · S. Aynharan · A. S. Samsudin
Ionic Materials Team, Faculty of Industrial Sciences and Technology, Universiti Malaysia Pahang Al-Sultan Abdullah, Gambang Kuantan, Pahang, Malaysia

A. F. Fuzlin
Centre for Advanced Intelligent Materials, Universiti Malaysia Pahang Al-Sultan Abdullah, Gambang Kuantan, Malaysia

S. Aynharan
Department of Physics, Faculty of Science, Eastern University, Chenkalady, Sri Lanka

M. Diantoro
Department of Physics, Faculty of Mathematics and Natural Science, Universitas Negeri Malang, Malang, Indonesia

Center of Advanced Materials for Renewable Energy, Universitas Negeri Malang, Malang, Indonesia

A. S. Samsudin (✉)
Centre for Automotive Engineering, Universiti Malaysia Pahang Al-Sultan Abdullah, Pekan, Pahang, Malaysia
e-mail: ahmadsalihin@umpsa.edu.my

L. Ravindran et al. (eds.), *Macro, Micro and Nanocomposites from Sustainable Sources*, Advanced Structured Materials 244, https://doi.org/10.1007/978-981-95-2469-3_9

9.1 Introduction

One of the greatest challenges of the twenty-first century is powering the world, and energy consumption has affected our health, welfare, economy, and society. Energy and water are the two most essential commodities for development. As a result of the rapid increase in human population, the widespread influence of globalization, the excessive consumption of energy and water resources, and the unstable state of the world economy in recent decades, the world is currently facing a multitude of environmental, financial, and social challenges that have affected nearly every corner of the globe. For a long time, our demands were met by the fossil fuels. Nevertheless, the current limiting availability of these resources indicates that these energy sources are non-renewable and have detrimental environmental impacts. The most significant repercussions include escalating fuel prices, nuclear incidents, global warming, and a dramatic annual increase in global carbon dioxide (CO_2) emissions (Lai and Locatelli 2021; Agency 2020). Contemporary energy sources, including nuclear and renewable resources, are gradually supplanting traditional energy sources. As population growth escalated in industrialized nations, per capita consumption has risen. Rapid lifestyle alterations result in heightened energy consumption. As a result, the energy sector is endeavoring to transition from fossil fuel-dependent energy generation to alternative renewable energy sources (RESs) including solar, wind, geothermal, hydropower, and biomass energy. Consequently, the transition from fossil and conventional fuels has become imperative for the contemporary world and its technological advancement.

The transition of the energy sector is a key focus of sustainable development. The insufficient utilization, manufacturing, and utilization of energy resources result in substantial harm to the environment. Figure 9.1 illustrates a forecasted trend in worldwide energy consumption by region (Amir et al. 2023). The International Energy Agency (IEA) reports that energy consumption accounts for a significant 75.6% (37.6 $GtCO_2eq$) of global anthropogenic greenhouse gas (GHG) emissions (Rothausen and Conway 2011). The primary sub-sectors of the energy sector contributing to GHG emissions are heat and electricity generation (31.8%), transportation (17%), and manufacturing and construction (12.7%) (Wali et al. 2024). As a result, global efforts have been initiated to establish and execute cross-sector activities in the energy sector, encompassing both production and consumption. Many countries have recognized the importance of sustainable energy transition and prioritized it in their national strategic documents. They have also emphasized the need for appropriate regulations, financing methods, implementation monitoring, and result evaluation (Radovanović et al. 2021).

Energy storage systems (ESS) are one of the best strategies to promptly satisfy electrical demand when the supply is from non-dispatchable sources. Moreover, ESS are transformative innovations that will influence power markets by providing numerous advantages for power systems, including improved reliability and power quality; an increased proportion of RESs; mitigation of intermittency issues associated with RESs; enhanced system stability; load management and peak shaving;

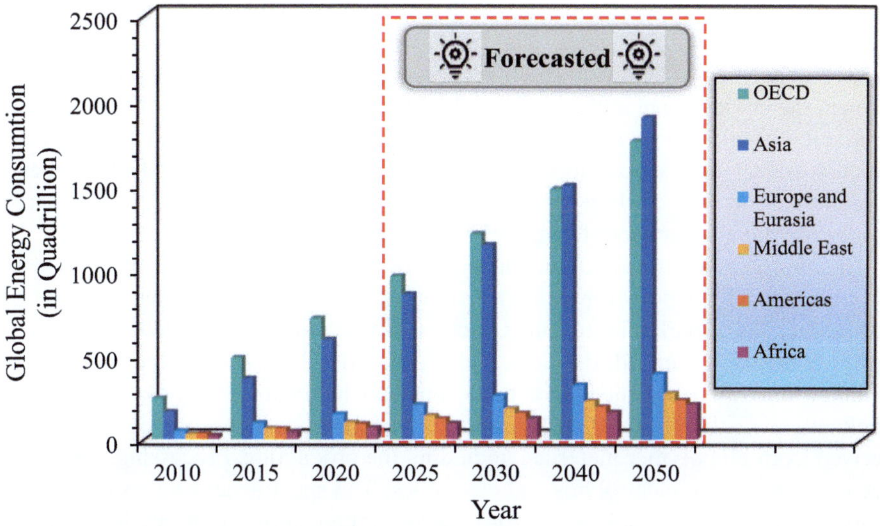

Fig. 9.1 Global consumption of energy by region until the year 2050

energy management and reduction of overall operational costs; support for spinning reserves; and alleviation of transmission congestion. Battery, fuel cell, and super-capacitors are prevalent energy storage devices. Modern technology has made it possible to transform electrical energy into forms that may be stored and then released back into the grid as and when it is required. Power grid reliability, efficiency, safety, security, and stability can all be enhanced with electricity storage, which also helps to lower the cost of electricity supply (Mostafa et al. 2020). ESS can be categorized based on their roles, storage duration, and the type of stored energy, as no single technology excels in all scenarios (Ming et al. 2019). For example, large-scale mechanical energy storage systems can transfer substantial quantities of electricity across different time periods, whereas batteries respond swiftly to signals. Consequently, a multitude of energy storage technologies with varying attributes are and will be required to serve diverse applications.

For instance, modern life relies on batteries, but the lead-acid battery was discovered 150 years ago (Ryś et al. 2024). However, to this day, lead-acid batteries are the standard for automobile starter batteries. Nonetheless, numerous issues exist regarding battery components. They predominantly depend on non-renewable sources, face difficulties in recycling, and lack regional resource availability, resulting in a significant CO_2 footprint. These challenges must be considered within the framework of a consistently rising proportion of electric vehicles, which is amplifying the demand specifically for lithium-ion batteries, alongside a swift reduction in costs that is likely to soon surpass the \$100/kWh benchmark (Schlemmer et al. 2021). An additional potential alternative for new energy material uses is biopolymers. Recent years have seen an increased demand for environmentally acceptable materials, resulting in the development of biopolymers, which are polymeric molecules formed by covalently bonded monomeric units. The transition to green energy is essential due to the prolonged use of synthetic fossil-fuel-based polymers, which has progressively harmed the environment and expedited climate change. The adverse effects of these synthetic polymers stem from their non-biodegradable properties and the synthesis process, which entails hazardous substances or produces toxic byproducts. Renewable raw resources, including organic waste from human and animal activities, serve as a beneficial alternative to fossil fuel feedstocks.

As concerns about environmental pollution intensify, biodegradable polymers have become a focal point for addressing issues of white pollution. Their widespread availability and remarkable attributes, such as affordability, biodegradability, renewability, non-toxicity, and desirable textural properties have galvanized researchers to delve into their chemical structures, physicochemical properties, and diverse applications (Zafar et al. 2024). This chapter explores energy storage components derived from biopolymer materials, including the production, processing, and characterization of both electrolytes and electrodes. The specific advantages offered by biopolymer-based materials for various energy storage applications are highlighted. To support further research and development, key challenges and perspectives are analyzed, along with proposed strategies to overcome these obstacles.

9.2 Biopolymers and Composites: Macro to Nano

Biopolymers are renewable, often biodegradable materials synthesized by biological systems like plants, microorganisms, and animals (Shaikh and Shaikh 2022). The prefix "bio" in "biopolymers" denotes that these polymers are derived from organic materials from living organisms. The "bio" prefix indicates their organic origin. Notable examples include anionic polysaccharides like alginate and carrageenan, extracted from seaweed (Awad et al. 2017), and chitosan, found in the exoskeletons of insects and crustaceans, as well as fungi and algae (Abd El Aty et al. 2017). The increasing focus on the environmental impact of fossil fuels drives the use of biopolymers. While petroleum is a crucial energy source and raw material, its environmental concerns have spurred interest in more sustainable alternatives (Nagalakshmaiah 2019).

The use of petroleum-based systems is increasingly deemed unsustainable due to their non-degradability and harmful environmental effects. As fossil fuel resources deplete, there is a growing focus on developing bio-based materials to reduce reliance on non-renewable resources (Trache et al. 2017). Biopolymers are gaining popularity as sustainable and environmentally friendly substitutes (Muneer et al. 2021). They are increasingly used in various applications, including bioplastics, agriculture, personal hygiene products, and medical products (Das et al. 2023; Sadasivuni et al. 2020). Biopolymers can be categorized into two types based on their: (1) polymer degradation and (2) source of origin, as depicted in Fig. 9.2.

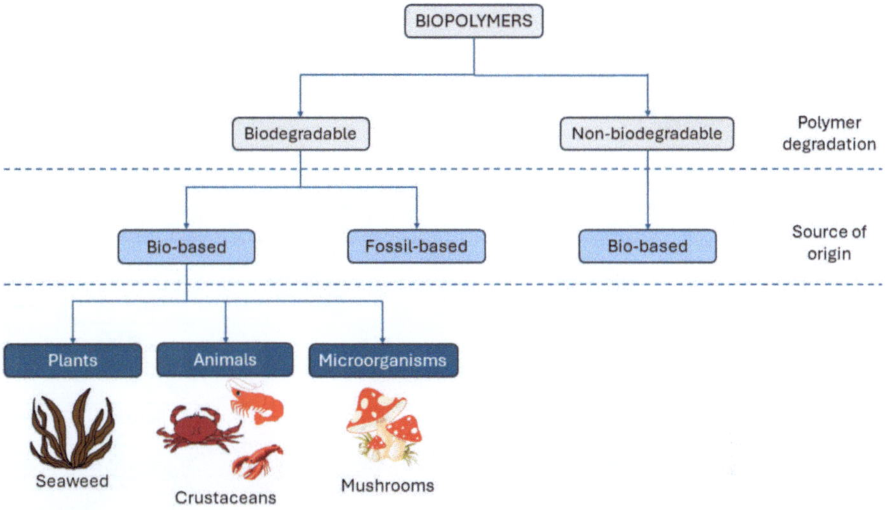

Fig. 9.2 Classification of biopolymers based on its polymer degradation and source of origin

Polymer composites combine different components to create a uniform material with unique properties (Pappu et al. 2016). Incorporating composites into polymers enhances mechanical properties like strength and stability, which is essential as biopolymers typically perform lower than synthetic materials. Adding micro/nanoscale reinforcement leads to a new class of biopolymer composites that merge the eco-friendly nature of biopolymers with the strength of petroleum-based materials. Biopolymer composites (as shown in Fig. 9.3) can be classified based on their reinforcement mechanisms (Swain 2018): (1) particle reinforcement composites, (2) structural or sandwich composites, and (3) fiber reinforcement composites, which provide exceptional strength.

Bio-nanocomposites, a class of composites that involve the integration of natural polymers to form biodegradable bio-composites (Sumrith et al. 2019), exhibit a diverse array of potential applications. For instance, in food packaging, suitable materials and techniques can effectively impede the undesired permeation of oxygen and moisture, preserving the food's physical and chemical properties. Bio-nanocomposites integrate two phases at the nanoscale (<100 nm) which these smaller particles being more advantageous to enhance the effectiveness of composites, making micro- or nanosized particles crucial in their development.

Bio-composites exhibit diverse chemical properties largely influenced by the biopolymer matrix and reinforcement type. The matrix, made of polysaccharides like cellulose or proteins such as gelatin, affects chemical stability, biodegradability, and environmental interactions like moisture and pH. Cellulose-based bio-composites have significant chemical resistance due to strong hydrogen bonding within cellulose chains (Chen et al. 2023; Wohlert 2022). Hydrogen bonding significantly enhances the stability of these bio-composites in diverse chemical environments. Additionally,

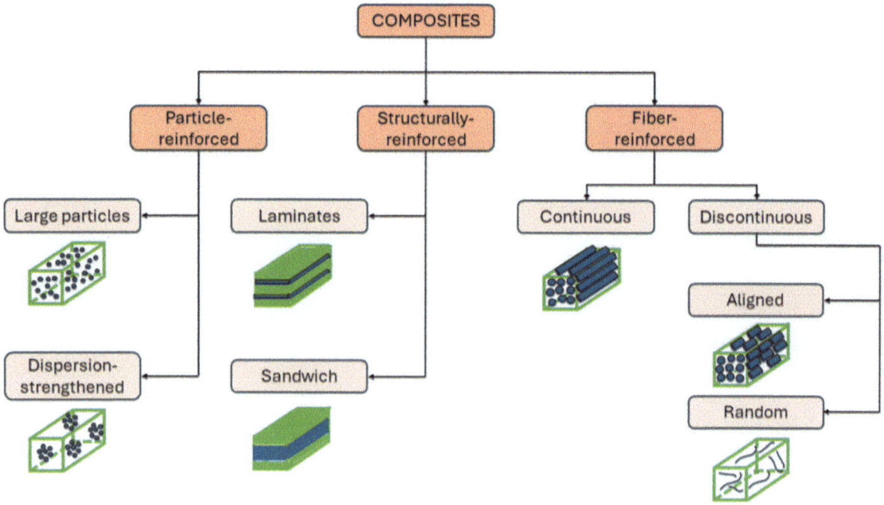

Fig. 9.3 Classification of composites

chemical modifications, such as grafting or cross-linking, can enhance properties, making these composites suitable for drug delivery systems by improving stability, reducing toxicity, and enhancing biodegradability (Mohammed et al. 2021). These adjustments increase the versatility of bio-composites for applications like packaging and biomedical devices.

The addition of organic or inorganic fillers significantly alters the chemical properties of bio-composites. Based on Fig. 9.4, nanofillers like rapheme oxide (GO), silica, or nanoclays can introduce new functional groups and improve chemical interactions within the composite. For example, hydroxyl groups in nanocellulose can form strong hydrogen bonds with the biopolymer matrix, enhancing mechanical strength and chemical resistance (Randhawa et al. 2022). Ensuring compatibility between the filler and matrix is essential for uniform dispersion and to prevent phase separation, which negatively affects properties. Research shows that surface modifications, such as functionalizing GO, can improve dispersion and stability (Zainal Abidin et al. 2018). The chemical properties of bio-composites also influence degradation, which is crucial for environmental sustainability. Biodegradable bio-composites, such as those made from poly(lactic acid) (PLA) or poly(hydroxyalkanoates) (PHA), can degrade through hydrolysis or enzymatic processes. Factors affecting degradation rates include the matrix composition, presence of hydrolytically labile bonds, and filler properties. For instance, adding natural fibers like jute to a PLA matrix can enhance biodegradation due to the fibers' hydrophilic nature, which facilitates water absorption and hydrolytic cleavage of ester bonds in PLA (Sanivada et al. 2020; Ejaz et al. 2020).

The physical properties of bio-composites are primarily influenced by the biopolymer matrix and the composition and structure of the reinforcement. Density is a key factor affecting the strength-to-weight ratio, making bio-composites suitable for

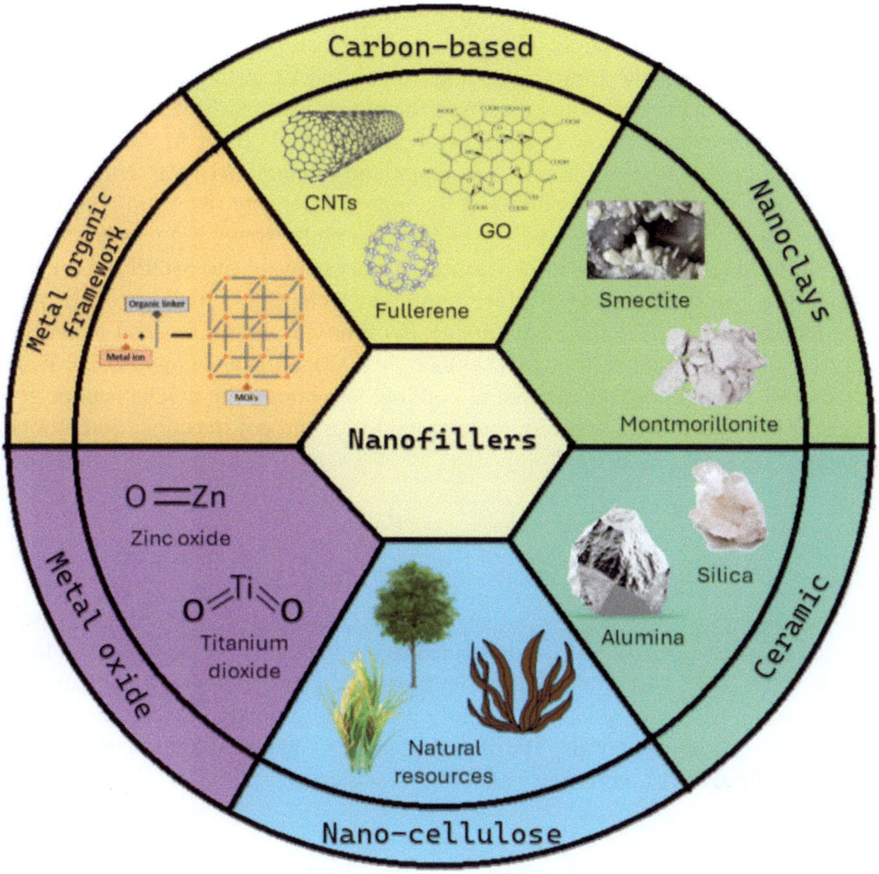

Fig. 9.4 Examples of nanofillers

lightweight applications like automotive and aerospace components. Due to the use of natural fibers or biodegradable polymers, bio-composites typically have lower densities than conventional composites, which enhances their appeal for weight-sensitive applications (Gholampour and Ozbakkaloglu 2020). For example, bio-composites reinforced with plant fibers like hemp or flax offer reduced density while maintaining good mechanical properties for lighter automotive interiors (Pokharel et al. 2022; Mahmud et al. 2021). Thermal stability is another critical property determined by the matrix and reinforcement materials. Techniques such as thermogravimetric analysis (TGA) and differential scanning calorimetry (DSC) are commonly used to assess thermal stability. Studies show that adding nanofillers like silica or montmorillonite can improve thermal stability by raising decomposition temperatures and enhancing resistance to thermal degradation (Majka 2016; Rodrigues et al. 2020). Incorporating montmorillonite into a chitosan matrix notably enhances the thermal stability, making the composite more suitable for high-temperature applications.

Monitoring the thermal properties of bio-composites is essential for applications like packaging and construction, where temperatures can vary or be high. Mechanical properties such as tensile strength, modulus, and impact resistance are also vital for evaluating bio-composites' suitability in different applications. Several factors influence these properties (Tănase et al. 2014; Aggarwal et al. 2020; Mohammed et al. 2022), including: (1) type of matrix, (2) nature and orientation of the reinforcement material, and (3) strength of bonding at the matrix-filler interface. Bio-composites reinforced with natural fibers like sisal or kenaf exhibit improved tensile strength and modulus due to the fibers' load-bearing capacity (Jawaid and Khalil 2011; Faruk et al. 2012; Yan et al. 2014). Surface treatment techniques, such as alkali treatment, enhance fiber adhesion to the matrix by removing impurities and increasing fiber roughness (Gholampour and Ozbakkaloglu 2020; Sahu and Gupta 2020; Wang et al. 2018a). This improves interfacial bonding and boosts the mechanical performance of bio-composites, making them suitable for the construction, automotive, and consumer goods industries.

9.2.1 Fabrication and Processing Techniques

Fabricating macro, micro, and nanocomposites involves various techniques tailored to their scales and desired properties. The fabrication and processing techniques of polymer composite have advanced recently due to the growing demand for sustainable and eco-friendly materials. The impact of processing techniques on the ultimate properties of polymer composites cannot be overstated.

9.2.1.1 Solution Casting Method

Solution casting is a popular method for fabricating bio-composites due to its simplicity, adaptability, and ability to produce uniform films with controlled properties (Sumrith et al. 2019; Bajpai 2017). This process involves dissolving or dispersing the biopolymer and fillers in a suitable solvent to create a uniform solution. The solvent must effectively dissolve the biopolymer, preserve its properties, and be compatible with the filler, commonly using water, ethanol, or organic solvents. The homogeneous solution is then applied to a flat surface, such as glass or Teflon, and allowed to evaporate under controlled temperature and humidity, forming a thin bio-composite layer for further testing.

Solution casting ensures uniform filler distribution in bio-composites, providing consistent mechanical, thermal, and barrier properties. Research shows that dispersing nanoparticles like GO or zinc oxide (ZnO) in cellulose-based bio-composites via solution casting improves mechanical strength and thermal stability (Zhang and Yang 2017). Additionally, it allows for the integration of functional fillers, such as plasticizers or antimicrobial agents, to customize properties. For instance, adding plasticizers during solution casting improves the flexibility and elongation at break of biopolymer films, making them suitable for packaging applications (Abdullah et al. 2022).

Drying conditions during solution casting, including temperature, humidity, and solvent evaporation rate, significantly impact the bio-composite film's properties, such as crystallinity and morphology. Rapid drying may create pores or cracks, damaging mechanical properties and barrier performance, while slower drying improves polymer chain and filler arrangement, enhancing density and strength (Rasib et al. 2021). Post-drying treatments like annealing or cross-linking can further improve the bio-composite properties, resulting in notable enhancements in both thermal stability and mechanical performance (Chen et al. 2020).

Despite its advantages, solution casting has limitations. The solvent must dissolve the polymer without degrading it or the nanoparticles. Additionally, the drying process can be time-consuming and needs careful temperature and humidity control to avoid defects. This method is generally limited to producing thin films and coatings, which may not suit all applications. However, ongoing research aims to address challenges by exploring alternative solvents, drying techniques, and combining solution casting with other methods. Despite its limitations, solution casting remains fundamental in polymer composites for its simplicity, versatility, and control over material properties.

9.2.1.2 In Situ *Polymerization*

In situ polymerization creates bio-composites by forming polymer matrices with reinforcing agents or fillers, enhancing fabrication versatility. The process begins with dissolving or dispersing monomers and initiators in a solvent, followed by adding the filler material. Polymerization is then initiated using thermal, chemical, or photochemical methods, resulting in a polymer network that encapsulates the filler particles (Kashfipour et al. 2018). This method ensures strong interfacial interaction between the polymer matrix and filler which is crucial for enhancing the mechanical, thermal, and barrier properties of the bio-composite.

In situ polymerization effectively disperses fillers within the polymer matrix, ensuring high filler dispersion. Unlike adding fillers to a pre-formed matrix, this technique allows the polymer to develop near the filler particles, promoting even distribution and reducing agglomeration. This is crucial for nanoscale fillers like rapheme oxide (GO), carbon nanotubes (CNTs), or nanoclays, as it significantly affects the composite's properties. For instance, in situ polymerization of polylactic acid (PLA) with nanoclay enhances tensile strength, thermal stability, and barrier properties due to the uniform distribution of nanoclay particles (Bikiaris et al. 2023). This method also enables the integration of various fillers, creating hybrid composites and broadening material properties.

The in situ polymerization method is highly adaptable, allowing for the production of bio-composites with specific properties. Researchers can adjust factors like monomers, fillers, and polymerization conditions to tailor bio-composites to precise requirements. For instance, modifying polymerization temperature, time, and initiator concentration can regulate the molecular weight, crystallinity, and crosslink density of the polymer matrix, affecting the mechanical and thermal properties of the bio-composite. In situ polymerization has successfully created various bio-composites, combining natural polymers like cellulose, chitosan, and starch with fillers such as natural fibers or nanoparticles. This versatility makes it ideal for applications ranging from packaging materials to biomedical devices.

Numerous studies demonstrate that in situ polymerization effectively creates high-performance polymer composites. For instance, polyamide-6/clay nanocomposites show significantly improved tensile strength and thermal stability compared to pure polyamide-6 (Sun et al. 2020). Similarly, research on polystyrene/montmorillonite nanocomposites revealed enhanced mechanical properties and reduced flammability, making them suitable for construction and automotive applications (Essawy et al. 2004). These examples highlight in situ polymerization as a versatile and effective method for producing advanced polymer composites with tailored properties for various applications.

9.2.1.3 Electrospinning

Electrospinning is an efficient technique for creating bio-composites with nanofiber structures, offering exceptional characteristics like high surface area and adjustable fiber diameters. The process involves applying a high-voltage electric field to a polymer solution, expelling it through a spinneret to produce fine fibers collected on a substrate. Fiber size, ranging from nanometers to micrometers, is influenced by polymer concentration, viscosity, voltage, and distance to the collector. This method allows precise control over fiber morphology, enhancing mechanical, thermal, and biological properties (Rodríguez-Tobías et al. 2019). Additionally, electrospinning enables the incorporation of various fillers or functional agents, broadening its applications.

A key advantage of electrospinning is its ability to incorporate various fillers, such as nanoparticles and nanoclays, into the polymer matrix during fiber formation, enhancing the mechanical, thermal, and electrical properties of the composites. For example, adding CNTs or rapheme to the polymer solution significantly improves electrical conductivity and mechanical strength (Venkatesan et al. 2021). The process parameters—like polymer concentration, viscosity, applied voltage, and needle–collector distance—can be precisely controlled to customize fiber diameter and morphology for specific applications (Seehra 2017). Electrospinning is particularly beneficial for high surface area and porosity applications, including filtration, tissue engineering, and energy storage. In filtration, electrospun nanofibers create membranes that capture particles at the nanoscale. In tissue engineering, the scaffolds promote cell attachment and proliferation (Jamil et al. 2024). Additionally, electrospun fibers are used in energy storage devices, where the high surface area facilitates efficient charge storage and transfer. The versatility and scalability of electrospinning make it a promising method for producing advanced polymer composites (Sharma 2023).

A study on PVA/GO nanocomposites fabricated via electrospinning showed significant improvements in mechanical strength and thermal stability compared to pure PVA fibers (Ahmadi Bonakdar and Rodrigue 2024). Another research on PCL/CNTs composites showed improved electrical conductivity and mechanical properties, making them ideal for flexible electronics and sensors (Zhang et al. 2021). Additionally, electrospun chitosan/PEO nanofibers with silver nanoparticles showed excellent antimicrobial properties, making them promising for wound dressings and biomedical devices.

9.2.2 *Biopolymer-Based Electrolytes*

Biopolymer electrolytes can be structured at macro, micro, and nanoscales to optimize performance for specific applications. Macro-structured electrolytes emphasize mechanical strength and stability, such as cellulose-based electrolytes in lithium-ion batteries that maintain electrode integrity and prevent dendrite formation (Zhang et al. 2022). At the microstructural level, porous networks in polymers enhance ion transport by increasing the surface area for movement, boosting ionic conductivity in energy storage devices (Qiu et al. 2024). Nanostructured electrolytes utilize nanoscale properties to significantly improve ionic conductivity by incorporating materials like GO or CNTs into the biopolymer matrices. These nanofillers create additional ion pathways, reducing resistance. For example, chitosan-based nanocomposite electrolytes with GO exhibit significant improvements in ionic conductivity and electrochemical stability, making them suitable for high-performance batteries and supercapacitors (Suneetha et al. 2019). Combining macro-, micro-, and nanostructures in one electrolyte system provides synergistic benefits, balancing mechanical strength with enhanced ionic transport. Recent research focuses on designing hierarchical structures in biopolymer electrolytes to achieve high performance and sustainability in next-generation energy storage devices.

Ionic conductivity and stability are essential for biopolymer electrolytes in energy storage, as they affect ion movement, power output, and efficiency. High ionic conductivity is challenging due to the insulating nature of biopolymers. Researchers enhance conductivity by doping with ionic salts, adding plasticizers, and incorporating nanomaterials to improve ionic pathways. Electrochemical stability is vital for longevity, requiring electrolytes to maintain conductivity across various temperatures and during prolonged cycling. For instance, pectin-based electrolytes with metal salts show improved conductivity and stable performance in capacitors over multiple cycles (Wilson and Botte 2024). Additionally, chemical stability prevents degradation and harmful byproducts during operation. However, enhancing conductivity often reduces stability. Recent studies aim to balance formulations to optimize both properties. For example, carboxymethyl cellulose-based biopolymer electrolytes, doped with various salts and plasticizers, effectively balance conductivity and stability, making them suitable for lithium-ion and zinc-ion batteries (Dueramae et al. 2020; Gupta and Varshney 2019).

Several case studies on biopolymer electrolytes in energy storage devices reveal their potential and challenges. One study examined carboxymethyl cellulose (CMC) as an electrolyte in lithium-ion batteries, finding that doping with ammonium bromide (NH_4Br) significantly enhanced ionic conductivity and improved battery performance while maintaining stability over multiple cycles (Samsudin et al. 2014). Another study focused on pectin-based hydrogels, which were doped with NaCl and KCl to enhance ionic conductivity. This hydrogel exhibited excellent mechanical properties and stability, showcasing its potential as a sustainable energy storage solution from agricultural waste (Wilson and Botte 2024). Additionally, chitosan-based electrolytes enhanced with rapheme oxide (GO) demonstrated significant improvements in ionic conductivity and thermal stability, making them suitable for high-performance batteries. These studies highlight the promise of combining biopolymers with advanced nanomaterials to meet the demands of modern energy storage devices (Hassan et al. 2023).

9.2.3 Biopolymer-Based Electrodes

The synthesis and modification of macro, micro, and nanostructured electrodes are critical steps in the development of advanced energy storage devices. Macrostructured electrodes are typically synthesized using conventional methods such as casting, extrusion, or compression molding. These methods are designed to create electrodes with uniform thickness and robust mechanical properties, which are essential for maintaining structural integrity during battery operation. However, macrostructured electrodes may suffer from limitations in terms of ion diffusion and electrochemical performance due to their relatively low surface area (Zhang et al. 2017). To address these limitations, researchers have been exploring ways to introduce micro and nanoscale features into the electrode structure. Micro-structured electrodes are engineered to have increased surface area and porosity, which can significantly enhance ion diffusion and charge storage capacity. These structures are often synthesized using techniques such as electrospinning, freeze-drying, or templating, which allow for precise control over the pore size and distribution within the electrode material. The introduction of micro-scale porosity not only improves ionic conductivity but also provides additional active sites for electrochemical reactions, leading to enhanced performance in devices such as supercapacitors and batteries (Qiu et al. 2024). Additionally, modifications such as the incorporation of conductive polymers or the use of hybrid materials can further enhance the electrochemical properties of micro-structured electrodes.

Nanostructured electrodes, on the other hand, take advantage of the unique properties that emerge at the nanoscale, such as increased surface area, quantum effects, and enhanced electrical conductivity. These electrodes are typically synthesized using techniques such as chemical vapor deposition (CVD), sol–gel processes, or hydrothermal synthesis. Nanostructuring can involve the incorporation of nanomaterials like rapheme, carbon nanotubes, or metal oxides, which provide superior electrochemical performance due to their high surface-to-volume ratio and excellent conductivity. The synthesis of nanostructured electrodes often requires careful control over the size, shape, and distribution of nanoparticles to achieve the desired electrochemical properties. These electrodes are particularly promising for use in high-performance energy storage devices, where both high-power density and energy density are required. The structural and electrochemical properties of electrodes are key determinants of their performance in energy storage devices. Structurally, electrodes must possess sufficient mechanical strength and flexibility to withstand the stresses associated with charge and discharge cycles. This is particularly important for flexible and wearable energy storage devices, where the electrodes must maintain their integrity under repeated bending and deformation. The structure of the electrode, including its porosity, surface area, and the distribution of active materials, directly influences its electrochemical performance. For instance, highly porous electrodes with a large surface area can facilitate faster ion diffusion and provide more active sites for electrochemical reactions, leading to higher charge storage capacity and faster charging times (Rahman et al. 2015). Electrochemical properties, such as ionic conductivity, electrical conductivity, and charge storage capacity, are crucial for determining the efficiency and effectiveness of an electrode. Ionic conductivity is influenced by the electrode's ability to facilitate the movement of ions within its structure, which is critical for achieving high power density. Electrical conductivity, on the other hand, ensures that electrons can move efficiently through the electrode material, reducing resistance and improving overall device performance. Materials with high intrinsic electrical conductivity, such as rapheme (Ibrahim et al. 2021) and carbon nanotubes (Wang et al. 2022a), are often incorporated into electrode structures to enhance their electrochemical properties. Additionally, the stability of these properties under different operational conditions, such as varying temperatures or prolonged cycling, is essential for the long-term performance of energy storage devices.

Modifying the structural and electrochemical properties of electrodes through techniques such as doping, surface functionalization, or the incorporation of nanomaterials can lead to significant improvements in performance. For example, doping electrodes with metal ions or incorporating nanostructured materials can enhance both ionic and electrical conductivity, leading to better performance in applications such as supercapacitors and batteries (Kacica and Biswas 2020). Additionally, surface functionalization of electrodes with chemical groups that improve ion adsorption or electron transfer can further enhance their electrochemical properties. These modifications are crucial for optimizing electrode performance and ensuring the efficient operation of energy storage devices. The integration of structured electrodes into energy storage devices is a critical step that involves not only the physical assembly of the device but also the optimization of electrode interactions with other components, such as electrolytes and current collectors. In lithium-ion batteries, for instance, the integration process must ensure that the electrode materials are compatible with the electrolyte to prevent unwanted side reactions and ensure efficient ion transport. The physical interface between the electrode and electrolyte plays a crucial role in determining the overall performance of the device, as it affects both the ionic conductivity and the stability of the electrode. Proper integration techniques, such as coating or layering, are often employed to optimize these interfaces and enhance the overall efficiency of the energy storage device.

For supercapacitors, the integration process often involves aligning the structured electrodes to maximize surface contact with the electrolyte, which is essential for achieving high capacitance and energy density. The alignment and connection of the electrodes to the current collectors must also be optimized to minimize resistance and ensure efficient electron transport. In some cases, hybrid integration approaches are used, where different types of electrodes (e.g., one with high energy density and another with high power density) are combined within the same device to achieve a balance between performance metrics (Ren et al. 2021). This hybrid integration can lead to devices that offer both rapid charging capabilities and long-term energy storage, making them suitable for a wide range of applications. The challenges associated with the integration of structured electrodes into energy storage devices include ensuring mechanical stability, maintaining high ionic and electrical conductivity, and preventing degradation over time. Advanced integration techniques, such as the use of binder-free electrodes or the development of solid-state electrolytes, are being explored to overcome these challenges. Binder-free electrodes, for example, eliminate the need for additional materials that can introduce resistance, thereby improving the overall conductivity and performance of the device (Kang et al. 2020). Solid-state electrolytes, which replace liquid electrolytes, offer improved safety and stability, particularly in high-energy–density applications. The successful integration of structured electrodes into energy storage devices is crucial for the development of next-generation technologies that offer improved performance, safety, and sustainability.

9.2.4 Characterization Techniques

9.2.4.1 Microstructural Analysis

The analysis of the microstructure of bio-composites is crucial for understanding their properties and performance. Advanced techniques like Field Emission Scanning Electron Microscopy (FESEM), Energy Dispersive X-ray Spectroscopy (EDX), and Transmission Electron Microscopy (TEM) provide insights into the crystalline structure, surface morphology, elemental composition, and internal structure of the materials. Through careful analysis of these factors, researchers can determine the impact of reinforcing agents on the polymer matrix's composition and characteristics. These performance evaluations are crucial for enhancing the design and manufacturing processes of bio-composites. This ensures that bio-composites meet the specific mechanical, thermal, and functional criteria required for their diverse applications.

Field Emission Scanning Electron Microscopy (FESEM) Analysis

FESEM is an effective analytical technique for studying the surface morphology and microstructure of bio-nanocomposites. High-resolution FESEM images allow researchers to examine the dispersion of nanofillers within the polymer matrix and assess interfacial bonding and overall structural integrity. This analysis is crucial for understanding the material's mechanical properties and thermal stability. A key application of FESEM is studying nanofiller distribution, as balanced dispersion is essential for enhancing mechanical properties. FESEM images provide insights into whether nanofillers are evenly dispersed or clustered, which can affect performance. Well-dispersed composites yield a consistent structure and improved stress transfer, while inadequate dispersion leads to agglomeration, weakening the composite's mechanical properties.

FESEM provides insights into interfacial bonding between nanofillers and the polymer matrix, which is crucial for effective stress transfer and improved mechanical strength. The images reveal the integration of nanofillers and can identify voids indicating inadequate bonding. FESEM also evaluates the effectiveness of surface modifications in enhancing adhesion. Additionally, it examines fracture surfaces after mechanical testing, revealing failure mechanisms; a rough surface suggests good energy absorption, while a smooth surface may indicate brittleness. FESEM can also analyze thermal stability by observing changes during heating, helping researchers understand how nanofillers affect structural integrity and contribute to flame retardancy during thermal degradation.

The FESEM analysis of PVA-based films by Perumal, Sellamuthu (Perumal et al. 2018), reveals key insights into the microstructure and compatibility of materials. Based on Fig. 9.5, the PVA/chitosan film showed a consistent microstructure with no signs of phase separation or fractures, indicating strong interaction between the matrices. Incorporating CNCs did not significantly change the fractured surfaces; the morphology remained undulating without noticeable clumps, suggesting an even distribution of CNCs. This uniformity indicates that incorporating CNCs could enhance the properties of the bio-nanocomposite films, potentially leading to a more robust material.

In another study by Govindan et al. (Govindan et al. 2012), FESEM analysis examined the surface morphology of the chitosan-silver (CS-Ag) nanocomposite. The CS appeared as leaf-like bundles, while the Ag nanoparticles were uniformly spherical. In the CS-Ag composite, the FESEM image from Fig. 9.6 showed Ag nanoparticles surrounded by the CS matrix, forming aggregates with low dispersibility but high stability. Improved dispersion and reduced agglomeration were achieved by dissolving CS for an extended period before adding the Ag nanoparticles. This

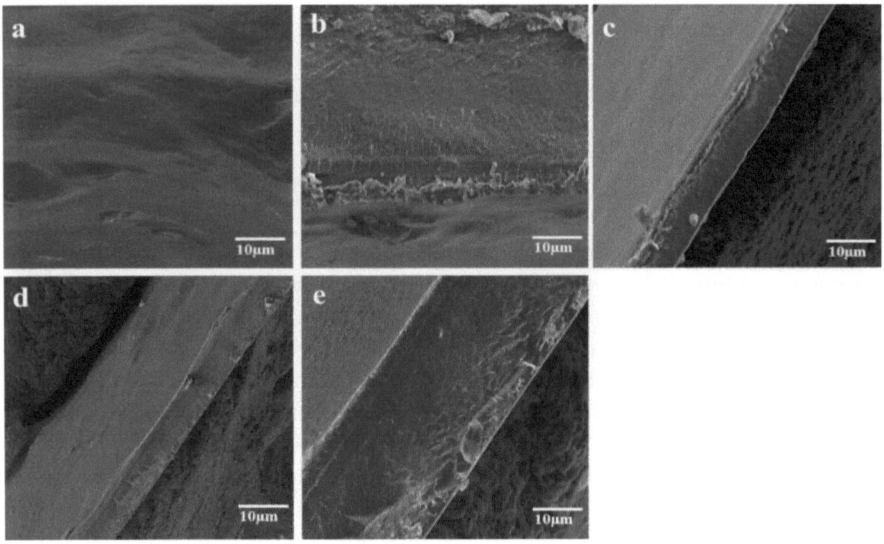

Fig. 9.5 FESEM images of the PVA/CS/CNC bio-composite films **a** PVA, **b** PVA/CS, **c** PVA/CS/CNC-1, **d** PVA/CS/CNC-3, and **e** PVA/CS/CNC-5. Reprinted with permission from Perumal et al. (2018), Copyright 2018, Elsevier

resulted in a well-dispersed composite with silver nanoparticles effectively embedded in the CS matrix, maintaining nanosized particle dimensions.

Energy-Dispersive x-ray Spectroscopy (EDX) Analysis

Energy-dispersive X-ray spectroscopy (EDX) is a key analytical technique for studying the elemental composition of bio-nanocomposites. By detecting X-rays emitted during electron beam bombardment, EDX provides quantitative and qualitative insights into present elements. This technique is crucial for analyzing the distribution, concentration, and interaction of fillers and matrices, which impact material properties. Researchers often use EDX alongside scanning electron microscopy (SEM) to assess surface morphology and elemental composition simultaneously. For example, EDX can confirm the presence of zinc and oxide in chitosan-based bio-nanocomposites with ZnO nanoparticles (Bharathi et al. 2019). Quantitative EDX data helps determine the proportion of elements, ensuring material purity and effective integration of nanofillers. Discrepancies in EDX spectra may indicate poor dispersion or clumping, highlighting areas for processing improvements.

EDX is essential for understanding the interactions between bio-based polymers and nanofillers at the interface. It reveals how elements from nanofillers interact with the polymer matrix, affecting the mechanical, thermal, and barrier properties of the bio-nanocomposite. For instance, EDX analysis of a starch-clay nanocomposite shows the elemental composition of clay layers and their interaction with starch

Fig. 9.6 FESEM images of **a** chitosan, **b** Ag nanoparticles, **c** chitosan-Ag nanocomposite, and **d** synthesized chitosan-Ag nanocomposite. Reprinted with permission from Govindan et al. (2012), Copyright 2012, Nature

molecules (Rosendo-González et al. 2019). This information links structural modifications to improved mechanical strength or thermal stability. EDX also assesses the uniformity of bio-nanocomposites, identifying clusters or agglomerates with elevated element levels, which provides feedback on the dispersion process and guides adjustments for better structural properties.

In a study by Ahmad (2017) on cellulose-poly(pyrrole) reinforced with GO (C/PPyGO), EDX analysis assessed the elemental composition of the bio-nanocomposites. The results from Fig. 9.7 revealed higher carbon content in C/PpyGO compared to the pure biodegradable polymer (C/Ppy), indicating successful incorporation of carbon-rich GO. The presence of nitrogen and sulfur suggests effective integration of the poly(pyrrole) component, enhancing material properties. Overall, EDX analysis offers valuable insights into elemental composition and nanofiller dispersion, improving material properties for applications ranging from eco-friendly packaging to medical materials.

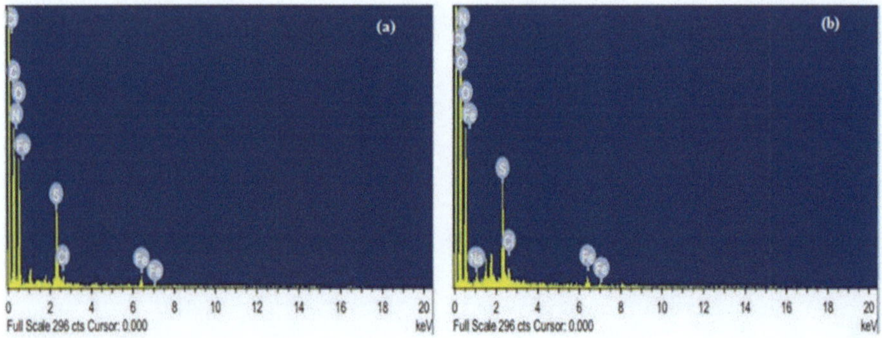

Fig. 9.7 **a** EDX images of the as-prepared materials (C/Ppy) and **b** bio-nanocomposite (C/Ppy-GO). Reprinted with permission from Ahmad et al. (2017), Copyright 2017, Royal Society of Chemistry

Transmission Electron Microscopy (TEM) Analysis

Transmission electron microscopy (TEM) is a valuable tool for examining the internal structure and dispersion of nanofillers within bio-nanocomposites at the nanoscale. It captures high-resolution images that reveal critical information about material composition and structure. TEM is essential for understanding the interaction between the biopolymer matrix and nanofillers, which affects the material's mechanical, thermal, and electrical properties. TEM analysis provides insights into nanofiller dispersion; evenly distributed nanofillers enhance composite properties, while clustered particles can create weaknesses. Well-dispersed systems show particles nicely separated by the polymer matrix, while poorly dispersed systems may have large aggregates.

TEM is commonly used to analyze the interfacial bonding between nanofillers and the biopolymer matrix, enhancing the mechanical properties of bio-nanocomposites through strong interactions. It reveals gaps or voids that may indicate compatibility issues. For instance, TEM images of chitosan-graphene oxide nanocomposites show well-separated rapheme oxide layers in the chitosan matrix, indicating strong interaction. TEM also provides insights into the nanostructure and morphology of materials, revealing structures like the needle-like or rod-shaped CNCs embedded in the polymer matrix (Pal et al. 2017), as illustrated in Fig. 9.8. Additionally, it often combines with diffraction techniques to obtain crystallographic data, crucial for understanding how crystalline phases influence the properties of bio-nanocomposites.

In a study by Patel (2014), TEM was used to evaluate the interaction and dispersion of gold nanoparticles (Au-NPs) in chitosan bio-nanocomposites. Based on Fig. 9.9, the analysis showed that the Au-NPs are spherical within the chitosan matrix, with high-resolution TEM images revealing lattice fringes at 0.239 nm and a

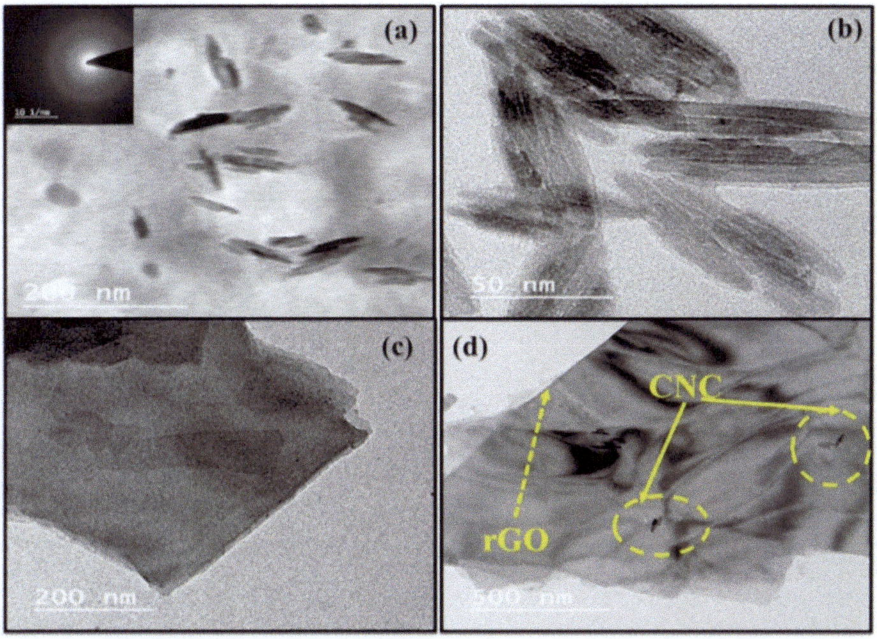

Fig. 9.8 TEM images of **a** pristine CNC, **b** PLA/1% CNC (Sample B), **c** PLA/0.5% rGO (Sample E), and **d** PLA/1% CNC/0.5% rGO (Sample H). Reprinted with permission from Pal et al. (2017), Copyright 2017, Elsevier

face-centered cubic structure confirmed by the SAED pattern. The narrow size distribution and absence of agglomerates suggest effective prevention of Au-NP agglomeration by the chitosan matrix. TEM also examines failure mechanisms, highlighting how nanofillers can bridge or deflect cracks, enhancing material durability. Overall, TEM analysis provides valuable insights into dispersion, interfacial bonding, and crystallinity, crucial for optimizing bio-nanocomposites' mechanical, thermal, and functional properties.

9.2.4.2 Electrochemical Characterization

Electrochemical Impedance Spectroscopy (EIS) Analysis

EIS effectively examines the electrical properties of bio-nanocomposites by applying an alternating current (AC) signal and measuring the material's response across frequencies. This technique provides insights into conductivity, dielectric properties, and electrochemical processes within the composite. EIS assesses electrical conductivity and charge transport mechanisms, showing how incorporating nanofillers enhances conductivity by improving charge carrier pathways. Impedance spectra

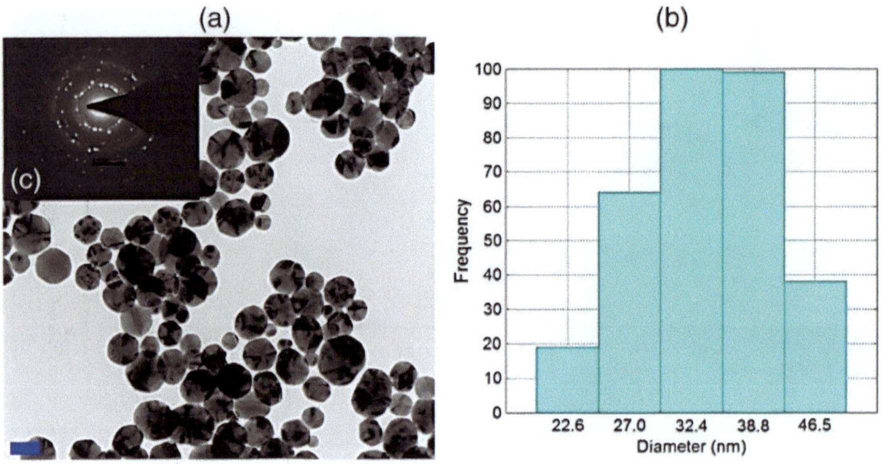

Fig. 9.9 **a** TEM micrograph of chitosan-Au nanocomposites, **b** size distribution histogram of chitosan-Au nanocomposites in aqueous solution, and **c** selected area electron diffraction (SAED) chitosan stabilized gold nanoparticles. Reprinted with permission from Patel et al. (2014), Copyright 2014, Elsevier

indicate resistance and capacitance at the polymer–nanofiller interface, where well-dispersed nanofillers reduce resistance and increase capacitance, leading to better adhesion and overall performance.

Studying the dielectric properties of bio-nanocomposites using EIS is essential. By analyzing frequency-dependent impedance data, researchers gain insights into the dielectric constant and loss, which are important for applications like energy storage and sensors. EIS helps identify relaxation phenomena from dipole reorientation or charge carrier movement, revealing dynamics of the polymer–nanofiller system. Equivalent circuit models depict the composite as a network of resistors and capacitors, providing measurements for parameters like bulk resistance and capacitance. This analysis enhances understanding of the material's electrochemical properties and guides the design of bio-nanocomposites for specific applications.

Islam and Mollik 2020) studied ion diffusion in electrode materials for energy storage using plasticized starch-GO (PS/GO) bio-nanocomposites through EIS analysis. The Nyquist plot in Fig. 9.10 compares pure plasticized starch, GO, and PS/GO composite electrodes at various AC frequencies. A smaller semicircle at low frequency indicates a lower charge-transfer barrier due to polar functional groups and defects in GO. The PS/GO (0.5%) composite shows the smallest semicircle radius, suggesting that GO incorporation reduces resistivity and enhances charge transport. Analyzing bio-nanocomposites is vital for developing materials with high ionic conductivity for batteries, supercapacitors, and sensors. Integrating EIS with techniques like cyclic voltammetry will facilitate creating high-performance materials.

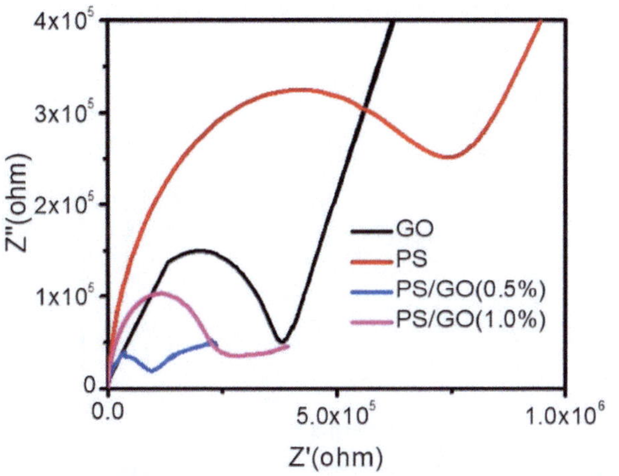

Fig. 9.10 Nyquist plots of pure PS, GO, and PS/GO electrodes at room temperature. Reprinted with permission from Islam and Mollik (2020), Copyright 2020, Elsevier

Cyclic Voltammetry (CV) Analysis

Cyclic voltammetry (CV) is an essential electrochemical method for characterizing the redox behavior and stability of bio-composites used in energy storage, sensing, and catalysis. It involves sweeping the electrode potential and measuring the resulting current to gain insights into electrochemical processes. For example, a study of polyaniline/chitosan bio-composites showed distinct redox peaks associated with polyaniline oxidation and reduction (Fig. 9.11) (Torvi et al. 2018). The shape and intensity of these peaks provide information about reaction kinetics, reversibility, capacitance, and energy storage capabilities.

CV is commonly used to study the interaction between polymer matrices and integrated fillers, which significantly impacts the electrochemical properties of bio-composites. For instance, studies on GO-based bio-composites show that including GO enhances electrical conductivity and electrochemical activity, evidenced by higher peak current densities in chitosan/GO composites (Arvand et al. 2017). CV also assesses durability through repeated cycling; consistent redox peak positions and intensities indicate excellent electrochemical stability, while deviations may suggest degradation or irreversible changes in the composite structure.

CV is crucial for exploring bio-composites in sensing technologies. By analyzing the redox behavior of specific analytes, researchers can develop sensitive and selective sensors. For instance, a CV study of reduced GO (rGO) in the PS/rGO nanocomposite showed that rGO distorts the CV curves due to its pseudo capacitance, leading to improved electrochemical stability (Mollik et al. 2021). The quasi-rectangular CV curves of the PS/rGO composite had a larger area than pure PS, indicating enhanced capacitive performance. This highlights the role of rGO nanofillers in improving carrier transport within the PS network, making CV essential for advanced material development.

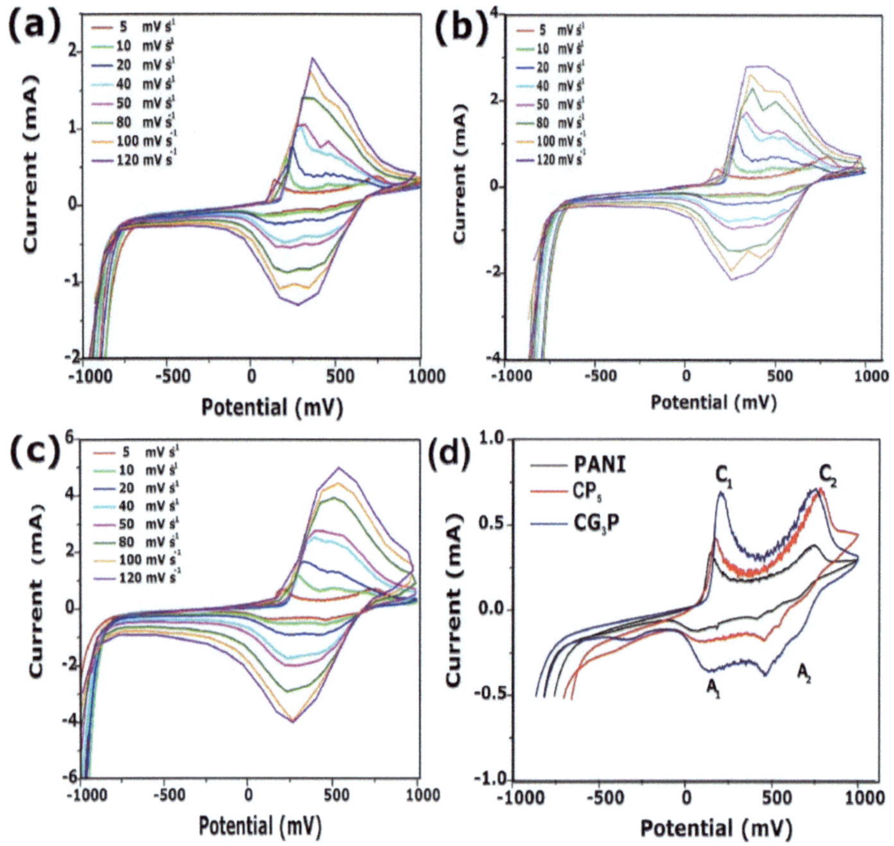

Fig. 9.11 Cyclic voltammogram of **a** PANI, **b** CP_5, and **c** CG_3P measured in three-electrode system with a potential range of −1 V to + 1 V in 1 M H_2SO_4 electrolyte solution at different scan rate of 5–120 mV s^{-1}; **d** cyclic voltammograms of PANI, CP5, and CG3P measured at a fixed scan rate of 5 mV s^{-1}. Reprinted with permission from Torvi et al. (2018), Copyright 2018, Elsevier

Galvanostatic Charge–discharge (GCD) Analysis

The galvanostatic charge–discharge (GCD) method is essential for assessing the electrochemical properties of bio-composites in energy storage systems like supercapacitors and batteries. This technique applies a constant electrical current while recording voltage changes over time, indicating how well a bio-composite can store and release energy. GCD tests provide insights into performance parameters such as specific capacitance, energy density, and cycling stability. For example, adding conductive fillers like CNTs to gelatin-based bio-composites significantly enhances their specific capacitance and energy density by increasing charge storage surface area and improving electron transport pathways (Alam et al. 2023).

Understanding GCD analysis is crucial for evaluating the rate capability of bio-composites, reflecting their ability to deliver energy at varying current densities.

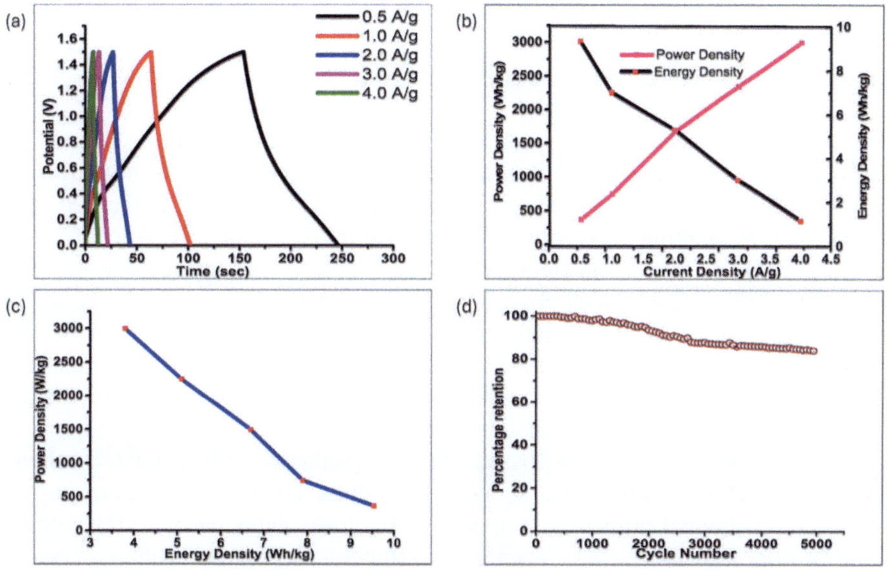

Fig. 9.12 **a** GCD curves of fabricated device, **b** power and energy density graphs as function of current density, **c** graph showing relationship between power and energy densities, and **d** graph of retention capacity. Reprinted with permission from Nayak et al. (2022a), Copyright 2022, Elsevier

Materials with excellent rate capability maintain significant capacitance at high current densities. As illustrated in Fig. 9.12, Nayak et al. (2022a) demonstrated that supercapacitors reinforced with zinc oxide and cobalt oxide achieved energy densities from 3.8 Wh kg^{-1} to 9.54 Wh kg^{-1} and power densities from 375.3 W kg^{-1} to 3000 W kg^{-1}. The device showed strong lifecycle stability, retaining 84.8% capacitance after 5000 cycles, attributed to the bio-composite's organized structure, which enables effective ion diffusion and charge transfer, highlighting its potential for energy storage applications.

Measuring a bio-composite's capacitance stability over multiple charge–discharge cycles is crucial in GCD analysis. Materials that retain capacitance demonstrate excellent cycling stability and durability, often due to strong interactions between the polymer matrix and conductive fillers. GCD analysis also reveals self-discharge behavior; a low self-discharge rate indicates high efficiency and charge retention, making the material desirable for energy storage applications. Understanding GCD characterization is essential for optimizing bio-composite compositions and enhancing charge storage capabilities to ensure the reliability of energy storage devices.

9.2.4.3 Thermal Properties

Differential Scanning Calorimetry (DSC) Analysis

Differential scanning calorimetry (DSC) is a key thermal analysis technique for investigating the thermal properties of bio-composites. It measures heat transfer during controlled temperature changes, providing insights into critical thermal transitions such as melting temperature (T_m), glass transition temperature (T_g), and crystallinity (T_c). These parameters are essential for understanding material stability. DSC analyzes phase transitions, characterizing thermal stability and melting behavior by measuring heat flow. For instance, it determines the melting temperature of a polymer matrix, crucial for processing, and identifies T_g, which indicates the temperature range where a polymer becomes flexible. This information is vital for predicting material performance under thermal conditions.

In polylactic acid (PLA)-based bio-composites, Differential Scanning Calorimetry (DSC) analysis has shown how natural fibers like cellulose impact the material's thermal properties (Fig. 9.13) (Bhiogade and Kannan 2021). Adding cellulose fibers affects PLA's glass transition temperature (T_g) and promotes crystallization, leading to increased crystallinity. Higher crystallinity is linked to improved thermal stability, making the composite suitable for high-temperature applications. Enhancing rigidity and stability are crucial for durability and resistance to thermal deformation. Overall, DSC is vital for optimizing bio-composite formulations, revealing how additives influence the thermal behavior of the base polymer for desired properties and performance.

DSC is crucial for assessing compatibility and interaction between the polymer matrix and fillers in bio-composites. In chitosan-based composites, DSC reveals changes in glass transition temperature (Tg) with fillers like graphene oxide or silica nanoparticles, often correlating with improved mechanical properties and thermal stability as fillers restrict polymer chain movement. DSC thermograms provide insights into interactions like cross-linking and degradation. Additionally, it studies thermal degradation behavior, essential for high-temperature applications. For instance, in starch and biodegradable polymer composites, DSC identifies degradation temperatures, showing how additives enhance thermal stability. Overall, DSC is essential for designing and improving bio-composites for various industrial applications.

Thermogravimetric Analysis (TGA) Analysis

Thermogravimetric Analysis (TGA) is essential for assessing the thermal stability and composition of bio-composites by analyzing weight fluctuations with temperature. For instance, in cellulose nanowhiskers (CNW)-reinforced polylactic acid (PLA) composites (Liu et al. 2018), based on Fig. 9.14a and b, TGA and DTGA curves show degradation starting around 260 °C, with over 96% weight loss at 350–360 °C. The initial decomposition temperature is influenced by PLA's molecular weight and crystallinity. The PLA/CNWs composite has higher residual weight due

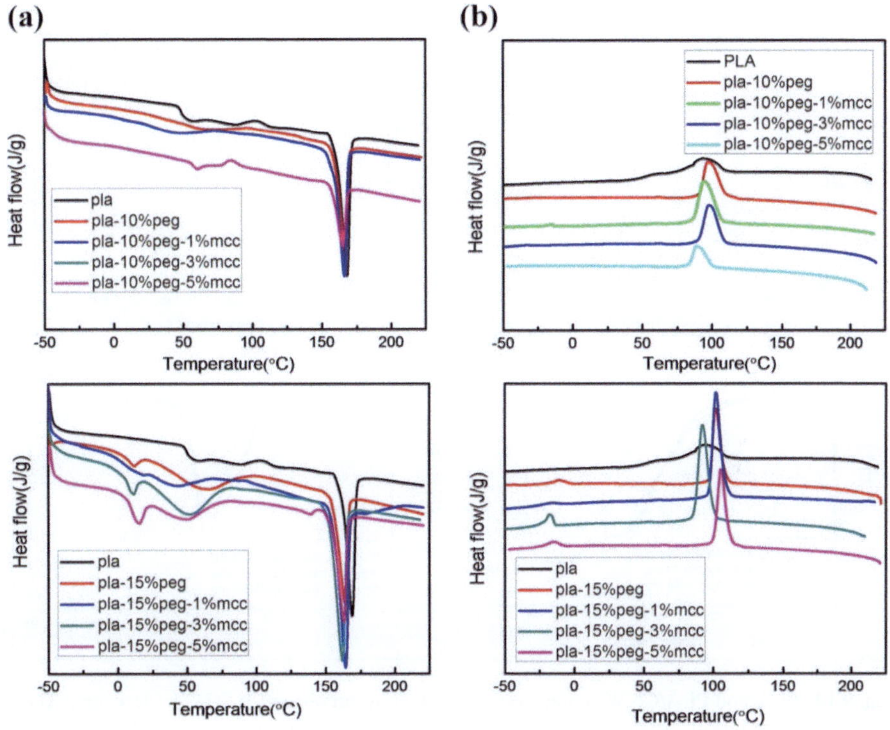

Fig. 9.13 DSC curves of PLA-based composites (MCC) in **a** heating and **b** cooling mode. Reprinted with permission from Bhiogade and Kannan (2021), Copyright 2021, Sage Publications

to CNWs' stability above 310 °C. Below 300 °C, the composite nanofibers lose more weight than neat PLA, but above 300 °C, it shows improved thermal stability with lower peak intensity and shifts in DTGA curves.

In a study by Rhim et al. (2014), TGA and DTGA depicted in Fig. 9.14c and d evaluated the thermal stability of agar and agar/silver nanoparticles (AgNPs) composite films. The thermograms show that both agar and agar/AgNPs films decompose in multiple steps, with a significant event around 300 °C due to agar breakdown. The higher residual percentages in the composite films are attributed to AgNPs. The thermal degradation of AgNPs-loaded agar films occurs slightly later than pure agar, indicating enhanced thermal stability due to the heat-resistant metallic silver, highlighting the role of AgNPs in improving the composite's thermal stability.

The addition of fillers like rapheme oxide or carbon nanotubes significantly enhances the thermal properties of bio-nanocomposites. TGA analysis indicates that these fillers improve thermal stability by restricting polymer chain movement and slowing degradation, broadening the applications of bio-nanocomposites in high-temperature environments. Additionally, TGA helps analyze the composition of bio-nanocomposites, determining the proportions of organic and inorganic components. By examining weight loss at various temperatures, researchers can assess the makeup

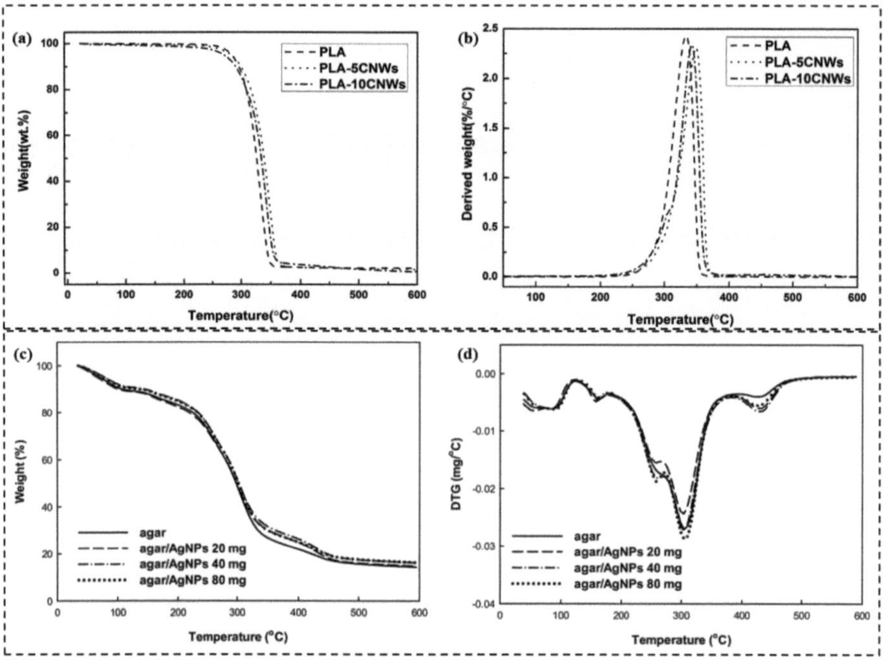

Fig. 9.14 PLA and PLA/CNWs composite mats: **a** TGA curves and **b** DTGA (derivative TGA) curves. Reprinted with permission from Liu et al. (2018), Copyright 2018, MDPI; **c** TGA and **d** DTGA curves of agar and agar/AgNPs composite films. Reprinted with permission from Rhim et al. (2014), Copyright 2014, Elsevier

of polymers, fillers, and residual ash content, which is crucial for ensuring the quality of bio-nanocomposites in industrial applications.

9.3 Recent Development of Biopolymers in Applications of Energy Storage Devices

As environmental challenges posed by synthetic compounds continue to intensify, there has been a significant shift toward the use of eco-friendly, biomass-derived biopolymers and other naturally occurring materials. These sustainable alternatives have garnered widespread attention and keen interest as potential replacements for conventional petroleum-based products. Energy storage technologies, including batteries and supercapacitors, are poised to play a crucial role in the transition to zero or low greenhouse gas emissions. By linking renewable energy generation with efficient energy storage systems, these technologies will be instrumental in facilitating a sustainable power shift.

9.3.1 Supercapacitors

A capacitor is a passive electrical component used to store and release small amounts of energy. Supercapacitors (SCs), an advanced version, are receiving increasing attention for their potential in both low-power electronics (like wearables) and high-power applications (such as military technologies). Their performance is influenced by the interaction between electrode and electrolyte materials (Kumar et al. 2022):

Electrical double layer capacitor (EDLCs)—Store energy through the separation of charges at the interface between the electrode and electrolyte without involving charge transfer.

Pesudocapacitors—Use redox reactions on the electrode surface, offering higher energy density but lower power density compared to EDLCs.

Hybrid/Asymmetric SCs—These are a combination of both EDLC and pseudocapacitor principles, which provide both high power density and energy density, making them versatile for various applications.

Research has focused on using natural polymers for sustainable supercapacitor materials, such as cellulose, starch, lignin, and alginate. Several techniques, including pyrolysis, electrospinning, hydrothermal carbonization, microwave-assisted activation, sol–gel processing, and templating (Appiah et al. 2022), have been employed to develop biopolymer-based materials. These methods help create porous carbon structures that improve surface area, ion transport, and overall supercapacitor performance.

Alginate

Alginate, a naturally occurring polyanionic biopolymer sourced from marine-derived materials such as sodium and potassium alginate, has garnered significant interest in the field of energy storage. In one study, Wei et al. reported the synthesis of oxygen-rich nitrogen-doped porous carbon by carbonizing ammonium alginate aerogels, followed by activation in air. The resulting material (illustrated in Fig. 9.15) exhibited an excellent specific capacitance of 204 F g^{-1} at 1 A g^{-1}, demonstrating strong rate capability and a 91.2% retention in capacitance after 10,000 charge–discharge cycles (Wei et al. 2022).

In related work, Wang et al. (2018b) presented a green synthesis route for macro-, meso-, and microporous carbon aerogels via a sol–gel technique. The aerogels featured a 3D macroporous structure composed of core regions 2–5 μm in diameter with ultrathin walls (<100 nm), allowing efficient ion diffusion by minimizing transport pathways. High-resolution TEM analysis confirmed the presence of micropores and semi-graphitized regions. Another approach introduced a scalable method to produce three-dimensional interconnected porous carbon (IPC) networks from potassium alginate through self-activation. The resulting IPC-800 material, rich in oxygen functionalities and possessing a large surface area, delivered outstanding electrochemical performance. Specifically, the IPC-800 electrode achieved a capacitance of 279 F g^{-1} at 1 A g^{-1}, maintained 200 F g^{-1} at a high current of 50 A g^{-1},

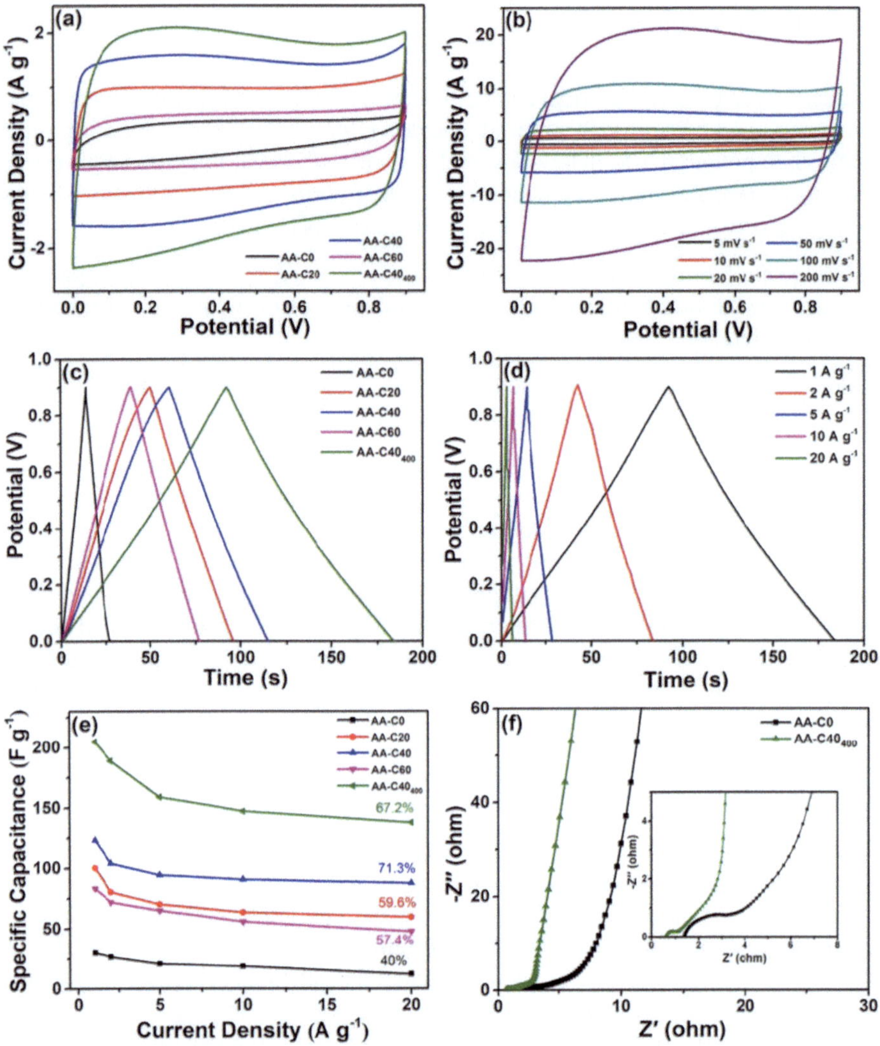

Fig. 9.15 Electrochemical behavior of carbon materials derived from AA evaluated in a two-electrode configuration. **a** CV profiles of different samples recorded at 20 mVs^{-1}. **b** CV curves of AA-C40400 at varying scan rates. **c** Galvanostatic charge–discharge (GCD) plots of all samples at a current density of 1 Ag^{-1}. **d** GCD responses of AA-C40400 across a range of current densities from 1 to 20 Ag^{-1}. **e** Specific capacitance trends of the samples as a function of current density. **f** Nyquist impedance spectra of AA-C0 and AA-C40400. Adapted with permission from Wei et al. (2022), Copyright 2022, Elsevier

and retained 96.6% capacitance after 10,000 cycles in 6 M KOH. Furthermore, the IPC-800-based symmetric supercapacitor demonstrated an energy density of 16.9 Wh kg^{-1} and excellent long-term cycling stability in a 1 M Na_2SO_4 electrolyte (Sun et al. 2019).

Xia (2019) reported an environmentally friendly synthesis approach for producing oxygen-doped porous carbon electrodes, in which oxygen-rich sodium alginate was used as the carbon precursor through a controlled carbonization process. By precisely tuning the carbonization temperature and carefully adjusting the molar ratio between α-L-guluronic acid and β-D-mannuronic acid segments in sodium alginate, an optimal synergy was established among the resulting carbon material's morphology, oxygen content, and electrical conductivity. The derived carbon electrodes exhibited a remarkable specific capacitance of 424.6 F g^{-1} in 6 M KOH at a current density of 1 A g^{-1}, along with excellent cycling durability—retaining more than 90% of their original capacitance after 20,000 cycles. This superior electrochemical behavior is attributed to their distinctive hierarchical porous structure (macro-, meso-, and micropores) and the abundance of oxygen-containing functionalities embedded within the carbonized sodium alginate aerogel matrix (Xia et al. 2019).

Cellulose

Cellulose is becoming popular as a sustainable material for supercapacitors due to its abundance, biodegradability, and ability to form nanostructures with a large surface area. Although its low conductivity is a challenge, it can be improved by incorporating conductive materials like graphene or carbon nanotubes. Cellulose-based materials also work well with various electrolytes and can be made into electrodes or separator membranes, making them versatile for eco-friendly supercapacitor applications. Liew et al. (2013) made notable progress in this field by creating a nanocomposite of cellulose nanocrystals (CNXLs) and polypyrrole (PPY). This composite demonstrated better capacitance and pore architecture compared to traditional electrodes, achieving a specific capacitance of 240 F g^{-1}. The efficiency increased with film thickness, reaching 1.54 F cm^{-2}, though thicker films faced electrolyte diffusion limitations, indicating areas for future improvement. Additionally, Hao et al. (2017) developed a silica-assisted method for creating 3D meso-microporous carbon nanofiber networks from bacterial cellulose (BC) through pyrolysis. These carbon nanofibers, with a large surface area of 624 m^2 g^{-1} and a mesoporous structure, were used in a binder-free electrode (CN-BC) that showed a specific capacitance of 302 F g^{-1} at 0.5 A g^{-1}, along with excellent high-rate performance and stability in a 6 M KOH electrolyte. In addition, Mo et al. (2018) developed a dense graphene-like porous carbon (GPC) with a hierarchical pore structure which synthesized from sheet cellulose, derived from the milling of bleached kraft pulp. This innovative pore structure resulted in an impressive specific surface area of up to 2045 m^2 g^{-1}, providing numerous storage sites and channels for electrolyte species. In a 6 M KOH aqueous electrolyte, the material demonstrated a gravimetric specific capacitance of 353 F/g and a volumetric specific capacitance of 309.7 F cm^{-3} at 1 A g^{-1}. Additionally, as shown in the Fig. 9.16, when used in a symmetrical supercapacitor with an ionic

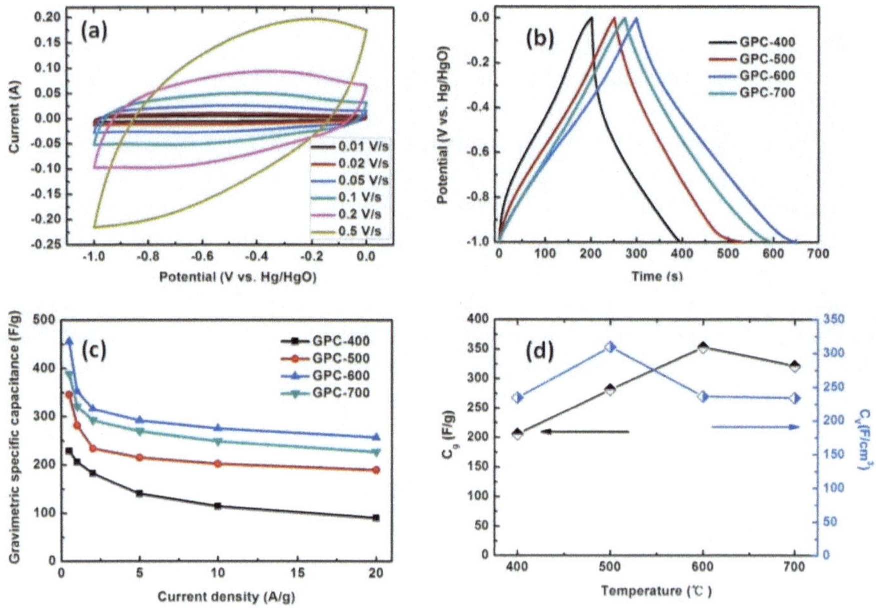

Fig. 9.16 Electrochemical evaluation of GPC using a three-electrode configuration in 6 M KOH electrolyte. **a** Cyclic voltammetry profiles of GPC-600 at different scan rates. **b** Charge–discharge curves obtained at a constant current of 1 A g^{-1}. **c** Variation of specific capacitance with respect to current density. **d** Gravimetric and volumetric capacitance values plotted as a function of temperature at 1 A g^{-1}. Adapted with permission from Mo et al. (2018), Copyright 2018, Elsevier

liquid electrolyte in acetonitrile, it achieved a gravimetric energy density of 120.1 Wh kg^{-1} and a volumetric energy density of 80.4 Wh L^{-1}.

Chitosan

Chitosan, a biopolymer derived from chitin, has gained attention as a promising material for supercapacitors due to its abundance, biodegradability, and adaptable chemical properties. When processed into carbon-based materials or used as a matrix to integrate conductive nanoparticles, chitosan significantly boosts supercapacitor performance. Its nitrogen content allows for doping, which enhances conductivity and surface wettability, facilitating better charge storage and faster ion diffusion. Additionally, chitosan-based electrodes can be designed with hierarchical porosity, increasing surface area and improving capacitance, rate capability, and cycling stability, making it an eco-friendly option for high-performance supercapacitors. Ba et al. (2018) developed nitrogen-doped hierarchical porous carbon (NHPC) materials using a chitosan–polyethylene glycol (PEG) blend as illustrated in Fig. 9.17. The optimal NHPC, made with a 3:2 chitosan-to-PEG ratio, exhibited a high surface area of 2269 m^2 g^{-1}, moderate nitrogen doping (3.22%), and a well-defined pore structure. This material delivered excellent electrochemical performance, achieving

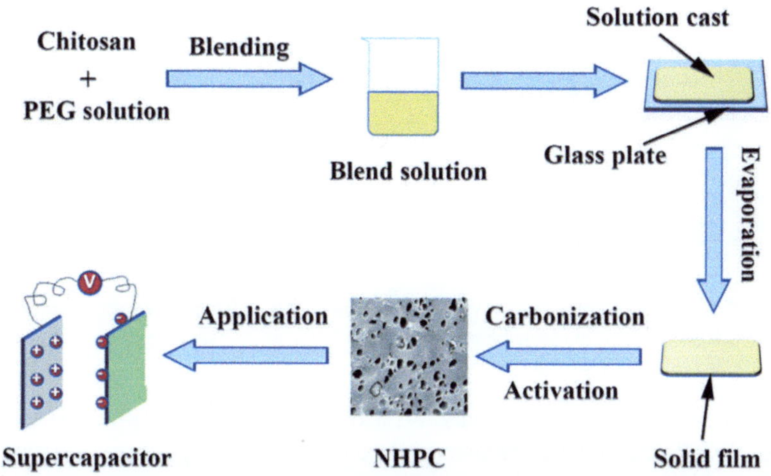

Fig. 9.17 Scheme illustration of the preparation of NHPC materials. Reprinted with permission from Ba et al. (2018), Copyright 2018, Royal Society of Chemistry

a specific capacitance of 356 F g^{-1} in 1 M H_2SO_4 and 271 F g^{-1} in 2 M KOH, with good retention even at higher current densities.

In another study, Nayak et al. (2022b) used a green synthesis method to produce an Ag-Zirconia nanocomposite, which showed a specific capacitance of 256 F g^{-1}, higher than pure Zirconia's 193 F g^{-1}. A supercapacitor device made with this nanocomposite and activated carbon demonstrated an energy density of 31.94 Wh kg^{-1}, excellent power density, and stability, retaining 89% capacitance after 2500 cycles.

Starch

Polysaccharides and similar biopolymers like starch offer significant potential as green electrode materials in supercapacitor applications. By leveraging their natural properties, they can be engineered to provide high-performance, eco-friendly alternatives to conventional electrode materials, supporting the transition to more sustainable energy storage technologies. In a recent groundbreaking study by Zhai et al. (2024), starch was innovatively used as a precursor for the synthesis of carbon aerogels. The process was notably simplified through the use of KOH, which enabled the conversion of starch suspension into hydrogels at room temperature. Based on Fig. 9.18, this method facilitated the molecular integration of KOH with starch, ensuring its optimal utilization during the activation phase. The resulting starch-based carbon aerogels exhibited an impressive specific surface area of 1349 m^2/g, with 43.7% of this surface area consisting of micropores. These aerogels demonstrated exceptional electrochemical performance as electrode materials for supercapacitors, achieving a specific capacitance of 211.5 F/g at 0.5 A/g and 138.5 F/g at 10 A/g, making them suitable for high-current applications. In a symmetrical supercapacitor configuration, the materials exhibited an energy density of 11.3 Wh/kg at a power density

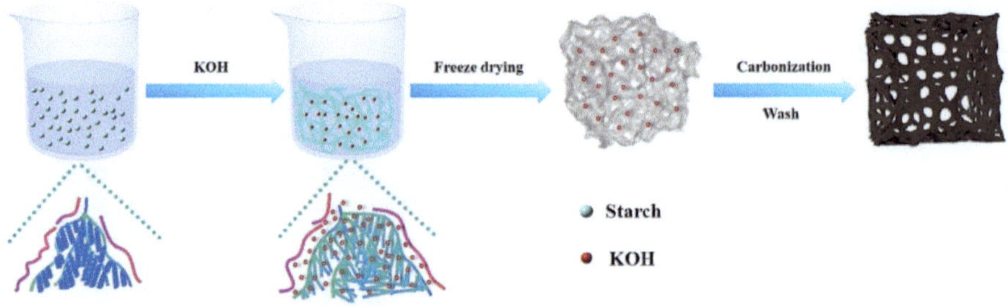

Fig. 9.18 Scheme illustration of the synthesis process of KS-F. Reprinted with permission from Zhai et al. (2024), Copyright 2024, Elsevier

of 0.5 kW/kg, and they maintained 98.91% of their capacitance after 10,000 cycles (Zhai et al. 2024). Table 9.1 enumerates the electrochemical performance metrics of various biopolymeric materials.

9.3.2 Rechargeable Batteries

Rechargeable batteries, also known as secondary cells or storage batteries, are electrochemical devices that can be charged, discharged, and recharged multiple times. Unlike disposable batteries, which are used once and discarded, rechargeable batteries offer a sustainable solution for powering a wide range of devices. There are several types of rechargeable batteries, each with distinct characteristics (Fichtner et al. 2022; Zhu et al. 2024):

1. *Lead-Acid Batteries*: The oldest type, known for their robustness and low cost. Commonly used in vehicles and backup power systems, they have a lifespan of 3 to 5 years but contain toxic lead, requiring proper disposal.
2. *Nickel–Cadmium (NiCd) Batteries*: These batteries can deliver high currents and are suitable for high-drain devices like power tools. However, they have a low capacity and are less favored today due to the toxicity of cadmium.
3. *Nickel-Metal Hydride (NiMH) Batteries*: NiMH batteries have improved capacity compared to NiCd and are widely used in consumer electronics. They are less toxic and have a lower self-discharge rate than older technologies.
4. *Lithium-Ion (Li-ion) Batteries*: Known for their high energy density and lightweight, Li-ion batteries are prevalent in portable electronics and electric vehicles. They have a longer lifespan and better performance than previous types but can be more expensive.
5. *Lithium Iron Phosphate ($LiFePO_4$) and Lithium Polymer (LiPo) Batteries*: They are variations of lithium technology, offering different advantages in terms of safety and energy density.

Table 9.1 Exploring the Physicochemical and Electrochemical Properties of Bio-based Polymer for Supercapacitors

Biopolymer	Synthesis method	Material obtained	Specific surface area (m^2 g^{-1})	C_s (F g^{-1})	Retention (%)	References
Lignin	Spray drying; KOH activation	Porous carbon	1590	340 @ 0.5 A/g	99.6 (10,000 cycles)	Appiah et al. (2022)
Alginate	Microwave-assisted method	Nanospheres	–	1329 @ 2 mA	91 (1000 cycles)	Jeyabanu et al. (2019)
	In situ hard template	Porous carbon	1838	350 @ 1 A/g	90.4 (5000 cycles)	Zhou et al. (2020a)
Cellulose	Crosslinking sol–gel method	Co-doped porous carbon	1501	358 @ 1 A/g	97.8 (10,000 cycles)	Lin et al. (2021)
	Molten salt method	Carbon nanofiber	899	150 @ 1 A/g	96 (10,000 cycles)	Zhong et al. (2022)
	Ice-templating carbonization	Porous carbon monoliths	724	210 @ 1 A/g	~ 100 (10,000 cycles)	Lee et al. (2020)
Agar	Carbonization-activation	Co-doped activated carbon	2134	313.7 @ 0.5 A/g	~100 (50,000 cycles)	Guo et al. (2021)
Chitosan	Microwave-assisted method	N-doped porous carbon	1170	435 @ 0.2 A/g	~100	Li et al. (2017)
	Hydrothermal	Chitosan/rGO composite	67	274 @ 0.5 A/g	92.57 (10,000 cycles)	Gao et al. (2021)
	Hydrothermal	Co-doped porous carbon	3646	586 @ 1 A/g	97 (10,000 cycles)	Xin et al. (2021)
Pectin	In situ template	3D porous carbon	2928	338 @ 1 A/g	85 (5000 cycles)	Zhou et al. (2020b)
Gelatin	Cross linking; activation	Porous carbon	1609	392 @ 1 A/g	83 (10,000 cycles)	Wang et al. (2022b)

(continued)

Table 9.1 (continued)

Biopolymer	Synthesis method	Material obtained	Specific surface area (m^2 g^{-1})	C_s (F g^{-1})	Retention (%)	References
Collagen	Feasible and simple approach	Mn-doped N-containing carbon	–	264.9 @ 1 A/g	81.4 (6,000 cycles)	Lei et al. (2018)
	Sol–gel, freeze-drying, carbonization, and KOH activation	Nitrogen-doped carbon aerogels	–	183.5 @ 0.5 A/g	90.2 (5,000 cycles)	Yu et al. (2018)
Chitin	Hydroxyapatite template (HAP)	N-doped carbon material (NC-HAP-700)	–	–	92 (5,000 cycles)	Chu et al. (2020)
	Pre-carbonization in air, followed by KOH activation	Cicada slough-derived carbon (CSC-2)	1243–2217	266.5 @ 0.5 A/g	92 (5,000 cycles)	Zhou et al. (2017)
Starch	TiO_2 activated starch template	A-TiO_2	=	388 @ –	99 (20,000 cycles)	Saliu et al. (2022)
Potato starch	Hydrothermal method and high-temperature carbonization	Uniform and monodisperse carbon microspheres (MCMSs)	456	61 @ 1 A/g	61 (4,000 cycles)	Ruibin et al. (2015)

Bio-based polymers are gaining traction as potential alternatives to conventional liquid electrolytes, particularly in rechargeable batteries. Researchers have increasingly focused on the development of these sustainable polymers to mitigate dependence on traditional fossil fuel-derived materials. Their intrinsic biodegradability and bio-compatibility make bio-based polymers particularly appealing, as they not only promote environmental sustainability but also enhance the functionality of embedded biomolecules, thereby elevating the performance of the overall material (Thomas et al. 2020). Other promising directions for biopolymer-based rechargeable batteries include the development of composite polymer electrolytes and hybrid inorganic–organic electrolytes (Boaretto et al. 2020; Wang et al. 2020).

Cellulose

Cellulose is categorized into several forms, such as bacterial nanocrystalline cellulose, nanofibrillated cellulose, and biodegradable, eco-friendly nanocomposites, each exhibiting markedly improved characteristics. Bacterial cellulose presents several unique advantages compared to conventional materials, including its porous structure, high specific surface area, excellent degradability, capability for large-scale production, and abundant –OH groups. These characteristics have spurred considerable research interest in developing multifunctional 3D carbonaceous materials for use in environmentally friendly rechargeable batteries (50). A recent study by Yanilmaz (2024) to enhance electrochemical performance, cellulose-based polymers were integrated with SnO_2, leading to the creation of binder-free, centrifugally spun carbon-coated SnO_2/carbon nanofibers (C@SnO_2/CNFs) as anodes in sodium-ion batteries. These anodes featured a highly porous architecture with unique sponge-like pore structures and SnO_2 nanoparticles, which enabled rapid Na-ion transport, enhanced electrolyte accessibility, and provided strong mechanical stability. Consequently, C@SnO_2/CNFs exhibited remarkable electrochemical performance in Fig. 9.19, including a high reversible capacity of 660 mAh/g at 50 mA/g, excellent rate capability, and outstanding cycling stability. Notably, the specific capacity was maintained at approximately 390 mAh/g after 2000 cycles at 0.2 A/g, underscoring the material's durability and superior performance (Yanilmaz et al. 2024).

Collagen

The abundant availability of collagen presents unique potential for its incorporation into rechargeable battery systems. Various research efforts have focused on transforming collagen derived from waste materials into functional components for electrochemical applications. As an example, Rubio et al. systematically investigated shrimp waste (SW) and two derived composites—ST-RT (synthesized at ambient temperature) and ST-CT (produced via carbothermal reduction). Among these, the ST-CT sample showed superior electrochemical behavior in sodium half-cells, delivering a reversible capacity of 162 mAh g^{-1} and maintaining 60% capacity over 200 cycles under varying current rates. Impedance spectroscopy analysis revealed a low total resistance (~0.14 $\Omega\ g^{-1}$), reflecting efficient charge transport and minimal interfacial resistance. Furthermore, a sodium-ion full cell, constructed using a $Na_2V_2(PO_4)_2$ cathode paired with the ST-CT anode, achieved reversible

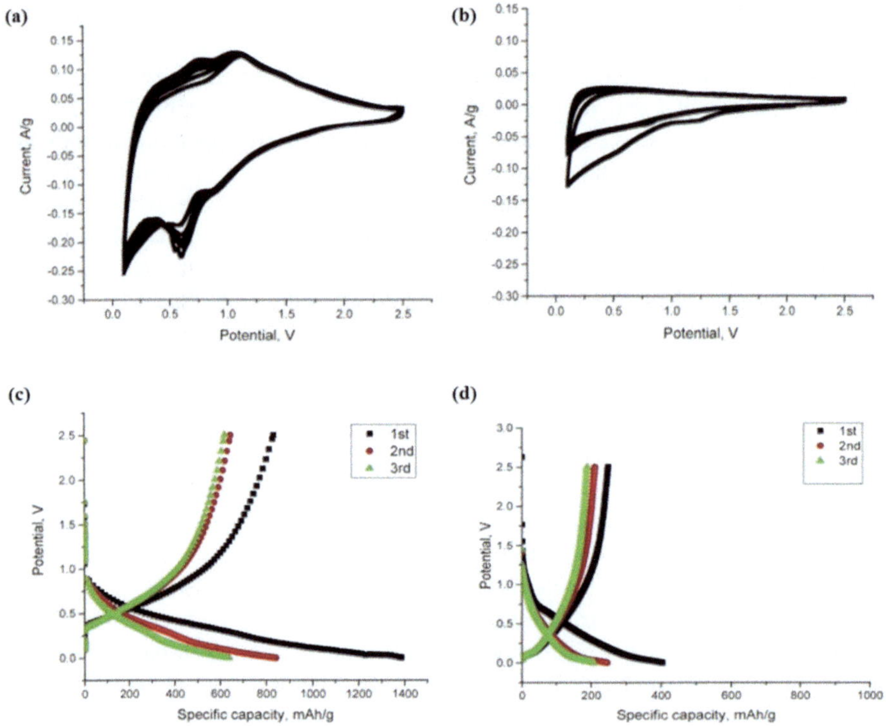

Fig. 9.19 CV curves of **a** C@SnO_2/CNFs, **b** C@CNFs, and first discharge charge curves of **c** C@SnO_2/CNFs and **d** C@CNFs. Reprinted with permission from Yanilmaz et al. (2024), Copyright 2024, Elsevier

charge/discharge capacities of 160/126 mAh g^{-1}, with an average operating voltage of approximately 2.5 V. As shown in Fig. 9.20, this configuration resulted in an impressive energy density of 310 Wh kg^{-1} (Rubio et al. 2022).

Lignin

Lignin, recognized as the second most abundant biopolymer on Earth, is inherently rich in redox-active phenolic groups, positioning it as a strong candidate for use in energy storage systems. Due to its low cost and widespread availability, lignin is considered one of the most promising natural materials for developing sustainable energy technologies. Leveraging lignin in the design of energy storage devices presents a substantial opportunity to advance green and eco-friendly innovations. In this context, a novel composite material composed of cellulose/lignin-based carbon nanobelt (CNB) aerogels integrated with Vse_2 nanosheets (Vse_2@CNB) is introduced as an anode for potassium-ion batteries (PIBs). This material was synthesized for the first time through a selenization process using $VO(acac)_2$, followed by in situ polymerization and coating with CNB aerogels. Recent developments in carbon

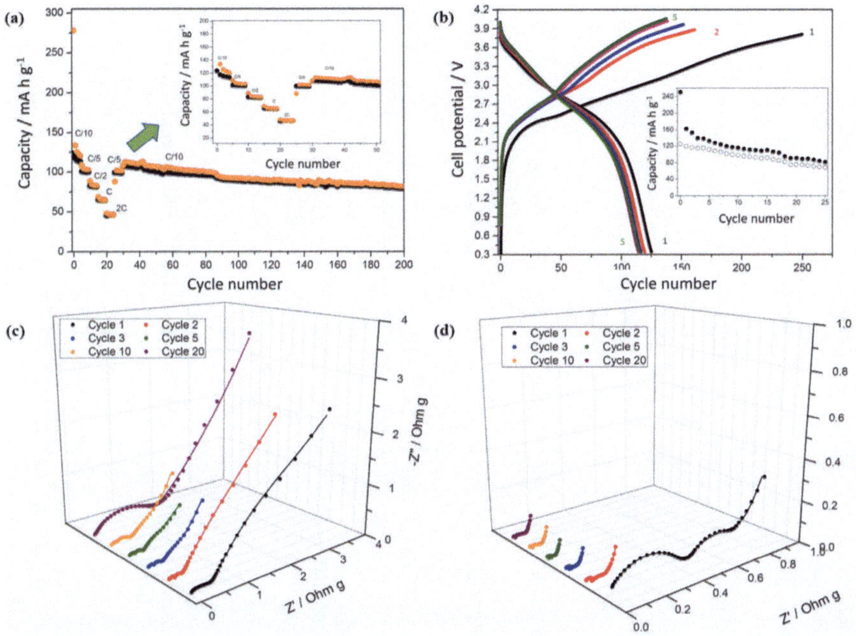

Fig. 9.20 **a** Cycling performance of the ST-CT electrode in sodium half-cells across various current rates ranging from C/10 to 2C, followed by a return to C/10. **b** Galvanostatic charge–discharge profiles and corresponding capacity retention (shown in the inset) of the full Na-ion cell composed of ST-CT/1M $NaPF_6$-DGM/C@$Na_2V_2(PO_4)_2$, tested at C/10. **c** and **d** Electrochemical impedance spectra (Nyquist plots) for ST-RT and ST-CT electrodes measured using a three-electrode setup from the 1st to the 20th cycle. Adapted with permission from Rubio et al. (2022), Copyright 2022, Elsevier

aerogel-based composites have shown their effectiveness in enhancing electrode properties for energy storage. The interconnected CNB aerogels, characterized by their high surface area and porous nature, contribute to improved electrical conductivity and mechanical integrity, thereby enhancing potassium storage capabilities. The Vse_2@CNB anode achieved a remarkable reversible capacity of 227.4 mAh g^{-1} at 100 mA g^{-1} after 100 cycles and maintained a capacity of 125 mAh g^{-1} even after 1000 cycles at 500 mA g^{-1}, accompanied by a high Coulombic efficiency of 99%. This work signifies meaningful progress toward the development of efficient and sustainable anode materials for future PIB technologies (Feng et al. 2022). Table 9.2 illustrates the electrochemical performance metrics of various biopolymeric materials.

Table 9.2 Exploring the physicochemical and electrochemical properties of bio-based polymer for batteries

Bio-polymers	Function	Initial reversible capacity	Coulombic efficiency	Cycling stability	References
Silk & nitrogen-doped carbon nanosheets	Anode in Li-ion	1913 $mA{\cdot}h{\cdot}g^{-1}$ at 0.1 $A{\cdot}g^{-1}$	49.2% at 0.1 $A{\cdot}g^{-1}$	9% loss after 10,000 cycles	Hou et al. (2015)
Silk-derived N/P co-doped porous carbon mixed with sulfur	Cathode in Li/s	888.5 $mA{\cdot}h{\cdot}g^{-1}$ at 1.0 C	97.6% at 1.0 C	0.032% loss per cycle over 500 cycles at 1.0	Dong et al. (2015)
Silk II–silk fibroin	Coating for Zn anode	189 $mA{\cdot}h{\cdot}g^{-1}$ at 0.1 $A{\cdot}g^{1}$	As high as 99.7%	As long as 3300 h at 10 $mA{\cdot}cm^{-2}$ and 10 $mA{\cdot}h{\cdot}cm^{-2}$	Lu et al. (2022)
Keratin-derived carbon added to $TiNb_2O_7$	Anode in Li-ion	356 $mA{\cdot}h{\cdot}g^{-1}$ at 0.1 C	55.0% at 0.1 C	85% retention after 50 cycles at 1 C	Thiyagarajan et al. (2021)
Keratin-derived carbon combined with α-Fe_2O_3 nanoparticles	Anode in Li-ion	1690 $mA{\cdot}h{\cdot}g^{-1}$ at 0.2 C	75% at 0.2 C	Capacity of 1000 $mA{\cdot}h{\cdot}g^{-1}$ at 0.2 C after 200 cycles	Song et al. (2019)
Collagenous and carbon combined with sulfur	Cathode in Li/S battery	1265 $mA{\cdot}h{\cdot}g^{-1}$	–	Capacity of 643	Wei et al. (2011)
Mesoporous cellulose nanocrystal membrane	Membrane for Li-ion	122 $mA{\cdot}h{\cdot}g^{-1}$ at C/2	Nearly 100% after initial decay	Retention above 90% up to C	Goncalves et al. (2019)
Bacterial cellulose combined with Al_2O_3 in a membrane	Separator for Li-ion	161 $mA{\cdot}h{\cdot}g^{-1}$ at 0.2 C	–	89% retention	Xu et al. (2017)
Agarose	Binder for $LiMn_2O_4$ cathode	101 $mA{\cdot}h{\cdot}g^{-1}$ at 0.05 C	~ 96.2% at 0.05 C	~ 100% retention after 400 cycles at 0.2 C	Hwang et al. (2016)
Lignocellulose	Electrolyte for LiS	1186.3 $mA{\cdot}h{\cdot}g^{-1}$ at 20 $mA{\cdot}g^{-1}$	–	55.1% capacity retention after 100 cycles at 20 $mA{\cdot}g^{-1}$	Song et al. (2018)

(continued)

Table 9.2 (continued)

Bio-polymers	Function	Initial reversible capacity	Coulombic efficiency	Cycling stability	References
Gelatin–silk protein film	Electrolyte for Zinc-ion batteries	311.7 $mA\cdot h\cdot g^{-1}$	Greater than 90% over 100 cycles	Greater than 90% over 100 cycles	Zhou et al. (2021)
Carrageenan and wool keratin	Electrolyte for Zn-Ion Batteries	271.6 $mA\cdot h\cdot g^{-1}$ at 0.1 $A\cdot g^{1}$	~ 98%	96% capacity retention after 500 cycles	Shao et al. (2022)
Chitosan and poly(vinyl alcohol	Electrolytic Manganese Dioxide Battery (EMD)	310 $mA\cdot h\cdot g^{-1}$ at 0.1 $A\cdot g^{1}$	96.5% at 0.5 $A\cdot g^{-1}$	~70% capacity retention after 300 cycles	Poosapati et al. (2021)
Chitosan-zinc electrolyte	Electrolyte for Zinc metal batteries	208 $mA\cdot h\cdot g^{-1}$	99.7%	Greater than 400 cycles	Wu et al. (2022)
Chitosan	Electrolyte for Separator for Zn batteries	323 $mA\cdot h\cdot g^{-1}$ at 0.1 $A\cdot g^{1}$	99.6% at 1 $mA\cdot cm^{-2}$ and 1 $mA\cdot h\cdot cm^{-2}$	98.4% retention over 1000 cycles	Yang et al. (2022)

9.3.3 Fuel Cells

Fuel cells are emerging as eco-friendly technologies for energy conversion, offering a cleaner alternative to traditional combustion-based power generation. They operate through electrochemical reactions that convert chemical energy directly into electrical energy, providing continuous power as long as fuel (e.g., hydrogen) and an oxidant are supplied. Unlike batteries, which store and deplete energy, fuel cells can function indefinitely with a constant fuel source.

Fuel cells are classified by the electrolyte used, which influences their efficiency, operating temperature, and applications (Ye et al. 2012):

1. *Proton Exchange Membrane Fuel Cells (PEMFCs)*—Operate at low temperatures (60–80 °C) with high power density, ideal for automotive and portable applications, though platinum catalysts make them costly.
2. *Solid Oxide Fuel Cells (SOFCs)*—High-temperature devices (600–1000 °C) with fuel flexibility and efficiency, suited for stationary and combined heat and power (CHP) applications.
3. *Alkaline Fuel Cells (AFCs)*—Use aqueous potassium hydroxide electrolyte, efficient in hydrogen and oxygen use but sensitive to CO_2 contamination, making them popular in space and military applications.
4. *Phosphoric Acid Fuel Cells (PAFCs)*—Operate at moderate temperatures (150–200 °C), commonly used in stationary power and large vehicles, offering good fuel tolerance but lower power density.
5. *Molten Carbonate Fuel Cells (MCFCs)*—Use molten carbonate salts, operate at high temperatures (600–700 °C), and are suited for industrial and CHP systems with high efficiency, though material stability is a challenge.

Biopolymers are gaining interest as sustainable materials for fuel cells, offering biodegradability and low toxicity. These natural polymers can be modified chemically or physically to enhance their properties for fuel cell use. Chitosan, in particular, has been widely studied as a solid polymer electrolyte in various fuel cells, including hydrogen-polymer electrolyte fuel cells (PEFCs), direct methanol fuel cells (DMFCs), and alkaline fuel cells (AFCs), due to its renewable and adaptable characteristics.

Chitosan

A pure chitosan membrane functions as an insulator under standard conditions but can support ionic conduction when doped with salts such as lithium nitrate (Li^+-based complexes) or proton-contributing compounds (H^+-based complexes). Strategies involving ionic and covalent cross-linking have been proposed to disrupt the crystalline regions within the chitosan matrix. Lowering crystallinity enhances the membrane's ability to absorb water, thereby creating a microstructure more conducive to ion mobility transport (Soontarapa and Intra 2006). In a study by Mukoma et al. (2004) cross-linked chitosan membranes were assessed for their potential as substitutes in proton exchange membrane fuel cells (PEMFCs). While these membranes demonstrated superior water absorption compared to Nafion 117 at room temperature, their practical use in high-temperature fuel cells was limited due to thermal instability. Therefore, their functionality remains suitable only under moderate temperature conditions. The blending of polymers aims to integrate the strengths of individual materials into a unified, near-homogeneous composite, although this goal is not always fully achieved. Chitosan has been commonly blended with other polymers such as poly(vinyl alcohol) (PVA), poly(acrylic acid) (PAA), poly(vinyl pyrrolidone) (PVP), nylon, alginate, and cellulose to improve overall performance and compensate for its weaknesses (Ikram et al. 2017).

Cellulose

The surface of cellulose is often tailored through modification with various organic and inorganic compounds to introduce functional groups such as sulfate esters, acetate units, succinate monoesters, aliphatic ethers, silyl ethers, carboxymethyl derivatives, and urethane linkages. Another common approach involves oxidation using a TEMPO-mediated process in the presence of hypochlorite, which introduces carboxylic acid functionalities (Moon et al. 2011). In one study, Lam et al. (2009) utilized cellulose acetate (CA) to fabricate a single-electrode configuration for direct methanol fuel cells (DMFCs). This innovative design eliminates the need for a separate polymer electrolyte membrane (PEM), integrating both the anode and cathode into a single structure. The resultant CA-based DMFC, with a compact thickness of 3.88×10^{-2} cm, achieved a 104% improvement in volumetric specific power density compared to conventional two-electrode DMFC systems operated passively at room temperature and ambient pressure (Lam et al. 2009). Harun et al. developed a solid polymer electrolyte using CA as the primary polymer, ammonium tetrafluoroborate (NH_4BF_4) as the ionic dopant, and polyethylene glycol (PEG600) as a plasticizing agent, employing a solution casting method. The highest proton conductivity for the CA-NH_4BF_4 system was recorded at 2.18×10^{-4} mS cm^{-1}, which increased significantly to 1.41×10^{-2} mS upon addition of 30 wt% PEG600.

Alginate

Alginate can be cross-linked effectively using bifunctional compounds such as glutaraldehyde, 1,6-hexanediamine, and others. However, similar to chitosan, alginate exhibits limitations in practical use due to high solubility in water and inadequate mechanical integrity. Several strategies have been explored to enhance its performance, including covalent and ionic cross-linking, as well as non-covalent interactions (Li et al. 2012). A unique membrane formulation was developed using pure alginate, calcium chloride ($CaCl_2$) as the ionic cross-linker, and glycerol serving as the plasticizer. This formulation demonstrated impressive proton conductivity of 10.1 mS cm^{-1} and remarkably low methanol permeability at 2.0×10^{-7} cm^2 s^{-1}. Additionally, the combination of $CaCl_2$ and glycerol significantly improved the mechanical properties of the alginate matrix (Shaari et al. 2018).

9.4 Challenges and Future Directions

The rapid advancement of new technologies is pushing many industries to adopt more environmentally responsible practices, especially in materials selection for energy storage devices. As the world moves toward generating 70% of power from renewable sources, energy storage plays a critical role in stabilizing supply and demand. Innovations in electrode and electrolyte materials have gained significant attention, leading to the development of sustainable, flexible, and lightweight materials. Natural biopolymers are becoming a key area of research, as the energy sector aims to reduce reliance on synthetic polymers that are non-biodegradable and non-renewable. Alongside this shift, biopolymers show great potential for use in flexible, wearable technologies, but production methods still require modernization to meet industrial demands.

Reinforcing biopolymers with fibers or nanoparticles has been a viable strategy to enhance their mechanical resilience, making them more durable for industrial applications. However, challenges remain in ensuring their processability and electrochemical performance, especially for energy storage devices like batteries and supercapacitors. The combination of biopolymers with advanced materials like graphene could offer a path forward, enhancing the energy storage capacity while maintaining the sustainability of the materials. Moreover, bio-based polymers from food waste are being investigated due to their global abundance and potential for large-scale application, but their full life cycle must be environmentally sound to avoid the production of harmful byproducts.

Electrochemical performance remains a critical challenge for biopolymer-based energy storage systems, particularly in achieving the desired energy and power densities. The ionic mobility within biopolymer-based electrolytes is still limited due to inadequate pore structures. The microstructure of these materials must be carefully engineered to improve ion diffusion, and the synthesis of carbon-based materials from biopolymers introduces further challenges in consistency. Advanced characterization methods such as TEM, AFM, SEM, and XRD offer valuable insights into the structural and electrochemical properties of these materials, guiding future development.

Finally, a deeper understanding of the interactions between electrolyte ions and biopolymer chains is needed to optimize ionic conductivity. Factors such as ion size, valence, hydration, and polymer molecular weight all influence the effectiveness of biopolymers in energy storage applications. The incorporation of electrochemically active elements into biopolymers also presents opportunities for improving mechanical stability and flexibility. By focusing on these challenges and refining the design of bio-based polymer materials, researchers can accelerate the transition from lab-scale innovations to large-scale, high-performance energy storage solutions.

References

Abd El Aty AA et al (2017) Covalent immobilization of Alternaria tenuissima KM651985 laccase and some applied aspects. Biocatalysis and Agricultural Biotechnology. 9 p. 74–81

Abdullah, et al., Biopolymer-based functional films for packaging applications: A review. Frontiers in Nutrition, 2022. 9: p 1000116

Agency IE (2020) Global energy review 2020: the impacts of the Covid-19 crisis on global energy demand and CO_2 emissions. OECD Publishing

Aggarwal J et al (2020) The realm of biopolymers and their usage: an overview. J Environ Treat Tech 8(2):1005–1016

Ahmad N et al (2017) Novel bio-nanocomposite materials for enhanced biodegradability and photocatalytic activity. New J Chem 41(18):10198–10207

Ahmadi Bonakdar M, Rodrigue D (2024) Electrospinning: Processes, structures, and materials. Macromol, 4(1): p. 58–103

Alam RB et al (2023) Improved electrochemical performance of multi-walled carbon nanotube reinforced gelatin biopolymer for transient energy storage applications. PLoS ONE 18(11):e0288113

Amir M et al (2023) Energy storage technologies: An integrated survey of developments, global economical/environmental effects, optimal scheduling model, and sustainable adaption policies. J Energy Storage 72:108694

Appiah ES et al (2022) Biopolymers-derived materials for supercapacitors: recent trends, challenges, and future prospects. Molecules 27(19):6556

Arvand M et al (2017) A new electrochemical sensing platform based on binary composite of graphene oxide-chitosan for sensitive rutin determination. Food Anal Methods 10:2332–2345

Awad GE et al (2017) A novel alginate–CMC gel beads for efficient covalent inulinase immobilization. Colloid Polym Sci 295:495–506

Ba Y et al (2018) Nitrogen-doped hierarchical porous carbon derived from a chitosan/polyethylene glycol blend for high performance supercapacitors. RSC Adv 8(13):7072–7079

Bajpai PK et al (2017) Processing and characterization of bio-composites. Handbook of ecomaterials, p 1–18

Bharathi D et al (2019) Preparation of chitosan coated zinc oxide nanocomposite for enhanced antibacterial and photocatalytic activity: as a bionanocomposite. Int J Biol Macromol 129:989–996

Bhiogade A, Kannan M (2021) Studies on thermal and degradation kinetics of cellulose micro/nanoparticle filled polylactic acid (PLA) based nanocomposites. Polymers and Polymer Composites, 29(9_suppl): p S85-S98

Bikiaris ND et al (2023) Recent advances in the investigation of poly (lactic acid)(PLA) nanocomposites: incorporation of various nanofillers and their properties and applications. Polymers 15(5):1196

Boaretto N et al (2020) Polymer electrolytes for rechargeable batteries: from nanocomposite to nanohybrid. J Electrochem Soc 167(7):070524

Chen J et al (2020) Effect of thermal annealing on crystal structure and properties of PLLA/PCL blend. J Polym Res 27:1–11

Chen Z et al (2023) Advances and applications of cellulose bio-composites in biodegradable materials. J Polym Environ 31(6):2273–2284

Chu M et al (2020) N-doped carbon derived from the monomer of chitin for high-performance supercapacitor. Appl Surf Sci 517:146140

Das A et al (2023) A comprehensive review on recent advances in preparation, physicochemical characterization, and bioengineering applications of biopolymers. Polym Bull 80(7):7247–7312

Dong H et al (2015) Facile synthesis of α-Fe2O3 nanoparticles on porous human hair-derived carbon as improved anode materials for lithium ion batteries. J Power Sources 300:104–111

Dueramae I et al (2020) Properties enhancement of carboxymethyl cellulose with thermo-responsive polymer as solid polymer electrolyte for zinc ion battery. Sci Rep 10(1):12587

Ejaz M et al (2020) Mechanical and biodegradable properties of jute/flax reinforced PLA composites. Fibers and Polymers 21:2635–2641

Essawy HA et al (2004) Polystyrene/montmorillonite nanocomposites prepared by in situ intercalative polymerization: influence of the surfactant type. Macromol Chem Phys 205(17):2366–2370

Faruk O et al (2012) Biocomposites reinforced with natural fibers: 2000–2010. Prog Polym Sci 37(11):1552–1596

Feng Y et al (2022) Cellulose/lignin-based carbon nanobelt aerogels coating on VSe2 nanosheets as anode for high performance potassium-ion batteries. J Power Sources 548:232033

Fichtner M et al (2022) Rechargeable batteries of the future—the state of the art from a BATTERY 2030+ perspective. Adv Energy Mater 12(17):2102904

Gao M et al (2021) Sandwich construction of chitosan/reduced graphene oxide composite as additive-free electrode material for high-performance supercapacitors. Carbohyd Polym 255:117397

Gholampour A, Ozbakkaloglu T (2020) A review of natural fiber composites: Properties, modification and processing techniques, characterization, applications. J Mater Sci 55(3):829–892

Goncalves R et al (2019) Mesoporous cellulose nanocrystal membranes as battery separators for environmentally safer lithium-ion batteries. ACS Applied Energy Materials 2(5):3749–3761

Govindan S et al (2012) Synthesis and characterization of chitosan–silver nanocomposite. Appl Nanosci 2:299–303

Guo Y et al (2021) Agar-based porous electrode and electrolyte for flexible symmetric supercapacitors with ultrahigh energy density. J Power Sources 507:230252

Gupta S, Varshney PK (2019) Effect of plasticizer on the conductivity of carboxymethyl cellulose-based solid polymer electrolyte. Polymer Bulletin p 1–10

Hao X et al (2017) Bacterial-cellulose-derived interconnected meso-microporous carbon nanofiber networks as binder-free electrodes for high-performance supercapacitors. J Power Sources 352:34–41

Hassan MM et al (2023) Novel nanoarchitecture of 3D ion transfer channel containing nanocomposite solid polymer electrolyte membrane based on holey graphene oxide and chitosan biopolymer. Chem Eng J 466:143159

Hou J et al (2015) Hierarchical porous nitrogen-doped carbon nanosheets derived from silk for ultrahigh-capacity battery anodes and supercapacitors. ACS Nano 9(3):2556–2564

Hwang G et al (2016) Multifunctional natural agarose as an alternative material for high-performance rechargeable lithium-ion batteries. Green Chem 18(9):2710–2716

Ibrahim A et al (2021) Graphene-based nanocomposites: Synthesis, mechanical properties, and characterizations. Polymers 13(17):2869

Ikram S et al (2017) Chitosan-based polymer electrolyte membranes for fuel cell applications. Organic-Inorganic composite polymer electrolyte membranes: preparation, properties, and fuel cell applications, p 381–398

Islam MR, Mollik SI (2020) Enhanced electrochemical performance of flexible and eco-friendly starch/graphene oxide nanocomposite. Heliyon, 6(10)

Jamil H et al (2024) Recent advances in polymer nanocomposites: unveiling the frontier of shape memory and self-healing properties—a comprehensive review. Molecules 29(6):1267

Jawaid M, Khalil HA (2011) Cellulosic/synthetic fibre reinforced polymer hybrid composites: A review. Carbohyd Polym 86(1):1–18

Jeyabanu K et al (2019) Preparation and characterization studies of La doped CuS nanospheres by microwave irradiation for high performance supercapacitors. Physica B 573:92–101

Kacica CT, Biswas P (2020) Improved conductivity and ionic mobility in nanostructured thin films via aliovalent doping for ultra-high rate energy storage. Nanoscale Advances 2(5):2160–2169

Kang Y et al (2020) Binder-free electrodes and their application for Li-ion batteries. Nanoscale Res Lett 15:1–19

Kashfipour MA et al (2018) A review on the role of interface in mechanical, thermal, and electrical properties of polymer composites. Adv Compos Hybrid Mater 1:415–439

Kumar N et al (2022) Recent advanced supercapacitor: a review of storage mechanisms, electrode materials, modification, and perspectives. Nanomaterials 12(20):3708
Lai CS, Locatelli G (2021) Are energy policies for supporting low-carbon power generation killing energy storage? J Clean Prod 280:124626
Lam A et al (2009) A novel single electrode supported direct methanol fuel cell. Electrochem Commun 11(7):1530–1534
Lee B-M et al (2020) Eco-friendly fabrication of porous carbon monoliths from water-soluble carboxymethyl cellulose for supercapacitor applications. J Ind Eng Chem 82:367–373
Lei J et al (2018) Novel collagen waste derived Mn-doped nitrogen-containing carbon for supercapacitors. Electrochim Acta 285:292–300
Li B et al (2017) Nitrogen doped and hierarchically porous carbons derived from chitosan hydrogel via rapid microwave carbonization for high-performance supercapacitors. Carbon 122:592–603
Liew SY et al (2013) High total-electrode and mass-specific capacitance cellulose nanocrystal-polypyrrole nanocomposites for supercapacitors. RSC Adv 3(24):9158–9162
Lin S et al (2021) N/O co-doped hierarchically porous carbon with three-dimensional conductive network for high-performance supercapacitors. J Alloy Compd 873:159705
Liu W et al (2018) Polylactic acid (PLA)/cellulose nanowhiskers (CNWs) composite nanofibers: microstructural and properties analysis. J Compos Sci 2(1):4
Lu J et al (2022) Silk Fibroin Coating Enables Dendrite-free Zinc Anode for Long-Life Aqueous Zinc-Ion Batteries. Chemsuschem 15(15):e202200656
Mahmud S et al (2021) Comprehensive review on plant fiber-reinforced polymeric biocomposites. J Mater Sci 56:7231–7264
Majka TM et al (2016) Thermal stability and degradation of polymer nanocomposites. Polymer Nanocomposites: Electrical and Thermal Properties, p 167–190
Ming J et al (2019) Zinc-ion batteries: Materials, mechanisms, and applications. Mater Sci Eng R Rep 135:58–84
Mo R-J et al (2018) Graphene-like porous carbon from sheet cellulose as electrodes for supercapacitors. Chem Eng J 346:104–112
Mohammed ASA et al (2021) Polysaccharides; classification, chemical properties, and future perspective applications in fields of pharmacology and biological medicine (a review of current applications and upcoming potentialities). J Polym Environ 29:2359–2371
Mohammed MM et al (2022) Interfacial bonding mechanisms of natural fibre-matrix composites: an overview. BioResources 17(4):7031
Mollik SI et al (2021) Improved thermal, mechanical, and electrochemical performance of bio-degradable starch/reduced graphene oxide nanocomposites. AIP Advances, 11(9)
Moon RJ et al (2011) Cellulose nanomaterials review: structure, properties and nanocomposites. Chem Soc Rev 40(7):3941–3994
Mostafa MH et al (2020) Techno-economic assessment of energy storage systems using annualized life cycle cost of storage (LCCOS) and levelized cost of energy (LCOE) metrics. J Energy Storage 29:101345
Mukoma P et al (2004) Synthesis and characterization of cross-linked chitosan membranes for application as alternative proton exchange membrane materials in fuel cells. J Power Sources 136(1):16–23
Muneer F et al (2021) Bioplastics from biopolymers: an eco-friendly and sustainable solution of plastic pollution. Polym Sci, Ser C 63:47–63
Nagalakshmaiah M et al (2019) Biocomposites: Present trends and challenges for the future. Green composites for automotive applications, p 197–215
Nayak S et al (2022a) Biosynthesis of zinc oxide-cobalt oxide nanocomposite as electrode material and its performance evaluation for the sustainable hybrid supercapacitor energy storage devices. Chem Phys Lett 806:140058
Nayak S et al (2022b) Green synthesis of Silver-Zirconia composite using chitosan biopolymer binder for fabrication of electrode materials in supercapattery application for sustainable energy storage. Curr Res Green Sustain Chem 5:100292

Pal N et al (2017) Combined effect of cellulose nanocrystal and reduced graphene oxide into polylactic acid matrix nanocomposite as a scaffold and its anti-bacterial activity. Int J Biol Macromol 95:94–105

Pappu A et al (2016) Facile extraction, processing and characterization of biorenewable sisal fibers for multifunctional applications. J Macromol Sci, Part A 53(7):424–432

Patel NG et al (2014) Fabrication, nanomechanical characterization, and cytocompatibility of gold-reinforced chitosan bio-nanocomposites. Mater Sci Eng, C 44:336–344

Perumal AB et al (2018) Development of polyvinyl alcohol/chitosan bio-nanocomposite films reinforced with cellulose nanocrystals isolated from rice straw. Appl Surf Sci 449:591–602

Pokharel A et al (2022) Biobased polymer composites: a review. J Compos Sci 6(9):255

Poosapati A et al (2021) Rechargeable zinc-electrolytic manganese dioxide (EMD) battery with a flexible chitosan-alkaline electrolyte. ACS Applied Energy Materials 4(4):4248–4258

Qiu J et al (2024) Insights into nano-and micro-structured scaffolds for advanced electrochemical energy storage. Nano-Micro Letters 16(1):130

Radovanović M et al (2021) Sustainable energy transition in Central Asia: status and challenges. Energy, Sustain Soc 11:1–13

Rahman MA et al (2015) A review on porous negative electrodes for high performance lithium-ion batteries. J Porous Mater 22:1313–1343

Randhawa A et al (2022) A review of properties of nanocellulose, its synthesis, and potential in biomedical applications. Appl Sci 12(14):7090

Rasib S et al (2021) Effect of waste fillers addition on properties of high-density polyethylene composites: mechanical properties, burning rate, and water absorption. Polym Bull 78:6777–6795

Ren K et al (2021) Recent developments of transition metal compounds-carbon hybrid electrodes for high energy/power supercapacitors. Nano-Micro Letters 13(1):129

Rhim J-W et al (2014) Preparation and characterization of bio-nanocomposite films of agar and silver nanoparticles: laser ablation method. Carbohyd Polym 103:456–465

Rodrigues C et al (2020) Mechanical, thermal and antimicrobial properties of chitosan-based-nanocomposite with potential applications for food packaging. J Polym Environ 28:1216–1236

Rodríguez-Tobías H et al (2019) Comprehensive review on electrospinning techniques as versatile approaches toward antimicrobial biopolymeric composite fibers. Mater Sci Eng, C 101:306–322

Rosendo-González V et al (2019) Synthesis and characterization of clay nanocomposites based on starch. MRS Advances 4(59):3243–3249

Rothausen SG, Conway D (2011) Greenhouse-gas emissions from energy use in the water sector. Nat Clim Chang 1(4):210–219

Rubio S et al (2022) Marine shrimp/tin waste as a negative electrode for rechargeable sodium-ion batteries. J Clean Prod 359:131994

Ruibin Q et al (2015) Monodisperse carbon microspheres derived from potato starch for asymmetric supercapacitors. Electrochim Acta 167:303–310

Ryś PA, et al (2024) Trends and prospects in lead-acid battery developments. J Power Technol. 104(1)

Sadasivuni KK et al (2020) Recent advances in mechanical properties of biopolymer composites: a review. Polym Compos 41(1):32–59

Sahu P, Gupta M (2020) A review on the properties of natural fibres and its bio-composites: Effect of alkali treatment. Proceedings of the institution of mechanical engineers, Part l: journal of materials: design and applications 234(1):198–217

Saliu O et al (2022) Starch built TiO2 nanoarchitecture with mixed anatase and rutile phase for high energy density supercapacitor electrode. J Energy Storage 49:104155

Samsudin AS et al (2014) Biopolymer materials based carboxymethyl cellulose as a proton conducting biopolymer electrolyte for application in rechargeable proton battery. Electrochim Acta 129:1–13

Sanivada UK et al (2020) PLA composites reinforced with flax and jute fibers—A review of recent trends, processing parameters and mechanical properties. Polymers 12(10):2373

Schlemmer W et al (2021) Polysaccharides for sustainable energy storage—A review. Carbohyd Polym 265:118063
Seehra M (2017) Nanostructured materials: Fabrication to applications. BoD–Books on Demand
Shaari N et al (2018) Enhanced mechanical flexibility and performance of sodium alginate polymer electrolyte bio-membrane for application in direct methanol fuel cell. J Appl Polym Sci 135(37):46666
Shaikh VA, Shaikh IV (2022) Polysaccharides: Sustainable Green-Composite Biomaterials. Encyclopedia of Green Materials. Springer, pp 1–11
Shao Y et al (2022) Regulating Interfacial Ion Migration via Wool Keratin Mediated Biogel Electrolyte toward Robust Flexible Zn-Ion Batteries. Small 18(10):2107163
Sharma M et al (2023) Recent advances in nanofillers for multidisciplinary applications of polymer nanocomposites. Handbook of Nanofillers, p 1–20
Song A et al (2018) Novel lignocellulose based gel polymer electrolyte with higher comprehensive performances for rechargeable lithium–sulfur battery. J Membr Sci 556:203–213
Song Z et al (2019) Synergistic confining polysulfides by rational design a N/P co-doped carbon as sulfur host and functional interlayer for high-performance lithium–sulfur batteries. J Power Sources 421:23–31
Soontarapa K, Intra U (2006) Chitosan-based fuel cell membranes. Chem Eng Commun 193(7):855–868
Sumrith N et al (2019) Biopolymers-based nanocomposites: Properties and applications. Bio-based Polymers and Nanocomposites: Preparation, Processing, Properties & Performance, p 255–272
Sun S et al (2019) Facile synthesis of three-dimensional interconnected porous carbon derived from potassium alginate for high performance supercapacitor. J Alloy Compd 803:401–406
Sun Y et al (2020) In-situ Polymerization of exfoliated structure PA6/organo-clay nanocomposites. Rev Adv Mater Sci 59(1):434–440
Suneetha RB et al (2019) Synthesis, structural and electrochemical characterization of Zn doped iron oxide/grapheneoxide/chitosan nanocomposite for supercapacitor application. Vacuum 164:396–404
Swain SK et al (2018) Functional biopolymer composites. Funct Biopolym, p 159–182
Tănase EE et al (2014) Biopolymers based on renewable resources-a review
Thiyagarajan GB et al (2021) TiNb2O7-Keratin derived carbon nanocomposites as novel anode materials for high-capacity lithium-ion batteries. Open Ceramics 6:100131
Thomas P et al (2020) Comprehensive review on nanocellulose: Recent developments, challenges and future prospects. J Mech Behav Biomed Mater 110:103884
Torvi A et al (2018) Development of supercapacitor systems based on binary and ternary nanocomposites using chitosan, graphene and polyaniline. Chem Data Collect 17:459–471
Trache D et al (2017) Recent progress in cellulose nanocrystals: sources and production. Nanoscale 9(5):1763–1786
Venkatesan D et al (2021) Synthesis, Properties, and Applications of Polymer Nanocomposite Matrices. Handbook of Polymer and Ceramic Nanotechnology. Springer, pp 465–485
Wali SB et al (2024) Grid-connected lithium-ion battery energy storage system towards sustainable energy: A patent landscape analysis and technology updates. J Energy Storage 77:109986
Wang F et al (2018a) Changes in the morphological–mechanical properties and thermal stability of bamboo fibers during the processing of alkaline treatment. Polym Compos 39(S3):E1421–E1428
Wang B et al (2018b) Alginate-based hierarchical porous carbon aerogel for high-performance supercapacitors. J Alloy Compd 749:517–522
Wang H et al (2020) Reviewing the current status and development of polymer electrolytes for solid-state lithium batteries. Energy Storage Materials 33:188–215
Wang S et al (2022a) Carbon nanotubes/reduced graphene oxide composites as electrode materials for supercapacitors. Appl Phys A 128(1):81
Wang G et al (2022b) Gelatin-derived honeycomb like porous carbon for high mass loading supercapacitors. Journal of Energy Storage 45:103525

Wei S et al (2011) Pig bone derived hierarchical porous carbon and its enhanced cycling performance of lithium–sulfur batteries. Energy Environ Sci 4(3):736–740
Wei K et al (2022) Oxygenated N-doped porous carbon derived from ammonium alginate: Facile synthesis and superior electrochemical performance for supercapacitor. J Energy Storage 51:104342
Wilson NW, Botte GG (2024) Novel biopolymer pectin-based hydrogel electrolytes for sustainable energy storage. Materials Advances
Wohlert M et al (2022) Cellulose and the role of hydrogen bonds: not in charge of everything. Cellulose, p 1–23
Wu M et al (2022) A sustainable chitosan-zinc electrolyte for high-rate zinc-metal batteries. Matter 5(10):3402–3416
Xia L et al (2019) Hierarchical macro-/meso-/microporous oxygen-doped carbon derived from sodium alginate: A cost-effective biomass material for binder-free supercapacitors. Mater des 182:108048
Xin X et al (2021) N, P-codoped porous carbon derived from chitosan with hierarchical N-enriched structure and ultra-high specific surface Area toward high-performance supercapacitors. J Mater Sci Technol 88:45–55
Xu Q et al (2017) A bacterial cellulose/Al 2 O 3 nanofibrous composite membrane for a lithium-ion battery separator. Cellulose 24:1889–1899
Yan L et al (2014) Flax fibre and its composites–A review. Compos B Eng 56:296–317
Yang X et al (2022) Chitosan modified filter paper separators with specific ion adsorption to inhibit side reactions and induce uniform Zn deposition for aqueous Zn batteries. Chem Eng J 450:137902
Yanilmaz M et al (2024) High-performance cellulose-based nanostructured carbons for rechargeable batteries. J Alloy Compd 1005:175961
Ye Y-S et al (2012) Water soluble polymers as proton exchange membranes for fuel cells. Polymers 4(2):913–963
Yu M et al (2018) Magnetic N-doped carbon aerogel from sodium carboxymethyl cellulose/collagen composite aerogel for dye adsorption and electrochemical supercapacitor. Int J Biol Macromol 115:185–193
Zafar S et al (2024) Recent developments and future prospective of polyurethane coatings for corrosion protection– a focused review. Eur Polym J, p 113421
Zainal Abidin AS et al (2018) Surface functionalization of graphene oxide with octadecylamine for improved thermal and mechanical properties in polybutylene succinate nanocomposite. Polym Bull 75:3499–3522
Zhai Z et al (2024) Starch-based carbon aerogels prepared by an innovative KOH activation method for supercapacitors. Int J Biol Macromol 257:128587
Zhang Z, Yang R (2017) Novel nanocomposites based on hydroxyethyl cellulose and graphene oxide. Fibers and Polymers 18:334–341
Zhang Y et al (2017) Review of macroporous materials as electrochemical supercapacitor electrodes. J Mater Sci 52:11201–11228
Zhang Q et al (2021) High-temperature polymers with record-high breakdown strength enabled by rationally designed chain-packing behavior in blends. Matter 4(7):2448–2459
Zhang L et al (2022) Cellulose-Based electrolytes for advanced lithium-ion batteries: recent advances and future perspectives. ChemNanoMat 8(8):e202200142
Zhong Y et al (2022) Carbon nanofibers derived from cellulose via molten-salt method as supercapacitor electrode. Int J Biol Macromol 207:541–548
Zhou J et al (2017) Chitin based heteroatom-doped porous carbon as electrode materials for supercapacitors. Carbohyd Polym 173:321–329
Zhou Y et al (2020a) In-situ template cooperated with thiourea to prepare oxygen/nitrogen co-doped porous carbons with adjustable pore structure for supercapacitors. Renewable Energy 153:1005–1015

Zhou Y et al (2020b) In-situ template cooperated with urea to construct pectin-derived hierarchical porous carbon with optimized pore structure for supercapacitor. Electrochim Acta 355:136801
Zhou J et al (2021) Humidity-sensitive, shape-controllable, and transient zinc-ion batteries based on plasticizing gelatin-silk protein electrolytes. Materials Today Energy 21:100712
Zhu J et al (2024) Low-Temperature-tolerant aqueous proton battery with porous Ti3C2Tx MXene electrode and phosphoric acid electrolyte. Batteries 10(6):207

Chapter 10
Self-Healing Biopolymer: A Promising Material for Sustainable Engineering Applications

F. S. Hosseini, K. Alasvand, M. Shariatmadar, S. Sohrabi, and B. Ramezanzadeh

Abstract This chapter introduces the concept of self-healing biomaterials with the prime rationale of autonomous restoration in case any damage occurs, hence prolonging lifetime or efficiency. These materials, inspired by biological systems, use natural and synthetic biopolymers to restore integrity after mechanical wear. This chapter outlines various biopolymers, including cellulose, chitosan, and hyaluronic acid, combined with synthetic ones, each possessing either intrinsic or extrinsic self-healing mechanisms. Such functionalization involves treatments with bio-chemicals, the addition of nanomaterials, and biomimetic strategies, given applications related to aerospace, medicine, and manufacturing industries. Advanced nanotechnology integration and responsive bonding mechanisms provide enhanced scalability, durability, and environmental compatibility—a pathway toward broader applications in many industries.

10.1 Introduction

The term "self-healing materials" generally refers to systems capable of restoring their original properties after mechanical damage (Joy et al. 2020). This innovative concept, inspired by biological processes like wound healing, has quickly gained traction in materials science. These materials can autonomously detect and repair minor issues such as micro-cracks before they escalate into larger failures. By maintaining the structural integrity of products without external intervention, self-healing materials help reduce repair costs and downtime. Industries such as aerospace and automotive, where even small defects can lead to major issues, stand to benefit greatly. Moreover, by increasing product longevity and minimizing waste, these materials align with sustainable manufacturing practices (Jayabalakrishnan et al. 2020).

F. S. Hosseini · K. Alasvand · M. Shariatmadar · S. Sohrabi · B. Ramezanzadeh (✉)
Department of Surface Coating and Corrosion, Institute for Color Science and Technology, Tehran, Iran
e-mail: ramezanzadeh-bh@icrc.ac.ir

L. Ravindran et al. (eds.), *Macro, Micro and Nanocomposites from Sustainable Sources*, Advanced Structured Materials 244, https://doi.org/10.1007/978-981-95-2469-3_10

Self-healing materials draw inspiration from natural repair mechanisms observed in biological systems (Cremaldi and Bhushan 2018). Biopolymers like cellulose and chitosan are vital to sustainable engineering because of their renewable sources and biodegradable properties. These natural polymers exhibit the ability to form dynamic bonds, allowing them to heal through reversible chemical interactions, which reduces environmental impact and waste. In parallel, advancements in synthetic polymers have introduced self-healing capabilities through mechanisms like covalent bond reformation, supramolecular chemistry, and shape-memory effects. Synthetic polymers, including polyurethanes and epoxy-based systems, utilize engineered molecular structures and embedded nanoparticles to mimic natural self-healing processes, offering greater versatility and tailored properties for diverse applications (Wang and Urban 2020).

Within the Polymers category, self-healing mechanisms are generally divided into two main types: intrinsic and extrinsic, each with its own approach to restoring integrity after damage. Self-healing polymers utilize intrinsic and extrinsic mechanisms to restore their integrity after damage (Irzhak et al. 2022). Intrinsic healing involves the polymer's inherent ability to reform broken bonds or structures through reversible interactions like dynamic covalent bonds or supramolecular interactions, triggered by external stimuli (e.g., heat or light). Conversely, extrinsic healing depends on embedded systems like microcapsules or vascular networks that release healing agents when damage occurs to fill the cracks. Both approaches have shown effectiveness in applications like aerospace, electronics, and biomedical fields (Kahar et al. 2021).

Despite notable advancements in self-healing materials, key challenges remain in achieving consistent scalability and stability under various conditions (Islam and Bhat 2021). Addressing issues such as optimizing healing efficiency while preserving the mechanical strength and durability of these materials is crucial for future development (Dallaev 2024). Recent innovations have focused on integrating advanced nanotechnology and dynamic bonding mechanisms, enhancing responsiveness and adaptability in self-healing systems. These efforts are illustrating the way of broader commercialization, particularly in sectors where reliability and extended lifespan are crucial (Islam and Bhat 2021; Dallaev 2024).

10.2 Types of Biopolymers for Self-Healing Applications

10.2.1 Natural Biopolymers

Natural biopolymers have gained significant attention in self-healing materials due to their biodegradability, biocompatibility, and renewable nature. These polymers are derived from natural sources, making them environmentally friendly and ideal for potential use in biomedical fields. Their self-healing capabilities often rely on reversible interactions like hydrogen bonding, ionic crosslinking, and dynamic covalent bonding. Cellulose, hyaluronic acid, chitosan, starch, and sodium alginate are among the most thoroughly studied natural biopolymers, each providing distinct properties that aid in creating sustainable and efficient self-healing systems.

10.2.1.1 Cellulose

One of the predominant polysaccharides found in nature is cellulose, sourced from diverse origins, comprising the cell walls of wood and plants, certain bacteria, algae, and tunicates—the only known animals that possess cellulose (Seddiqi et al. 2021). The structure of cellulose is made up of a straight chain, where the basic repeating unit consists of two anhydroglucose molecules ($C_6H_{10}O_5$). These molecules are linked by an oxygen atom, forming a covalent bond between the carbon atoms (Santos et al. 2021). This unique structure of cellulose has made it a strong, fibrous polymer that does not dissolve in water and is crucial for providing structural support to plant cell walls. In addition, this polymer is natural and is biodegradable, renewable, and compatible with biological systems (George and Sabapathi 2015). These properties have made cellulose a widely used material in various industries. Cellulose-based materials have long been utilized in traditional industries like paper and textile manufacturing. In recent years, cellulose-based materials have found broader applications in the biomedical sector, driven by their biocompatibility, ability to degrade naturally, and minimal toxicity (Hasanin 2022). These characteristics make cellulose environmentally friendly and cost-effective for modern medical uses. Additionally, cellulose can be chemically treated to create valuable products. It has significant roles in healthcare and disease control, with new applications emerging in fields like tissue engineering, wound care, and drug delivery (Seddiqi et al. 2021).

In addition to extensive use in medical and industrial fields, one of the remarkable features of cellulose that has recently gained more attention is its ability to be treated and combined with other substrates to create self-healing composites (Tian and He 2016). Due to its adaptable chemical structure and biocompatibility, cellulose has emerged as an ideal material for designing and producing self-healing biocomposites. As a polymer derived from D-glucose units, cellulose has shown great promise in self-healing hydrogels (Kundu et al. 2022). Through chemical modifications, cellulose derivatives like carboxymethyl cellulose (CMC) introduce functional groups such as hydroxyl and carboxyl, which enable the formation of self-healing networks via hydrogen bonds, metal–ligand interactions, and dynamic covalent bonds. These hydrogels demonstrate excellent mechanical properties and rapid healing capabilities, rendering them appropriate for biomedical uses due to their biocompatibility and biodegradability. However, the high crystallinity and limited solubility of cellulose pose challenges in processing and integrating with other polymers (Wang et al. 2023a).

10.2.1.2 Chitosan

This polymer is classified as a polysaccharide that is naturally sourced from chitin, which is deacetylated through chemical, enzymatic, or microwave-assisted processes (Chatterjee et al. 2021). It is the only natural polycation, with its charge density dependent on the degree of acetylation and the surrounding pH. Chitosan's solubility is influenced by both its molecular weight and acetylation level; while low molecular weight chitosan is soluble across a broad range of pH levels, including neutral, high molecular weight chitosan dissolves only in acidic solutions. This limitation has prompted the creation of chitosan derivatives with enhanced solubility for broader applications (Aranaz et al. 2021). Chitosan consists of the acetylated unit that is connected by β-(1 → 4)-glycosidic bonds to the deacetylated unit (Chatterjee et al. 2021).

Chitosan is a naturally derived polymer known for its biocompatibility, biodegradability, and non-toxicity (Shukla et al. 2013). It can form various physical forms such as hydrogels, nanoparticles, and films, making it versatile for different applications. Additionally, chitosan's ability to form chelates and its favorable mechanical properties are further enhanced by chemical modifications like crosslinking and grafting, which allow for tailored applications in fields like tissue engineering, drug delivery, and environmental protection (Kluczka 2024).

Chitosan is widely recognized for its biocompatibility, antimicrobial effects, and hemostatic characteristics, rendering it an excellent option for wound healing applications. Chitosan can form self-repairing hydrogels by reversible crosslinks, involving both dynamic covalent bonds like Schiff base, boronate esters, and acyl hydrazone bonds, as well as physical interactions comprising hydrogen bonding, electrostatic, and ionic interactions (Zhang et al. 2024a). In addition to hydrogels, chitosan-based self-healing substances have been broadly applied in tissue engineering, drug delivery, and anti-corrosion. These materials repair themselves via dynamic covalent and non-covalent interactions, which allow for autonomous restoration after damage. Consequently, chitosan-based hydrogels have been developed to manage complex conditions in both biomedical and industrial fields (Ding et al. 2018).

Chitosan is commonly incorporated in fields like biotechnology, medicine, and wastewater treatment, but its poor solubility and weak mechanical properties limit its applications in biomedical fields. Modifications have been developed to enhance solubility and create derivatives with improved physicochemical properties (Wang et al. 2020). Chitosan-based nanocomposites exhibit better mechanical strength, thermal stability, and antimicrobial properties compared to pure chitosan. However, despite advancements, challenges like poor scaffold strength in tissue engineering and potential cytotoxicity remain, requiring further research to fully realize its clinical potential (Yadav et al. 2023).

10.2.1.3 Hyaluronic Acid

This anionic polysaccharide is renowned for its outstanding biocompatibility, biodegradability, and moisturizing properties. Hyaluronic acid (HA) is essential in controlling angiogenesis, inflammation, and cellular behavior, which makes it highly versatile and valuable in biomedical applications (Luo et al. 2023). HA has a high molecular weight linear polysaccharide composed of D-glucuronic acid and N-acetylglucosamine units (Moreira et al. 2024).

HA possesses remarkable viscoelasticity, exceptional moisture retention, high biocompatibility, and hygroscopic characteristics (Kim et al. 2017). Even at low concentrations (0.1 wt.%), HA can produce a high viscosity, making it a proper lubricant, shock absorber, joint stabilizer, and regulator of water balance (Gupta et al. 2019).

Hyaluronic acid is a versatile glycosaminoglycan that plays key roles in the extracellular matrix, tissue repair, epithelial tissue maintenance, and neural tissue function (Dovedytis et al. 2020). Self-healing materials made from hyaluronic acid are commonly used in targeted drug delivery, protein and cell transport, electronics, biosensors, and various other biomedical applications (Kikani et al. 2023). Hyaluronic acid-based self-healing hydrogels primarily rely on Dynamic covalent bonding, like imine (Schiff base) bonds, which are created through reactions between amine and aldehyde groups. These bonds allow the substance to recover its structure autonomously after damage (Li et al. 2020). Additionally, non-covalent interactions, including hydrogen bonding, ionic bonds, and host–guest engagements, contribute to the flexibility and resilience of the hydrogel. This combination of covalent and non-covalent mechanisms enhances its applications, particularly in biomedical fields like wound healing, where the hydrogel mimics the extracellular matrix and supports tissue regeneration (Anupama Devi et al. 2021).

One of the main challenges of using hyaluronic acid in biomedical applications is its low mechanical stability. While HA forms hydrogels easily, these materials often lack sufficient mechanical strength, especially under stress, which limits their use in structural applications. Additionally, HA is subject to rapid degradation in biological environments, primarily due to enzymatic activity and oxidative damage caused by reactive oxygen species (ROS). This rapid turnover can hinder long-term applications, such as in tissue scaffolding. Chemical modifications like crosslinking or combining HA with other polymers are often necessary to address these issues, improve its mechanical properties, and prolong its stability (Aranaz et al. 2021; Shukla et al. 2013; Kluczka 2024).

10.2.1.4 Sodium Alginate

Sodium alginate (SA), often referred to as kelp gum, is a polysaccharide derived from marine plants like kelp and Sargassum. After removing components such as iodine and mannitol (Zhou et al. 2021). Sodium alginate is a hydrophilic, negatively charged linear polysaccharide made up of α-D-mannuronic acid and β-L-guluronic acid segments. When exposed to multivalent cations, it quickly forms hydrogels through metal–ligand interactions at room temperature, making it valuable for wound dressings, drug delivery, and tissue engineering (Wang et al. 2023a).

SA is a water-soluble and non-toxic polysaccharide known for its gelation properties, especially when interacting with calcium ions to form "egg-box" structures. The fraction of mannuronic acid (M) to guluronic acid (G) units in alginate affects its properties: a higher G content provides stronger and more stable gels, while more M units enhance elasticity (Wang and Lu 2023). Incorporating reversible chemical or physical interactions into the crosslinking network is a key method to enable self-healing in hydrogels (Wu et al. 2023). Dynamic bonds, such as imine bonds formed by the Schiff base reaction between aldehyde and amino groups, allow the network to recover and rebuild over time (Malik et al. 2022). A study demonstrated that oxidizing alginate by 50% and crosslinking it with acrylamide-modified chitosan resulted in a hydrogel capable of self-healing within an hour. However, increasing the concentration of imine bonds may affect mechanical performance and potentially introduce toxicity. Adjustments in crosslinker structure can also enhance self-healing by boosting intermolecular interactions (Wang et al. 2022).

Sodium alginate is broadly employed in both traditional and modern applications owing to its biocompatibility and gel-forming capability. Traditionally, it has been generally utilized in the food industry as a thickening agent, stabilizer, and emulsifier, especially in products like ice cream, sauces, and jellies (Abka-khajouei et al. 2022). In the biomedical field, alginate hydrogels are currently employed in wound dressings due to their moisture-retentive properties, promoting healing by maintaining a moist environment (Aderibigbe and Buyana 2018). Furthermore, sodium alginate has found applications in drug delivery systems, where it is used to encapsulate drugs for controlled release and in cell encapsulation for tissue engineering. Its ability to form stable gels in the presence of calcium ions also makes it ideal for tissue scaffolding in regenerative medicine (Draget and Taylor 2011).

Sodium alginate faces several challenges, including its low mechanical strength, restricting its use in load-bearing implementations, and its unpredictable degradation rate in biomedical contexts, which can hinder long-term applications. Additionally, some chemical modifications may lead to potential cytotoxicity, making it essential to find safer alternatives. However, future research is focused on improving alginate's properties through more biocompatible crosslinking agents, integrating nanotechnology to enhance drug delivery, and expanding its use in environmental and biomedical fields, including tissue scaffolding and sustainable packaging (Rahman et al. 2024; Adamiak and Sionkowska 2023).

10.2.1.5 Starch

Starch is a naturally abundant product found in various plant tissues, including leaves, seeds, fruits, and tubers. It serves as a primary energy storage form, converting solar energy into a chemical form that supports life on Earth (Villa Zabala 2020). Starch consists of two polysaccharides: amylose, which has linear α-(1 $\rightarrow$ 4) connectors, and amylopectin, which contains both α-(1 → 4) and approximately 5% α-(1 → 6) segment linkages. These components form water-insoluble granules with partial crystallinity, whose characteristics hinge on the biological origin of the starch (Robyt 2008).

Starch exhibits a range of important physicochemical properties, such as swelling power, gelatinization, and rheological behavior, which vary depending on its botanical source. When heated with water, starch undergoes gelatinization, a process where its granules swell and lose their crystalline structure, making it an ideal thickening agent and stabilizer. Additionally, starch's ability to form gels and retain water is crucial in both food and non-food applications (Zhu and Xie 2018).

Starch-based self-healing hydrogels utilize reversible bonding comprising hydrogen bonding, ionic crosslinking, and host–guest interactions to restore their structure after mechanical damage. These materials can incorporate dynamic covalent bonds, like those formed through Schiff base reactions, where imine bonds enable the hydrogel to autonomously repair its network (Rumon et al. 2022). Despite their promise in biomedical fields like wound healing, tissue engineering, and drug delivery, starch-based hydrogels confront challenges in terms of mechanical strength and stability in aqueous environments (Ding et al. 2023). Researchers are working on overcoming these limitations by incorporating hydrophobic interactions and combining starch with other polymers to strengthen mechanical characteristics and self-healing capabilities (Rumon et al. 2022; Ding et al. 2023).

10.2.1.6 Other Natural Biopolymers with Self-Healing Properties

Pectin, a polysaccharide commonly found in the cell walls of plants, especially fruits, exhibits self-healing properties in hydrogels through its capacity to create ionic crosslinks with divalent cations, including calcium (Celus et al. 2018). This makes it useful in biomedical usages, including drug delivery and wound healing (Bostancı et al. 2022). Additionally, gelatin, a protein derived from collagen, is often used in self-healing materials when combined with other polymers (Sisso et al. 2020). Its application is in tissue engineering, where it helps create frameworks that can support tissue growth and repair. It is also used in regenerative medicine. While gelatin's mechanical properties are weaker, it is frequently modified to enhance its self-healing capacity (Balavigneswaran and Muthuvijayan 2021).

10.2.2 Synthetic Biopolymers

Synthetic biopolymers provide notable benefits in self-healing materials, primarily because of their ability to be customized, their enhanced mechanical strength, and their chemical stability. Unlike natural biopolymers, synthetic polymers can be designed with precise regulation of molecular structure, allowing for tailored properties to suit specific applications. These polymers are widely used in industries such as coatings, foams, protective films, and medical devices. Self-healing in synthetic biopolymers is often achieved through dynamic covalent bonds, reversible physical interactions, or responsive behavior to environmental stimuli. Key examples include polyurethane, polyethylene glycol (PEG), silicone polymers, and polycaprolactone (PCL), which are integral to developing robust self-healing systems for industrial and biomedical use.

10.2.2.1 Polyurethane

Polyurethanes are a highly versatile and customizable class of polymers with unique properties, including high tensile strength, resistance to abrasion and fatigue, and flexibility at low temperatures (Engels et al. 2013). Their adaptability has made polyurethanes widely used across various industries, from apparel and automotive to construction and medical fields. Polyurethanes, in particular, can be engineered to possess excellent biocompatibility and hemocompatibility, rendering them well-suited biomaterials for diverse medical applications, such as drug delivery systems. The capability to tune the structure of polyurethanes facilitates management of their characteristics, resulting in highly functional materials for specific applications (Wienen et al. 2023).

Polyurethanes are composed of alternating hard segments (HS), derived from isocyanates (either aliphatic or aromatic), and soft segments (SS), derived from macrodiols such as polyester, polyether, or polycarbonate. The hard segment aids in stiffness and abrasion resistance, while the soft segment provides flexibility and elastomeric properties. The balance between these two segments enables the customization of polyurethane's mechanical and chemical properties to suit targeted uses (Patti et al. 2021).

Polyurethanes with dynamic oxime–urethane bonds demonstrate excellent self-healing properties without the need for external stimuli. These materials can autonomously repair themselves after mechanical damage, making them well-suited for biomedical applications. Their self-healing capacity is particularly useful in clinical settings where reducing procedural complexity is critical, such as in nerve repair and bone immobilization. This ability to restore mechanical integrity ensures longer-lasting and more reliable performance (Jiang et al. 2021).

Polyurethanes are commonly employed in self-healing materials owing to their tunability and strong mechanical properties. Recent advances focus on combining dynamic covalent bonds with other self-healing mechanisms, including hydrogen bonding and metal coordination. This combination improves both the efficiency and the mechanical strength of polyurethane-based materials, making them ideal for applications like coatings, elastomers, and biomedical devices. These materials offer enhanced durability and functionality in industries such as automotive, construction, and healthcare (An et al. 2023).

Polyurethanes face several challenges in self-healing applications, primarily balancing mechanical strength with healing efficiency. Dynamic bonds, like disulfides or Diels–Alder reactions, enable self-repair but can compromise the material's overall durability, making it prone to deformation under stress. Additionally, many polyurethane systems require elevated temperatures to initiate healing, limiting their usability in everyday applications. Moreover, the need for external stimuli, like heat or UV light, adds complexity and reduces their practicality in real-world settings. Further research is necessary to develop systems that can self-heal autonomously at room temperature while maintaining strong mechanical properties (Aguirresarobe et al. 2021).

10.2.2.2 Polyethylene Glycol

Polyethylene glycol (PEG) is a synthetic polymer known for its biocompatibility, water solubility, and affordability. These properties make PEG highly suitable for usages comprising drug delivery systems and wound dressings, where its safety and functionality are essential (Zahedi et al. 2017).

PEGs are linear polymers formed through a base-catalyzed reaction that adds ethylene oxide units to an ethylene core. Their molecular weight varies widely, ranging from 200 to several million. PEGs with lower molecular weights are liquids, while higher molecular weight PEGs are solids. These polymers are nonvolatile, water-soluble, and miscible with a broad range of solvents, rendering them versatile in various applications (Smyth et al. 1950).

PEG enhances self-healing capabilities in hydrogels by forming hydrogen bonds and dynamic covalent links. These interactions enable PEG-based hydrogels to restore their structure after damage, effectively repairing any disruptions. The quick recovery of PEG, particularly following mechanical stress, makes it exceptionally well-suited for tissue engineering applications (Yasar et al. 2024). Moreover, PEG's hydrophilic properties contribute to the flexibility and moisture retention of the hydrogel, which is crucial for supporting tissue regeneration and ensuring efficient drug delivery (Sun et al. 2023). Additionally, when PEG forms cross-linked networks with other polymers, it boosts the hydrogel's mechanical strength without compromising its self-healing abilities (Kizilel et al. 2009). Despite these promising self-healing capabilities, PEG-based materials face several challenges in practical applications. PEG-based hydrogels face mechanical limitations in self-healing materials, particularly in load-bearing applications where their strength may be insufficient (Wen et al. 2020). Despite its hydrophilic nature aiding in tissue regeneration, PEG can suffer from excessive swelling, which compromises structural integrity (Papavasiliou et al. 2012). However, ongoing research addresses these issues by combining PEG with other polymers to improve its mechanical durability and self-healing capabilities (Wen et al. 2020; Zengin et al. 2021).

10.2.2.3 Polycaprolactone

This substance is a commercially available synthetic polymer known for its controlled biodegradability and versatile mechanical properties. Its slow degradation rate, which can extend over 3–4 years in water, makes it ideal for biomedical usages like drug delivery and tissue engineering. Due to its superior viscoelastic characteristics, polycaprolactone (PCL) can be easily fabricated into diverse structures, comprising scaffolds and implantable devices, offering wide-ranging applications in bioengineering and green chemistry (King et al. 2021). PCL is a semi-crystalline polyester with an extended chain structure featuring five non-polar methylene segments along with a polar ester segment. It undergoes bulk degradation through hydrolysis after complete water penetration, which contributes to its slow degradation rate. PCL also has a low melting point, is highly malleable, and is FDA-approved, making it ideal for biomedical applications and environmental remediation (King et al. 2021).

PCL has a melting point between 60- 65°C and a low glass transition temperature of around -60°C. Its solubility in organic solvents allows for easy blending with other polymers, increasing its versatility in biomedical applications. However, its hydrophobic nature can limit cell adhesion, requiring surface modifications to enhance its interaction with biological tissues (Mateos-Timoneda 2009).

PCL exhibits self-healing properties when combined with other polymers, utilizing various mechanisms. PCL blended with natural rubber demonstrates self-healing through supramolecular hydrogen bonding, achieving up to 79.9% healing efficiency when assisted by shape memory effects (Lai et al. 2020). PCL networks with disulfide bonds exhibit self-healing, shape memory, processability, and degradability characteristics (Zhang et al. 2018). A novel approach called "shape memory assisted self-healing" (SMASH) combines shape memory and self-healing characteristics in PCL blends, enabling rebounding with a simple thermal trigger (Gnedenkov et al. 2023). These interactions allow PCL-based hydrogels to recover after mechanical damage, making them effective in wound healing and tissue scaffolding (Gnedenkov et al. 2023). PCL faces several challenges in self-healing applications (Jiménez-Suárez et al. 2020). Additionally, mechanical strength is often insufficient for load-bearing applications, necessitating blending with stronger polymers (Liang et al. 2024). Furthermore, PCL's hydrophobic nature limits cell adhesion, hindering its effectiveness in tissue engineering unless surface modifications are applied (Jeon et al. 2014; Chen et al. 2019a).

10.2.2.4 Polyvinyl Alcohol

Polyvinyl alcohol (PVA) is a biodegradable, semi-crystalline synthetic polymer that is extensively utilized in the biotechnology field, comprising tissue regeneration and drug delivery systems (Feldman 2020). The polymer's linear carbon chain with hydroxyl functional groups allows for variations in solubility and crystallinity depending on the level of hydrolysis, enabling the formation of flexible films with increased permeability to oxygen and water vapor. This characteristic accelerates wound healing. Additionally, PVA easily blends with other polymers like chitosan, enhancing its biodegradability, biocompatibility, and antimicrobial properties (Gherman et al. 2023).

PVA with a chemical formula of $(C_2H_4O)n$, where n ranges between 500 and 5000, has a molecular weight of approximately 20,000 to 200,000 Da. This structure contributes to its unique solubility and semi-crystalline characteristics (Benetti et al. 2021). Properties give PVA its hydrophilic nature, enabling it to form films that are flexible, stable, and adaptable for use in biomedical and industrial fields (Gaaz et al. 2015).

PVA hydrogels exhibit self-healing properties through a mechanism that relies on physical cross-linking and hydrogen bonding within the polymer network. The hydroxyl groups in PVA facilitate these bonds, enabling the material to repair itself autonomously without external intervention. The ability to restore structural integrity is affected by factors like polymer concentration and the number of freezing/thawing cycles during preparation. These characteristics make PVA-based hydrogels highly suitable for biomedical applications, owing to their biocompatibility and capacity for healing at room temperature (Zhang et al. 2012a).

Despite its versatility, PVA's low mechanical strength and poor water resistance limit its use in demanding applications (Li et al. 2010). Its sensitivity to moisture can lead to swelling and loss of mechanical integrity in humid conditions (Liu et al. 2022). To overcome these challenges, combining PVA with inorganic materials like silica has been shown to significantly enhance its mechanical strength and water resistance, broadening its potential applications in medical materials (Li et al. 2010).

10.2.2.5 Other Synthetic Biopolymers with Self-Healing Properties

Several synthetic biopolymers demonstrate remarkable self-healing capabilities that make them suitable for various advanced applications. Poly(ethylene oxide) (PEO) can form self-healing hydrogels through reversible noncovalent and dynamic covalent interactions (Li et al. 2010). Recent advancements have contributed to the advancement of self-healing Polydimethylsiloxane (PDMS) elastomers, incorporating reversible dynamic bonds into the polymer network (Zhang et al. 2017a).

10.2.3 Biocomposites with Self-Healing Properties

Self-healing biocomposites have attracted considerable interest in recent years because of their biocompatibility, self-repair capabilities, and environmental sustainability. These materials combine natural polymers like cellulose, starch, and chitosan with biopolymers or synthetic polymers, offering optimized mechanical properties and biodegradability. By utilizing these polymers, biocomposites are developed with high resistance and the capability to regain their original properties after minor damage. The internal structure and interactions within these biocomposites are vital in achieving self-healing characteristics, making them applicable in areas like packaging, tissue engineering, and construction materials (Kumar et al. 2023; Rendón-Villalobos et al. 2016).

The development of self-healing biocomposites has emerged as a significant advancement in material science, driven by the need for sustainable, resilient, and biocompatible solutions. These materials utilize a blend of synthetic and natural polymers to achieve optimized environmental sustainability and mechanical performance (Kumar et al. 2023).

Cellulose-based biocomposites, which encompass nanocellulose, cellulose nanofibrils (CNF), and cellulose nanocrystals (CNC), demonstrate superior self-healing properties owing to their ability to establish strong interfacial bonds and enhance structural integrity (Khalid et al. 2024). Similarly, starch-based biocomposites are designed by blending starch with polymers like PLA or chitosan, thereby improving flexibility and enabling the restoration of properties after minor damages (Jayarathna et al. 2022). Chitosan-based biocomposites leverage the natural binding abilities of chitosan, allowing them to achieve a balance of rigidity and elasticity that supports self-repair in tissue engineering applications (Liu et al. 2011). Finally, polyester-based biocomposites, especially those reinforced with PLA and cellulose or chitosan, offer improved resistance to wear and demonstrate the ability to recover from stress and damage (Tripathi et al. 2021). These biocomposites are utilized in numerous areas, including packaging, construction, and biomedical engineering, where durability and environmental sustainability are crucial (Lavania et al. 2023).

10.3 Mechanisms of Self-Healing in Biopolymers

10.3.1 Overview of Healing Mechanisms

One of their fundamental ideas is the self-healing materials' capacity to fix damage independently without outside assistance. Biocomposites and polymer materials are being researched for their potential to cure themselves. Numerous polymer types, including elastomers, thermoplastics, and thermostats, offer enormous potential for use in self-healing materials with various self-healing mechanisms. The damage volume may be impacted by loading circumstances, shape, and damaged polymer structure (Jadoun 2024). The two main sorts of self-healing processes in biopolymers are extrinsic and intrinsic. Extrinsic agents, which contain blood vessel systems or alternative substances capable of being triggered when the substance is damaged for restoration, require an extraneous agent to start the restorative process. Materials can autonomously restore without the aid of external agents thanks to the intrinsic agents' dynamic cross-linking mechanism and ability to mend themselves without any outside assistance (Jadoun 2024; Zhang and Rong 2013; Wang et al. 2014; Everitt et al. 2015). Additionally, depending on the stimulus needs and the type of self-healing technique, self-healing systems can be classified into two categories, namely automated and non-automated procedures (Jadoun 2024; Tan et al. 2021).

Autonomic systems start the self-healing process in the event of damage, but non-autonomous systems need outside stimuli like heat, light, or chemicals to do so (Hager et al. 2010).

10.3.2 Intrinsic Healing Mechanisms

Materials with intrinsic self-healing abilities can mend damage without requiring additional repair agents, as they initiate the healing process independently. Compared to extrinsic self-healing materials, this method is more robust and dependable due to reproducible healing actions in these materials (Jadoun 2024; Narayan et al. 2022; Chen et al. 2017; Garcia and Fischer 2014). Through a momentary increase in mobility, these materials can restore damage by returning to the affected region. The process of restoring chemical bond strength or recovery relies on the specific molecular structures and functionality of polymers that sustain more harm (such as temperature) to recover (Garcia and Fischer 2014). These materials repair damage through the innate reversibility of physical or chemical bonds at their structural design sites, such as the swelling of shape-retaining polymers, the freezing and the fusion of thermoplastic substances, and the increase in the viscosity of pH-sensitive micro-gels. These materials undergo a polymer phase when exposed to external stimuli like heat (Jadoun 2024; Yuan et al. 2008). Reversible chemical events, including metal–ligand, supramolecular, and reversible covalent bonds, underpin most of the widespread intrinsic self-healing mechanisms. The material may mend the damage by breaking and reforming these links. Reversible covalent bonds, for instance, may be present in the chains of a self-healing polymer. These connections can be broken and reformed in reaction to harm, so restoring the substance. Reversible physical bonds, comprising van der Waals forces or hydrogen bonds, provide the foundation of another kind of intrinsic self-healing mechanism. When damage occurs, these relationships may be broken up, causing the material to release healing molecules that can fix it. However, intrinsic self-healing also has drawbacks, including the possibility of sluggish healing rates and minimal therapeutic efficacy. Furthermore, adding reversible physical interactions or dynamic chemical bonds might diminish the material and reduce its structural unity (Jadoun 2024; Narayan et al. 2022).

The self-reactivity mechanism, reversible covalent or non-covalent bonds, are incorporated as crosslinks in the polymer matrix as part of the chemistry that makes self-healing polymers possible (Li and Zuo 2019).

10.3.2.1 Reversible Chemistry

Reversible covalent bonds in intrinsic self-healing mechanisms include Diels–Alder reactions, [2 + 2] ring formation, acyl hydrazone bonds, trithiocarbonate, imine, boronate, oxime, and disulfide bonds. These reversible bonds and interactions are broken and reversibly modified, enabling polymers to exhibit reproducible self-healing properties (Anupama Devi et al. 2021; Li and Zuo 2019). Reversible grafts react to external triggers like chemicals, temperature, and light, enabling multiple healing cycles of a gap (Li and Zuo 2019).

The Diels–Alder reaction does not require a catalyst or coupling reagent and is ideal for creating self-healing hydrogels (Anupama Devi et al. 2021). Chen and colleagues produced a transparent organic polymer material that is a hard solid at ambient temperature but self-heals upon heating. With heat, the cross-links within the polymer break and can be reformed after cooling, effectively curing the material. They used Diels–Alder reactions to create these reversible cross-links in polymeric materials. The specific DN used by them is norbornene and dienophile, maleic anhydride (Chen et al. 1979). In 2004, a polymer material was created by Chung and colleagues whose cracks healed upon exposure to UV light. This material contained cyclohexene and benzene monomers, which underwent reversible [2 + 2] cycloaddition reactions after damage and formed a four-membered ring. As a result, they form cross-links that bridge the crack and restore the material's integrity by maintaining the structural properties (Chung et al. 2004). Rapid and straightforward reaction circumstances, accessibility of raw materials and reagents, and thermal reversibility are the advantages of Diels–Alder reactions (Wang et al. 2023a).

Another self-healing process initiated by exposure to UV light is the reversible reaction of trithiocarbonates (TTC). Amamoto et al. fabricated covalently bonded polymers through reversible fragmentation chain transfer (RAFT) polymerization. When such trithiocarbonate (TTC) units in cross-linked polymers are damaged, they deform under UV irradiation, allowing the material to repair itself (Amamoto et al. 2011).

Acylhydrazone bonds form when ketone or aldehyde groups engage with hydrazine groups, creating a reversible covalent connection. These bonds can be formed through hydrolysis or exchange reactions at acidic pH. Self-healing hydrogels, featuring acylhydrazone grafts, have been developed for their potential in wound healing applications and tissue engineering (Anupama Devi et al. 2021). The researchers synthesized polymer gels using covalent cross-linking methods that could reversibly switch between sol (liquid) and gel (solid) states through acyl hydrazone linkages as dynamic covalent bonds in the polymer matrix. When the gel is exposed to certain circumstances, such as alterations in temperature, pH, or the introduction of particular chemicals, the acyl hydrazone bonds can fracture and re-establish, enabling materials to self-heal (Deng et al. 2010). Chen and colleagues created an injectable hydrogel with remarkable characteristics such as dynamic covalent self-healing, sustained drug release, and antibacterial properties. Following injection or injury, because of the dynamic nature of imine and acyl hydrazone linkages, shattered hydrogel segments exhibit excellent self-repair capability to produce a unified hydrogel. Also, bFGF@PLGA microspheres, as the basic fibroblast growth factor, and chlorhexidine acetate (CHA) as an antimicrobial agent, were embedded in hydrogels (Fig. 10.1). This bFGF@PLGA/CHA/hydrogel could maintain morphology

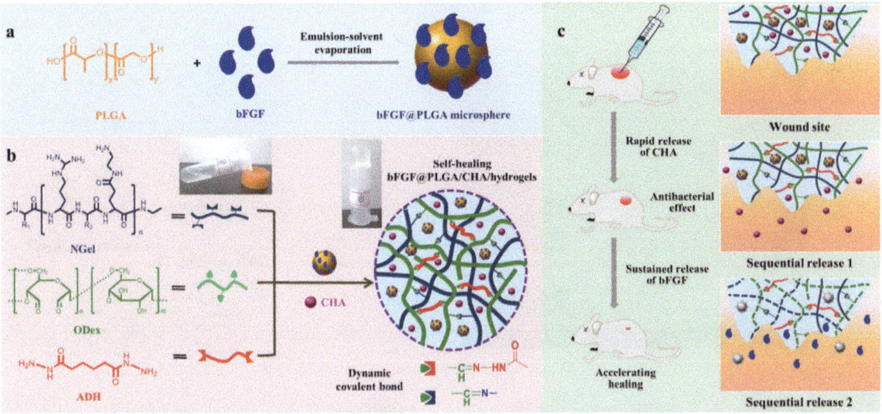

Fig. 10.1 Schematic illustration of bFGF@PLGA/CHA/hydrogels with self-healing and sequential release properties for inhibiting bacterial infection and accelerating healing in wound repair. **a** Fabrication of bFGF@PLGA microspheres. **b** Preparation of bFGF@PLGA/CHA/hydrogels. This injectable and self-healing hydrogel was constructed via the dynamic reaction of NGel, ODex, and ADH, which was attributed to dynamic imine and acyl hydrazone bonds. bFGF@PLGA microspheres and CHA were simultaneously embedded into the hydrogels. **c** Illustration of sequential drug release procedure of bFGF@PLGA/CHA/hydrogels. Once injected into the wound site, the hydrogels can effectively prevent infection at the early stages due to the rapid release of CHA, and then further accelerate cell proliferation and wound healing through the sustained release of bFGF (Chen et al. 2019b)

stability after swelling. After 3 days of incubation, the cell viability of neat hydrogels exceeded 95%, whereas bFGF@PLGA/CHA/hydrogels with varying amounts of amino gelatin surpassed 135%.

According to Fig. 10.2f, with increasing time from 1 to 5 days, while the cell viability of bFGF@PLGA/CHA/hydrogel illustrated a notable rise, the blank hydrogel group showed no significant difference (Chen et al. 2019b). Also, in pH = 7.4, a self-healing injectable hydrogel based on alginate and PEG was developed by Xie et al., which was fabricated through a thiol-Michael addition reaction between the quinone group of oxidized dopamine in the dopamine-modified alginate (SA-DA) and the thiol group in the 4-arm PEG-SH. This hydrogel exhibited excellent tissue adhesion and rapid self-healing by forming strong covalent bonds without external stimuli. The presence of dopamine also enhanced the adhesion and stability of the gel in humid environments (Xie et al. 2025).

Recent research emphasizes the importance of dynamic covalent bonds, such as imine bonds, in photocurable, bio-based self-healing polyurethane materials (Tang et al. 2025). Compared to imine linkages, acyl hydrazone linkages are particularly susceptible to degradation in mildly acidic environments (pH~4–7) or at elevated temperatures (Wang et al. 2023a).

Disulfide bonds are formed through a thiol/disulfide interchange mechanism in a non-acidic environment. These bonds undergo an exchange reaction that increases the

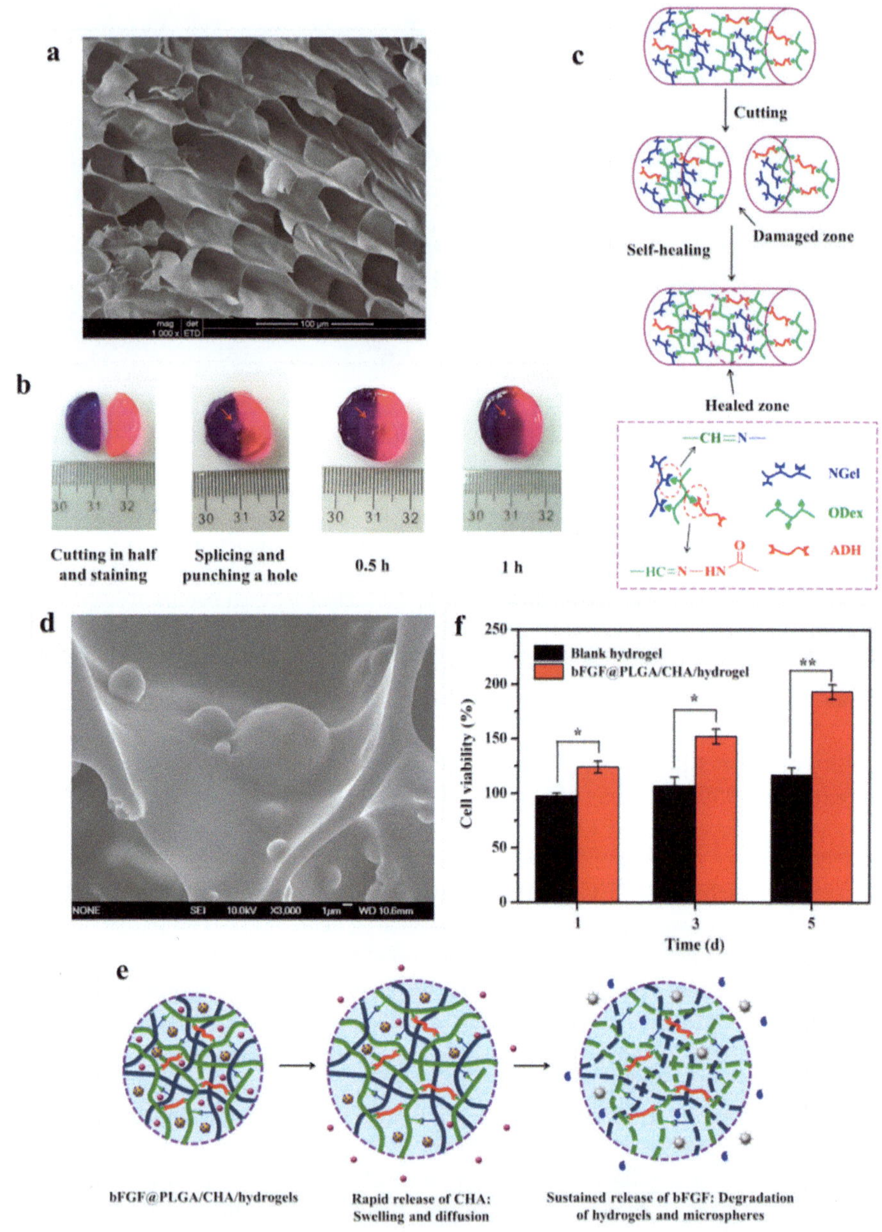

Fig. 10.2 SEM image of NGel-ODex-ADH hydrogels **a**, the appearance of self-healing process between two different colored semicircle hydrogels. The middle of the united gel is punched a hole, and the red arrow points to the hole site **b**. Schematic illustration of the self-healing mechanism of NGel-ODex-ADH hydrogel **c**, SEM image of bFGF@PLGA/CHA/hydrogels **d**, schematic representation of the sequential delivery of bFGF@PLGA/CHA/hydrogels **e**, cell proliferation of 3T3 cells co-cultured with blank hydrogels and bFGF@PLGA/CHA/hydrogels at 1, 3, and 5 d **f** (Chen et al. 2019b)

self-healing ability of composite hydrogels (Anupama Devi et al. 2021). The incorporation of disulfide bonds into rubber through covalent bonding was able to show auto-repair of cuts at moderate thermal conditions, leading to the complete restoration of mechanical properties. Breaking disulfide bonds produces sulfenyl radicals that can attack neighboring disulfides and facilitate the healing process (Canadell et al. 2011).

Among other dynamic covalent interactions for creating self-healing hydrogels are imine bonds, which are also known as Schiff base bonds and create crosslinks between amine groups and aldehyde or ketone polymers in physiological conditions. These grafts have a fast reaction rate under mild conditions, which makes them suitable for tissue engineering applications (Anupama Devi et al. 2021). In 2025, a biobased hydrogel from soy protein and vanillin featuring reversible physical and chemical cross-links, including dynamic Schiff base bonds, was developed that enables rapid self-healing within 10 s. These bonds rearrange after damage, restoring the polymer network, while hydrogen and van der Waals interactions enhance structural stability. The hydrogel demonstrated high mechanical strength, antibacterial and antioxidant properties, effectively stopped hepatic bleeding, and significantly accelerated diabetic wound healing in animal models (Wang et al. 2025). Ding et al. have prepared hydrogels from collagen (COL), chitosan (CS), and polyethylene glycol-modified with dibenzaldehyde (PEG 2000), which are cross-linked through dynamic imine linkages. This hydrogel responds to changes in pH and strain and remains stable in different temperature conditions. It has also shown good performance in wound healing and hemostatic ability (Ding et al. 2020). Jungprasertchai et al. developed a lyophilized carboxymethyl chitosan-starch foam for application in hemostatic healing dressing. They use carboxymethyl chitosan (CMC), starch, and glyoxal (as a cross-linking agent). This modified PEG, serving as a telechelic binder, demonstrates automatic recovery abilities in under 15 min upon crack initiation. This foam showed rapid water absorption and retention and maintained good mechanical properties even in a humid environment. This dressing helps to stop bleeding while still allowing blood to clot at the wound site (Anupama Devi et al. 2021; Jungprasertchai et al. 2022). Generally, temperature and pH significantly affect imine bonds. This, in turn, limits the restoration efficiency of hydrogels in environments with high acidity or alkalinity and at extreme temperatures (Wang et al. 2023a).

Other reversible covalent bonds are boronate bonds, which arise from the reaction of diols with boronic acid (Anupama Devi et al. 2021). He and his team created a pH-responsive self-repair hydrogel by the formation of boronate-catechol assemblies. To synthesize the hydrogel, they used 1,3-benzene diboronic acid (BDBA) as a linker and 4-arm PEG-catechol, a polymer with four catechol groups, in PBS buffer solution. This hydrogel exhibited pH-responsive characteristics and established a stable network in alkaline conditions by forming ester borate bonds among boronic acid and catechol groups. At neutral or acidic pH, these bonds dissociate and cause the hydrogel to change from a gel state to a liquid state (He et al. 2011). Dynamic borate bonds in self-healing injectable hydrogels based on plant polysaccharides also provide healing capabilities, which are formed between the hydroxyl groups of okra polysaccharide (OP) and xyloglucan (XG), and while having inherent antioxidant activity, they accelerate the wound healing process by enhancing epithelialization and angiogenesis. This hydrogel was able to neutralize 73.9% of DPPH free radicals and accelerate the wound healing process by reducing inflammation, increasing collagen deposition, and stimulating angiogenesis (Liu et al. 2024).

Oxime linkages are ideal ways to synthesize hydrogels that have high reaction efficiency and environmental reaction conditions. This bond forms through the interaction of aldehyde or ketone groups with hydroxylamine. Usually, protein-polymer conjugation uses these oxime bond formations (Anupama Devi et al. 2021).

The reversible nature of crosslinks in polymer networks, often achieved through dynamic covalent or non-covalent interactions, enables self-healing in polysaccharide-based hydrogels to mimic biological tissues (Elangwe et al. 2025). Intrinsic self-healing relies on reversible interactions that can be covalent or non-covalent (Jain et al. 2025). In general, self-healing polymers are based on non-covalent factors, including hydrogen bonds, ionic interactions, metal–ligand bonds, π-π interactions, and guest–host interactions.

10.3.2.2 Viscoelastic Recovery

The viscoelastic mechanism inherent in self-healing polymers is a combination of viscous and elastic properties that enables the material to undergo deformation when stressed and revert to its original form once the applied stress is removed. In the domain of self-repairing polymers employing metal–ligand bonds, the viscoelastic response involves the reversible shearing of metal–ligand interactions, which increases the mobility of the polymer and facilitates dynamic movement and self-healing of damage (Li and Zuo 2019).

Tunn et al. focused on the advancement of self-healing hydrogels using nature-inspired strategies. Their goal was to create self-healing hydrogels with tunable mechanical properties by histidine-metal coordination compound in the coil structure. The hydrogels were founded on PEG as core material and histidine (His) and zinc ion (Zn^{2+}) as cross-linking agents. They incorporated histidine-metal coordination sites into coiled-coil (CC) heterodimeric peptides. They synthesized and bioengineered. As a result, α-helical cross-linked grains, Zn^{2+}, are created. The hydrogels showed self-healing properties with the addition of Zn^{2+} ions, which allowed reversible switching between viscoelastic and elastic states. The dynamic nature of these reversible crosslinks affects the viscoelastic characteristics of these hydrogels. Viscoelasticity allows the polymer to flow and fill cracks (viscous behavior) while maintaining structural integrity (elastic behavior). When the material is damaged, the cross-links break and the material flows into the crack. After modification of the crosslinks, the material regains its original structure and mechanical properties. By using intermolecular coordination bonds between metal and histidine as the second cross-linking method, these CC-forming peptides contribute to the creation of higher-order complexes and function as a non-covalent cross-linker for hydrogels based on polyethylene glycol. The reversible interactions between histidine's CC and Zn^{2+} lead to a hydrogel exhibiting dual self-healing modes. Reversible and dynamic tweaking of the hydrogel's viscoelastic characteristics (i.e., relaxation time) as a function of Zn^{2+} concentration is made possible by the extra cross-linking and self-healing mode (Li and Zuo 2019; Tunn et al. 2019). Enke et al. developed synthetic analogs using histidine-rich block copolymers via reversible addition-fragmentation chain transfer (RAFT) polymer formation to mimic hierarchical architectures and the self-healing ability of oyster filaments. Polystyrene is incorporated within the hard domain, while n-butyl acrylate (BA) and histidine unis serve as ligands in the soft domain. Various Zn^{2+} salts facilitate the joining of block copolymers. The mechanism of the self-repair capabilities of the resultant polymers is attributed to the rheological behavior of the rigid segment during thermal exposure (initial phase) and the increased polymeric chain mobility following the reversible severing of metal-binding interactions (subsequent phase) (Li and Zuo 2019; Enke et al. 2018).

In addition to ligand–metal interactions, hydrogen bonds and supramolecular interactions are also effective in the viscoelastic mechanism (Zhu et al. 2024). Zhang et al. developed a biocompatible and self-adhesive organohydrogel using a DLP printer, which has a network of mixed covalent and physical bonds. Viscoelastic recovery in this network structure helps to rearrange hydrogen bonds, which enhances mechanical properties, increases elasticity, and improves healability. The ability to quickly recover function after strain confirms the existence of this mechanism in the hydrogel. This hydrogel, made of methacrylic anhydride-grafted with kappa-carrageenan (MA-κ-CA) with a grafting rate of 6.25%, exhibited the highest tensile modulus (57.1 kPa), strain at break (1133%), and fracture energy (3.6 MJm^{-3}) (Zhang et al. n.d). Intending to develop an elastic, conductive, and self-healing hydrogel for wearable sensors and soft electronics, Khadka et al. synthesized a hydrogel from a combination of PVA and the electrically conductive polymer Poly(3,4-ethylenedioxythiophene) (PEDOT): poly(styrenesulfonate) (PSS) along with boric acid (BA) and ethylenediamine (EDA). The dynamic and reversible B–O, B–N, C–N, and hydrogen bonds provide fast healing capability without external stimuli, in less than 5 min with an efficiency of 80%. In the test of 1000 cycles of stretching and recovery of this hydrogel, no performance degradation was observed (Fig. 10.3) (Khadka et al. 2024).

10.3.2.3 Shape-Memory Polymers (SMP)

When a material can stabilize a transient shape and revert to its permanently "remembered" shape in response to an external stimulus, it is said to possess shape memory qualities. These materials often have two different kinds of cross-linking: an irreversible cross-linking that stabilizes the permanent shape and a reversible "cross-linking" that maintains the temporary shape. By employing various techniques to turn the contacts on and off, metal–ligand bonds can also be utilized as temporary alteration components for SMPs. If the distribution of metal ions can be regulated, the kinetics of metal–ligand binding association and dissociation, which is significantly reliant on the metallic cation, can be effective in developing spatially selective flexibility (Zou et al. 2017).

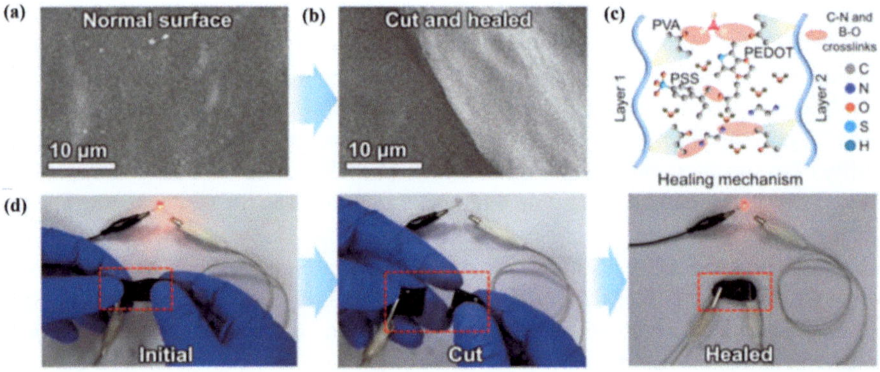

Fig. 10.3 **a** Surface of Case 3 sample before healing; **b** after healing. **c** Schematic of self-healing mechanism. **d** Conductivity before and after healing. Case 3 (highest PEDOT:PSS content) shows the lowest sheet resistance (288 $\Omega.sq^{-1}$), the highest conductivity (0.5 S/m), the highest resistance change (15.6%) at 25% strain, and the lowest charge transfer resistance (7 kΩ) due to dense PEDOT: PSS networks in PVA matrix (Khadka et al. 2024)

A polyester polymer functionalized with isonicotinate (PIE) with pyrazinamide side chains was created by Wang and colleagues. When combined with Cu^{2+} ions, pyrazinamide can create a physical cross-link with superior shape-retentive and strong self-repairing abilities in gentle conditions. The pyrazinamide groups on the polymer's side chains within this polymer network, an effect akin to a ligand, is produced with the adsorbed Cu^{2+} cations, which gives the polymer its shape memory function. Rebuilding the coordination linkages and rearranging the polymeric molecular chains on the wounded surfaces are the primary causes of the self-healing characteristic. When exposed to an external stimulus, like heat or a change in pH, a distorted material can regain its previous shape (Wang et al. 2015a). Referred to as shape memory-assisted self-healing (SMASH) materials, the shape-retaining capability can be utilized to augment the self-repair abilities of substances. Shape recovery is the first step in the self-healing process of SMASH materials, which brings the cracks closer together. The repair process then occurs in the damaged region (Li and Zuo 2019). Luo et al. prepared coatings with thermoplastic PCL fibers randomly dispersed in a shape memory epoxy resin. These coatings exhibit self-healing capabilities, while their shape memory function allows the coatings to return to their original shape after deformation. PCL fibers contain both amorphous and crystalline regions. The crystalline domains function as physical network junctions, aiding the material in retaining its original shape. Also, when the coating deforms, the PCL fibers also deform, and upon heating, the crystalline regions melt, allowing the fibers to return to their original shape. The self-healing capability is facilitated by dynamic interactions in the epoxy matrix (Luo and Mather 2013).

Tandon et al. investigated the self-healing mechanism of PCL/TPU composite fibers produced by FDM and MEW 3D printing methods. The healing is activated by heating below the melting point of PCL (~90°C), which causes chain movement in the

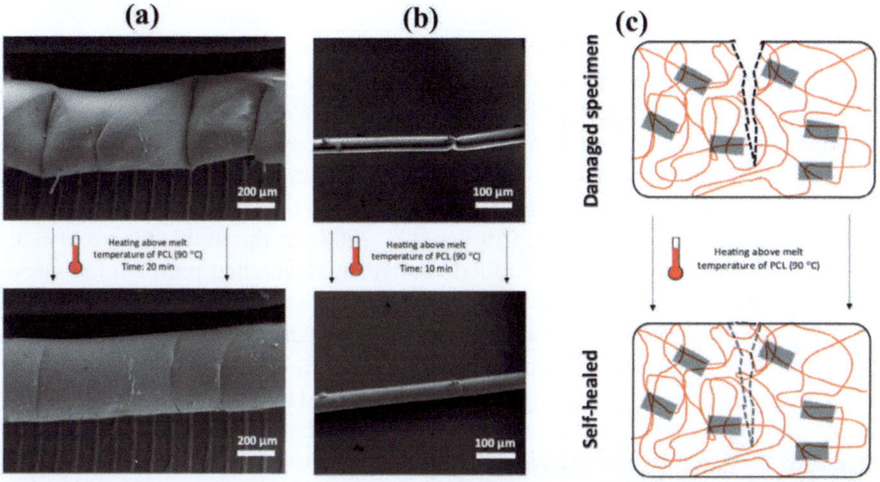

Fig. 10.4 Self-healing property of the fibers. SEM image showing **a** FDM fiber demonstrating healing when kept at a temperature of 90 °C for 20 min, **b** MEW fiber demonstrating disappearance of indents after 10 min of heat treatment, and **c** schematic of the self-healing mechanism (Tandon et al. 2025)

damaged area of the polymer and repairs the surface with the help of intermolecular forces. After cooling, the recrystallization of PCL stabilizes the structure. TPU plays the role of a receptive support, but must be dependent on the thermoplasticity and shape memory of PCL. This process is carried out without chemical reactions, based solely on physical and thermal properties. Figure 10.4 shows the results of scanning electron microscopy (SEM) images and pre- and post-healing imaging studies along with a schematic of the repair, which indicates the high effectiveness of this mechanism in modifying the damaged surface and returning to its original state afterwards (Tandon et al. 2025).

Similarly, shape memory vitrimer polymers with self-healing and flame resistance use dynamic imine and phosphate bonds. At ~100 °C, bond exchange enables fast network rearrangement and damage repair within minutes. They maintain shape memory even after flame exposure, making them promising for safety, engineering, and electronics (Razzaq et al. 2025).

10.3.2.4 Supramolecular Chemistry Approaches

Crosslinks in polymers are produced by non-bonding interactions like van der Waals forces, hydrogen interactions, π-π stacking, water-repellent interactions, metal-complex interactions, and electrostatic interactions. Larger, more ordered structures are created through non-covalent interactions between molecules known as supramolecular interactions (Anupama Devi et al. 2021; Li and Zuo 2019). Hydrogen bonds are highly directional, meaning they form specific angles and orientations that help the molecules accurately assemble into larger structures. These bonds are reversible and allow dynamic assembly and disassembly of supramolecular structures. In polymer networks, hydrogen atoms engage in reversible cross-linking with atoms of high electronegativity, like fluorine, oxygen, and nitrogen, demonstrating increased bond strength and improved self-repair abilities. Hydrogen bonds are relatively strong compared to other non-covalent interactions; however, compared to hydrogels prepared using ionic and covalent interactions, they are relatively less stable (Anupama Devi et al. 2021; Cordier et al. 2008). In the study of Cordier et al., the supramolecular network was formed via hydrogen bonding between fatty acids and urea compounds, which allows the elastomer to repair itself at room temperature and exhibit reversible properties because the hydrogen bonds can be broken and be modified and enabling materials to be deformed several times (Cordier et al. 2008). Since individual hydrogen bonds are weak, rigid hydrogen-bonded hydrogels consistently have robust interactions due to extended hydrogen-bonding sequences. Therefore, they aren't easily recovered after breaking and have limited self-repair efficiency. This self-repair process relies on the level of protonation and the surrounding environment, altering the reactive group's polarities. As a result, the self-healing abilities are constrained in strong acidic or basic environments (Wang et al. 2023a).

These intelligent hydrogels can automatically mend themselves when they experience tears or damage. Their durability and stability are enhanced by their ability to be shaped, injected, and stretched, maintaining their initial mechanical properties (Anupama Devi et al. 2021). Hydrophobic interactions occur in them when the hydrophobic regions of the polymer chains accumulate to minimize their contact with water. These interactions help to form a strong network structure. These hydrophobic associations lead to the formation of finely divided domains in the hydrogel. These domains perform physical cross-links and increase the mechanical strength and toughness of hydrogels by more efficiently dissipating energy during deformation. Jiang et al. prepared hydrophobic crosslinking hydrogels (HA-gels) via copolymerization within micelles. It also showed that hydrogels with a semi-penetrating network structure exhibit higher mechanical strength than those with a single network structure (Fig. 10.5) (Jiang et al. 2010). This interaction is utilized in the design of self-repairing physical hydrogels via the self-assembly process of hydrophobic monomers in an aqueous solution. Micellar aggregation is a commonly employed technique for crafting self-healing hydrogels through hydrophobic grafting. This process involves essential elements like hydrophilic and hydrophobic units, electrolytes, and surfactants (Anupama Devi et al. 2021).

Typically, hydrogels formed through non-covalent interactions exhibit high flexibility and self-healing properties because they can easily break and reform crosslinks. In contrast, hydrogels produced from covalent linkages tend to be much more stable (Anupama Devi et al. 2021). Hydrophobic interaction strength is size-dependent, and by altering the size of hydrophobic chains and substituents, the mechanical characteristics and self-repair capabilities of the cross-linked hydrogel can be fine-tuned (Wang et al. 2023a).

Aromatic rings engage in π–π stacking bonds, where the electron-rich (donor) and electron-deficient (acceptor) regions of the molecules align and overlap. The

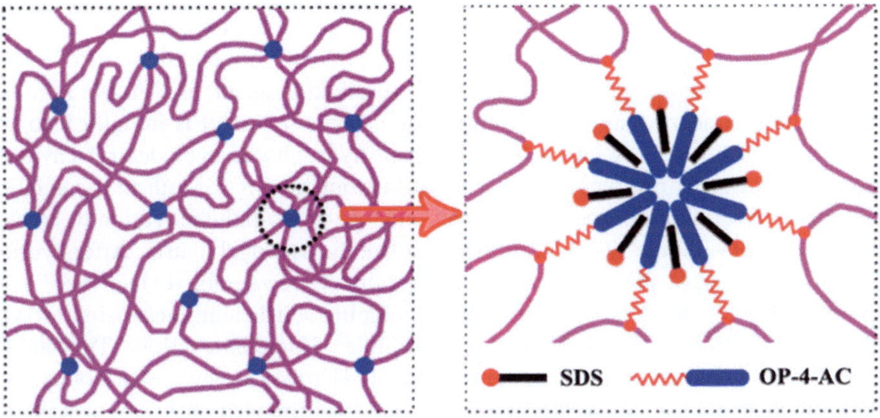

Fig. 10.5 Schematic illustration Schematic illustration of the structural model with associating networks in the HA-gels (Jiang et al. 2010)

research work of Burattini and colleagues explores the advancement of a supramolecular polymer network showing self-repair properties due to π-π. This supramolecular polymer system exhibits enhanced mechanical properties and easy healing properties. In the polymer system, the terminal pyrenyl groups of polyamide are introduced through synergistic π-π electronic interaction in the chain folds of polyamide. When materials are damaged, the π-π interactions are broken, and upon recovery, these interactions can be repaired and effectively heal the material. The non-covalent character of the π-π stacking connection makes the healing process reversible, meaning that it can be initiated by temperature changes (Burattini et al. 2009). In 2012, researchers centered on growing a new class of substances that combine robustness and self-repair capabilities. They developed supramolecular polymer nanocomposites by incorporating cellulose nanocrystals (CNC) into a curable supramolecular polymer matrix. These nanocomposites were prepared through solvent casting and then compression molding. Nanocomposites showed increased mechanical properties with different weight percentages of CNCs (from 1.25 to 0.20 weight percent). The addition of CNC significantly improved the strength and toughness of the material. When nanocomposites are damaged, they can recover their mechanical properties after exposure to certain conditions, such as heating or exposure to certain solvents. Here, the stacked π-π interactions act as transient cross-links that can be recovered under thermal or water solvent conditions. It also promotes the self-assembly of polymer chains, resulting in a strong and well-organized network (Fox et al. 2012).

Specific complementary interactions in supramolecular materials with host–guest dynamics are realized through various guest groups, including azobenzene, adamantane, cholic acid, ferrocene, and cholesterol. Hydrogels formed through these interactions show promising potential in biomedical research (Anupama Devi et al. 2021). Jalavandi and colleagues developed a shear-thinning hydrogel that can break apart when shear forces are applied and repair itself after the shear forces are removed. They produced a combination of weak guest–host interactions and secondary photocrosslinking. They used cyclodextrin-modified alginate as the hydrogel base, methacrylated gelatin as the crosslinking agent, and visible light to initiate the secondary photocrosslinking process. The hydrogel is initially formed through weak host–guest interactions between hydrophobic cyclodextrin cavities and gelatin aromatic residues (amino acids with aromatic rings). Methacrylic gelatin has photo-crosslinking properties that can undergo photo-polymerization to create secondary cross-linking sites and further stabilize the hydrogel matrix. After hydrogels are subjected to shear forces, they are effectively reassembled (Jalalvandi and Shavandi 2019). Zhang et al. focused on developing self-healing gels using crown ether-based host–guest bonds, which is an example of ionic interactions. They engineered a polymer featuring a pendant dibenzo[24]crown-8 component connected via bisammonium salts to create supramolecular gels founded on host–guest interactions. These polymer networks have self-healing properties, meaning they can repair themselves automatically if damaged, and the gels also respond to stimuli like changes in pH values. The crown ether units act as hosts that form complexes with bisammonium salts that are guests. These interactions are reversible and dynamic and allow materials to self-heal. When the gel is damaged, host–guest interactions at the site of damage are disrupted, and upon exposure to appropriate conditions (such as changes in pH), these interactions can be modified, leading to a self-healing process (Zhang et al. 2012b). Thanks to metal cations and reactive organic ligands, self-repair polymers utilizing coordination bonds can serve multiple purposes, such as reacting to various stimuli, dielectric, light-emitting, shape-memorable, magnetically responsive, and catalysis. Self-healing polymers are created by cross-linking different coordinating links. Ligands can be added to the polymer chain as pendant groups or conjugated to the existing polymer backbone. By altering the metal ions, the bond strength can be changed. Numerous ligands, including carboxylate, pyridinyl, histidine, and catechol, are useful for extracting self-healing polymers and coordinating with metal ions (Li and Zuo 2019). Qin et al. detailed a general in situ polymerization method for creating anisotropic hydrogels. These hydrogels are composed of highly regular layered networks connected by metal nanostructured assemblies. They have self-healing capabilities facilitated by the dynamic coordination of thiolate and silver metal under near-infrared (NIR) irradiation and low pH conditions. Also, they show significant anisotropic performances in terms of mechanical, optical, deflation, and inflation behaviors (Qin et al. 2019).

The association between oppositely charged polymers leads to cross-linking within polymer solutions throughout hydrogel synthesis. Ionic interaction between free ions and polymers results in a reversible response as a result of the free ions' movement to uncrosslinked or damaged regions of polymer chains, which triggers self-repair (Anupama Devi et al. 2021). In the research work of Liu et al., the aim was to provide an effective mechanism for energy dissipation, increased toughness, and self-healing ability of hydrogel by creating a multi-linked network (MBN) with hierarchical cross-links. They used poly(acrylic acid) (PAA) as the primary polymer used to create the hydrogel and iron(III) chloride ($FeCl_3$) to create ionic cross-links with the carboxyl groups in PAA. Poly-acrylic acid hydrogels originate from double-bond networks comprising ionic cross-links and hydrogen bonds, which were synthesized and exhibited excellent mechanical characteristics. Also, hydrogels showed complete self-repair after 48 h at 50°C (ying Liu et al. 2017). Bai et al. focused on the development of a new type of carboxy betaine acrylamide hydrogel with unique self-healing properties. They use zwitterionic polymers that have positive and negative charges to facilitate self-healing. The hydrogel showed spontaneous and time-independent self-healing properties under physiological conditions driven by the zwitterionic fusion mechanism, where cationic and anionic groups rapidly combine to form zwitterionic pairs through electrostatic attraction. These hydrogels can also withstand repeated mechanical stress without significant degradation (Bai et al. 2014). Self-healing ionic hydrogels often have a longer lifespan since they can be repaired several times. Furthermore, intrinsic conductivity is produced by the hydrogel's freely moving counterions. However, the structural robustness and self-repair capabilities of ionically bonded hydrogels depend on the ionic concentration of the surrounding fluid (Wang et al. 2023a).

In recent years, self-healing networks, engineered using supramolecular strategies like metal coordination and hydrogen bonding, have garnered significant interest because of their potential to increase the endurance of materials, enhance safety, and also guarantee stability (Thangavel et al. 2019). Xiong et al. reported that CO_2-switchable self-healing energy hydrogels based on ionic liquids formed through supramolecular interactions between ionic liquids and water at room temperature. When CO_2 gas is passed through them, diamine anions or primary/secondary amines and protons react with it to form carbamates or bicarbonates, respectively. Thus, hydrogen bonds between anions and water molecules and ionic bonds between anions and ions occur, and the gel structure is converted to a sol solution state. Then, by heating or passing N_2 gas, the initial gel structure is restored (Xiong et al. 2025).

10.3.3 Extrinsic Healing Mechanisms

A category of materials known as exogenous self-repair materials contains external agents, like vascular networks or microcapsules, that can be released when the healing material is damaged. These agents, including monomers, adhesives, or various activated materials, subsequently restore damaged materials (Jadoun 2024). Therefore, the restoring agent is incorporated within the system as a separate, distinct phase in the form of capsules, fibers, and nanocarriers (Garcia and Fischer 2014). Extrinsic restorative agents, like nanocapsules and microcapsules, that are loaded and subsequently embedded with substances that promote auto-repair, are the foundation of extrinsic self-healing composites. The capillary effect causes the brittle capsules to release the material onto the fracture after it ruptures, which causes the cracks to mend. External self-healing materials boast advantages like activation in response to various forms of damage, superior restoration efficiency, and reasonable cost. However, they require a high concentration of healing agents and have the potential for agent leakage (Jadoun 2024).

Romero et al. focused on creating a self-healing polymer system that works effectively at low and ultralow temperatures, which relies on the Ring-Opening Metathesis Polymerization of 5-ethylidene norbornene/dicyclopentadiene monomers using a ruthenium-based third-generation Grubbs catalyst. This catalyst was microencapsulated in polyurea containers and embedded within different epoxy matrices (Fig. 10.6). The resulting system demonstrates over 100% self-healing efficiency at low temperatures, not only repairing itself but also improving its mechanical properties beyond pre-damage levels. Upon material damage, the microcapsules discharge the healing agent, which subsequently reacts with the polymer matrix to mend the damage (Romero-Sabat et al. 2021).

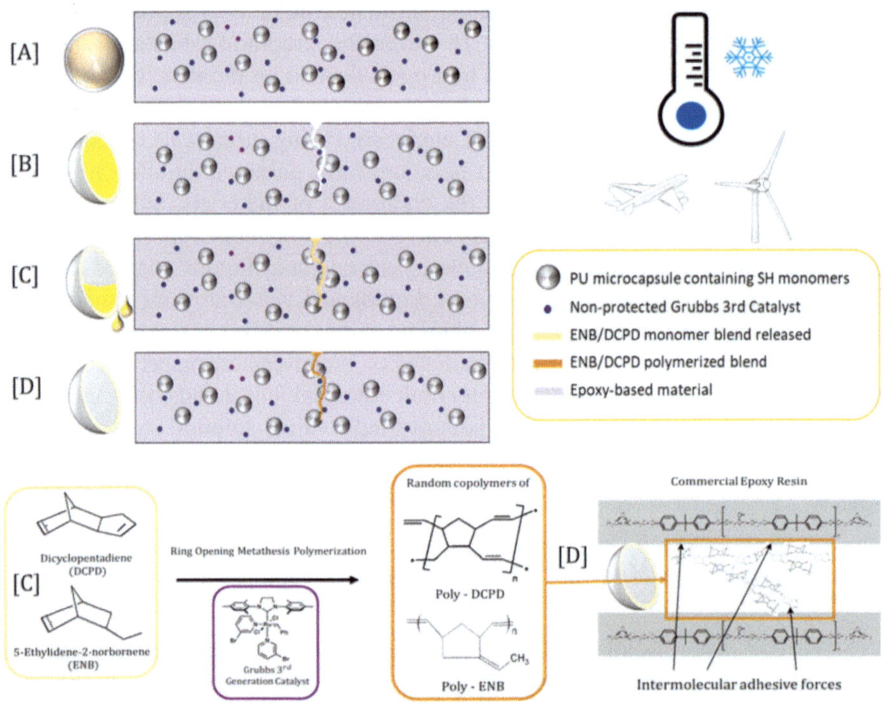

Fig. 10.6 Scheme of the extrinsic ROMP-based self-healing system developed in this work for epoxy-based materials used in highly demanding fields efficient at ultra-low temperatures. Strained cyclic olefins with no bulky groups around the double bonds such as 5-Ethylidiene-2-norbornene (ENB) and Dicyclopentadiene (DCPD) are microencapsulated and dispersed in epoxy-based materials. When a crack appears (due to erosion or external damage, for example), it breaks the composite along with the capsules, and the monomers are released from their vessels and reacts at low temperature (−20 °C) in presence of a Ru-based catalyst (3rd Generation Grubbs Catalyst) previously embedded into the material to generate a crosslinked random copolymer of poly-ENB/DCPD, generating an adhesive layer and attaching both sides of the damaged composite. With this, the material is healed and its operating shelf-life is increased (Romero-Sabat et al. 2021)

The primary limitation of intrinsic self-healing lies in the constraints related to multiple repairs at previously damaged and healed locations (Garcia and Fischer 2014).

10.3.3.1 Microcapsule-Based Healing

Self-healing microcapsules are tiny capsules embedded in the substance and encapsulate repairing agents like catalysts, resins, or monomers. Furthermore, several catalysts and healing agents have been produced, including polydimethylsiloxane, epoxy resins, linseed oil, and isocyanates (Jadoun 2024).

Upon sustaining damage, the microcapsules rupture, releasing repair material within the affected area. These agents then undergo polymerization, creating new bonds to repair the damage. Research focused on spherical capsules containing healing agents has progressed in liquid encapsulation primarily in four sections:

1. Improvements in encapsulation methods.
2. Choosing suitable restorative agent-crosslinker/catalyst combinations for diverse frameworks.
3. Materials used in encapsulating shells, development of therapeutic agents that do not require crosslinkers or catalysts, and.
4. Implementation The concept of encapsulation for more practical research (Garcia and Fischer 2014).

In spherical capsules, three primary approaches leveraging embedded liquid repair substances:

1. In dual-agent systems, the repairing substance is encapsulated while the catalyst is distributed within the substrate. Upon encountering the damaged region, the repair agent hardens, fusing the opposing sides of the crack and acting as an adhesive. Similarly, dual encapsulation systems may be employed, with a uniform capsule containing the healing agent and the other containing a hardener.
2. An adaptive system interacting with its surroundings, like the silyl ester method for corrosion prevention, reacts with moisture and the underlying metal to create a protective barrier layer.
3. Solvent-Like matrix-reactive agent encapsulated in a thermoplastic, which heals the damage with sufficient mobility (Garcia and Fischer 2014).

White and his research team report a unique polymeric material that shows the potential to self-repair cracks through the incorporation of a microencapsulated healing agent. After crack formation, dicyclopentadiene is formed at the crack position and, upon contact with the catalyst, transforms into polydicyclopentadiene. When a fracture appears, a dicyclopentadiene monomer gets released and absorbed into the microcapsules within the matrix through capillary movement (White et al. 2001). In another work, this author and colleagues focused on the development of dicyclopentadiene (DCPD)–containing microcapsules using in situ polymerization for self-healing applications in epoxy resin. They used urea and formaldehyde as poly(urea formaldehyde) (PUF) shell-forming monomers and an oil-in-water emulsion system to facilitate the in situ polymerization process. When the epoxy was damaged, DCPD was released from the microcapsules, polymerized, and repaired the cracks in the epoxy matrix. In this process, DCPD is dispersed in an aqueous phase, and urea and formaldehyde are polymerized at the interface to form a PUF shell around the DCPD droplets (Brown et al. 2003).

In addition to spherical capsules containing therapeutic agents, fibers have also been used in self-healing, but in this case, it is more likely that healing will cause damage.

This method is intended for direct application to fine composites through three established techniques utilizing hollow fibers: 1. hollow fibers, 2. segmented segments, and 3. vascular systems (Sect. 3.3.2).

The most common ones are single hollow fibers and vascular systems, where vascular systems are networks of fibers capable of transporting therapeutic agents (Garcia and Fischer 2014). The inclusion of nanoparticles such as MWCNT increases the efficiency of the restoration process and even allows better mechanical recovery and electrical characteristics (Kosarli et al. 2021). Certain self-healing nanocomposites are created via hydrogen interactions between graphene oxide and amine. Liu et al. use graphene oxide (GO) as an activator and multifunctional networking center and polyacrylamide (PAM) as the polymer matrix that forms the hydrogel. Also, graphene peroxide acts as a cross-linker in the hydrogel network and creates a strong network with PAM chains. These robust interactions between graphene oxide sheets and polyacrylamide chains were crucial for enhancing the mechanical (Liu et al. 2013). The segmented fiber method harnesses the benefits of hollow capsules, known for their high breakage and release probability, while also incorporating spherical and elongated microcapsules for targeted release and repeated repairs within the unified structure. Unlike split fibers, hollow fibers limit the healing of multiple injuries in the same area due to blockage or widespread diffusion of the healing agent. But in split fibers, due to the limitation of the amount of released repair agent, the damage that can be repaired is less than the remaining strategies. Conversely, vascular systems can trigger multiple healing events. However, their complexity in implementation and susceptibility to blockages under certain conditions pose significant challenges (Garcia and Fischer 2014).

10.3.3.2 Vascular Networks

Vascular self-healing involves integrating a network of vessels into a substance, enabling the discharge of restorative compounds to mend the damage. This process represents an extrinsic self-healing mechanism. The vascular network can contain healing agents like adhesives, monomers, or resins and is made to resemble the blood vessels of living things. The vascular network in the affected area bursts when the matrix is damaged, releasing the healing agent, which then works to heal the damage. The ability to precisely administer the repairing substance to the damaged location results in quick and efficient healing, making this mechanism extremely successful (Jadoun 2024). Huang and colleagues investigated the capability for self-repair mechanisms in cementitious substances using the healing agent saturated calcium hydroxide solution ($Ca(OH)_2$) using two different approaches of the capsule and vascular system. The efficiency of self-healing using capsules was low due to the rapid absorption of the healing agent by the bulk matrix. However, increasing the dosage and size of the capsules improves the efficacy. However, in the vascular system, the efficiency is much higher due to the continuous supply of $Ca(OH)_2$ solution to the cracks (Huang et al. 2014). Similarly, in the study by Zhang et al., using 3D printing of volatile inks containing GO, methyl palmitate (MPA), and dodecyldimethylammonium bromide (DDAB), physical networks are embedded in the concrete matrix, which remain as hollow pathways after hydration. These networks activate carbonation reactions by infiltrating CO_2 into the cracks, which leads to crack healing and strength recovery through the formation of $CaCO_3$ crystals. Figures 10.7 and 10.8 show SEM images and schematics of the network formation process, respectively, along with surface healing and an increase in compressive strength of up to 28.1%. This system can be used to construct self-healing concrete by intelligently designing the ink and exploiting the interfacial stability (Zhang et al. 2024b).

It has been discovered that healing catalysts, which are chemically broken down into networks via capillaries or hollow channels, can remarkably regenerate repetitive injury events in vascular repair (Jadoun 2024). Qamar et al. set out to pioneer a new technique for vascular self-repair by embedding porous thermoplastic networks, 3D-printed into a thermoset matrix. This strategy leverages the unique chemical properties of thermoplastic materials, enabling repair agents to flow into and mend cracks (Qamar and Trask 2017). Long-term storage of healing agents in the vascular network and their availability are among the advantages of this system. Furthermore, the network of vessels can sequentially disperse several restorative agents, enabling more intricate and comprehensive repairs (Jadoun 2024). To create core–shell nanofiber coatings, Lee et al. used a double emulsion electrospinning technique consisting of dimethyl siloxane as the healing agent, dimethyl-methyl hydrogen-siloxane as the shell, and poly(dimethyl siloxane) (PDMS) as the outer matrix (Lee et al. 2014).

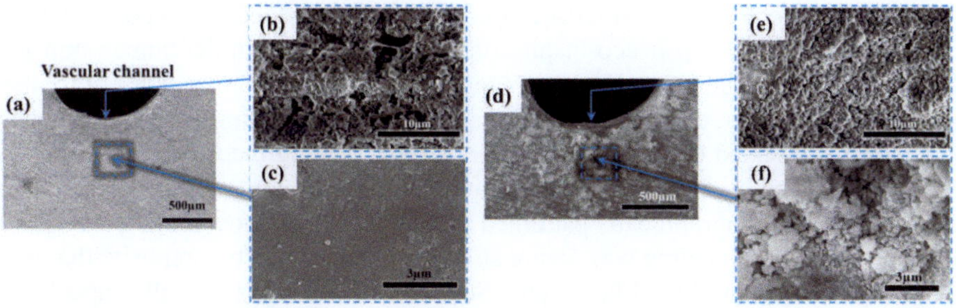

Fig. 10.7 SEM images comparing carbonated and non-carbonated specimens: **a–c** Magnified images of non-carbonated specimens at the vascular channel surface and interior location; **d–f** Magnified images of carbonated specimens at the vascular channel surface and interior location (Zhang et al. 2024b)

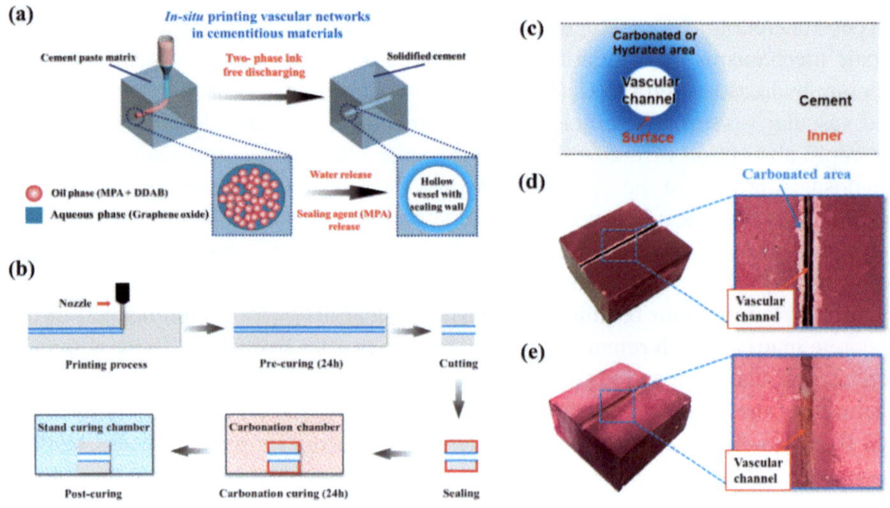

Fig. 10.8 In-situ printing and formation of a vascular channel. **a** Schematic illustration of in-situ printing of fugitive ink and the formation of a hollow vascular channel. **b** Schematic representation of the carbonation process. **c** Sampling locations within the cement paste specimen. **d** Carbonated cement paste specimen evaluated using the phenolphthalein test. **e** Reference cement paste specimen without carbonation curing, also tested with phenolphthalein (Zhang et al. 2024b). The phenolphthalein test is used to determine the depth of carbonation in cement paste by applying a phenolphthalein solution to a freshly cut surface. In carbonated regions, the pH is reduced due to the reaction of $Ca(OH)_2$ with CO_2, forming $CaCO_3$, which prevents the phenolphthalein from turning pink—these areas remain colorless, indicating carbonation has occurred

Some nanocomposites, composed of healing polymer compounds and nanofillers like cellulose nanocrystals in various ratios, showcase therapeutic behaviors akin to a range of biological healing processes through circulatory and vascular designs. To ensure the ongoing delivery of healing agents with controlled flow, a recirculation system can coat and replenish the healing agent. Toohey et al. created 3D vascular self-healing materials where healing agents are routinely and automatically dispensed to defects via a 3D microvascular network embedded in an epoxy coating. Their goal was to develop an eco-inspired coating design that mimics human skin and allows continuous delivery of healing agents into polymer coating cracks (Toohey et al. 2007). However, the potential of these designs to heal is diminished by the need to minimize the fixed catalyst and provide various healing mediators inside them. To address these issues, Hansen et al. created a two-part epoxy curing solution that retains the photolithographically patterned areas inside the microvascular embedding network. Better self-healing was demonstrated by the sandwich composite that was created using a 1D or 2D fluidic system. Self-healing materials have the capacity to lower maintenance costs in the aerospace sector by fixing minor dings or scratches in an aircraft's wings or fuselage. However, among its drawbacks are the high cost of manufacturing, the potential for material deterioration in the vascular network, and the potential for structural integrity to be compromised (Jadoun 2024).

10.3.4 Combination Mechanisms

To circumvent the drawbacks of a fully reversible system made up of relatively weak crosslinks, hybrid architectures with both reversible and irreversible crosslinks may be employed in the creation of intrinsically autonomous repair polymer systems (Garcia and Fischer 2014). To create self-healing hydrogels from biopolymers like chitosan, cellulose, gelatin, sodium alginate, and their derivatives, imine bonds and acyl hydrazones are typically employed to establish ionic, hydrophobic, and hydrogen interactions, and dynamic covalent bonds (Fig. 10.9) (Wang et al. 2023a). Creating self-healing hydrogels with hydrophobic interactions often involves intricate processes and specialized materials. Consequently, these hydrogels are typically combined with other dynamic bonds, like hydrogen and ionic bonds (Wang et al. 2023a). Takafuji et al. investigate the auto-repair hydrogels supramolecular assembly of chiral nanofibers G-Py^+ (glutamine-derived lipids containing pyridinium head groups) with S-PEGn-S (polyethylene glycol sulfate), which have self-healing properties. Hydrogel shows hydrophilicity at ambient temperature. G-Py^+ forms nanotube aggregates in water, which are stabilized by hydrogen bonds. Also, S-PEGn-S has terminal anionic sulfate groups that interact electrostatically with G-Py^+ cationic pyridinium groups. The chiral nature of the G-Py^+ aggregates helps maintain the structural integrity and orientation of the nanofibrous assemblies, even after the addition of S-PEGn-S. Therefore, in addition to stabilizing the structure, chirality also facilitates the dynamic coupling necessary for effective self-healing (Takafuji et al. 2024). Xue et al. investigated hierarchical reversible cross-linked polymer hydrogels with dual-shape memory behavior in response to pH and Ca^{2+} ions. This hydrogel was composed of poly(vinyl alcohol) (PVA) and borax as a cross-linker that interacts with PVA to form borate bonds. They utilized Poly(1-allyl-2-thiourea-co-acrylamide) as a penetrating polymer in the borax/PVA framework to create a semi-impregnated network (semi-IPN). This hydrogel was subjected to a freeze–thaw process to produce PVA crystal domains, which acted as additional cross-links, and thus exhibited proper mechanical properties, as well as a rapid self-healing ability of more than 90%, showing its initial strength percentage within 10 min. This backbone self-healing was driven by reversible borate interactions and plentiful hydrogen bridges in polymeric matrices via freeze–thaw iterations. Besides acting as crosslinking agents and forming reversible ionic bonds with polymer chains, Ca^{2+} ions enhance the shape memory characteristics of the hydrogel (Xue et al. 2021).

10.4 Functionalization of Biopolymers for Enhanced Self-Healing

Biopolymers are increasingly recognized as sustainable alternatives to fossil-based materials, offering the potential to reduce environmental impact while providing functional properties for diverse applications. Functionalization techniques are

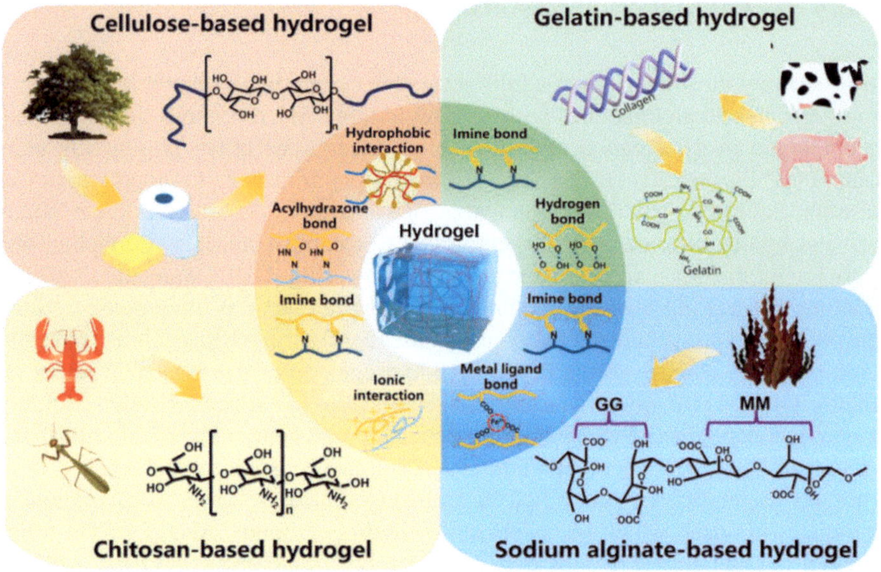

Fig. 10.9 The schematic of the main interactions of self-healing hydrogels from different biopolymers (Wang et al. 2023a)

being explored to enhance the intrinsic properties of biopolymers, particularly for self-healing applications, to compete effectively with petroleum-derived materials (George et al. 2020; Choi et al. 2023). Functionalization introduces specific chemical groups or structures that improve the material's ability to heal damage autonomously, hence extending its lifespan and expanding its applicability. The benefits of functionalizing biopolymers for self-healing purposes include:

- **Enhanced Self-Healing Efficiency:** Functionalization introduces dynamic chemical bonds or cross-links, comprising covalent, ionic, or hydrogen interactions, which enable rapid and efficient self-repair mechanisms. These interactions allow biopolymers to restore their structural integrity after damage, even under mild conditions (Yang and Urban 2018).
- **Stimuli-Responsive Behavior:** By incorporating functional groups that are sensitive to external stimuli (e.g., temperature, pH, moisture, or light), biopolymers can achieve self-healing in response to specific environmental changes. This stimuli-responsiveness enables targeted healing processes, making the materials suitable for applications in smart coatings, biomedical devices, and wearable electronics (Wei et al. 2017; de las et al. 2005).
- **Improved Mechanical Performance:** The introduction of functional groups that form reversible bonds or dynamic cross-links improves the mechanical characteristics of biopolymers, including tensile strength, toughness, and flexibility. This makes the materials more resilient to mechanical stress and capable of maintaining performance across repeated healing cycles (Neal et al. 2015; Wang et al. 2021).

- **Enhanced Adhesion and Compatibility:** Functionalization can improve the adhesive properties of biopolymers, enabling better bonding to various substrates or reinforcing agents. This is crucial for applications such as protective coatings, adhesives, and composite materials where strong interfacial adhesion is required. Furthermore, tailored functionalization can enhance compatibility with other materials, making biopolymers suitable for hybrid systems and multifunctional applications (Wang et al. 2017; Montanari et al. 2020).
- **Biocompatibility and Environmental Sustainability:** Functionalized biopolymers can retain or enhance their biocompatibility by using naturally derived or non-toxic functional groups, rendering them perfect for biomedical fields. Additionally, the sustainable nature of biopolymers contributes to reducing the environmental footprint related to conventional substances (Troy et al. 2021).

The strategies for functionalizing biopolymers to enhance self-repairing can be broadly grouped into three main types: chemical modifications, nanomaterial integration, and biomimetic or bioinspired approaches. Each of these methods provides unique mechanisms for improving self-repair efficiency and expanding the material's range of applications.

10.4.1 Chemical Modifications

Such chemical modifications are of high importance and serve to enhance the self-healing capabilities of biopolymers in advanced applications such as biomedical devices, coatings, and energy storage. Dynamic covalent bonds introduced through Schiff base reactions, disulfide bonds, and Diels–Alder chemistry allow for reversible interactions that enable self-repair. For instance, Schiff base reactions between aldehyde and amino groups in hydroxyl-rich polymers such as PVA produce self-healing hydrogels capable of dynamic reforming of cross-links, hence allowing for effective healing (Takács et al. 2022). Another widely used approach involves forming disulfide bonds, where thiol-modified biopolymers, including gelatin or chitosan, can undergo reversible disulfide-thiol exchange, hence allowing for fast recoveries of damaged materials. This class of materials is particularly well-suited for applications with repeated healing cycles, such as coatings and wearable electronics (Canadell et al. 2011; Yimyai et al. 2022; Cho et al. 2022).

Most biopolymer functionalization via dynamic covalent chemistry is conducted with the help of boronic ester linkages. The boronic esters, either as a home functional or a hetero functional cross-linker, revert dynamically with the diols or polyols, such as in the polysaccharides-alginate-releasing self-healing hydrogels with adjustable mechanical characteristics for applications in tissue engineering (Aeridou et al. 2020; Pettignano et al. 2017; Hong et al. 2018).

Physical cross-linking via hydrogen bonding and ionic interactions is an essential methodology for the modification of biopolymers (Zhang et al. 2017b). As an example, the ionic cross-linking of chitosan with multivalent anions like sulfate or phosphate shows an improvement in self-healing and mechanical properties that make these materials suitable for use as wound dressing (Mndlovu et al. 2020). Hydrogen-bonding interactions in biopolymers such as gelatin and poly(vinyl alcohol) can be considerably reinforced by incorporating ureido-pyrimidinone groups, forming more hydrogen bonds, improving the self-healing capability of the substance (Liu et al. 2023; Zhang et al. 2019).

Functionalization using click chemistry methods, such as azide-alkyne cycloaddition, often provides biopolymers with precise functional groups, illustrating the way for the fabrication of substances with controllable self-healing and mechanical characteristics (Sumerlin and Vogt 2010; Nimmo and Shoichet 2011; Fournier et al. 2007). Modifying biopolymers like cellulose and chitosan enables the development of smart hydrogels and shape-memory materials (Muir and Burdick 2021). These chemical modifications offer versatile tools for designing biopolymers with tailored self-healing capabilities, thereby expanding their potential in biomedical, environmental, and industrial applications. The functional scope and healing efficiency of bio-based materials continue to evolve, driven by ongoing exploration of novel dynamic chemistries and cross-linking techniques.

10.4.2 Nanomaterial Integration

Integrating nanomaterials into biopolymers is a key strategy for improving self-healing properties through mechanical reinforcement, enhanced thermal stability, and functional capabilities. Besides graphene, carbon nanotubes (CNTs), and nano clays, other nanomaterials such as nanocellulose, metallic nanoparticles, and MXenes are gaining attention for their potential to enhance self-healing in biopolymer-based systems (Alcântara and Darder 2018; Jamróz et al. 2019; Solangi et al. 2024; Kolya and Kang 2023). Graphene oxide (GO) and reduced graphene oxide (rGO) improve the self-healing efficiency of biopolymer composites by facilitating conductive networks that activate thermal self-repair mechanisms and increase mechanical robustness (Lee and Cha 2020; Du et al. 2018). When used with biopolymers like PVA and chitosan, graphene enhances mechanical strength and electrical conductivity, making these composites suitable for flexible electronics and wearable sensors (Li et al. 2012).

CNTs contribute to self-healing in biopolymers by forming networks that enable energy dissipation and mechanical reinforcement. Their ability to improve thermal conductivity is valuable for self-healing substances utilized in electronic devices where thermal management is vital (Kolya and Kang 2023). The surface modification of CNTs with functional groups (e.g., carboxyl or amine groups) facilitates better dispersion in biopolymer matrices, thus enhancing the interfacial bonding and self-healing properties (Makvandi et al. 2021).

Nanoclays, such as montmorillonite and bentonite, improve the thermal and barrier properties of biopolymer composites by creating tortuous pathways that hinder crack propagation and moisture ingress. The functionalization of nano clays with surfactants or ionic liquids can further enhance their compatibility with hydrophilic biopolymers, improving their dispersion and mechanical properties (Mapossa et al. 2023).

Nanocellulose, a renewable nanomaterial derived from plant fibers, has emerged as an effective filler for enhancing the self-healing properties of biopolymers due to its high aspect ratio, mechanical strength, and biodegradability. It can form hydrogen-bonding networks with biopolymers such as alginate and gelatin, enabling reversible cross-linking and self-healing behavior. Applications of nanocellulose-enhanced biopolymer composites include biomedical devices, wound dressings, and eco-friendly packaging (Aigaje et al. 2023; Gumrah Dumanli 2017).

Metallic nanoparticles have been incorporated into biopolymers to improve antibacterial properties, thermal conductivity, and UV resistance while enhancing self-healing through photo-thermal effects or catalysis, mainly for silver, zinc oxide, and gold particles. For instance, zinc oxide nanoparticles can activate self-healing through light-induced thermal effects in gelatin-based hydrogels, making them suitable for biomedical applications like skin regeneration (Mollajavadi et al. 2023; Farazin et al. 2023; Mao et al. 2017; Dzulkharnien and Rohani 2022).

MXenes, a novel class of two-dimensional materials, exhibit exceptional electrical conductivity, hydrophilicity, and mechanical properties, making them promising candidates for enhancing self-healing in biopolymers. Their ability to form strong interfacial interactions with polymer matrices enables the advancement of conductive, flexible, and self-healing materials for applications such as wearable electronics and energy storage (Zhu et al. 2022; Wan et al. 2023; Liao et al. 2019).

Despite the significant potential of nanomaterial-enhanced biopolymers, challenges such as uniform dispersion, achieving optimal filler content, maintaining biodegradability, and nanosafety concerns need to be addressed. Ongoing research aims to develop advanced surface modification techniques and scalable synthesis methods to resolve these issues and develop the practical applications of nanocomposite self-healing substances (Anwer et al. 2023).

10.4.3 Biomimetic and Bioinspired Approaches

Biomimetic and bioinspired functionalization of biopolymers represents an approach to utilizing design principles of nature in the advancement of materials able to mimic both the structure and function of biological structures. The methods allow the development of advanced self-healing materials applied for biomedical, environmental, and industrial usages. Anisotropic biopolymer structures, which have their properties dependent on a given direction, are under extensive study for biomedical applications such as in tissue scaffolding, an application that utilizes nanocellulose due to its outstanding strength, flexibility, and biocompatibility. Its unique ability to form aligned fiber networks that replicate the extracellular matrix makes it particularly suitable for neural and musculoskeletal tissue engineering (Lu et al. 2023).

Lignin, a plant-derived polyphenol, is emerging as a functional component in biomimetic materials, acting as a natural antioxidant and reinforcing agent. When incorporated into biopolymers like gelatin or poly(lactic acid), lignin improves the mechanical strength, biodegradability, and thermal resilience, making the composites ideal for sustainable packaging and biomedical applications (Mennani et al. 2023). Other plant-based biopolymers, like cellulose, chitosan, and starch, are actively researched as eco-friendly alternatives to fossil-based polymers owing to their inherent biodegradability, renewability, and potential for chemical modification (Singh et al. 2023). These materials can be functionalized to incorporate bioinspired motifs such as adhesive proteins found in mussels or structural elements from nacre, thus improving their self-healing and mechanical performance.

The use of metal–ligand coordination in biomimetic approaches has shown promise for creating self-repairing materials with strong mechanical strength and dynamic responsiveness. Bioinspired functional groups such as catechol, found in mussel adhesive proteins, and carboxylates are capable of reversibly binding to metal ions like Fe^{3+}, Zn^{2+}, or Ca^{2+}, forming dynamic networks that facilitate self-repair. Biopolymers like hyaluronic acid and gelatin can be functionalized with these groups, resulting in injectable hydrogels and bioadhesives with potential applications in wound healing, tissue regeneration, and drug delivery (Shi et al. 2019; Lee et al. 2016). Additionally, incorporating other bioinspired cross-linking strategies, such as reversible disulfide bonds or hydrogen-bonded networks, can further improve the self-healing efficiency of these materials (Wang et al. 2015b).

Bioinspired approaches also draw from the hierarchical structures found in natural materials like bone, nacre, and wood. Mimicking these structures in biopolymers enables the development of tough and lightweight composites that exhibit self-healing properties. For example, layer-by-layer assembly and freeze-casting techniques can be employed to replicate the micro- and nanoscale structures of these natural materials, resulting in composites that combine toughness with flexibility. While challenges remain in precisely mimicking these intricate architectures, advancements in molecular design, functionalization techniques, and analytical methods are paving the way for new applications in healthcare, environmental sustainability, and high-performance materials (Lu et al. 2023).

10.5 Self-Healing Biopolymers (SHBP) in Sustainable Engineering

The major applications of SHBP fall into three categories: construction, corrosion protection coating, and biomedical fields.

10.5.1 Application in Construction Materials

The utilization of these SHBP over cracks can extend the durability of buildings by retarding the penetration of moisture (Choi et al. 2023). Arroyave et al. reported monitoring of cementitious cracks' healing with an SHBP composite for 28 days. A reduction in crack width is observed when applying a dosage of 2.5 wt. % of biopolymer/Silicate-based resin composite material (Arroyave et al. 2023).

A recent study asserts that SHBPs used in construction engineering are not limited to lignin, cellulose, and starch. Chitosan and proteins used in biomedical applications are also used in construction engineering (Philip and Reshmy 2024).

10.5.2 Use in Coatings and Corrosion Protection

A protective coating is obligatory in industries such as automobiles and architecture (Choi et al. 2023). Surface modification and coating can enhance hydrophobicity to increase anti-corrosion properties (George et al. 2022). For marine equipment, due to the possibility of microalgae adhesion, anti-corrosion coatings should be strengthened with antifouling properties (Wang et al. 2024).

The most recurrent coating methods for the deposition and adhesion of SHBP onto substrates are dip coating, spin coating, and electrospinning (Song et al. 2020).

10.5.3 Applications in Biomedical Engineering

Self-healing hydrogels can be used as scaffolds for research and development on wound healing composites (Wang et al. 2023b). Wound healing may take > 55 days up to < 85 days with no use of SHBP. However, the use of an ideal dressing should protect the wound bed from moisture loss, repair cracks, and reduce the time to 42 days (Mehvari et al. 2024). Moreover, applications of SHBP in biomedical engineering can create new opportunities for drug delivery (Machado et al. 2020). M. Chen et al. have designed a composite SHPB hydrogel derived from chitosan and hydroxyapatite/gelatin for drug delivery and bone repair with a more than 65% reduction in swelling rate (Chen et al. 2022).

Dental adhesive polymeric composite is susceptible to bacterial infection due to breakage; however, the formulation of SHBP can be reinforced by antibacterial agents to avoid this issue (Pathan and Shende 2021).

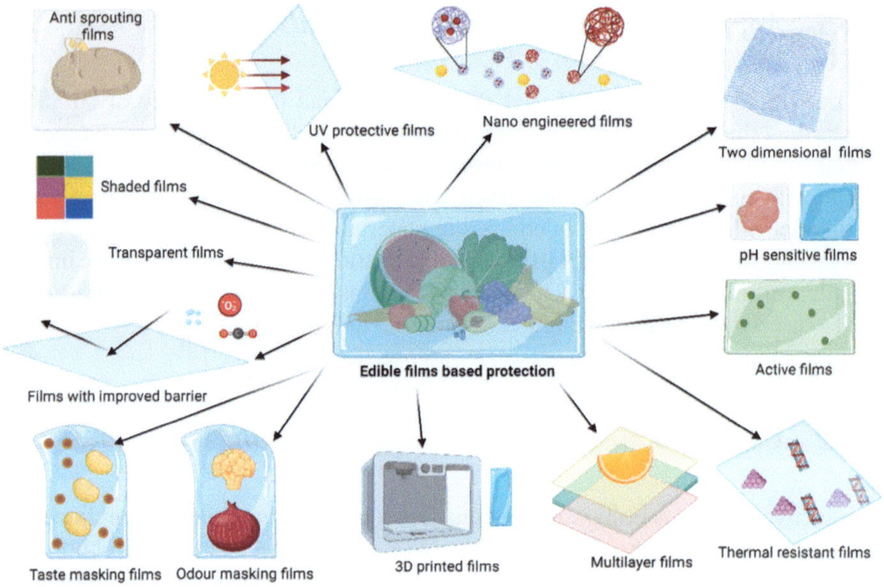

Fig. 10.10 Different goals in the SHBP used for food packing (Shah et al. 2024)

10.5.4 Self-Healing in Packaging and Consumer Products

Using plastics in packaging has severe environmental issues owing to their non-biodegradability. To avoid this problem, investigation in the field of biocompatible and biodegradable biopolymer films focuses on increasing the protection of food products against oxidative or microbial decay, heat, light, and the bioavailability of vitamins and odor (Fig. 10.10) (Shah et al. 2024). Generally, SHBP materials that are reported in food packaging applications are produced from the hybridization of chitosan with sodium alginate or polyacrylamide and gelatin extract with green tea (Gopalakrishnan and Mishra 2024).

10.6 Environmental and Economic Benefits

10.6.1 Biodegradability

The massive occurrence of packaging in daily life is inevitable. The non-biodegradability of many petrochemical industry–produced plastics can cause serious environmental problems (Ibrahim et al. 2022). Recently, biodegradable packaging of pharmaceuticals, food, and cosmetics has been reported (Getahun et al. 2024;

Dini 2024; Saha et al. 2020). Moreover, there is a growing interest in research in the field of biomedical applications of biodegradable biopolymers (Getahun et al. 2024).

10.6.2 Cost-Effectiveness in Sustainable Development

The use of abundant lignocellulosic agricultural byproducts in sustainable engineering applications can not only reduce expenses but also establish associated businesses, which can accelerate the development of rural places (Machado et al. 2020).

10.6.3 Reducing Waste and Energy Consumption

The biodegradable nature of SHBP moderates the burden of plastic waste (Choi et al. 2023). Recently, the replacement of fossil fuel-derived hydrogel with SHBP hydrogels as renewable electrolyte materials has been a hot topic (Xu et al. 2022). Cellulose-based foams and aerogel composites with high porosity and stability have an excellent potential for energy storage and environmental concerns (Liu et al. 2021). The production of supercapacitors and batteries from biodegradable material, such as SHBP, allows safe application and low or zero disposal (Sundharamurthi et al. 2021).

10.7 Challenges and Future Directions

10.7.1 Limitations in Current Biopolymer Self-Healing Systems

The self-healing superhydrophobic surfaces are non-autonomous, which means they require high temperature or light activation for restoration. Moreover, the research on this topic is limited to the microscale, and repairing at the mm or cm scale is challenging (Ma et al. 2024). The compatibility of SHBP with extrusion and other film-making processes, which result in a uniform film on a commercial scale, is still challenging (Gopalakrishnan and Mishra 2024). There is a lack of standards to ensure the prevention of allergic or harmful interactions of SHBP packaging with human beings.

10.7.2 Scaling and Commercialization Barriers

Depending on the types and applications of DHBP, their industrialization faces some challenges. For cellulose-based SHBP, the barrier to commercialization is the drying process, which requires high-temperature and high-pressure equipment, and the current procedure is not efficient (Liu et al. 2021). For bacterial-based biopolymers, the upscaling process poses a challenge in terms of homogeneity. Thus, the quality of the product may vary with light, temperature, and ecological sources. On the other hand, by the utilization of artificial light, a further cost burden is imposed (Fradinho et al. 2021). For microbial biopolymers, the cost of production is a significant challenge (Sharma et al. 2024). For SHBP, a certificate should ensure their safety (Moradali and Rehm 2020).

10.7.3 Future Trends

To optimize the manufacturing process and application, the utilization of artificial intelligence (AI) is highly recommended, which can contribute to more competent production of SHBP (Song et al. 2020). Starch and cellulose-based SHBP have an ascending trend in the construction market, owing to the development of plastering machines in areas that rely on conventional construction methods (Philip and Reshmy 2024). The emergence of 3-D printing technology can extend the biomedical applications of SHBP (Diba et al. 2018). Future studies can focus on the interaction of SHBP with customer satisfaction surveillance studies.

10.8 Conclusion

Self-healing biopolymers are an innovative, sustainable engineering approach regarding their capacity to replicate the natural, complex process of healing, restoring structural integrity, and prolonging the material's lifespan. Biopolymers like cellulose, chitosan, and hyaluronic acid, incorporated with synthetic polymers like polyurethane and polyethylene glycol, provided the possibility for materials design capable of autonomous repair following mechanical damage, whereby costs associated with maintenance and material waste could be reduced. These are done for biomedical engineering, packaging, construction, and environmental protection by solutions that balance functionality and sustainability. Self-healing in such SHBP systems is successful through various intrinsic and extrinsic mechanisms, from reversible covalent bonding and supramolecular interactions to microencapsulation and vascular networks. For the period under review, their mechanical strength, scalability, and long-term stability remain challenges that need attention. For instance, different self-healing systems require external stimuli, like heat or light, which reduces their practicality for daily life. Besides, consistency in healing efficiencies with the preservation of the mechanical properties of materials is another critical issue for wider commercialization.

Functionalization of biopolymers through chemical modifications, integration of nanomaterials, and biomimetic approaches has considerably expanded the scope and efficiency of self-healing materials. Innovation in dynamic covalent chemistry, nanotechnology, and bio-inspired design has allowed the possibility for SHBP to be tailored for specific requirements such as stimuli-responsiveness and improved mechanical performance. From another perspective, self-healing biocomposites could be a harmless alternative to existing materials; thus, they have huge potential in applications where biodegradability and environmental sustainability are of concern. In the future, overcoming present limitations in scaling barriers will be required before wider utilization of SHBP can be envisioned. Future research should be channeled toward developing self-healing systems performing under environmental conditions, improving the compatibility of the material with industrial manufacturing techniques, and employing state-of-the-art technologies to optimize material design. As development in this area continues, SHBP could provide enabling technology that makes a qualitative leap toward durable, cost-effective, and sustainable solutions for diverse material applications.

References

Abka-khajouei R, Tounsi L, Shahabi N, Patel AK, Abdelkafi S, Michaud P (2022) Structures, properties and applications of alginates, mar drugs 20. https://doi.org/10.3390/md20060364

Adamiak K, Sionkowska A (2023) State of innovation in alginate-based materials, mar drugs 21. https://doi.org/10.3390/md21060353

Aderibigbe BA, Buyana B (2018) Alginate in wound dressings, Pharmaceutics 10. https://doi.org/10.3390/pharmaceutics10020042

Aeridou E, Díaz Díaz D, Alemán C, Pérez-Madrigal MM (2020) Advanced functional hydrogel biomaterials based on dynamic B–O bonds and polysaccharide building blocks, biomacromolecules 21. 3984–3996. https://doi.org/10.1021/acs.biomac.0c01139.

Aguirresarobe RH, Nevejans S, Reck B, Irusta L, Sardon H, Asua JM, Ballard N (2021) Healable and self-healing polyurethanes using dynamic chemistry, Prog Polym Sci 114. https://doi.org/10.1016/j.progpolymsci.2021.101362

Aigaje E, Riofrio A, Baykara H (2023) Processing, properties, modifications, and environmental impact of nanocellulose/biopolymer composites: a review. Polymers (Basel) 15:1219. https://doi.org/10.3390/polym15051219

Alcântara ACS, Darder M (2018) Building up functional bionanocomposites from the assembly of clays and biopolymers. Chem Rec 18:696–712. https://doi.org/10.1002/tcr.201700076

Amamoto Y, Kamada J, Otsuka H, Takahara A, Matyjaszewski K (2011) Repeatable photoinduced self-healing of covalently cross-linked polymers through reshuffling of trithiocarbonate units. Angew Chem 123:1698–1701. https://doi.org/10.1002/ange.201003888

An ZW, Xue R, Ye K, Zhao H, Liu Y, Li P, Chen ZM, Huang CX, Hu GH (2023) Recent advances in self-healing polyurethane based on dynamic covalent bonds combined with other self-healing methods. Nanoscale 15:6505–6520. https://doi.org/10.1039/d2nr07110j

Anupama Devi VK, Shyam R, Palaniappan A, Jaiswal AK, Oh TH, Nathanael AJ (2021) Self-healing hydrogels: Preparation, mechanism and advancement in biomedical applications, Polymers (Basel) 13. https://doi.org/10.3390/polym13213782

Anwer AH, Ahtesham A, Shoeb M, Mashkoor F, Ansari MZ, Zhu S, Jeong C (2023) State-of-the-art advances in nanocomposite and bio-nanocomposite polymeric materials: A comprehensive review. Adv Colloid Interface Sci 318:102955. https://doi.org/10.1016/j.cis.2023.102955

Aranaz I, Alcántara AR, Civera MC, Arias C, Elorza B, Caballero AH, Acosta N (2021) Chitosan: An overview of its properties and applications, Polymers (Basel) 13. https://doi.org/10.3390/polym13193256

Arroyave S, Asensio E, Perilla JE, Narváez-Rincón PC, Cadavid A, Guerrero A (2023) Evaluation and characterization of autonomous self-healing cementitious materials with low carbon footprint using hybrid organic/inorganic microcapsules. Mater Today Proc. https://doi.org/10.1016/j.matpr.2023.03.673

Bai T, Liu S, Sun F, Sinclair A, Zhang L, Shao Q, Jiang S (2014) Zwitterionic fusion in hydrogels and spontaneous and time-independent self-healing under physiological conditions. Biomaterials 35:3926–3933. https://doi.org/10.1016/j.biomaterials.2014.01.077

Balavigneswaran CK, Muthuvijayan V (2021) Nanohybrid-reinforced gelatin-ureidopyrimidinone-based self-healing injectable hydrogels for tissue engineering applications. ACS Appl Bio Mater 4:5362–5377. https://doi.org/10.1021/acsabm.1c00458

Benetti C, Colombo P, Wong TW (2021) In vitro and in vivo particle coating for oral targeting and drug delivery, handbook of modern coating technologies: fabrication methods and functional properties 231–258. https://doi.org/10.1016/B978-0-444-63240-1.00009-7

Bostancı NS, Büyüksungur S, Hasirci N, Tezcaner A (2022) Potential of pectin for biomedical applications: a comprehensive review. J Biomater Sci Polym Ed 33:1866–1900. https://doi.org/10.1080/09205063.2022.2088525

Brown EN, Kessler MR, Sottos NR, White SR (2003) In situ poly(urea-formaldehyde) microencapsulation of dicyclopentadiene. J Microencapsul 20:719–730. https://doi.org/10.3109/02652040309178083

Burattini S, Colquhoun HM, Fox JD, Friedmann D, Greenland BW, Harris PJF, Hayes W, MacKay ME, Rowan SJ (2009) A self-repairing, supramolecular polymer system: Healability as a consequence of donor-acceptor π-π Stacking interactions, Chemical Communications. 6717–6719. https://doi.org/10.1039/b910648k

de las C, Alarcón H, Pennadam S, Alexander C (2005) Stimuli responsive polymers for biomedical applications, Chem Soc Rev 34 276–285. https://doi.org/10.1039/B406727D

Canadell J, Goossens H, Klumperman B (2011) Self-healing materials based on disulfide links. Macromolecules 44:2536–2541. https://doi.org/10.1021/ma2001492

Celus M, Kyomugasho C, Van Loey AM, Grauwet T, Hendrickx ME (2018) Influence of pectin structural properties on interactions with divalent cations and its associated functionalities. Compr Rev Food Sci Food Saf 17:1576–1594. https://doi.org/10.1111/1541-4337.12394

Chatterjee R, Maity M, Hasnain MS, Nayak AK (2021) Chitosan: source, chemistry, and properties, Chitosan in Drug Delivery 1–22. https://doi.org/10.1016/B978-0-12-819336-5.00001-7

Chen X, Dam MA, Ono K, Mal A, Shen H, Nutt SR, Sheran K, Wudl F (1979) A thermally remendable cross-linked polymeric material. Science 295(2002):1698–1702. https://doi.org/10.1126/science.1065879

Chen M, Tian J, Liu Y, Cao H, Li R, Wang J, Wu J, Zhang Q (2019b) Dynamic covalent constructed self-healing hydrogel for sequential delivery of antibacterial agent and growth factor in wound healing. Chem Eng J 373:413–424. https://doi.org/10.1016/j.cej.2019.05.043

Chen D, Wang D, Yang Y, Huang Q, Zhu S, Zheng Z (2017) Self-healing materials for next-generation energy harvesting and storage devices, Adv Energy Mater 7. https://doi.org/10.1002/aenm.201700890

Chen X, Lin Z, Feng Y, Tan H, Xu X, Luo J, Li J (2019) Zwitterionic PMCP-Modified polycaprolactone surface for tissue engineering: antifouling, cell adhesion promotion, and osteogenic differentiation properties, small 15. https://doi.org/10.1002/smll.201903784

Chen M, Tan H, Xu W, Wang Z, Zhang J, Li S, Zhou T, li J, Niu X (2022) A self-healing, magnetic and injectable biopolymer hydrogel generated by dual cross-linking for drug delivery and bone repair, Acta Biomater 153 159–177. https://doi.org/10.1016/j.actbio.2022.09.036

Cho DH, Cho KG, An S, Kim MS, Oh HW, Yeo J, Yoo WC, Hong K, Kim M, Lee KH (2022) Self-healable, stretchable, and nonvolatile solid polymer electrolytes for sustainable energy storage and sensing applications. Energy Storage Mater 45:323–331. https://doi.org/10.1016/j.ensm.2021.11.047

Choi K, Noh A, Kim J, Hong PH, Ko MJ, Hong SW (2023) Properties and applications of self-healing polymeric materials: a review. Polymers (Basel) 15:4408. https://doi.org/10.3390/polym15224408

Chung CM, Roh YS, Cho SY, Kim JG (2004) Crack healing in polymeric materials via photochemical [2+2] cycloaddition. Chem Mater 16:3982–3984. https://doi.org/10.1021/cm049394+

Cordier P, Tournilhac F, Soulié-Ziakovic C, Leibler L (2008) Self-healing and thermoreversible rubber from supramolecular assembly. Nature 451:977–980. https://doi.org/10.1038/nature06669

Cremaldi JC, Bhushan B (2018) Bioinspired self-healing materials: lessons from nature. Beilstein J Nanotechnol 9:907–935. https://doi.org/10.3762/bjnano.9.85

Dallaev R (2024) Advances in materials with self-healing properties: a brief review, materials 17. https://doi.org/10.3390/ma17102464

Deng G, Tang C, Li F, Jiang H, Chen Y (2010) Covalent cross-linked polymer gels with reversible sol-gel transition and self-healing properties. Macromol 43:1191–1194. https://doi.org/10.1021/ma9022197

Diba M, Spaans S, Ning K, Ippel BD, Yang F, Loomans B, Dankers PYW, Leeuwenburgh SCG (2018) Self-Healing biomaterials: from molecular concepts to clinical applications. Adv Mater Interfaces 5. https://doi.org/10.1002/admi.201800118

Ding F, Li H, Du Y, Shi X (2018) Recent advances in chitosan-based self-healing materials. Res Chem Intermed 44:4827–4840. https://doi.org/10.1007/s11164-018-3339-7

Ding C, Tian M, Feng R, Dang Y, Zhang M (2020) Novel Self-Healing hydrogel with injectable, pH-Responsive. Strain-Sensitive, Promoting Wound-Healing, and Hemostatic Properties Based on Collagen and Chitosan, ACS Biomater Sci Eng 6:3855–3867. https://doi.org/10.1021/acsbiomaterials.0c00588

Ding X, Fan L, Wang L, Zhou M, Wang Y, Zhao Y (2023) Designing self-healing hydrogels for biomedical applications. Mater Horiz 10:3929–3947. https://doi.org/10.1039/d3mh00891f

Dini I (2024) “Edible Beauty”: The Evolution of Environmentally Friendly Cosmetics and Packaging. Antioxidants 13:742. https://doi.org/10.3390/antiox13060742

Dovedytis M, Liu ZJ, Bartlett S (2020) Hyaluronic acid and its biomedical applications: A review. Eng Regen 1:102–113. https://doi.org/10.1016/j.engreg.2020.10.001

Draget KI, Taylor C (2011) Chemical, physical and biological properties of alginates and their biomedical implications. Food Hydrocoll 25:251–256. https://doi.org/10.1016/j.foodhyd.2009.10.007

Du Y, Li D, Liu L, Gai G (2018) Recent achievements of self-healing graphene/polymer composites. Polymers (Basel) 10:114. https://doi.org/10.3390/polym10020114

Dzulkharnien NSF, Rohani R (2022) A review on current designation of metallic nanocomposite hydrogel in biomedical applications. Nanomaterials 12:1629. https://doi.org/10.3390/nano12101629

Elangwe CN, Samuilova EO, Uspenskaya MV, Olekhnovich RO (2025) Review of self-healing polysaccharide-based hydrogels in tissue regeneration: characterization methods and applications. Int J Biol Macromol 311:144149. https://doi.org/10.1016/j.ijbiomac.2025.144149

Engels HW, Pirkl HG, Albers R, Albach RW, Krause J, Hoffmann A, Casselmann H, Dormish J (2013) Polyurethanes: Versatile materials and sustainable problem solvers for today’s challenges. Angew Chem-Int Ed 52:9422–9441. https://doi.org/10.1002/anie.201302766

Enke M, Bose RK, Zechel S, Vitz J, Deubler R, Garcia SJ, Van Der Zwaag S, Schacher FH, Hager MD, Schubert US (2018) A translation of the structure of mussel byssal threads into synthetic materials by the utilization of histidine-rich block copolymers. Polym Chem 9:3543–3551. https://doi.org/10.1039/c8py00663f

Everitt DT, Luterbacher R, Coope TS, Trask RS, Wass DF, Bond IP (2015) Optimisation of epoxy blends for use in extrinsic self-healing fibre-reinforced composites. Polymer (Guildf) 69:283–292. https://doi.org/10.1016/j.polymer.2015.02.047

Farazin A, Mohammadimehr M, Naeimi H (2023) Flexible self-healing nanocomposite based gelatin/tannic acid/acrylic acid reinforced with zinc oxide nanoparticles and hollow silver nanoparticles based on porous silica for rapid wound healing. Int J Biol Macromol 241:124572. https://doi.org/10.1016/j.ijbiomac.2023.124572

Feldman D (2020) Poly(Vinyl alcohol) recent contributions to engineering and medicine. J Compos Sci 4. https://doi.org/10.3390/jcs4040175

Fournier D, Hoogenboom R, Schubert US (2007) Clicking polymers: a straightforward approach to novel macromolecular architectures. Chem Soc Rev 36:1369. https://doi.org/10.1039/b700809k

Fox J, Wie JJ, Greenland BW, Burattini S, Hayes W, Colquhoun HM, MacKay ME, Rowan SJ (2012) High-strength, healable, supramolecular polymer nanocomposites. J Am Chem Soc 134:5362–5368. https://doi.org/10.1021/ja300050x

Fradinho J, Allegue LD, Ventura M, Melero JA, Reis MAM, Puyol D (2021) Up-scale challenges on biopolymer production from waste streams by Purple Phototrophic Bacteria mixed cultures: A critical review. Bioresour Technol 327:124820. https://doi.org/10.1016/j.biortech.2021.124820

Gaaz TS, Sulong AB, Akhtar MN, Kadhum AAH, Mohamad AB, Al-Amiery AA, McPhee DJ (2015) Properties and applications of polyvinyl alcohol, halloysite nanotubes and their nanocomposites. Molecules 20:22833–22847. https://doi.org/10.3390/molecules201219884

Garcia SJ, Fischer HR (2014) Self-healing polymer systems: Properties, synthesis and applications. Smart Polym Their Appl. 271–298. https://doi.org/10.1533/9780857097026.1.271

George J, Sabapathi SN (2015) Cellulose nanocrystals: Synthesis, functional properties, and applications. Nanotechnol Sci Appl 8:45–54. https://doi.org/10.2147/NSA.S64386

George A, Sanjay MR, Srisuk R, Parameswaranpillai J, Siengchin S (2020) A comprehensive review on chemical properties and applications of biopolymers and their composites. Int J Biol Macromol 154:329–338. https://doi.org/10.1016/j.ijbiomac.2020.03.120

George JS, Vijayan P, Hoang AT, Kalarikkal N, Nguyen-Tri P, Thomas S (2022) Recent advances in bio-inspired multifunctional coatings for corrosion protection, Prog Org Coat 168 106858. https://doi.org/10.1016/j.porgcoat.2022.106858

Getahun MJ, Kassie BB, Alemu TS (2024) Recent advances in biopolymer synthesis, properties, & commercial applications: a review. Process Biochem 145:261–287. https://doi.org/10.1016/j.procbio.2024.06.034

Gherman SP, Biliuță G, Bele A, Ipate AM, Baron RI, Ochiuz L, pac AF Ș, Zavastin DE (2023) Biomaterials based on chitosan and polyvinyl alcohol as a drug delivery system with wound-healing effects, Gels 9. https://doi.org/10.3390/gels9020122

Gnedenkov AS, Sinebryukhov SL, Filonina VS, Ustinov AY, Sukhoverkhov SV, Gnedenkov SV (2023) New polycaprolactone-containing self-healing coating design for enhance corrosion resistance of the magnesium and its alloys. Polymers (Basel) 15. https://doi.org/10.3390/polym15010202.

Gopalakrishnan K, Mishra P (2024) Self-Healing polymer a dynamic solution in food industry: a comprehensive review. Food Biophys 19:1–17. https://doi.org/10.1007/s11483-023-09780-z

Gumrah Dumanli A (2017) Nanocellulose and its composites for biomedical applications, Curr Med Chem 24 512–528. https://doi.org/10.2174/0929867323666161014124008

Gupta RC, Lall R, Srivastava A, Sinha A (2019) Hyaluronic acid: Molecular mechanisms and therapeutic trajectory, Front Vet Sci 6. https://doi.org/10.3389/fvets.2019.00192

Hager MD, Greil P, Leyens C, Van Der Zwaag S, Schubert US (2010) Self-healing materials. Adv Mater 22:5424–5430. https://doi.org/10.1002/adma.201003036

Hasanin MS (2022) Cellulose-Based biomaterials: chemistry and biomedical applications, starch/staerke 74. https://doi.org/10.1002/star.202200060

He L, Fullenkamp DE, Rivera JG, Messersmith PB (2011) PH responsive self-healing hydrogels formed by boronate-catechol complexation. Chem Commun 47:7497–7499. https://doi.org/10.1039/c1cc11928a

Hong SH, Kim S, Park JP, Shin M, Kim K, Ryu JH, Lee H (2018) Dynamic bonds between boronic acid and alginate: hydrogels with stretchable, self-healing, stimuli-responsive, remoldable, and adhesive properties. Biomacromol 19:2053–2061. https://doi.org/10.1021/acs.biomac.8b00144

Huang H, Ye G, Shui Z (2014) Feasibility of self-healing in cementitious materials—By using capsules or a vascular system? Constr Build Mater 63:108–118. https://doi.org/10.1016/j.conbuildmat.2014.04.028

Ibrahim ID, Hamam Y, Sadiku ER, Ndambuki JM, Kupolati WK, Jamiru T, Eze AA, Snyman J (2022) Need for Sustainable Packaging: An Overview. Polymers (Basel) 14:4430. https://doi.org/10.3390/polym14204430

Irzhak VI, Uflyand IE, Dzhardimalieva GI (2022) Self-Healing of polymers and polymer composites, polymers (Basel) 14. https://doi.org/10.3390/polym14245404

Islam S, Bhat G (2021) Progress and challenges in self-healing composite materials. Mater Adv 2:1896–1926. https://doi.org/10.1039/d0ma00873g

Jadoun S (2024) Synthesis, mechanism, and applications of self-healing materials. Biomed Mater Devices 2:225–240. https://doi.org/10.1007/s44174-023-00107-7

Jain A, Goswami S, Banerjee S (2025) Advances in Self-healing polymers: mechanisms, applications, and future perspectives. Macromol Chem Phys. https://doi.org/10.1002/macp.202500159

Jalalvandi E, Shavandi A (2019) Shear thinning/self-healing hydrogel based on natural polymers with secondary photocrosslinking for biomedical applications. J Mech Behav Biomed Mater 90:191–201. https://doi.org/10.1016/j.jmbbm.2018.10.009

Jamróz E, Kulawik P, Kopel P (2019) The effect of nanofillers on the functional properties of biopolymer-based films: a review. Polymers (Basel) 11:675. https://doi.org/10.3390/polym11040675

Jayabalakrishnan D, Naga Muruga DB, Bhaskar K, Pavan P, Balaji K, Rajakumar PS, Priya C, Deepa RAB, Sendilvelan S, Prabhahar M (2020) Self-Healing materials–A review, Mater Today Proc 45. 7195–7199. https://doi.org/10.1016/j.matpr.2021.02.415

Jayarathna S, Andersson M, Andersson R (2022) Recent advances in starch-based blends and composites for bioplastics applications, Polymers (Basel) 14. https://doi.org/10.3390/polym14214557

Jeon H, Lee H, Kim G (2014) A surface-modified poly(ε-caprolactone) scaffold comprising variable nanosized surface-roughness using a plasma treatment. Tissue Eng Part C Methods 20:951–963. https://doi.org/10.1089/ten.tec.2013.0701

Jiang G, Liu C, Liu X, Chen Q, Zhang G, Yang M, Liu F (2010) Network structure and compositional effects on tensile mechanical properties of hydrophobic association hydrogels with high mechanical strength. Polymer (Guildf) 51:1507–1515. https://doi.org/10.1016/j.polymer.2010.01.061

Jiang C, Zhang L, Yang Q, Huang S, Shi H, Long Q, Qian B, Liu Z, Guan Q, Liu M, Yang R, Zhao Q, You Z, Ye X (2021) Self-healing polyurethane-elastomer with mechanical tunability for multiple biomedical applications in vivo, Nat Commun 12. https://doi.org/10.1038/s41467-021-24680-x

Jiménez-Suárez A, Del Rosario G, Sánchez-Romate XX, Prolongo SG (2020) Influence of morphology on the healing mechanism of PCL/epoxy blends, Materials 13. https://doi.org/10.3390/MA13081941

Joy J, George E, Anas S, Thomas S (2020) Applications of self-healing polymeric systems, Self-Healing Polymer-Based systems. 495–513. https://doi.org/10.1016/B978-0-12-818450-9.00016-7

Jungprasertchai N, Chuysinuan P, Ekabutr P, Niamlang P, Supaphol P (2022) Freeze-Dried carboxymethyl chitosan/starch foam for use as a haemostatic wound dressing. J Polym Environ 30:1106–1117. https://doi.org/10.1007/s10924-021-02260-w

Kahar NNFNMN, Osman AF, Alosime E, Arsat N, Azman NAM, Syamsir A, Itam Z, Hamid ZAA (2021) The versatility of polymeric materials as self-healing agents for various types of applications: A review, Polymers (Basel) 13. https://doi.org/10.3390/polym13081194

Khadka A, Pradhan S, Samuel E, Joshi B, Gao H, Aldalbahi A, Periyasami G, Lee H-S, Yoon SS (2024) Rapidly self-healing, highly conductive, stretchable, body-attachable hydrogel sensor for soft electronics. Compos Commun 52:102158. https://doi.org/10.1016/j.coco.2024.102158

Khalid MY, Arif ZU, Al Rashid A, Bukhari SMZS, Hossain M, Koç M (2024) Shape-memory and self-healing properties of sustainable cellulosic nanofibers-based hybrid materials for novel applications, Giant 19. https://doi.org/10.1016/j.giant.2024.100299

Kikani T, Dave S, Thakore S (2023) Functionalization of hyaluronic acid for development of self-healing hydrogels for biomedical applications: A review, Int J Biol Macromol 242. https://doi.org/10.1016/j.ijbiomac.2023.124950

Kim H, Jeong H, Han S, Beack S, Hwang BW, Shin M, Oh SS, Hahn SK (2017) Hyaluronate and its derivatives for customized biomedical applications. Biomaterials 123:155–171. https://doi.org/10.1016/j.biomaterials.2017.01.029

King NS, Luster-Teasley S, Clark CJ (2021) Preliminary analyses of controlled release of potassium permanganate encapsulated in polycaprolactone. J Water Resour Prot 13:32–43. https://doi.org/10.4236/jwarp.2021.131003

Kizilel S, Pérez-Luna VH, Teymour F (2009) Modeling of PEG hydrogel membranes for biomedical applications. Macromol React Eng 3:271–287. https://doi.org/10.1002/mren.200900005

Kluczka J (2024) Chitosan: structural and chemical modification, properties, and application, Int J Mol Sci 25. https://doi.org/10.3390/ijms25010554

Kolya H, Kang C-W (2023) Next-Generation water treatment: exploring the potential of biopolymer-based nanocomposites in adsorption and membrane filtration. Polymers (Basel) 15:3421. https://doi.org/10.3390/polym15163421

Kosarli M, Foteinidis G, Tsirka K, Bekas DG, Paipetis AS (2021) Concurrent recovery of mechanical and electrical properties in nanomodified capsule-based self-healing epoxies, Polymer (Guildf) 227. https://doi.org/10.1016/j.polymer.2021.123843

Kumar A, Mishra RK, Verma K, Aldosari SM, Maity CK, Verma S, Patel R, Thakur VK (2023) A comprehensive review of various biopolymer composites and their applications: From biocompatibility to self-healing. Mater Today Sustain 23. https://doi.org/10.1016/j.mtsust.2023.100431.

Kundu R, Mahada P, Chhirang B, Das B (2022) Cellulose hydrogels: green and sustainable soft biomaterials, current research in green and sustainable chemistry 5. https://doi.org/10.1016/j.crgsc.2021.100252

Lai SM, Liu JL, Huang YH (2020) Preparation of Self-healing Natural Rubber/Polycaprolactone (NR/PCL) Blends. J Macromol Sci Part B Phys 59:587–607. https://doi.org/10.1080/00222348.2020.1757218

Lavania S, Mehta J, Bhardwaj P, Tripathi A, Gupta N, Gupta P (2023) Biocomposites: prospects and manifold applications for human and environmental sustainability. ECS J Solid State Sci Technol 12:037002. https://doi.org/10.1149/2162-8777/acbe17

Lee W-J, Cha S-H (2020) Improvement of mechanical and self-healing properties for polymethacrylate derivatives containing maleimide modified graphene oxide. Polymers (Basel) 12:603. https://doi.org/10.3390/polym12030603

Lee MW, An S, Lee C, Liou M, Yarin AL, Yoon SS (2014) Self-healing transparent core-shell nanofiber coatings for anti-corrosive protection. J Mater Chem A Mater 2:7045–7053. https://doi.org/10.1039/c4ta00623b

Lee J, Chang K, Kim S, Gite V, Chung H, Sohn D (2016) Phase controllable hyaluronic acid hydrogel with Iron(III) Ion-Catechol induced dual cross-linking by utilizing the gap of gelation kinetics. Macromolecules 49:7450–7459. https://doi.org/10.1021/acs.macromol.6b01198

Li J, Suo J, Deng R (2010) Structure, mechanical, and swelling behaviors of poly(vinyl alcohol)/SiO2 hybrid membranes. J Reinf Plast Compos 29:618–629. https://doi.org/10.1177/0731684408100696

Li Y, Yu T, Yang T, Zheng L, Liao K (2012) Bio-Inspired nacre-like composite films based on graphene with superior mechanical, electrical, and biocompatible properties. Adv Mater 24:3426–3431. https://doi.org/10.1002/adma.201200452

Cheng-Hui Li, Jing-Lin Zuo (2019) Self-Healing polymers based on coordination bonds, Adv Mater 32

Li S, Pei M, Wan T, Yang H, Gu S, Tao Y, Liu X, Zhou Y, Xu W, Xiao P (2020) Self-healing hyaluronic acid hydrogels based on dynamic Schiff base linkages as biomaterials, Carbohydr Polym 250. https://doi.org/10.1016/j.carbpol.2020.116922

Liang H-Y, Lee W-K, Hsu J-T, Shih J-Y, Ma T-L, Vo TTT, Lee C-W, Cheng M-T, Lee I-T (2024) Polycaprolactone in bone tissue engineering: a comprehensive review of innovations in scaffold fabrication and surface modifications. J Funct Biomater 15:243. https://doi.org/10.3390/jfb15090243

Liao H, Guo X, Wan P, Yu G (2019) Conductive MXene Nanocomposite Organohydrogel for Flexible, Healable, Low-Temperature Tolerant Strain Sensors, Adv Funct Mater 29 (2019). https://doi.org/10.1002/adfm.201904507

Liu X, Ma L, Mao Z, Gao C (2011) Chitosan-based biomaterials for tissue repair and regeneration. Adv Polym Sci 244:81–128. https://doi.org/10.1007/12_2011_118

Liu J, Song G, He C, Wang H (2013) Self-healing in tough graphene oxide composite hydrogels. Macromol Rapid Commun 34:1002–1007. https://doi.org/10.1002/marc.201300242

Liu H, Du H, Zheng T, Liu K, Ji X, Xu T, Zhang X, Si C (2021) Cellulose based composite foams and aerogels for advanced energy storage devices. Chem Eng J 426:130817. https://doi.org/10.1016/j.cej.2021.130817

Liu X, Gallavardin T, Burel F, Vuluga D (2023) Influence of quadruple hydrogen bonding on polyvinyl butyral resin properties. Polym Degrad Stab 208:110243. https://doi.org/10.1016/j.polymdegradstab.2022.110243

Liu Y, Teng J, Huang R, Zhao W, Yang D, Ma Y, Wei H, Chen H, Zhang J, Chen J (2024) Injectable plant-derived polysaccharide hydrogels with intrinsic antioxidant bioactivity accelerate wound healing by promoting epithelialization and angiogenesis. Int J Biol Macromol 266:131170. https://doi.org/10.1016/j.ijbiomac.2024.131170

Liu B, Zhang J, Guo H (2022) Research progress of polyvinyl alcohol water-resistant film materials, membranes (Basel) 12. https://doi.org/10.3390/membranes12030347

Lu X, Jiao H, Shi Y, Li Y, Zhang H, Fu Y, Guo J, Wang Q, Liu X, Zhou M, Ullah MW, Sun J, Liu J (2023) Fabrication of bio-inspired anisotropic structures from biopolymers for biomedical applications: A review. Carbohydr Polym 308:120669. https://doi.org/10.1016/j.carbpol.2023.120669

Luo Y, Tan J, Zhou Y, Guo Y, Liao X, He L, Li D, Li X, Liu Y (2023) From crosslinking strategies to biomedical applications of hyaluronic acid-based hydrogels: A review, Int J Biol Macromol 231. https://doi.org/10.1016/j.ijbiomac.2023.123308

Luo X, Mather PT (2013) Shape memory assisted self-healing coating. ACS Macro Lett 2:152–156. https://doi.org/10.1021/mz400017x

Ma Q, Yang Q, Zhang J, Ren F, Xia C, Chen F (2024) Anti-corrosion properties of bio-inspired surfaces: a systematic review of recent research developments. Mater Adv 5:2689–2718. https://doi.org/10.1039/D3MA01058A

Machado G, Santos F, Lourega R, Mattia J, Faria D, Eichler P, Auler A (2020) Biopolymers from lignocellulosic biomass, in: lignocellulosic biorefining technologies, Wiley, pp 125–158. https://doi.org/10.1002/9781119568858.ch7

Makvandi P, Iftekhar S, Pizzetti F, Zarepour A, Zare EN, Ashrafizadeh M, Agarwal T, Padil VVT, Mohammadinejad R, Sillanpaa M, Maiti TK, Perale G, Zarrabi A, Rossi F (2021) Functionalization of polymers and nanomaterials for water treatment, food packaging, textile and biomedical applications: a review. Environ Chem Lett 19:583–611. https://doi.org/10.1007/s10311-020-01089-4

Malik US, Niazi MBK, Jahan Z, Zafar MI, Vo DVN, Sher F (2022) Nano-structured dynamic Schiff base cues as robust self-healing polymers for biomedical and tissue engineering applications: a review. Environ Chem Lett 20:495–517. https://doi.org/10.1007/s10311-021-01337-1

Mao C, Xiang Y, Liu X, Cui Z, Yang X, Yeung KWK, Pan H, Wang X, Chu PK, Wu S (2017) Photo-Inspired antibacterial activity and wound healing acceleration by hydrogel embedded with Ag/Ag@AgCl/ZnO nanostructures. ACS Nano 11:9010–9021. https://doi.org/10.1021/acsnano.7b03513

Mapossa AB, da Silva Júnior AH, de Oliveira CRS, Mhike W (2023) Thermal, morphological and mechanical properties of multifunctional composites based on biodegradable polymers/bentonite clay: a review, Polymers (Basel) 15. 3443. https://doi.org/10.3390/polym15163443

Mateos-Timoneda MA (2009) Polymers for bone repair. Bone Repair Biomater 231–251. https://doi.org/10.1533/9781845696610.2.231

Mehvari F, Ramezanzade V, An J, Kim J, Dinari M, Seung Kim J (2024) Biopolymer-based hydrogels for biomedical applications: Bioactivity and wound healing properties, Coord Chem Rev 518 216093. https://doi.org/10.1016/j.ccr.2024.216093

Mennani M, Kasbaji M, Ait Benhamou A, Boussetta A, Mekkaoui AA, Grimi N, Moubarik A (2023) Current approaches, emerging developments and functional prospects for lignin-based catalysts—a review, Green Chemistry 25 (2023) 2896–2929. https://doi.org/10.1039/D3GC00072A

Mndlovu H, du Toit LC, Kumar P, Choonara YE, Marimuthu T, Kondiah PPD, Pillay V (2020) Bioplatform fabrication approaches affecting chitosan-based interpolymer complex properties and performance as wound dressings. Molecules 25:222. https://doi.org/10.3390/molecules25010222

Mollajavadi MY, Tarigheh FF, Eslami-Farsani R (2023) <scp>Self-healing</scp> polymers containing nanomaterials for biomedical engineering applications: A review. Polym Compos 44:6869–6889. https://doi.org/10.1002/pc.27603

Montanari C, Olsén P, Berglund LA (2020) Interface tailoring by a versatile functionalization platform for nanostructured wood biocomposites. Green Chem 22:8012–8023. https://doi.org/10.1039/D0GC02768E

Moradali MF, Rehm BHA (2020) Bacterial biopolymers: from pathogenesis to advanced materials. Nat Rev Microbiol 18:195–210. https://doi.org/10.1038/s41579-019-0313-3

Moreira TD, Martins VB, da Silva Júnior AH, Sayer C, de Araújo PHH, Immich APS (2024) New insights into biomaterials for wound dressings and care: Challenges and trends, Prog Org Coat 187. https://doi.org/10.1016/j.porgcoat.2023.108118

Muir VG, Burdick JA (2021) Chemically modified biopolymers for the formation of biomedical hydrogels. Chem Rev 121:10908–10949. https://doi.org/10.1021/acs.chemrev.0c00923

Narayan R, Laberty-Robert C, Pelta J, Tarascon JM, Dominko R (2022) Self-Healing: an emerging technology for next-generation smart batteries, Adv Energy Mater 12. https://doi.org/10.1002/aenm.202102652

Neal JA, Mozhdehi D, Guan Z (2015) Enhancing mechanical performance of a covalent self-healing material by sacrificial noncovalent bonds. J Am Chem Soc 137:4846–4850. https://doi.org/10.1021/jacs.5b01601

Nimmo CM, Shoichet MS (2011) Regenerative biomaterials that "click": simple, aqueous-based protocols for hydrogel synthesis, surface immobilization, and 3D patterning. Bioconjug Chem 22:2199–2209. https://doi.org/10.1021/bc200281k

Papavasiliou G, Sokic S, Turturro M (2012) Synthetic peg hydrogels as extracellular matrix mimics for tissue engineering applications. Biotechnol-Mol Stud Nov Appl Improv Qual Hum Life. https://doi.org/10.5772/31695

Pathan N, Shende P (2021) Strategic conceptualization and potential of self-healing polymers in biomedical field. Mater Sci Eng, C 125:112099. https://doi.org/10.1016/j.msec.2021.112099

Patti A, Costa F, Perrotti M, Barbarino D, Acierno D (2021) Polyurethane impregnation for improving the mechanical and the water resistance of polypropylene-based textiles, Materials 14. https://doi.org/10.3390/ma14081951

Pettignano A, Grijalvo S, Häring M, Eritja R, Tanchoux N, Quignard F, Díaz Díaz D (2017) Boronic acid-modified alginate enables direct formation of injectable, self-healing and multistimuli-responsive hydrogels. Chem Commun 53 3350–3353. https://doi.org/10.1039/C7CC00765E

Philip E, Reshmy R (2024) Applications of biopolymers in construction and civil engineering, in: applications of biopolymers in science. Biotechnology, and Engineering, Wiley, pp 265–289. https://doi.org/10.1002/9781119783473.ch11

Qamar IPS, Trask RS (2017) Development of multi-dimensional 3D printed vascular networks for self-healing materials, In: ASME 2017 conference on smart materials, adaptive structures and intelligent systems, SMASIS 2017 1. https://doi.org/10.1115/SMASIS2017-3829

Qin H, Zhang T, Li N, Cong HP, Yu SH (2019) Anisotropic and self-healing hydrogels with multi-responsive actuating capability, Nat Commun 10. https://doi.org/10.1038/s41467-019-10243-8

Rahman MM, Shahid MA, Hossain MT, Sheikh MS, Rahman MS, Uddin N, Rahim A, Khan RA, Hossain I (2024) Sources, extractions, and applications of alginate: a review, discover applied sciences 6. https://doi.org/10.1007/s42452-024-06151-2

Razzaq MY, Shinde KS, Rupp H, Balk M, Schadewald A (2025) Vitrimeric shape-memory polymers with intrinsic flame retardancy and self-healing capabilities. Macromol Rapid Commun. https://doi.org/10.1002/marc.202401101

Rendón-Villalobos R, Ortíz-Sánchez A, Tovar-Sánchez E, Flores-Huicochea E (2016) The role of biopolymers in obtaining environmentally friendly materials. Compos Renew Sustain Mater. https://doi.org/10.5772/65265

Robyt JF (2008) Starch: structure, properties, chemistry, and enzymology, glycoscience (2008) 1437–1472. https://doi.org/10.1007/978-3-540-30429-6_35

Romero-Sabat G, Gago-Benedí E, Roa Rovira JJ, González-Gálvez D, Mateo A, Medel S, Tolentino Chivite A (2021) Development of a highly efficient extrinsic and autonomous self-healing polymeric system at low and ultra-low temperatures for high-performance applications. Compos Part A Appl Sci Manuf 145. https://doi.org/10.1016/j.compositesa.2021.106335

Rumon MMH, Akib AA, Sultana F, Moniruzzaman M, Niloy MS, Shakil MS, Roy CK (2022) Self-Healing hydrogels: development, biomedical applications, and challenges, polymers (Basel) 14. https://doi.org/10.3390/polym14214539

Saha T, Hoque ME, Mahbub T (2020) Biopolymers for sustainable packaging in food, cosmetics, and pharmaceuticals. In: advanced processing, properties, and applications of starch and other bio-based polymers, Elsevier, pp 197–214. https://doi.org/10.1016/B978-0-12-819661-8.00013-5

Santos RF, Ribeiro JCL, Franco de Carvalho JM, Magalhães WLE, Pedroti LG, Nalon GH, de Lima GES (2021) Nanofibrillated cellulose and its applications in cement-based composites: A review, Constr Build Mater 288. https://doi.org/10.1016/j.conbuildmat.2021.123122

Seddiqi H, Oliaei E, Honarkar H, Jin J, Geonzon LC, Bacabac RG, Klein-Nulend J (2021) Cellulose and its derivatives: towards biomedical applications. Cellulose 28:1893–1931. https://doi.org/10.1007/s10570-020-03674-w

Shah YA, Bhatia S, Al-Harrasi A, Khan TS (2024) Advancements in the biopolymer films for food packaging applications: a short review. Biotechnol Sustain Mater 1:2. https://doi.org/10.1186/s44316-024-00002-1

Sharma M, Tellili N, Kacem I, Rouissi T (2024) Microbial biopolymers: from production to environmental applications—a review. Appl Sci 14:5081. https://doi.org/10.3390/app14125081

Shi L, Ding P, Wang Y, Zhang Y, Ossipov D, Hilborn J (2019) Self-Healing polymeric hydrogel formed by metal–ligand coordination assembly: design, fabrication, and biomedical applications. Macromol Rapid Commun 40. https://doi.org/10.1002/marc.201800837

Shukla SK, Mishra AK, Arotiba OA, Mamba BB (2013) Chitosan-based nanomaterials: A state-of-the-art review. Int J Biol Macromol 59:46–58. https://doi.org/10.1016/j.ijbiomac.2013.04.043

Singh NK, Baranwal J, Pati S, Barse B, Khan RH, Kumar A (2023) Application of plant products in the synthesis and functionalisation of biopolymers. Int J Biol Macromol 237:124174. https://doi.org/10.1016/j.ijbiomac.2023.124174

Sisso AM, Boit MO, DeForest CA (2020) Self-healing injectable gelatin hydrogels for localized therapeutic cell delivery. J Biomed Mater Res A 108:1112–1121. https://doi.org/10.1002/jbm.a.36886

Smyth HF, Carpenter CP, Weil CS (1950) The toxicology of the polyethylene glycols. J Am Pharm Assoc Am Pharm Assoc (Baltim) 39:349–354. https://doi.org/10.1002/jps.3030390615

Solangi NH, Karri RR, Mubarak NM, Mazari SA (2024) Mechanism of polymer composite-based nanomaterial for biomedical applications. Advanced Industrial and Engineering Polymer Research 7:1–19. https://doi.org/10.1016/j.aiepr.2023.09.002

Song J, Winkeljann B, Lieleg O (2020) Biopolymer-Based coatings: promising strategies to improve the biocompatibility and functionality of materials used in biomedical engineering, Adv Mater Interfaces 7. https://doi.org/10.1002/admi.202000850

Sumerlin BS, Vogt AP (2010) Macromolecular engineering through click chemistry and other efficient transformations. Macromolecules 43:1–13. https://doi.org/10.1021/ma901447e

Sun S, Cui Y, Yuan B, Dou M, Wang G, Xu H, Wang J, Yin W, Wu D, Peng C (2023) Drug delivery systems based on polyethylene glycol hydrogels for enhanced bone regeneration. Front Bioeng Biotechnol 11. https://doi.org/10.3389/fbioe.2023.1117647

Sundharamurthi S, Sreeraj TR, Amalraj A, Gopi S (2021) Biopolymer composites in supercapacitor and energy storage devices, in: Biopolymers and Their Industrial Applications, Elsevier. pp 305–329. https://doi.org/10.1016/B978-0-12-819240-5.00013-4

Takács T, Abdelghafour MM, Lamch Ł, Szenti I, Sebők D, Janovák L, Kukovecz Á (2022) Facile modification of hydroxyl group containing macromolecules provides autonomously self-healing polymers through the formation of dynamic Schiff base linkages. Eur Polym J 168:111086. https://doi.org/10.1016/j.eurpolymj.2022.111086

Takafuji M, Kawamoto K, Hano N, Otsuki M, Ihara H (2024) Nanofibrous chiral supramolecular assembly-derived self-healing hydrogels with polyethylene glycol. Nanoscale Adv 6:3850–3856. https://doi.org/10.1039/d4na00353e

Tan YJ, Susanto GJ, Anwar Ali HP, Tee BCK (2021) Progress and roadmap for intelligent self-healing materials in autonomous robotics. Adv Mater 33. https://doi.org/10.1002/adma.202002800

Tandon B, Sabahi N, Farsi R, Kangur T, Boero G, Bertsch A, Li X, Brugger J (2025) Performance comparison of shape memory polymer structures printed by fused deposition modeling and melt electrowriting, Adv Mater Technol 10. https://doi.org/10.1002/admt.202400466

Tang X, Huang Y, Guo X, Xu C, Yuan T (2025) Light cured Bio-Based Self-Healing polyurethane coating based on imine bond. J Appl Polym Sci. https://doi.org/10.1002/app.57323

Thangavel G, Tan MWM, Lee PS (2019) Advances in self-healing supramolecular soft materials and nanocomposites, Nano Converg 6. https://doi.org/10.1186/s40580-019-0199-9

Tian H, He J (2016) Cellulose as a scaffold for self-assembly: from basic research to real applications. Langmuir 32:12269–12282. https://doi.org/10.1021/acs.langmuir.6b02033

Toohey KS, Sottos NR, Lewis JA, Moore JS, White SR (2007) Self-healing materials with microvascular networks. Nat Mater 6:581–585. https://doi.org/10.1038/nmat1934

Tripathi N, Misra M, Mohanty AK (2021) Durable polylactic acid (PLA)-Based sustainable engineered blends and biocomposites: recent developments, challenges, and opportunities. ACS Engineering Au 1:7–38. https://doi.org/10.1021/acsengineeringau.1c00011

Troy E, Tilbury MA, Power AM, Wall JG (2021) Nature-based biomaterials and their application in biomedicine. Polymers (Basel) 13:3321. https://doi.org/10.3390/polym13193321

Tunn I, Harrington MJ, Blank KG (2019) Bioinspired Histidine-Zn2+ Coordination for tuning the mechanical properties of self-healing coiled coil cross-linked hydrogels. Biomim 4. https://doi.org/10.3390/biomimetics4010025

Villa Zabala CC (2020) An overview on starch structure and chemical nature. 3–9. https://doi.org/10.1007/978-3-030-42542-5_2

Wan H, Chen Y, Tao Y, Chen P, Wang S, Jiang X, Lu A (2023) MXene-Mediated cellulose conductive hydrogel with ultrastretchability and self-healing ability. ACS Nano 17:20699–20710. https://doi.org/10.1021/acsnano.3c08859

Wang Y, Lu Y (2023) Sodium alginate-based functional materials toward sustainable applications: water treatment and energy storage. Ind Eng Chem Res 62:11279–11304. https://doi.org/10.1021/acs.iecr.3c01082

Wang S, Urban MW (2020) Self-healing polymers. Nat Rev Mater 5:562–583. https://doi.org/10.1038/s41578-020-0202-4

Wang X, Li Y, Sun J (2014). Bioinspired Self-Healing Coatings. https://doi.org/10.1142/9789814354936_0016

Wang L, Di S, Wang W, Zhou S (2015a) Self-healing and shape memory capabilities of copper-coordination polymer network. RSC Adv 5:28896–28900. https://doi.org/10.1039/c4ra16833j

Wang W, Xu Y, Li A, Li T, Liu M, von Klitzing R, Ober CK, Kayitmazer AB, Li L, Guo X (2015b) Zinc induced polyelectrolyte coacervate bioadhesive and its transition to a self-healing hydrogel. RSC Adv 5:66871–66878. https://doi.org/10.1039/C5RA11915D

Wang Z, Kang H, Zhang W, Zhang S, Li J (2017) Improvement of interfacial adhesion by bio-inspired catechol-functionalized soy protein with versatile reactivity: preparation of fully utilizable soy-based film. Polymers (Basel) 9:95. https://doi.org/10.3390/polym9030095

Wang L, Yang K, Li X, Zhang X, Zhang D, Wang L-N, Lee C-S (2021) A double-crosslinked self-healing antibacterial hydrogel with enhanced mechanical performance for wound treatment. Acta Biomater 124:139–152. https://doi.org/10.1016/j.actbio.2021.01.038

Wang X, Zhang HJ, Yang Y, Chen Y, Zhu X, You X (2023b) Biopolymer-based self-healing hydrogels: A short review. Giant 16:100188. https://doi.org/10.1016/j.giant.2023.100188

Wang Z, Jin Z, Liu H, Liu R, Zhang Y, Yin Y, Liu H, Yuan X, Fan S, Liu H (2024) Insights into the hydrophobic coating with integrated high-efficiency anti-corrosion, anti-biofouling and self-healing properties based on anti-bacterial nano LDH materials. Corros Sci 231:111995. https://doi.org/10.1016/j.corsci.2024.111995

Wang H, Lu B, Zhou J, Lai J, Zheng X, Guo S-Z, Zhang L-M (2025) Biobased physicochemical reversible dual-cross-linked hydrogel: self-healing, antibacterial, antioxidant, and hemostatic properties for diabetic wound healing. Biomacromol 26:2637–2653. https://doi.org/10.1021/acs.biomac.5c00087

Wang W, Meng Q, Li Q, Liu J, Zhou M, Jin Z, Zhao K (2020) Chitosan derivatives and their application in biomedicine, Int J Mol Sci 21. https://doi.org/10.3390/ijms21020487

Wang D, Li Y, Zhang H, Ren Z, Fan K, Cheng J, Zhang J, Gao F (2022) The design of rapid self-healing alginate hydrogel with dendritic crosslinking network, molecules 27. https://doi.org/10.3390/molecules27217367

Wang X, Zhang HJ, Yang Y, Chen Y, Zhu X, You X (2023) Biopolymer-based self-healing hydrogels: A short review, Giant 16. https://doi.org/10.1016/j.giant.2023.100188

Wei M, Gao Y, Li X, Serpe MJ (2017) Stimuli-responsive polymers and their applications. Polym Chem 8:127–143. https://doi.org/10.1039/C6PY01585A

Wen J, Zhang X, Pan M, Yuan J, Jia Z, Zhu L (2020) A robust tough and multifunctional polyurethane/tannic acid hydrogel fabricated by physical-chemical dual crosslinking. Polymers (Basel) 12. https://doi.org/10.3390/POLYM12010239

White SR, Sottos NR, Geubelle PH, Moore JS, Kessler MR, Sriram SR, Brown EN, Viswanathan S (2001) Autonomic healing of polymer composites. Nature 409:794–797. https://doi.org/10.1038/35057232

Wienen D, Gries T, Cooper SL, Heath DE (2023) An overview of polyurethane biomaterials and their use in drug delivery. J Control Release 363:376–388. https://doi.org/10.1016/j.jconrel.2023.09.036

Wu M, Han L, Yan B, Zeng H (2023) Self-healing hydrogels based on reversible noncovalent and dynamic covalent interactions: A short review, Supramolecular Materials 2. https://doi.org/10.1016/j.supmat.2023.100045

ying Liu X, Zhong M, kuan Shi F, Xu H, ming Xie X (2017) Multi-bond network hydrogels with robust mechanical and self-healable properties. Chin J Polym Sci (English Edition) 35 1253–1267. https://doi.org/10.1007/s10118-017-1971-0

Xie Y, Li G, Wu J, Zhu J, Cai X, Zhao P, Zhang D, Zhong Y (2025) Injectable self-healing alginate/PEG hydrogels cross-linked via thiol-Michael addition bonds for hemostasis and wound healing. Carbohydr Polym 348:122864. https://doi.org/10.1016/j.carbpol.2024.122864

Xiong D, Ma H, He J, Han Y, Ma W, Wu C, Shi Y, Li Z, Wang H, Zhao Y, Wang J (2025) CO2-switchable self-healing of ionic liquids-based hydrogels. J Colloid Interface Sci 693:137594. https://doi.org/10.1016/j.jcis.2025.137594

Xu T, Liu K, Sheng N, Zhang M, Liu W, Liu H, Dai L, Zhang X, Si C, Du H, Zhang K (2022) Biopolymer-based hydrogel electrolytes for advanced energy storage/conversion devices: Properties, applications, and perspectives. Energy Storage Mater 48:244–262. https://doi.org/10.1016/j.ensm.2022.03.013

Xue S, Wu Y, Liu G, Guo M, Liu Y, Zhang T, Wang Z (2021) Hierarchically reversible crosslinking polymeric hydrogels with highly efficient self-healing, robust mechanical properties, and double-driven shape memory behavior. J Mater Chem A Mater 9:5730–5739. https://doi.org/10.1039/d0ta10850b

Yadav M, Kaushik B, Rao GK, Srivastava CM, Vaya D (2023) Advances and challenges in the use of chitosan and its derivatives in biomedical fields: A review, carbohydrate polymer technologies and applications 5. https://doi.org/10.1016/j.carpta.2023.100323

Yang Y, Urban MW (2018) Self-Healing of Polymers via Supramolecular Chemistry, Adv Mater Interfaces 5. https://doi.org/10.1002/admi.201800384

Yasar M, Oktay B, Gurkan AC, Apohan NK (2024) Self-healing and cell adhesion properties of dynamic and photo-crosslinking PEG-based dual networks soft hydrogels. J Polym Sci. https://doi.org/10.1002/pol.20240198

Yimyai T, Pena-Francesch A, Crespy D (2022) Transparent and self-healing elastomers for reconfigurable 3D materials. Macromol Rapid Commun 43. https://doi.org/10.1002/marc.202200554

Yuan YC, Yin T, Rong MZ, Zhang MQ (2008) Self healing in polymers and polymer composites. Concepts, Realization and Outlook: a Review, Express Polym Lett 2:238–250. https://doi.org/10.3144/expresspolymlett.2008.29

Zahedi E, Ansari S, Wu BM, Bencharit S, Moshaverinia A (2017) Hydrogels in craniofacial tissue engineering. Biomater Oral Dent Tissue Eng. 47–64. https://doi.org/10.1016/B978-0-08-100961-1.00004-9

Zengin A, Castro JPO, Habibovic P, Van Rijt SH (2021) Injectable, self-healing mesoporous silica nanocomposite hydrogels with improved mechanical properties. Nanoscale 13:1144–1154. https://doi.org/10.1039/d0nr07406c

Zhang MQ, Rong MZ (2013) Intrinsic self-healing of covalent polymers through bond reconnection towards strength restoration. Polym Chem 4:4878–4884. https://doi.org/10.1039/c3py00005b

Zhang H, Xia H, Zhao Y (2012a) Poly(vinyl alcohol) hydrogel can autonomously self-heal. ACS Macro Lett 1:1233–1236. https://doi.org/10.1021/mz300451r

Zhang M, Xu D, Yan X, Chen J, Dong S, Zheng B, Huang F (2012b) Self-Healing supramolecular gels formed by crown ether based host-guest interactions. Angew Chem 124:7117–7121. https://doi.org/10.1002/ange.201203063

Zhang C, Yang Z, Duong NT, Li X, Nishiyama Y, Wu Q, Zhang R, Sun P (2019) Using dynamic bonds to enhance the mechanical performance: from microscopic molecular interactions to macroscopic properties. Macromolecules 52:5014–5025. https://doi.org/10.1021/acs.macromol.9b00503

Zhang Y, Lu P, Fang G, Dong B, Hong S, Wang Y, Li J, Fan S (2024b) Enhancing mechanical properties of concrete with 3D printed vascular networks via carbonation strengthening. Cem Concr Compos 154:105791. https://doi.org/10.1016/j.cemconcomp.2024.105791

Zhang B, Zhang P, Zhang H, Yan C, Zheng Z, Wu B, Yu Y (2017) A transparent, highly stretchable, autonomous Self-Healing Poly(dimethyl siloxane) elastomer, macromol rapid commun 38. https://doi.org/10.1002/marc.201700110

Zhang G, Lv L, Deng Y, Wang C (2017) Self-Healing gelatin hydrogels cross-linked by combining multiple hydrogen bonding and ionic coordination, macromol rapid commun 38. https://doi.org/10.1002/marc.201700018

Zhang L, Qiu T, Zhu Z, Guo L, Li X (2018) Self-Healing polycaprolactone networks through thermo-induced reversible disulfide bond formation. Macromol Rapid Commun 39. https://doi.org/10.1002/marc.201800121

Zhang X, Liang Y, Huang S, Guo B (2024) Chitosan-based self-healing hydrogel dressing for wound healing, Adv Colloid Interface Sci 332. https://doi.org/10.1016/j.cis.2024.103267

Zhang Z, Song K, Hirose K, He J, Li Q, Li Y, Pan Y, Adel M, Zhuang R, Iwai S, Fath El-Bab AMR, Fang H, Yang Z, Umezu S (n.d) A highly sensitive, self-adhesive, biocompatible DLP 3D printed organohydrogel for flexible sensors and wearable devices

Zhou X, Liu L, Zeng X (2021) Research progress on the utilisation of embedding technology and suitable delivery systems for improving the bioavailability of nattokinase: a review, food structure 30. https://doi.org/10.1016/j.foostr.2021.100219

Zhu H, Dai W, Wang L, Yao C, Wang C, Gu B, Li D, He J (2022) Electroactive oxidized Alginate/Gelatin/MXene (Ti3C2Tx) composite hydrogel with improved biocompatibility and self-healing property. Polymers (Basel) 14:3908. https://doi.org/10.3390/polym14183908

Zhu F, Xie Q (2018) Structure and physicochemical properties of starch, physical modifications of starch 1–14. https://doi.org/10.1007/978-981-13-0725-6_1

Zhu P, Li Z, Pang J, He P, Zhang S (2024) Latest developments and trends in electronic skin devices. Soft Sci 4. https://doi.org/10.20517/ss.2024.05

Zou W, Dong J, Luo Y, Zhao Q, Xie T (2017) Dynamic covalent polymer networks: from old chemistry to modern day innovations. Adv Mater 29. https://doi.org/10.1002/adma.201606100

Chapter 11
Shape Memory Biopolymers: Design, Synthesis, and Applications

Bantamlak Birlie Kassie, Dehenenet Flatie Tassaw, and Tekalgn Mamay Daget

Abstract Shape memory biopolymers (SMBs) are smart materials that revert to their original shape when exposed to stimuli like temperature, light, pH, or moisture. This chapter examines the design, synthesis, and applications of SMBs, emphasizing their versatility and potential in various fields. The fundamental principles of shape memory behavior in biopolymers are discussed, focusing on molecular architecture and cross-linking methods that enhance or modulate these properties. The chapter covers polymerization techniques, including triple helix formation, ionic interactions, and cross-linking, addressing the challenges in synthesizing environmentally friendly SMBs. Applications of SMBs are explored across biomedical, environmental, and other industrial fields. Biomedical uses include drug delivery systems, tissue engineering scaffolds, wound healing, self-healing materials, biosensors, responsive hydrogels, and smart sutures. Environmental applications highlight biodegradable responsive packaging and smart textiles, while industrial applications cover actuators, sensors, 3D printing, and advanced technologies. The chapter concludes by discussing challenges in the development and commercialization of SMBs, such as scalability, cost-effectiveness, and the need for robust materials. Future research directions and technological advancements are outlined, aiming to overcome these challenges and promote broader adoption of SMBs. This chapter provides a comprehensive overview of SMBs' current state, design, synthesis, and application potential, serving as a valuable resource for researchers and professionals.

Keywords Biopolymer synthesis · Biomedical applications · Shape memory biopolymers · Smart materials · Stimuli-responsive polymers

B. B. Kassie · D. F. Tassaw · T. M. Daget
Textile Faculty, Ethiopian Institute of Textile and Fashion Technology, Bahir Dar University, P.O.box 1037, Bahir Dar, Ethiopia

B. B. Kassie (✉) · T. M. Daget
Textile Science and Engineering Research Center, Ethiopian Institute of Textile and Fashion Technology, Bahir Dar University, P.O.box 1037, Bahir Dar, Ethiopia
e-mail: bantamlakbirlie@gmail.com

D. F. Tassaw
Institute of Architecture of Textiles, Faculty of Material Technologies and Textile Design, Lodz University of Technology, 116 Zeromskiego Str., 90-924 Lodz, Poland

L. Ravindran et al. (eds.), *Macro, Micro and Nanocomposites from Sustainable Sources*, Advanced Structured Materials 244, https://doi.org/10.1007/978-981-95-2469-3_11

11.1 Introduction

In the pursuit of sustainable and high-performance materials, shape memory biopolymers (SMBs) have emerged as a transformative innovation, bridging the gap between functionality and environmental responsibility (Basak and Bandyopadhyay 2024). In recent years, the field of materials science has witnessed a significant shift towards the development of sustainable and functional materials that not only meet performance criteria but also address environmental concerns (Ashby and Johnson 2013). Among these innovative materials, shape memory polymers (SMPs) have emerged as a prominent category, particularly in the realm of biopolymers (Pilate et al. 2016). Shape memory biopolymers have garnered increasing attention due to their exceptional capability to undergo programmed shape transformations in response to external stimuli, such as electric and magnetic fields, pressure (Fang et al. 2015), radiation and laser heating (Pilate et al. 2016), heat (Lendlein et al. 2019), temperature (Liu et al. 2015), light, pH (Guo et al. 2014), optical and microwave (Du et al. 2015), absorbed gas molecules, or moisture (Toit et al. 2016; Leng et al. 2011; Pisani et al. 2022). Their unique properties, combined with their renewable nature, position them as viable candidates for a wide range of applications, particularly in the biomedical field (Fernandes et al. 2019), as well as in environmental, camouflage, packaging, aerospace, electronic information, sensors, self-assembling structures, and smart textiles (Uyan and Celiktas 2023). The increasing demand for adaptive and sustainable materials has driven significant advancements in SMB research, leading to the exploration of their vast potential across multiple disciplines.

The concept of shape memory behavior was first introduced in metallic alloys, where certain compositions exhibited the ability to "remember" a specific shape after being deformed (Kousbroek 2018). This phenomenon, known as the shape memory effect (SME), has since been harnessed and adapted for biopolymers, leading to a new frontier in material design (Ahn et al. 2010). The flexibility of polymer chemistry allows for the tuning of properties through the careful selection of monomers, crosslinking agents, and processing conditions (Schnepp 2013). Biopolymer-based SMBs, derived from renewable sources, further enhance this tunability while simultaneously addressing sustainability concerns. This tunability is further enhanced in biopolymers derived from natural resources, such as polysaccharides (Balart et al. 2021), proteins (Tănase et al. 2014), cellulose (Motaung and Linganiso 2018), starch (Shahabi-Ghahfarrokhi et al. 2020), collagen (Fassini et al. 2021), chitin (Ali et al. 2024), Polyhydroxyalkanoates (PHAs) (Tan et al. 2014), keratin (Saha et al. 2019), xanthan gum (Sulaiman et al. 2022), alginate (Pettignano et al. 2019), and agar which not only possess intrinsic biodegradability but also exhibit biocompatibility a crucial requirement for medical, bioelectronics, responsive packaging, smart textiles applications. By integrating sustainable sourcing with advanced material engineering, SMBs are redefining the landscape of functional materials. As shown in Fig. 11.1 shape memory biopolymers respond to various stimuli and are used in biomedical, environmental, and other industrial applications.

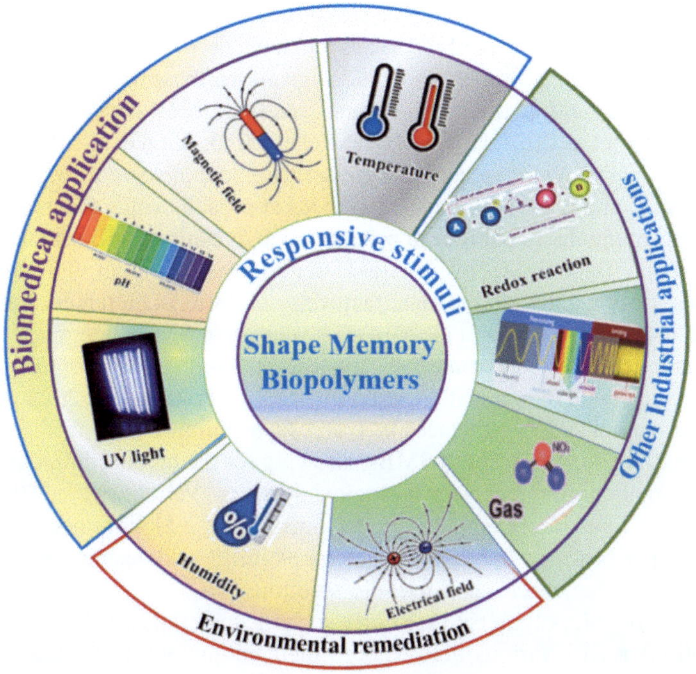

Fig. 11.1 Responsive stimuli for shape memory biopolymers and their multifunctional applications

Shape memory biopolymers are synthesized through various methods, often involving polymerization and crosslinking techniques. These materials typically combine a biocompatible polymer matrix, such as polylactic acid (PLA) or polycaprolactone (PCL), with responsive components that enable shape recovery upon external stimuli (Khalid et al. 2024). The synthesis process may include electrospinning, solvent casting, or 3D printing, depending on the desired properties (Delaey et al. 2020). Advancements in these techniques have enabled precise control over microstructural properties, thereby enhancing the performance of SMB-based materials. Recent advancements in processing techniques, such as 3D printing and electrospinning, have enabled the fabrication of complex structures that leverage the shape memory biopolymer effects (Olawumi et al. 2024; Sikhosana et al. 2021; Vach Agocsova et al. 2023). Crosslinking agents, like UV light or chemical reagents, are used to stabilize the network, ensuring that the material retains its shape memory capabilities (Pisani et al. 2022; Kirillova and Ionov 2019). This has opened avenues for the development of responsive materials that can change their shape under physiological conditions, thus enhancing their functionality in medical devices and drug delivery systems. These improvements have paved the way for next-generation biomaterials that seamlessly integrate into biomedical and industrial applications.

The design and synthesis of SMBs are driven by the need to create biocompatible materials that can recover their shape in response to stimuli like heat. For instance, polylactic acid (PLA) (Molina et al. 2023) is often blended with thermoplastic starch (TPS) and oligomeric lactic acid (OLA) as a plasticizer to enhance flexibility while preserving shape memory properties. This design allows these biopolymers to recover their shape at body-compatible temperatures, making them suitable for resorbable implants (Sessini et al. 2024; Navarro-Baena et al. 2016; Zhou et al. 2021). A significant advancement in synthesis has been the development of solvent-free methods for producing segmented polyurethane SMBs. This approach avoids the use of harmful solvents, improving the scalability of production and resulting in biostable materials with high shape fixity and recovery rates. Such innovations align with the growing need for sustainable and scalable manufacturing solutions in biomaterial production (Ramezani et al. 2024; Zhang et al. 2014). The design of SMBs integrates principles from polymer chemistry and materials science, enabling the development of smart materials that respond dynamically to their environments (Luo et al. 2023; Korde and Kandasubramanian 2020). Advances in synthesis techniques have further enhanced the functionality and versatility of SMBs, allowing for the incorporation of bioactive components that can interact with biological systems (Valioglu et al. 2024; Basak 2021; Mbeh and do Nascimento 2015). As a result, these biopolymers are not only designed for performance but also for biocompatibility and sustainability, aligning with the growing demand for eco-friendly solutions.

Shape memory biopolymers (SMBs) are versatile materials that have a wide range of applications across various fields due to their unique ability to return to a predefined shape upon external stimuli. In the biomedical sector, SMBs can be utilized in self-expanding stents, where the polymer is deformed and then allowed to return to its original shape upon reaching body temperature, facilitating minimally invasive procedures. They are also employed in surgical sutures, dressings, and implantable devices, where their ability to respond to environmental stimuli can lead to enhanced healing and patient outcomes (Laasch 2012). Additionally, SMBs play a significant role in developing scaffolds for tissue engineering, enhancing cell adhesion and facilitating regeneration (Ramaraju et al. 2020). In dental applications, they are used in orthodontic devices that adapt to tooth movements, improving patient comfort (Bruni et al. 2019). Furthermore, SMBs function as coagulants and flocculants in wastewater treatment, adsorbing heavy metals and organic pollutants while contributing to biodegradable materials that help reduce plastic waste (Gupta et al. 2021; Peterson et al. 2017; Nagahama et al. 2009; Ruiz-Rubio et al. 2020). This convergence of sustainability and performance underscores the transformative potential of SMBs across industries. Beyond healthcare, SMBs demonstrate their versatility in textiles, packaging, and actuators. In the realm of smart textiles, they are integrated into fabrics that can adapt to changes in temperature and moisture, enhancing user comfort (Failed 2017). In packaging, they provide smart sealing mechanisms that preserve products and reduce waste (Talegaonkar et al. 2017). In robotics, SMBs facilitate the development of soft actuators, enhancing the flexibility of robotic systems (Schönfeld et al. 2021). They are incorporated into dynamic seals and lightweight structures in the aerospace and automotive industries, improving performance and energy efficiency (Harun-Ur-Rashid and Imran 2024; Singh et al. 2024). SMBs also find applications in consumer products, such as hair styling tools that adjust to different hair types (Thadepalli 2022), and in smart home devices, creating blinds that respond to sunlight and temperature changes (Leeladhar and Singh 2018). Their exploration in electronics includes self-healing circuits and flexible sensors, showcasing their significant potential for innovation and practical applications (Ghosh 2009; Zheng et al. 2016). However, challenges remain in their design and implementation, particularly regarding the balance between mechanical performance, shape memory characteristics, and biocompatibility of SMBs. Addressing these challenges requires a collaborative effort across multiple disciplines, fostering interdisciplinary innovation to fully unlock the potential of SMBs.

This chapter explores the multifaceted world of shape memory biopolymers, emphasizing their design principles and synthesis methods, along with diverse applications categorized into biomedical, environmental, industrial, and consumer sectors. In the biomedical field, shape memory biopolymers are utilized in drug delivery systems, tissue engineering scaffolds, wound healing, self-healing materials, biosensors, responsive hydrogels, minimally invasive medical devices, and smart sutures and staples. These materials offer enhanced functionality and adaptability, making them ideal for various medical applications. In the environmental sector, they contribute to sustainable food packaging and intelligent textiles for adaptive clothing, promoting eco-friendly practices. In industrial applications, shape memory biopolymers are increasingly used in sports equipment, enhancing performance and comfort through their responsive properties. They also play a role in energy harvesting, where mechanical energy is converted into usable power, and in 3D printing, enabling the creation of customized components with complex geometries and functionalities. Moreover, these biopolymers are integrated into consumer products, providing innovative features for everyday items, and are being explored in robotics, where they contribute to the development of soft actuators and adaptive structures that mimic natural movements. As interdisciplinary collaboration continues to drive innovation, the full potential of these remarkable materials is yet to be realized, promising a new era of smart, eco-friendly solutions for global challenges.

11.2 Molecular Mechanism and Synthesis of Shape-Memory Biopolymer

There are three main classes of biopolymers classified according to the monomeric units used and the structure of the biopolymer formed (Pattanashetti et al. 2017).

i. Polynucleotides (RNA and DNA)—these are long polymers composed of 13 or more nucleotide monomers.
ii. Polypeptides—these are short polymers of amino acids.
iii. Polysaccharides—these are linear bonded polymeric carbohydrate structures.

Biopolymers exhibit chemical structures and compositions that closely resemble the macromolecules found in the native extracellular environment (Yates and Barlow

2013). These materials, often categorized as soft, elastic, and gel-like, possess characteristics that lie between solids and liquids. In living organisms, biopolymers are considered smart and adaptable materials, as their structures are continuously modified either by enzymes during various life stages or in response to environmental stimuli (Galaev and Mattiasson 1999).

The shape-memory phenomenon in polymers stems from a dual-segment system, where one segment is highly elastic, while the other can significantly decrease its stiffness when triggered by a specific stimulus (Lendlein and Langer 2002). The shape-memory mechanism is fundamentally based on entropic elasticity, which allows the polymer to recover its original shape after deformation by utilizing the stored elastic energy (Behl et al. 2009; Leng and Du 2010). In general, shape-memory polymers (SMPs) consist of two active phases: the fixity phase and the switch phase. The fixity phase is responsible for memorizing the polymer's original shape, while the switch phase allows the temporary shape to be fixed. The switch phase also has a transition temperature (T_{trans}), which triggers the recovery of the primary shape. This transition temperature can either be the melting temperature (Tm) or the glass transition temperature (Tg), depending on whether the switch phase is crystalline or amorphous in nature (Peponi et al. 2014).

- When the polymer is synthesized, its chains adopt a coil conformation, which is the most favorable arrangement in terms of entropy. This coiled structure represents the polymer's natural, high-entropy state (Lendlein and Kelch 2002).
- The fixity phase prevents the polymer chains from slipping by stabilizing them through intermolecular or intramolecular bonds. This ensures that the polymer maintains its fixed, temporary shape until the switch phase is activated (black dots in Fig. 11.2).
- The shape memory effect is enabled by the switching chains. Below the transition temperature (Ttrans) of the switch phase (shown as light grey in Fig. 11.2), the chain movements are restricted. However, when the temperature rises above T_{trans}, the chains become mobile. In this state, if stress is applied, the chains undergo elastic deformation, leading to a reduction in the entropy of their conformation. Once the stress is released, the polymer is able to return to its original shape, restoring its initial conformation.
- Conversely, when the temperature is below the transition temperature (T_{trans}), the switching phase (depicted in light grey in Fig. 11.2) stabilizes the chains, maintaining the temporary shape. In this state, it is possible to release the stress without altering the fixed configuration of the polymer.
- Finally, when the temperature is raised above T_{trans}, the polymer will revert to its original shape, as this conformation is entropically more favorable.

The natural organization of biopolymers, which is often sensitive to various stimuli, can be leveraged for the reversible formation of temporary crosslinks, either through physical or chemical means (Lendlein and Kelch 2002). Controlling self-organization phenomena in biopolymer-based shape-memory polymers (SMPs) is challenging. As a result, it is essential to design a suitable polymer network architecture, where crosslink density significantly impacts swellability,

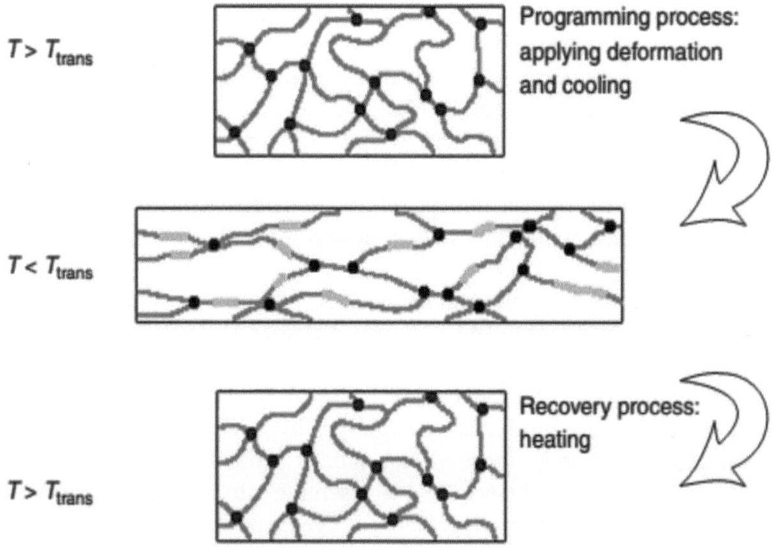

Fig. 11.2 Shape memory effect in polymers

mechanical strength, and elasticity. However, the crosslink density should not be excessively high, as this can hinder self-organization due to steric effects. Additionally, the strength and quantity of non-covalent interactions that facilitate self-organization must also be taken into account (Löwenberg et al. 2017). In addition to the concept of natural self-organization, conventional molecular switches, including thermo-sensitive hydrophobic interactions and pH-sensitive mechanisms, have also been utilized in biopolymer-based shape-memory hydrogels (SMHs) (Izawa and Kadokawa 2010).

Biopolymer-based Shape memory materials can be equipped with a shape switching function when facilitating, e.g., triple helix formation in proteins or ionic interactions in polysaccharides (Izawa and Kadokawa 2010). Polysaccharides like alginate, chitosan and carrageenan are anionic biopolymers derived from plants and animals (Liang et al. 2019).

11.2.1 Triple Helix Formation for Shape Memory Effect (SME) in Protein-Based Biopolymer

The shape memory effect in biopolymers was achieved by using recombinant telechelic polypeptides, which feature collagen or gelatin-like end blocks and a random coil-like middle block. This effect was induced through the chemical cross-linking of lysine residues present in the random coil (Skrzeszewska et al. 2011). The adoption of preferred conformations in protein polymer is sequence (primary structure)-dependent (Munoz and Cerminara 2016), and is therefore inherently included information. The conformation is associated with the biological function of the protein, such as specificity in biological interactions (Hiramatsu et al. 2015) or its mechanical performance (Lv et al. 2010). Since the conformations are determined by non-covalent interactions, they can be modified and potentially switched "on" or "off" by altering the environment, such as through changes in pH, the presence and concentration of ions or specific molecules, or temperature. This versatility has inspired the use of biopolymers as switching segments in shape-memory polymers (SMPs).

The tropocollagen, or "collagen molecule," is a rod-like subunit measuring approximately 300 nm in length and 1.5 nm in diameter. It consists of three polypeptide strands, each forming a left-handed helix. These three left-handed helices are twisted together into a right-handed coiled coil, creating a triple helix that is a cooperative quaternary structure stabilized by numerous hydrogen bonds. These hydrogen bonds typically begin to dissociate when heated to around 35–37 °C (Bella 2016). Tropocollagen subunits spontaneously self-assemble into larger arrays in the extracellular spaces of tissues, featuring regularly staggered ends. Within the triple helices, some covalent cross-linking occurs, and there is also a variable degree of covalent cross-linking between tropocollagen helices. This variation leads to the formation of different types of collagen found in various mature tissues (Parvizi 2010).

Biotechnologically produced ABA triblock peptides with collagen-like A blocks and lysine-containing B blocks that were covalently crosslinked with glutaraldehyde (Skrzeszewska et al. 2011) showed a shape-memory effect after programming using the triple helix formation as temporary net points as shown in Fig. 11.3. This strategy, however, did not allow easy tailoring of the mechanical properties of the network. An interpenetrating double network consisting of a covalent polymeth-acrylate network and a physical gelatin network (Chen et al. 2019) likewise exploited the triple helix association, but required the synthetic polymer network for the shape-memory effect.

Biopolymer-based stimuli-sensitive switchable matrices demonstrate a shape-memory function that can be alternately activated by two different input signals, such as changes in environmental salt concentration or temperature. This functionality is achieved through the implementation of a shape-memory mechanism in hydrogels, which relies on the coil-to-helix transition of protein chains within gelatin-based networks. The gelatin-based covalent network system is formed by reacting glycidyl methacrylated gelatin (GMA-gelatin) with oligo (ethylene glycol) (OEG) α, ω-dithiols (as illustrated in Fig. 11.4a). Utilizing a bifunctional crosslinker to crosslink gelatin has been shown to be advantageous for effectively forming a polymer network. By adjusting the length and rigidity of the crosslinker, the properties of the network can be customized without changing the underlying crosslinking chemistry. In GMA-gelatin OEG networks, the cross-linking is relatively loose, enabling the gelatin chains to retain sufficient flexibility to adopt triple helices under favorable environmental conditions (Löwenberg et al. 2020). Lowering the aging temperatures promote helix formation but reduce stability, while slight temperature increases partially melt junction zones, leading to their reformation with varying lengths and stabilities (Sugiyama 2004) but low concentrations of salt facilitate the

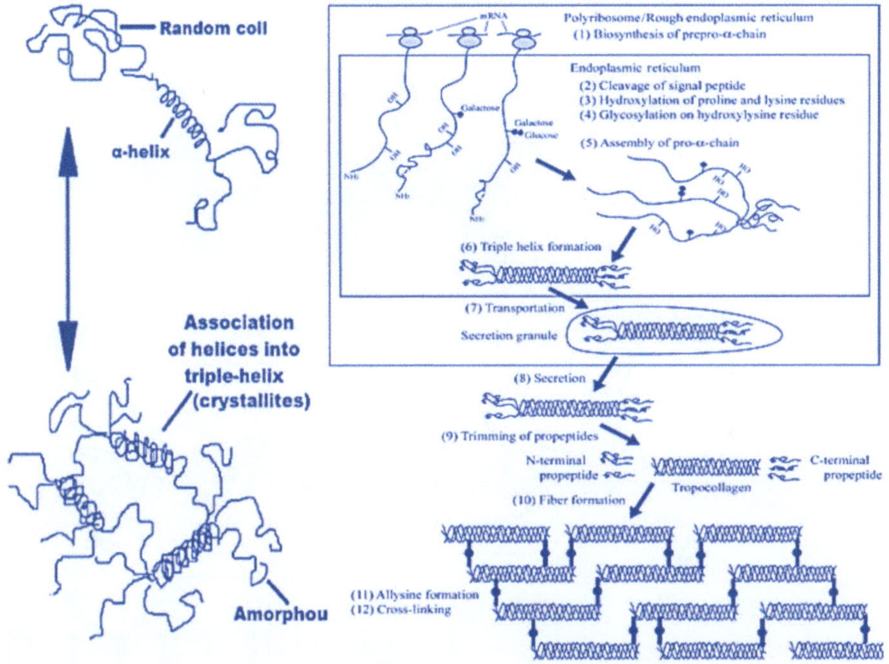

Fig. 11.3 Triple helix formation of protein-based biopolymers

coil-to-helix transition of gelatin, while higher concentrations inhibit it. In contrast, NaCl and $CrCl_3$ consistently decrease both triple helix content and storage modulus, highlighting the significant role of salt concentration and type in modulating gelatin's structural properties (Qiao et al. 2018) also the higher salt content disrupts the spatial structure of gelatin and inhibits hydrogen bond formation (Xiang et al. 2018).

For example, a thermally induced shape-memory effect (SME) in peptide-derived hydrogels with collagen-like transient nodes utilizes hydrogen bonding for helix formation and the fixation of a temporary shape upon cooling. The recovery of the permanent shape is triggered by heating, which causes the melting of triple helices due to hydrogen bond cleavage. The thermally induced shape-memory behavior of the gelatin-based biopolymer network is synthesized by first functionalizing gelatin with double bonds through a reaction with glycidyl methacrylate, followed by crosslinking the gelatin solutions in water via thiol-Michael addition with OEG α, ω-dithiols. In this system, triple helices are employed to fix the temporary shapes in gelatin-based interpenetrating double networks that incorporate graphene oxide (Neffe et al. 2021) as shown in Fig. 11.5.

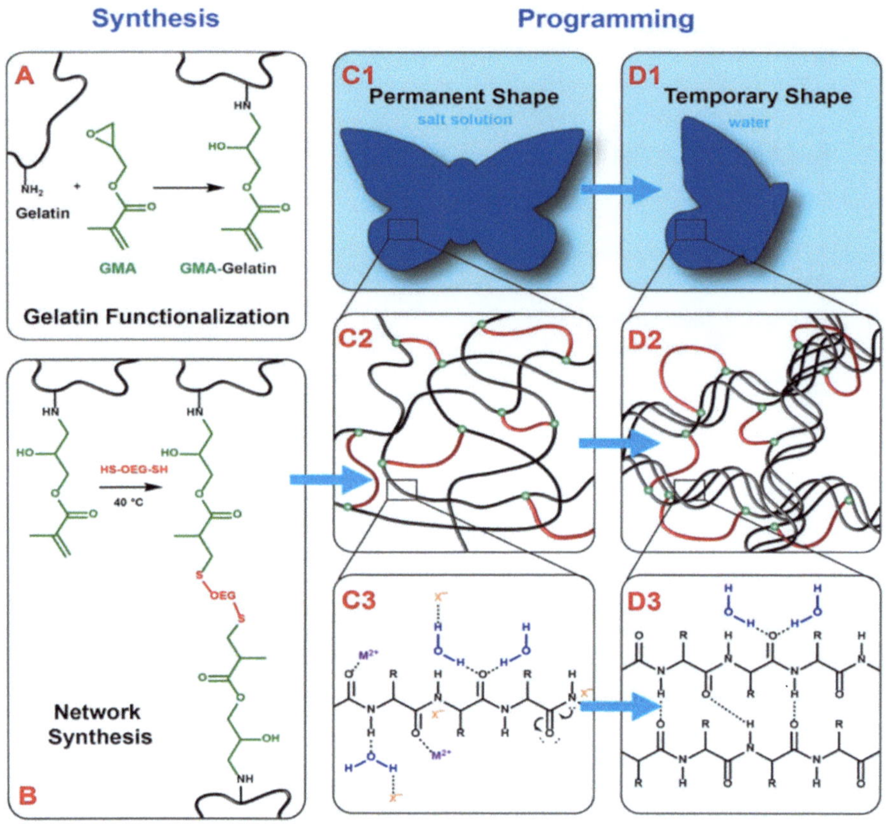

Fig. 11.4 Schematic overview about the synthesis and programming steps for implementing an SME into gelatin-based hydrogels: **a** Gelatin functionalization with GMA. **b** Synthesis of gelatin-based networks by thiol-ene Michael-addition; gelatin (black), GMA (green), and OEG dithiol (red). **c** Network in the permanent shape: **c1** macroscopic view, **c2** schematic view with a random coil orientation of the gelatin main-chain (black), crosslinked with OEG (red) where the covalent netpoints (green) were previously formed by the thiol-ene reaction, **c3** expected mechanism of ion coordination along the gelatin backbone inhibiting the gelatin-gelatin interaction. **d** Network in the temporary shape after the programming process (deformation): **d1** macroscopic view, **d2** schematic view with the gelatin main-chain (black) in a helical conformation acting as temporary netpoints, crosslinked with OEG (red) where the covalent netpoints (green) were previously formed by the thiol-ene reaction, **d3** expected mechanism of gelatin-gelatin intramolecular interactions allowing the helicalization to fix the temporary shape

11.2.2 Ionic Interactions for Shape Memory Effect (SME) in Polysaccharides-Based Biopolymer

The complex carbohydrates found in polysaccharides exhibit a wide range of chemical structures, compositions, and ionic properties. Due to their inherent functional

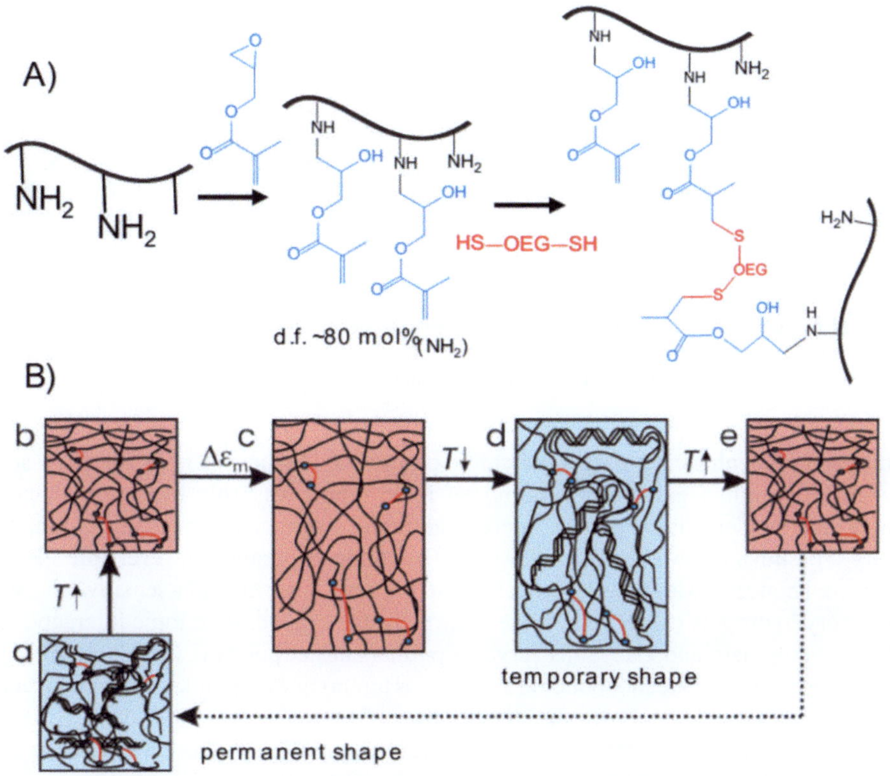

Fig. 11.5 **a** Network formation is achieved in two steps: (i) functionalization of gelatin with glycidyl methacrylate; and (ii) crosslinking with an oligo (ethylene glycol) (OEG) α, ω-dithiol. **b** Molecular principle of the envisioned thermally-induced shape-memory effect of the investigated gelatin-based networks. The networks after synthesis at room temperature contain triple helices. **a** Heating leads to dissociation of the triple helices (**b**). Deformation at Thigh (**c**) and subsequent cooling (**d**) fixes the temporary shape of the material by triple helices acting as temporary net points. **a**, **d** is the programming of the material. Subsequent heating leads to the dissociation of the temporary net points and readoption of the permanent shape, the shape-memory effect (**e**). Returning to room temperature re-establishes status (**a**) without further shape change

groups, polysaccharides are well-suited for engineering biopolymers, such as hydrogels, through various crosslinking techniques, which can be either non-covalent or covalent (Failed 2024). Among natural polysaccharides, chitosan stands out as a fundamental polymer primarily due to its abundance of amino groups distributed throughout its structural chains, which impart a positive charge to the molecule (Rasool et al. 2024). The formation of shape-memory hydrogel using chitosan through ionic interactions involves gradual interactions between molybdate polyoxyanions and the residues on chitosan under acidic conditions, contributing to the gelation process (Sacco et al. 2018). As shown in Fig. 11.6, xanthan gum forms a network with ionic liquid, creating imidazolium-carboxylate pairs and alkyl chain

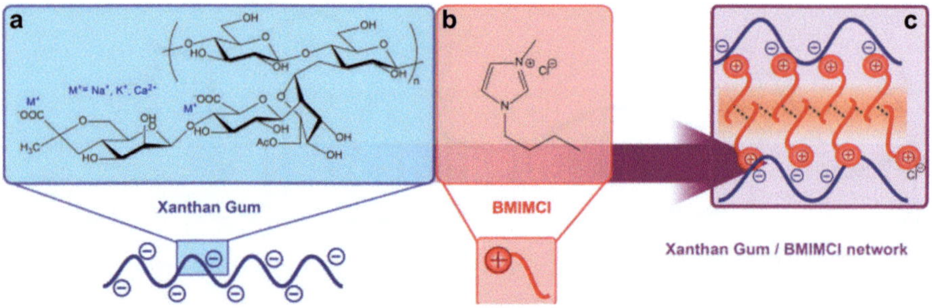

Fig. 11.6 Schematic illustration of network formation between xanthan gum (**a**) and an ionic liquid (**b**). The resulting network (**c**) consists of imidazolium-carboxylate pairs and intermolecular interactions between the alkyl chains of the ionic liquid

interactions, enhancing shape-switching properties. As an example for polysaccharides, xanthan gum incorporating an ionic liquid was able to switch shapes using thermo-sensitive regularly ordered intermolecular interactions acting as temporary crosslinks (Izawa and Kadokawa 2010). The incorporation of reversible PBA-catechol bonds into alginate hydrogels led to the development of pH-sensitive shape-memory hydrogels (SMHs). In these hydrogels, the egg-box-like ionic interactions between alginate and Ca^{2+} ions serve as permanent net points (Li et al. 2016). At alkaline pH, PBA-catechol bonds are formed as temporary crosslinks, while recovery is triggered at acidic pH by the dissociation of these bonds. Additionally, these hydrogels exhibit multi-stimuli responsiveness, as the PBA-diol ester bonds can be reversibly cleaved by the presence of sugars (Li et al. 2016).

The ionic cross-linking of multilayered chitosan (CHI)–alginate (ALG) films led to freestanding membranes with multiple interesting properties, such as: improved mechanical strength, calcium-induced adhesion and shape memory ability (Silva et al. 2016). A common approach used to stiff these multilayers consists in the post-assembly treatment with cross-linking agents such as glutaraldehyde, 1-ethyl-3-(3-dimethylaminopropyl)-carbodiimide in combination with N-hydroxysulfo-succinimide (EDC/NHs) and genipin (Chaubaroux et al. 2012). A membrane crosslinked with genipin and $CaCl_2$, combining covalent and ionic cross-linking. The genipin cross-linking has the ability to obtain aminecontaining polymers, generating a greenish color on the samples (Chen et al. 2005). An ionic cross-linking was also performed due to the ability of ALG to have reversible gelling properties upon ionic interactions with divalent ions, such as calcium and the carboxylic acid moieties on the guluronic acid residues of ALG (Lee and Mooney 2012; Yang et al. 2014). $CaCl_2$ was used to induce ionic cross-linking in control membranes, namely CHI–ALG membranes with and without previous genipin cross-linking. As shown in Fig. 11.7, electrostatic interactions drive the production of CHI-ALG membranes, illustrating multilayer rearrangements via covalent and ionic cross-linking processes.

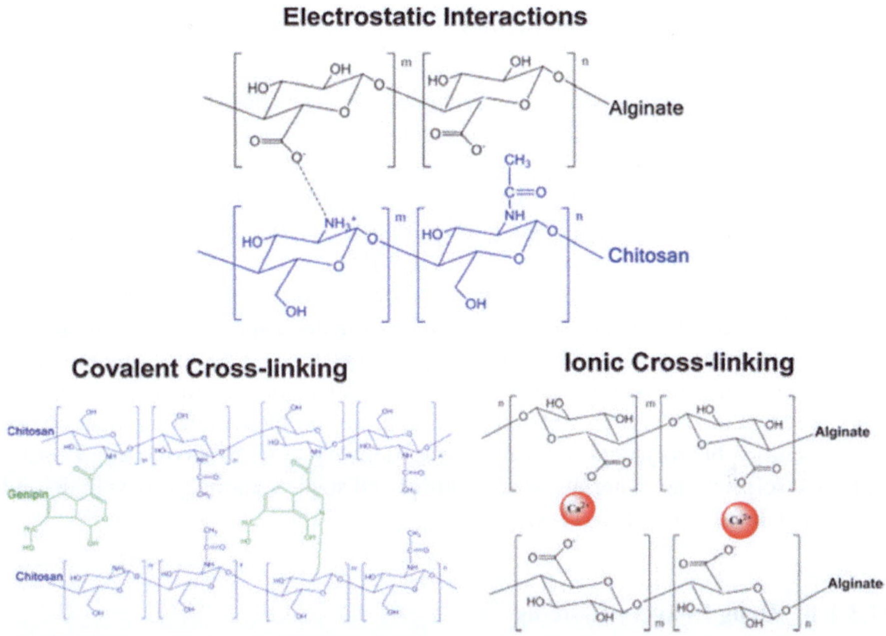

Fig. 11.7 The schematic illustration shows the electrostatic interactions behind the production of CHI–ALG membranes as well as their multilayer rearrangements upon covalent and ionic cross-linking

11.3 Applications of Shape Memory Biopolymers

Polymers play a crucial role in our daily lives due to their unique properties, making them suitable for a variety of applications. In particular, biopolymers are increasingly recognized for their biodegradability, biocompatibility, renewability, and cost-effectiveness (Pattanashetti et al. 2017). Shape-memory polymers (SMP) are a unique class of polymers capable of transitioning between a permanent shape and a temporary one, allowing them to be used in a wide range of applications (Delaey et al. 2020). Recently, Shape-Memory Polymers (SMPs) have gained considerable attention in biomedical applications, including tissue engineering, drug delivery, endovascular surgery, orthodontics, and orthopedics, attracting interest from both academia and industry (Sabahi et al. 2020). Furthermore, these shape memory bio polymers are widely used in industrial and environmental applications.

11.3.1 Biomedical Applications

Shape-memory polymers (SMPs) can remember their original shape after deformation when triggered by specific stimuli, such as heat, light, or moisture, without requiring external assistance once the stimulus is removed (Pisani et al. 2022). From a biomedical perspective, they must meet stringent requirements to function as biomaterials, including being non-toxic, hemocompatible, non-carcinogenic, non-inflammatory, and non-allergenic (Failed 2015a). Additionally, they should not release harmful residues. Furthermore, these biomaterials must be biocompatible, ensuring they produce the desired effects in their intended applications without triggering an immune response. Notably, biopolymers tend to be less inflammatory compared to synthetic polymers, making them advantageous in biomedical contexts (Singhwane et al. 2023). This chapter will explore the biomedical applications of shape-memory biopolymers, including drug delivery systems, tissue engineering scaffolds, self-healing materials, smart sutures and staples, cardiac valve repair, and other specialized fields (Fig. 11.8).

11.3.1.1 Drug Delivery Systems

Shape memory polymers are particularly valuable as drug carriers, especially for drugs that have low solubility or stability in the bloodstream, as they can enhance the bioavailability of such drugs. As shown in Fig. 11.9 these polymers can also provide controlled, sustained drug release, which is beneficial when a drug needs to be delivered over an extended period. Smart polymers, a new category of materials, offer innovative approaches to drug delivery systems. However, their suitability for use as matrices in drug delivery depends on key factors such as drug loading capacity, behavior in aqueous environments, release patterns, and degradation characteristics (Sezer 2016; Yu and Yu 2015).

Biomedical nanotechnology has made remarkable progresses in transforming traditional drug delivery systems into smart, stimuli-responsive platforms using shape memory biopolymers. These advanced systems, designed to respond to specific internal or external triggers, offer significant advantages such as enhanced drug targeting efficiency and reduced side effects key to improving patient compliance. Wide range of smart drug delivery systems have been developed, including polymeric nanoparticles, liposomes, metals/metal oxides, and exosomes, each showcasing unique capabilities (Liu et al. 2016).

Shape memory hydrogels are increasingly utilized in biomedical engineering applications, particularly in drug delivery, due to their adaptability, excellent biocompatibility, and potential for biodegradability. Their wet and soft properties closely resemble those of human tissues, making them highly suitable for these functions (Fig. 11.10) (Korde and Kandasubramanian 2020).

With its enhanced potential for drug delivery, bacterial nanocellulose (BNC) becomes even more effective when modified with hydrophilic additives such as

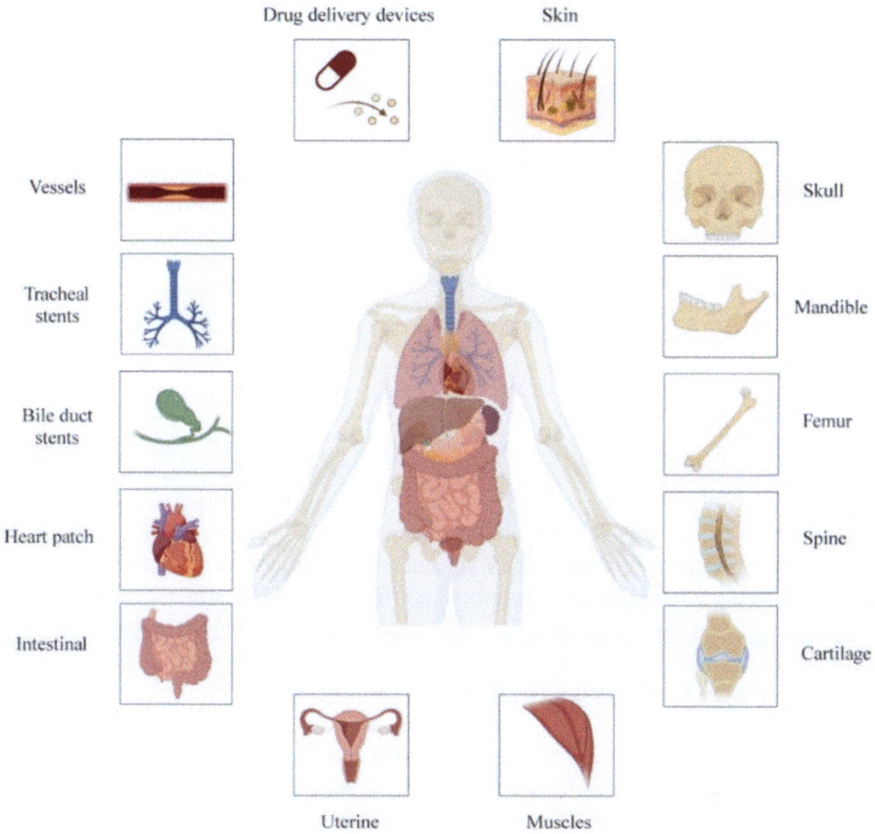

Fig. 11.8 Bio medical applications of shape memory biopolymers (Wu et al. 2024)

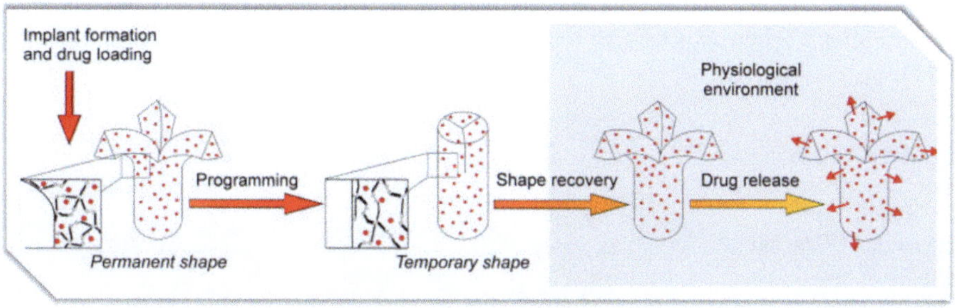

Fig. 11.9 Schematic diagram drug delivery system using shape memory biopolymers (Wischke et al. 2009)

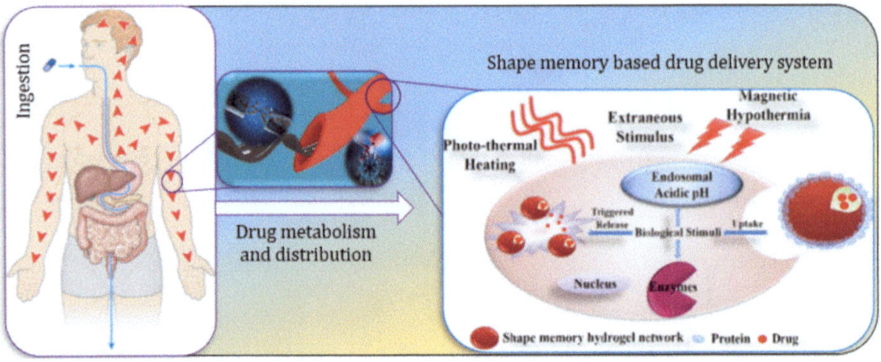

Fig. 11.10 Action of shape memory biopolymers in drug delivery system (Korde and Kandasubramanian 2020)

magnesium chloride, glucose, and sucrose. This modification resolves bacterial nanocellulose's poor re-swelling (shape memory) effect, allowing for easy storage and insertion into body cavities. Once inserted, the material re-swells to fill spaces, making it highly adaptable for biomedical applications (Müller et al. 2014).

11.3.1.2 Tissue Engineering Scaffolds

Smart tissue engineering scaffolds made from shape memory polymers are gaining attention in tissue engineering due to their unique ability to change shape (Zare et al. 2021). This shape actuating feature is being leveraged to enhance the delivery of scaffolds or to fill tissue defects more effectively. Such smart scaffolds can deploy, expand, or fit themselves automatically (Fig. 11.11). Typically, these smart scaffolds are either thermo-responsive or hydro-responsive, meaning they return to their original shape when exposed to heat or moisture, respectively (Pfau and Grunlan 2021).

Fig. 11.11 Shape memory biopolymers based smart scaffolds in tissue engineering (Pfau and Grunlan 2021)

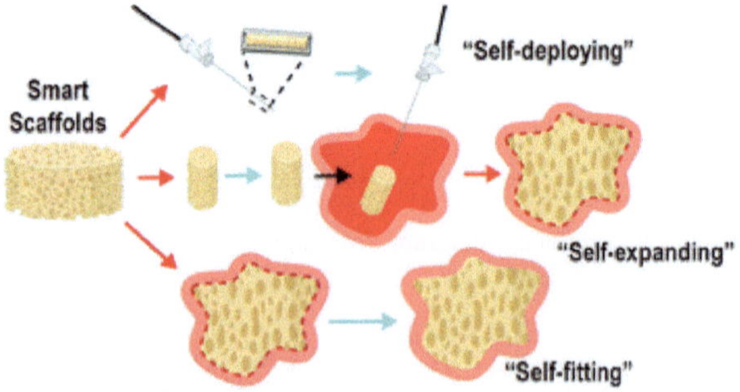

Most smart shape memory biopolymers based scaffolds have been directed towards regenerating bone, cartilage, and cardiovascular tissues, skin, skeletal tissue, and nerves (Fig. 11.12) (Najjari et al. 2022; Mantha et al. 2019). A vast variety of smart scaffolds can be prepared with properties targeted for a specific tissue application. Smart scaffold compositions span across several distinct classes of shape memory bio polymers, affording further tunability of properties using numerous approaches (Kirillova et al. 2021). Shape memory bio polymers are increasingly used in cardiovascular stents due to their cost-effectiveness and ability to reconfigure shape, making them ideal for angioplasty. Different types, including thermo-induced and bio-absorbable shape memory biopolymers, enhance the effectiveness of stent application. Additionally, these materials are used for heart valve replacements, aiming to overcome limitations of animal-derived or natural polymer valves. The functional scaffolds, created via melt electro writing, are designed to mimic the microstructure of heart valves and handle blood flow pressure (Holman et al. 2021; Solazzo et al. 2019; Saidy et al. 2019).

Nerve guidance conduits (NGCs), made from shape memory biopolymers, are used for nerve repair. Key advancements include the use of cellulose/soy protein isolate membrane nerve guidance conduits filled with hydrogels to bridge nerve defects and agarose gel combined with lipid microtubules to create anisotropic gels

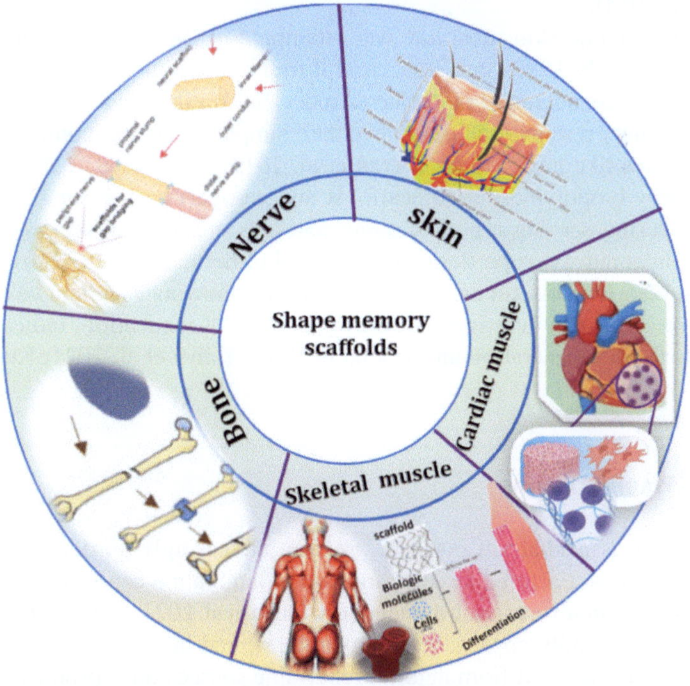

Fig. 11.12 Applications of shape memory biopolymers in medical Scaffold

with nerve guidance conduits gradient distribution as fillers (Luo et al. 2021). Adaptive microporous hydrogels, composed of micro-scale building blocks embedded nerve guidance conduits, have been developed as injectable materials that maintain their structure for localized bioactive agent delivery. Injectable hydrogels, often integrated with microspheres or bioactive molecules, mimic neural tissue mechanics, regulate cell fate, and promote peripheral nerve injury repair (Qi et al. 2022; Hsu et al. 2019).

Minimally invasive scaffold implantation techniques are used to reduce trauma and accelerate healing in muscle tissue (Carnes and Pins 2020). A shape-memory alginate scaffold, designed to be highly porous and compressible, has been used for delivery to injured tibialis anterior muscle via minimally invasive surgery. This scaffold offers key properties for muscle repair: shape-memory for easy delivery, high porosity for dense cell loading and a supportive microenvironment, and preprogrammed degradation that aligns with the muscle healing process, eliminating the need for surgical removal and minimizing inflammation. The hydration-mediated shape memory allows the scaffold to be compressed and delivered through a catheter, reducing invasiveness and preserving local physiological conditions for better regeneration. These scaffolds also support continuous cell migration, and their anti-necrosis, anti-apoptosis, and anti-inflammatory effects help prevent fibroblast infiltration at injured sites, promoting functional muscle recovery in severe injury cases (Wang et al. 2014).

The first synthetic skin substitute was attempted in 1962, but it wasn't until the late 1970s and early 1980s that a successful trial marked the modern era of tissue engineering and regenerative medicine (Archer 2023). Skin and bone regeneration have been primary research focuses due to their simpler structures compared to more complex tissues like heart valves (Perán et al. 2013; Jeong et al. 2017). Recently, shape memory biopolymer-based scaffolds and aerogels have been developed for tissue engineering, serving as both dressing materials and artificial skin by mimicking the extracellular matrix properties of native skin. Furthermore, scaffolds from gelatin, cellulose acetate, and elastin using electrospinning, showing excellent adhesion and proliferation of human keratinocytes during a week of incubation, indicating their potential for skin regeneration and wound healing (Yahya et al. 2021; Khalili et al. 2019).

11.3.1.3 Wound Healing

Wound healing is a vital yet intricate process, increasingly burdened by the rise of chronic diseases. Factors like infection and inflammation can further complicate healing, resulting in severe tissue damage (Raziyeva et al. 2021). Therefore, exploring innovative, eco-friendly approaches to enhance wound healing is critically important. Biopolymers derived from natural, renewable sources offer promising benefits for skin regeneration. They have lower infection risks, are biodegradable through enzyme-controlled processes, exhibit excellent biocompatibility, and possess high chemical versatility. These attributes make them an attractive option for advancing

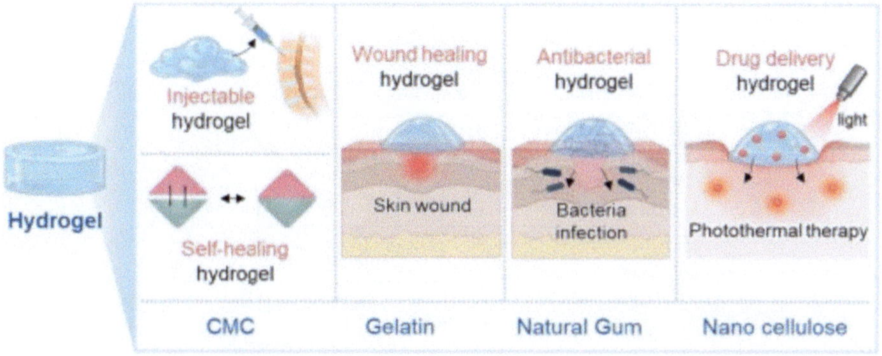

Fig. 11.13 Schematic wound healing principles of hydrogel shape memory biopolymer (Mehvari et al. 2024)

wound care (Moholkar et al. 2021). Shape-memory hydrogels are biodegradable and biocompatible materials made from recombinant telechelic polypeptides, featuring collagen-like end blocks and a random coil-like middle block. The programmed shape is achieved through chemical cross-linking of lysine residues in the random coil. These soft networks can stretch up to 200% and be pinned in a temporary shape by lowering the temperature, allowing the collagen-like ends to form physical nodes. At room temperature, the hydrogel maintains its deformed shape for days but reverts to its original shape within minutes when heated to 50 °C or higher. The shape-memory behavior arises from the reversible association of collagen-like blocks into triple helical nodes, and the rate of shape recovery can be adjusted by altering temperature or the hydrogel's physicochemical properties. As shown in Fig. 11.13, shape memory biopolymer hydrogels also exhibit excellent thermomechanical memory, returning to their original shape even after multiple cycles, and offer the additional advantage of being soft, biodegradable, and biocompatible (Skrzeszewska et al. 2011).

Furthermore, carboxymethyl chitosan, with abundant hydrogen bonding and coordination-active groups, acts as a natural ligand for metal ion coordination, giving it great potential for wound healing applications (Soares et al. 2022). In double network hydrogels, the interpenetrated carboxymethyl chitosan retains its strong chelating ability, forming metal-coordinated gels through spontaneous complexation with metal ions such as Cu^{2+}, Zn^{2+}, Ni^{2+}, Co^{2+}, Fe^{3+}, and Cr^{3+} (Yan et al. 2021; Zhuo et al. 2024). The double network gel features two independent and reversible physical cross-links, enabling multifunctionality such as pH-responsive shape memory, multi-stage self-healing, and antibacterial properties (Li et al. 2023a). This coordination-driven double network gel offers valuable insight for developing multifunctional biopolymer hydrogels, with promising applications in nanocomponent assembly, biomedical systems, and soft electronic biosensing. Notably, the double network gel demonstrates a unique multi-phase self-healing ability, where two gel pieces self-repair within minutes to form a monolithic gel strong enough to support itself when suspended in air (Yan et al. 2020).

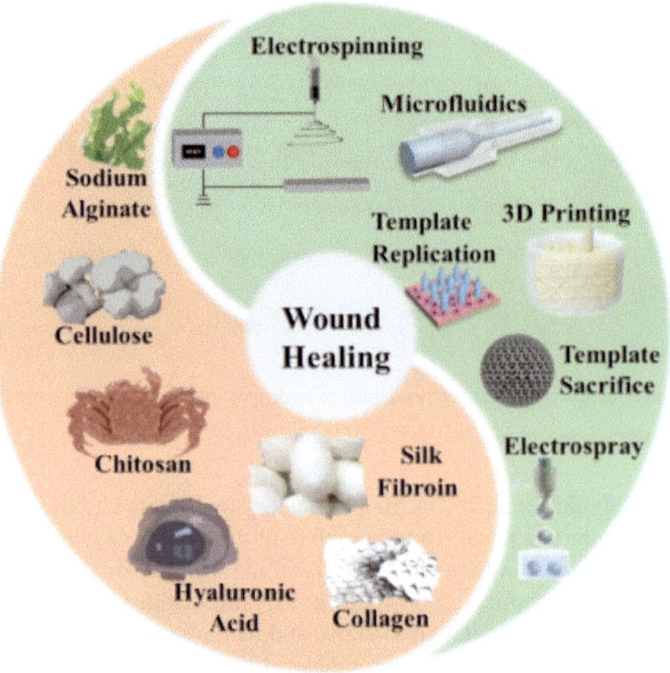

Fig. 11.14 Biopolymers and innovative techniques for wound healing applications (Zhang et al. 2024a)

Moreover, as illustrated in Fig. 11.14, biomaterials such as sodium alginate, nanocellulose, chitosan, collagen, and silk fibroin are commonly used for wound healing applications. These materials can be developed using various advanced techniques, including 3D printing, electrospinning, electrospray, and template replication. These methods enable precise control over the structure and properties of the biomaterials, enhancing their effectiveness in promoting wound healing.

11.3.1.4 Clot Removal Devices

Stroke is a prevalent condition primarily caused by atherosclerosis in nerves and blood vessels (Murphy and Werring 2020). When blood vessels narrow due to hemagglutination and clots, blood flow can be obstructed, leading to a stroke (Meglio et al. 2019). The resulting ischemia can cause irreversible brain damage within a short time and can be life-threatening if untreated (Lee et al. 2018). In addition to medication, non-drug treatments are being explored for stroke patients, offering the potential for more rapid restoration of blood flow compared to traditional drug therapies. The innovative shape memory biopolymer is used to remove clots in stroke patients by first guiding the device into the thrombus area via a tube, puncturing the clot, and then

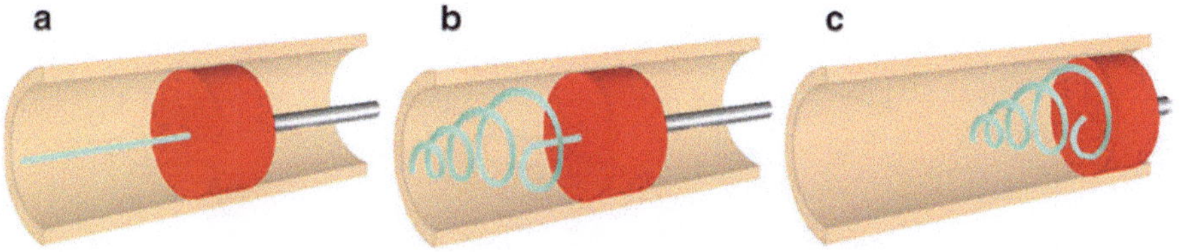

Fig. 11.15 Operating principle of thrombectomy device based on shape memory polymer. **a** The clot is first punctured by device **b** then device is activated to form a coil shape **c** both the device and clot is pulled to remove simultaneously (Zhao et al. 2019; Small et al. 2005)

reverting to a coiled shape to collect and extract it (Fig. 11.15). Recent advancements include the incorporation of laser-driven shape memory biopolymers for increased stiffness and the development of a hybrid device that combines the shape memory polymers with nickel-titanium alloy, resulting in a stronger restoring force.

In another application, shape memory biopolymers are utilized to address vascular restenosis, a condition where the arterial lumen narrows again after a vascular intervention aimed at treating ischemic lesions. Biodegradable shape memory bio polymers, especially those based on poly(lactic) acid (PLA) and poly-caprolactam (PCL), are being used into novel porous vascular scaffolds to effectively address this issue (Failed 2009). Unlike non-biodegradable shape memory alloys, which require secondary surgeries for scaffold and balloon removal-adding pain and burden for patients (Fig. 11.16). These biodegradable materials eliminate the need for such procedures (Zhang et al. 2023). Poly(lactic) acid (PLA) is an optically active material that can undergo hydrolysis by hydrolase enzymes, breaking the polymer down into its monomers. These monomers are further metabolized into carbon dioxide and water. Its composites are widely used in clinical settings, and their degradation rate can be controlled by varying the composition ratio of different components (Zhao et al. 2023).

11.3.1.5 Biosensors

Bioinspired actuators (biosensors) with stimuli-responsive and deformable properties are crucial for applications in artificial tissues, medical devices, diagnostics, and intelligent biosensors. They require biocompatibility, controlled deformability, biodegradability, mechanical durability, and stable reversibility (Zhang et al. 2022). The bionic actuator system composed of genetically engineered silk-elastin-like protein hydrogels and wood-derived cellulose nanofibers, responding to temperature and ionic strength using has been explored using eco-friendly methods. These actuators can perform programmed site-selective actuation and fold into three-dimensional (3D) origami-like shapes, demonstrating reversible deformation and complex spatial transformations, including a biomimetic flower design (Shi et al. 2019; Ge et al. 2021). By incorporating principles of shape-memory biopolymers, these entirely

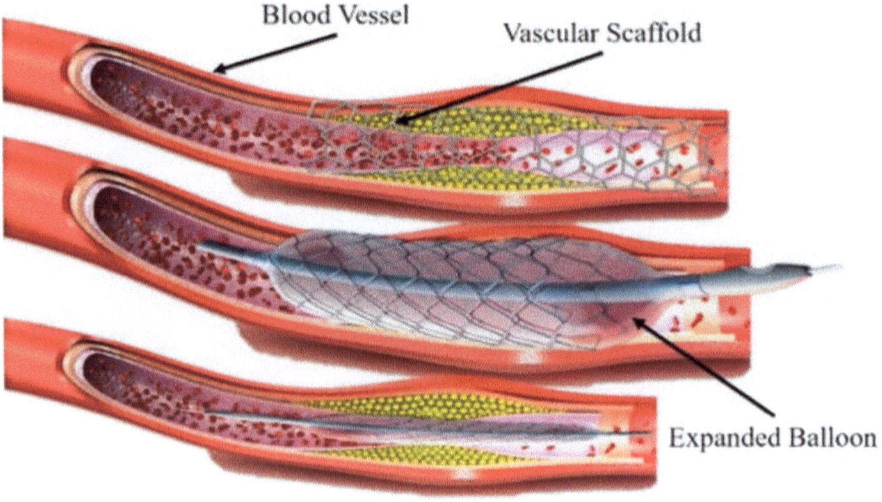

Fig. 11.16 Shape memory biopolymers balloon-assisted vascular implantation (Zhang et al. 2023)

biocompatible and biodegradable actuators enhance adaptability and responsiveness in dynamic biological environments, offering promising avenues for in vivo biomedicine, soft robotics, and bionic research (Wang et al. 2020). An example of shape memory biopolymer biosensor the biomimetic skin-like materials that sense external stimuli can be enhanced by integrating shape-memory biopolymers, improving the functionality of biosensors. And it is reported that the mechanical and sensory capabilities of natural skin while utilizing shape-memory properties for better responsiveness (Youn et al. 2024). Incorporating these advanced materials into biosensors enhances their sensitivity and adaptability, enabling smart microrobots to effectively detect and respond to biological signals in real time, thereby advancing biomedical applications (Neettiyath and Pumera 2024; Zhang et al. 2024b).

Moreover, wearable sensors provide a cutting-edge method for diagnosing diseases by continuously monitoring patients' physiological and pathophysiological data in real-time. Although this technology is still in its early stages, significant progress has been made in developing flexible, polymer-based sensors. Among these, bio-based polymers are gaining increasing attention due to their advantages, including biocompatibility, biodegradability, eco-friendliness, and cost-effectiveness (Shajari et al. 2023). Key natural materials like cellulose, chitosan, silk, gelatin, and alginate have proven effective in enhancing the sensing functionality of these wearable devices. Various fabrication techniques, such as films, hydrogels, aerogels, 3D layers, and textiles, have been explored to develop efficient sensor frameworks (Cui et al. 2020).

Looking ahead, the integration of sustainable, green materials and miniaturized sensor designs offers exciting opportunities for the next generation of wearable sensors, paving the way for more advanced and environmentally friendly technologies (Haick and Tang 2021; Jin et al. 2020).

11.3.1.6 Smart Sutures and Staples

Sutures play a key role in wound healing, and recent advancements have focused on developing materials to improve tissue approximation and wound closure (Dennis et al. 2016). Innovations include barbed, antimicrobial, and drug-eluting sutures, with biostable and bioresorbable materials gaining attention for their superior tissue support. Biodegradable polymeric sutures are also widely explored due to their effectiveness (Joseph et al. 2017). Additionally, stimuli-responsive sutures made from shape-memory biopolymers have been developed, capable of reacting to environmental changes such as electric or magnetic fields. These materials are promising for applications like cell adhesion, nerve regeneration, drug delivery, and controlled degradation in biomedical settings (Das et al. 2020).

Shape-memory biopolymers are suitable enough for biomedical applications due to their lightweight nature, affordability, ease of processing, high recoverable strains, customizable thermo-mechanical properties, and responsiveness to various stimuli, including remote actuation. These materials are particularly promising for active medical devices, including self-tightening sutures, surgical fasteners, intraluminal stents for vascular, urogenital, and gastrointestinal applications, anchoring cannulas, orthopedic fixations, expandable implants for reconstructive or aesthetic surgery, and craniofacial meshes and plates (Strandman and Zhu 2015). The shape-memory property of polymeric materials enables minimally invasive surgery by allowing smaller incisions for compressed medical devices, thus reducing surgical stress. Degradable shape-memory biopolymers, which can be loaded with drugs, are suitable for applications as injectable and implantable self-anchoring rods for targeted fixation and localized drug release (Sun et al. 2012).

Additionally, shape-memory biopolymers are effective in wound closure devices, such as self-tightening sutures and surgical fasteners, which facilitate tissue regeneration and minimize scarring (Narasimhan et al. 2023). Smart degradable sutures have been developed that can tie themselves, aligning with minimally invasive techniques. A notable advancement is the self-untie property in self-anchoring sutures, featuring filaments with retainers that engage and secure tissue without knots, adapting their shape in response to specific stimuli (Feng et al. 2009).

11.3.2 Environmental Applications

11.3.2.1 Biodegradable and Responsive Packaging

Shape memory biopolymers in biodegradable packaging enable smart and sustainable applications (Mishra et al. 2024; Liu et al. 2007). These materials revert to a preset shape when activated by heat or moisture, facilitating self-folding and adjustable designs, which can be utilized in practical packaging solutions such as self-sealing lids, re-closable pouches, or tamper-evident seals. In nature, biopolymers like proteins and nucleic acids fold into complex 3D structures, while lipid bilayers transition from flat sheets to spheres. Their biodegradability, combined with their ability to reduce environmental impact, offers an eco-friendly alternative to conventional plastics, making them ideal for food, medical, and consumer goods packaging. These materials not only enhance functionality but also promote sustainability (Kisak et al. 2004; Fernandes and Gracias 2012; Yaakob et al. 2024; Failed 2015b). Nanocomposite packaging materials derived from biopolymers offer significant potential for advanced smart food packaging with bio-functional properties. Incorporating nanomaterials enhances thermal, mechanical, and gas barrier properties while ensuring biodegradability, resulting in improved barrier performance, increased mechanical strength, and antibacterial activity (Mishra et al. 2024; Yaakob et al. 2024).

11.3.2.2 Smart Textiles

Shape memory biopolymers have promising applications in various areas of smart textiles. In healthcare, they can be used in adaptive garments that adjust to body temperature and movements, providing enhanced comfort for patients (Sokolowski et al. 2007). In sportswear, these materials can optimize fit and performance by responding to physical activity (Hassabo et al. 2023). Additionally, they are ideal for fashion, enabling dynamic designs that change shape or texture (Zarek et al. 2016). With their biodegradable properties, these biopolymers also promote sustainability, making them a valuable choice in eco-friendly textile innovations (Gowri et al. 2022). Stimuli-responsive hydrogel-treated textiles in skin-care products offer moisturizing, whitening, and anti-aging benefits. A facial mask made from thermoresponsive polymeric gel adjusts to body temperature, enabling controlled release of nutrients and vitamins like vitamin C by varying temperature (Failed 2017). Deodorant fabrics utilize stimuli-responsive biopolymeric hydrogels to release agents at specific temperatures. Cross-linked with functional monomers and β-cyclodextrin, these fabrics enhance controlled release properties and can incorporate various hydrophobic deodorant moieties (Pathak 2012).

11.3.3 *Industrial Applications*

11.3.3.1 Actuators and Sensors

Shape memory polymers (SMPs) are one candidate for actuation material in the field of shape control. There is a significant interest in these polymeric actuators as artificial muscles as well as devices with multiple degrees of freedom for endoscopic surgery application (Pilate et al. 2016). The two-way shape-memory effect occurs on both heating and cooling step in the presence or absence of external stress (Qin and Mather 2009). It was found that reversible chain conformations were responsible for this kind of actuation (Raquez et al. 2011). The key to realize two-way shape memory effect (SME) is the formation of internal stress within shape memory polymers (Chen et al. 2008; Stroganov et al. 2015). For soft sensory actuators, work has been progressed in the way of embracing sensors in or on actuators, in which the actuators can get real-time signal feedback by their sensing abilities.

11.3.3.2 3D Printing and Additive Manufacturing

Additive manufacturing (AM), or 3D printing, has arisen as a booming technology for the development of more creative devices and potential functional applications in biomedical, electronics, sensors, and tissue engineering through the pioneering combination of shape memory polymers (SMPs) (Verma and Verma 2022). A challenge remains in developing 3D printed SM devices that have quick responses, strength, precision, and excellent durability, despite the fact that combining "3D printing technology" with SM polymers allows for additional shape-controllable changes to meet real-world necessities such as 3D configuration restoration and target deformation (Huang et al. 2019). "Fused deposition modelling (FDM)" is the widely employed "3D printing technology". FDM builds up the part "layer by layer by extruding a thermoplastic filament". The invention of a printable SMP by FDM can bring in SMPs to a plethora of new possible applications (Rivero et al. 2014; Tabatabaei et al. 2021).

11.4 Challenges and Future Prospects

Shape memory biopolymers have experienced significant advancements, leading to enhanced multifunctionality and performance. These improvements have expanded their applications in various commercial and advanced applications (Hu et al. 2012). This book chapter discusses various types of shame memory biopolymers and highlighting the advantages of exploiting shape memory effects in innovative polymers. Recent advancements in synthesis and fabrication have broadened their application scope, with different application areas. Shape memory biopolymers show significant promise in biomedical applications due to their biocompatibility and mechanical properties. However, challenges such as limited clinical applications, the need for material optimization, integration with biological agents, and precise control of stimuli responsiveness hinder their full potential. Most current research is still in the preclinical stage, focusing on cellular experiments and animal models, particularly for irregular bone repair (Li et al. 2023b). Looking ahead, advancements in 4D printing technology and multi-material printing could revolutionize the creation of adaptable medical devices and custom implants. Future research will also emphasize enhancing sustainability and biodegradability, as well as exploring new material combinations to improve functionality. By addressing these challenges, shape memory biopolymers could significantly impact tissue engineering and regenerative medicine (Khalid et al. 2022).

Future research must address critical challenges to fully harness the potential of SMBs. Emphasis should be placed on synthesizing and reinforcing two-way shape memory biopolymers for greater applicability in soft robotics and flexible electronics, as these materials can achieve recovery stress levels similar to shape memory alloys. Additionally, precise control over shape recovery behavior, particularly the intermediate shapes during recovery, presents exciting opportunities to develop SMPs with single transition phases. Exploring new materials during polymer synthesis and investigating non-reversible and a thermal chemical reaction may also uncover novel shape memory phenomena, further advancing the field.

11.5 Conclusion

Shape memory biopolymers (SMBs) signify a groundbreaking evolution in materials science, integrating the advanced features of smart materials with the eco-friendly characteristics of biopolymers. These unique materials can return to a pre-defined shape upon exposure to external stimuli such as temperature, light, pH, or moisture, opening up diverse applications, particularly in the biomedical field. This chapter offers a comprehensive overview of the design and synthesis of SMBs, emphasizing their molecular architecture and the critical cross-linking methods that enhance their shape memory behavior. Techniques such as the formation of triple helices in protein-based biopolymers and ionic interactions in polysaccharide-based polymers are fundamental in optimizing their functionality. However, challenges such as scalability and cost-effectiveness in the synthesis of environmentally friendly SMBs remain significant barriers that need addressing for broader commercial adoption.

The potential applications of SMBs are vast, especially within biomedicine. They play pivotal roles in drug delivery systems, where they enable controlled release of therapeutic agents, and in tissue engineering, where they provide scaffolding that supports cell growth and regeneration. Furthermore, innovations in wound healing technologies and biosensors highlight their capacity to improve patient outcomes. Beyond the biomedical realm, SMBs offer promising solutions in environmental applications, such as biodegradable packaging and smart textiles, significantly reducing ecological footprints. In industrial sectors, their use in actuators, sensors, and 3D printing enhances operational efficiency and functionality. While SMBs hold immense promise, ongoing research and development are crucial to overcoming existing challenges, including the need for robust materials and scalable production techniques. This chapter lays the groundwork for future investigations into the design, synthesis, and application of SMBs, underscoring their potential to drive advancements across multiple fields while promoting sustainability.

References

Ahn S-K, Deshmukh P, Kasi RM (2010) Shape memory behavior of side-chain liquid crystalline polymer networks triggered by dual transition temperatures. Macromolecules 43(17):7330–7340

Ali G, Sharma M, Salama E-S, Ling Z, Li X (2024) Applications of chitin and chitosan as natural biopolymer: potential sources, pretreatments, and degradation pathways. Biomass Convers Biorefinery 14(4):4567–4581

Archer J (2023) The Seldinger technique: a revolution in medicine. ANZ J Surg 93(S1):184–214

Ashby MF, Johnson K (2013) Materials and design: the art and science of material selection in product design. Butterworth-Heinemann

Balart R, Garcia-Garcia D, Fombuena V, Quiles-Carrillo L, Arrieta MP (2021) Biopolymers from natural resources 13:2532

Basak S (2021) Redesigning the modern applied medical sciences and engineering with shape memory polymers. Adv Compos Hybrid Mater 4:223–234

Basak S, Bandyopadhyay A (2024) Solvent-responsive shape memory polymers for biomedical applications. CRC Press

Behl M, Zotzmann J, Lendlein A (2009) Shape-memory polymers and shape-changing polymers. Shape-Memory Polym 1–40

Bella J (2016) Collagen structure: new tricks from a very old dog. Biochem J 473(8):1001–1025

Bruni A, Serra FG, Deregibus A, Castroflorio T (2019) Shape-memory polymers in dentistry: systematic review and patent landscape report. Materials 12(14):2216

Carnes ME, Pins GD (2020) Skeletal muscle tissue engineering: biomaterials-based strategies for the treatment of volumetric muscle loss. Bioengineering 7(3):85

Chaubaroux C et al (2012) Collagen-based fibrillar multilayer films cross-linked by a natural agent. Biomacromol 13(7):2128–2135

Chen H, Ouyang W, Lawuyi B, Martoni C, Prakash S (2005) Reaction of chitosan with genipin and its fluorogenic attributes for potential microcapsule membrane characterization. J Biomed Mater Res Part a: off J Soc Biomater Japan Soc Biomater Aust Soc Biomater Korean Soc Biomater 75(4):917–927

Chen S, Hu J, Zhuo H, Zhu Y (2008) Two-way shape memory effect in polymer laminates. Mater Lett 62(25):4088–4090

Chen F, Yang K, Zhao D, Yang H (2019) Thermal-and salt-activated shape memory hydrogels based on a gelatin/polyacrylamide double network. RSC Adv 9(32):18619–18626

Cui C, Fu Q, Meng L, Hao S, Dai R, Yang J (2020) Recent progress in natural biopolymers conductive hydrogels for flexible wearable sensors and energy devices: materials, structures, and performance. ACS Appl Bio Mater 4(1):85–121

Das D, Mulchandani N, Kumar A, Katiyar V (2020) Fabrication of stimuli-responsive polymers and their composites: candidates for resorbable sutures. Adv Sustain Polym Synth Fab Charact 121–144

Delaey J, Dubruel P, Van Vlierberghe S (2020) Shape-memory polymers for biomedical applications. Adv Func Mater 30(44):1909047

Dennis C, Sethu S, Nayak S, Mohan L, Morsi Y, Manivasagam G (2016) Suture materials—current and emerging trends. J Biomed Mater Res, Part A 104(6):1544–1559

Di Meglio L et al (2019) Acute ischemic stroke thrombi have an outer shell that impairs fibrinolysis. Neurology 93(18):e1686–e1698

Du H, Song Z, Wang J, Liang Z, Shen Y, You F (2015) Microwave-induced shape-memory effect of silicon carbide/poly (vinyl alcohol) composite. Sens Actuators, A 228:1–8

du Toit LC, Marimuthu T, Kumar P, Choonara YE, Pillay V (2016) Customized shape memory biopolymers. Unfold Biopolym Landsc Front Biomater 2:92e123

Fang Y, Ni Y, Leo S-Y, Taylor C, Basile V, Jiang P (2015) Reconfigurable photonic crystals enabled by pressure-responsive shape-memory polymers. Nat Commun 6(1):7416

Fassini D et al (2021) Diverse and productive source of biopolymer inspiration: marine collagens. Biomacromol 22(5):1815–1834

Feng Y, Behl M, Kelch S, Lendlein A (2009) Biodegradable multiblock copolymers based on oligodepsipeptides with shape-memory properties. Macromol Biosci 9(1):45–54

Fernandes R, Gracias DH (2012) Self-folding polymeric containers for encapsulation and delivery of drugs. Adv Drug Deliv Rev 64(14):1579–1589

Fernandes C, Heggannavar GB, Kariduraganavar MY, Mitchell GR, Alves N, Morouço P (2019) Smart materials for biomedical applications: the usefulness of shape-memory polymers. Appl Mech Mater 890:237–247

Galaev IY, Mattiasson B (1999) 'Smart'polymers and what they could do in biotechnology and medicine. Trends Biotechnol 17(8):335–340

Ge Y et al (2021) Programmable humidity-responsive actuation of polymer films enabled by combining shape memory property and surface-tunable hygroscopicity. ACS Appl Mater Interfaces 13(32):38773–38782

Ghosh SK (2009) Self-healing materials: fundamentals, design strategies, and applications. Wiley Online Library

Gowri M, Latha N, Rajan M (2022) Fiber-and fabric-reinforced shape-memory polymers. In: Shape memory composites based on polymers and metals for 4D printing: processes, applications and challenges. Springer, pp 267–286

Guo W et al (2014) pH-stimulated DNA hydrogels exhibiting shape-memory properties. Adv Mater (Deerfield Beach, Fla.) 27(1):73–78

Gupta A et al (2021) A review of adsorbents for heavy metal decontamination: growing approach to wastewater treatment. Materials 14(16):4702

Haick H, Tang N (2021) Artificial intelligence in medical sensors for clinical decisions. ACS Nano 15(3):3557–3567

Harun-Ur-Rashid M, Imran AB (2024) Emerging trends in engineering polymers: a paradigm shift in material engineering

Hassabo AG et al (2023) Applications of nanotechnology in the creation of smart sportswear for enhanced sports performance: efficiency and comfort. J Text Color Polym Sci 20(1):11–28

Hiramatsu N, Chiang W-C, Kurt TD, Sigurdson CJ, Lin JH (2015) Multiple mechanisms of unfolded protein response–induced cell death. Am J Pathol 185(7):1800–1808

Holman H, Kavarana MN, Rajab TK (2021) Smart materials in cardiovascular implants: shape memory alloys and shape memory polymers. Artif Organs 45(5):454–463

Hsu RS et al (2019) Adaptable microporous hydrogels of propagating NGF-gradient by injectable building blocks for accelerated axonal outgrowth. Adv Sci 6(16):1900520

Hu J, Zhu Y, Huang H, Lu J (2012) Recent advances in shape–memory polymers: Structure, mechanism, functionality, modeling and applications. Prog Polym Sci 37(12):1720–1763

Huang H-J, Tsai Y-L, Lin S-H, Hsu S-H (2019) Smart polymers for cell therapy and precision medicine. J Biomed Sci 26:1–11

Izawa H, Kadokawa J-I (2010) Preparation and characterizations of functional ionic liquid-gel and hydrogel materials of xanthan gum. J Mater Chem 20(25):5235–5241

Jeong K-H, Park D, Lee Y-C (2017) Polymer-based hydrogel scaffolds for skin tissue engineering applications: a mini-review. J Polym Res 24:1–10
Jin X, Liu C, Xu T, Su L, Zhang X (2020) Artificial intelligence biosensors: challenges and prospects. Biosens Bioelectron 165:112412
Joseph B, George A, Gopi S, Kalarikkal N, Thomas S (2017) Polymer sutures for simultaneous wound healing and drug delivery–a review. Int J Pharm 524(1–2):454–466
Khalid MY, Arif ZU, Noroozi R, Zolfagharian A, Bodaghi M (2022) 4D printing of shape memory polymer composites: a review on fabrication techniques, applications, and future perspectives. J Manuf Process 81:759–797
Khalid MY, Arif ZU, Al Rashid A, Bukhari SMZS, Hossain M, Koç M (2024) Shape-memory and self-healing properties of sustainable cellulosic nanofibers-based hybrid materials for novel applications. Giant 100299
Khalili S, Khorasani SN, Razavi SM, Hashemibeni B, Tamayol A (2019) Nanofibrous scaffolds with biomimetic composition for skin regeneration. Appl Biochem Biotechnol 187:1193–1203
Kirillova A, Ionov L (2019) Shape-changing polymers for biomedical applications. J Mater Chem B 7(10):1597–1624
Kirillova A et al (2021) Fabrication of biomedical scaffolds using biodegradable polymers. Chem Rev 121(18):11238–11304
Kisak E, Coldren B, Evans C, Boyer C, Zasadzinski J (2004) The vesosome—a multicompartment drug delivery vehicle. Curr Med Chem 11(2):199–219
Korde JM, Kandasubramanian B (2020) Naturally biomimicked smart shape memory hydrogels for biomedical functions. Chem Eng J 379:122430
Kousbroek R (2018) Shape memory alloys. In: Metal and ceramic biomaterials. CRC Press, pp 63–90
Laasch HU (2012) Current designs of self-expanding stents. In: Self-expandable stents in the gastrointestinal tract. Springer, pp 51–69
Lee KY, Mooney DJ (2012) Alginate: properties and biomedical applications. Prog Polym Sci 37(1):106–126
Lee RH et al (2018) Cerebral ischemia and neuroregeneration. Neural Regen Res 13(3):373–385
Lendlein A, Kelch S (2002) Shape-memory polymers. Angew Chem Int Ed 41(12):2034–2057
Lendlein A, Langer R (2002) Biodegradable, elastic shape-memory polymers for potential biomedical applications. Science 296(5573):1673–1676
Lendlein A, Balk M, Tarazona NA, Gould OE (2019) Bioperspectives for shape-memory polymers as shape programmable, active materials. Biomacromol 20(10):3627–3640
Leng J, Lan X, Liu Y, Du S (2011) Shape-memory polymers and their composites: stimulus methods and applications. Prog Mater Sci 56(7):1077–1135
Leng J, Du S (2010) Shape-memory polymers and multifunctional composites. CRC Press (2010)
Li Z et al (2016) Mussel-inspired multifunctional supramolecular hydrogels with self-healing, shape memory and adhesive properties. Polym Chem 7(34):5343–5346
Li X, Li M, Tang L, Shi D, Lam E, Bae J (2023a) 3D shape morphing of stimuli-responsive composite hydrogels. Mater Chem Front 7(23):5989–6034
Li T, Chen L, Yuan Y, Shi R (2023b) The current status, prospects, and challenges of shape memory polymers application in bone tissue engineering. Polymers 15(3):556
Liang R, Wang L, Yu H, Khan A, Amin BU, Khan RU (2019) Molecular design, synthesis and biomedical applications of stimuli-responsive shape memory hydrogels. Eur Polymer J 114:380–396
Liu C, Qin H, Mather P (2007) Review of progress in shape-memory polymers. J Mater Chem 17(16):1543–1558
Liu Z, Jiao D, Zhang Z (2015) Remarkable shape memory effect of a natural biopolymer in aqueous environment. Biomaterials 65:13–21
Liu D, Yang F, Xiong F, Gu N (2016) The smart drug delivery system and its clinical potential. Theranostics 6(9):1306

Löwenberg C, Balk M, Wischke C, Behl M, Lendlein A (2017) Shape-memory hydrogels: evolution of structural principles to enable shape switching of hydrophilic polymer networks. Acc Chem Res 50(4):723–732
Löwenberg C, Julich-Gruner KK, Neffe AT, Behl M, Lendlein A (2020) Salt-induced shape-memory effect in gelatin-based hydrogels. Biomacromol 21(6):2024–2031
Luo L et al (2021) Application of bioactive hydrogels combined with dental pulp stem cells for the repair of large gap peripheral nerve injuries. Bioact Mater 6(3):638–654
Luo K, Xu Z-Y, Zhan R, Yang K-K (2023) Creating shape memory polymers from biobased resources. Sustain Polym Energy 1(2):10009
Lv S, Dudek DM, Cao Y, Balamurali M, Gosline J, Li H (2010) Designed biomaterials to mimic the mechanical properties of muscles. Nature 465(7294):69–73
Mantha S et al (2019) Smart hydrogels in tissue engineering and regenerative medicine. Materials 12(20):3323
Mbeh D, do Nascimento R (2015) Biocompatibility of shape-memory polymers for biomedical applications. In: Shape memory polymers for biomedical applications. Elsevier, pp 77–95
Mehvari F, Ramezanzade V, An J, Kim J, Dinari M, Kim JS (2024) Biopolymer-based hydrogels for biomedical applications: bioactivity and wound healing properties. Coord Chem Rev 518:216093
Mishra B et al. (2024) Recent advances in sustainable biopolymer-based nanocomposites for smart food packaging: a review. Int J Biol Macromol 135583
Moholkar DN, Sadalage PS, Peixoto D, Paiva-Santos AC, Pawar KD (2021) Recent advances in biopolymer-based formulations for wound healing applications. Eur Polym J 160:110784
Molina BG, Ocón G, Silva FM, Iribarren JI, Armelin E, Alemán C (2023) Thermally-induced shape memory behavior of polylactic acid/polycaprolactone blends. Eur Polym J 196:112230
Motaung TE, Linganiso LZ (2018) Critical review on agrowaste cellulose applications for biopolymers. Int J Plast Technol 22(2):185–216
Müller A et al (2014) Bacterial nanocellulose with a shape-memory effect as potential drug delivery system. RSC Adv 4(100):57173–57184
Munoz V, Cerminara M (2016) When fast is better: protein folding fundamentals and mechanisms from ultrafast approaches. Biochem J 473(17):2545–2559
Murphy SJ, Werring DJ (2020) Stroke: causes and clinical features. Medicine 48(9):561–566
Nagahama K, Ueda Y, Ouchi T, Ohya Y (2009) Biodegradable shape-memory polymers exhibiting sharp thermal transitions and controlled drug release. Biomacromol 10(7):1789–1794
Najjari A et al (2022) Smart piezoelectric biomaterials for tissue engineering and regenerative medicine: a review. Biomed Eng Biomedizinische Technik 67(2):71–88
Narasimhan AK, Rahul TS, Krishnan S (2023) Revisiting the properties of suture materials: an overview. Adv Technol Polym Mater Surgical Sutures 199–235
Navarro-Baena I, Sessini V, Dominici F, Torre L, Kenny JM, Peponi L (2016) Design of biodegradable blends based on PLA and PCL: from morphological, thermal and mechanical studies to shape memory behavior. Polym Degrad Stab 132:97–108
Neettiyath A, Pumera M (2024) Micro/nanorobots for advanced light-based biosensing and imaging. Adv Funct Mater 2415875
Neffe AT, Löwenberg C, Julich-Gruner KK, Behl M, Lendlein A (2021) Thermally-induced shape-memory behavior of degradable gelatin-based networks. Int J Mol Sci 22(11):5892
Olawumi MA, Omigbodun FT, Oladapo BI (2024) Integration of sustainable and net-zero concepts in shape-memory polymer composites to enhance environmental performance. Biomimetics 9(9):530
Parvizi J (2010) High yield orthopaedics E-Book. Elsevier Health Sciences
Pathak H (2012) Smart' textile materials: a paradigm shift. J Text Sci Eng Ineering 2:107
Pattanashetti NA, Heggannavar GB, Kariduraganavar MY (2017) Smart biopolymers and their biomedical applications. Procedia Manuf 12:263–279
Peponi L, Navarro-Baena I, Kenny JM (2014) Shape memory polymers: properties, synthesis and applications. In: Smart polymers and their applications. Elsevier, pp 204–236

Perán M, García MA, Lopez-Ruiz E, Jiménez G, Marchal JA (2013) How can nanotechnology help to repair the body? advances in cardiac, skin, bone, cartilage and nerve tissue regeneration. Materials 6(4):1333–1359

Peterson GI, Dobrynin AV, Becker ML (2017) Biodegradable shape memory polymers in medicine. Adv Healthc Mater 6(21):1700694

Pettignano A, Aguilera DA, Tanchoux N, Bernardi L, Quignard F (2019) Alginate: a versatile biopolymer for functional advanced materials for catalysis. In: Studies in surface science and catalysis, vol 178. Elsevier, pp 357–375

Pfau MR, Grunlan MA (2021) Smart scaffolds: shape memory polymers (SMPs) in tissue engineering. J Mater Chem B 9(21):4287–4297

Pilate F, Toncheva A, Dubois P, Raquez J-M (2016) Shape-memory polymers for multiple applications in the materials world. Eur Polym J 80:268–294

Pisani S, Genta I, Modena T, Dorati R, Benazzo M, Conti B (2022) Shape-memory polymers hallmarks and their biomedical applications in the form of nanofibers. Int J Mol Sci 23(3):1290

Qi Z et al (2022) A dual-drug enhanced injectable hydrogel incorporated with neural stem cells for combination therapy in spinal cord injury. Chem Eng J 427:130906

Qiao C, Zhang J, Ma X, Liu W, Liu Q (2018) Effect of salt on the coil-helix transition of gelatin at early stages: optical rotation, rheology and DSC studies. Int J Biol Macromol 107:1074–1079

Qin H, Mather PT (2009) Combined one-way and two-way shape memory in a glass-forming nematic network. Macromolecules 42(1):273–280

Ramaraju H, Akman RE, Safranski DL, Hollister SJ (2020) Designing biodegradable shape memory polymers for tissue repair. Adv Func Mater 30(44):2002014

Ramezani M, Getya D, Gitsov I, Monroe MBB (2024) Solvent-free synthesis of biostable segmented polyurethane shape memory polymers for biomedical applications. J Mater Chem B 12(5):1217–1231

Raquez JM et al (2011) Design of cross-linked semicrystalline poly (ε-caprolactone)-based networks with one-way and two-way shape-memory properties through Diels–Alder reactions. Chem A Eur J 17(36):10135–10143

Rasool A et al (2024) Chitosan-based smart polymeric hydrogels and their prospective applications in biomedicine. Starch-Stärke 76(1–2):2100150

Raziyeva K, Kim Y, Zharkinbekov Z, Kassymbek K, Jimi S, Saparov A (2021) Immunology of acute and chronic wound healing. Biomolecules 11(5):700

Raghavendra GM, Varaprasad K, Jayaramudu T (2015) Biomaterials: design, development and biomedical applications, nanotechnology applications for tissue engineering. William Andrew Publishing Norwich, NY, USA

Raturi LP, Singh J (2018) Sunlight-driven eco-friendly smart curtain based on infrared responsive graphene oxide-polymer photoactuators. Sci Repor 8(1):3687

Rivero G, Nguyen L-TT, Hillewaere XK, Du Prez FE (2014) One-pot thermo-remendable shape memory polyurethanes. Macromolecules 47(6):2010–2018

Rumon MHM et al (2024) Polysaccharide-based hydrogels for advanced biomedical engineering applications. ACS Polym Au

Ruiz-Rubio L, Pérez-Álvarez L, Vilas-Vilela JL (2020) Biodegradable shape-memory polymers. Shape Memory Polym Blends Compos Adv Appl 219–236

Sabahi N, Chen W, Wang C-H, Kruzic JJ, Li X (2020) A review on additive manufacturing of shape-memory materials for biomedical applications. Jom 72:1229–1253

Sacco P, Furlani F, De Marzo G, Marsich E, Paoletti S, Donati I (2018) Concepts for developing physical gels of chitosan and of chitosan derivatives. Gels 4(3):67

Saha S, Arshad M, Zubair M, Ullah A (2019) Keratin as a biopolymer. In: Keratin as a protein biopolymer: extraction from waste biomass and applications, pp 163–185

Saidy NT et al (2019) Biologically inspired scaffolds for heart valve tissue engineering via melt electrowriting. Small 15(24):1900873

Schnepp Z (2013) Biopolymers as a flexible resource for nanochemistry. Angew Chem Int Ed 52(4):1096–1108

Schönfeld D, Chalissery D, Wenz F, Specht M, Eberl C, Pretsch T (2021) Actuating shape memory polymer for thermoresponsive soft robotic gripper and programmable materials. Molecules 26(3):522
Sessini V et al (2024) Thermally-activated shape memory behavior of biodegradable blends based on plasticized PLA and thermoplastic starch. Polymers 16(8):1107
Sezer AD (2016) Smart drug delivery system. BoD–Books on Demand
Shahabi-Ghahfarrokhi I, Almasi H, Babaei-Ghazvini A (2020) Characteristics of biopolymers from natural resources. In: Processing and development of polysaccharide-based biopolymers for packaging applications. Elsevier, pp 49–95
Shajari S, Kuruvinashetti K, Komeili A, Sundararaj U (2023) The emergence of AI-based wearable sensors for digital health technology: a review. Sensors 23(23):9498
Shi Q, Liu H, Tang D, Li Y, Li X, Xu F (2019) Bioactuators based on stimulus-responsive hydrogels and their emerging biomedical applications. NPG Asia Mater 11(1):64
Sikhosana S, Gumede T, Malebo N, Ogundeji A (2021) Poly (lactic acid) and its composites as functional materials for 3-d scaffolds in biomedical applications: a mini-review of recent trends. eXPRESS Polym Lett 15(6):568–580
Silva JM, Caridade SG, Reis RL, Mano JF (2016) Polysaccharide-based freestanding multilayered membranes exhibiting reversible switchable properties. Soft Matter 12(4):1200–1209
Singh AB, Khandelwal C, Dangayach GS (2024) Revolutionizing healthcare materials: Innovations in processing, advancements, and challenges for enhanced medical device integration and performance. J Micromanuf 25165984241256234
Singhwane A et al (2023) Smart shape memory polymers and their significance in biomedical applications: a review. Polym Adv Technol 34(11):3552–3568
Skrzeszewska PJ, Jong LN, de Wolf FA, Cohen Stuart MA, van der Gucht J (2011) Shape-memory effects in biopolymer networks with collagen-like transient nodes. Biomacromolecules 12(6):2285–2292
Small W IV, Wilson TS, Benett WJ, Loge JM, Maitland DJ (2005) Laser-activated shape memory polymer intravascular thrombectomy device. Opt Express 13(20):8204–8213
Snoeck D (2015) Self-healing and microstructure of cementitious materials with microfibres and superabsorbent polymers. Ghent University
Soares AMBF, Carvalho CLC, de Andrade Rodrigues G, Luz RA, Gerôncio ET Cantanhêde W (2022) Supramolecular electrochemistry: recent trends and perspectives. Adv Bioelectrochem Vol 1: Surf, Electron Transf Techn 89–113
Sokolowski W, Metcalfe A, Hayashi S, Raymond J (2007) Medical applications of shape memory polymers. Biomed Mater 2(1):S23
Solazzo M, O'Brien FJ, Nicolosi V, Monaghan MG (2019) The rationale and emergence of electroconductive biomaterial scaffolds in cardiac tissue engineering. APL Bioeng 3(4)
Sternberg K (2009) Current requirements for polymeric biomaterials in otolaryngology. In: GMS current topics in otorhinolaryngology, head and neck surgery, vol 8
Strandman S, Zhu X (2015) Biodegradable shape-memory polymers for biomedical applications. In: Shape memory polymers for biomedical applications. Elsevier, pp 219–245
Stroganov V, Al-Hussein M, Sommer J-U, Janke A, Zakharchenko S, Ionov L (2015) Reversible thermosensitive biodegradable polymeric actuators based on confined crystallization. Nano Lett 15(3):1786–1790
Sugiyama N (2004) Influence of thermal history on the stability of gelatin gels. Shizuokaken Kougyogijyutsu Senta Kenkyu Houkoku 49:78–80
Sulaiman H, Taha MR, Abd Rahman N, Taib AM (2022) Performance of soil stabilized with biopolymer materials–xanthan gum and guar gum. Phys Chem Earth, Parts A/B/C 128:103276
Sun L et al (2012) Stimulus-responsive shape memory materials: a review. Mater des 33:577–640
Tabatabaei M, Taleghani AD, Li G, Zhang T (2021) Shape memory polymers as lost circulation materials for sealing wide-opened natural fractures. SPE Drill Complet 36(04):931–942
Talegaonkar S, Sharma H, Pandey S, Mishra PK, Wimmer R (2017) Bionanocomposites: smart biodegradable packaging material for food preservation. In: Food packaging. Elsevier, pp 79–110

Tan G-YA et al (2014) Start a research on biopolymer polyhydroxyalkanoate (PHA): a review. Polymers 6(3):706–754
Tănase EE, Râpă M, Popa O (2014) Biopolymers based on renewable resources—a review
Thakur S (2017) Shape memory polymers for smart textile applications. Text Adv Appl 10
Thadepalli S (2022) Review of multifarious applications of polymers in medical and health care textiles. Mater Today: Proc 55:330–336
Uyan M, Celiktas MS (2023) Evaluation of the bio-based materials utilization in shape memory polymer composites production. Eur Polym J 195:112196
Vach Agocsova S et al (2023) Resorbable biomaterials used for 3D scaffolds in tissue engineering: a review. Materials 16(12):4267
Valioglu F et al (2024) Recent advances in shape memory scaffolds and regenerative outcomes. Biomed Eng Lett 1–23
Verma S, Verma VK (2022) Shape memory polymers for additive manufacturing: an overview. Mater Today: Proceed 57:2077–2081
Wang L, Cao L, Shansky J, Wang Z, Mooney D, Vandenburgh H (2014) Minimally invasive approach to the repair of injured skeletal muscle with a shape-memory scaffold. Mol Ther 22(8):1441–1449
Wang Y et al (2020) Stimuli-responsive composite biopolymer actuators with selective spatial deformation behavior. Proc Natl Acad Sci 117(25):14602–14608
Wischke C, Neffe AT, Steuer S, Lendlein A (2009) Evaluation of a degradable shape-memory polymer network as matrix for controlled drug release. J Control Release 138(3):243–250
Wu H et al (2024) Current status and challenges of shape memory scaffolds in biomedical applications. MedComm–Biomater Appl 3(3)e95
Xiang LIU, Cheng-qiang Z, Yi-qun H, Yu-xia FAN, Ke-qiang LAI (2018) Effect of sodium chloride content on gel properties and structure of tilapia fish skin gelatin. Sci Technol Food Ind 39(20):18–24. https://doi.org/10.13386/j.issn1002-0306.2018.20.004
Yaakob MNA, Roslan R, Rahim MHA (2024) Rheologic and shape memory properties of natural biopolymers. In: Handbook of natural polymers, vol 2. Elsevier, pp 501–522
Yahya EB et al (2021) Insights into the role of biopolymer aerogel scaffolds in tissue engineering and regenerative medicine. Polymers 13(10):1612
Yan K et al (2020) A multifunctional metal-biopolymer coordinated double network hydrogel combined with multi-stimulus responsiveness, self-healing, shape memory and antibacterial properties. Biomater Sci 8(11):3193–3201
Yan K et al (2021) Interpenetrating polysaccharide-based hydrogel: a dynamically responsive versatile medium for precisely controlled synthesis of nanometals. Mater Sci Eng, C 127:112211
Yang J-A, Yeom J, Hwang BW, Hoffman AS, Hahn SK (2014) In situ-forming injectable hydrogels for regenerative medicine. Prog Polym Sci 39(12):1973–1986
Yates MR, Barlow CY (2013) Life cycle assessments of biodegradable, commercial biopolymers—a critical review. Resour Conserv Recycl 78:54–66
Youn S, Ki M-R, Abdelhamid MA, Pack S-P (2024) Biomimetic materials for skin tissue regeneration and electronic skin. Biomimetics 9(5):278
Yu L, Yu H (2015) Light-powered tumbler movement of graphene oxide/polymer nanocomposites. ACS Appl Mater Interfaces 7(6):3834–3839
Zare M, Davoodi P, Ramakrishna S (2021) Electrospun shape memory polymer micro-/nanofibers and tailoring their roles for biomedical applications. Nanomaterials 11(4):933
Zarek M et al (2016) 4D printing shape memory polymers for dynamic jewellery and fashionwear. Virtual Phys Prototyp 11(4):263–270
Zhang L et al (2014) Bio-based shape memory polyurethanes (Bio-SMPUs) with short side chains in the soft segment. J Mater Chem A 2(29):11490–11498
Zhang Z, Wang Y, Wang Q, Shang L (2022) Smart film actuators for biomedical applications. Small 18(36):2105116

Zhang L, Hanif M, Li J, Shah AH, Hussain W, Zhang G (2023) Fused deposition modeling and characterization of heat shape memory poly (lactic) acid-based porous vascular scaffold. Polymers 15(2):390
Zhang H, Lin X, Cao X, Wang Y, Wang J, Zhao Y (2024a) Developing natural polymers for skin wound healing. Bioactive Mater 33:355–376
Zhang H, Tang J, Cao H, Wang C, Shen C, Liu J (2024b) Review of the applications of micro/nanorobots in biomedicine. ACS Appl Nano Mater 7(15):17151–17192
Zhao W, Liu L, Zhang F, Leng J, Liu Y (2019) Shape memory polymers and their composites in biomedical applications. Mater Sci Eng, C 97:864–883
Zhao W, Yue C, Liu L, Liu Y, Leng J (2023) Research progress of shape memory polymer and 4D printing in biomedical application. Adv Healthc Mater 12(16):2201975
Zheng N, Huang Y, Zhao Q, Feng X, Xie T (2016) Shape memory polymers for flexible electronics. Sci Sinica Phys Mech Astron 46(4):044602
Zhou H et al (2021) The preparation and characterization of biodegradable PCL/PLA shape memory blends. J Macromol Sci Part A 58(10):669–676
Zhuo S, Liang Y, Wu Z, Zhao X, Han Y, Guo B (2024) Supramolecular hydrogels for wound repair and hemostasis. Mater Horiz

Chapter 12
Biopolymers: The Next Frontier in Therapeutics

Nazim Hussain, Hira Ijaz, Zainab Amjad, Rabia Arooj, Waqas Ahmed Khan, Hafiza Ayesha Malik, Ayesha Munawar, and Umme Laila

Abstract Biopolymers are biodegradable polymers which are synthesized by living organisms and they have found remarkably interesting potential as therapeutic materials in the recent past based on their renewable nature, compatibility with biological systems, degradability, and lack of toxicity. They are extracted from monomers like saccharides, amino acids, and nucleic acids. Biopolymers represent potential applications in wound healing, tissue engineering, and drug delivery systems. This chapter focuses on the basic characteristics of biopolymers and divides them into natural, synthetic, and semi-synthetic ones, such as alginate, chitosan, poly lactic acid, and poly glycolic acid. Biopolymers are employed frequently in scaffolds for tissue engineering, as well as therapies like gene delivery and drug delivery systems based in hydrogel and nanoparticles. There are also highlighted impacts of biopolymers, for instance, the fact that they reduce toxicity level in comparison with non-biodegradable polymers so that they are suitable in various advanced therapeutic applications. Current challenges and limitations with ongoing clinical trials of biopolymers in therapeutics are also discussed in this chapter. Novel technologies of three-dimensional bioprinting and nanotechnology are also explained to illustrate the advancement of application of biopolymers in regenerative medicine and genetic therapies. These advancements present potential for creating new varieties of medical treatments which are safer, more effective and easier to implement in the future.

Graphical Abstract

N. Hussain (✉) · H. Ijaz · Z. Amjad · R. Arooj · H. A. Malik · A. Munawar · U. Laila
Center for Applied Molecular Biology (CAMB), University of the Punjab, Lahore, Pakistan
e-mail: nazim.camb@pu.edu.pk

W. A. Khan
Department of Biotechnology, University of Sargodha, Sargodha, Pakistan
e-mail: waqas.ahmad@uos.edu.pk

L. Ravindran et al. (eds.), *Macro, Micro and Nanocomposites from Sustainable Sources*,
Advanced Structured Materials 244, https://doi.org/10.1007/978-981-95-2469-3_12

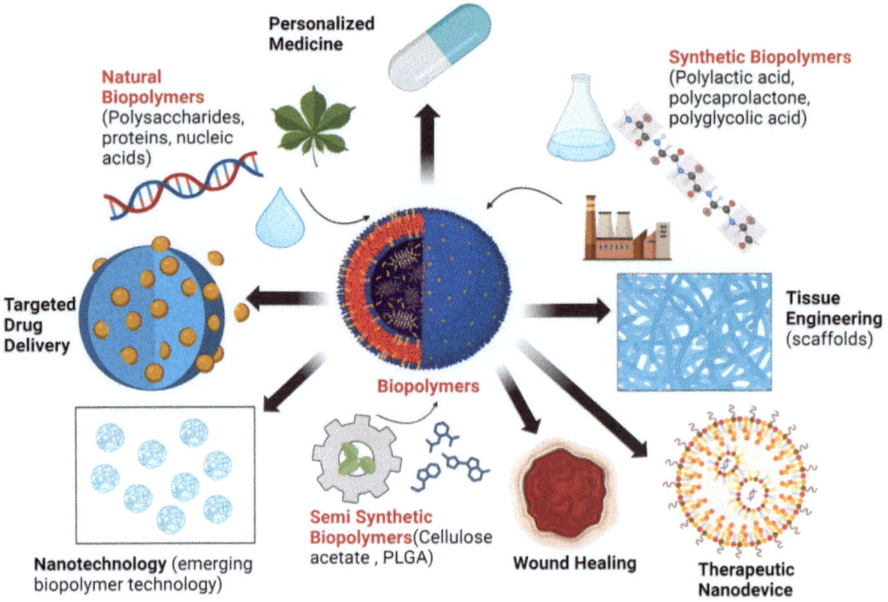

12.1 Introduction to Biopolymers

Biopolymers are a class of polymers where the prefix "bio" refers to living organisms. Living organisms produce biopolymers with either branched or linear structure which is composed of monomeric units such as saccharides, amino acids, and nucleic acids (Khalil et al. 2023). Biopolymers are derived from renewable sources and these are inexpensive, frequently found in nature, biocompatible, and biodegradable without leaving any harmful residues. These qualities of biopolymers make them an exceptional choice for therapeutic applications (Babu et al. 2013). Biopolymers are used as hydrogels in wound healing and films in liver and cardiac tissue engineering due to their excellent attributes of non-toxicity and biodegradability (Basu et al. 2017). Biopolymers interact with biological systems in a highly specific way and mimic natural processes which provide therapeutic effects with minimal side effects. Biopolymers are categorized in to three types based on their composition and origin including natural biopolymers, synthetic biopolymers, and semisynthetic biopolymers (Kalia and Avérous 2011).

Natural biopolymers include alginate, agar, carrageenan, starch, chitosan, protein, gluten, collagen, fibrin, gelatin, etc., and these are used in drug delivery systems, wound healing, and tissue engineering. The polymeric material is utilized in many applications as film coating agents, emulsifying agents, disintegrating agents, binding agents, viscosity modifiers, drug release matrix, and bioadhesives (Nitta and Numata 2013). Synthetic biopolymers include poly lactic acid (PLA), poly glycolic acid

(PGA), polycaprolactone (PCL), polybutylene succinate (PBS) and poly vinyl alcohol (PVC), etc., and they are extensively used in therapeutic applications. These synthetic biopolymers and their derivatives have excellent attributes of low cytotoxicity, biocompatibility, and biodegradability which makes them an exceptional use for the fabrication of tissue scaffolds as they mimic the natural processes of extracellular matrix of host (Gunatillake et al. 2003). The purpose of biopolymer-based scaffolds is regeneration of damaged organs and tissues in the host body. These scaffolds disintegrate and proliferate efficiently for the engineering of tissues like vascular, skin, bones, cartilage, etc., and cells and organs in the body and enhance body's natural healing process. Non-biocompatible polymers implants produce high toxicity, immunological activity, and severe inflammation in the body. The use of biopolymer-based scaffolds help in the minimization of toxic effects of their conventional counterparts.

Biopolymers are engineered into hydrogels, nanoparticles and microspheres to provide sustainable release of therapeutic agents in drug delivery process (Fazal et al. 2023). In tissue engineering, biopolymers are used as scaffolds for tissue regeneration and healing process. In gene therapy, biopolymers are used as effective carriers for delivery of therapeutic agents to treat various genetic disorders. Biopolymers like chitosan and collagen are used for wound healing where they are used as dressings to facilitate the migration of cells crucial for closure of wound (Khademhosseini and Langer 2016). Recent advancements in 3D bioprinting and nanotechnology have emerged as significant tools in tissue engineering, drug delivery, and regenerative medicine. As researchers continue to explore the biopolymers, their role in shaping the future includes the development of more effective, sustainable, and safer methods in the field of therapeutics (Khang 2017).

12.2 Types of Biopolymers

Biopolymers are classified into various categories on the basis of their origin, biological composition, and source. Each category of biopolymer exhibits unique attributes that make it suitable for therapeutic applications.

12.2.1 Nucleic Acids

Nucleic acids are linear biopolymers that which consist of small molecules of monomer units called nucleotides linked in a sequence (Biswas 2022). DNA, RNA (including mRNA, siRNA, miRNA) are the types of nucleic acids potentially used in the field of therapeutics. Nucleic acids are used in gene therapy to treat various genetic disorders. CRISPR-Cas 9 gene editing is used to treat Transthyretin amyloidosis which is a life threatening disease which is caused by progressive accumulation of TTR protein in heart and nerve tissues (Gillmore et al. 2021).

mRNA-based therapeutics are used in regenerative medicine, cancer treatment, and vaccination of various infectious diseases (Gupta et al. 2021). RNAi and siRNA-based therapeutics are used to treat cancer as they can inhibit the expression of any cancer related genes (Traber and Yu 2023). Antisense oligonucleotides are also used to treat various infectious diseases, cancer, and genetic disorders as they are used to inhibit or alter the activity of defective genes (Ramasamy et al. 2022).

12.2.2 Proteins

Proteins are made up of amino acids which are linked together by peptide bonds to make a polypeptide. Proteins are used in therapeutics due to their high biological activity and ability to form gels and fibers (Biswas 2022). Collagen due to its excellent biocompatibility is used in drug delivery systems in the form of hydrogels, coating, film, and micro particle. It is also used in natural wound healing process and in tissue engineering as scaffold to provide a suitable environment for the repair and regeneration of tissue (Vashist et al. 2014). Gelatin is also used in therapeutics with a binding protein to deliver therapeutic agents in the target cell. It is also used to create microsphere and nanoparticle to release specific therapeutic agent at the targeted site (Smart 2004). Fibrin and silk fibroin have excellent biocompatibility and biodegradability properties and they are used in sutures and wound dressings to promote wound healing (Mazurek et al. 2022).

12.2.3 Polysaccharides

Polysaccharides consist of linear or branched long chains of monosaccharide molecules linked together by glycosidic bonds. Polysaccharides include alginate, hyaluronic acid, starch, glycogen, cellulose, and chitin. Alginate polysaccharide, chitosan, and cellulose are excellent candidates for drug delivery system as they have the ability to encapsulate the therapeutic agents and their effective release in the host. Hyaluronic acid-based hydrogels are used in wound dressing that helps in accelerated wound healing. Carboxymethyl cellulose derivatives and chitosan-based biopolymers are used in tissue engineering as scaffolds to make porous structures for cell growth and skin, bone and cartilage tissue regeneration (Biswas 2022).

12.2.4 Polyesters

Polyesters are a group of synthetic biopolymers that are produced by the process of polymerization and they have ester functional group. Polyglycolic acid (PGA), polylactic acid (PLA), polycaprolactone (PCL), polyvinyl alcohol (PVA), polyacrylamide

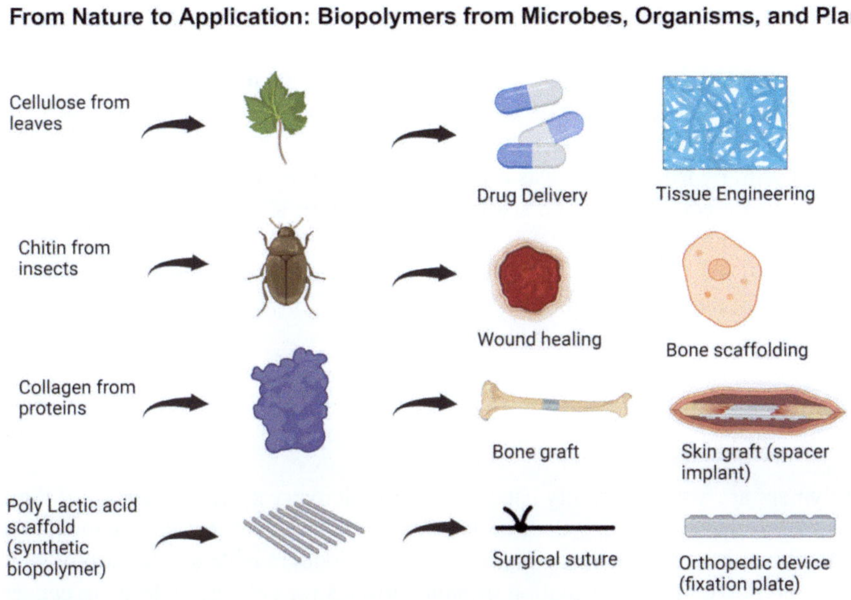

Fig. 12.1 Sources of biopolymers and their potential therapeutic uses

(PAM), and polyethylene oxide (PEO) are biopolymers used in tissue engineering as scaffolds due to their excellent properties of biodegradability, biocompatibility, and low cytotoxicity (Vedadghavami et al. 2017).

Figure 12.1 shows the different sources of biopolymers like cellulose, chitin, collagen, and poly lactic acid scaffold (synthetic polymer) with their biomedical applications in drug delivery, tissue engineering, wound dressings, bone and skin grafts, surgical sutures, and orthopedic devices.

12.3 Applications of Biopolymers in Therapeutics

12.3.1 Tissue Engineering and Regenerative Medicine

In tissue engineering, biopolymers are used as matrices for cell adhesion and the formation of tissues. An example is the application of chitosan, which is a chitin derivative, for skin tissue engineering. Chitosan-based scaffolds have revealed improved wound healing characteristics because of biocompatibility, antibacterial effects, and high cell adhesion and proliferation (Ahmed and Ikram 2016). The recent study tested the application of a chitosan-based hydrogel for the effective treatment of chronic wounds. The hydrogel promoted more skin cell migration and the formation of an extracellular matrix to bring about more rapid and effective wound healing (Wang et al. 2023). Also, there was also found a decrease in inflammation and enhanced skin surface roughness in the patient after treatment. Another success story includes the use of collagen-based biopolymers in the field of bone tissue engineering (Nijsure and Kishore 2017). One of the major proteins in humans, collagen, has bridged the gap with respect to bone regeneration. In clinical trials, collagen combined with hydroxyapatite was effective in promoting the bone regeneration of critical-sized defects. Due to the scaffold's capacity to align with the structure of the natural bone extracellular matrix, the scaffold was able to encourage new bone formation and the recovery of the entire defective region within six months after implantation in most of the patients (Shineh et al. 2023).

12.3.2 Drug Delivery

Biopolymers are also commonly used in the drug delivery systems because of their effectiveness in encapsulating drug molecules and subsequently releasing them in a controlled manner. Polylactic-co-glycolic acid, a biodegradable synthetic polymer, has been employed in encapsulation of nanoparticles for delivering drugs in cancer therapy (Perinelli et al. 2019). In a successful case study, PLGA nanoparticles were synthesized that encapsulated the anticancer drug doxorubicin (Pieper et al. 2019). These nanoparticles were trialed in a group of patients diagnosed with metastatic breast cancer. The analysis revealed enhanced drug effectiveness compared to traditional approaches with fewer side effects like cardiotoxicity that patients can experience with doxorubicin (Smith et al. 2010). The longer circulation time of PLGA nanoparticles due to the controlled release profile provided increased drug concentrations at the tumor site while reducing the systemic levels of the drug.

Likewise, alginate has been applied in OSD systems for oral drug delivery because of its muco adhesive characteristic and tendency to shelter delicate drugs from the severe pH of the GI tract (Sobieraj 2024). Freezable microspheres that have been made from alginate encapsulated insulin have been of value to a diabetic patient by constructing non-invasive insulin delivery systems. Another study conducted which involve a Phase II trial showed positive outcomes where adopting, alginate-insulin microsphere had enhanced glycemic control with reduced side effects in patients. This biopolymer-based system for delivering insulin may be a new technique to overcome the set back of injectable insulin that could have enhanced patient compliance and quality of life (Lopez-Mendez et al. 2021).

Figure 12.2 depicts a number of nanotechnology-based drug delivery systems such as nanogels, liposomes, micelles, dendrimers, polymeric nanoparticles, nanotubes, polymersomes, and quantum dots. Both systems are intended to improve specific delivery and effectiveness of therapeutic agents in human body and serve as effective methods of targeted drug delivery.

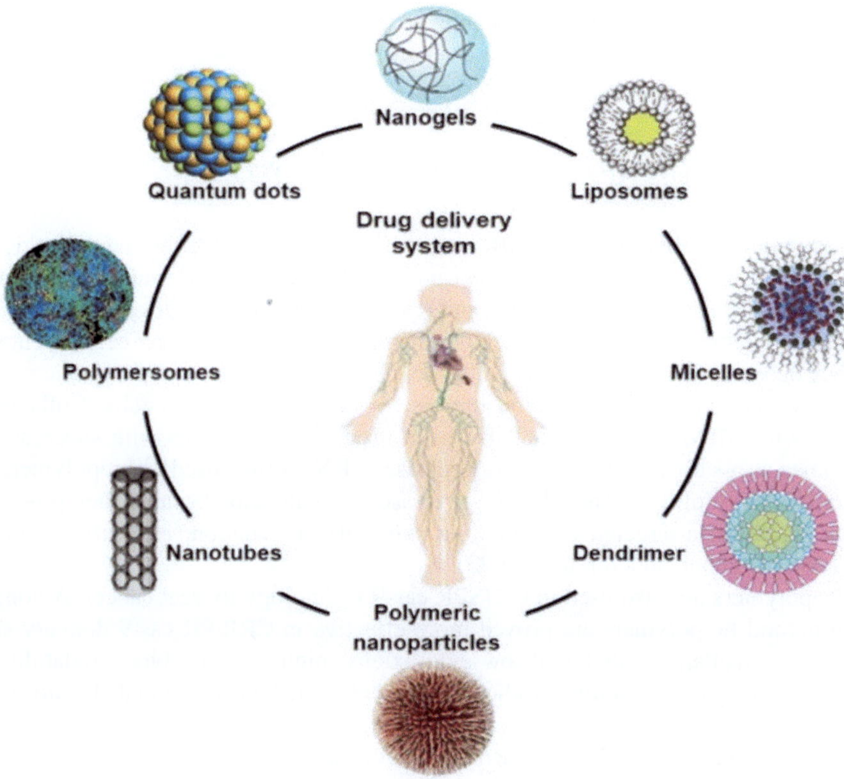

Fig. 12.2 Different drug carriers for drug delivery system

12.3.3 Wound Healing

Biopolymers have also found applications in wound healing and regeneration medicine in particular. The natural protein-based biopolymer, silk fibroin, is gradually proving useful in managing major burn and long-term wounds (Hassan et al. 2024). Surgical intervention was not needed in a case study conducted where the application of silk fibroin dressings was tested for second-degree burns. By enhancing the moisture content, a biopolymer dressing facilitated early formation of a confluent layer of epithelial cells, leading to almost complete reduction of scar tissue. Silk fibroin-based dressings provided improved healing and reduced pain for patients during clinical trials compared to traditional dressings (Indrakumar et al. 2023).

In relation to regenerative medicine, PDO is a successful example of a synthetic biopolymer that can be used as a suture and could also promote tissue growth. The clinical trial on the application of PDO in soft tissue rejuvenation after reconstructive surgeries has been proved successful. This research proved that PDO sutures not only provided mechanical "fixation" in the form of tissue approximation but also promoted the manufacture of new collagen, speeding up the process of wound recuperation and enhancing surgical outcomes (Contreras et al. 2023).

12.3.4 Gene Therapy and Nucleic Acid Delivery

Biopolymers are effectively used as therapeutic agents in delivering nucleic acids (DNA, RNA, mRNA, and siRNA) in gene therapy. These therapeutic nucleic acids are used in mRNA vaccines, gene editing, and RNA interference technology. Natural and synthetic polymers are effectively used as therapeutic agents in cancer cells through siRNA delivery. Biopolymers like cellulose, chitosan, alginate, and polylactic acid have been effectively used in solid tumors by siRNA delivery to cells. Collagen loaded with siRNA formulations have been used in solid tumors with successful outcomes as a safe and sustained drug release. siRNA drugs used as biopolymeric materials are helpful in controlled drug release internalization because they protect drugs against non-targeted drug binding sites, pH degradation, and extracellular trafficking (Subhan and Torchilin 2023).

Biopolymers are also used in CRISPR cas-9 technology to treat cancer. Among them natural biopolymers are proved more effective in CRISPR cas-9 delivery as they have excellent attributes of low cytotoxicity, high stability, biodegradability, and excellent biocompatibility. Studies have demonstrated great potential of chitosan, hyaluronic acid, cyclodextrin, protamine, and oligonucleotide-based drug delivery in CRISPR cas-9 genome editing in the treatment of cancer (Lin and Wang 2023). Poly-lactic glycolic acid (PLGA) and chitosan-based polymeric nanoparticles are effectively used in the treatment of cystic fibrosis and other genetic disorders (Fernández Fernández et al. 2018). Protein biopolymers have been used to synthesize mRNA with its potential use in immunotherapy, vaccination, and nanotechnology (Urbina et al. 2020).

12.4 Advantages of Biopolymers in Therapeutics Over Conventional Materials

12.4.1 Biocompatibility and Biodegradability

Biopolymers have been used effectively in therapeutic applications because of their excellent attributes of biocompatibility, non-immunogenicity, and biodegradability. Poly glutamic acid composites (PGA) are used as nanoparticles, nanofibers, and hydrogels in successful medical applications such as cancer therapy, vaccine adjuvants, and tissue engineering (Park et al. 2021). Poly lactic acid being highly biodegradable is used in medical devices made of copolymers of PLA-PGA that are used for effective release of drugs in cells. PGA and PLGA are highly biocompatible and biodegradable they are effectively used in drug delivery and tissue engineering like bone fixation devices. Polycaprolactone is also highly biodegradable and is used in sutures, wound dressing, drug delivery systems, and tissue engineering. Polyesteramides are biocompatible and biodegradable biopolymers used in hydrogels, drug delivery systems, and tissue regeneration (Azimi et al. 2014). Chitosan, hyaluronic acid, and alginate are also highly suitable for drug delivery systems, and tissue regeneration because they have natural origin which makes them non-toxic and reduced risk of inflammatory response as compared to the conventional materials (Mukhtar et al. 2021).

12.4.2 Enhanced Drug Loading and Release Profiles

Chitosan/alginate biopolymers-based nanocapsules have proved high potential for sustainable release profile and on demand drug delivery with their applications in implantable biosensors (Xu et al. 2019). Polylactic acid (PLA), polyglycolic acid (PGA), and polycaprolactone (PCL) are highly used as therapeutic agents of drug loading and delivery in a fairly controlled manner (Dorj et al. 2014). Natural biopolymers like silk fibroin, cellulose, alginate, gelatin, and hyaluronic acid are highly used in therapeutic applications because they have the ability to be chemically modified for localized drug delivery (Wu et al. 2022). Chitosan and PLGA-based biopolymers are effectively used in a coating to promote accelerated osteoblast adhesion and proliferation without causing any bacterial colonization (Kumeria et al. 2015). Bacterial nanocellulose is used as a potential biopolymer for drug delivery system of proteins due its attractive properties of hydrophilicity, biocompatibility, and controlled drug loading (Müller et al. 2013). Chitosan, starch, and β-cyclodextrin (β-CD) have been used as potential biopolymer nanocomposites of drug delivery for ibuprofen drug. This revealed excellent loading capacity and sustainable release profiles of these carbohydrate biopolymers. While conventional materials do not have the equilibrium between the drug loading and controlled release for effective delivery (Abukhadra et al. 2020).

12.4.3 Natural Interaction with Biological Systems

Many biopolymers are extensively used in tissue engineering and regenerative medicine as they mimic the natural processes of extracellular matrix in human tissues. Alginate is a natural polysaccharide and is used in hydrogels, wound healing, tissue engineering, and drug delivery due to its excellent property of mimicking natural processes of host tissue especially in the creation of environment similar to extracellular matrix (ECM) (Sahoo and Biswal 2021). Chitosan-alginate co polysaccharide-based hydrogels have been effectively used in tissue engineering applications as this

PEG penetrated co polysaccharide has the excellent property of mimicking ECM of host tissue (Radhakrishnan et al. 2015). Alginate microfibers are widely used in 3D culture, cell encapsulation, cell therapy, and regenerative medicine as their delivery scaffolds mimic tissues effectively (Xie et al. 2022). Collagen and hyaluronic acid hydrogels are extensively used in therapeutics for creating in vitro tissue models and tissue engineering as both of them represent structural and tissue integrity being a component of extracellular matrix in many tissues (Xu et al. 2021). Conventional materials lack these properties of natural interaction, cell growth, and proliferation which limit their applications in regenerative medicine.

12.4.4 Non-toxic Degradation Products

Biodegradable synthetic polymers are effective in drug delivery as compared to non-biodegradable polymers as they produce non-toxic or less toxic byproducts. Poly lactic acid (PLA) is most remarkable in cancer drug delivery as it produces non-toxic degradation products which makes it suitable in various biomedical applications (Mundel et al. 2022). Polyhydroxyalkanoate-based biopolymers also degrade in to non-toxic byproducts and they are highly used in surgical implants, sutures, tissue engineering scaffolds, and controlled drug delivery (Ansari et al. 2021). PHA is also used to treat many health issues in orthopedics, heart and wound management, artificial membranes for kidneys and load bearing cartilage as it has excellent properties of high immunotolerance and low cytotoxicity (Mukheem et al. 2018). Chitosan is known to produce less toxic degradation products and its biopolymers are extensively used in carcinoma therapy, oral drug delivery, gene delivery and colon targeted delivery. Chitosan has shown good anticancer activity as it has inhibitory effects on growth of tumor cells and tumor metastasis (Sousa Victor et al. 2020).

Figure 12.3 explains the advantages of biopolymers including biocompatibility, targeted drug delivery, natural integration, low toxic effect, reduced immunogenicity, customizability, regulatory compliance, personalized medicine, and biodegradability.

12.5 Current Challenges and Limitations

12.5.1 Stability and Storage Issues

Biopolymers are polymers derived from biological sources. They are extensively used in medical applications due to their availability, low cost, operation at elevated temperature, non-toxicity, biocompatibility, biodegradability, characteristics like high adsorption capacity and ease of surface modification (Das et al. 2023). They have been examined suitable in biomedical, pharmaceutical, drug delivery, tissue engineering, and medications. Stability and storage are important issues relevant

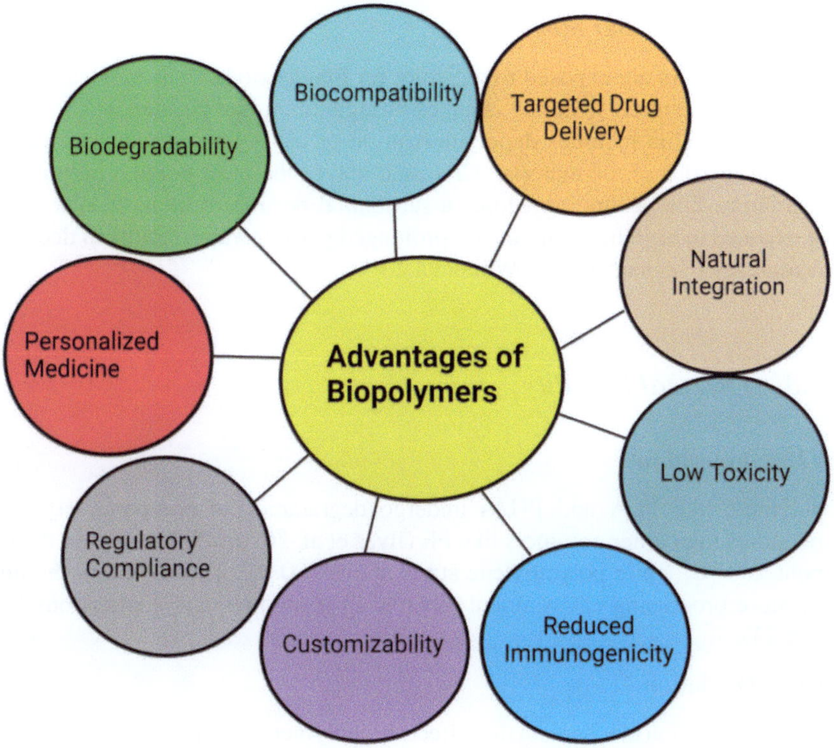

Fig. 12.3 Advantages of biopolymers in therapeutics over conventional materials

to biopolymers as they are natural materials and more prone to degradation in comparison with synthetic polymers (Mtibe et al. 2021). The subsequent subtopics discuss concerns arising from moisture sensitivity and thermal stability, microbial deterioration, and methods of preservation.

12.5.2 Moisture Sensitivity

Hydrolytic Degradation

Starch and proteins are examples of biopolymers that undergo hydrolytic degradation if exposed to moisture. Working as a plasticizer, water lowers the glass transition temperature (Tg) and enhances the mobility of polymer chains in the system. This results in a substantial loss of mechanical characteristics like tensile strength and elongation at break. Starch based films experienced a loss in tensile strength by up to 50% when stored under conditions of relative humidity (RH) greater than 70% (Thakur et al. 2019).

Alteration in Molecular Structure

When biopolymers are exposed to moisture for long periods, their molecular structure cannot be restored since the changes brought about are permanent (Soldo and Miletic 2022). This includes depolymerization or cross-linking which cause loss of some properties of the material. Chitosan, one of the most popular biopolymers used in biomedical technology, loses a substantial portion of its molecular weight when exposed to high humidity over a prolonged period, which results in decreased bioavailability (Szymańska and Winnicka 2015).

12.5.3 Thermal Stability

Low Decomposition

Biopolymers like PLA and PHAs undergo degradation at comparatively lower temperatures than other polymers like PE (Ilyas et al. 2020). PLA begins to degrade at about 180 °C, while polyethylene starts above 300 °C. This is due to the high-temperature processing environment inherent in many practical applications (Taib et al. 2023).

Thermal Oxidation

Another degradation mechanism that affects biopolymers is thermal oxidation, which results from the exposure of the biopolymers to high temperatures with formation of free radicals (Dintcheva and D'Anna 2019). These radicals can cause a breakdown of the polymer backbone by chain scission, involving the loss of mechanical properties and discoloration. Excessive heat damages the molecular weight of PLA and causes it to turn yellow, which affects the appearance and mechanical properties of the final product (Ho et al. 2015).

12.5.4 Microbial Degradation

Biodegradation of Biopolymers

Biopolymers that are exposed to microbial action are more affected by microbial degradation especially if they are stored in the humid region or areas with biological activity (Siracusa 2019). Among the various PHAs, polyhydroxybutyrate (PHB) is more vulnerable toward biodegradation in soil and get decomposed within three to six months by the action of microbes. This makes it unsuitable for use in applications where the coating is required to last longer, or where it is subject to regular physical use (Prudnikova et al. 2024).

Enzyme Degradation

Some enzymes like amylase acting on starch or lipase on PHAs are capable of hydrolyzing biopolymers. In storage environment where microbes are present, these enzymes can enhance the degradation process. This is illustrated in films made from PHAs, where it has been observed that mechanical properties deteriorate as the material degrades under the action of lipase-producing bacteria (Prakash et al. 2020).

12.6 Preservation Strategies

12.6.1 Antioxidants

Antioxidant additives that can address this issue include butylated hydroxytoluene (BHT), which can prevent oxidative degradation of the biopolymers such as PLA (Sivakanthan et al. 2020). However, the efficiency of these additives depends upon their migration rates and compatibility with the polymer matrix, and sometimes it results in undesirable modification of the material properties.

12.6.2 Plasticizers

Glycerol and sorbitol act as the biopolymers additives. They can be used to reduce the temperature at which the biopolymers become brittle or rigid by decreasing the Tg (Arfat 2017). Thus, plasticizers can negatively affect biopolymers by decreasing their barrier properties and making them more permeable to moisture, which explains the trade-off between flexibility and moisture resistance. The combination of glycerol with films formed from starch enhances flexibility but at the same time, it encourages water intake at rates of 30% (Basiak et al. 2018).

12.6.3 Antimicrobial Agents

To avoid microbial degradation, the nanocomposite must contain antimicrobial agents such as silver nanoparticles (Azizi-Lalabadi et al. 2021). These agents can be expensive, and as such dispose potential environmental and health risks due to toxic properties. The use of antimicrobial agents also fosters resistance to microbes in the long run as they become resistant to the antimicrobial agents used on them.

12.7 Scalability and Production Costs for Biopolymers

There are issues with providing raw materials for biopolymer production, increased production costs, and technological issues when scaling up the production of biopolymers for therapeutic applications (Gahlawat et al. 2020). They greatly determine the economics of the biopolymers and their market acceptance.

12.7.1 Raw Material Availability

Food Resources

Lactic acid production from corn, sugarcane, and soybeans for biopolymers and bioplastics compete directly with food and feed production. It was estimated that about 2.5 tons of corn is used to produce 1 ton of PLA (Groot and Borén 2010). This competition affects prices of raw materials and could be unethical in a situation where some parts of the world are struggling to feed themselves.

Feedstock Sustainability

Regarding the agricultural feedstocks, there are several drawback of biopolymers: land use, water consumption, usage of fertilizer and pesticides (Puglia et al. 2021). But the environmental consequences or the environmental conditions under which biopolymers are derived from these feedstocks may be high thus eliminating the biodegradability factor.

12.7.2 High Production Cost

Fermentation Process

There are high costs involved in the production of biopolymers such as PLA and PHAs by fermentation which include high capital and energy costs (Kourmentza et al. 2017). The cost of controlling the conditions required for microbial growth, such as temperature, pH, and aeration, additives to cost of production. PHAs are synthesized via a two-stage fermentation process and the downstream processing, which includes solvent extraction and precipitation, costs time and money and can reach up to half of the production cost (Yadav et al. 2020).

Downstreaming and Purification

The process of purification and extraction of biopolymers involves the use of costly solvents and tend to involve energy demanding drying procedures (Flórez et al. 2023). These processes are biopolymer production. For example, the process of isolating PLA from fermentation broth includes the use of solvents such as chloroform, which is expensive and has detrimental effects on the environment and the health of the workers (Calleja Luque 2020).

12.7.3 Technological Barriers

Low Yield and Production

Biopolymer production processes may have lower yields and productivity than synthetic polymer production. The practical conversion efficiency of synthesizing PHA is less than 80% of the actual achievable, which is less compared with the petrochemical based polymers (Adeleye et al. 2020). This low yield means that the required feedstock quantity is high and it takes longer to process which are costly.

Lack of Processing Technologies

The processing technologies for biopolymers are still in their development stage of development compared to synthetic polymers. For example, materials such as PLA exhibit different melt flow and thermal characteristics to conventional polymers and consequently, extrusion and injection molding processes often have to be adapted (Kumaran et al. 2020). The absence of mature processing technologies also acts as a hindrance to the scale up of production of biopolymer and raise issues of capital intensity.

Capital Investment

One of the key challenges to scale up is the high capital cost needed for facilities to produce biopolymers. This investment also encompasses the costs incurred for performing research and development activities, construction of a pilot plant, and compliance with applicable laws and regulations (Nouri 2007). The investment in commercial-scale PLA production facility may reach $100 million, thus is a huge capital investment for the companies especially during the early stages of the industry development.

Market Demand

Biopolymers are comparatively less used as compared to traditional plastics which also hinders the chances of attaining economy of scale (Kabir et al. 2020). The price sensitivity of consumers and the industries it indirectly affects also worsens this problem due to the high cost of biopolymers relative to other materials. For example, the price of PLA is generally two to three times that of polyethylene, so it is more costly and not suitable for many low-cost applications.

12.8 Regulatory Hurdles

The current regulation of biopolymers can be described as convoluted and there are major issues within this area that may serve as barriers to marketization. Other challenges are regulatory issues which include safety and environmental impacts, labeling and certifications, trade barriers and patents.

12.8.1 Safety and Environmental Assessment

Rigorous Testing

The use of biopolymers does not guarantee the biopolymers non-toxicity and its effect to the environment. This comprises of toxicity tests, biodegradation studies, and life cycle assessment (LCA) (Kabir et al. 2020). For instance, for a biopolymer to be certified for composting, it has to disintegrate fully within a given time of 12 weeks under industrial composting environment of 58 °C (Wojtowicz 2009). These testing requirements entail significant time and cost, which would contribute to either delays in market entry or expensive development costs.

Uncertain Environmental Impact

The lifetime behavior of biopolymers is still unclear, especially regarding the formation of microplastics and its stability in the environment. For example, PLA, which is one type of microplastics, can break down into microplastic particles under specific circumstances and become hazardous to marine organisms and the environment (Tamayo-Belda et al. 2022). This can result in more rigorous regulation and may limit the application of biopolymers at some extent.

Labeling and Certifications

It is critical to use appropriate labels for biopolymer products to educate the consumers as well as to meet the legal requirements. However, due to these complexities, there is often confusion in some labeling claims such as biodegradable, bio-based, and the overall impact on the environment is expensive to comply with. Mislabeling or misleading claims could lead to punitive repercussions and detrimental impacts on business image. Products intended for global markets in the sector of biopolymers require product standards that can meet regional and even international standards such as the ASTM, ISO, and the EN standards (Wilde 2020). Gaining a certification for biodegradability, compostability or bio-based content can be difficult because of the strict guidelines and differences in standards in the global market. Biopolymer that has been certified under the ASTM D6400 test in the United States will not qualify for the EN 13432 in Europe, which requires the biopolymer to undergo further certification.

12.9 Ongoing Clinical Trials

Cancer Treatment

The most promising field of biopolymer application is cancer treatment. PLGA nanoparticles have demonstrated applicability as carriers for chemotherapeutic agents where a targeted release and consequently fewer side effects are possible. Another active clinical trial is the study of PLGA nanoparticles incorporated with paclitaxel, an anti-cancerous agent, particularly in the treatment of ovarian cancer (Sharma et al. 2022). The key objectives of the trial are to: (i) Determine safety and tolerability; (ii) Evaluate effectiveness; and (iii) Examine pharmacokinetics of the nanoparticle-based drug delivery system. First set of results have been very encouraging with patients experiencing lesser SAE and tumor size.

Wound Treatment

Another area of interest is the application of hydrogels based on chitosan for wound treatment in patients with diabetes. The use of a chitosan hydrogel dressing in the management of diabetic foot ulcer has been assessed in a Phase III clinical trial (Djavid et al. 2020). Early observations suggest that treated wounds heal faster and are less prone to infection when using the biopolymer dressing. This trial is especially significant due to increasing worldwide prevalence of diabetes and its comorbidities, including chronic ulcers.

Cardiovascular Disease

Biopolymers are also considered for cardiovascular usage. Another active clinical trial focuses on the application of alginate scaffolds in heart muscle repair after myocardial infraction. The goals of the study include determining the safety and effectiveness of injecting alginate scaffolds into the injured heart tissue to facilitate tissue remodeling and reverse the negative changes in cardiac function (Ruvinov and Cohen 2016). Initial outcomes indicate that the biopolymer scaffold has a positive effect on cell migration and angiogenesis thus improving cardiac function in animal models. Currently, human trials are going on to evaluate its applicability for broader clinical practice.

12.10 Future Directions in Clinical Research

Personalized Drug

Personalized drug delivery systems mark one of the leading trends in the field of biopolymers. To improve over other forms of drug delivery systems, the current advancements in polymers for drug delivery are seeking to develop biopolymer delivery systems that conform to the target genetic makeup, current disease state, or metabolic rate of the patient. Upcoming clinical trials will study the usage of 3D printed biopolymer scaffolds, which may be customizable to conform to the patient's body shape (Weems et al. 2020). Such scaffolds could not only provide site-specific drug delivery but also harbor tissues for tissue engineering in the field of personalized medicine.

Genetic Therapy

Another exciting prospect in the field is the utilization of gene therapy alongside biopolymers. Biopolymers can also be designed to encapsulate and transport genetic components and factors such as CRISPR-Cas9 agents to targeted tissues within the body. Active pre-clinical research is being conducted to deliver the gene editing tools through the biopolymer-based nanoparticles to treat genetic diseases which include cystic fibrosis and muscular dystrophy (Santos-Carballal et al. 2018). Subsequent clinical studies are anticipated on the biopolymer-based gene therapeutic systems on these aspects of safety and efficacy.

Wound Healing

In the field of wound healing, it is anticipated that future clinical studies will bring out smart biopolymer dressings that will be sensitive to variations in the surrounding environment of the wound (Farahani and Shafiee 2021). Such dressings would be able to deliver antimicrobial agents or growth factors on a controlled basis in the event of sepsis or tissue injury. Furthermore, biodegradable biopolymers that have the ability to actively stimulate tissue healing through the delivery of bioactive signals is studied in preliminary works. Clinical trials in this area will likely examine the efficacy of these bioactive materials to treat chronic wounds and burns.

Chronic Disease Management

Biopolymers are expected to have a notable role in the treatment of such chronic ailments, including diabetes, and cardiovascular ailments. Some of the current research includes the creation of new insulin delivery systems using biopolymers to reduce injection frequency (Shah et al. 2019). Also under development are biopolymer-based stents which are intended for long-term implantation in cardiovascular patients with minimal risks of restenosis. Subsequent clinical trials are expected to address the question of whether these biopolymer-based devices cause adverse effects and are as effective in the long run.

12.11 Future Trends and Innovations

There is a trend in the advancement of new technologies for biopolymer that addresses on property improvement, industrial applicability, and affordability of the biopolymers along with environmental problems related to the formation and degradation of the biopolymers. Several trends are emerging particularly in the use of unexplored feedstock, synthesis of new biopolymers, and formation of novel biopolymers with enhanced properties by incorporating additives.

12.11.1 Emerging Biopolymer Technologies

Application of Waste Biomass

One of the emerging trends in biopolymer technology is the utilization of waste biomass to produce the reinforcing materials. Components derived from the conventional feedstocks including corn starch and sugarcane are often in conflict with the food supply chain, making it unethical and unsustainable (Ghatak 2011). While waste biomass derived from the agricultural and food processing industries is the most sustainable and cheaper form of biomass, through transforming organic waste into useful polymers the product recovery process is economically advantageous and ecologically sensitive. For instance, polyhydroxyalkanoates (PHAs), a category of biodegradable polymers, are currently being manufactured from different types of feeds such as food waste, sugar beet pulp, or wastewater (Pradhan et al. 2020). New light is another company that has benefited from the utilization of greenhouse gases through converting methane and carbon dioxide to PHAs biopolymeric materials. These PHAs are being used in the drug delivery systems for their biodegradibility and biocompatibility. Likewise, there has been a paradigm shift in the fabrication of polylactic acid (PLA) where the feedstock of choice is lignocellulosic biomass including agricultural waste. For instance, the combination of biorefinery processes, which concern the transformation of biomass into bio-based chemicals and polymers, is driving up the creation of more sustainable biopolymers. This trend shows the possibility of developing a cyclic bioeconomy where the waste products are further used as secondary resources that do not harm the environment.

Application of Biopolymer Composite

Two more significant innovations are bio-blends and biopolymer composites, which include blending biopolymers with reinforcements. Some natural polymers like cellulose, starch, and chitosan are used to enhance biopolymers and their composites to provide better mechanical properties, thermal stability, and increased durability. Cellulose nanofibers (CNFs), for instance, have gained enormous popularity as reinforcement material for biopolymers (Djafari Petroudy et al. 2021). Ideally when PLA or PHAs are combined with cellulose nanofibers, the mechanical properties of the resultant composite would be improved while biodegradability would not be compromised. The PLA and PCL-based synthetic biopolymers are used in tissue engineering and regenerative medicine.

Self-healing Biopolymers

Such recent areas of study include self-healing biopolymers that are capable of restoring themselves when under threat of destruction or degradation, thus enhancing their durability and usability. Self-healing materials are quite relevant in biomedical field because material failure can be very expensive and risky. Self-healing biopolymers are also convenient to the sustainability of products given that they may not require constant replacement (Kumar et al. 2023). For instance, hydrogels prepared from gelatin with the ability to repair itself automatically are considered for use in wound healing. These hydrogels are capable of undergoing dynamic cross-linking in the event of injury to enhance tissue repair and wound healing (Jacob et al. 2021).

Integration with Nanotechnology and Bioengineering

Biopolymers' compatibility with nanotechnology and bioengineering is seen as an emerging field, which offers significant development potential to improve the performance and broaden uses of biopolymers. Biopolymers that are made at the nanoscale level and those that also contain biological particles can have better properties such as delivering medicine to specific areas of the body, better mechanical strength, and activity (Udayakumar et al. 2021). This integration is predicted to transform areas like medicine, environmental conservation, and even industrial production.

Nanotechnological Biopolymers

Nanotechnology provides an opportunity to introduce appropriate changes into biopolymers at the molecular level, and it allows for the creation of materials with a defined set of characteristics. One potential application of biopolymers is in the creation of nanoparticle systems for drug delivery applications. For instance, poly (lactic-co-glycolic acid) or PLGA nanoparticles can be encapsulated in biopolymer matrices to achieve targeted drug delivery with consistent release rates. It has been particularly applied in cancer treatment because of the ability to increase the concentration of the drug in the affected area but with limited effects on the rest of the body. In new emerging technologies for biomedical applications silica-based nanoparticles are used with a range of functional biopolymers for the development of therapeutic nanodevices (Gándara et al. 2023).

Figure 12.4 explains the modification of silica in different forms of polymers. For loading polymer, RNA, proteins, and peptides can be used for pore functionalization, expansion, and surface functionalization. The therapeutic usage of modified silica and biopolymer is present in the form of vaccine, anticancer drugs, and antibacterial drugs.

Scaffold Biopolymers

In the case of tissue engineering and the use of biopolymers, bioengineering techniques are being used to create innovative biomimetic scaffolding for regenerative therapies. Natural and biodegradable materials like collagen, alginate, and fibrin are often employed for the synthesis of scaffolds that give a matrix-like structure where cells can proliferate and tissues can be regenerated (Donnaloja et al. 2020). New advances in bioengineering have allowed researchers to enhance 3D-printed biopolymer scaffolds with a high degree of architectural complexity. This technology makes it possible to design structures to suit the need of specific clients and promote the growth of certain tissue types like bones, cartilage, and skin. The additional incorporation of these various bioactive molecules including growth factors and peptides into the mentioned scaffolds further boosts its capabilities of regeneration (Hassan et al. 2019). Scientists have come up with alginate-based scaffolds for cartilage repair in osteoarthritis patients. These scaffolds are expected to exhibit a controlled bioactivity that encourages stem cells to develop into chondrocytes for cartilage repair.

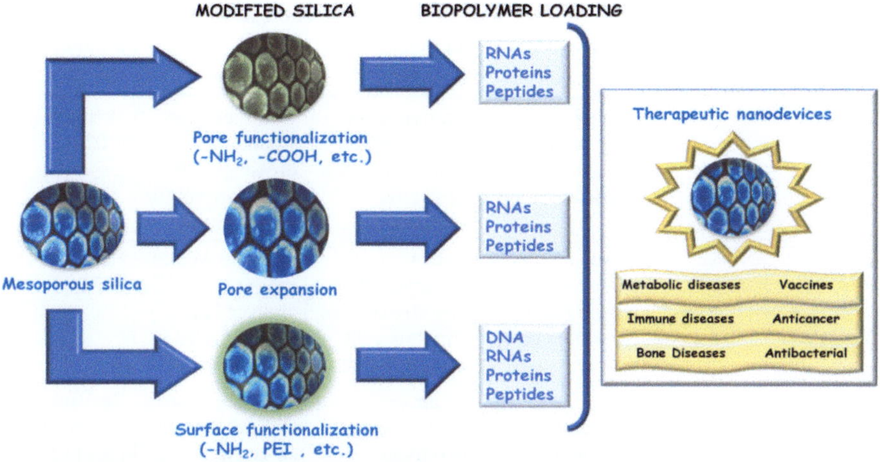

Fig. 12.4 Schematic representation of modified silica for biopolymer loading

The use of bioengineering and biopolymers in regenerative medicine is one of the most effective and efficient methods that can be used in treatment of various diseases and injuries such as organ failure and trauma (Table 12.1).

Conclusion

This chapter critically evaluated the structural diversity, synthesis methods, and the potential for applications of biopolymers while noting that the use of biopolymers continues to gain traction in sustainable material science. Although biopolymers like polylactic acid (PLA), polyhydroxyalkanoates (PHAs), and chitosan have distinct environmental and biomedical merits, (biodegradability, biocompatibility, and renewability), its commercial prospects are drowned in high production costs, fluctuating mechanical competencies, and scalability challenges. Particularly, microbial fermentation and enzymatic polymerization, though environmentally friendly, must be optimized to be able to rival the yield and continuity of petrochemical polymer production. In addition, majority of biopolymers exhibit hydrophilicity as well as poor thermal resistance which limit them from industrial packaging and high temperature applications. The chapter also indicated a damning void on lifecycle evaluations and end of life management strategies for biopolymer-based products, which called for standardization and regulatory accord. New development in genetic engineering, nanocomposite formulations, and polymer blending present promising avenues of overcoming these limitations, but translation research requires interdisciplinary collaboration. In moving forward, the biotechnology segment needs to prioritize making improvements to strain engineering, feedstock efficiency, as well as coming up with immutable closed-loop production models in pursuit of economic and ecological sustainability. Therefore, biopolymers offer great promise but the contingency of future success will rely on focused innovation and systems-based thinking based in both scientific and industrial realities.

Table 12.1 Mechanism of action and therapeutic applications of biopolymers

Type of biopolymer	Therapeutic application	Mechanism of action	Specific examples	References
Chitosan	Wound healing, drug delivery, antimicrobial applications	Promotes cell migration and adhesion; forms protective barrier	Chitosan-based wound dressings, nano-carriers for drug delivery	Verma et al. (2023)
Alginate	Tissue engineering, wound dressing, controlled drug delivery	Forms hydrogels that absorb exudates, supports cell encapsulation	Alginate wound dressings, alginate beads for drug release	Hegde, et al. 2022)
Collagen	Skin regeneration, bone grafts, tissue scaffolds	Promotes cellular growth and tissue remodeling; biocompatible scaffold	Collagen-based skin substitutes, collagen sponges for bone repair	Rezvani Ghomi et al. (2021)
Hyaluronic acid	Joint lubrication, ophthalmic surgery, skin care, drug delivery	Retains water, aids in tissue hydration and wound healing	Hyaluronic acid injections for osteoarthritis, eye surgeries	Huynh and Priefer (2020)
Gelatin	Drug delivery, tissue scaffolding, regenerative medicine	Biodegradable matrix for cell adhesion and proliferation	Gelatin sponges in drug delivery, 3D gelatin scaffolds	Singaravelu et al. (2016)
Polylactic acid (PLA)	Biodegradable sutures, drug delivery, tissue engineering scaffolds	Degrades into lactic acid, supporting tissue repair and drug release	PLA sutures, microspheres for controlled drug delivery	Liu et al. (2019)
Polyhydroxyalkanoates (PHA)	Tissue engineering, drug delivery, biodegradable implants	Biocompatible and biodegradable; forms scaffolds for tissue growth	PHA-based implants, drug delivery carriers	Li et al. (2017)
Starch-based polymers	Drug delivery, wound healing	Biodegradable with controlled drug release	Starch-based drug carriers, wound dressings	Torres et al. (2013)
Dextran	Drug delivery, anti-thrombotic therapy, tissue engineering	Enhances vascular permeability and drug solubility	Dextran-coated drug nanoparticles, dextran hydrogels	Zeng et al. (2022)

(continued)

Table 12.1 (continued)

Type of biopolymer	Therapeutic application	Mechanism of action	Specific examples	References
Silk fibroin	Tissue scaffolds, drug delivery, bone regeneration	Strong biocompatible fibers that support tissue regrowth	Silk fibroin scaffolds for bone and cartilage repair	Zhou et al. (2022)
Cellulose derivatives	Drug delivery, wound healing, tissue engineering	Forms hydrogels for drug encapsulation and controlled drug release	Hydroxypropyl cellulose in wound care, cellulose-based hydrogels	Alven and Aderibigbe (2020)
Pectin	Wound healing, drug delivery, colon-targeted therapies	Forms gels that support wound healing and controlled drug release	Pectin-based wound dressings, pectin hydrogels for drug delivery	Han et al. (2022)

Future Prospects

The biopolymer therapeutics path will be defined by new delivery system breakthroughs; smart wound dressings; regenerative medicine. With clinical trials already showing success of using PLGA nanoparticles for target cancer therapy and chitosan-based hydrogels in diabetic wound healing, future research should be concentrated around improving stability of biopolymer, loading capacity, even responsiveness to physiologic stimuli. Incorporation of 3D bioprinting technologies will enable patient-customized scaffold design using materials like alginate and collagen to a larger extent particularly in cartilage and cardiac tissue regeneration. In addition, the biopolymers, including nanotechnology-enabled ones, such as siRNA-loaded chitosan or CRISPR-Cas9 delivery using alginate nanoparticles, are likely to drive non-viral gene therapies in such diseases as cystic fibrosis and various cancers. Overcoming the existing short comings (hydrolytic degradation, microbial susceptibility, and thermal instability) will necessitate molecular adjustments, cross-linking approaches, and addition of stabilizers. In addition, the utilization of agricultural waste, as well as greenhouse gases, will align with economic and environmental goals. Clear regulation especially on standardization of biodegradation and clinical safety will be important in ensuring commercial success from laboratory success. We consider that the potential future of therapeutic biopolymers is their convergence with precision medicine, smart materials, and green manufacturing for safer patient-specific, and scalable healthcare solutions.

References

Abukhadra MR et al (2020) Insight into the role of integrated carbohydrate polymers (starch, chitosan, and β-cyclodextrin) with mesoporous silica as carriers for ibuprofen drug; equilibrium and pharmacokinetic properties. Int J Biol Macromol 156:537–547

Adeleye AT et al (2020) Sustainable synthesis and applications of polyhydroxyalkanoates (PHAs) from biomass. Process Biochem 96:174–193

Ahmed S, Ikram S (2016) Chitosan based scaffolds and their applications in wound healing. Achiev Life Sci 10(1):27–37

Alven S, Aderibigbe BA (2020) Chitosan and cellulose-based hydrogels for wound management 21(24): 9656

Ansari S et al (2021) Biomedical applications of environmental friendly poly-hydroxyalkanoates. Int J Biol Macromol 183:549–563

Arfat YA (2017) Plasticizers for biopolymer films. Glass transition and phase transitions in food and biological materials, 159–182

Azimi B et al (2014) Poly (lactide-co-glycolide) fiber: an overview. J Eng Fibers Fabr 9(1):155892501400900100

Azizi-Lalabadi M, Garavand F, Jafari SM (2021) Incorporation of silver nanoparticles into active antimicrobial nanocomposites: release behavior, analyzing techniques, applications and safety issues. Adv Coll Interface Sci 293:102440

Babu RP, O'connor K, Seeram R (2013) Current progress on bio-based polymers and their future trends. Prog Biomater 2:1–16

Basiak E, Lenart A, Debeaufort F (2018) How glycerol and water contents affect the structural and functional properties of starch-based edible films. Polymers 10(4):412

Basu P et al (2017) PEO–CMC blend nanofibers fabrication by electrospinning for soft tissue engineering applications. Mater Lett 195:10–13

Biswas MC et al (2022) Recent advancement of biopolymers and their potential biomedical applications. J Polym Environ 1–24

Calleja Luque S (2020) State of the art in the purification of PHB's from biomass and new harmless extraction methods

Contreras C, Ariza-Donado A, Ariza-Fontalvo A (2023) Using PDO threads: a scarcely studied rejuvenation technique. Case report and systematic review. J Cosmet Dermatol 22(8):2158–2165

Das A et al (2023) A comprehensive review on recent advances in preparation, physicochemical characterization, and bioengineering applications of biopolymers. Polym Bull 80(7):7247–7312

de Sousa Victor R et al (2020) A review on Chitosan's uses as biomaterial: tissue engineering, drug delivery systems and cancer treatment. Materials 13(21):4995

de Wilde B (2020) International and national norms on biodegradability and certification procedures. In: Handbook of biodegradable polymers, vol 2

Dintcheva NT, D'Anna F (2019) Anti-/pro-oxidant behavior of naturally occurring molecules in polymers and biopolymers: a brief review. ACS Sustain Chem Eng 7(15):12656–12670

Djafari SR et al (2021) Recent advances in cellulose nanofibers preparation through energy-efficient approaches: a review. Energies 14(20):6792

Djavid GE et al (2020) Application of a collagen matrix dressing on a neuropathic diabetic foot ulcer: a randomised control trial. J Wound Care 29(Sup3):S13–S18

Donnaloja F et al (2020) Natural and synthetic polymers for bone scaffolds optimization. Polymers 12(4):905

Dorj B et al (2014) A novel therapeutic design of microporous-structured biopolymer scaffolds for drug loading and delivery. Acta Biomater 10(3):1238–1250

Farahani M, Shafiee A (2021) Wound healing: from passive to smart dressings. Adv Healthcare Mater 10(16):2100477

Fazal T et al (2023) Recent developments in natural biopolymer based drug delivery systems. RSC Adv 13(33):23087–23121

Fernández E et al (2018) Biopolymer-based nanoparticles for cystic fibrosis lung gene therapy studies. Materials 11(1):122

Flórez M, Cazón P, Vázquez M (2023) Selected biopolymers' processing and their applications: a review. Polymers 15(3):641

Gahlawat G, Kumari P, Bhagat NR (2020) Technological advances in the production of polyhydroxyalkanoate biopolymers. Curr Sustain/Renew Energy Rep 7:73–83

Gándara Z, Rubio N, Castillo RR (2023) Delivery of therapeutic biopolymers employing silica-based nanosystems. Pharmaceutics 15(2):351

Ghatak HR (2011) Biorefineries from the perspective of sustainability: feedstocks, products, and processes. Renew Sustain Energy Rev 15(8):4042–4052

Gillmore JD et al (2021) CRISPR-Cas9 in vivo gene editing for transthyretin amyloidosis. N Engl J Med 385(6):493–502

Groot WJ, Borén T (2010) Life cycle assessment of the manufacture of lactide and PLA biopolymers from sugarcane in Thailand. Int J Life Cycle Assess 15:970–984

Gunatillake PA, Adhikari R, Gadegaard N (2003) Biodegradable synthetic polymers for tissue engineering. Eur Cell Mater 5(1):1–16

Gupta A et al (2021) Nucleic acid delivery for therapeutic applications. Adv Drug Deliv Rev 178:113834

Han SS et al (2022) Pectin based hydrogels for drug delivery applications: a mini review 8(12):834

Hassan MA et al (2024) Advancements in silk fibroin and silk sericin-based biomaterial applications for cancer therapy and wound dressing formulation: a comprehensive review. Int J Pharm 124494

Hassan M et al (2019) 3D printing of biopolymer nanocomposites for tissue engineering: nanomaterials, processing and structure-function relation. Eur Polymer J 121:109340

Hegde V et al (2022) Alginate based polymeric systems for drug delivery, antibacterial/microbial, and wound dressing applications 33:104813

Ho M-P et al (2015) Improvement on the properties of polylactic acid (PLA) using bamboo charcoal particles. Compos B Eng 81:14–25

Huynh A, Priefer RJC (2020) Hyaluronic acid applications in ophthalmology, rheumatology, and dermatology 489:107950

Ilyas R et al (2020) Properties and characterization of PLA, PHA, and other types of biopolymer composites. Advanced processing, properties, and applications of starch and other bio-based polymers. Elsevier, pp 111–138

Indrakumar S et al (2023) Photopolymerized silk fibroin gel for advanced burn wound care. Int J Biol Macromol 233:123569

Jacob S et al (2021) Emerging role of hydrogels in drug delivery systems, tissue engineering and wound management. Pharmaceutics 13(3):357

Kabir E et al (2020) Prospects of biopolymer technology as an alternative option for non-degradable plastics and sustainable management of plastic wastes. J Clean Prod 258:120536

Kalia S, Avérous L (2011) Biopolymers: biomedical and environmental applications. John Wiley & Sons

Khademhosseini A, Langer R (2016) A decade of progress in tissue engineering. Nat Protoc 11(10):1775–1781

Khalil HA et al (2023) Biopolymers based aerogels: a review on revolutionary solutions for smart therapeutics delivery. Prog Mater Sci 131:101014

Khang G (2017) Handbook of intelligent scaffolds for tissue engineering and regenerative medicine. CRC Press

Kourmentza C et al (2017) Recent advances and challenges towards sustainable polyhydroxyalkanoate (PHA) production. Bioengineering 4(2):55

Kumar A et al (2023) A comprehensive review of various biopolymer composites and their applications: from biocompatibility to self-healing. Mater Today Sustain 23:100431

Kumaran SK et al (2020) Biopolymers and natural polymers. Polymer science and nanotechnology. Elsevier, pp 245–256

Kumeria T et al (2015) Advanced biopolymer-coated drug-releasing titania nanotubes (TNTs) implants with simultaneously enhanced osteoblast adhesion and antibacterial properties. Colloids Surf, B 130:255–263
Li Z, Loh XJ (2017) Nanobiotechnology. Recent advances of using polyhydroxyalkanoate-based nanovehicles as therapeutic delivery carriers 9(3):e1429
Lin M, Wang X (2023) Natural biopolymer-based delivery of CRISPR/Cas9 for cancer treatment. Pharmaceutics 16(1):62
Liu S et al (2019) Preparation and properties of poly (lactic acid)(PLA) suture loaded with PLA microspheres enclosed drugs (PM-Ds)
Lopez-Mendez TB et al (2021) Cell microencapsulation technologies for sustained drug delivery: clinical trials and companies. Drug Discov Today 26(3):852–861
Mazurek Ł et al (2022) Silk fibroin biomaterials and their beneficial role in skin wound healing. Biomolecules 12(12):1852
Mtibe A et al (2021) Synthetic biopolymers and their composites: advantages and limitations—an overview. Macromol Rapid Commun 42(15):2100130
Mukheem A et al (2018) Bioplastic Polyhydroxyalkanoate (PHA): recent advances in modification and medical applications. Prepr Org 10
Mukhtar M et al (2021) Chitosan biopolymer, its derivatives and potential applications in nano-therapeutics: a comprehensive review. Eur Polymer J 160:110767
Müller A et al (2013) The biopolymer bacterial nanocellulose as drug delivery system: investigation of drug loading and release using the model protein albumin. J Pharm Sci 102(2):579–592
Mundel R, Thakur T, Chatterjee M (2022) Emerging uses of PLA–PEG copolymer in cancer drug delivery. 3 Biotech 12(2):41
Nijsure MP, Kishore V (2017) Collagen-based scaffolds for bone tissue engineering applications. Orthop Biomater Adv Appl 187–224
Nitta SK, Numata K (2013) Biopolymer-based nanoparticles for drug/gene delivery and tissue engineering. Int J Mol Sci 14(1):1629–1654
Nouri S (2007) Environmental assessment of emerging technologies: the case of biopolymers. Chalmers Tekniska Hogskola (Sweden)
Park S-B et al (2021) Poly (glutamic acid): production, composites, and medical applications of the next-generation biopolymer. Prog Polym Sci 113:101341
Perinelli DR et al (2019) PEGylated polylactide (PLA) and poly (lactic-co-glycolic acid)(PLGA) copolymers for the design of drug delivery systems. J Pharm Investig 49:443–458
Pieper S et al (2019) Incorporation of doxorubicin in different polymer nanoparticles and their anticancer activity. Beilstein J Nanotechnol 10(1):2062–2072
Pradhan S, Dikshit PK, Moholkar VS (2020) Production, characterization, and applications of biodegradable polymer: polyhydroxyalkanoates. Advances in sustainable polymers: synthesis, fabrication and characterization. Springer, pp 51–94
Prakash O, Nimonkar Y, Desai D (2020) A recent overview of microbes and microbiome preservation. Ind J Microbiol 60(3):297–309
Prudnikova SV et al (2024) Biodegradation of microbial plastic poly (3-hydroxybutyrate) in soil ecosystems at different latitudes. Giant 18:100288
Puglia D et al (2021) The opportunity of valorizing agricultural waste, through its conversion into biostimulants, biofertilizers, and biopolymers. Sustainability 13(5):2710
Radhakrishnan A, Jose GM, Kurup M (2015) PEG-penetrated chitosan–alginate co-polysaccharide-based partially and fully cross-linked hydrogels as ECM mimic for tissue engineering applications. Prog Biomater 4:101–112
Ramasamy T et al (2022) Nano drug delivery systems for antisense oligonucleotides (ASO) therapeutics. J Control Release 352:861–878
Rezvani E et al (2021) Collagen-based biomaterials for biomedical applications 109(12):1986–1999
Ruvinov E, Cohen S (2016) Alginate biomaterial for the treatment of myocardial infarction: progress, translational strategies, and clinical outlook: from ocean algae to patient bedside. Adv Drug Deliv Rev 96:54–76

Sahoo DR, Biswal T (2021) Alginate and its application to tissue engineering. SN Appl Sci 3(1):30
Santos-Carballal B, Fernández Fernández E, Goycoolea FM (2018) Chitosan in non-viral gene delivery: role of structure, characterization methods, and insights in cancer and rare diseases therapies. Polymers 10(4):444
Shah SA et al (2019) Biopolymer-based biomaterials for accelerated diabetic wound healing: a critical review. Int J Biol Macromol 139:975–993
Sharma T et al (2022) Advancements in clinical translation of flavonoid nanoparticles for cancer treatment. OpenNano 8:100074
Shineh G et al (2023) Functional approaches in promoting vascularization and angiogenesis in bone critical-sized defects via delivery of cells, growth factors, drugs, and particles. J Funct Biomater 14(2):99
Singaravelu S et al (2016) Biomimetic interconnected porous keratin–fibrin–gelatin 3D sponge for tissue engineering application 86:810–819
Siracusa V (2019) Microbial degradation of synthetic biopolymers waste. Polymers 11(6):1066
Sivakanthan S et al (2020) Antioxidant and antimicrobial applications of biopolymers: a review. Food Res Int 136:109327
Smart JD (2004) Recent developments in the use of bioadhesive systems for delivery of drugs to the oral cavity. Crit Rev™ Ther Drug Carr Syst 21(4)
Smith LA et al (2010) Cardiotoxicity of anthracycline agents for the treatment of cancer: systematic review and meta-analysis of randomised controlled trials. BMC Cancer 10:1–14
Sobieraj J et al (2024) How biodegradable polymers can be effective drug delivery systems for cannabinoids? prospectives and challenges. Int J Nanomed 4607–4649
Soldo A, Miletic M (2022) Durability against wetting-drying cycles of sustainable biopolymer-treated soil. Polymers 14(19):4247
Subhan MA, Torchilin VP (2023) Biopolymer-based nanosystems for siRNA drug delivery to solid tumors including breast cancer. Pharmaceutics 15(1):153
Szymańska E, Winnicka K (2015) Stability of chitosan—a challenge for pharmaceutical and biomedical applications. Mar Drugs 13(4):1819–1846
Taib N-AAB et al (2023) A review on poly lactic acid (PLA) as a biodegradable polymer. Polym Bull 80(2):1179–1213
Tamayo-Belda M et al (2022) Identification and toxicity towards aquatic primary producers of the smallest fractions released from hydrolytic degradation of polycaprolactone microplastics. Chemosphere 303:134966
Thakur R et al (2019) Starch-based films: major factors affecting their properties. Int J Biol Macromol 132:1079–1089
Torres FG, Commeaux S, Troncoso OP (2013) Starch-based biomaterials for wound-dressing applications 65(7–8):543–551
Traber GM, Yu A-M (2023) RNAi-based therapeutics and novel RNA bioengineering technologies. J Pharmacol Exp Ther 384(1):133–154
Udayakumar GP et al (2021) Biopolymers and composites: properties, characterization and their applications in food, medical and pharmaceutical industries. J Environ Chem Eng 9(4):105322
Urbina F, Morales-Pison S, Maldonado E (2020) Enzymatic protein biopolymers as a tool to synthetize eukaryotic messenger ribonucleic acid (mRNA) with uses in vaccination, immunotherapy and nanotechnology. Polymers 12(8):1633
Vashist A et al (2014) Recent advances in hydrogel based drug delivery systems for the human body. J Mater Chem B 2(2):147–166
Vedadghavami A et al (2017) Manufacturing of hydrogel biomaterials with controlled mechanical properties for tissue engineering applications. Acta Biomater 62:42–63
Verma D et al (2023) Sustainable functionalized chitosan based nano-composites for wound dressings applications: a review 116580
Wang X et al (2023) Chitosan-based hydrogels for infected wound treatment. Macromol Biosci 23(9):2300094

Weems AC et al (2020) 3D printing for the clinic: examining contemporary polymeric biomaterials and their clinical utility. Biomacromol 21(3):1037–1059
Wojtowicz A (2009) Biodegradability and compostability of biopolymers. Thermoplast Starch: Green Mater Various Ind 55–76
Wu J et al (2022) Localized, on-demand, sustained drug delivery from biopolymer-based materials. Expert Opin Drug Deliv 19(10):1317–1335
Xie Y et al (2022) Alginate microfibers as therapeutic delivery scaffolds and tissue mimics. Exp Biol Med 247(23):2103–2118
Xu L et al (2019) Electrostatically assembled multilayered films of biopolymer enhanced nanocapsules for on-demand drug release. ACS Appl Bio Mater 2(8):3429–3438
Xu Q et al (2021) Collagen-and hyaluronic acid-based hydrogels and their biomedical applications. Mater Sci Eng R Rep 146:100641
Yadav B et al (2020) A review on recovery of proteins from industrial wastewaters with special emphasis on PHA production process: sustainable circular bioeconomy process development. Biores Technol 317:124006
Zeng N et al (2022) Synthesis of magnetic/pH dual responsive dextran hydrogels as stimuli-sensitive drug carriers 520:108632
Zhou Z et al (2022) Silk fibroin-based biomaterials for cartilage/osteochondral repair 12(11):5103

Chapter 13
Sustainable Polymer Composites for the Next Generation Aerospace Applications

Abir Abdessalem, Samia Nouira, Joseph Fitoussi, and Mohammadali Shirinbayan

Abstract Increasing concern for reducing the weight of aircraft and aerospace components has led to the use of environmentally friendly composite materials. These materials can be recycled as they are manufactured using either natural fibers reinforcement or natural matrix, or sometimes both. However, synthetic materials cannot be recycled. This transition to sustainable materials is essential to reduce the carbon footprint in the world. Environmentally friendly materials are known to have less negative impact on the atmosphere and the environment. Thus, the study of their durability is of great interest, especially under environmental conditions (moisture, UV, thermal oxidation, etc.) and mechanical loads. This chapter is dedicated to the study of next generation polymer composites for aerospace applications. The aim is to propose a comprehensive experimental approach with the aim of understanding all the physical phenomena induced by aging and involved in damage and fracture mechanisms. To fully utilize these materials, it is essential to understand the relationships between microstructure(s) and macroscopic properties. Therefore, this study focuses primarily on the mastery of scale transitions. Subsequently, multi-scale experimental approaches are proposed with the aim of developing micromechanical models that can be used to predict the life-cycle behavior and damage of composite materials.

A. Abdessalem
Ecole Nationale d'Ingénieurs de Sousse (ENISo), Laboratoire LMS—Université de Sousse, Pôle Technologique de Sousse, Sousse, Tunisie
e-mail: abir.abdessalem@eniso.u-sousse.tn

S. Nouira · J. Fitoussi · M. Shirinbayan (✉)
Arts et Metiers Institute of Technology, CNRS, CNAM, PIMM, Paris, France
e-mail: mohammadali.shirinbayan@ensam.eu

S. Nouira
e-mail: samia.nouira@univ-lorraine.fr

J. Fitoussi
e-mail: joseph.fitoussi@ensam.eu

S. Nouira
Université de Lorraine, Laboratoire LEM3 UMR 7239, 57045 Metz, France

L. Ravindran et al. (eds.), *Macro, Micro and Nanocomposites from Sustainable Sources*, Advanced Structured Materials 244, https://doi.org/10.1007/978-981-95-2469-3_13

The influence of aging phenomena on the deformation and damage mechanisms of materials will also be addressed.

Keywords Green composites · Predictive methodology · Environmental aging

13.1 Introduction

In the twenty-first century, the aeronautical community is confronted with the imperative to expedite the development of novel, sustainable aircraft. Hi-Rate Composite Aircraft Manufacturing (HiCAM), a recent initiative by NASA, aims to address this challenge. HiCAM's research focuses on the accelerated manufacturing of many aircraft components, such as wings and fuselages, using sustainable composite materials, with the objective of achieving a manufacturing rate that is comparable to that of aircraft composed of metallic materials (https://www.militaryaerospace.com/commercial-aerospace/article/14291126/nasa-names-awardees-for-sustainable-aviation-component-technology) HiCAM is a part of the "Sustainable Flight National Partnership (SFNP)," through which NASA and Airbus are collaborating to make the aerospace field more sustainable through the use of green technologies. Composite materials can be made from a customized mixture of materials such as plastics, graphite, and ceramics. The construction of aircraft parts made of composites can significantly reduce the weight of an aircraft, which in turn makes it more fuel efficient (https://www.airbus.com/en/newsroom/sto-ries/2024-06-developing-bio-based-composites-that-are-fit-to-fly).

Research in biomaterials with self-healing, shape memory, or environmentally responsive properties is growing (Simões 2024). For instance, the development of advanced composites that incorporate phase change materials to manage heat can improve the energy efficiency of vehicles and aircraft (Liu et al. 2022; Chen et al. 2023). These materials possess the capacity to absorb, store, and release thermal energy, thereby enhancing performance in a variety of applications.

In this review, we will deal with polymer fiber reinforced composites, where a polymer resin, which is homogeneous and isotropic, is typically used as the matrix material. Composites are constructed to combine the properties of their constituents in order to obtain a superior end product. Such materials can be manufactured in a few ways, depending on the matrix and also the orientation of the reinforcement. Synthetic fibers are generally used in these composites thanks to their controllable properties compared to those of natural fibers.

Fiber-reinforced composites are materials that consist of bundles of fibers incorporated in the matrix material. They are known as the brightest materials of this century. Their durability in different service environments can be modified by the response of their constituents (i.e., fiber or polymer matrix) and the interphase between the fiber and polymer matrix. Their sensitivity to degradation depends on the climate environment in which the material is exposed during its service life. Thus, environmental conditions can include high and low temperatures, moisture, and ultraviolet radiation (UV).

Furthermore, the influence of the aging process on the material damage mechanism should be studied carefully. As it is known, the degradation processes generated by aging of polymeric materials by photo-irradiation, photo-oxidation, thermal degradation, and also hydrolysis processes lead to changes in the physical, chemical, and mechanical properties of the polymer composites (Tamboura et al. 2022; Abdessalem et al. 2019).

However, fully synthetic composites may have a negative impact on the environment because they are not recyclable at the end of their life as a consequence of dissimilarity between the constituents (fiber, matrix, and sometimes fillers). Such materials can highly contribute to global warming and boost toxic environmental effects. Meanwhile, natural fibers are biodegradable at the end of their life. Furthermore, if the matrix is also recyclable, the fully composite can be decomposed after its end-of-service. The use of a "green" polymer matrix, generally based on renewable materials, minimizes human dependency on carbon-based fuels. Thus, the development of ecological composites with lower environmental effects is encouraged.

Despite their impressive advantages, fiber-reinforced polymers (FRPs) have been used in recent decades, although they are highly sensitive to debonding and fracture propagation at the fiber-matrix interface. In addition, the complex short fiber orientation created by the polymer flow during the injection and molding process generates a strong anisotropy of these materials (Abdessalem et al. 2021). Consequently, life cycle assessment considering the anisotropy induced by fiber orientation is of major interest to improve the durability of FRPs. Many researchers have used micro-computer-tomography analysis and numerical simulations to analyze the dispersive behavior of FRPs. For example, Vincent et al. (2005) proposed in their studies an injection molding analysis model by calculating the fiber orientation tensors from two-dimensional optical image data. At the same time, different studies have been carried out on the durability characteristics of FRPs under complex loading conditions. Gude et al. (2006) Recently, a failure criterion was proposed using Wohler curves obtained from multi-axial fatigue experiments on composite-based polymers, CFRP. In addition, Tanaka et al. (2016) studied the notch effect and anisotropy, simultaneously, by using a single-edge notched plate sample in different directions.

Thus, fatigue is one of the most common and critical loading conditions for composite-based polymers. Despite many studies conducted in this field (see, e.g., Zuo et al. 2021), fatigue behavior remains one of the major challenges for fiber-reinforced polymers (FRPs) due to their complex microstructure and architecture, high sensitivity to environmental factors, and the interactions occurring across macro-, micro-, and nano-scales.

Predicting the fatigue behavior of FRPs is an important challenge. Many studies have been carried out to understand the fatigue behavior of composites. For example, Belingardi et al. (2006) evaluated the bending fatigue performance of glass and carbon-based epoxy composite. They found huge differences in damage initiation and stiffness degradation compared to non-hybrid composites. Researches fulfilled by Seghini (2020) on the tensile-fatigue behavior of flax and basalt fibers-based epoxy composites, demonstrated the positive influence of hybridization. At the same time, Barouni et al. (2022) came to a similar conclusion for hybrid flax-glass–epoxy composites. The fatigue damage mechanisms of short fiber reinforced PP, PA6, and SMC composites have also been investigated using scanning electron microscope (SEM) images (Tamboura et al. 2022; Abdessalem et al. 2021). PA6-based composites showed a local ductile fracture behavior at room temperature with limited decohesion fiber/matrix. Moreover, cohesive deformation appears around the fibers due to the non-negligeable interfacial bonding. At the same time, the PP fibers reinforced polymer showed crack initiation and propagation due to fiber pullout. It was concluded that, the ductile fracture behavior can be reduced with the decrease in fiber volume fraction (Tanaka et al. 2014).

In overall these studies, the fatigue behavior was evaluated based on the classical S–N curves approach. Thus, the development of a new approach providing a fast assessment is essential. Therefore, in this review, we propose a new fast predictive methodology based on quasi-static loads to obtain fatigue behavior.

13.2 Green Composites for Aerospace Applications

Aerospace applications are one of the most complex areas where a wide variety of plastics and composites are used. The aircraft industry is, as well, looking toward the use of bio-fiber reinforced composites for its components. For example, the hemp fiber reinforced epoxy matrix material has been successfully proven to replace the steel racks used in the electronic parts of a helicopter. Composites made of flax and glass fibers and combined with phenolic resins show good potential to be used in small drones' bodies. The commercial aircraft industry used flax, jute, hemp, and sisal fiber composites in interior panels as they have very good properties and can compete with synthetic fibers in specific strength and modulus. Table 13.1 lists the mechanical properties of synthetic and natural fibers (Wallenberger and Weston 2004).

a. **Stability Logic in Sustainable Polymer Composites**

The stability of plastic composites in the field of aeronautics refers to their ability to ensure mechanical integrity, thermal resistance, and durability over time under severe environmental conditions. Ensuring environmental sustainability, fuel efficiency, and structural reliability are critical factors that must be taken into account when assessing the stability of polymer composites in the field of aerospace.

The integration of bio-based and recyclable materials in sustainable polymer composites is a challenge to reduce environmental impacts without losing stability. The following aspects of such stability logic are of particular significance:

- Material degradation resistance (e.g., relative humidity, oxidation, thermal cycling, Ultra-Violet exposure).

Table 13.1 Comparison of the mechanical properties of natural and synthetic fibers

Fibre	Density (g/cm^3)	Elongation (%)	Tensile strength (MPa)	Young's modulus (GPa)
Cotton	1.5–1.6	3.0–10.0	287–597	5.5–12.6
Jute	1.3–1.46	1.5–1.8	393–800	10–30
Flax	1.4–1.5	1.2–3.2	345–150	27.6–80
Hemp	1.48	1.6	550–900	70
Ramie	1.5	2.0–3.8	220–938	44–128
Sisal	1.33–1.5	2.0–14	400–700	9.0–38.0
Coir	1.2	15.0–30.0	175–220	4.0–6.0
Softwood craft	1.5	–	1000	40.0
E-glass	2.5	2.5–3.0	2000–3500	70.0
S-glass	2.5	2.8	4570	86.0
Aramid (normal)	1.4	3.3–3.7	3000–3150	63.0–67.0
Carbon (standard)	1.4	1.4–1.8	4000	230.0–240.0

- Mechanical stability under dynamic, impact, and fatigue loading conditions.
- Chemical durability when exposed to hydraulic fluids, aviation fuels, and environmental factors.
- Moisture resistance is crucial to prevent the early failure of components in harsh and aggressive aerospace environments.
- Innovations in nano-material reinforcement, intelligent composite systems, and self-healing polymers to strengthen the sustainability and the stability of next-generation aerospace materials.

b. **Environmental Impact of Polymer Composites in Aerospace**

Biopolymer composites have become increasingly important due to their environmental compatibility.

Conventional aerospace composite materials have significant environmental issues resulting from high energy consumption and end-of-life issues. The focus on sustainable polymer composites allows us to address these issues through:

- Bio-based material (e.g., soy-derived, polylactic acid (PLA) matrix, or lignin-based).
- Natural fiber reinforcements (e.g., sisal, hemp, flax).
- Eco-friendly manufacturing techniques (e.g., additive manufacturing, reducing the Volatile-Organic-Compound (VOC) resin infusion).
- Decrease the carbon footprint when using light-weight materials that enhance fuel efficiency.
- Replace petroleum-based composites with sustainable materials. The aerospace industry can notably decrease greenhouse gas emissions, unsafe waste generation, and non-renewable resource consumption.

c. **Recycling Strategies for Aerospace Polymer Composites**

The recycling of composites in the aerospace industries is still a major engineering challenge, mainly because of the cross-linked properties of thermoset polymers as well as interphase adhesion between fibers and matrix. Nevertheless, new strategies for recycling materials are becoming available, including:

- Mechanical recycling: This method involves the grinding of composites into reusable particles.
- Thermal recycling: This method is based on composite Pyrolysis to recover reinforcement and generate organic fuels.
- Chemical recycling is used to separate components (matrix, fibers, and fillers) through depolymerization and solvolysis.
- Hybrid recycling method: This approach combines both chemical and mechanical processes to enhance material recovery.
- The uses of recyclable resins (e.g., polyamide (PA), polyether ether ketone (PEEK), and polyetherimide (PEI)) provide new perspectives for recycling and environmental economies in aerospace manufacture.

d. **Life Cycle Assessment (LCA) for Sustainable Aerospace Composites**

The use of LCA is a guaranteed method to evaluate the impact of aerospace components on the environment during their life cycle. For sustainable composites, a comprehensive Life Cycle Assessment includes:

- Material Extraction and Production
- Manufacturing and Processing
- Use Phase and Performance
- Life Cycle Management.

By including Life Cycle Assessment (LCA) data in the design of aerospace materials, manufacturers will be able to enhance material selection and improve sustainability criteria that satisfy global environmental regulations such as the Carbon Reduction Strategy. Finally, in aerospace applications, the introduction of sustainable composites depends on their environmental impact, stability, recyclability, as well as life-cycle performance.

However, their performance and durability are highly influenced by aging and mechanical stress.

In fact, the performance of bio-composites in aerospace-environment conditions is strongly impacted by their inherent properties and the physicochemical changes they undergo as they age. These biopolymer-based composites are influenced by moisture diffusion, chemical degradation due to oxidation, and ultraviolet exposure, resulting in physical modifications (which affect the composite's physical properties).

e. **Influence of Weather Aging on Green Composites Properties**

This section critically reviews and summarizes the influence of aging on the properties of biopolymer composites. Photodegradation of natural composites is a very complex process, and the penetration of ultraviolet radiation is one of the important factors that can deeply affect the mechanical properties and the durability of such materials. Many studies have focused on UV-induced degradation of mechanical behavior in bio-composites (Badji et al. 2018), precisely in the correlation between chemical process and loss of properties. The degradation seems to be proportional to natural fiber content and varies according to the type of matrix. Recently, natural fiber reinforced polypropylene (PP) was found to be very resistant to weathering compared to the low- and high-density polyethylene (LDPE and HDPE), and polyvinylchloride (PVC). Furthermore, mechanical properties' loss was also found to be dependent on composite manufacturing process, e.g., injection molding provides greater resistance than thermo-compression molding due to better fiber distribution in the matrix. Other investigations suggested that the decrease in mechanical properties is largely dependent on the chemical composition of the fibers, particularly the cellulose and lignin content. In this context, Peng et al. (2015) showed that the plastic stiffness was increased thanks to the removal of lignin from wood fibers. However, fibers rich in cellulose were found to be more sensitive to UV, which can dramatically increase microcrack formation. This suggests that for some applications, the addition of lignin may slow the loss of function of the biomaterial. More recently, pine fiber reinforced polypropylene was found to be more resistant to dry UV than flax fiber reinforced polypropylene due to the presence of more lignin (Nasri et al. 2022). The effects of photooxidation on the chemical and mechanical properties of bio-composites have been largely studied and reviewed.

Degradation of natural composites has been found to be more significant in a wet environment than in a dry one because of the increase in chemical reactivity. The diffusion of moisture inside the composite and its kinetics depends on the degradation mechanisms. Thus, to study the humid aging of composites in a limited time, an accelerated method is required.

Many diffusion models can be used to explain water uptake (Carter and Kibler 1978; Fick 1995). The simplest is the Fickian diffusion (Fick 1995). Berges et al. (2016) studied the moisture uptake behavior of unidirectional (UD) flax fiber reinforced epoxy. They concluded that the diffusion kinetics of unidirection flax/epoxy composites followed a one-dimensional Fickian behavior when subjected to hygrothermal aging at 70 °C and 85% RH (relative humidity). Chilali et al. (2018), in their studies, they elaborate a comparison between two composites: flax fiber reinforced thermoplastic and flax fiber reinforced thermoset composites when immersed in water for different periods of time. The water absorption of the flax composites was the same in the two materials.

The results show that at the beginning of moisture absorption, the amount of water increased linearly, then, when the saturation is reached the diffusion slowed down. Despite the hydrophobic character of the matrix, the hydrophilic fibers show an important absorption of atmosphere moisture, with a high probable degradation of the fiber and also the fiber-matrix interface. Consequently, a significant decrease in bio-composite mechanical properties was recorded. Such results have also been observed for hybrid-natural-fiber bio-composites made of an epoxy resin reinforced with kenaf and sisal fibers. Furthermore, the aging impact on aged and non-aged sisal fiber reinforced polypropylene have been studied in the literature (Mukhopadhyay et al. 2008). The results explain that aged-fibers composites have higher mechanical properties than non-aged fiber-based composites. Fiber strength (better in fresh fibers) and good interfacial adhesion (necessary for stress transfer) are the reasons behind this.

There are also several works in the literature on the effects of aging on flax fiber reinforced bio-composites. Moudood et al. (2019) have correlated humid aging with a decrease in tensile stiffness and strength by 57 and 9%, respectively. The effect of water aging on the mechanical properties of green polyethylene reinforced with flax, coconut, and basalt composites was also investigated. In particular, Bajwa et al. (2015) showed that the treatment of bio-flax fibers allows better durability of the green composite. At the same time, Lu et al. (2016) showed that natural fibers treatments boost the performance and the thermal stability of their bio-composites made from jute fibers which are treated with a low concentrated alkaline solution.

Obviously, these changes affect the life-cycle performance: the longer the material remains in the environment, the more it degrades. Several works were carried out on mechanical characterization of bio-polymer-based composites, but few studies were interested to evaluate fatigue life-cycle of these materials thoroughly analyzing the damage mechanisms.

In the literature, several reported studies were concentrated on the development of appropriate manufacturing processes and the investigation of mechanical characterization, basically by considering the static loading under service conditions. Although cyclic-loads are the most common service conditions, very limited works were conducted on the plant fiber reinforced bio-composites behavior under fatigue and dynamic loading. More specifically, these studies have shown that fatigue performance is enhanced with both the volume fraction of long plant-fibers and the adhesion between the fiber and the matrix. Moreover, the fatigue resistance of these materials was marked by good fatigue ratios, commonly close to 0.5, which is mainly related to the quality of the bio-composites (adhesion, void fraction, etc.) as well as to the environmental factors (humidity, etc.) that can affect the fiber-matrix adhesion. Gassan et al. (2002) showed that fiber-matrix adhesion had a significant effect on the fatigue behavior of polyester composites reinforced with flax and jute fibers. El-Sawi et al. (2014) examined the fatigue damage mechanisms of UD flax composites were and crack propagation by infrared thermography. They concluded that natural fiber reinforced plastics with stronger fibers and adhesions or higher fiber fractions had higher critical loads in damage propagation and higher failure loads.

Liang et al. (2014) studied the flax/epoxy composite under tension-tension fatigue loading with different stacking order and then compared the obtained results with similar epoxy reinforced glass fiber composites. The results presented an excellent fit to the Wöhler model and the values of fatigue strength in flax-epoxy were close to those of glass–epoxy below 64 MPa of stress level. At the same time, Yuanjian et al. (2007) compared the fatigue properties of a hemp/polyester composite and a glass/polyester composite. They found that the hemp fiber composites exhibit higher fatigue ratios despite their lower ultimate fatigue strength.

Bensadoun et al. (2016) have shown recently that bio-composites with different flax fiber architectures can improve stiffness and strength (orientation of the fibers in the load direction), as well as ameliorate fatigue life, slow down the damage initiation and its propagation rate. However, in Campilho (2015) the authors concluded that, in order to obtain higher mechanical behaviors, plant fibers should contain high fracture of cellulose.

In Shah (2016), it was observed that the stiffness of jute, flax, and hemp fiber composites decreased significantly (up to 50%) at low applied cycles and stabilized later. However, the same fatigue tests applied to glass fiber composites (GFCs) have shown that the resulting stiffness of natural fiber composites is lower than that of GFCs. Understanding the natural FRP's mechanical behavior requires a deep study of the effect of loading frequency and dynamic response, which requires a good comprehension of their viscoelastic behavior.

In this context, dynamic mechanical analysis (DMA) is emerging as a powerful technique for such analysis in a wide range of applications. Based on such technique, it is possible to obtain information on temperature and time-dependent behavior, as well as aging damage, glass transition temperature, cross-linking and phase mixing degrees, crystallinity proportion, interfacial adhesion, among others. More recently, the use of the DMA technique was applied to characterize the mechanical properties of non-homogeneous materials as well as the crack healing.

Many researchers have investigated the dynamic mechanical properties of non-synthetic fiber reinforced composites. Ray et al. (2002) examined the dynamic mechanical behavior of treated and non-treated jute fiber/vinyl ester composites. The primary results show a significant decrease in the storage modulus (E') with the increase in temperature. The tan delta of the composites was also reduced with the presence of jute fibers. This can be explained by the lower interfacial stress transfer. In another study, the DMA properties of plant flax fiber reinforced polylactic acid (PLA), polypropylene, and epoxy-based composites were reported to be better than those of glass and carbon composites. The best compromise between damping and stiffness was obtained in material with flax fiber reinforced with semi-crystalline natural PLA. The dynamic mechanical stiffness analysis was performed on cellulose microfibers (CMF) reinforced poly (ethylene-co-vinyl acetate) (EVA) composites. The findings suggest that stiffness and damping decrease with CMF load, while the (E') increases by 7.5%. However, the modulus decreases with higher CMF filler content. Kumar et al. (2014) evaluated the damping properties of treated and non-treated coconut fibers. The results showed lower values of tan δ but higher values in E' of the treated composites.

This was attributed to the increased interfacial cohesion between the coconut sheath fibers and the epoxy matrix. Consequently, surface treatments of organic fibers have a significant effect on fiber-matrix interfacial bonding, as it results in an effective load transfer. Limited knowledge of in-service performance and degradation characteristics, especially for new materials like green composites, is a barrier to evaluating the environmental impact of material lifetimes. The behavior of a material during its service life decreases continuously due to degradation mechanisms. Therefore, service life performance has to be considered in the Life Cycle Assessment (LCA) to guarantee a complete comprehension of durability.

13.3 Weather Aging

Carbon fiber reinforced polymers (CFRPs) are a family of advanced composite materials used in aerospace applications, consisting of carbon fiber reinforcing matrix. When exposed to the weather environment, they are subjected to aging mechanisms. The principal mechanism of photooxidative aging and thermal oxidation of these materials is the free radical reaction mechanism, which involves chain growth and chain breakage. (CFRPs) combined with water and oxygen molecules in humid environment under Ultra-Violet radiation result in the formation of oxidation products that contain carbon radicals. As aging progresses, branching, cross-linking, and molecular chain scission occur.

The long-term service life of composites at high temperatures is affected by thermal and oxidative deterioration of the materials. Despite the fact that the fibers are relatively stable in the composites, the matrix as well as the fiber/matrix interface can be easily damaged at elevated temperatures. Thus, it is important to understand how the materials will perform over the intended service life when considering composite materials for longtime applications.

It was convenient by many authors that the incorporation of antioxidants is the most and effective way to prevent polymers from thermo-oxidative aging (Ding et al. 2022; Fusco et al. 2007). However, conventional low molecular weight antioxidants are found to migrate to outer surface of polymer materials as they have low thermal stability. These changes result in a significant decrease of antioxidant activity (Jomova et al. 2024). To solve the migration problem, different studies have proposed to immobilize antioxidants on the inorganic filler surfaces (silica, titanium dioxide). The bounded antioxidants can increase the compatibility of the polymer composite to inorganic fillers, leading thus to the increase in antioxidative performance and mechanical properties. Another effective method is to increase the molecular weight of the added antioxidant by grafting them onto the polymer (natural or synthetic). Wang et al. (2023) prepared macro-molecular antioxidants by grafting phenolic antioxidants to polybutadiene using isophorone-diisocyanate as a bridging agent. The resulting macromolecular antioxidant shows better resistance to thermo-oxidative aging.

Another type of aging was largely discussed in literature which is Ultra-Violet radiations.

In a recent study, Shi et al. (2022) showed that increasing UV irradiation exposure, increases progressively the damage on the specimen surfaces compared to the non-aged specimen which surfaces were entirely covered with the resin. The authors found that short exposure to UV (10 days) form fine cracks at the fiber-matrix junction. Moreover, after 40 days of irradiation, more obvious damage was registered in the matrix: significant holes appeared on the surface of the tested sample and the number of bonded fibers increased.

Upon 80 days of aging, the voids became larger, and many fibers were fully exposed, the fractures in the matrix started to expand inward, and rapidly, the matrix was delaminated, explaining the loss of the sample's quality in later aging time.

UV aging has an evident damaging effect on carbon composites, the principal mechanism is as described as follows: UV-rays are very energetic and active than visible light, and the energy of ultraviolet rays can lead to the breakage of several covalent bonds. Therefore, under UV-rays, the polymer macromolecular chains will break, and the resulting small molecular chains will cover a large area and cause small cracks. In addition, the polymer and the fiber have not the same coefficients of thermal expansion. Consequently, the resin deformation is greater at high temperatures than that of the fiber, which results in thermal stress at the junction of fiber/matrix and gradually breaks the deformation coordination relationship between the fiber and the polymer, causing the fiber and the polymer to disassociate from each other, further aggravating the expansion of cracks.

UV aging has an evident damaging effect on carbon composites, the principal mechanism is as described as follows: UV-rays are very energetic and active than visible light, and the energy of ultraviolet rays can lead to the breakage of several covalent bonds. Therefore, under UV-rays, the polymer macromolecular chains will break, and the resulting small molecular chains will cover a large area and cause small cracks. In addition, the polymer and the fiber have not the same coefficients of thermal expansion. As a consequence, the resin deformation is greater at high temperatures than that of the fiber, which results in thermal stress at the junction of fiber/matrix and gradually breaks the deformation coordination relationship between the fiber and the polymer, causing the fiber and the polymer to disassociate from each other, further aggravating the expansion of cracks.

These changes lead to fiber/matrix debonding, which affects the mechanical behavior of the material (Dever et al. 2001). It is, therefore, important to study the mechanical performance of the aged samples, and with the use of nanoindentation techniques, it is possible to characterize the material surface mechanical properties. Pecora et al. (2021) carried out nanoindentation tests, before and after thermal aging, on epoxy matrix reinforced by 3D interlocked carbon fiber composites. They concluded that both the elastic and plastic behaviors were influenced by thermal aging; however, the viscoelastic properties were not affected. Babu et al. (2018) performed also nanoindentation tests on epoxy-based carbon fibers composites before and after exposure to UV-rays. They reported that UV-rays caused an elevation of the modulus close to the fibers, which was associated to high cross-linking densities in these areas. Most nanoindentation characterizations techniques have dealt with the composite matrix response under UV aging without detailed investigations on the reinforcing and interfacial phases.

In addition to UV irradiation and thermal-oxidative aging, humid aging has attracted significant attention and has been thoroughly investigated over the past few decades.

In fact, unsaturated polyester reinforced by glass fiber composites are materials widely used in various fields. However, prolonged exposure to environmental conditions like moisture and temperature can cause damage. In recent years, many studies have been made in order to better assess the long-term impact of these factors and to understand the possible degradation mechanisms. To explain, in polymeric materials, moisture can easily diffuse either in singular areas (defects due to elaboration process) or in the matrix macromolecular networks. This leads to many changes in the properties according to the physical and chemical mechanisms occurring in the material. At the beginning, physical aging seems to be reversible damage leading, thus, to network plasticization, which causes a decrease in the glass transition temperature (Tg) and so far, the properties of the material. These phenomenon can be reversed after water desorption and a return to unaged properties can be almost achieved. Unfortunately, water aging causes irreversible damage: hydrolysis and crack propagation. In this study, we focus on physical and chemical degradation due to moisture absorption.

An original method was proposed in recent years to improve humid aging resistance consists of incorporating nanofillers in the polymer (Wang et al. 2020). Different authors have found that the mechanical, thermomechanical, and electrical properties of the modified polymer were greatly improved after nanofillers incorporation.

In this context, Shettar et al. (2019) investigated the durability of epoxy composites exposed to 50 °C at 80% of relative humidity and found that the incorporation of nanoclay reduced water absorption. They concluded that flexural strength of aged nonoclay/epoxy-based composite is better than non-aged material.

Several studies were conducted to study the kinetics of water uptake in composites. The moisture uptake behavior can typically be described by Fick or by Carter and Kibler's theory (known as Langmuir-type diffusion). Fick's theory describes moisture diffusion in homogeneous materials. However, Langmuir can predict water diffusion in heterogeneous materials. In their theory, Carter and Kibler assume the presence of two different types of water molecules in the composite known as "mobile" and "bound" water which are, respectively, the free (non-hydrogen bound) molecules, and the hydrogen bound molecules.

The Langmuir diffusion theory is given by the following equations:

$$\frac{n(t)}{M} = \frac{\beta}{\beta+\gamma}\left\{1 - \frac{8}{\pi^2}e^{-kt}\right\} \tag{13.1}$$

$$\frac{N(t)}{M} = \frac{\gamma\beta}{\beta+\gamma}e^{-\beta t}\left\{\frac{1}{\beta}\left(e^{\beta t}-1\right) - \frac{8}{\pi^2 k}\left(e^{-kt}-1\right)\right\}$$

where N(t) and n(t) are, respectively, the bound and free water molecules, M is the number of diffused water molecules at a given time t ($M(t) = n(t) + N(t)$), γ and β are the probabilities of a water molecule to be hydrogen bounded and free forms, respectively. In this context, Abdessalem et al. (2019) showed in a recent study that water diffusion in Sheet Molding Compound (SMC) composites follows Langmuir

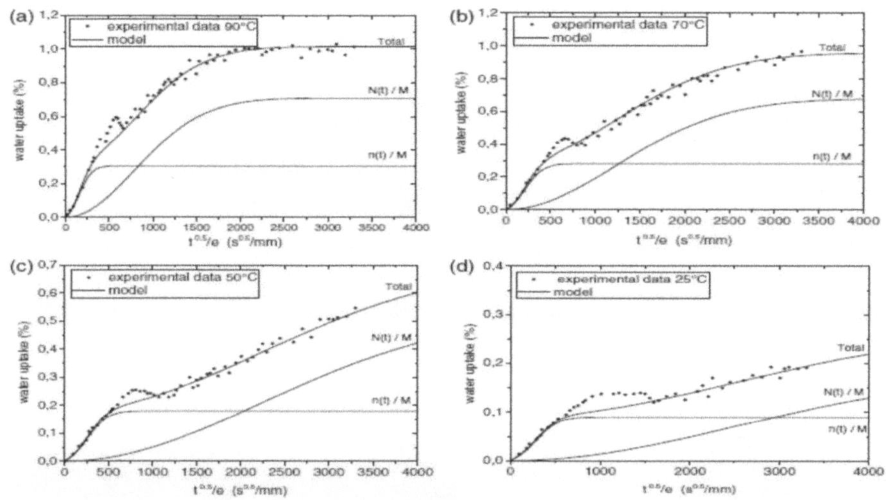

Fig. 13.1 Experimental and modeled gravimetry tests for moisture diffusion in 100% relative humidity (RH) at **a** 90 °C, **b** 70 °C, **c** 50 °C, and **d** 25 °C (•) experimental, (−) model

theory (see Fig. 13.1). They deduced that the diffusion depends on several parameters, in particular the volumetric fraction of fillers (chalk particles) and heterogeneities (voids due to the molding process).

Despite the high correlation between the gravimetric values and the model results, a small deviation was observed, which was related to the leakage of chalk particles, as the material is highly charged with $CaCO_3$.

In this context, moisture diffusion can cause permanent damage. Water transport can be encouraged by diffusion in the matrix which contains imperfections (microspaces, pores or cracks) as well as by capillarity along the fiber/matrix interface (Tamboura et al. 2022). Microcracking and delamination are commonly the pre-dominant forms of damage to take place in such materials when exposed to hydrothermal aging as shown in Fig. 13.2 (Abdessalem et al. 2021).

Finally to summarize, the kinetics of mobile (free) water diffusion is affected by the presence of heterogeneities and micro-cracks in the material. To explain, water absorption is governed by diffusion and capillarity in micro-pore spaces (Langmuir diffusion). A progressive increase in the water content in these areas leads to a gradual increase in the osmotic pressure in these regions, resulting in swelling zones (bubbles in Fig. 13.3).

When the critical osmotic pressure is exceeded, crack propagation is induced, and hence an increase in water absorption.

Once cracking has occurred, the evolution of micro-cracks is not reparable and causes the failure of the material. Micro-cracks are caused by an osmosis mechanism (Tcharkhtchi et al. 2000) and can be facilitated by the presence of imperfections such as micro-voids. Other types of damage can be found in thermoset plastic composites: blistering. In fact, when composites are conditioned, the absorbed moisture can

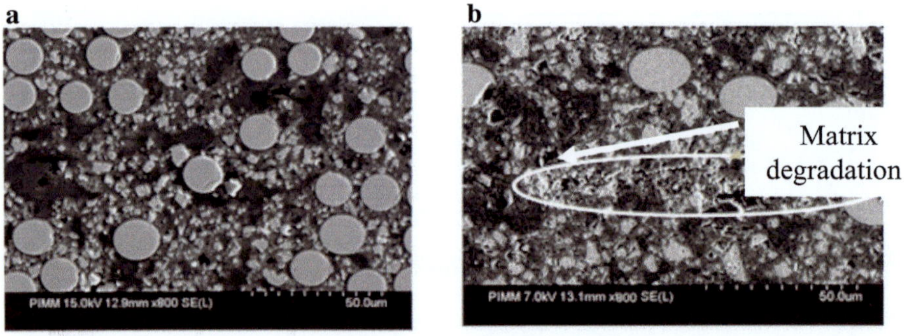

Fig. 13.2 Matrix degradation at 90 °C: **a** before and **b** after 15% of total water uptake

Fig. 13.3 Blistering zones after 30% of total absorbed water at **a** 25 °C, **b** 50 °C, **c** 70 °C, and **d** 90 °C

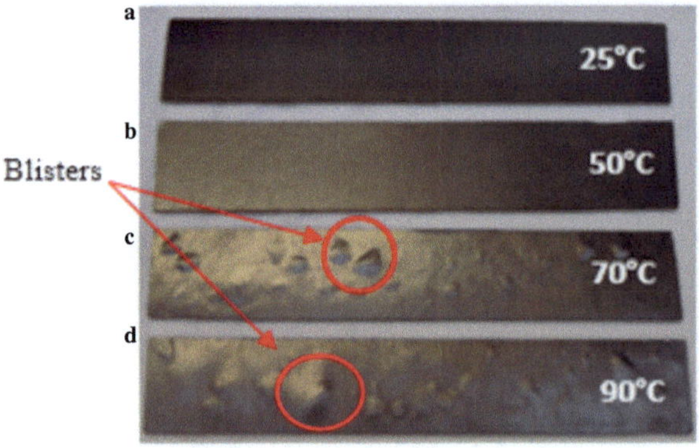

induce an internal pressure known as "osmotic pressure." When the osmotic pressure is greater than the matrix strength, blisters in the form of bubbles at the external surfaces appear (see Fig. 13.3).

On the other hand, since polymer materials are highly sensitive to moisture due to the molecular polarity, the entrance of water molecules into the network is facilitated by their hydrophilic character. Therefore, since water molecules can interact with the hydrophilic groups of the matrix, water uptake is enhanced by the formation of hydrogen bonds with the polar chemical groups. Desorption tests made by Zhou and Lucas (1993) have shown that these bonds can take two forms, based on the activation energy: "Type I" characterized by the creation of a single hydrogen-bond with a low activation energy (10 kcal/mol) and the "Type II" is marked by the formation of a multiple hydrogen-bonds (forming a bridge) with a high activation energy (15 kcal/mol).

To understand and assess the durability of composite materials, the damage process should be well characterized. Many techniques were employed to qualify and quantify internal degradation in polymer composites. Hsu et al. (2000) have used an instrumented impact testing method, called "tap test," to evaluate damage in composite structures used in aircraft. Kinra et al. (2006) examined matrix microcracks in laminated composites by means of an ultrasonic backscattering technique. Lastly, since damage in composites is dispersed because of their complex morphology, X-ray micro-computed tomography (XμCT) has been recognized as an efficient technique for damage analysis in such materials. The three-dimensional map obtained by this technique allows the visualization and the quantification of the inner microstructure, such as the volume fraction of each constituent and their spatial distribution.

In this context, Schilling et al. (2005) used X-ray tomography to study internal delamination, flaws and microcracks in graphite and glass reinforced epoxy composites after mechanical loads. Awaja et al. (2009) used X-ray tomography to evaluate the impact of accelerated thermal degradation and microcracking in epoxy-based composites. Saucedo-Mora et al. (2016) examined the damage evolution of ceramic matrix reinforced by ceramic composites using volume correlation to measure deformation and to continuously follow crack evolution. Arif et al. (2014) characterized the fatigue damage mechanisms of composites made of polyamide 66 and short glass fiber following the injection molding process to well understand the induced fatigue damage. Further, Abdessalem et al. (2021) studied aging of synthetic SMC composite when immersed in tempered water. As water aging is a slow phenomenon, they used different temperature to fasten the degradation. In order to characterize fracture nucleation and propagation, the authors used in-situ microscopic observation tests. They found that the first occurring failure is fiber/matrix decohesion due to local concentrated stress at the interfaces. The damage mechanism seems to be the same for aged and non-aged composites as shown in Fig. 13.4. Besides, they concluded that the increase in water uptake ratio increases the interfacial damage and so far, favors the hole degradation of the material.

In the following section, we plan to discuss the different proposed models in the literature. These models are used to well predict the remaining life cycle of composite material in service environment which can make timely risk assessment for many fields, namely aerospace field. This is a practical method to ensure a safe production in the aeronautic field in order to guarantee a stable operation of composite materials.

13.4 New Fast Methodology for Predicting Composites' End Life Cycle: Application to SMC Composite

Many studies are available in the literature on the development of predictive mechanical characterization of composites-based polymer matrix.

Recent researches showed that the mechanical response of both A-SMC and SMC composites is strongly influenced by microstructure variability and strain rate (Fitoussi et al. 2013, 2005; Shirinbayan et al. 2017). In this context, a micromechanical proposed recently by Fitoussi (1998) was carried out to predict the tensile test curves and the loss of stiffness caused by damage in SMC composites. The

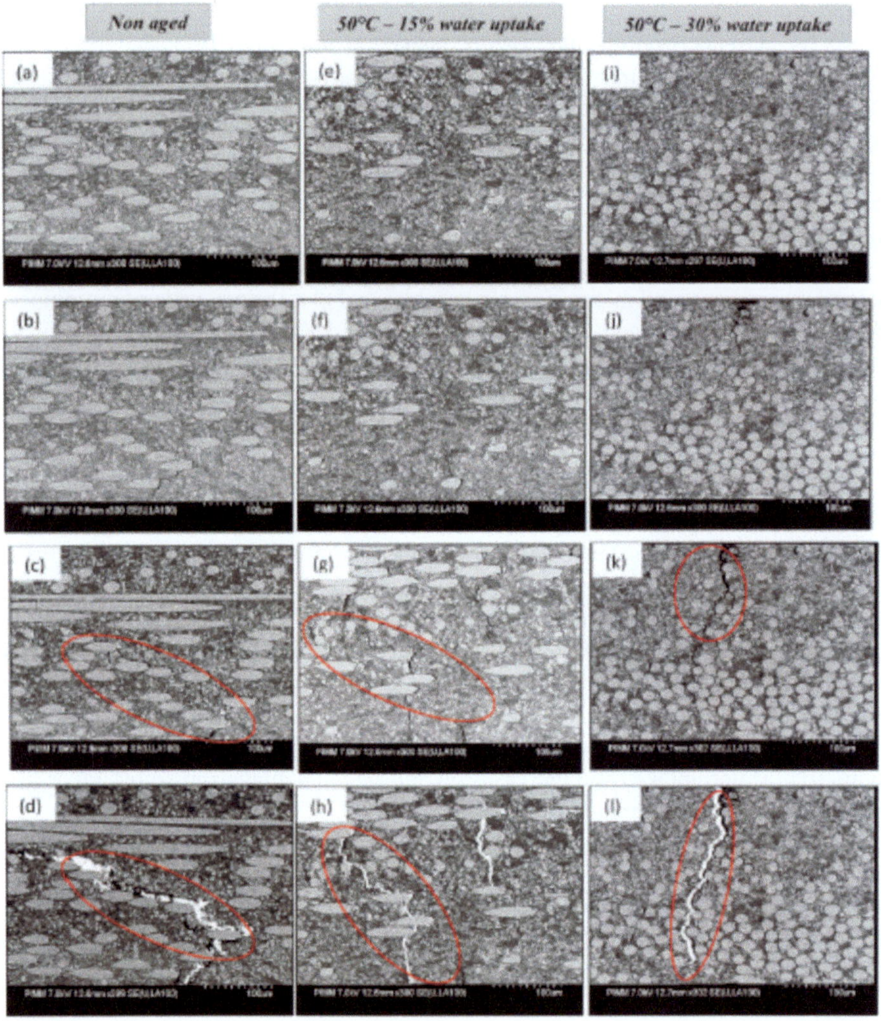

Fig. 13.4 In-situ tests for aged and non-aged samples

proposed model is essentially based on a statistical local interface failure criterion introduced in the Mori and Tanaka model. Meraghni et Benzaggah (1995) introduced this type of micromechanical model in a finite element analysis to study the damage of composite structures. This approach has been verified in a bending torsion loading test. Jendli et al. (2005) applied this approach in the case of high strain rate loading interrupted tensile tests. The authors used a multi-scale experimental approach to evaluate a multi-scale visco-damage model. Shirinbayan et al. (2017) used the described method on advanced SMC composites (A-SMC) to analyze damage mechanisms as a function of fibers orientation and strain velocity to provide the necessary database to build a predictive model of the micro-mechanical response of A-SMC under crash tests.

Many authors have proposed different models to describe the damage behavior of composites to be implanted in numerical simulations (Kästner et al. 2012; Kweon and Benzerga 2013). When describing the mechanical behavior of composite material, two predictive models can be distinguished: phenomenological and micromechanical approaches based specifically on damage variables. Indeed, the micromechanical approach describes the evolution of the microcrack fraction during loading. However, the phenomenological approach focuses on the macroscopic consequences of local damage, for example, the loss of stiffness.

Here, the macroscopic scale is equivalent to the Representative Volume Element (RVE) of the material. To facilitate micromechanical analysis, the damage variable is, in some cases, non-local. Therefore, the cracks parameters are homogenized. This approach is frequently used in the case of mixed damage models which combines micromechanical and phenomenological characteristics. On the other hand, the damage variable is usually independent to the physical reality when using energetic approaches (Nedjar 2001). The damage models can describe softening deformation in some cases. Mixed models are usually associated with a failure criterion. In phenomenological models, the laws of mechanical behavior are deduced from thermodynamic properties and the description of mechanical degradation is more finite. In numerical simulations, it is commonly easy to use phenomenological failure models. However, micromechanical failure model requires the microscopic data.

In micromechanical models, the damage parameter has a physical meaning related to the crack volume ratio, the fracture dissipated energy and crack parameters. Furthermore, micromechanical modeling is considered as the most appropriate approach used to predict the damage behavior of composites, especially when considering the material anisotropy. In the above-described models, failure can be considered through several local damage criteria, which determine the initiation of microcracks at a local scale. It can be noted that, regardless the used numerical approach, a deep knowledge of damage mechanisms at a local scale and their kinetics is always required to well identify failure variables (Jendli et al. 2004).

Understanding the mechanical behavior of SMC composites needs to consider several failure modes, namely fatigue behavior as it is the most important failure mode to be considered in the design of SMC aerospace components.

Indeed, the lack of sufficient knowledge of this failure mode results in specifications given by the manufacturers that often require clarification. Besides, the available experimental information generally used to carry out pre-design are still limited especially for new material formulations. Thus, including material formulation becomes a must to optimize development times.

In this way, many iterations are usually required to find the right formulations for each new application. It is, therefore, essential to develop a rapid methodology allowing the pre-dimensioning of structures in a short time by using the right material formulation for each new aerospace component.

This methodology should be pragmatic for aerospace industries purposes. When the components are subjected to fatigue loading, an experimental campaign is required to establish the SN curves which is a critical issue as the tests are very long costly. Furthermore, short fiber reinforced composites are very complex materials because of their heterogeneous structure, anisotropy, and failure development compared to homogeneous materials.

In fact, the mechanical properties of these materials have been found to be strongly influenced by their composition: matrix (thermoset/thermoplastic), and fibers (geometry, volume fraction and spatial distribution).

The diversity of architectures in polymer-based composite materials and their high complexity has their own specificities for each application. This large variety of microstructures leads to non-negligeable mechanical scattering properties, particularly under fatigue loading. Thus, predicting the thresholds properties and kinetics and their impact on the residual mechanical properties, as well as the lifetime of the materials become an interesting challenge. The scattered data on composite Wohler curves have been highlighted by Barnard et al. (1985). In their study, they acknowledge that the damage mechanism changes are due to microstructural variation and failure during the different cycles of fatigue process. Then they concluded that there is a strong relationship between the statistical variation in strength and the scatter recorded in the composite Wohler curves. In the case of SMC materials, this effect would be expected to be greatly increased due to the high heterogeneities of the microstructure induced by the molding process (Abdessalem et al. 2021). Indeed, even for a given formulation, many microstructures can be present in aerospace components. Therefore, the predictive model should take into account microstructure variation as an input data.

Moreover, given the complexity of the local phenomena that occur when loading, the proposed approach must be founded on a deep knowledge of these mechanisms. Different studies have shown that local damage should be a central part of the design methodology when studying short fiber reinforced composites (e.g., Sheet Molding Compound) (Shirinbayan et al. 2015).

On one hand, when using thermoplastic polymers-based composite, the non-linear characteristics due to polymer visco-plasticity at low frequency and ambient temperature do not need to be considered in the model. On the other hand, the effects of environmental and loading conditions (stress ratio, temperature, frequency) are of major importance for the fatigue behavior.

Therefore, the proposed methodology should be capable to predict the life-cycle of these composite materials within these different complex conditions. The fatigue failure behavior of composite-based polymer materials seems to be very complex. In particular, three categories of predictive fatigue models applied to PFCs exists in literature (Laribi et al. 2018): phenomenological model based on residual strength/stiffness variations, micro-mechanical model based on microscopic failure mechanism description and empirical model based on S–N curves.

Considerable efforts have been made to develop new methods that can significantly reduce the cost and time required to characterize fatigue behavior.

In their study, Jegou et al. (2013) introduced a new methodology which focuses on temperature measurements. The energetic fatigue criterion, obtained from the measurement series, is based on only one microstructure in order to be able to predict the S–N curve from the heat build-up one. The overall duration tests are less than 48 h. In another recent study conducted on natural rubber compound filled with carbon black, Marco et al. (2017) proposed a comparable approach. The authors related the fatigue life cycle to crack evolution by using interrupted fatigue tests, and also the energy dissipation.

The proposed approach gives an easy prediction of S–N curves based on temperature measures and an energetic criteria. It has been tested on a large range of thermoset and thermoplastic polymers.

Jain et al. (2016) suggested an approach which focuses on a micromechanical model that relates the macroscopic damage state under quasi-static loading to the state reached at a given state of fatigue curve. The authors have also presented a methodology that allows the prediction of fatigue behaviors for other composite structures on the basis of the knowledge of the so-called "master curve," which represents the Wohler curve obtained for a specific structure used as a reference. Nevertheless, the proposed methodology needs experimental finding. In fact, for the number of cycles to fatigue failure, the gap between the stiffness loss curves of all microstructures is not important and can be accepted as equal from a statistical point of view.

In the case of Sheet Molding Compound Composites, experimental results show that, before failure, the stiffness loss kinetics and also its critical level are strongly dependent on the microstructure. In recent research conducted by LARIBI, a hybrid modeling approach was presented. In order to consider simultaneously the microstructure and the damage effect, the proposed model was based on a Mori and Tanaka predictive approach which is based on the introduction of a failure criterion at a microstructural scale. The model was developed using simple tensile loading for a given microstructure. It is also capable to predict the mechanical response under monotonic loads for any other microstructure.

Many authors have highlighted an existing relationship between monotonic and fatigue behavior when studying composite materials (Laribi et al. 2018; Jain et al. 2016). In addition, many fatigue prediction models focus on residual monotonic strength (Fitoussi et al. 2013). In fact, the authors concluded that the same failure mechanisms were registered in both monotonic and fatigue loading.

It can therefore be assumed that a particular local damaged in microstructure, generally described by a microcrack content, can be obtained under different loading history (e.g., fatigue and monotonic test creep and thermal strain loading) and that the local damage is related to macroscopic characteristics (e.g., loss of stiffness).

Recently, Laribi et al. (2018) have recently introduced a state equation that correlates the local damage with the macroscopic loss of stiffness under monotonic loads. The authors concluded that the proposed equation is also suitable for fatigue loading. It is therefore possible to assume the presence of a multi-scale relationship between local and macroscopic failure that is not related to the macroscopic loading history. If microscopic and macroscopic failure are described by relative values, this intrinsic relationship remains independent of microstructure. To explain, we consider two composite samples subjected to the same relative loss of stiffness under monotonic loading and under fatigue loading separately. The two materials develop the same local damage rate. Furthermore, Laribi et al. (2018) found that ultimate damage occurs when a critical level of local failure rate is achieved. In fact, it can be considered that failure starts when the resistant surface achieves a critical rate. Finally, microcracks coalesce to form a macroscopic crack leading to the breakdown of the structure.

Therefore, damage and failure under monotonic loading are intrinsically linked to fatigue life. The relationship between the first loss of stiffness (E_1/E_0) and the number of cycles to failure (N_R) for each fatigue test can be highlighted by plotting them on a graph. This relationship is emphasized for all working aging temperatures at different immersion periods (Sect. 13.5) and different microstructures (Laribi et al. 2018).

In the next section, the authors aim to investigate the different steps used in the described methodology to predict the (N_R) for aged composite materials at different water aging conditions.

13.5 Application: Predictive Methodology for Aged SMC Composites

Now, we have a relationship that allows prediction of fatigue life based on the applied stress. In the following paragraph, the experimental tests and protocols are explained.

In this context, several tests were carried out:

- Quasistatic tensile tests until failure were conducted. It is important to confirm that the samples with similar ultrasonic results record similar mechanical proprieties.
- Tension-tension fatigue loads at many levels of applied stress. Here, the working stress ratio were $R = 0.1$. In order to better determine E_0 and the residual stiffness after the first cycle, E_1, each fatigue test was preceded by a static load-unload-reloading elastic cycle.

The relationship between the first loss of stiffness (E_1/E_0) and the number of cycles to failure (N_R) for each fatigue test can be highlighted by plotting them on a graph. This relationship is emphasized for all working temperatures at different immersion periods and for different microstructure (Laribi et al. 2018).

The detailed predictive methodology was applied to aged SMC composites with temperature varying from 25 to 90 °C. The numerical procedure can be summarized graphically to establish the SN curve for any given microstructure. The steps are presented in Fig. 13.5 and described as follow:

Step 1: An experimental campaign is realized using loading–unloading tensile tests to establish the experimental reference curve; $(D, \sigma)_{Ref}$ curve where D is the macroscopic damage ($D = 1 - E/E_0$). This step should be carried out for a given state (non-aged, aged specimen at a given time and temperature).

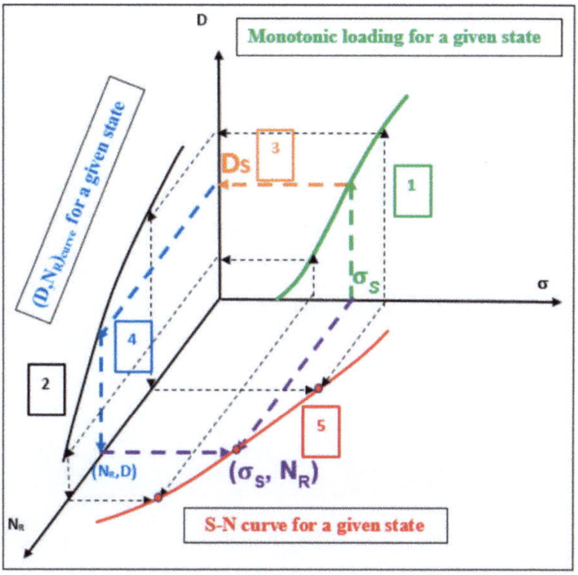

Fig. 13.5 A graphic summarizes of proposed predictive methodology

Step 2: A second experimental campaign is realized based on fatigue tests until-failure preceded-by a loading–unloading reloading quasi-static-test are carried out at different amplitudes in order to establish the $(D_1, N_R)_{Ref}$ curve.

Step 3: The value of the macroscopic damage D_S reached for σ_S is extracted from the target curve $(D, \sigma)_{Target}$.

Step 4: The number of cycles at break, N_r, corresponding to $D = D_S$ is identified on the curve $(D_1, N_R)_{Ref.}$

Step 5: Recording the couple of value (σ, N_R).

Step 6: Steps 1–5 are repeated-for another chosen-value of stress S as many times are needed—to establish the SN-curve of aged microstructure.

The given methodology was applied to the SMC composite aged at different temperatures time. The results are shown in Fig. 13.6.

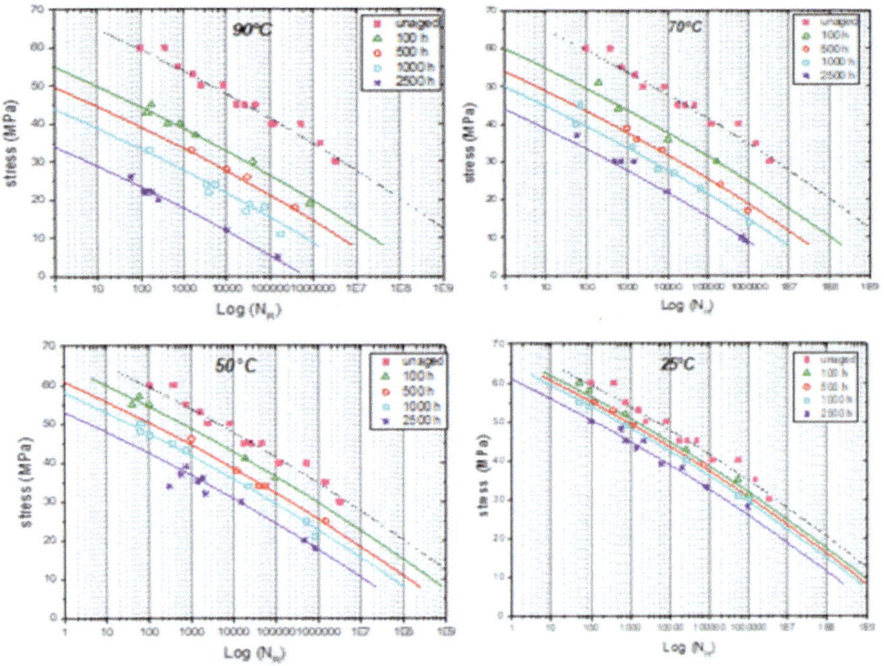

Fig. 13.6 Modeling of Wohler curves under fatigue loading using the predictive model

13.6 Conclusion

This review focuses on the potential of composites to transform the aerospace industry, where a good balance between high strength, lightweight properties, and multi-functionality is necessary. Key findings highlight the benefits of fiber reinforced composites, particularly those including natural fibers. These materials provide unique mechanical, thermal, and environmental properties that enable aerospace applications. Selecting the appropriate matrix and fiber combination, complemented by advanced interface engineering, is essential for maximizing material performance and durability under harsh service conditions. Composite manufacturing techniques have made significant advances. These methods enable the creation of complex, highly customized structures while improving manufacturing accuracy and reducing waste.

It is believed that to fully exploit these materials it is essential to gain an understanding of the relationships between microstructures and macroscopic properties.

In this context, original experimental methodologies have been developed for two different purposes. Firstly, these methodologies have been used for the statistical analysis of the microstructures and local physical phenomena that occur during the application of different types of thermomechanical stress. Secondly, these methods have been used for the analysis of macroscopic mechanical responses.

Therefore, these studies focus primarily on the control of scale transitions. Subsequently, multi-scale experimental approaches are proposed with the aim of developing micromechanical models that can be used to predict the behavior and damage of composite materials. The experimental results obtained will allow the identification and validation of the predictive models developed. The influence of aging phenomena on the deformation and damage mechanisms of materials have also been addressed.

The future directions in multifunctional composite design are poised to significantly influence various industries, particularly the aerospace and automotive industries. As the demand for materials that can perform multiple functions simultaneously increases, researchers and engineers focus on innovative approaches to enhance these composites' performance, sustainability, and applicability.

In fact, biocomposites represent an important step forward as they are in line with the current trend toward integrated and smart material systems. However, despite this progress, many challenges need to be resolved, especially in terms of scalability, recyclability, and costs. The development of bio-based and sustainable materials and successful recycling solutions is required to decrease environmental impacts as well as guarantee the long-term life of composite material technologies. In addition, studies on long-lasting raw materials as well as manufacturing processes are essential to further decrease the environmental impact of such materials.

References

Abdessalem A, Tamboura S, Fitoussi J, Daly HB, Tcharkhtchi A (2019) Bi-phasic water diffusion in sheet molding compound composite. J Appl Polym Sci 48381. https://doi.org/10.1002/app.48381

Abdessalem A, Tamboura S, Fitoussi J, Daly HB, Tcharkhtchi A, Meraghni F (2021) Microstructure investigation of hydrothermal damage of aged SMC composites using micro-computed tomography and scanning electron microscopy. Eng Fail Anal 121:105177. https://doi.org/10.1016/j.engfailanal.2020.105177

Arif MF, Meraghni F, Chemisky Y, Despringre N, Robert G (2014) In situ damage mechanisms investigation of PA66/GF30 composite: effect of relative humidity. Compos B Eng 58:487–495. https://doi.org/10.1016/j.compositesb.2013.11.001

Awaja F, Arhatari BD (2009) X-ray Micro Computed Tomography investigation of accelerated thermal degradation of epoxy resin/glass microsphere syntactic foam. Compos A Appl Sci Manuf 40:1217–1222. https://doi.org/10.1016/j.compositesa.2009.05.014

Babu LK, Mishra K, Singh RP (2018) Near–fiber effects of UV irradiation on the fiber–matrix interphase: a combined experimental and numerical investigation. Mater des 157:294–302. https://doi.org/10.1016/j.matdes.2018.07.050

Badji C, Beigbeder J, Garay H, Bergeret A, Bénézet J-C, Desauziers V (2018) Exterior and under glass natural weathering of hemp fibers reinforced polypropylene biocomposites: impact on mechanical, chemical, microstructural and visual aspect properties. Polym Degrad Stab 148:104–116. https://doi.org/10.1016/j.polymdegradstab.2017.12.015

Bajwa DS, Bajwa SG, Holt GA (2015) Impact of biofibers and coupling agents on the weathering characteristics of composites. Polym Degrad Stab 120:212–219. https://doi.org/10.1016/j.polymdegradstab.2015.06.015

Barnard PM, Butler RJ, Curtis PT (1985) Fatigue scatter of UD glass epoxy, a fact or fiction? In: Marshall IH (ed) Composite structures 3. Springer, Netherlands, Dordrecht, pp 69–82

Barouni A, Lupton C, Jiang C, Saifullah A, Giasin K, Zhang Z, Dhakal HN (2022) Investigation into the fatigue properties of flax fibre epoxy composites and hybrid composites based on flax and glass fibres. Compos Struct 281:115046. https://doi.org/10.1016/j.compstruct.2021.115046

Belingardi et al (2006) Bending fatigue behavior of glass–carbon/epoxy hybrid composites. Compos Sci Technol 66:222–232 (2006). https://doi.org/10.1016/j.compscitech.2005.04.031

Bensadoun F, Vallons KAM, Lessard LB, Verpoest I, Van Vuure AW (2016) Fatigue behaviour assessment of flax–epoxy composites. Compos A Appl Sci Manuf 82:253–266. https://doi.org/10.1016/j.compositesa.2015.11.003

Berges M, Léger R, Placet V, Person V, Corn S, Gabrion X, Rousseau J, Ramasso E, Ienny P, Fontaine S (2016) Influence of moisture uptake on the static, cyclic and dynamic behaviour of unidirectional flax fibre-reinforced epoxy laminates. Compos A Appl Sci Manuf 88:165–177. https://doi.org/10.1016/j.compositesa.2016.05.029

Campilho RDSG (ed) (2015) Natural fiber composites. CRC Press, Boca Raton

Carter HG, Kibler KG (1978) Langmuir-type model for anomalous moisture diffusion in composite resins. J Compos Mater 12:118–131. https://doi.org/10.1177/002199837801200201

Chen X, Xu J, Li Y, Gao Y, Wang G (2023) Integrating multiple energy storage in 1D–2D bridged array carbon-based phase change materials. SusMat 3:510–521. https://doi.org/10.1002/sus2.147

Chilali A, Zouari W, Assarar M, Kebir H, Ayad R (2018) Effect of water ageing on the load-unload cyclic behaviour of flax fibre-reinforced thermoplastic and thermosetting composites. Compos Struct 183:309–319. https://doi.org/10.1016/j.compstruct.2017.03.077

Developing bio-based composites that are fit to fly | Airbus. https://www.airbus.com/en/newsroom/stories/2024-06-developing-bio-based-composites-that-are-fit-to-fly

Dever JA, Messer R, Powers C, Townsend J, Wooldridge E (2001) Effects of vacuum ultraviolet radiation on thin polyimide films. High Perform Polym 13:S391–S399. https://doi.org/10.1088/0954-0083/13/3/331

Ding P, Zou Y, He J, Sun Y, Liu F (2022) Effects of a novel chitosan based macromolecule antioxidant COS-GMMP on the thermo-oxidative aging of styrene-butadiene rubber/silica composites. Polym Degrad Stab 195:109813. https://doi.org/10.1016/j.polymdegradstab.2021.109813

El Sawi I, Fawaz Z, Zitoune R, Bougherara H (2014) An investigation of the damage mechanisms and fatigue life diagrams of flax fiber-reinforced polymer laminates. J Mater Sci 49:2338–2346. https://doi.org/10.1007/s10853-013-7934-0

Fick A (1995) On liquid diffusion. J Membr Sci 100:33–38. https://doi.org/10.1016/0376-7388(94)00230-V

Fitoussi J, Meraghni F, Jendli Z, Hug G, Baptiste D (2005) Experimental methodology for high strain-rates tensile behaviour analysis of polymer matrix composites. Compos Sci Technol 65:2174–2188. https://doi.org/10.1016/j.compscitech.2005.05.001

Fitoussi J, Bocquet M, Meraghni F (2013) Effect of the matrix behavior on the damage of ethylene–propylene glass fiber reinforced composite subjected to high strain rate tension. Compos B Eng 45:1181–1191. https://doi.org/10.1016/j.compositesb.2012.06.011

Fitoussi J, Guo G, Baptiste D (1998) A statistical micromechanical model of anisotropic damage for S.M.C. composites. Compos Sci Technol 58:759–763 (1998). https://doi.org/10.1016/S0266-3538(97)00163-2

Fusco D, Colloca G, Monaco MRL, Cesari M (2007) Effects of antioxidant supplementation on the aging process. Clin Interv Aging 2:377–387. https://doi.org/10.2147/cia.S12159918

Gassan J (2002) A study of fibre and interface parameters affecting the fatigue behaviour of natural fibre composites. Compos A Appl Sci Manuf 33:369–374. https://doi.org/10.1016/S1359-835X(01)00116-6

Gude M, Hufenbach W, Koch I, Protz R (2006) Fatigue failure criteria and degradation rules for composites under multiaxial loadings. Mech Compos Mater 42:443–450. https://doi.org/10.1007/s11029-006-0054-z

Hsu DK, Barnard DJ, Peters JJ, Dayal V (2000) Physical basis of tap test as a quantitative imaging tool for composite structures on aircraft. AIP Conf Proc 509:1857–1864. https://doi.org/10.1063/1.1306256

Jain A, Veas JM, Straesser S, Van Paepegem W, Verpoest I, Lomov SV (2016) The Master SN curve approach—a hybrid multi-scale fatigue simulation of short fiber reinforced composites. Compos A Appl Sci Manuf 91:510–518. https://doi.org/10.1016/j.compositesa.2015.11.038

Jegou L, Marco Y, Le Saux V, Calloch S (2013) Fast prediction of the Wöhler curve from heat build-up measurements on short fiber reinforced plastic. Int J Fatigue 47:259–267. https://doi.org/10.1016/j.ijfatigue.2012.09.007

Jendli Z, Meraghni F, Fitoussi J, Baptiste D (2004) Micromechanical analysis of strain rate effect on damage evolution in sheet molding compound composites. Compos A Appl Sci Manuf 35:779–785. https://doi.org/10.1016/j.compositesa.2004.01.020

Jendli Z, Fitoussi J, Meraghni F, Baptiste D (2005) Anisotropic strain rate effects on the fibre–matrix interface decohesion in sheet moulding compound composites. Compos Sci Technol 65:387–393. https://doi.org/10.1016/j.compscitech.2004.09.027

Jomova K, Alomar SY, Alwasel SH, Nepovimova E, Kuca K, Valko M (2024) Several lines of antioxidant defense against oxidative stress: antioxidant enzymes, nanomaterials with multiple enzyme-mimicking activities, and low-molecular-weight antioxidants. Arch Toxicol 98:1323–1367. https://doi.org/10.1007/s00204-024-03696-4

Kästner M, Obst M, Brummund J, Thielsch K, Ulbricht V (2012) Inelastic material behavior of polymers—experimental characterization, formulation and implementation of a material model. Mech Mater 52:40–57. https://doi.org/10.1016/j.mechmat.2012.04.011

Kinra VK, Ganpatye AS, Maslov K (2006) Ultrasonic ply-by-ply detection of matrix cracks in laminated composites. J Nondestruct Eval 25:37–49. https://doi.org/10.1007/s10921-006-0001-x

Kweon S, Benzerga AA (2013) Finite element implementation of a macromolecular viscoplastic polymer model. Int J Numer Meth Eng 94:895–919. https://doi.org/10.1002/nme.4474

Laribi MA, Tamboura S, Fitoussi J, Tié Bi R, Tcharkhtchi A, Ben Dali H (2018) Fast fatigue life prediction of short fiber reinforced composites using a new hybrid damage approach: Application to SMC. Compos B Eng 139:155–162. https://doi.org/10.1016/j.compositesb.2017.11.063

Liang S, Gning P-B, Guillaumat L (2014) Properties evolution of flax/epoxy composites under fatigue loading. Int J Fatigue 63:36–45. https://doi.org/10.1016/j.ijfatigue.2014.01.003

Liu P, Chen X, Li Y, Cheng P, Tang Z, Lv J, Aftab W, Wang G (2022) Aerogels meet phase change materials: fundamentals, advances, and beyond. ACS Nano 16:15586–15626. https://doi.org/10.1021/acsnano.2c05067

Lu T, Solis-Ramos E, Yi Y-B, Kumosa M (2016) Synergistic environmental degradation of glass reinforced polymer composites. Polym Degrad Stab 131:1–8. https://doi.org/10.1016/j.polymdegradstab.2016.06.025

Lucas JP, Zhou J (1993) The effects of sorbed moisture on resin-matrix composites. JOM 45:37–40. https://doi.org/10.1007/BF03222513

Marco Y, Huneau B, Masquelier I, Le Saux V, Charrier P (2017) Prediction of fatigue properties of natural rubber based on the descriptions of the cracks population and of the dissipated energy. Polym Test 59:67–74. https://doi.org/10.1016/j.polymertesting.2017.01.015

Meraghni F, Benzeggagh ML (1995) Micromechanical modelling of matrix degradation in randomly oriented discontinuous-fibre composites. Compos Sci Technol 55:171–186. https://doi.org/10.1016/0266-3538(95)00096-8

Moudood A, Rahman A, Khanlou HM, Hall W, Öchsner A, Francucci G (2019) Environmental effects on the durability and the mechanical performance of flax fiber/bio-epoxy composites. Compos B Eng 171:284–293. https://doi.org/10.1016/j.compositesb.2019.05.032

Mukhopadhyay et al (2008) Effect of ageing of sisal fibres on properties of sisal—polypropylene composites. Polym Degrad Stab 93:2048–2051. https://doi.org/10.1016/j.polymdegradstab.2008.02.018

NASA names awardees for sustainable aviation component technology. https://www.militaryaerospace.com/commercial-aerospace/article/14291126/nasa-names-awardees-for-sustainable-aviation-component-technology

Nasri K, Toubal L, Loranger É, Koffi D (2022) Influence of UV irradiation on mechanical properties and drop-weight impact performance of polypropylene biocomposites reinforced with short flax and pine fibers. Compos Part C: Open Access. 9:100296. https://doi.org/10.1016/j.jcomc.2022.100296

Nedjar B (2001) Elastoplastic-damage modelling including the gradient of damage: formulation and computational aspects. Int J Solids Struct 38:5421–5451. https://doi.org/10.1016/S0020-7683(00)00358-9

Pecora M, Smerdova O, Gigliotti M (2021) Gradients of cyclic indentation mechanical properties in PR520 epoxy and its 3D carbon fiber composite induced by aging at 150 °C. Polym Degrad Stab 193:109720. https://doi.org/10.1016/j.polymdegradstab.2021.109720

Peng Y, Liu R, Cao J (2015) Characterization of surface chemistry and crystallization behavior of polypropylene composites reinforced with wood flour, cellulose, and lignin during accelerated weathering. Appl Surf Sci 332:253–259. https://doi.org/10.1016/j.apsusc.2015.01.147

Ray D, Sarkar BK, Das S, Rana AK (2002) Dynamic mechanical and thermal analysis of vinylester-resin-matrix composites reinforced with untreated and alkali-treated jute fibres. Compos Sci Technol 62:911–917. https://doi.org/10.1016/S0266-3538(02)00005-2

Saucedo-Mora L, Lowe T, Zhao S, Lee PD, Mummery PM, Marrow TJ (2016) In situ observation of mechanical damage within a SiC-SiC ceramic matrix composite. J Nucl Mater 481:13–23. https://doi.org/10.1016/j.jnucmat.2016.09.007

Schilling PJ, Karedla BR, Tatiparthi AK, Verges MA, Herrington PD (2005) X-ray computed microtomography of internal damage in fiber reinforced polymer matrix composites. Compos Sci Technol 65:2071–2078. https://doi.org/10.1016/j.compscitech.2005.05.014

Seghini MC, Touchard F, Sarasini F, Chocinski-Arnault L, Ricciardi MR, Antonucci V, Tirillò J (2020) Fatigue behaviour of flax-basalt/epoxy hybrid composites in comparison with non-hybrid composites. Int J Fatigue 139:105800. https://doi.org/10.1016/j.ijfatigue.2020.105800

Shah DU (2016) Damage in biocomposites: stiffness evolution of aligned plant fibre composites during monotonic and cyclic fatigue loading. Compos A Appl Sci Manuf 83:160–168. https://doi.org/10.1016/j.compositesa.2015.09.008

Shettar M, Kini UA, Sharma S, Hiremath P, Gowrishankar MC (2019) Study on the mechanical properties of nanoclay-epoxy composites under different hygrothermal aging conditions. Mater Res Express 6:085333. https://doi.org/10.1088/2053-1591/ab2502

Shi Z, Zou C, Zhou F, Zhao J (2022) Analysis of the mechanical properties and damage mechanism of carbon fiber/epoxy composites under UV aging. Materials 15:2919. https://doi.org/10.3390/ma15082919

Shirinbayan M, Fitoussi J, Meraghni F, Surowiec B, Bocquet M, Tcharkhtchi A (2015) High strain rate visco-damageable behavior of advanced sheet molding compound (A-SMC) under tension. Compos B Eng 82:30–41. https://doi.org/10.1016/j.compositesb.2015.07.010

Shirinbayan M, Fitoussi J, Bocquet M, Meraghni F, Surowiec B, Tcharkhtchi A (2017) Multi-scale experimental investigation of the viscous nature of damage in Advanced Sheet Molding Compound (A-SMC) submitted to high strain rates. Compos B Eng 115:3–13. https://doi.org/10.1016/j.compositesb.2016.10.061

Simões S (2024) High-performance advanced composites in multifunctional material design: state of the art, challenges, and future directions. Materials 17:5997. https://doi.org/10.3390/ma17235997

Suresh Kumar SM, Duraibabu D, Subramanian K (2014) Studies on mechanical, thermal and dynamic mechanical properties of untreated (raw) and treated coconut sheath fiber reinforced epoxy composites. Mater des 59:63–69. https://doi.org/10.1016/j.matdes.2014.02.013

Tamboura S, Abdessalem A, Fitoussi J, Ben Daly H, Tcharkhtchi A (2022) On the mechanical properties and damage mechanisms of short fibers reinforced composite submitted to hydrothermal aging: application to sheet molding compound composite. Eng Fail Anal 131:105806. https://doi.org/10.1016/j.engfailanal.2021.105806

Tanaka K, Oharada K, Yamada D, Shimizu K (2016) Fatigue crack propagation in short-carbon-fiber reinforced plastics evaluated based on anisotropic fracture mechanics. Int J Fatigue 92:415–425. https://doi.org/10.1016/j.ijfatigue.2016.01.015

Tanaka et al (2014) Effect of fiber orientation on fatigue crack propagation in short-fiber reinforced plastics. Eng Fract Mech 123:44–58. https://doi.org/10.1016/j.engfracmech.2014.03.019

Tcharkhtchi A, Bronnec PY, Verdu J (2000) Water absorption characteristics of diglycidylether of butane diol–3,5-diethyl-2,4-diaminotoluene networks. Polymer 41:5777–5785. https://doi.org/10.1016/S0032-3861(99)00801-0

Vincent M, Giroud T, Clarke A, Eberhardt C (2005) Description and modeling of fiber orientation in injection molding of fiber reinforced thermoplastics. Polymer 46:6719–6725. https://doi.org/10.1016/j.polymer.2005.05.026

Wang Y, Chen G, Wan B, Cai G, Zhang Y (2020) Behavior of circular ice-filled self-luminous FRP tubular stub columns under axial compression. Constr Build Mater 232:117287. https://doi.org/10.1016/j.conbuildmat.2019.117287

Wang J, Huang S, Yan K, Shi J, Shi S, Jin Y, Yuan L (2023) Sustainable macromolecular antioxidants from eugenol with synergistically enhanced storage stability for active PVA packaging. React Funct Polym 191:105671. https://doi.org/10.1016/j.reactfunctpolym.2023.105671

Wallenberger FT, Weston NE (2004) Natural fibers, plastics and composites. Springer US, Boston, MA

Yuanjian T, Isaac DH (2007) Impact and fatigue behaviour of hemp fibre composites. Compos Sci Technol 67:3300–3307. https://doi.org/10.1016/j.compscitech.2007.03.039

Zuo P, Anastasios P (2021) Review of hybrid composites fatigue. Compos Struct 274:114358. https://doi.org/10.1016/j.compstruct.2021.114358

Chapter 14
Advanced Polymer Composites for Eco-Friendly Marine Applications

Muhammed Nayeem Mullungal and S. S. Shaju

Abstract Advanced polymer composites (APCs) are becoming increasingly popular in marine environments because they possess better mechanical qualities, are resistant to corrosion, and are lightweight. Compared to conventional materials such as steel and wood, advanced polymer composites (APCs) provide several important advantages. These composites are made up of high-performance polymer reinforced with natural or synthetic fibers. This study investigates the diverse applications of APCs in marine sectors, including boat and shipbuilding, offshore structures, underwater equipment, and maintenance systems. Their application in shipbuilding, where they are used in hulls and decks, improves fuel efficiency and reduces the weight of the structure. Wind turbine blades, floating platforms, and subsea pipelines are all examples of offshore infrastructure that can benefit from the increased endurance and performance that APCs provide. Underwater systems, such as remotely operated vehicles (ROVs) and sensors, benefit from their capacity to provide mechanical durability and long-term stability. The ability to recycle, the utilization of biodegradable polymers, and the reduction in emissions throughout the lifecycle are some of the environmental benefits that position APCs as sustainable solutions in the field of marine engineering. Even if these advantages are present, there are still obstacles to overcome in terms of recycling, the longevity of natural fibers, and the high production costs. In order to improve both performance and sustainability, future developments will concentrate on biodegradable composites, nanotechnology, and automated manufacturing techniques. APCs, in general, have the potential to change marine sectors by providing material solutions that are efficient, long-lasting, and kind to the environment while also being environmentally friendly.

M. N. Mullungal (✉)
Environmental Science Program, Department of Biological and Environmental Sciences, College of Arts and Sciences, Qatar University, Doha, Qatar
e-mail: mmullungal@qu.edu.qa

S. S. Shaju
Department of Chemical Oceanography, Cochin University of Science and Technology, Kochi, India

L. Ravindran et al. (eds.), *Macro, Micro and Nanocomposites from Sustainable Sources*, Advanced Structured Materials 244, https://doi.org/10.1007/978-981-95-2469-3_14

14.1 Advanced Polymer Composites (APCs)

High-performance polymers are combined with reinforcing fibers or fillers to produce advanced polymer composites, which have improved mechanical, thermal, and chemical properties. The combination results in materials that are strong, lightweight, and adaptable to various applications. This chapter provides detailed overview of APCs, their components, properties, and applications, particularly in marine contexts. APCs are gaining traction in eco-friendly marine applications due to their lightweight, corrosion-resistant, and high-strength properties (Nunes and D'Aloia 2024). These materials offer significant advantages over traditional materials like metals and wood, making them suitable for various marine structures, vessels, and components (Thakur et al. 2014).

The advanced polymer composite essentially consists of two component materials: (i) the matrix material or polymer, which is generally the low-strength and low-modulus component and (ii) the fiber, which is the relatively high-strength and high-modulus component (Vaziri and Wagner 2014). The polymer matrix serves as the continuous phase that binds the reinforcements, transfers loads (Chawla 2012), and protects them from environmental damage. Common polymer matrices include epoxy, polyester, vinyl ester, and thermoplastics like polyetheretherketone (PEEK) and polypropylene (PP). The reinforcing materials are usually fibers that provide mechanical strength and stiffness to the composite. These fibers can be either natural (e.g., flax, jute) (Faruk et al. 2012) or synthetic (e.g., glass, carbon, aramid) (Mallick 2007).

14.2 Thermosetting and Thermoplastic Composites

Polymer composites are broadly categorized into thermosetting and thermoplastic composites, based on the matrix type.

Thermosetting composites are made with matrices that undergo an irreversible chemical curing process, forming cross-linked networks. Epoxy and polyester are common examples. These materials offer excellent thermal and chemical resistance but cannot be reshaped or reprocessed once cured (Chawla 2012).

Thermoplastic composites, on the other hand, use matrices that soften upon heating and harden upon cooling, making them recyclable and more suitable for rapid manufacturing. Although traditionally less stiff than thermosets, modern thermoplastics like PEEK and polyphenylene sulfide (PPS) offer comparable performance in demanding applications (Friedrich and Almajid 2013).

14.2.1 Types of Reinforcements

14.2.1.1 Natural Fibers

Natural fibers such as flax, hemp, kenaf, and jute have gained attention due to their sustainability and environmental benefits (Saito and Nishiyama 2014). These fibers are renewable, biodegradable, lightweight, and offer relatively good specific strength and stiffness (Nair et al. 2022; Mohanty et al. 2005). They also contribute to a reduction in carbon footprint compared to synthetic alternatives. In addition, natural fiber-reinforced composites often show improved acoustic insulation and reduced tool wear during processing (Saba et al. 2015).

However, they are typically less durable and more susceptible to moisture absorption, which can lead to degradation over time. Surface treatments or hybridization with synthetic fibers can mitigate these issues while retaining environmental advantages.

14.2.1.2 Synthetic Fibers

Synthetic reinforcements include glass fibers, carbon fibers, and aramid fibers (e.g., Kevlar). These fibers are engineered to offer superior mechanical properties, including high tensile strength, durability, and resistance to chemicals and temperature extremes (Hull and Clyne 1996). Carbon fiber composites, for instance, are widely used in aerospace and high-performance automotive parts due to their excellent strength-to-weight ratio. However, synthetic fibers are often more expensive and less environmentally friendly compared to natural alternatives.

14.2.2 Manufacturing Techniques

A variety of manufacturing techniques are used to fabricate APCs, depending on the application, type of matrix, and reinforcement used (Rajak et al. 2017). Common processes include.

Hand Lay-Up: One of the simplest and most cost-effective methods, this involves manually placing fibers into a mold and applying resin with brushes or rollers. Though labor-intensive, it is ideal for low-volume production and prototyping (Vaziri and Wagner 2014).

Resin Infusion (e.g., Vacuum-Assisted Resin Transfer Molding—VARTM): This process uses vacuum pressure to infuse resin into a dry fiber lay-up enclosed in a mold. It provides better control over resin content and results in stronger, void-free laminates, making it popular in marine and wind energy sectors (Campbell 2010).

Automated Processes: Advanced techniques such as automated fiber placement (AFP) and pultrusion are used for high-volume or high-precision manufacturing. These methods offer improved repeatability, speed, and quality control. Thermoplastic composites particularly benefit from automation due to their shorter processing times and lack of curing requirements (Friedrich and Almajid 2013).

14.2.3 Properties of Advanced Polymer Composites

The performance of advanced polymer composites in marine environments is influenced by their mechanical, thermal, and degradation properties. Key performance characteristics include:

1. Mechanical Properties: Natural fiber-reinforced composites exhibit high tensile strength and stiffness, although these properties can be lower than those of synthetic fiber composites. For instance, composites reinforced with Posidonia oceanica fibers have shown a tensile modulus of up to 2.4 GPa (Ramesh et al. 2017; Seggiani et al. 2017, 2018).
2. Seawater Resistance: The resistance of composites to seawater aging is a critical factor. Thermoplastic composites have shown comparable fatigue performance to thermoset composites after seawater immersion, with minimal loss in mechanical properties (Stankovic et al. 2024; Devine et al. 2023).
3. Thermal Stability: Biopolymers like PHBV and PBSA have shown good thermal stability, although their properties can degrade with increasing fiber content. The addition of plasticizers like acetyl tributyl citrate (ATBC) has been shown to improve processability and thermal stability (Strangis et al. 2023; Seggiani et al. 2018).
4. Low Maintenance: Reduced need for maintenance compared to traditional materials, leading to long service life.

14.3 Significance of Advanced Polymer Composites in the Marine Environment

Advanced polymer composites (APCs) are increasingly used in marine applications due to their high strength-to-weight ratio, corrosion resistance, and durability. However, their environmental impact—particularly in marine ecosystems—must be carefully considered.

14.3.1 Sustainability in Marine Applications

Sustainability is a critical factor in the development of APCs for marine environments. Traditional composites often rely on petroleum-based resins and synthetic fibers, which contribute to pollution and persist in ecosystems. To mitigate this, researchers are exploring:

- Biodegradable Resins: Bio-based epoxy and polyester resins derived from plant oils (e.g., soybean, linseed) reduce dependency on fossil fuels and degrade more readily in marine environments (Reddy and Yang 2019; Puglia et al. 2020).
- Natural Fiber Reinforcements: Fibers such as flax, hemp, and jute are renewable and biodegradable, making them suitable for temporary marine structures like buoys and biodegradable fishing gear (Joshi et al. 2004).

However, natural fibers face challenges in marine use, such as moisture absorption and reduced mechanical strength compared to synthetic alternatives. Hybrid composites—combining natural fibers with minimal synthetic reinforcements—offer a balanced solution (Zhang and Li 2022; La Mantia and Morreale 2011).

14.3.2 Recycling and Waste Management

Marine pollution from discarded composites (e.g., boat hulls, fishing equipment) is a growing concern. APCs can be engineered for improved recyclability:

- Thermoplastic Composites: Polymers like polypropylene (PP) and polyethylene terephthalate (PET) can be remelted and reshaped, allowing for closed-loop recycling (Pickering et al. 2016). This is particularly useful for marine components such as panels and non-structural parts.
- Thermosetting Composites: Traditional thermosets (e.g., epoxy, vinyl ester) are harder to recycle, but emerging chemical recycling methods (e.g., solvolysis, pyrolysis) break them down into reusable materials (Yang et al. 2023).

Despite progress, recycling infrastructure for marine composites remains limited (Sadowski and Krzysztof 2020). Developing specialized marine waste collection and processing systems is essential to prevent composite debris from accumulating in oceans (Hopewell et al. 2009).

14.3.3 Reduced Carbon Footprint in Marine Vessels

The marine industry contributes significantly to global CO_2 emissions, and lightweight APCs can help mitigate this impact:

- Fuel Efficiency: Lightweight composites reduce vessel weight, leading to lower fuel consumption and emissions. For example, carbon fiber-reinforced polymers (CFRPs) in ship hulls can decrease fuel use by up to 20% (Das et al. 2016).
- Life Cycle Assessment (LCA): Studies show that while composite production has a higher initial carbon footprint than steel, the long-term fuel savings offset this over the vessel's operational life (Sutherland et al. 2018).

However, the energy-intensive production of synthetic fibers (e.g., carbon fiber) remains an environmental concern. Transitioning to low-carbon manufacturing and renewable energy sources can further improve sustainability (Marsh 2020).

14.4 Marine Applications of Advanced Polymer Composites

Advanced polymer composites (APCs) have emerged as a transformative material in marine applications, offering unmatched combination of strength, lightness, corrosion resistance, and design flexibility (Nielson and Pickering 2023; Siddique and Ochi 2020). The harsh and variable conditions of marine environments—such as saltwater corrosion, UV exposure, dynamic loading, and biological fouling—make APCs ideal alternatives to traditional materials like steel, aluminum, and wood. APCs are increasingly used in the construction of vessels, offshore structures, underwater equipment, and even in maintenance and repair, offering performance and environmental benefits across the lifecycle. The various types of APCs that are used in the marine environment are listed the Table 14.1.

Table 14.1 Comparison of advanced polymer composites for marine applications

Composite type	Key applications	Environmental benefits
Natural fiber-reinforced	Shipbuilding, offshore structures	Biodegradable, reduced carbon footprint
Biopolymer-based composites	Aquaculture tools, fishing gear	High biodegradation rates, reduced marine pollution
Hybrid fiber-reinforced	Marine renewable energy, tidal turbines	Balanced mechanical properties and sustainability
Thermoplastic composites	Underwater repair, ship components	Recyclable, improved seawater resistance
Nanocomposites	Advanced structural applications	Enhanced mechanical and thermal properties, biodegradability

14.4.1 Boat and Ship Building

One of the most widespread uses of APCs in the marine sector is in boat and ship construction, particularly for hulls, decks, bulkheads, and superstructures. The primary benefit of APCs in this context lies in their high strength-to-weight ratio, which significantly reduces the overall weight of marine vessels. For instance, fiber-reinforced polymer (FRP) composites—especially those reinforced with glass or carbon fibers—offer equivalent or better strength than steel while being up to 60% lighter (Campbell 2010).

- Weight Reduction: The use of composite sandwich structures (CSSs) allows for significant weight savings compared to traditional materials like steel and aluminum. For instance, the incorporation of fiber-reinforced plastics (FRPs) as load-bearing skins in CSSs provides a combination of lightness and strength, which is essential for high-performance vessels (Daniel et al. 2023).
- Fuel Efficiency: By reducing the overall weight of the vessel, APCs contribute to lower fuel consumption. This is particularly beneficial for commercial shipping, where operational costs are closely tied to fuel efficiency.
- Durability: APCs exhibit excellent resistance to corrosion and environmental degradation, which is vital for marine applications. This durability extends the lifespan of vessels, reducing maintenance costs over time (Dow and Bird 1994).
- Design Flexibility; APCs also offer greater design flexibility compared to metals, enabling manufacturers to produce complex, integrated shapes that reduce the number of parts and joints. This not only enhances structural integrity but also decreases production time and costs.

14.4.2 Marine Structures

14.4.2.1 Offshore Wind Farms

As the global demand for renewable energy grows, offshore wind farms have become increasingly common. These structures are subjected to extreme mechanical loads, wave impacts, saltwater corrosion, and temperature variations, making material selection critical. APCs, especially glass fiber-reinforced composites, are extensively used in wind turbine blades due to their high fatigue resistance, lightweight, and durability (Cousins et al. 2019).

Composite wind turbine blades can reach lengths of 100 m or more, and their reduced weight allows for larger blades with better energy capture without overburdening the support structure (Brøndsted et al. 2008). Furthermore, the non-conductive nature of many APCs helps reduce lightning strike risk, a significant concern in marine environments.

14.4.2.2 Floating Platforms and Pontoons

Composites are also used in floating structures, such as offshore oil platforms, pontoons, floating docks, and breakwaters. These structures require materials that are buoyant, strong, and resistant to fatigue from continuous wave action. APCs meet these requirements while offering excellent corrosion resistance, eliminating the need for regular painting or sacrificial anodes.

Sandwich structures made from polymer cores (e.g., PVC foam) and composite skins (e.g., glass fiber) are often employed to build lightweight, strong, and buoyant platforms (Daniel et al. 2023). These structures maintain dimensional stability over time and reduce overall life-cycle costs compared to steel alternatives.

14.4.2.3 Subsea Pipelines and Risers

In subsea applications, composite pipes are increasingly used as alternatives to metal piping, particularly in aggressive or deepwater environments. APCs offer superior corrosion resistance and can be designed to withstand high-pressure and high-temperature conditions, extending the life of the infrastructure while reducing maintenance needs (Hull and Clyne 1996).

14.4.2.4 Underwater Equipment

Remotely Operated Vehicles (ROVs)

ROVs are essential tools for underwater exploration, inspection, and maintenance tasks. The use of APCs in ROV frames, arms, and housing components allows for weight savings, increased maneuverability, and improved energy efficiency. For deep-sea ROVs, where high pressure and low temperatures prevail, carbon fiber-reinforced thermoplastics provide the mechanical resilience and fatigue strength needed for sustained operation (Strong 2008).

Moreover, APCs can be electrically non-conductive, which is advantageous for minimizing electromagnetic interference with onboard electronics and instruments.

Underwater Sensors and Casings

Advanced sensors used for monitoring oceanographic parameters (e.g., salinity, pressure, temperature) must be housed in robust, pressure-resistant, and corrosion-resistant materials. Composites offer a customizable solution, allowing tailored designs that integrate sensors and electronics within a lightweight, protective shell. In particular, polymer matrix composites with thermosetting epoxy or thermoplastic resins provide both mechanical protection and long-term stability under submersion.

Propellers and Thrusters

Although traditionally made from metals, propellers and thruster blades are increasingly being fabricated from composites, especially in small and mid-size vessels. Composite blades offer lower inertia, leading to faster acceleration and improved fuel efficiency. Their resistance to cavitation erosion and corrosion makes them ideal for long-term use in saltwater environments.

14.4.3 Aquaculture and Fishing Gear

Biodegradable composites are being developed for aquaculture tools and fishing gear to reduce marine pollution. For example, biodegradable fishing gear made from polybutylene succinate (PBS) and mussel-byssus-inspired fibers has shown promise in mitigating the environmental impact of abandoned fishing gear (Hong et al. 2024).

14.4.4 Repair and Maintenance

14.4.4.1 Retrofitting and Structural Reinforcement

One of the more strategic applications of APCs in marine environments is in retrofitting and repairing existing structures—such as aging ships, piers, jetties, and offshore platforms. Using composite wraps or laminates, engineers can restore structural integrity without dismantling large sections or using heavy equipment.

Fiber-reinforced polymer (FRP) wraps can be applied directly to damaged steel or concrete structures to increase load-bearing capacity, resist further corrosion, and prolong service life. This is especially useful for underwater or splash zone repairs, where conventional repair techniques are costly and time-consuming (Bae and Kim 2021).

FRP repair systems are lightweight, require minimal site preparation, and can be cured in place using ambient or low-temperature processes. For example, carbon fiber-reinforced epoxy wraps have been used to strengthen offshore risers and pilings exposed to dynamic loading and corrosion (Oliveux et al. 2001).

14.4.4.2 Corrosion Protection

APCs are inherently resistant to corrosion, and when used as surface coatings or barriers, they can protect steel and concrete substrates from environmental degradation. Composite materials used in these coatings are often enhanced with nanoparticles or hybrid fibers to improve thermal and chemical resistance.

In addition, composite grating, decking, and ladders are increasingly replacing metallic components in marine facilities to improve safety, reduce maintenance, and extend lifespan.

14.5 Performance Characteristics

The performance evaluation and testing of advanced polymer composites (APCs) in marine environments are crucial for ensuring their durability and effectiveness in harsh conditions. These composites are increasingly used in marine applications due to their favorable properties such as thermal stability, corrosion resistance, and high strength-to-weight ratios.

Testing standards, real-world examples, and regulatory compliance are key aspects of evaluating these materials for marine use.

Testing Standards

- **Mechanical Testing**: APCs are subjected to mechanical tests to assess their strength, fatigue resistance, and durability under marine conditions. These tests often include stress analysis and fatigue testing to simulate the stresses experienced in marine environments (Dow and Bird 1994; Pei et al. 2023).
- **Environmental Testing**: Environmental tests evaluate the impact of factors like seawater exposure, temperature variations, and UV radiation on APCs. For instance, Glass Fiber Reinforced Polymers (GFRPs) are tested for seawater absorption and thermal effects, which can affect their mechanical properties (Cavasin et al. 2019).
- **Long-Term Durability**: Studies on fiber-reinforced polymers (FRPs) focus on their long-term performance, particularly in prestressing applications for marine infrastructure, highlighting the importance of durability in harsh marine conditions (Tang and Wang 2023; Sen 2003).

Regulatory Compliance

Meeting maritime safety and environmental regulations is essential for the adoption of APCs:

- **Safety Standards**: Compliance with international maritime safety standards ensures that materials can withstand the rigors of marine environments, protecting both human life and marine ecosystems (Dow and Bird 1994).
- **Environmental Regulations**: Adhering to regulations helps mitigate the ecological impact of marine operations, promoting the use of sustainable materials and practices.
- **Certification (ISO 12215):** Mandates material testing for small marine vessels.

Real-world Case Studies

Real-world case studies illustrate the significant advancements in composite materials within marine applications. For instance, America's Cup yachts have utilized carbon fiber-reinforced hulls, achieving a remarkable 40% weight reduction and enhanced hydrodynamic efficiency, which directly correlates to improved speed and performance (Miller 2003; Hamilton and Patterson 1993). Offshore wind turbine blades made from glass fiber-reinforced polymer (GFRP) have undergone rigorous fatigue testing, demonstrating durability to withstand over 20 years of marine conditions, which has led to a 30% reduction in maintenance costs (Pimenta and Pinho 2011; Wu and Zhang 2020; Heres 2015). Additionally, the U.S. Navy's Littoral Combat Ship has validated carbon fiber-reinforced polymer (CFRP) hulls through extensive ballistic and fire resistance tests, ensuring compliance with stringent military standards (Heres 2015; Hamilton and Patterson 1993).

While these advancements showcase the benefits of composite materials, challenges remain, such as the high acquisition costs and the need for improved material characterization methods to enhance reliability and performance in various marine environments.

14.6 Challenges

Despite their advantages, advanced polymer composites (APCs) face several challenges in marine environments, limiting their widespread adoption.

Moisture Absorption: Natural fibers tend to absorb moisture, leading to a reduction in mechanical properties over time. This can be mitigated through surface treatments and the use of compatibilizers (Boccarusso et al. 2023).

Limited Durability: The biodegradability of natural fiber composites, while environmentally beneficial, can limit their long-term durability in marine environments. This requires careful balancing of biodegradation rates and mechanical performance (Salas et al. 2023; Seggiani et al. 2018).

High Material and Manufacturing Costs: Carbon and Aramid Fibers: Expensive compared to steel and aluminum, increasing upfront costs (Marsh 2020). Specialized Manufacturing Techniques like resin infusion and autoclave curing require significant investment (Campbell 2004).

Manufacturing Scalability: Hand lay-up and vacuum infusion are labor-intensive and slow for large marine structures (Potter and Liu 2021) lowering the production speed. Quality Control by ensuring uniform fiber distribution and resin curing in large components remains difficult (Gibson 2016).

Market Acceptance and Regulatory Hurdles: Conservative industry structure of the shipbuilders and offshore operators often prefer traditional materials due to familiarity (Greene 2019). Certification processes for the new composites are lengthy as they must undergo lengthy testing to meet maritime safety standards (IMO, DNV GL) (Sutherland et al. 2018).

Environmental and End-of-Life Concerns: Thermoset composites are hard to recycle, leading to waste accumulation (Yang et al. 2023). Non-biodegradable composites may contribute to marine pollution if not properly managed known as bio fueling (Chambers et al. 2006; Tang and Wang 2023).

14.7 Future Directions and Innovative Developments

The future of advanced polymer composites for marine applications hinges on overcoming existing challenges while embracing innovative materials and manufacturing techniques. As the maritime industry seeks to enhance performance and sustainability, the integration of advanced polymer composites presents a promising avenue. This response will explore the current challenges, innovative materials, and manufacturing techniques that are shaping the future of these composites in marine applications.

14.7.1 Innovative Materials

Recent advancements in materials science have led to the development of several innovative polymer composites suitable for marine applications:

Fiber-Reinforced Composites: These composites, especially with superhydrophobic coatings, exhibit excellent strength and corrosion resistance, making them ideal for marine structures.

Biodegradable Composites: Plant biomass-based composites offer eco-friendly alternatives, although their mechanical performance needs enhancement for effective marine use.

Microwave-Absorbing Composites: These materials provide dual functionality, essential for naval applications, by combining corrosion resistance with electromagnetic interference shielding.

Nanocomposites: The incorporation of nanomaterials, such as marine-derived polymers, to enhance mechanical and thermal properties while maintaining biodegradability (Yosri et al. 2024).

Sustainable Manufacturing: The development of eco-friendly manufacturing processes, such as sol–gel and hot pressing, to reduce the environmental impact of composite production (Yosri et al. 2024; Santhosh et al. 2024).

14.7.2 Advanced Manufacturing Techniques

The evolution of manufacturing techniques plays a crucial role in the development of advanced polymer composites:

Additive Manufacturing (AM): Techniques such as fused filament fabrication and selective laser sintering allow for the creation of complex geometries and tailored material properties, enhancing performance in marine applications (Rajapakse and Daniel 2022; Wang et al. 2023).

Concurrent Engineering: This approach integrates design and production processes, facilitating the development of optimized solutions that consider the entire lifecycle of marine structures (Shenoi et al. 2011; Jolly 2023).

Structural Health Monitoring: Advanced inspection techniques, including vibration-based damage identification, are essential for maintaining the integrity of marine structures over time (Shenoi et al. 2011).

In contrast to the optimistic outlook for polymer composites, there are concerns regarding their long-term sustainability and environmental impact. The challenge of plastic pollution remains a pressing issue, necessitating a focus on recycling and the development of renewable plastics to mitigate adverse effects on marine ecosystems (Dhandapani et al. 2023, 2024). In summary, while the future of advanced polymer composites in marine applications is promising, it is imperative to address the challenges of moisture resistance, safety, and environmental impact through innovative materials and manufacturing techniques. Continued research and development in these areas will be crucial for realizing the full potential of polymer composites in the maritime sector.

14.8 Conclusion

Advanced polymer composites (APCs) are revolutionizing marine engineering with their strength, lightweight, corrosion resistance, and design flexibility. By combining high-performance polymers with natural or synthetic fibers, APCs outperform steel and wood in harsh marine conditions. Their rising application in shipbuilding, offshore constructions, underwater equipment, and aquaculture technologies shows their versatility. APCs improve fuel efficiency, maintenance costs, and service life, which are crucial in maritime operations.

APC sustainability is growing, notably with biodegradable polymers and natural materials to reduce environmental effect. Thermoplastic composites are recyclable, while hybrid systems combine sustainability and mechanical efficiency. Despite their benefits, natural fibers absorb moisture, synthetic fibers are expensive and energy-intensive, and recycling infrastructure—especially for thermosets—needs improvement. These issues can be solved with microwave-absorbing composites, nanocomposites, and bio-based polymers and advanced manufacturing technologies including additive manufacturing and concurrent engineering. APCs are proven effective in maritime applications like America's Cup yachts and offshore wind turbines. Additional study on material durability, recyclability, and cost-effectiveness is needed to optimize APC potential. APCs will help create a more sustainable, efficient, and resilient maritime industry that meets performance and environmental goals amid global challenges through innovation.

References

Bae S-Y, Kim Y-H (2021) Study on structural design and analysis of 18 ft CFRP leisure boat. Int J Mod Phys B 35(14):2140036. https://doi.org/10.1142/S0217979221400361

Boccarusso L, Durante M, Cioffi S, Caruso A (2023) Moisture absorption and mechanical behavior of natural fiber-reinforced composites in marine environments. Compos B Eng 240:110015. https://doi.org/10.1016/j.compositesb.2022.110015

Brøndsted P, Lilholt H, Lystrup A (2008) Composite materials for wind turbine blades. Annu Rev Mater Res 38:505–538. https://doi.org/10.1146/annurev.matsci.38.060407.130434

Campbell FC (2004) Manufacturing processes for advanced composites. Elsevier

Campbell FC (2010) Structural composite materials. ASM International

Cavasin M, Boria S, Belingardi G (2019) Durability of GFRP composites in seawater: a review of current research and applications. Mar Struct 66:115–134. https://doi.org/10.1016/j.marstruc.2019.03.005

Chambers LD, Stokes KR, Walsh FC, Wood RJK (2006) Modern approaches to marine antifouling coatings. Biofouling 22(6):311–327. https://doi.org/10.1080/08927010600980223

Chawla KK (2012) Composite materials: science and engineering, 3rd edn. Springer

Cousins DS, Suzuki Y, Murray RE, Samaniuk JR, Stebner AP (2019) Recycling glass fiber thermoplastic composites from wind turbine blades. J Clean Prod 209:1252–1263. https://doi.org/10.1016/j.jclepro.2018.10.286

Daniel IM, Gdoutos EE, Rajapakse YDS (eds) (2023) Major accomplishments in composite materials and sandwich structures: an anthology of ONR sponsored research. Office of Naval Research

Das S, Graziano D, Upadhyayula VKK, Masanet E, Riddle M (2016) Life-cycle analysis and optimization of carbon fiber-reinforced polymer composites. Resour Conserv Recycl 107:1–12. https://doi.org/10.1016/j.resconrec.2015.11.002

Devine DM, Tawfick S, Friedrich K (2023) Aging performance of thermoplastic composites in seawater immersion. Polym Test 122:107659. https://doi.org/10.1016/j.polymertesting.2023.107659

Dhandapani A, Krishnasamy S, Kumar Thiagamani SM, Periasamy D, Muthukumar C, Kumaran Sundaresan T, Ali S, Kurniawan R (2023) Evolution, prospects, and predicaments of polymers in marine applications: a potential successor to traditional materials

Dhandapani S, Narayanasamy M, Krishnan KR (2024) Addressing marine plastic pollution through advanced polymer recycling. Environ Sci Pollut Res 31(5):4879–4891. https://doi.org/10.1007/s11356-023-29517-6

Dow NF, Bird RM (1994) Design and fabrication of composite marine structures. Compos Eng 4(8):847–861. https://doi.org/10.1016/0961-9526(94)90092-2

Faruk O, Bledzki AK, Fink HP, Sain M (2012) Biocomposites reinforced with natural fibers: 2000–2010. Prog Polym Sci 37(11):1552–1596. https://doi.org/10.1016/j.progpolymsci.2012.04.003

Friedrich K, Almajid AA (2013) Manufacturing aspects of advanced polymer composites for automotive applications. Appl Compos Mater 20(2):107–128. https://doi.org/10.1007/s10443-012-9263-9

Gibson RF (2016) Principles of composite material mechanics, 4th edn. CRC Press
Greene E (2019) Marine composites: design and performance. In: Mayer RM (ed) Marine applications of advanced fiber-reinforced composites. Woodhead Publishing, pp 1–30
Hamilton H, Patterson D (1993) Carbon fiber usage in the U.S. Navy's littoral combat ship. Naval Eng J 105(4):53–62. https://doi.org/10.1111/j.1559-3584.1993.tb04832.x
Heres M (2015) Offshore wind turbine blade design and testing. Renewable Energy World. https://www.renewableenergyworld.com
Hong J, Lee SH, Kim HJ (2024) Biodegradable composites for sustainable aquaculture: mussel-inspired materials and performance analysis. J Clean Prod 436:140261. https://doi.org/10.1016/j.jclepro.2024.140261
Hopewell J, Dvorak R, Kosior E (2009) Plastics recycling: challenges and opportunities. Philos Trans R Soc B: Biol Sci 364(1526):2115–2126. https://doi.org/10.1098/rstb.2008.0311
Hull D, Clyne TW (1996) An introduction to composite materials, 2nd edn. Cambridge University Press
Jolly S (2023) Engineering advancements in composite materials and their applications in engineering and technology. J Mater Sci Eng
Joshi SV, Drzal LT, Mohanty AK, Arora S (2004) Are natural fiber composites environmentally superior to glass fiber reinforced composites? Compos A Appl Sci Manuf 35(3):371–376. https://doi.org/10.1016/j.compositesa.2003.09.016
La Mantia FP, Morreale M (2011) Green composites: a brief review. Compos A Appl Sci Manuf 42(6):579–588. https://doi.org/10.1016/j.compositesa.2011.01.017
Mallick PK (2007) Fiber-reinforced composites: materials, manufacturing, and design, 3rd edn. CRC Press
Marsh G (2020) Composites and sustainability: carbon fiber's high cost and energy demand. Reinf Plast 64(4):118–123. https://doi.org/10.1016/j.repl.2020.03.006
Miller R (2003) Composite engineering in the America's Cup. J Compos Mater 37(22):2011–2020. https://doi.org/10.1177/0021998303037405
Mohanty AK, Misra M, Drzal LT (2005) Natural fibers, biopolymers, and biocomposites. CRC Press
Nguyen L, Lee J, Kim J (n.d.) Structural optimization of composite sandwich structures for marine applications. Compos Struct. https://doi.org/10.1016/j.compstruct.2023.115435
Nair SS, Kandasubramanian B, Baradwaj B (2022) Sustainable marine composites from renewable resources: a review. J Marine Sci Eng 10(6):754. https://doi.org/10.3390/jmse10060754
Nielson JA, Pickering KL (2023) Advancements in bio-based marine composites: challenges and prospects. Compos Sci Technol 225:109664. https://doi.org/10.1016/j.compscitech.2023.109664
Nunes S, D'Aloia A (2024) Eco-friendly composites for marine applications: a review of current materials and technologies. Mar Pollut Bull 194:115038. https://doi.org/10.1016/j.marpolbul.2024.115038
Oliveux B, Le Maout C, Pyl L (2001) Carbon-fiber wraps for the strengthening of offshore structures. J Compos Technol Res 23(1):32–40
Pei A, Hu X, Wang Y (2023) Effects of seawater aging on mechanical properties of carbon fiber composites. Compos B Eng 246:110376. https://doi.org/10.1016/j.compositesb.2023.110376
Pickering KL, Efendy MA, Le TM (2016) A review of recent developments in natural fibre composites and their mechanical performance. Comp Part A Appl Sci Manuf 83:98–112. https://doi.org/10.1016/j.compositesa.2015.08.038
Pimenta S, Pinho ST (2011) Mechanical behavior of glass fiber reinforced polymer composites subjected to seawater immersion. Compos Struct 94(4):1391–1400. https://doi.org/10.1016/j.compstruct.2011.01.019
Potter KD, Liu Z (2021) High-performance marine composites: design, fabrication, and testing. J Compos Mater 55(3):397–416. https://doi.org/10.1177/0021998320965430

Puglia D, Luzi F, Lilli M, Sbardella F, Pauselli M, Torre L, Benincasa P (2020) Straw fibres from barley hybrid lines and their reinforcement effect in polypropylene based composites. Ind Crops Prod 154:112736. ISSN 0926-6690. https://doi.org/10.1016/j.indcrop.2020.112736

Rajak DK, Mukhopadhyay S, Kumar V (2017) Manufacturing and sustainability of polymer composites: a review. Mater Today: Proc 4(2):1653–1660. https://doi.org/10.1016/j.matpr.2017.01.016

Rajapakse YDS, Daniel IM (2022) Recent advances in marine composite materials. In: Advances in composite materials for marine applications. Springer, pp 85–115

Ramesh M, Palanikumar K, Rajkumar K (2017) Mechanical behavior and durability of marine composites: a review. J Ind Text 47(7):1661–1682. https://doi.org/10.1177/1528083716673745

Reddy NM, Yang Y (2019) Biodegradable polymer composites for marine applications: progress and prospects. ACS Sustain Chem Eng 7(22):18929–18938. https://doi.org/10.1021/acssuschemeng.9b03821

Russell A, Smith B (2018) Recycling composite materials in the marine industry: challenges and opportunities. Mar Pollut Bull 130:231–239. https://doi.org/10.1016/j.marpolbul.2018.03.042

Saba N, Tahir PM, Jawaid M (2015) A review on potential materials for automotive applications. Renew Sustain Energy Rev 47:365–377. https://doi.org/10.1016/j.rser.2015.03.071

Sadowski Ł, Krzysztof P (2020) Environmental impact and recyclability of composite materials. J Clean Prod 245:118930. https://doi.org/10.1016/j.jclepro.2019.118930

Saito T, Nishiyama Y (2014) Cellulose nanofibers for marine applications. Mar Biotechnol 16(2):118–131. https://doi.org/10.1007/s10126-013-9543-8

Salas C, Naceur M, Zeghloul S (2023) Seawater aging effects on polymer composites: a review. Polym Degrad Stab 208:110134. https://doi.org/10.1016/j.polymdegradstab.2023.110134

Santhosh Kumar PK, Akbar H, Pisker B, Ramanathan HN (2024) Sustainable production practices and circular economy. pp 336–355. https://doi.org/10.4018/979-8-3693-3880-3.ch014

Seggiani M, Cinelli P, Mallegni N, Balestri E, Puccini M, Vitolo S, Lardicci C, Lazzeri A (2017) New bio-composites based on polyhydroxyalkanoates and posidonia oceanica fibres for applications in a marine environment. Materials 10:326

Seggiani M, Cinelli P, Balestri E, Mallegni N, Stefanelli, E, Rossi, A, Lardicci C, Lazzeri A (2018) Novel sustainable composites based on poly(hydroxybutyrate-co-hydroxyvalerate) and seagrass beach-CAST fibers: performance and degradability in marine environments. Materials 11:772

Sen R (2003) Durability of advanced composites in a marine environment. Int J Mater Prod Technol 19(1/2):118. https://doi.org/10.1504/ijmpt.2003.003548

Shenoi RA, Dulieu-Barton JM, Quinn S, Blake JIR, Boyd SW (2011) Composite materials for marine applications: key challenges for the future. In: Composite materials: a vision for the future. CRC Press

Siddique M, Ochi S (2020) Durability of polymer composites for marine applications: an overview. Mar Struct 71:102761. https://doi.org/10.1016/j.marstruc.2020.102761

Stankovic D, Obande W, Devine M, Bajpai A, Brádaigh CMÓ, Ray D (2024) Accelerated seawater ageing and fatigue performance of glass fibre reinforced thermoplastic composites for marine and tidal energy applications. Compos Part C Open Access 14:100470. https://doi.org/10.1016/J.JCOMC.2024.100470

Sutherland WJ, Sutherland GS et al (2018) Life cycle assessment of composite materials in the construction of ships. J Ship Res 38(2):75–92

Tang Q, Wang X (2023) Impact of seawater on polymer matrix composites for marine applications. Compos Sci Technol 237:109925. https://doi.org/10.1016/j.compscitech.2023.109925

Thakur VK, Thakur MK, Gupta RK (2014) Recent advances in green composites based on natural fibers. ACS Sustain Chem Eng 2(7):1775–1789. https://doi.org/10.1021/sc500294m

Vaziri R, Wagner HD (2014) Advanced fiber composites for marine applications. Mater Today 17(6):274–281. https://doi.org/10.1016/j.mattod.2014.04.011

Wang S, Wu Q, Li X (2023) Design and manufacturing of marine polymer composites: challenges and prospects. Compos Struct 304:116410. https://doi.org/10.1016/j.compstruct.2022.116410

Wu H, Zhang Q (2020) Effects of seawater aging on mechanical properties of glass fiber composites. J Compos Mater 54(12):1639–1651. https://doi.org/10.1177/0021998319878926

Yang J, Zhang S, Liu D (2023) Sustainable composite materials for marine applications: advances and challenges. Mater Today Commun 37:106425. https://doi.org/10.1016/j.mtcomm.2023.106425

Yosri N, Khalifa SAM, Attia NF, Du M, Yin L, Abolibda TZ, Zhai K, Guo Z, El-Seedi HR (2024) Sustainability in the green engineering of nanocomposites based on marine-derived polysaccharides and collagens: a review. Int J Biol Macromol 225:133249. https://doi.org/10.1016/j.ijbiomac.2024.133249

Zhang W, Li Y (2022) Marine environment effects on the durability of polymer composites. Polym Degrad Stab 198:109883. https://doi.org/10.1016/j.polymdegradstab.2022.109883

Chapter 15
Biopolymer-Based Therapeutics: Processing, Applications, and Future Prospects

Derniza-Elena Cozorici, Ionut-Cristian Radu, Erika Blanzeanu, and Catalin Zaharia

Abstract As therapeutic approaches evolve to meet the growing demand for more precise and effective treatments, biopolymers have emerged as versatile materials with great potential across a variety of medical applications. This chapter emphasizes the crucial role of biopolymeric materials in advancing therapeutic strategies, with a focus on areas like drug delivery systems, regenerative medicine, cancer therapy, and vaccine development. Due to their biocompatibility, biodegradability, and adaptability, engineered biopolymeric materials are promising alternatives to conventional therapeutic approaches by overcoming limitations such as inadequate drug targeting, low bioavailability, and adverse side effects. This chapter aims to explore the advancements of biopolymers in therapeutics, highlighting innovations such as targeted nanocarriers, smart drug delivery systems, and engineered tissues. Additionally, it reviews cutting-edge techniques like electrospinning and 3D bioprinting for processing biopolymers, while also addressing emerging areas such as artificial cells and organs-on-a-chip. This chapter shows the potential use of biopolymeric materials in modern therapeutics and underscores the need for continued research and innovation in this rapidly evolving field.

Keywords Biopolymers · Drug delivery · Therapeutics · Vaccines · Nanofibers · 3D bioprinting

D.-E. Cozorici · I.-C. Radu · E. Blanzeanu · C. Zaharia (✉)
Faculty of Chemical Engineering and Biotechnology, Department of Bioresources and Polymer Science, Advanced Polymer Materials Group, National University of Science and Technology POLITEHNICA Bucharest, Bucharest, Romania
e-mail: catalin.zaharia@upb.ro

D.-E. Cozorici
e-mail: derniza.cozorici@upb.ro

I.-C. Radu
e-mail: ionut_cristian.radu@upb.ro

E. Blanzeanu
e-mail: erika.blanzeanu@upb.ro

L. Ravindran et al. (eds.), *Macro, Micro and Nanocomposites from Sustainable Sources*, Advanced Structured Materials 244, https://doi.org/10.1007/978-981-95-2469-3_15

15.1 Introduction

As the therapeutic field continues to expand and evolve, driven by the pursuit of more effective and innovative treatment solutions for various diseases and conditions, the development of biomaterials has received a lot of attention. Different types of biomaterials have proven to be pivotal in exploring new approaches to disease management, targeted treatments, precise therapeutic options, and curative therapies. In this direction, due to their abundance, biodegradability, environmentally friendly characteristics, and biocompatibility, biopolymeric-based materials have gained significant attention. These properties have boosted their use in the biomedical field, making biopolymers an important class of biomaterials.

As a broad definition, biopolymers are macromolecular compounds with a high molecular weight, consisting of multiple monomeric units that are covalently bonded. Their defining characteristic is that the monomeric units are sourced from either living organisms or resources that are considered to be renewable, or, in the case of synthetic monomeric units, they are chemically synthesized in such a way that they can degrade naturally and leave no remnants that might threaten human health or the environment (Rahman et al. 2019; Baranwal et al. 2022; Upadhye et al. 2022). From packaging and textiles to agriculture and electronics, biopolymers offer a promising portfolio of sustainable material alternatives, aligning with the growing demand for eco-friendly solutions across various industries.

Biopolymeric materials can be classified based on several aspects, with the most general classification based on their origin, dividing biopolymers into two main categories: natural and synthetic biopolymers. Natural biopolymers are derived directly from living organisms, either from plant- or animal-based sources. On the other hand, synthetic biopolymers can be obtained by chemical synthesis, starting from renewable bio-based monomers, such as polylactic acid (PLA) derived from lactic acid, or from synthetic monomers, like polycaprolactone (PCL) synthesized from petroleum-derived ε-caprolactone. These materials are engineered to combine the advantageous properties of both natural biopolymers and synthetic polymers, overcoming a significant drawback of synthetic polymeric materials by offering an eco-friendly alternative (Sessini et al. 2023; Mtibe et al. 2021). Based on their origin, another class of biopolymers is microbial biopolymers, which are produced through fermentation by various microorganisms. The type of biopolymer produced depends on the microorganism involved, with one notable example being the class of polyhydroxyalkanoates (PHAs), which are synthesized through bacterial fermentation (Fatima 2024; Moradali and Rehm 2020; Li et al. 2016).

Biopolymers can also be classified based on their composition. Carbohydrate monomers serve as the building monomeric blocks for polysaccharide-based biopolymers, such as chitosan, alginate, and cellulose. Amino acid monomers are the building blocks of protein-based biopolymers, with collagen, elastin, and silk being common examples. Nucleotide monomers, on the other hand, form the basis of nucleic acid-based biopolymers, including deoxyribonucleic acid (DNA) and ribonucleic acid (RNA). Additional classification criteria include their structure, dividing biopolymers into linear, branched, or cross-linked types. Another important factor to consider is biodegradability, as biopolymers can either be biodegradable or non-biodegradable, along with other specific properties (Kalpana Manivannan et al. 2024; Baranwal et al. 2022; Hassan et al. 2019; Das et al. 2023).

There are some differences between natural and synthetic biopolymers in terms of their structural and functional aspects. While natural biopolymers are valued for their inherent biocompatibility and similarity to the extracellular matrix, their complex structures are harder to control. In contrast, synthetic biopolymers, with their simpler structures, offer greater flexibility for tailored modifications (Das et al. 2023; Chen et al. 2022). For instance, the specific characteristics of each type of biopolymer can either encourage or limit its use in a certain biomedical application, and multiple aspects should be considered based on the targeted application.

Literature extensively explores its diverse uses, highlighting a broad range of approaches. Research covers biopolymeric materials at all scales, from nano- to macro-, explored in various forms, including scaffolds, particles, membranes, fibers, and hydrogels, showcasing their versatility and potential to revolutionize medical treatments and technologies (Varma et al. 2021). Natural, synthetic, and microbial biopolymers—such as chitosan, cellulose, collagen, gelatin, keratin, silk fibroin, pectin, carrageenan, alginate, polylactic acid, polycaprolactone, polyethylene glycol, and polyhydroxybutyrate—are extensively studied in this context (Das et al. 2023; Balakrishnan et al. 2018; Nazeer et al. 2020; Gregory et al. 2022).

A key property of biopolymers for application in biomedicine is their ability to interact safely with living systems. Besides the biocompatibility aspect, which can be influenced by factors such as surface chemistry, topography, and charge, or by biodegradability, degradation rate, and the size or shape of the materials, biopolymers are also notable for their ability to have their properties tuned in ways that lead to advanced materials with adaptive functionalities, meeting desired performance criteria (Jurak et al. 2021).

In terms of valuable properties, some biopolymers, such as chitosan, inherently exhibit antibacterial activity, making them suitable for biomedical applications, particularly in wound healing for the development of wound dressings to prevent infection in open wounds (Li and Zhuang 2020; Torabi and Hassanzadeh-Tabrizi 2024). Other biopolymers, like hyaluronic acid have anti-inflammatory activity, further enhancing their potential for therapeutic uses (Hu et al. 2022). Beyond these intrinsic biological functions, advanced biopolymer-based materials are being engineered to possess properties such as self-healing, responsiveness to stimuli like temperature, pH, light, or magnetic and electric fields, and other tailored behaviors, falling into the category of smart biopolymers. This class of biomaterials allows for the development of multifunctional systems that can, for example, employ a controlled release mechanism triggered by a certain stimuli, enhancing the efficacy of drug delivery systems while simultaneously offering self-healing properties to improve wound healing (Pathan and Shende 2021; Jaiswal and Sherje 2024; Hu et al. 2024).

Improving the performance of biopolymeric materials often requires enhancing certain properties or introducing new functionalities. This is commonly achieved by blending the base biopolymer with others or by incorporating additional materials into the matrix to create composites. Biopolymer blends are typically produced by combining two or more biopolymers, often with the use of compatibilizers to enhance compatibility between the constituent polymers. The targeted properties improved through such blends include mechanical strength, rheological behavior, thermal stability, degradation rates, and, in some cases, even production costs (HPS et al. 2023; Fredi and Dorigato 2024). Additionally, biopolymer-based composites can be developed by introducing reinforcement agents or fillers, which represent the discontinuous phase of the composite system, into the continuous biopolymeric matrix. Natural fibers, such as cellulose, hemp, bamboo, and flax fibers, as well as bioceramics like calcium phosphates, clays, metallic particles, and metal oxides, are all potential candidates as reinforcement or filler materials. Reducing filler dimensions to the nanoscale enables the incorporation of various nanofillers such as carbon-based materials like graphene nanoplatelets and carbon nanotubes, silica nanoparticles, silver and gold nanoparticles, and diverse nanofibers for the development of advanced nanocomposites. The incorporation of these nanoscale reinforcements can significantly improve the thermal, mechanical, optical, and other properties of the biopolymer matrix, including imparting stimuli-responsive behavior (Udayakumar et al. 2021; Al Mahmud 2023; Sofini et al. 2024).

These strategies allow the development of new materials that combine the desired characteristics of different materials, resulting in improved performance and expanded application possibilities. In addition to serving as the primary material in a system, biopolymers can be used as coatings, fillers, or additives to enhance the biocompatibility or biological activity of other materials, thereby improving their suitability for therapeutic applications. The high number of biopolymer-based materials that can be developed with tailored properties highlights the versatility of this class of biomaterials. The wide-ranging potential of biopolymers is demonstrated by the extensive scientific interest they have attracted, as well as by the development of commercially available therapeutic products, emphasizing their relevance in the biomedical field (Biswas et al. 2022).

Selecting the appropriate biopolymer for a given therapeutic application involves considering several factors, highlighted in Fig. 15.1. This process starts with identifying the type of biopolymer suited for the intended use, followed by assessing the size and shape of the final system. Furthermore, it is essential to assess whether a blend or composite system is most suitable, and whether surface modifications are necessary to enable targeted delivery particularly in drug delivery applications or to improve interactions with biological environments. Additionally, a wide range of properties, including bioactivity, responsiveness to specific or multiple stimuli, appropriate degradation rates, and other specific requirements, can be customized based on the application. Manufacturing feasibility and cost considerations should also be considered (Kalpana Manivannan et al. 2024; Fazal et al. 2023).

This book chapter explores the revolutionary role of biopolymers in therapeutics, emphasizing the major progress in drug delivery systems, regenerative medicine, cancer therapy and vaccines. It presents a comparison to conventional therapeutic strategies, showing the improved efficacy and innovation brought by biopolymeric

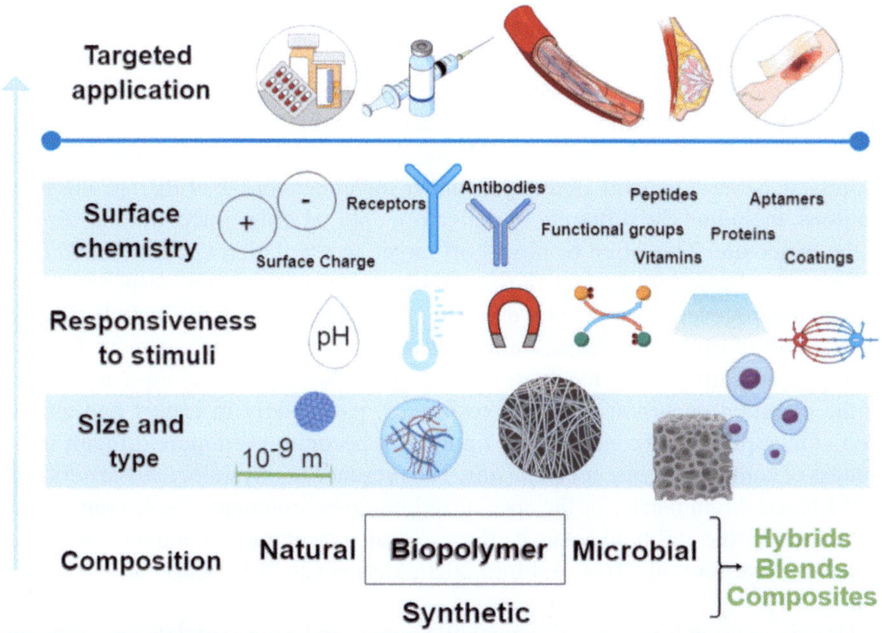

Fig. 15.1 Key considerations in designing biopolymer-based therapeutic materials

materials while discussing the limitations of traditional approaches. In response to the limitations of conventional therapeutic strategies, recent progress in biopolymer research has led to the development of innovative solutions aimed at improving drug targeting, enhancing selectivity and bioavailability, and reducing side effects. Targeted nanocarriers, smart drug delivery systems, artificially engineered tissues, advanced wound dressings, and artificial blood vessels, among other innovations, are all described as promising solutions to the existing limitations in conventional therapies. Recent advancements in the processing of biopolymeric materials are discussed, with a focus on techniques such as electrospinning and 3D printing, including the innovative approach of 3D bioprinting. Additionally, the use of biopolymers in the development of artificial cells and emerging research areas such as organs-on-a-chip and their potential applications in therapeutics are addressed.

15.2 The Role of Biopolymers in Therapeutic Applications

The field of therapeutics is vast, as it encompasses a wide range of approaches and technologies aimed at treating diseases and improving patient outcomes. This branch of medicine is evolving rapidly, driven by advancements in technology, new drug development strategies, and a deeper understanding of disease mechanisms.

Conventional therapy strategies face some major limits in a variety of medical areas, leading to a reassessment of therapeutic approaches aimed at more efficient treatments, fewer side effects, and solutions for previously untreatable human diseases. Limitations in therapeutic approaches can be identified at various levels, depending on the specific category and complexity of the medical conditions being addressed. Several general challenges hinder the effectiveness of therapeutic interventions, including the difficulty of delivering optimal drug concentrations directly to the target site. This often results in off-target accumulation and limited efficacy, further complicated by patient-to-patient variability in treatment response due to disease heterogeneity and genetic diversity. Additional limitations include the poor bioavailability of certain drugs, which often necessitates higher or more frequent dosing, potentially leading to adverse effects and reduced patient compliance as well as the common occurrence of drug resistance, particularly in cancer and antimicrobial therapies. Overcoming these limitations becomes even more difficult in the context of complex disease mechanisms. The presence of physiological barriers such as the blood–brain barrier or the specialized microenvironment of solid tumors can further hinder the ability to achieve effective drug concentrations at the intended site of action (Sharma et al. 2024; Ashford 2020; Vasan et al. 2019; Lai et al. 2022; Rizk et al. 2017).

Ongoing research into novel therapeutic agents and advanced delivery strategies seeks to enable more targeted and controlled drug administration. At the same time, efforts are being directed toward developing precision medicine approaches that account for patient-specific factors such as genetic variations, disease heterogeneity, and individual treatment responses. By tailoring therapies in a patient-specific way, treatment outcomes can be optimized, and unwanted side effects can be reduced. Additionally, efforts are being made to identify more efficient and non-invasive routes of administration, which could improve patient compliance (Bae and Park 2020).

The role of biopolymers in advancing various therapeutic areas will be explored, with a focus on their potential to overcome existing limitations and on the remarkable versatility of these polymeric biomaterials.

15.2.1 Advancements in Drug Delivery Systems

Progress in therapeutics has brought notable changes to medical treatment, with drug discovery and delivery methods playing a key role in driving these developments.

The limitations of conventional drug delivery methods, particularly the lack of control over drug release, distribution, and targeting, as seen in immediate-release formulations such as oral tablets, capsules, injectable solutions, topical preparations, and suppositories have prompted the development of advanced drug delivery systems. A drug delivery system refers to a formulation or device designed to improve the administration of a therapeutic agent by enhancing its efficacy and safety. This is achieved by controlling the rate, timing, and site of drug release within the biological system (Jain 2020; Tiwari et al. 2012; Ezike et al. 2023). Drug delivery systems have evolved over time, influenced by advancements in materials science, fabrication techniques, and growing knowledge of pharmacokinetics. This evolution is generally grouped into three main generations, based on improvements in design, technological sophistication, and the ability to control drug release and targeting (Ezike et al. 2023).

In the early stages, drug delivery systems primarily focused on developing oral and transdermal formulations. Efforts were made to develop delayed-release, prolonged-release, and sustained-release options aimed at optimizing therapeutic effects while minimizing adverse reactions. This first generation of drug delivery systems was based on four mechanisms of drug release: diffusion-controlled, dissolution-controlled, osmosis-controlled, and ion exchange-controlled release, all with an emphasis on maintaining constant drug plasma concentrations. The drug delivery systems developed in this category were designed at the macroscale. The second generation of drug delivery systems explored more advanced carriers and approaches to deliver therapeutics, focusing on both micro- and nanoscale systems designed for site-specific delivery. This targeted approach aimed to enhance the local concentration of the drug at the site of action while minimizing exposure to healthy tissues, thereby improving overall therapeutic efficacy. In this context, various materials such as microparticles and hydrogels as potential carriers were developed, and novel systems were introduced, including smart drug delivery platforms based on intelligent polymers capable of releasing therapeutics in response to external or internal stimuli. Additionally, techniques to improve the pharmacokinetics and safety of therapeutic agents were explored, and early progress in this generation of drug delivery systems led to the development of nanocarriers, such as nanoparticles, which introduced the concept of nanomedicine. Another important aspect of research for this generation has been the development of methods for delivering macromolecules such as peptides, proteins, and nucleic acids, in addition to traditional small molecule drugs. The third generation, which is still under continuous development, is expanding the portfolio of nano-based drug delivery systems by focusing on active targeting and long-term delivery. It aims to overcome the limitations of previous systems through the development of novel strategies, resulting from a collaborative effort across multiple disciplines (Ezike et al. 2023; Park et al. 2022; Vargason et al. 2021; Jain et al. 2018a).

The use of biopolymers in advancing drug delivery systems has led to several notable achievements, from their role as excipients in pharmaceutical formulations to more systematic applications in developing sustained or controlled release systems. Additionally, biopolymers have been employed to enhance drug solubility and bioavailability through drug-polymer complexes and conjugates.

Biopolymers serve various roles as excipients in drug formulations. For example, cellulose and its derivatives, such as microcrystalline cellulose and hydroxypropyl cellulose, act as adsorbents, glidants, suspending agents, diluents, and disintegrants, improving both stability and tablet disintegration. Another biopolymer, gelatin, functions as a coating agent, film-former, gelling agent, suspending agent, binder, and viscosity enhancer. Similarly, alginate provides stability, acts as a suspending agent and sustained release adjuvant, and serves as a binder, disintegrant, and viscosity enhancer. Additionally, starch is primarily used in solid oral dosage forms as a binder, diluent, and disintegrant to ensure consistency and facilitate proper tablet disintegration into small fragments upon contact with a liquid medium (Rowe et al. 2006).

Besides their use as excipients, several notable synthetic biopolymers have emerged on the list of Food and Drug Administration (FDA)-approved drug delivery systems. The initial use of poly (lactic-co-glycolic acid) (PLGA) as a synthetic biopolymer matrix in drug delivery systems can be considered a milestone event due to its role in enhancing the controlled release of therapeutics. The first significant milestone can be traced back to the FDA's approval of Lupron Depot in 1989. This PLGA-based delivery system was the first long-acting injectable formulation of leuprolide acetate, providing a controlled release for the treatment of prostate cancer over a period of one month. Over the years, PLGA has been used in a variety of FDA-approved formulations, marking the continued progress of PLGA-based injectable formulations in the form of microparticles, in situ forming implants, and solid implants administered via subcutaneous, intramuscular, intra-articular, and intravitreal injections. Several FDA-approved drugs utilize PLGA-based drug delivery systems, showcasing the versatility and effectiveness of this synthetic biopolymer. For instance, Risperdal Consta is approved for the treatment of schizophrenia and bipolar I disorder, using PLGA microspheres to enable biweekly intramuscular injections. Bydureon, approved for type 2 diabetes, also employs PLGA microspheres for weekly subcutaneous injections. Additionally, ZILRETTA is approved for osteoarthritis and utilizes PLGA microspheres for intra-articular injections (Paik et al. 2019; Ma et al. 2022).

Another key milestone in the development of drug delivery systems, underscoring the importance of biopolymers, was the emergence of polymer-drug conjugates. This strategy aimed to enhance the pharmacokinetics and pharmacodynamics of protein-based therapeutics by improving their stability, solubility, and circulation half-life, ultimately leading to improved therapeutic efficacy and reduced side effects. Polyethylene glycol (PEG) is the most widely used synthetic biopolymer in this context. The covalent attachment of PEG to therapeutic proteins has resulted in the development of several FDA-approved PEGylated products for the treatment of various conditions, including hepatitis B and C, multiple sclerosis, acromegaly, hemophilia A, and infections during chemotherapy (Sanchez Armengol et al. 2022; Gao et al. 2024a).

Several biopolymers have also been utilized in the development of hydrogel-based drug delivery systems, including xanthan gum, gellan gum, hydroxypropyl methylcellulose, PEG, alginate, hyaluronic acid, polyvinyl alcohol (PVA), and PLGA. These biopolymers play a crucial role in the formation of hydrogels that can enhance drug stability, control release rates, and improve therapeutic efficacy. Commercial hydrogel-based drug delivery systems often incorporate multiple biopolymers as inactive ingredients to form a polymeric matrix. For instance, REGRANEX®, a topical treatment for diabetic foot ulcers, uses sodium carboxymethylcellulose as the matrix along with a recombinant human platelet-derived growth factor to promote wound healing. Other examples include Zilactin-B Gel®, an oral pain reliever that contains hydroxypropyl cellulose and PEG, while Timolol GFS®, an ophthalmic gel-forming solution for glaucoma treatment, incorporates xanthan gum (Raeisi and Farjadian 2024; Cascone and Lamberti 2020). Among the FDA-approved hydrogel-based formulations is OncoGel, a smart drug delivery system composed of a PLGA-PEG-PLGA triblock copolymer. This thermoresponsive injectable undergoes a sol-gel transition at body temperature, allowing for the controlled release of paclitaxel in the treatment of solid tumors (Elstad and Fowers 2009).

Reducing the size of drug carriers to nanometric dimensions represents another step forward in advancing the efficacy of drug delivery systems. In this context, nanoparticles are extensively studied within the field of nanomedicine, as their nanoscale dimensions result in a high surface area, which in turn enhances the stability and solubility of encapsulated molecules (Mitchell et al. 2021). By optimizing the size, shape, and surface properties of nanocarriers, these formulations can efficiently deliver therapeutics to targeted diseased cells, overcoming intracellular barriers to achieve the desired therapeutic effect while minimizing off-target impacts. Various nanocarriers have shown great promise, with several receiving FDA approval as drug delivery systems, including some that contain biopolymers in their formulation (Mitchell et al. 2021; Liu et al. 2024a; Foroozandeh and Aziz 2018; Thapa and Kim 2023). Polymeric nanomaterials are particularly attractive due to the ease with which their physicochemical properties, such as size, shape, architecture, charge, and surface functionality can be tuned. Biopolymeric nanostructures, including nanoparticles, nanofilms, and nanofibers, are being explored across a wide range of therapeutic areas (Beach et al. 2024; Singh et al. 2021).

The range of drug delivery systems and approaches used to administer therapeutics has expanded over time and is expected to continue advancing in the coming years, as innovations in biotechnology, nanotechnology, and materials science open new possibilities. The next generation of drug delivery systems is focused on enhancing targeting efficiency of therapeutic agents, addressing the challenges imposed by physiological barriers by exploring new routes and pathways, and advancing personalized therapies (Bae and Park 2020; Gao et al. 2023). Considering the progress of biopolymers in this direction, along with advancements in nanotechnology, advanced fabrication techniques, and cell-based technologies, it can be predicted that biopolymers will be part of this progress. Patient compliance and convenience are also targeted objectives in the future of therapeutics. The development of long-acting drug delivery systems, particularly for chronic disease management, aims to reduce the frequency of drug administration while maintaining therapeutic drug levels. These advancements include effective oral delivery systems, transdermal patches, implants, inhalers, single-injection multi-dose vaccines, cell-based immunotherapies, and targeted smart systems, all designed to improve patient compliance to treatment regimens (Jindal et al. 2023; Baryakova et al. 2023).

15.2.2 Regenerative Medicine and Tissue Engineering

Regenerative medicine is a rapidly advancing field that has the potential to transform healthcare by offering innovative solutions for a wide range of conditions. This interdisciplinary area focuses on the restoration, repair, or replacement of damaged tissues and organs affected by various factors such as age, disease, trauma, and congenital abnormalities. As a broad field, regenerative medicine integrates knowledge and techniques from multiple disciplines and encompasses several subdisciplines, with tissue engineering and cell therapies being key categories. These areas are supported by ongoing research and advancements in cell technology and manipulation, with stem cells playing a pivotal role. Key concentrations in this field include the manipulation of various cell types, the use of growth factors, and the development of engineered biomaterials. Together, these components pave the way for novel technologies and therapeutic strategies that enhance tissue regeneration and healing. The regulatory landscape for regenerative medicine is important for ensuring safety and efficacy and in this direction FDA regulates a variety of products in this category, including cell therapies, therapeutic tissue engineering products, combination products that incorporate these therapies, certain gene therapy products, and human cell and tissue products (Altyar et al. 2023; Dzobo et al. 2018; https://www.cancer.gov/research/progress/250-years-milestones).

Biopolymers have made significant strides in regenerative medicine, first demonstrated by the transition from traditional plasters and bandages to advanced wound dressings—such as films, sponges, hydrogels, and fibers—that not only improve the healing process but also reduce the risk of infection. Currently, several biopolymer-based wound dressings are available on the market, made from alginate, gelatin, hyaluronic acid, collagen, and even compositions that contain silver to promote antimicrobial effects. Furthermore, ongoing research is investigating smart dressings with stimuli-responsive drug delivery and integrated multiplexed sensors for real-time wound monitoring and therapeutic functions. These innovations enable personalized treatment strategies for complex wounds, including non-healing ones (Torabi and Hassanzadeh-Tabrizi 2024; Nguyen et al. 2023; Horta-Velázquez et al. 2024; Dong et al. 2024). More advanced achievements include the use of biopolymers in artificial grafts, such as vascular, tendon, and ligament grafts, with ongoing research aimed at enhancing their functionality (Francesco et al. 2023; Shiroud Heidari et al. 2023).

As research advances, the applications of regenerative medicine are expected to expand, with biopolymers playing an important role in this progress. Biopolymeric materials remain essential for developing scaffolds that mimic the extracellular matrix, providing a 3D framework for cells and bioactive molecules, such as growth factors. These scaffolds help create engineered substrates that replicate the targeted tissue microenvironment and support tissue formation. They can be designed to degrade at a rate that aligns with tissue regeneration, offering temporary support until new tissue is fully formed. Additionally, ongoing research is exploring the use of biopolymers for targeted delivery of drugs, growth factors, or genes, enabling controlled, sustained, and localized release of therapeutics. Efforts are also underway to integrate biopolymer-based materials with other technologies, such as electrospinning and 3D bioprinting, which are further discussed in this chapter. These advancements aim to create functional materials and develop models of physiological barriers, diseases, and drug testing platforms, all contributing to the progress in therapeutic innovations (Kalpana Manivannan et al. 2024; Sanchez Armengol et al. 2024; Capella-Monsonís et al. 2024).

15.2.3 Biopolymers in Cancer Therapy

Cancer treatment has made considerable progress over the years. While traditional methods such as surgery, radiotherapy, and chemotherapy remain central, they have been steadily refined and adapted to improve outcomes. Meanwhile, the focus is increasingly shifting toward more targeted, precision-based strategies that aim to improve effectiveness and reduce the side effects of treatments. However, according to the World Health Organization (WHO), the global burden of cancer is expected to rise, with over 35 million new cases predicted by 2050, a staggering 77% increase from the estimated 20 million cases in 2022. This emphasizes the critical need for continued research into treatment options and comprehensive cancer management strategies, beginning with prevention, early detection, and extending through to palliative care (https://www.cancer.gov/research/progress/250-years-milestones)

Since the late 1990s, several key advancements have significantly improved cancer treatments. Notable among these are the development of targeted therapies, marked by FDA-approved monoclonal antibodies that specifically target cancer cells. Additionally, immunotherapy has seen remarkable progress with the introduction of immune checkpoint inhibitors, such as pembrolizumab, which enhance the immune system's ability to fight cancer. Vaccines like Gardasil and Cervarix have also been approved by FDA to prevent cervical cancer caused by human papillomavirus (HPV). Furthermore, the first FDA-approved cancer treatment vaccine, Sipuleucel-T, is a personalized therapy made from a patient's own immune cells. It is designed to treat metastatic or symptomatic prostate cancer that no longer responds to hormonal therapy. Building on these advancements, there has also been a collaborative effort to improve therapeutic efficiency through innovative drug delivery strategies. In addition to new approved anticancer drugs, formulations based on liposomes, polymeric approaches like micelles and polymer-protein conjugates, as well as PEGylated systems, have been approved. Other noteworthy approaches include antibody-drug conjugates and metallic nanoparticles. Strategies like these have received approval from the FDA or the European Medicines Agency (EMA) for the treatment of various types of cancer (https://www.cancer.gov/research/progress/250-years-milestones; Mohite et al. 2024; Anassi and Ndefo 2011; Mansouri et al. 2024; Karahmet Sher et al. 2024).

Future innovations in cancer treatment, as can be predicted from current research efforts, are centered on a multifaceted approach to address the complexities of this disease. With a primary focus on balancing therapeutic efficacy and safety, further research into targeted drug delivery systems, immunotherapies, cell therapies, and combination treatments is essential to fully realize their potential. Additionally, breakthroughs in genomics, transcriptomics, and proteomics are paving the way for precision medicine, which will tailor treatments to individual genetic, environmental, and lifestyle factors, leading to more personalized and effective outcomes (Al Meslamani 2023; Tsimberidou et al. 2020).

Regarding the role of biopolymers in the future of cancer treatment options, the focus is specifically on their function as material carriers for delivering anti-cancer agents. In addition to their primary role, biopolymers can also serve as secondary components in various therapeutic strategies, enhancing the efficacy of existing treatments. Various approaches concerning drug vehicles, targeting strategies, and release mechanisms are explored in the context of engineering novel drug delivery systems. Table 15.1 presents recent examples from the literature that illustrate ongoing research in this field.

Recent research outlined in the literature highlights several trends in biopolymer-based drug delivery systems for cancer treatment. These developments are contributing to the advancement of cancer therapies by improving precision, therapeutic efficiency, and patient outcomes. Increasing attention is being directed toward hybrid and composite systems, which demonstrate superior performance compared to conventional single-component approaches. Nano-sized particles are also a central area of interest due to their potential for highly precise drug delivery. Meanwhile, smart systems are being designed to enable targeted, controlled release, in response to specific stimuli, thereby optimizing therapeutic efficacy. Combining the advantages of biopolymeric nanocarriers with those of stimuli-responsive systems, which can release encapsulated drugs in response to specific internal signals such as changes in pH, temperature, or the presence of certain enzymes, or external triggers such as magnetic fields, temperature, ultrasound, or light, offers a promising platform for precise cancer treatment. This approach can significantly reduce the severe side effects and drug resistance commonly associated with conventional chemotherapy. Additionally, functionalization with specific cancer-targeting moieties is gaining increasing prominence, alongside the growing trend of combining multiple therapeutic approaches for their synergistic effects (Xiao et al. 2023; Sun et al. 2023). These strategies lead to a future of more personalized, efficient, and minimally invasive cancer treatments, driven by advanced drug delivery systems.

15.2.4 Biopolymers in Vaccine Development

Vaccines have long been recognized for their crucial role in preventing infectious diseases, but their impact in therapeutics is expanding with the emergence of therapeutic vaccines. Unlike traditional prophylactic vaccines, which are designed to prevent a disease, therapeutic vaccines offer a groundbreaking approach by treating existing diseases through the modulation of the immune response. Conventional

Table 15.1 Summary of biopolymer-based systems explored for cancer treatment

Biopolymer	System type	Specific properties	Therapeutic agent	Potential application in cancer treatment	References
Chitosan	Injectable hydrogel with 2D peptide nanosheet-drug conjugates	pH-responsive drug release	Doxorubicin	Promising strategy for enhancing targeted cancer treatment	Luan et al. (2024)
Hyaluronic acid	Hyaluronic acid-conjugated nanodrug system	Targeted drug delivery	Methotrexate 5-fluorouracil	Dual nanodrug delivery system for selective targeting of cancer cells and tumor tissues	Shao et al. (2024)
PLGA chitosan	PLGA nanoparticles with folic acid-linked chitosan	Antioxidant, anti-angiogenic, and selective cytotoxicity	Methylurolitin-A	Targeted anti-cancer system for human colon cancer treatment	Alhazami et al. (2024)
Zein starch	Nanocomposite of co-biopolymers and graphitic carbon nitride nanosheets	pH-sensitivity	5-fluorouracil	Targeted therapy for brain cancer	Pourmadadi et al. (2024)
Hyaluronic acid	Polymer conjugated nanoprodrug	CD44 receptor targeting, acid-sensitive release, autophagy inhibition	Doxorubicin primaquine	Enhanced selectivity and efficacy in triple-negative breast cancer treatment	Wu et al. (2024a)
PLGA chitosan	Chitosan folate-targeted PLGA nanoparticles	Folate receptor targeting	Palbociclib	Targeted breast cancer therapy	Mehata et al. (2024)
Chitosan PEG	Hydrogel-forming chitosan-based microneedles chemo wrap patch with PEGylated liposomes	Multiple approaches to enhance drug-skin affinity	5-fluorouracil	Treatment for pre-carcinogenic and carcinogenic skin lesions	Suriyaamporn et al. (2024)

(continued)

Table 15.1 (continued)

Biopolymer	System type	Specific properties	Therapeutic agent	Potential application in cancer treatment	References
Sodium alginate	Injectable hydrogel	Temperature and pH responsiveness	5-fluorouracil	Controlled and sustained delivery for breast cancer	Pansuriya et al. (2024)
Hyaluronic acid	Injectable hydrogel with polydopamine	Synergistic photothermal and chemotherapeutic effects	Paclitaxel	Breast cancer treatment	Gao et al. (2024b)
Cellulose chitosan	Interpenetrating network hydrogel	Temperature and pH responsiveness	5-fluorouracil	Controlled-release drug delivery system	Zhang et al. (2025)
Hyaluronic acid	Soluble microneedle patch	Photosensitivity	5-aminolevulinic acid	Oral mucosal drug delivery system for photodynamic therapy in oral leukoplakia treatment	Sang et al. (2025)
Sodium alginate PCL PEG	Injectable multifunctional hydrogel incorporating micelles	Combined chemo-immunotherapy	Doxorubicin	Colorectal cancer treatment	Yang et al. (2024a)
PLGA chitosan	PLGA/MXene nanofibers-CS aerogel composite	Temperature and pH responsiveness	Doxorubicin	Chemo-photothermal therapy to inhibit tumor growth, prevent recurrence, and control bleeding in skin cancer surgery	Fu et al. (2025)

vaccines are primarily developed to prevent diseases by inducing immunity against specific pathogens. This is achieved by administering an antigen to healthy individuals, which triggers the body's immune response and builds targeted protection. Such vaccines are particularly effective against communicable diseases like HIV, hepatitis A, B, and C, and measles. In contrast, therapeutic vaccines are designed to enhance the immune system's ability to combat existing diseases. Their mechanism of action involves stimulating cytotoxic T cells and B cells to produce disease-specific antibodies, effectively targeting and fighting the disease. Therapeutic vaccines are being developed to address a wide range of diseases, primarily focusing on infectious diseases and chronic noncommunicable conditions. The most common types of diseases targeted by these vaccines include cancers, cardiovascular diseases, pulmonary diseases, diabetes, Parkinson's disease, autoimmune diseases, and others (Tian et al. 2022; Li et al. 2023).

Nowadays, several biopolymers are commonly used as stabilizers in vaccine formulations, including dextran in Rotarix®, approved for rotavirus gastroenteritis prevention (https://www.fda.gov/media/75726/download), gelatin in Varivax®, a varicella vaccine (https://www.fda.gov/media/76904/download), and FluMist®, a nasal influenza vaccine (https://www.fda.gov/vaccines-blood-biologics/vaccines/flumist), and albumin in Pentacel®, a combination vaccine for diphtheria, tetanus, pertussis, poliomyelitis, and invasive disease due to *Haemophilus influenzae* type b (https://www.fda.gov/media/74385/download?).

With ongoing progress in key fields such as virology, immunology, and molecular biology, current strategies in vaccine development include nucleic acid-based vaccines, therapeutic viral vector vaccines, and non-viral vaccine technologies, such as virus-like particles administered intranasally. These strategies also involve enhancing efficacy through the addition of adjuvants, using engineered carriers for improved precision and targeted delivery, and exploring minimally invasive, needle-free administration routes like transdermal or intranasal delivery (Poria et al. 2024; Weerarathna et al. 2024; Wang et al. 2023).

In this context, biopolymers can be utilized in vaccines as adjuvants, non-immunogenic materials that stimulate the immune system, and as carriers for vaccine delivery. As adjuvants, biopolymers enhance immune responses by directly activating immune cells through pattern recognition receptors or by modulating the immune response to favor either humoral or cell-mediated immunity. As carriers, they protect the vaccine from enzymatic degradation, preserving its structural integrity and ensuring efficient delivery (Bose et al. 2019; Wibowo et al. 2021; Nevagi et al. 2019).

Table 15.2 presents various strategies currently being explored for using biopolymers in vaccine development. These include nanoscale carriers, with a focus on nanoparticles, and microscale approaches such as hydrogels and microparticles. The development of such carriers aims to enhance antigen stability, enable controlled release, and boost immune activation, thereby improving overall vaccine efficacy. Additionally, biopolymers are being investigated for use in the development of microneedle patches, an innovative, painless, and self-administered platform for vaccine delivery, as well as in intranasal administration, another minimally invasive route where chitosan and its derivatives are commonly studied as adjuvants (Rai et al. 2024; Kumar et al. 2024).

15.3 New Frontiers in Biopolymer Therapeutics: Processing Innovations and Conceptual Advancements

Although considerable progress has been made in developing therapeutic strategies by exploring various types of materials and methods to deliver therapeutic agents as efficiently as possible, there remains an urgent need for innovative approaches and advanced tools to expand the understanding of disease mechanisms and to evaluate newly developed therapeutic agents. This includes the need for suitable materials for emerging technologies such as electrospinning and 3D bioprinting. The growing use of biopolymers in artificial cells and organ-on-a-chip technologies highlights their potential to address current limitations. When integrated with advanced fabrication techniques, these platforms offer promising pathways toward more effective therapeutic solutions.

15.3.1 Processing Innovations in Biopolymers

15.3.1.1 Electrospinning

Biopolymeric fibers are widely utilized in the biomedical field due to their versatility and ability to be engineered with specific functionalities for applications such as drug delivery, tissue engineering, and wound healing. Special emphasis is placed on nanofibers, which offer unique properties, such as high surface area, high porosity that allows for large drug loading and tissue scaffolding, and the possibility of functionalization that enhances their performance in various biomedical applications (Mwiiri et al. 2020). Electrospinning is one of the most established and widely employed techniques for producing fibers ranging in size from the nanometer to the micrometer scale (Mooss et al. 2024; Hiwrale et al. 2023; Al-Enizi et al. 2018).

The fundamental principle of electrospinning is the use of electrostatic forces to generate and expand fibers from especially polymeric solutions, suspensions, or melts. Key components of an electrospinning setup include a high-voltage power source, a spinneret, and a collector. The high-voltage power source is essential for creating the electric field that drives fiber formation. The spinneret, often a syringe equipped with a needle, dispenses the polymeric solution at a controlled rate. Positioned at a specific distance from the needle tip, the collector, typically a conductive surface, serves as the target for the deposited fibers. The collector can be either

Table 15.2 Biopolymers explored in different vaccine development strategies

Vaccine strategy	Biopolymer	Role of biopolymer	Administration route	Potential application	References
T cell-based intranasal vaccine	Polyhydroxybutyrate	Antigen carrier	Intranasal	Broadly protective influenza vaccine	Liu et al. (2024b)
Mucosal vaccination for respiratory pathogens	N-dihydrogalactochitosan	Mucosal adjuvant	Intransal	Protection against SARS-CoV-2 and other respiratory viruses by inducing localized and systemic immunity	Hoover et al. (2024)
mRNA nasal vaccine delivery	Polyethylene glycol, gelatin	Thermoresponsive matrix for controlled gelation and sustained release of mRNA	Intranasal	Nasal gene delivery platform that overcomes issues of enzymatic degradation	Kopilovic et al. (n.d.)
Cancer vaccine targeting tumor-specific neoantigens	Polyhydroxybutyrate	Delivery platform for neoantigens	Intraperitoneal	Targeted cancer immunotherapy platform with potential for metastasis treatment and personalized vaccines	Pounraj et al. (2024)
Microneedle platform for delivery of peptide-coated viral cancer vaccines	Polyvinylpyrrolidone	Base material for the microneedle patch	Transdermal	Cancer immunotherapy	D'Amico et al. (2025)
Microneedle patch for nanovaccine delivery	Methacrylated gelatin	Base material for the microneedle patch	Transdermal	Postoperative treatment of malignant melanoma and skin tissue regeneration	Chen et al. (2024)
Microneedle patch for nanovaccine delivery	Methacrylated gelatin	Base material for the microneedle patch	Subcutaneous	Treatment of gliomas through enhanced immunotherapy to induce strong anti-tumor immune responses and promote immune memory	Qin et al. (2024)

(continued)

Table 15.2 (continued)

Vaccine strategy	Biopolymer	Role of biopolymer	Administration route	Potential application	References
mRNA vaccine delivery	Silk fibroin, gelatin	Hydrogel matrix for RNA molecules	Injectable	Tailored RNA delivery systems for therapeutics	Mahaling et al. (2024)
Multi-epitope protein vaccine	Alginate	Nanocarrier for the vaccine protein	Oral	Oral vaccine for the infectious disease Shigellosis	Hashemi et al. (2024)
Multifunctional platform for injectable chemo-immunotherapy and personalized tumor vaccine	Alginate	Forms injectable hydrogel depot	Intratumoral	Treatment of colorectal cancer through enhanced drug penetration and immune activation	Yang et al. (2024a)
Therapeutic vaccination against cutaneous leishmaniasis, a skin infection	Poly (lactic-co-glycolic acid)	Nanoparticles co-delivering leishmania antigens and toll-like receptor (TLR) agonists	Subcutaneous	Treatment of cutaneous leishmaniasis	Katebi et al. (2024)
Controlled-release subunit vaccines	Poly (lactic-co-glycolic acid)	Carrier for antigens and immune potentiators	Injectable	Development of vaccines for infectious diseases	Cassaidy et al. (2024)
pH-responsive nanovaccine platform for co-delivering peptide antigens and adjuvants	Polyethylene glycol	Hydrophilic block in nanoparticle carrier formation	Subcutaneous	Therapeutic cancer vaccines targeting neoantigens	Baljon et al. (2024)

stationary or rotating, influencing the alignment and morphology of the resulting fibers. The mechanism of fiber formation begins with an electric field established between the needle tip and the collector by applying a specified voltage. Due to the surface tension of the solution, droplets form at the spinneret and acquire an induced charge on their surface. When the electrostatic force becomes equivalent to the surface tension of the solution, the droplet deforms into a conical shape known as a Taylor cone. As the applied voltage increases, the electrostatic force eventually overcomes the surface tension, leading to the formation of a jet. This jet travels toward the collector, undergoing stretching and whipping caused by the electric field. During this process, solvent evaporation occurs, resulting in solid fibers being deposited on the collector (Xue et al. 2019). There are several variables that need to be adjusted in order to obtain fibers of desired dimensions with a constant diameter, characterized by a smooth structure free of beads. Environmental parameters, such as temperature, atmospheric pressure, and humidity, as well as variables regarding the polymeric solution—including the concentration and molecular weight of the polymer, viscosity, surface tension, conductivity, the choice of solvent, and the flow rate of the solution—should be considered. Additionally, setup parameters such as the distance between the needle and the collector, the needle diameter, and the applied voltage play crucial roles in this process (Abdulhussain et al. 2023; Haider et al. 2018).

Although the basic principle of electrospinning might be regarded as relatively simple, significant ongoing efforts have driven advancements, leading to the development of more advanced setups. Progress in electrospinning principles and technology focused on obtaining fibers with controlled dimensions and morphologies, including multicomponent fibers. As a result of this progress, fibers with a wide range of functionalities can be developed through the continuous optimization of parameters.

Electrospinning has evolved from mono-axial techniques to more complex multi-axial approaches, such as co-axial electrospinning, which enables the development of fibers with core-shell or hollow structures, and tri-axial electrospinning, which can produce intricate trilayer structures by varying the physicochemical properties of each layer. Additionally, portable electrospinning devices have been developed for on-site fiber production, with potential applications in wound management. More recent systems can combine electrospinning with 3D printing, enabling the development of 3D fibrous scaffolds with tailored properties, which could serve as potential solutions in tissue engineering. Besides needle-based techniques, needle-less approaches have been developed to overcome limitations such as needle clogging and low fiber production capacity. These methods generally operate using the surface of an open vessel filled with the prepared solution, where fibers are generated by either stationary or rotating spinnerets (Keirouz et al. 2023; Govind Kumar et al. 2022; Ejiohuo 2023).

Drug-loaded biopolymer nanofibers based on gelatin, chitosan, collagen, alginate, silk fibroin, zein, polycaprolactone, polyvinylpyrrolidone, and polyvinyl alcohol are extensively explored, covering a wide range of therapeutic applications and underscoring ongoing research into hybrid, composite, and functional materials. Several promising applications can be identified, spanning from tissue regeneration, including skin, bone, cartilage, and nerve tissue engineering, to drug delivery systems for various therapeutic agents, including chemotherapeutic drugs for localized cancer therapy (Patel et al. 2025; Tören et al. 2024). Smart wound dressings designed to react to specific stimuli, such as temperature and pH changes, to provide dynamic treatment options and assess wound status, as well as multifunctional electrospun nanofiber scaffolds engineered to achieve both antitumor therapy and tissue repair as a postoperative treatment for tumors, are just some examples of how biopolymer nanofibers could advance therapeutic approaches (Qu et al. 2024; Kuang et al. 2024).

In this direction, nanofibers made from silk fibroin and fibrin have recently been explored for the development of a scaffold that mimics the natural structure and function of blood vessels as a potential candidate for artificial blood vessels (Yang et al. 2024b). Another study presents biodegradable electrospun fibrous scaffolds made from PCL and PVA nanofibers, designed to create effective tissue engineering scaffolds that replicate the properties of the extracellular matrix (Mitropoulou et al. 2024).

In cancer therapy, PCL and gelatin were utilized to develop drug-loaded multifunctional bilayer nanofiber scaffolds for postoperative wound care in patients with cutaneous squamous cell carcinoma (Zhou et al. 2024; Song et al. 2024). Additionally, other studies have explored the development of composite nanofibers, such as electrospun nanofibers made from casein and PVA, reinforced with silver nanoparticles for antimicrobial wound dressings to promote wound healing (George et al. 2024). These investigations underscore the ongoing research into hybrid, composite, and functional materials in various therapeutic applications.

15.3.1.2 3D Bioprinting

As 3D printing continues to grow and find applications across a variety of industries, the shift toward 3D bioprinting marks a significant step in the evolution of additive manufacturing, particularly in biomedical applications (Elhadad et al. 2023). 3D bioprinting applies the principles of 3D printing to create tissue-engineered constructs by precisely depositing living cells and biomaterials, with the ultimate goal of fabricating functional biological tissues (Tripathi et al. 2023).

The materials specifically designed for use in 3D bioprinting are called bioinks. Bioinks are typically composed of three key elements: a polymeric component, cells, and bioactive molecules. The polymeric component, often a hydrogel, provides a biocompatible matrix that supports cell viability and functionality. Integrated within this structure are living cells, which contribute to the bioink's capacity to replicate tissue-specific properties. Additionally, bioactive molecules, such as growth factors, are incorporated to promote cell adhesion, proliferation, and differentiation, enhancing the overall performance of the bioink formulation. Depending on the presence of the polymeric component, bioink-based materials are classified as either scaffold-based or scaffold-free bioinks. In scaffold-based bioinks, cells are typically embedded in hydrogel formulations. These hydrogels protect the cells from stressors during the 3D bioprinting process, support cell proliferation and growth, and facilitate tissue formation. In contrast, scaffold-free bioinks use cell aggregates as building blocks, which can take the form of mono- or multicellular structures such as cell sheets, pellets, spheroids, or tissue strands. Ink-based materials can serve different roles based on their functionality. Structural bioinks provide the necessary mechanical support for 3D bioprinted constructs. Functional bioinks are designed to enhance biological activity by incorporating specific bioactive molecules that promote cell differentiation and propagation. Supportive bioinks offer temporary mechanical or structural stability during the printing process. Finally, sacrificial bioinks are employed to form temporary structures that are later removed, allowing for the creation of hollow channels within the printed constructs (Khoshnood and Zamanian 2020; Vanaei et al. 2021; Raees et al. 2023).

To ensure the success of 3D bioprinted constructs for a given application, the biocompatible matrix within bioinks must exhibit key properties that support both the viability and functionality of the embedded cells. Biocompatibility is essential, meaning that the matrix in which cells are embedded should be non-toxic and non-immunogenic to support cell viability and function. The ideal ink-based material should also possess a certain level of porosity, promoting interconnectivity within the printed structure, needed for the transport of nutrients and oxygen to the interior of the construct while allowing for the removal of cellular waste. Further, pH neutrality, insolubility in cell culture medium, and the ability to promote cell adhesion are crucial characteristics to recreate the physiological environment, while also providing structural integrity, preventing degradation in culture conditions (Raees et al. 2023; Nagaraja et al. 2024). Ensuring appropriate mechanical properties to maintain the structural integrity, stability, shape, and function of the printed construct is essential. Additionally, printability, which is influenced by the composition, rheological properties, gelation kinetics, and surface tension of the selected materials, must also be carefully considered (Bartolo et al. 2022). Besides this, depending on the targeted application the ink formulation should be biodegradable or not. For some applications, biodegradable matrices can be designed to degrade into non-toxic components at a rate suitable for tissue regeneration, supported by the encapsulated cells, while also releasing specific therapeutic agents. In contrast, non-biodegradable formulations may be used in cases where permanent support structures are needed (Yilmaz et al. 2021).

Such considerations are of major importance, as 3D bioprinted constructs are designed to uphold their shape, structural strength, and integrity during the printing process. They should also sustain their 3D configuration in vitro for a certain period and, once implanted, either degrade appropriately or remain stable, depending on the therapeutic goal.

Selecting an appropriate bioink formulation is a critical step in ensuring successful 3D bioprinting outcomes, as it directly influences cell viability, construct stability, and overall functionality of the 3D bioprinted tissue. An important factor to consider is the type of 3D bioprinting technology, whether extrusion-based, droplet-based, or laser-based, as each method requires specific bioink properties to ensure optimal performance. The formulation of the bioink should also be tailored to the specific requirements of the target tissue, as different tissues demand distinct conditions for cell proliferation, differentiation, and mechanical support. The complexity of the construct also determines the appropriate bioink characteristics. Simple geometries might tolerate a wider range of bioinks, while complex geometries or non-planar constructs may require bioinks with specific viscoelastic properties to prevent collapse or deformation during 3D printing. Another important aspect to be considered is the crosslinking method essential for stabilizing the 3D bioprinted constructs and providing suitable mechanical properties. The choice of crosslinking method, whether ionic, photo, or thermal, will depend on the required stiffness and stability of the construct as well as the compatibility with encapsulated cells. Additionally, the transparency of the ink-based material is an important factor for monitoring and analysis, as opaque bioinks can obstruct the visual assessment of cell distribution, morphology, and interactions (Jose et al. 2021; Chen et al. 2023; Williams et al. 2018).

In the research driven to develop bioink formulations both natural and synthetic biopolymers are regarded as benchmark material. Commercially available bioinks are typically formulated using biopolymers such as gelatin, collagen, alginate, chitosan, and fibrin to meet the requirements of various 3D bioprinting technologies and crosslinking mechanisms. These bioinks are often modified, such as through the methacrylation of gelatin for ultraviolet (UV) crosslinking, which makes it suitable for this specific type of crosslinking technique. They can also be customized for targeted tissue development by incorporating growth factors or recreating specific biomechanical characteristics. Notable examples of such bioinks are those developed by Rokit Healthcare, Cellink Global, and Bio Inx.

Although each type of biopolymer offers distinct advantages and ensures high cell viability, challenges still exist regarding structural stability, mechanical properties, and the fidelity of 3D bioprinted constructs. Natural biopolymers are highly biocompatible and effectively mimic biological environments, while synthetic biopolymers provide enhanced tunability and stability. To overcome the limitations of basic formulations, strategies such as copolymerizing synthetic biopolymers with natural ones, blending various biopolymers, and incorporating fillers to create composite or hybrid bioinks are being explored to develop advanced formulations with enhanced properties (Khoeini et al. 2021; Sadeghianmaryan et al. 2024; Schwab et al. 2020).

Incorporating micro- and nanosized fillers into ink-based matrix can enhance their mechanical properties, adjust rheological behavior, influence the degradation rate of the bioink and boost cellular activity. The literature reports the use of various fillers, such as silica nanoparticles, graphene oxide, cellulose nanofibers, and nanoclays like laponite and montmorillonite, to improve the mechanical and rheological properties. In addition, bioactive fillers, such as bioactive glass, hydroxyapatite, and bioactive ions like magnesium (Mg^{2+}), calcium (Ca^{2+}), and zinc (Zn^{2+}), further promote cellular activity and tissue regeneration by enhancing cell-material interactions (Chandra et al. 2024; Heid and Boccaccini 2020; Jo et al. 2024).

As an important extension of 3D bioprinting, the concept of 4D bioprinting has emerged, focusing on smart bioink formulations that can change over time in response to specific stimuli. This innovative approach allows for the development of programmable bioinks capable of undergoing controlled changes in shape and function in response to various stimuli. The integration of nanofillers into the biopolymeric matrix enhances these stimuli-responsive properties. For instance, iron oxide nanoparticles can respond to magnetic fields, gold nanorods can provide photo-responsive behavior, and silver nanowires or carbon-based nanomaterials can react to electrical stimuli. This progress opens new possibilities for creating dynamic and functional tissue constructs (Rahimnejad et al. 2024; Lai et al. 2024; Rasouli et al. 2024).

Trying to keep the balance between printability and cell viability, efforts are also focused on developing multicomponent bioinks suitable for creating complex tissue models that closely mimic native structures (Cui et al. 2020; Kim et al. 2020).

The impact of 3D bioprinted constructs can be categorized into two primary areas. First, these constructs can restore the function of diseased or damaged tissues by serving as scaffolds for the delivery of therapeutic agents, making them applicable across various fields of regenerative medicine. Additionally, they hold the potential to replace entire tissues or organs, representing a significant long-term objective of 3D bioprinting (Huang et al. 2024). Secondly, 3D bioprinted constructs can serve as platforms for evaluating drug efficacy and modeling disease mechanisms, facilitating the exploration of personalized therapeutic approaches (Wu et al. 2024b). This dual functionality underscores the versatility and potential of 3D bioprinting in both clinical applications and biomedical research.

The potential of 3D bioprinting for developing drug delivery systems has also been documented in the literature. For instance, a study by Mahdizadeh et al. (2024) demonstrated the use of a 3D-bioprinted gelatin–alginate construct to incorporate niosome vesicles loaded with letrozole, creating a pH-responsive drug delivery system designed for targeted breast cancer therapy (Mahdizadeh et al. 2024).

Integration of 3D bioprinting and multi-algorithm machine learning is also an exciting direction. Tang et al. explored this integration to enhance the understanding and assessment of glioma treatment responses and the characteristics of the tumor microenvironment through an innovative combination. For this purpose, patient-derived glioma tissues that closely mimic the molecular properties and drug responses of actual tumors were 3D bioprinted using modified gelatin and hyaluronic acid-based bioink. By utilizing a multi-algorithm machine learning strategy to develop robust gene expression predictors, the investigation identifies promising drugs and compounds for glioma treatment and uncovers specific characteristics of the tumor microenvironment, including immunosuppressive and angiogenic features associated with myeloid cell infiltration (Tang et al. 2024).

Both electrospinning and 3D bioprinting are essential processing techniques that can contribute to the development of biopolymer-based materials for therapeutics, either by fabricating scaffolds for the delivery of therapeutic agents or by creating tissue constructs for research purposes, as described in Fig. 15.2.

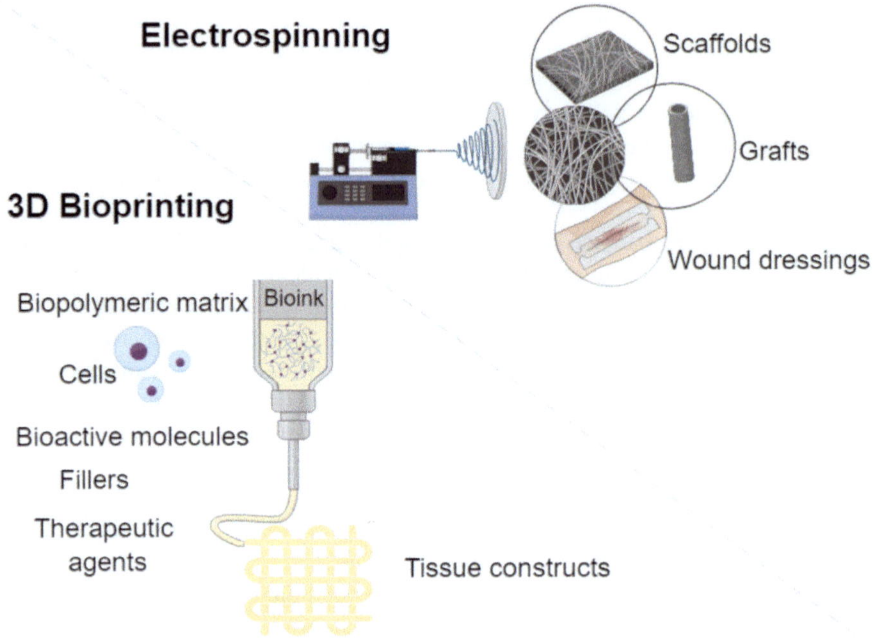

Fig. 15.2 Electrospinning and 3D bioprinting: Cutting-edge techniques for tissue engineering and regenerative medicine

15.3.2 Biopolymers in Artificial Cell Development

The concept of artificial cells centers on the development of cell-like structures that replicate key functions of natural cells, providing valuable insights into their characteristics and behavior. Artificial cells can be obtained through various methods, typically following either a top-down or bottom-up approach, each with its own set of advantages and challenges. The top-down approach modifies existing cells to produce minimal, functional versions, whereas the bottom-up approach assembles individual components to emulate the structure and function of living cells. The development of artificial cells is a complex, multidisciplinary effort that integrates expertise from various fields and considers several aspects necessary for replicating the fundamental attributes that define living cells (Ugya et al. 2022; Xu et al. 2023; Chang 2000; Waeterschoot et al. 2504).

Inspired by the intricate compartmentalization found in living cells, artificial cells can showcase a remarkable variety of architectures, ranging from simple single compartments to complex multiscale structures known as compartments-in-compartments. In the construction of these compartments, building blocks based on lipids, polymers, peptides, proteins, or their combinations are utilized, and they can range in size from nanometers to micrometers. Compartments based on polymeric systems are reported to exhibit superior physical stability and versatility in chemical design compared to phospholipid bilayers (Palivan et al. 2024).

Common structural scaffolds used in artificial cells construction include vesicles and hydrogels, where biopolymers such as chitosan, glucomannans, cellulose, alginate, and hyaluronic acid can serve as structural components. These constructs can be engineered to encapsulate, protect, and control the release of therapeutic agents. Therefore, in addition to modeling specific cellular behaviors, artificial cells can also be designed to function as drug delivery systems (Westensee et al. 2024a; Jiang et al. 2022).

In addition, the study by Westensee et al. designed and developed alginate-based artificial cells with minimal liver-like function, functionalized with metalloporphyrins to mimic the catalytic activity of CYP1A2, a major liver enzyme responsible for metabolizing various drugs and endogenous compounds. The study demonstrated the successful integration of functionalized hydrogel-based artificial cells with mammalian cells in 3D bioprinted constructs. These artificial cells were shown to support and enhance the enzymatic activity of HepG2 cells, highlighting the potential of artificial cells for tissue engineering in the development of hybrid biological-artificial tissues (Westensee et al. 2024b).

15.3.3 Emerging Research: Organs-on-a-Chip

In addition to developing materials such as drug delivery systems, wound dressings, scaffolds, and implants based on biopolymers for therapeutic applications, it is essential to have tools for testing new drugs that provide reliable predictions of therapeutic efficacy and safety to accelerate advancements in the therapeutic field.

Despite considering conventional in vitro models such as 2D or 3D cell cultures, there are persistent general limitations in accurately predicting the effects of a tested drug for a specific condition. One limitation that is observed is the inability to replicate dynamic physiological conditions such as fluid flow and mechanical forces, resulting in an inaccurate representation of in vivo microenvironments. Another significant challenge lies in the replication of organ-level functions. Although more advanced in vitro models such as organoids aim to mimic organ structures, achieving the full functionality of entire organs remains a challenging task. Additionally, in vitro models lack a comprehensive representation of the immune system, which poses challenges in accurately studying immune responses (Białkowska et al. 2020; Hofer and Lutolf 2021; Cardoso et al. 2023). Although animal models have long been an established method for drug testing in the pharmaceutical sector, their use in clinical research is hindered by limited relevance to human biology, as well as high costs, time consumption, and ethical concerns. Exploring alternative approaches to animal models has been a subject of continuous investigation in recent years and is expected to be further encouraged in the coming years as a result of the of the FDA Modernization Act 2.0. This legislation, adopted on December 29, 2022, permits the use of specific alternatives to animal testing, contradicting the Federal Food, Drug, and Cosmetic Act of 1938, which previously required animal testing in all new drug development protocols (Zushin et al. 2023).

In response to the limitations of traditional drug testing approaches, organ-on-a-chip technology presents a promising opportunity to study human physiology, explore novel in vitro disease models, and replicate tissues and complex cell–cell interactions through engineered microfluidic platforms, with the aim of accelerating drug development. Organ-on-a-chip devices act as platforms for simulating the in vivo environment of living tissues and organs, providing more physiologically appropriate in vitro organ models by utilizing the principles of microfluidics. Microfluidic technology has proven to be an exceptional tool for simulating human organs by cultivating living cells in micrometer-scale chambers that are continuously perfused to replicate the physiological processes of tissues and organs (Azizipour et al. 2020; Bhatia and Ingber 2014).

The basic idea is to accurately replicate essential features of tissue and organ physiology, allowing scientists to study disease progression, assess drug responses, and develop personalized therapies. Current studies are focused on creating models that closely mimic human physiology, leading to the development of advanced devices such as lung-on-a-chip, liver-on-a-chip, kidney-on-a-chip, and heart-on-a-chip, which provide more accurate platforms for research and drug testing (Jain et al. 2018b; Dauth et al. 2017; Deng et al. 2020; Yin et al. 2020; Liu et al. 2020).

To develop effective treatments, it is essential to first comprehend the complex mechanisms underlying various diseases. Understanding different organ dysfunctions is also valuable for designing models for complex disorders. This approach is referred to as disease-on-a-chip. Microfluidic devices with enormous medicinal potential might give a better understanding of disease mechanisms in a more precise and controlled way. Multiple conditions impacting the functionality of various organs or tissues have been reported in the literature in this context utilizing microengineered platforms. Epilepsy (Liu et al. 2019), lung inflammation (Os et al. 2023), ischemic stroke (Wevers et al. 2021), nonalcoholic fatty liver disease (Du et al. 2021), and intestinal inflammation (Gjorevski et al. 2020) are just a few examples of disorders that have been recreated. Additionally, other models have been developed to investigate the complex tumor microenvironment and cancer-related pathophysiological events, aiming to stimulate the development of novel therapeutic options and ultimately improve cancer survival rates (Cauli et al. 2023; Baka et al. 2023).

Another fascinating area of research in organ-on-a-chip platforms is their potential to revolutionize personalized medicine. This can be achieved by facilitating the testing of pharmaceuticals and therapies on cells or tissues that are specific to individual patients. Consequently, this advancement can result in more precise predictions regarding the effectiveness of drugs and their associated side effects. The application of personalized platforms can extend drug discovery, toxicology testing, and the modeling of rare diseases, predicting individual responses to drugs and optimizing therapy effectiveness while minimizing adverse effects (Berg et al. 2019; Ingber 2022).

Selecting appropriate materials is a critical factor in the development process of organ-on-a-chip platforms. These microfluidic platforms can be complex devices with multiple components, one of the most crucial being the microfluidic channels. These channels are created from materials such as glass, thermoplastics, elastomers, or hydrogels, using various microfabrication techniques like photolithography and soft lithography, or advanced 3D printing technologies, including stereolithography, fused deposition modeling, and 3D bioprinting (Cho et al. 2023; Leung et al. 2022). Biopolymers used in the development of bulk microfluidic devices include PLA, PCL, poly (ethylene glycol) (PEG), and hydrogels based on gelatin, alginate, and hyaluronic acid. While hydrogels are less commonly involved in the fabrication of the primary organ-on-a-chip device, they are widely employed as scaffolding materials, filling the channels of prefabricated microfluidic devices to recreate the desired microenvironment for cells. Biopolymer-based hydrogels can be designed to mimic the native extracellular matrix environment, and several properties can be tailored to influence cell behavior and organization for specific applications. In this context, cells can be embedded in collagen, fibrin, alginate, or PEG matrices, which provide structural support, promote cell adhesion, and guide differentiation, depending on the properties and composition of the hydrogel matrix (Campbell et al. 2021; Terrell et al. 2020; Kutluk et al. 2023).

Besides serving as a 3D matrix, biopolymers are also considered for use as coatings on surfaces that come into contact with cells to ensure biocompatibility and enhance cell adhesion (Shakeri et al. 2023). Additionally, platforms frequently include porous membranes that support cell growth and facilitate compartmental separation within the microfluidic system. These membranes enable the regulated flow of gases and essential nutrients between distinct compartments, enhancing biomimicry by replicating the barrier functions and selective permeability of biological membranes, thereby allowing for a more accurate simulation of physiological processes. Organ-on-a-chip models often utilize semipermeable thin films made from biopolymers like collagen, PLGA, chitosan, or polycaprolactone, each characterized by a specific porosity tailored for optimal performance (Corral-Nájera et al. 2023). Figure 15.3 illustrates the organizational level of an organ-on-a-chip platform, which additionally integrates biosensors, attached pumps for fluid control, and external read-out equipment for monitoring.

Examples from the literature highlight the successful incorporation of various biopolymers into organ-on-a-chip platforms for a wide range of biomedical applications. Many studies report the use of biopolymers, and some notable examples include the use of a collagen membrane in a lung airway-on-a-chip model. This setup facilitates the study of respiratory disease mechanisms and therapeutic testing, replicating a lung epithelial barrier with bidirectional airflow and a glycocalyx to closely mimic in vivo conditions (Park et al. 2023). Another example involves fibrin hydrogel serving as a scaffold in a heart-on-chip platform, which models human myocardium. This system has proven to be a valuable preclinical tool for cardiac safety screening, enhancing research on cardiovascular health and drug testing (Visone et al. 2022). Additionally, collagen hydrogel is employed as a matrix in ischemia-on-chip models, enabling the development of more physiologically relevant co-culture systems. These models are particularly useful for studying processes such as angiogenesis, wound

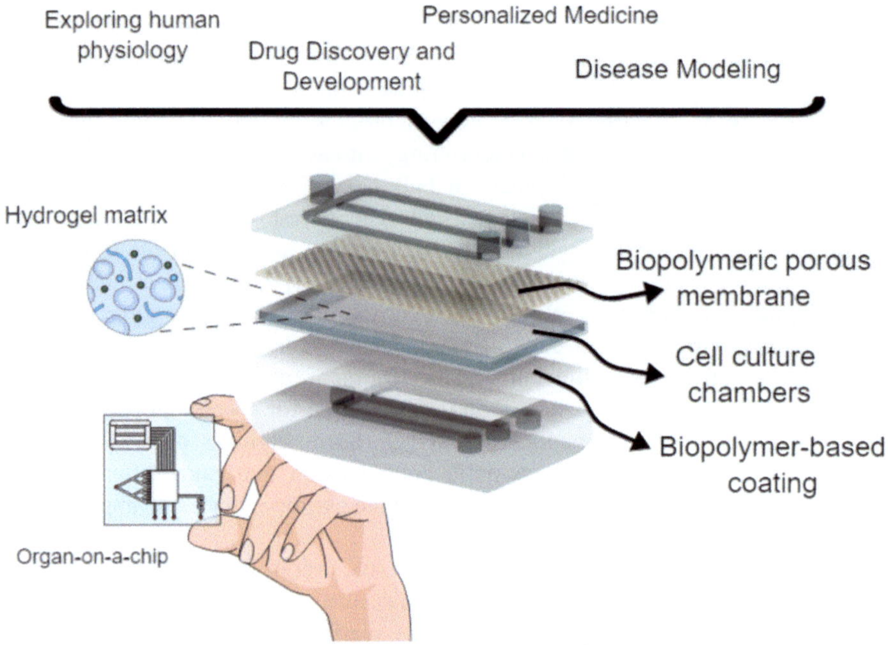

Fig. 15.3 Integration of biopolymeric materials in an organ-on-a-chip platform

healing, tumor microenvironments, and ischemic conditions (González-Lana et al. 2023).

15.4 Conclusions

The future of therapeutics can be significantly influenced by the use of biopolymers, with emerging technologies and novel biopolymer formulations paving the way for new approaches in disease prevention, targeted treatments, personalized therapeutic options, and curative therapies.

Looking to the future of therapeutics, there is a significant shift from conventional approaches toward precision medicine, where therapies are tailored to individual genetic profiles and health conditions, enabling targeted treatments that improve patient outcomes and minimize side effects.

As a general observation, there is a common trend in therapeutic approaches to employ composite and hybrid materials with multifunctional properties. Following this trend, biopolymers are proving to be promising materials for advancing the therapeutic field, not only by acting as carriers of therapeutic agents and scaffolds in regenerative medicine but also by providing reliable in vitro models for drug testing and disease modeling to speed up the drug development process while reducing reliance on animal testing.

Moreover, the integration of biopolymers with advanced technologies like electrospinning, 3D bioprinting, and machine learning is expected to bring about significant progress in the therapeutic field.

Although challenges persist at different levels, these developments have the potential to transform current treatment paradigms and promote interdisciplinary collaboration in the quest for innovative solutions to address unmet medical needs.

Acknowledgements This work was supported by a grant from the Ministry of Research, Innovation and Digitization, CNCS/CCCDI–UEFISCDI, Project Number PN-III-P4-ID-PCE-2020-1448, within PNCDI III.

References

Abdulhussain R, Adebisi A, Conway BR, Asare-Addo K (2023) Electrospun nanofibers: exploring process parameters, polymer selection, and recent applications in pharmaceuticals and drug delivery. J Drug Deliv Sci Technol 90:105156. https://doi.org/10.1016/j.jddst.2023.105156

Al Mahmud MZ (2023) Exploring the versatile applications of biocomposites in the medical field. Bioprinting. 36:e00319. https://doi.org/10.1016/j.bprint.2023.e00319

Al Meslamani AZ (2023) The future of precision medicine in oncology. Expert Rev Precis Med Drug Dev 8(1):43–47. https://doi.org/10.1080/23808993.2023.2292988

Al-Enizi AM, Zagho MM, Elzatahry AA (2018) Polymer-based electrospun nanofibers for biomedical applications. Nanomaterials (Basel) 8(4). https://doi.org/10.3390/nano8040259

Alhazami AJZ, Pouresmaeil V, Homayouni TM (2024) The Methylurolithin A-loaded PLGA-folate-chitosan nanoparticles (Mu-PFCNPs), as the novel safe selective anti-colon cancer drug delivery system. J Cluster Sci 35(6):1719–1730. https://doi.org/10.1007/s10876-024-02618-9

Altyar AE, El-Sayed A, Abdeen A, Piscopo M, Mousa SA, Najda A et al (2023) Future regenerative medicine developments and their therapeutic applications. Biomed Pharmacother 158:114131. https://doi.org/10.1016/j.biopha.2022.114131

Anassi E, Ndefo UA (2011) Sipuleucel-T (provenge) injection: the first immunotherapy agent (vaccine) for hormone-refractory prostate cancer. Pharm Therapeutics 36(4):197–202

Ashford M (2020) Drug delivery—The increasing momentum. Drug Deliv Transl Res 10(6):1888–1894. https://doi.org/10.1007/s13346-020-00858-6

Azizipour N, Avazpour R, Rosenzweig DH, Sawan M, Ajji A (2020) Evolution of biochip technology: a review from lab-on-a-chip to organ-on-a-chip. Micromachines 11(6):599

Bae YH, Park K (2020) Advanced drug delivery 2020 and beyond: perspectives on the future. Adv Drug Deliv Rev 158:4–16. https://doi.org/10.1016/j.addr.2020.06.018

Baka Z, Stiefel M, Figarol A, Godier C, Mallick A, Joubert O et al (2022) Cancer-on-chip technology: current applications in major cancer types, challenges and future prospects. Progr Biomed Eng 4. https://doi.org/10.1088/2516-1091/ac8259

Balakrishnan P, Geethamma VG, Sreekala MS, Thomas S (2018) Polymeric biomaterials: State-of-the-art and new challenges. In: Thomas S, Balakrishnan P, Sreekala MS (eds) Fundamental biomaterials: polymers. Woodhead Publishing, pp 1–20

Baljon JJ, Kwiatkowski AJ, Pagendarm HM, Stone PT, Kumar A, Bharti V et al (2024) A cancer nanovaccine for co-delivery of peptide neoantigens and optimized combinations of STING and TLR4 agonists. ACS Nano 18(9):6845–6862. https://doi.org/10.1021/acsnano.3c04471

Baranwal J, Barse B, Fais A, Delogu GL, Kumar A (2022) Biopolymer: a sustainable material for food and medical applications. Polymers (Basel) 14(5). https://doi.org/10.3390/polym14050983

Bartolo P, Malshe A, Ferraris E, Koc B (2022) 3D bioprinting: materials, processes, and applications. CIRP Ann 71(2):577–597. https://doi.org/10.1016/j.cirp.2022.06.001

Baryakova TH, Pogostin BH, Langer R, McHugh KJ (2023) Overcoming barriers to patient adherence: the case for developing innovative drug delivery systems. Nat Rev Drug Discov 22(5):387–409. https://doi.org/10.1038/s41573-023-00670-0

Beach MA, Nayanathara U, Gao Y, Zhang C, Xiong Y, Wang Y et al (2024) Polymeric nanoparticles for drug delivery. Chem Rev 124(9):5505–5616. https://doi.org/10.1021/acs.chemrev.3c00705

Bhatia SN, Ingber DE (2014) Microfluidic organs-on-chips. Nat Biotechnol 32(8):760–772. https://doi.org/10.1038/nbt.2989

Białkowska K, Komorowski P, Bryszewska M, Miłowska K (2020) Spheroids as a type of three-dimensional cell cultures—Examples of methods of preparation and the most important application. Int J Mol Sci 21(17):6225

Biswas MC, Jony B, Nandy PK, Chowdhury RA, Halder S, Kumar D et al (2022) Recent advancement of biopolymers and their potential biomedical applications. J Polym Environ 30(1):51–74. https://doi.org/10.1007/s10924-021-02199-y

Bose RJC, Kim M, Chang JH, Paulmurugan R, Moon JJ, Koh W-G et al (2019) Biodegradable polymers for modern vaccine development. J Ind Eng Chem 77:12–24. https://doi.org/10.1016/j.jiec.2019.04.044

Campbell SB, Wu Q, Yazbeck J, Liu C, Okhovatian S, Radisic M (2021) Beyond polydimethylsiloxane: alternative materials for fabrication of organ-on-a-chip devices and microphysiological systems. ACS Biomater Sci Eng 7(7):2880–2899. https://doi.org/10.1021/acsbiomaterials.0c00640

Capella-Monsonís H, Crum RJ, Hussey GS, Badylak SF (2024) Advances, challenges, and future directions in the clinical translation of ECM biomaterials for regenerative medicine applications. Adv Drug Deliv Rev 211:115347. https://doi.org/10.1016/j.addr.2024.115347

Cardoso BD, Castanheira EMS, Lanceros-Méndez S, Cardoso VF (2023) Recent advances on cell culture platforms for in vitro drug screening and cell therapies: from conventional to microfluidic strategies. Adv Healthc Mater 12(18):2202936. https://doi.org/10.1002/adhm.202202936

Cascone S, Lamberti G (2020) Hydrogel-based commercial products for biomedical applications: a review. Int J Pharm 573:118803. https://doi.org/10.1016/j.ijpharm.2019.118803

Cassaidy BJ, Moser BA, Solanki A, Chen Q, Shen J, Gotsis K et al (2024) Immune potentiation of PLGA controlled-release vaccines for improved immunological outcomes. ACS Omega 9(10):11608–11614. https://doi.org/10.1021/acsomega.3c06552

Cauli E, Polidoro MA, Marzorati S, Bernardi C, Rasponi M, Lleo A (2023) Cancer-on-chip: a 3D model for the study of the tumor microenvironment. J Biol Eng 17(1):53. https://doi.org/10.1186/s13036-023-00372-6

Chandra DK, Reis RL, Kundu SC, Kumar A, Mahapatra C (2024) Nanomaterials-based hybrid bioink platforms in advancing 3d bioprinting technologies for regenerative medicine. ACS Biomater Sci Eng 10(7):4145–4174. https://doi.org/10.1021/acsbiomaterials.4c00166

Chang TM (2000) Artificial cell biotechnology for medical applications. Blood Purif 18(2):91–96. https://doi.org/10.1159/000014430

Chen W-H, Chen Q-W, Chen Q, Cui C, Duan S, Kang Y et al (2022) Biomedical polymers: synthesis, properties, and applications. Sci China Chem 65(6):1010–1075. https://doi.org/10.1007/s11426-022-1243-5

Chen XB, Fazel Anvari-Yazdi A, Duan X, Zimmerling A, Gharraei R, Sharma NK et al (2023) Biomaterials/bioinks and extrusion bioprinting. Bioact Mater 28:511–536. https://doi.org/10.1016/j.bioactmat.2023.06.006

Chen Z, Guo Z, Hu T, Huang B, Zheng Q, Du X et al (2024) Double-layered microneedle patch loaded with bioinspired nano-vaccine for melanoma treatment and wound healing. Int J Biol Macromol 262:129961. https://doi.org/10.1016/j.ijbiomac.2024.129961

Cho S, Lee S, Ahn SI (2023) Design and engineering of organ-on-a-chip. Biomed Eng Lett 13(2):97–109. https://doi.org/10.1007/s13534-022-00258-4

Corral-Nájera K, Chauhan G, Serna-Saldívar SO, Martínez-Chapa SO, Aeinehvand MM (2023) Polymeric and biological membranes for organ-on-a-chip devices. Microsyst Nanoeng 9(1):107. https://doi.org/10.1038/s41378-023-00579-z

Cui X, Li J, Hartanto Y, Durham M, Tang J, Zhang H et al (2020) Advances in extrusion 3D bioprinting: a focus on multicomponent hydrogel-based bioinks. Adv Healthc Mater 9(15):1901648. https://doi.org/10.1002/adhm.201901648

D'Amico C, Fusciello M, Hamdan F, D'Alessio F, Bottega P, Saklauskaite M et al (2025) Transdermal delivery of PeptiCRAd cancer vaccine using microneedle patches. Bioact Mater 45:115–127. https://doi.org/10.1016/j.bioactmat.2024.11.006

Das A, Ringu T, Ghosh S, Pramanik N (2023) A comprehensive review on recent advances in preparation, physicochemical characterization, and bioengineering applications of biopolymers. Polym Bull 80(7):7247–7312. https://doi.org/10.1007/s00289-022-04443-4

Dauth S, Maoz BM, Sheehy SP, Hemphill MA, Murty T, Macedonia MK et al (2017) Neurons derived from different brain regions are inherently different in vitro: a novel multiregional brain-on-a-chip. J Neurophysiol 117(3):1320–1341. https://doi.org/10.1152/jn.00575.2016

Deng J, Cong Y, Han X, Wei W, Lu Y, Liu T et al (2020) A liver-on-a-chip for hepatoprotective activity assessment. Biomicrofluidics 14(6):064107. https://doi.org/10.1063/5.0024767

Di Francesco D, Pigliafreddo A, Casarella S, Di Nunno L, Mantovani D, Boccafoschi F (2023) Biological materials for tissue-engineered vascular grafts: overview of recent advancements. Biomolecules 13(9):1389

Dong Y, Fu S, Yu J, Li X, Ding B (2024) Emerging smart micro/nanofiber-based materials for next-generation wound dressings. Adv Func Mater 34(9):2311199. https://doi.org/10.1002/adfm.202311199

Du K, Li S, Li C, Li P, Miao C, Luo T et al (2021) Modeling nonalcoholic fatty liver disease on a liver lobule chip with dual blood supply. Acta Biomater 134:228–239. https://doi.org/10.1016/j.actbio.2021.07.013

Dzobo K, Thomford NE, Senthebane DA, Shipanga H, Rowe A, Dandara C et al (2018) Advances in regenerative medicine and tissue engineering: innovation and transformation of medicine. Stem Cells Int 2018(1):2495848. https://doi.org/10.1155/2018/2495848

Ejiohuo O (2023) A perspective on the synergistic use of 3D printing and electrospinning to improve nanomaterials for biomedical applications. Nano Trends 4:100025. https://doi.org/10.1016/j.nwnano.2023.100025

Elhadad AA, Rosa-Sainz A, Cañete R, Peralta E, Begines B, Balbuena M et al (2023) Applications and multidisciplinary perspective on 3D printing techniques: recent developments and future trends. Mater Sci Eng R Rep 156:100760. https://doi.org/10.1016/j.mser.2023.100760

Elstad NL, Fowers KD (2009) OncoGel (ReGel/paclitaxel)—Clinical applications for a novel paclitaxel delivery system. Adv Drug Deliv Rev 61(10):785–794. https://doi.org/10.1016/j.addr.2009.04.010

Ezike TC, Okpala US, Onoja UL, Nwike CP, Ezeako EC, Okpara OJ et al (2023) Advances in drug delivery systems, challenges and future directions. Heliyon 9(6):e17488. https://doi.org/10.1016/j.heliyon.2023.e17488

Fatima SN (2024) Microbial biopolymers: a sustainable alternative to traditional petroleum-based polymers for a greener future. Mater Today Commun 40:109846. https://doi.org/10.1016/j.mtcomm.2024.109846

Fazal T, Murtaza BN, Shah M, Iqbal S, Rehman M-U, Jaber F et al (2023) Recent developments in natural biopolymer based drug delivery systems. RSC Adv 13(33):23087–23121. https://doi.org/10.1039/D3RA03369D

Foroozandeh P, Aziz AA (2018) Insight into cellular uptake and intracellular trafficking of nanoparticles. Nanoscale Res Lett 13(1):339. https://doi.org/10.1186/s11671-018-2728-6

Fredi G, Dorigato A (2024) Compatibilization of biopolymer blends: a review. Adv Ind Eng Polym Res 7(4):373–404. https://doi.org/10.1016/j.aiepr.2023.11.002

Fu Y, Li C, Chen C, An Q, Zhang W, Jiang Y et al (2025) Hemostatic nanofibers/chitosan composite aerogel for potential chemo-photothermal therapy. J Mol Struct 1319:139477. https://doi.org/10.1016/j.molstruc.2024.139477

Gao J, Karp JM, Langer R, Joshi N (2023) The future of drug delivery. Chem Mater 35(2):359–363. https://doi.org/10.1021/acs.chemmater.2c03003

Gao Y, Joshi M, Zhao Z, Mitragotri S (2024a) PEGylated therapeutics in the clinic. Bioeng Transl Med 9(1):e10600. https://doi.org/10.1002/btm2.10600

Gao H, Li H, Shao S, Tan L, Wang Y, Li D et al (2024b) Self-enhanced PTX@HSA-HA loaded functionalized injectable hydrogel for effective local chemo-photothermal therapy in breast cancer. Carbohydr Polym 345:122569. https://doi.org/10.1016/j.carbpol.2024.122569

George N, Chakraborty S, Mary NL, Suguna L (2024) Incorporating silver nanoparticles into electrospun nanofibers of casein/polyvinyl alcohol to develop scaffolds for tissue engineering. Int J Biol Macromol 267:131501. https://doi.org/10.1016/j.ijbiomac.2024.131501

Gjorevski N, Avignon B, Gérard R, Cabon L, Roth AB, Bscheider M et al (2020) Neutrophilic infiltration in organ-on-a-chip model of tissue inflammation. Lab Chip 20(18):3365–3374. https://doi.org/10.1039/D0LC00417K

González-Lana S, Randelovic T, Ciriza J, López-Valdeolivas M, Monge R, Sánchez-Somolinos C et al (2023) Surface modifications of COP-based microfluidic devices for improved immobilisation of hydrogel proteins: long-term 3D culture with contractile cell types and ischaemia model. Lab Chip 23(10):2434–2446. https://doi.org/10.1039/D3LC00075C

Govind Kumar S, Nirmala Rachel J (2022) Electrospinning: the technique and applications. In: Maaz K, Samson Jerold Samuel C (eds) Recent developments in nanofibers research. IntechOpen, Rijeka

Gregory DA, Taylor CS, Fricker ATR, Asare E, Tetali SSV, Haycock JW et al (2022) Polyhydroxyalkanoates and their advances for biomedical applications. Trends Mol Med 28(4):331–342. https://doi.org/10.1016/j.molmed.2022.01.007

HPS AK, Fazita M, Nurazzi N (2023) Biopolymer composites, pp 1–104

Haider A, Haider S, Kang I-K (2018) A comprehensive review summarizing the effect of electrospinning parameters and potential applications of nanofibers in biomedical and biotechnology. Arab J Chem 11(8):1165–1188. https://doi.org/10.1016/j.arabjc.2015.11.015

Hashemi P, Osanloo M, Farjadfar A, Nasiri-Ghiri M, Zarenezhad E, Mahmoodi S (2024) A multi-epitope protein vaccine encapsulated in alginate nanoparticles as a candidate vaccine against Shigella sonnei. Sci Rep 14(1):22484. https://doi.org/10.1038/s41598-024-73105-4

Hassan M, Bai J, Dou D-Q (2019) Biopolymers; definition, classification and applications. Egypt J Chem. https://doi.org/10.21608/EJCHEM.2019.6967.1580

Heid S, Boccaccini AR (2020) Advancing bioinks for 3D bioprinting using reactive fillers: a review. Acta Biomater 113:1–22. https://doi.org/10.1016/j.actbio.2020.06.040

Hiwrale A, Bharati S, Pingale P, Rajput A (2023) Nanofibers: a current era in drug delivery system. Heliyon 9(9):e18917. https://doi.org/10.1016/j.heliyon.2023.e18917

Hofer M, Lutolf MP (2021) Engineering organoids. Nat Rev Mater 6(5):402–420. https://doi.org/10.1038/s41578-021-00279-y

Hoover AR, More S, Liu K, West CL, Valerio TI, Furrer CL et al (2024) N-dihydrogalactochitosan serves as an effective mucosal adjuvant for intranasal vaccine in combination with recombinant viral proteins against respiratory infection. Acta Biomater 175:279–292. https://doi.org/10.1016/j.actbio.2023.12.039

Horta-Velázquez A, Mota-Morales JD, Morales-Narváez E (2024) Next-generation of smart dressings: integrating multiplexed sensors and theranostic functions. Int J Biol Macromol 254:127737. https://doi.org/10.1016/j.ijbiomac.2023.127737

https://www.cancer.gov/research/progress/250-years-milestones

https://www.fda.gov/media/74385/download?

https://www.fda.gov/media/75726/download

https://www.fda.gov/media/76904/download

https://www.fda.gov/vaccines-blood-biologics/vaccines/flumist

Hu L, Nomura S, Sato Y, Takagi K, Ishii T, Honma Y et al (2022) Anti-inflammatory effects of differential molecular weight Hyaluronic acids on UVB-induced calprotectin-mediated keratinocyte inflammation. J Dermatol Sci 107(1):24–31. https://doi.org/10.1016/j.jdermsci.2022.06.001

Hu Z, Zhang H, Li Z, Zhao T, Gu Z, Yuan Q et al (2024) Multifunctional photothermal hydrogels: design principles, various functions, and promising biological applications. Chin Chem Lett 35(10):109527. https://doi.org/10.1016/j.cclet.2024.109527

Huang G, Zhao Y, Chen D, Wei L, Hu Z, Li J et al (2024) Applications, advancements, and challenges of 3D bioprinting in organ transplantation. Biomater Sci 12(6):1425–1448. https://doi.org/10.1039/D3BM01934A

Ingber DE (2022) Human organs-on-chips for disease modelling, drug development and personalized medicine. Nat Rev Genet 23(8):467–491. https://doi.org/10.1038/s41576-022-00466-9

Jain A, Kumari R, Tiwari A, Verma A, Tripathi A, Shrivastava A et al (2018a) Nanocarrier based advances in drug delivery to tumor: an overview. Curr Drug Targets 19(13):1498–1518. https://doi.org/10.2174/1389450119666180131105822

Jain A, Barrile R, van der Meer AD, Mammoto A, Mammoto T, De Ceunynck K et al (2018b) Primary human lung alveolus-on-a-chip model of intravascular thrombosis for assessment of therapeutics. Clin Pharmacol Ther 103(2):332–340. https://doi.org/10.1002/cpt.742

Jain KK (2020) An overview of drug delivery systems. In: Jain KK (ed) Drug delivery systems. Springer, New York, NY, pp 1–54

Jaiswal R, Sherje AP (2024) Recent advances in biopolymer-based smart hydrogel for wound healing. J Drug Deliv Sci Technol 99:105990. https://doi.org/10.1016/j.jddst.2024.105990

Jiang W, Wu Z, Gao Z, Wan M, Zhou M, Mao C et al (2022) Artificial cells: past, present and future. ACS Nano 16(10):15705–15733. https://doi.org/10.1021/acsnano.2c06104

Jindal AB, Bhide AR, Salave S, Rana D, Benival D (2023) Long-acting parenteral drug delivery systems for the treatment of chronic diseases. Adv Drug Deliv Rev 198:114862. https://doi.org/10.1016/j.addr.2023.114862

Jo HJ, Kang MS, Kim JM, Raja IS, Jang HJ, Park KM et al (2024) Bioactive ions-loaded bioinks primed for 3D printing of artificial tissues. Biomed Mater Dev 2(2):811–833. https://doi.org/10.1007/s44174-023-00151-3

Jose J, Peter A, Thajudeen KY, Gomes Pereira MDL, VP A, Bhat SG et al (2024) Recent advances in the design and development of bioink formulations for various biomedical applications. Results Eng 22:102060. https://doi.org/10.1016/j.rineng.2024.102060

Jurak M, Wiącek AE, Ładniak A, Przykaza K, Szafran K (2021) What affects the biocompatibility of polymers? Adv Coll Interface Sci 294:102451. https://doi.org/10.1016/j.cis.2021.102451

Kalpana Manivannan R, Sharma N, Kumar V, Jayaraj I, Vimal S, Umesh M (2024) A comprehensive review on natural macromolecular biopolymers for biomedical applications: recent advancements, current challenges, and future outlooks. Carbohydr Polym Technol Appl 8:100536. https://doi.org/10.1016/j.carpta.2024.100536

Karahmet Sher E, Alebić M, Marković Boras M, Boškailo E, Karahmet Farhat E, Karahmet A et al (2024) Nanotechnology in medicine revolutionizing drug delivery for cancer and viral infection treatments. Int J Pharm 660:124345. https://doi.org/10.1016/j.ijpharm.2024.124345

Katebi A, Riazi-rad F, Varshochian R, Ajdary S (2024) PLGA nanoparticle-delivered Leishmania antigen and TLR agonists as a therapeutic vaccine against cutaneous leishmaniasis in BALB/c mice. Int Immunopharmacol 138:112538. https://doi.org/10.1016/j.intimp.2024.112538

Keirouz A, Wang Z, Reddy VS, Nagy ZK, Vass P, Buzgo M et al (2023) The history of electrospinning: past, present, and future developments. Adv Maters Technol 8(11):2201723. https://doi.org/10.1002/admt.202201723

Khoeini R, Nosrati H, Akbarzadeh A, Eftekhari A, Kavetskyy T, Khalilov R et al (2021) Natural and synthetic bioinks for 3D bioprinting. Adv NanoBiomed Res 1(8):2000097. https://doi.org/10.1002/anbr.202000097

Khoshnood N, Zamanian A (2020) A comprehensive review on scaffold-free bioinks for bioprinting. Bioprinting 19:e00088. https://doi.org/10.1016/j.bprint.2020.e00088

Kim D, Kim M, Lee J, Jang J (2022) Review on multicomponent hydrogel bioinks based on natural biomaterials for bioprinting 3D liver tissues. Front Bioeng Biotechnol 10. https://doi.org/10.3389/fbioe.2022.764682

Kopilovic B, Laroui N, Coutinho JAP, Pichon C, Freire MG (n/a) Nasal delivery of mRNA using thermoresponsive gellable aqueous biphasic systems. Small 2409527. https://doi.org/10.1002/smll.202409527

Kuang G, Lin X, Li J, Sun W, Zhang Q, Zhao Y (2024) Electrospun nanofibers-derived functional scaffolds for cancer therapy. Chem Eng J 489:151253. https://doi.org/10.1016/j.cej.2024.151253

Kumar G, Virmani T, Misra SK, Sharma A, Pathak K (2024) Exploration of chitosan and its modified derivatives as vaccine adjuvant: a review. Carbohydr Polym Technol Appl 8:100537. https://doi.org/10.1016/j.carpta.2024.100537

Kutluk H, Bastounis EE, Constantinou I (2023) Integration of extracellular matrices into organ-on-chip systems. Adv Healthc Mater 12(20):2203256. https://doi.org/10.1002/adhm.202203256

Lai CKC, Ng RWY, Leung SSY, Hui M, Ip M (2022) Overcoming the rising incidence and evolving mechanisms of antibiotic resistance by novel drug delivery approaches—An overview. Adv Drug Deliv Rev 181:114078. https://doi.org/10.1016/j.addr.2021.114078

Lai J, Liu Y, Lu G, Yung P, Wang X, Tuan RS et al (2024) 4D bioprinting of programmed dynamic tissues. Bioact Mater 37:348–377. https://doi.org/10.1016/j.bioactmat.2024.03.033

Leung CM, de Haan P, Ronaldson-Bouchard K, Kim G-A, Ko J, Rho HS et al (2022) A guide to the organ-on-a-chip. Nate Rev Methods Prim 2(1):33. https://doi.org/10.1038/s43586-022-00118-6

Li J, Zhuang S (2020) Antibacterial activity of chitosan and its derivatives and their interaction mechanism with bacteria: current state and perspectives. Eur Polymer J 138:109984. https://doi.org/10.1016/j.eurpolymj.2020.109984

Li Z, Yang J, Loh XJ (2016) Polyhydroxyalkanoates: opening doors for a sustainable future. NPG Asia Mater 8(4):e265. https://doi.org/10.1038/am.2016.48

Li T, Qian C, Gu Y, Zhang J, Li S, Xia N (2023) Current progress in the development of prophylactic and therapeutic vaccines. Sci China Life Sci 66(4):679–710. https://doi.org/10.1007/s11427-022-2230-4

Liu J, Sternberg AR, Ghiasvand S, Berdichevsky Y (2019) Epilepsy-on-a-chip system for antiepileptic drug discovery. IEEE Trans Biomed Eng 66(5):1231–1241. https://doi.org/10.1109/tbme.2018.2871415

Liu H, Bolonduro OA, Hu N, Ju J, Rao AA, Duffy BM et al (2020) Heart-on-a-chip model with integrated extra- and intracellular bioelectronics for monitoring cardiac electrophysiology under acute hypoxia. Nano Lett 20(4):2585–2593. https://doi.org/10.1021/acs.nanolett.0c00076

Liu J, Cabral H, Mi P (2024a) Nanocarriers address intracellular barriers for efficient drug delivery, overcoming drug resistance, subcellular targeting and controlled release. Adv Drug Deliv Rev 207:115239. https://doi.org/10.1016/j.addr.2024.115239

Liu Z, Kabir MT, Chen S, Zhang H, Wakim LM, Rehm BHA (2024b) Intranasal epitope-polymer vaccine lodges resident memory T cells protecting against influenza virus. Adv Healthc Mater. 13(15):e2304188. https://doi.org/10.1002/adhm.202304188

Luan X, Hu H, Zhu D, He P, Sun Z, Xi Y et al (2024) Injectable chitosan hydrogels doped with 2D peptide nanosheet-drug conjugates for glutathione-responsive sustained drug delivery. Chem Eur J 30(28):e202400021. https://doi.org/10.1002/chem.202400021

Ma L, Zheng X, Lin R, Sun AR, Song J, Ye Z et al (2022) Knee osteoarthritis therapy: recent advances in intra-articular drug delivery systems. Drug des Devel Ther. 16:1311–1347. https://doi.org/10.2147/dddt.S357386

Mahaling B, Roy C, Ghosh S (2024) Silk–gelatin hybrid hydrogel: a potential carrier for RNA therapeutics. J Mater Chem B 12(25):6203–6220. https://doi.org/10.1039/D4TB00491D

Mahdizadeh N, Khorshid Shabestari M, Tafvizi F, Khodarahmi P (2024) Delivery of letrozole-encapsulated niosomes via a 3D bioprinting gelatin–alginate scaffold for potential breast cancer treatment. Cancer Nanotechnol 15(1):33. https://doi.org/10.1186/s12645-024-00271-5

Mansouri S, Alharbi Y, Alqahtani A (2024) Current status and prospects for improved targeted delivery approaches for cancer. Pathol Res Pract 253:154993. https://doi.org/10.1016/j.prp.2023.154993

Mehata AK, Bonlawar J, Tamang R, Malik AK, Setia A, Kumar S et al (2024) PLGA nanoplatform for the hypoxic tumor delivery: folate targeting, therapy, and ultrasound/photoacoustic imaging. ACS Appl Bio Mater 7(8):5754–5770. https://doi.org/10.1021/acsabm.4c00853

Mitchell MJ, Billingsley MM, Haley RM, Wechsler ME, Peppas NA, Langer R (2021) Engineering precision nanoparticles for drug delivery. Nat Rev Drug Discov 20(2):101–124. https://doi.org/10.1038/s41573-020-0090-8

Mitropoulou A, Markatos DN, Dimopoulos A, Marazioti A, Mikelis C-M, Mavrilas D (2024) Development and evaluation of biodegradable core-shell microfibrous and nanofibrous scaffolds for tissue engineering applications. J Mater Sci Mater Med 35(1):10. https://doi.org/10.1007/s10856-024-06777-z

Mohite P, Yadav V, Pandhare R, Maitra S, Saleh FM, Saleem RM et al (2024) Revolutionizing cancer treatment: unleashing the power of viral vaccines, monoclonal antibodies, and proteolysis-targeting chimeras in the new era of immunotherapy. ACS Omega 9(7):7277–7295. https://doi.org/10.1021/acsomega.3c06501

Mooss VA (2024) Different types of fiber in polymer matrices for biomedical engineering. In: Parameswaranpillai J, Ganguly S, Das P, Gopi JA (eds) Fiber and ceramic filler-based polymer composites for biomedical engineering. Springer Nature, Singapore, pp 45–61

Moradali MF, Rehm BHA (2020) Bacterial biopolymers: from pathogenesis to advanced materials. Nat Rev Microbiol 18(4):195–210. https://doi.org/10.1038/s41579-019-0313-3

Mtibe A, Motloung MP, Bandyopadhyay J, Ray SS (2021) Synthetic biopolymers and their composites: advantages and limitations—An overview. Macromol Rapid Commun 42(15):2100130. https://doi.org/10.1002/marc.202100130

Mwiiri FK, Daniels R (2020) Chapter 3—Electrospun nanofibers for biomedical applications. In: Shegokar R (ed) Delivery of drugs. Elsevier, pp 53–74

Nagaraja K, Dhokare P, Bhattacharyya A, Noh I (2024) Recent advances in 3D bioprinting of polysaccharide-based bioinks for fabrication of bioengineered tissues. Mol Syst Des Eng 9(10):977–999. https://doi.org/10.1039/D4ME00001C

Nazeer N, Ahmed M (2020) Chapter 13—Polymers in medicine. In: Narain R (ed) Polymer science and nanotechnology. Elsevier, pp 281–323

Nevagi RJ, Skwarczynski M, Toth I (2019) Polymers for subunit vaccine delivery. Eur Polymer J 114:397–410. https://doi.org/10.1016/j.eurpolymj.2019.03.009

Nguyen HM, Le Ngoc TT, Nguyen AT, Le Thien HN, Pham TT (2023) Biomedical materials for wound dressing: recent advances and applications. RSC Adv 13(8):5509–5528. https://doi.org/10.1039/D2RA07673J

Paik J, Duggan ST, Keam SJ (2019) Triamcinolone Acetonide extended-release: a review in osteoarthritis pain of the knee. Drugs 79(4):455–462. https://doi.org/10.1007/s40265-019-01083-3

Palivan CG, Heuberger L, Gaitzsch J, Voit B, Appelhans D, Borges Fernandes B et al (2024) Advancing artificial cells with functional compartmentalized polymeric systems—In Honor of Wolfgang Meier. Biomacromolecules 25(9):5454–5467. https://doi.org/10.1021/acs.biomac.4c00769

Pansuriya R, Patel T, Singh K, Al Ghamdi A, Kasoju N, Kumar A et al (2024) Self-healable, stimuli-responsive bio-ionic liquid and sodium alginate conjugated hydrogel with tunable Injectability and mechanical properties for the treatment of breast cancer. Int J Biol Macromol 277:134112. https://doi.org/10.1016/j.ijbiomac.2024.134112

Park H, Otte A, Park K (2022) Evolution of drug delivery systems: from 1950 to 2020 and beyond. J Control Release 342:53–65. https://doi.org/10.1016/j.jconrel.2021.12.030

Park S, Newton J, Hidjir T, Young EWK (2023) Bidirectional airflow in lung airway-on-a-chip with matrix-derived membrane elicits epithelial glycocalyx formation. Lab Chip 23(16):3671–3682. https://doi.org/10.1039/D3LC00259D

Patel DK, Won S-Y, Jung E, Han SS (2025) Recent progress in biopolymer-based electrospun nanofibers and their potential biomedical applications: a review. Int J Biol Macromol 293:139426. https://doi.org/10.1016/j.ijbiomac.2024.139426
Pathan N, Shende P (2021) Strategic conceptualization and potential of self-healing polymers in biomedical field. Mater Sci Eng C 125:112099. https://doi.org/10.1016/j.msec.2021.112099
Poria R, Kala D, Nagraik R, Dhir Y, Dhir S, Singh B et al (2024) Vaccine development: current trends and technologies. Life Sci 336:122331. https://doi.org/10.1016/j.lfs.2023.122331
Pounraj S, Chen S, Triscott H, Lam AK, Ma L, Rehm BHA (2024) Engineering neoantigens to form immunogenic biopolymer particles targeting metastatic breast cancer. Appl Mater Today 38:102238. https://doi.org/10.1016/j.apmt.2024.102238
Pourmadadi M, Poorkhalili P, Sorourian M, Sorourian G, Ghaderi R, Mehrabi MG et al (2024) The smart nanocarrier containing zein/starch co-biopolymers enhanced by graphitic carbon nitride; exploring opportunities in brain cancer treatment. Int J Biol Macromol 274:133275. https://doi.org/10.1016/j.ijbiomac.2024.133275
Qin D, Wenyong H, Dengke S, Longyi C, Zeyu Y, Boyang W et al (2024) GelMA microneedle-loaded bio-derived nanovaccine shows therapeutic potential for gliomas. Sci Technol Adv Mater 25(1):2426444. https://doi.org/10.1080/14686996.2024.2426444
Qu Z, Wang Y, Dong Y, Li X, Hao L, Sun L et al (2024) Intelligent electrospinning nanofibrous membranes for monitoring and promotion of wound healing. Mater Today Bio 26:101093. https://doi.org/10.1016/j.mtbio.2024.101093
Raees S, Ullah F, Javed F, Akil HM, Jadoon Khan M, Safdar M et al (2023) Classification, processing, and applications of bioink and 3D bioprinting: a detailed review. Int J Biol Macromol 232:123476. https://doi.org/10.1016/j.ijbiomac.2023.123476
Raeisi A, Farjadian F (2024) Commercial hydrogel product for drug delivery based on route of administration. Front Chem 12:1336717. https://doi.org/10.3389/fchem.2024.1336717
Rahimnejad M, Jahangiri S, Zirak Hassan Kiadeh S, Rezvaninejad S, Ahmadi Z, Ahmadi S et al (2024) Stimuli-responsive biomaterials: smart avenue toward 4D bioprinting. Crit Rev Biotechnol 44(5), 860–891. https://doi.org/10.1080/07388551.2023.2213398
Rahman M, Hasan MR (2019) Synthetic biopolymers. In: Jafar Mazumder MA, Sheardown H, Al-Ahmed A (eds) Functional biopolymers. Springer International Publishing, Cham, pp 1–43
Rai C-I, Kuo T-H, Chen Y-C (2024) Novel administration routes, delivery vectors, and application of vaccines based on biotechnologies: a review. Vaccines 12(9):1002
Rasouli R, Sweeney C, Frampton JP (2024) Heterogeneous and composite bioinks for 3D-bioprinting of complex tissue. Biomed Mater Dev. https://doi.org/10.1007/s44174-024-00171-7
Rizk M, Zou L, Savic R, Dooley K (2017) Importance of drug pharmacokinetics at the site of action. Clin Transl Sci 10(3):133–142. https://doi.org/10.1111/cts.12448
Rowe R, Sheskey P, Owen S (2006) Hand book of pharmaceutical excipients. American Pharmaceutical Association, Pharmaceutical Press, London, p 918
Sadeghianmaryan A, Ahmadian N, Wheatley S, Alizadeh Sardroud H, Nasrollah SAS, Naseri E et al (2024) Advancements in 3D-printable polysaccharides, proteins, and synthetic polymers for wound dressing and skin scaffolding—A review. Int J Biol Macromol 266:131207. https://doi.org/10.1016/j.ijbiomac.2024.131207
Sanchez Armengol E, Unterweger A, Laffleur F (2022) PEGylated drug delivery systems in the pharmaceutical field: past, present and future perspective. Drug Dev Ind Pharm 48(4):129–139. https://doi.org/10.1080/03639045.2022.2101062
Sanchez Armengol E, Hock N, Saribal S, To D, Summonte S, Veider F et al (2024) Unveiling the potential of biomaterials and their synergistic fusion in tissue engineering. Eur J Pharm Sci 196:106761. https://doi.org/10.1016/j.ejps.2024.106761
Sang Z, Zhu T, Qu X, Zhang Z, Wang W, Hao Y (2025) A hyaluronic acid-based dissolving microneedle patch loaded with 5-aminolevulinic acid for improved oral leukoplakia treatment. Colloids Surf B 245:114216. https://doi.org/10.1016/j.colsurfb.2024.114216

Schwab A, Levato R, D'Este M, Piluso S, Eglin D, Malda J (2020) Printability and Shape Fidelity of Bioinks in 3D Bioprinting. Chem Rev 120(19):11028–11055. https://doi.org/10.1021/acs.chemrev.0c00084

Sessini V, Ghosh S, Mosquera MEG (2023) Chapter 1—Introduction to biopolymer synthesis, properties, and emerging applications. In: Sessini V, Ghosh S, Mosquera MEG (eds) Biopolymers. Elsevier, pp 1–20

Shakeri A, Jarad NA, Khan S, F Didar T (2022) Bio-functionalization of microfluidic platforms made of thermoplastic materials: a review. Analytica Chimica Acta 1209:339283. https://doi.org/10.1016/j.aca.2021.339283

Shao W, Yang Y, Shen W, Ren L, WenwenWang, Zhu P (2024) Hyaluronic acid-conjugated methotrexate and 5-fluorouracil for targeted drug delivery. Int J Biol Macromol 273:132671. https://doi.org/10.1016/j.ijbiomac.2024.132671

Sharma KR, Malik A, Roof RA, Boyce JP, Verma SK (2024) New approaches for challenging therapeutic targets. Drug Discov Today 29(4):103942. https://doi.org/10.1016/j.drudis.2024.103942

Shiroud Heidari B, Ruan R, Vahabli E, Chen P, De-Juan-Pardo EM, Zheng M et al (2023) Natural, synthetic and commercially-available biopolymers used to regenerate tendons and ligaments. Bioact Mater 19:179–197. https://doi.org/10.1016/j.bioactmat.2022.04.003

Singh TV, Shagolsem LS (2021) Biopolymer based nano-structured materials and their applications. In: Swain BP (ed) Nanostructured materials and their applications. Springer, Singapore, pp 337–366

Sofini SP, Balasubramanian D, Girigoswami A, Girigoswami K (2024) Biomedical applications of natural and synthetic polymer based nanocomposites. J Biomater Sci Polym Ed 35(2):269–294. https://doi.org/10.1080/09205063.2023.2283910

Song Y, Hu Q, Liu S, Yao G, Zhang H (2024) Electrospinning drug-loaded polycaprolactone/polycaprolactone-gelatin multi-functional bilayer nanofibers composite scaffold for postoperative wound healing of cutaneous squamous cell carcinoma. Biomed Technol 8:65–80. https://doi.org/10.1016/j.bmt.2024.10.001

Sun L, Liu H, Ye Y, Lei Y, Islam R, Tan S et al (2023) Smart nanoparticles for cancer therapy. Signal Transduct Target Ther 8(1):418. https://doi.org/10.1038/s41392-023-01642-x

Suriyaamporn P, Dechsri K, Charoenying T, Ngawhirunpat T, Rojanarata T, Patrojanasophon P et al (2024) Multiple strategies approach: a novel crosslinked hydrogel forming chitosan-based microneedles chemowrap patch loaded with 5-fluorouracil liposomes for chronic wound cancer treatment. Int J Biol Macromol 279:134973. https://doi.org/10.1016/j.ijbiomac.2024.134973

Tang M, Jiang S, Huang X, Ji C, Gu Y, Qi Y et al (2024) Integration of 3D bioprinting and multi-algorithm machine learning identified glioma susceptibilities and microenvironment characteristics. Cell Discov 10(1):39. https://doi.org/10.1038/s41421-024-00650-7

Terrell JA, Jones CG, Kabandana GKM, Chen C (2020) From cells-on-a-chip to organs-on-a-chip: scaffolding materials for 3D cell culture in microfluidics. J Mater Chem B 8(31):6667–6685. https://doi.org/10.1039/D0TB00718H

Thapa RK, Kim JO (2023) Nanomedicine-based commercial formulations: current developments and future prospects. J Pharm Investig 53(1):19–33. https://doi.org/10.1007/s40005-022-00607-6

Tian Y, Hu D, Li Y, Yang L (2022) Development of therapeutic vaccines for the treatment of diseases. Mol Biomed 3(1):40. https://doi.org/10.1186/s43556-022-00098-9

Tiwari G, Tiwari R, Sriwastawa B, Bhati L, Pandey S, Pandey P et al (2012) Drug delivery systems: an updated review. Int J Pharm Investig 2(1):2–11. https://doi.org/10.4103/2230-973x.96920

Torabi S, Hassanzadeh-Tabrizi SA (2024) Effective antibacterial agents in modern wound dressings: a review. Biofouling 40(5–6):305–332. https://doi.org/10.1080/08927014.2024.2358913

Tören E, Buzgo M, Mazari AA, Khan MZ (2024) Recent advances in biopolymer based electrospun nanomaterials for drug delivery systems. Polym Adv Technol 35(3):e6309. https://doi.org/10.1002/pat.6309

Tripathi S, Mandal SS, Bauri S, Maiti P (2023) 3D bioprinting and its innovative approach for biomedical applications. MedComm. 4(1):e194. https://doi.org/10.1002/mco2.194

Tsimberidou AM, Fountzilas E, Nikanjam M, Kurzrock R (2020) Review of precision cancer medicine: evolution of the treatment paradigm. Cancer Treat Rev 86. https://doi.org/10.1016/j.ctrv.2020.102019

Udayakumar GP, Muthusamy S, Selvaganesh B, Sivarajasekar N, Rambabu K, Banat F et al (2021) Biopolymers and composites: properties, characterization and their applications in food, medical and pharmaceutical industries. J Environ Chem Eng 9(4):105322. https://doi.org/10.1016/j.jece.2021.105322

Ugya AY, Pohan L, Wang Q, Meguellati K (2022) Design of artificial cells: artificial biochemical systems, their thermodynamics and kinetics properties. Egypt J Basic Appl Sci 9(1):393–412. https://doi.org/10.1080/2314808X.2022.2104071

Upadhye M, Kuchekar M, Pujari R, Sable N (2022) Biopolymers: a comprehensive review. Open Access Res J Sci Technol 4:013–018. https://doi.org/10.53022/oarjst.2022.4.1.0070

van den Berg A, Mummery CL, Passier R, van der Meer AD (2019) Personalised organs-on-chips: functional testing for precision medicine. Lab Chip 19(2):198–205. https://doi.org/10.1039/C8LC00827B

van Os L, Yeoh J, Witz G, Ferrari D, Krebs P, Chandorkar Y et al (2023) Immune cell extravasation in an organ-on-chip to model lung inflammation. Eur J Pharm Sci 187:106485. https://doi.org/10.1016/j.ejps.2023.106485

Vanaei S, Parizi MS, Vanaei S, Salemizadehparizi F, Vanaei HR (2021) An overview on materials and techniques in 3D bioprinting toward biomedical application. Eng Regen 2:1–18. https://doi.org/10.1016/j.engreg.2020.12.001

Vargason AM, Anselmo AC, Mitragotri S (2021) The evolution of commercial drug delivery technologies. Nat Biomed Eng 5(9):951–967. https://doi.org/10.1038/s41551-021-00698-w

Varma K, Gopi S (2021) Chapter 7—Biopolymers and their role in medicinal and pharmaceutical applications. In: Thomas S, Gopi S, Amalraj A (eds) Biopolymers and their industrial applications. Elsevier, pp 175–191

Vasan N, Baselga J, Hyman DM (2019) A view on drug resistance in cancer. Nature 575(7782):299–309. https://doi.org/10.1038/s41586-019-1730-1

Visone R, Lozano-Juan F, Marzorati S, Rivolta MW, Pesenti E, Redaelli A et al (2022) Predicting human cardiac QT alterations and pro-arrhythmic effects of compounds with a 3D beating heart-on-chip platform. Toxicol Sci 191(1):47–60. https://doi.org/10.1093/toxsci/kfac108

Waeterschoot J, Gosselé W, Lemež Š, Casadevall i Solvas X (2024) Artificial cells for in vivo biomedical applications through red blood cell biomimicry. Nat Commun 15(1):2504. https://doi.org/10.1038/s41467-024-46732-8

Wang EY, Sarmadi M, Ying B, Jaklenec A, Langer R (2023) Recent advances in nano- and micro-scale carrier systems for controlled delivery of vaccines. Biomaterials 303:122345. https://doi.org/10.1016/j.biomaterials.2023.122345

Weerarathna IN, Doelakeh ES, Kiwanuka L, Kumar P, Arora S (2024) Prophylactic and therapeutic vaccine development: advancements and challenges. Mol Biomed 5(1):57. https://doi.org/10.1186/s43556-024-00222-x

Westensee IN, de Dios AP, Städler B (2024a) From single-compartment artificial cells to tissue-like materials. Adv Mater Technol 9(18):2301804. https://doi.org/10.1002/admt.202301804

Westensee IN, Paffen LJMM, Pendlmayr S, De Dios AP, Ramos Docampo MA, Städler B (2024b) Artificial cells and HepG2 cells in 3D-Bioprinted arrangements. Adv Healthc Mater 13(12):2303699. https://doi.org/10.1002/adhm.202303699

Wevers NR, Nair AL, Fowke TM, Pontier M, Kasi DG, Spijkers XM et al (2021) Modeling ischemic stroke in a triculture neurovascular unit on-a-chip. Fluids Barriers CNS 18(1):59. https://doi.org/10.1186/s12987-021-00294-9

Wibowo D, Jorritsma SHT, Gonzaga ZJ, Evert B, Chen S, Rehm BHA (2021) Polymeric nanoparticle vaccines to combat emerging and pandemic threats. Biomaterials 268:120597. https://doi.org/10.1016/j.biomaterials.2020.120597

Williams D, Thayer P, Martinez H, Gatenholm E, Khademhosseini A (2018) A perspective on the physical, mechanical and biological specifications of bioinks and the development of functional tissues in 3D bioprinting. Bioprinting 9:19–36. https://doi.org/10.1016/j.bprint.2018.02.003

Wu Y, Liu X, Yao C, Chen J, Wu X, Zhu M (2024a) Fluorescent hyaluronic acid nanoprodrug: a tumor-activated autophagy inhibitor for synergistic cancer therapy. Int J Biol Macromol 274:133360. https://doi.org/10.1016/j.ijbiomac.2024.133360

Wu X, Shi W, Liu X, Gu Z (2024b) Recent advances in 3D-printing-based organ-on-a-chip. EngMedicine 1(1):100003. https://doi.org/10.1016/j.engmed.2024.100003

Xiao R, Zhou G, Wen Y, Ye J, Li X, Wang X (2023) Recent advances on stimuli-responsive biopolymer-based nanocomposites for drug delivery. Compos B Eng 266:111018. https://doi.org/10.1016/j.compositesb.2023.111018

Xu Q, Zhang Z, Lui PPY, Lu L, Li X, Zhang X (2023) Preparation and biomedical applications of artificial cells. Mater Today Bio 23:100877. https://doi.org/10.1016/j.mtbio.2023.100877

Xue J, Wu T, Dai Y, Xia Y (2019) Electrospinning and electrospun nanofibers: methods, materials, and applications. Chem Rev 119(8):5298–5415. https://doi.org/10.1021/acs.chemrev.8b00593

Yang X, Huang C, Wang H, Yang K, Huang M, Zhang W et al (2024a) Multifunctional nanoparticle-loaded injectable alginate hydrogels with deep tumor penetration for enhanced chemo-immunotherapy of cancer. ACS Nano 18(28):18604–18621. https://doi.org/10.1021/acsnano.4c04766

Yang L, Wang X, Xiong M, Liu X, Luo S, Luo J et al (2024b) Electrospun silk fibroin/fibrin vascular scaffold with superior mechanical properties and biocompatibility for applications in tissue engineering. Sci Rep 14(1):3942. https://doi.org/10.1038/s41598-024-54638-0

Yilmaz B, Al Rashid A, Mou YA, Evis Z, Koç M (2021) Bioprinting: a review of processes, materials and applications. Bioprinting 23:e00148. https://doi.org/10.1016/j.bprint.2021.e00148

Yin L, Du G, Zhang B, Zhang H, Yin R, Zhang W et al (2020) Efficient drug screening and nephrotoxicity assessment on co-culture microfluidic kidney chip. Sci Rep 10(1):6568. https://doi.org/10.1038/s41598-020-63096-3

Zhang B, Hu C, Wang M, Wei H, Li S, Yu H et al (2025) Facile fabrication of a thermal/pH responsive IPN hydrogel drug carrier based on cellulose and chitosan through simultaneous dual-click strategy. J Colloid Interface Sci 678:827–841. https://doi.org/10.1016/j.jcis.2024.08.208

Zhou J, Chen Y, Liu Y, Huang T, Xing J, Ge R et al (2024) Electrospun medicated gelatin/polycaprolactone Janus fibers for photothermal-chem combined therapy of liver cancer. Int J Biol Macromol 269:132113. https://doi.org/10.1016/j.ijbiomac.2024.132113

Zushin P-JH, Mukherjee S, Wu JC (2023) FDA Modernization Act 2.0: transitioning beyond animal models with human cells, organoids, and AI/ML-based approaches. J Clin Investig 133(21). https://doi.org/10.1172/JCI175824

Chapter 16
Functional Bio-based Materials for Sustainable Wastewater Treatment

Karolina Labus, Katarzyna Czyżewska, Mateusz Jackowski, and Halina Maniak

Abstract Wastewater is an undesirable by-product of human activity that constitutes an enormous global problem. Considering domestic and industrial wastes, an ongoing excess in their production is observed, counted in billions of hectolitres. Moreover, the growing needs of the human population indicate that the amount of wastewater will constantly increase annually. The next aspect is economic growth in numerous countries that inflicts wastewater production, especially that coming from industry. The rising amount of wastewater creates new environmental threats that should be mitigated to maintain CO_2 emissions at an acceptable level. In the era of increasing ecological awareness, sustainable solutions are sought for effective wastewater management. One of the leading directions is the use of biomaterials for this purpose, both those specially functionalised for a given application, as well as those derived from waste biomass of industrial processing (e.g. lignocellulose, bran, chitin, collagen, gelatine, fruit pomace, brewer's grain, etc.). The most challenging is the development of new bio-based materials for wastewater treatment, especially for developing countries that require easy-to-use and cost-effective solutions to mitigate the challenges of developing economies and their impact on the environment. In this chapter, authors aim to overview different types of wastewater with specific properties and the possibilities of their effective processing using various pro-ecological solutions involving bio-based materials. First, the diversity of chemical composition of wastewater from different sectors is provided to demonstrate noticeable differences in the amount, type and concentration of the compounds they contain. Next, biomaterials with potential for use in sustainable wastewater treatment are classified

K. Labus (✉) · K. Czyżewska · M. Jackowski · H. Maniak
Department of Micro, Nano and Bioprocess Engineering, Faculty of Chemistry, Wrocław University of Science and Technology, Wrocław, Poland
e-mail: karolina.labus@pwr.edu.pl

K. Czyżewska
e-mail: katarzyna.czyzewska@pwr.edu.pl

M. Jackowski
e-mail: mateusz.jackowski@pwr.edu.pl

H. Maniak
e-mail: halina.maniak@pwr.edu.pl

L. Ravindran et al. (eds.), *Macro, Micro and Nanocomposites from Sustainable Sources*, Advanced Structured Materials 244, https://doi.org/10.1007/978-981-95-2469-3_16

and characterised. The main part of the chapter concerns the application of bio-based materials as high-performance pro-ecological tools in the monitoring, neutralisation, treatment, and valorisation of wastewater. Among others, the use of biomaterials in separation, sorption, detection, and conversion of hazardous compounds is discussed. The chapter also provides information on the latest innovative solutions and future prospects for sustainable wastewater treatment using bio-based materials.

16.1 Introduction

A critical challenge facing modern civilization is the increasingly limited availability of freshwater resources. With only 1% of the Earth's water accessible for human consumption, there is a growing emphasis on alternative water sources, such as recycled and reclaimed water, particularly for industrial and agricultural use (Rock et al. 2019). According to the United Nations World Water Development Report 2024, global freshwater use is increasing at a rate of just under 1% per year. While agriculture accounts for approximately 70% of total freshwater consumption, the industrial (20%) and domestic (10%) sectors are the primary drivers of water depletion, contributing significantly to the strain on water resources. This situation is further exacerbated by factors such as climate change, geopolitical unrest, pandemics, mass migration, hyperinflation, and other crises, which intensify the pressure on already scarce water supplies (UN Water red 2024).

Beyond the issue of water accessibility, a significant contemporary challenge lies in water pollution and wastewater generation, both of which severely impact global water resources. Water pollution resulting from human activities is primarily attributed to contaminants such as organic dyes, acids, antibiotics, inorganic salts, and heavy metals. Additionally, agricultural, textile, and pharmaceutical pollutants pose a significant threat to water quality (Annu et al. 2024). The exponential rise in wastewater generation presents a significant environmental threat, particularly where untreated wastewater is discharged. Insufficient regulatory enforcement and oversight have led to the treatment of only 8% of industrial wastewater in developing countries, intensifying the risks of ecosystem degradation. This issue is further compounded by data indicating an average annual wastewater production of 380 billion m^3, with projections suggesting a 51% increase by 2050, highlighting the growing challenge of managing wastewater effectively (Soo et al. 2024).

With growing ecological awareness, considerable focus is being placed on developing sustainable technologies for the neutralisation, management, and valorisation of wastewater. Information and communication technologies, such as artificial intelligence (AI), play a key role in this effort by providing data that support water management, earth observation, and the design of sensors based on low-cost technologies (UN Water red 2024). In recent years, a variety of strategies have been explored to effectively manage wastewater from diverse sources, including membrane-based processes, adsorption, coagulation, flocculation, precipitation, solvent extraction, and ion exchange. However, many of these methods require the use of hazardous chemical compounds, which pose environmental risks and increase the overall cost of wastewater treatment (Soo et al. 2024; Pandey 2020).

An alternative approach for wastewater treatment focuses on utilising bio-based materials, which can be employed either directly due to their inherent properties or after undergoing specific functionalisation processes. In this context, raw and waste materials, particularly from the agro-industrial sector—such as lignocellulose, chitin, and fruit pomace—play a pivotal role (Xiao et al. 2023). Biopolymers represent another crucial category, being abundant, biodegradable, non-toxic, and cost-effective, making them well suited for sustainable water treatment applications. Their production from renewable biomass offers a viable alternative to synthetic polymers derived from petroleum-based products, thereby reducing the overall carbon footprint. Furthermore, the integration of biopolymers aligns with the principles of a circular economy, promoting resource efficiency and waste minimisation (Elgarahy et al. 2023). Biopolymers also exhibit specific physicochemical properties that enhance their efficacy in wastewater remediation. One innovative solution is the development of biopolymeric nanocomposites, which combine the favourable properties of natural polymers with those of nanomaterials, yielding highly absorbent, selective, and efficient materials tailored for water purification and sustainable water management (Annu et al. 2024). Additionally, biopolymers are being increasingly utilized in the design of advanced wastewater treatment technologies, serving as coagulants, flocculants, adsorbents, membranes, and photocatalytic agents, thus significantly improving the efficiency and environmental sustainability of water treatment processes (Elgarahy et al. 2023).

This study provides a comprehensive review of recent advancements in wastewater treatment technologies employing biomaterials. It synthesises current data on the global wastewater crisis and the associated environmental risks. The review focuses on biomaterials' characteristics and potential applications in wastewater treatment, including their use as sorbents, membranes, sensors, and in biotransformation processes through enzymatic activity. The concluding section offers a critical evaluation of the efficacy of biomaterials in water purification, highlighting the challenges and opportunities posed by emerging biomaterial-based technologies.

16.2 Classification of Wastewater Types

The definition and classification of wastewater is a complex problem. They may be grouped by the source, matter state, chemical composition, or concentration. Moreover, every group of wastewater may be divided into subgroups. However, for this work, the classification given by Encyclopaedia Britannica will be applied (Fig. 16.1).

Wastewater can be defined as follows (Nathanson and Ambulkar n.d.): "Wastewater is the polluted form of water generated from rainwater runoff and human activities. It is also called sewage. It is typically categorised by how it is generated, specifically, as domestic sewage, industrial sewage, or storm sewage (stormwater)."

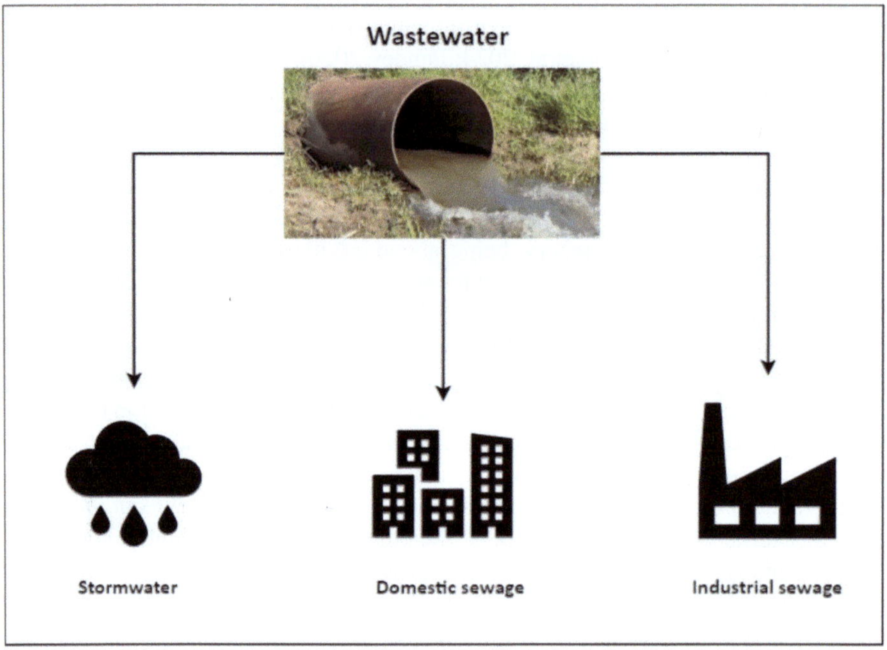

Fig. 16.1 Basic classification of wastewater types

Stormwater is wastewater that is an effect of rain or snow. Their amount is not constant and depends on the season, climate, geographical location, and landform. Therefore, its composition may vary even in the same place throughout the year. In general, rainwater is perceived as relatively clean. Nevertheless, it may contain compounds that cause acid rains, which are mainly a mixture of sulphuric and nitric acids, depending on the relative quantities of SOX, NOX, and CO_2 emissions (Chen et al. 2012). Moreover, in recent years, microplastic has been detected in rainwater, making it a more complex type of wastewater than previously thought (Parashar and Hait 2023). Another factor that inflicts the composition of stormwater is the area from which it is collected. For instance, stormwater from cities will be rich in microplastic, tire particles, dust, and chemical compounds used for city maintenance (Wilfong et al. 2023). Stormwater from rural areas will be rich in dust, and chemicals used in the agroindustry like fertilizers and pesticides (Taneez et al. 2023). Stormwater may be collected in the constructed wetlands, which is a method for water treatment from agriculture. The composition of agriculture wastewater may vary depending on climate, grown plants, and geological environment. Nevertheless, literature gives some examples of agro-industrial wastewaters that enter constructed wetlands BOD up to 8750 mg L^{-1}, total phosphorus up to 92 mg L^{-1}, total nitrogen up to 402 mg L^{-1}, ammonium up to 317 mg L^{-1}, *Escherichia coli* up to 7.52 × 1011 CFU/100 mL (Rozema et al. 2016).

Domestic sewage is wastewater derived from basic human activities like metabolism, hygienic purposes, cooking, and cleaning. It is often divided into two subgroups:

- **Greywater**—wastewater composed of weakly polluted water like water used for cooking, hand washing, and household maintenance. It is mainly composed of inorganic suspensions, detergents, dust, etc. The composition of greywater varies widely depending on factors such as water source, quality of water supply, water use, household activities, and socio-economic and cultural factors. Studies showed that the chemical oxygen demand of greywater varies from 35 to 66 mg L^{-1}, whereas biological oxygen demand varies from 18to 42 mg L^{-1} (Sievers et al. 2016).
- **Blackwater**—wastewater resulting from human and animal metabolism and represents fecal contamination. Moreover, chemicals like pharmaceuticals are often found in blackwater. This waste is more dangerous than greywater and requires more sophisticated treatment. For instance, the phosphorus level of wastewater collected in the dormitory reached 100 mg L^{-1} in unfiltered wastewater, where total nitrogen was observed at 460 mg L^{-1} (Sallanko and Sarpola 2007).

Unfortunately, black and grey wastewater are usually collected together, so domestic sewage is mostly a composition of both.

Industrial sewage is wastewater produced by industry. Since every branch of industry creates different types of wastewater, it is necessary to mention the exact kinds of industries and connected wastewater. Although it is rather impossible to describe all types of industry, some of them should be listed as representatives of specific branches that produce characteristic wastewater.

- **The mining industry** produces most of the raw materials (minerals) used by heavy industries. Mining operations generate large quantities of wastewater. In most cases, they represent low pH due to the accelerated oxidation of iron pyrite (FeS_2) and other sulfide minerals resulting from the exposure of these minerals to oxygen and water due to the mining and processing of metal ores and coals (Liu et al. 2012a). Coal mine wastewater represents conductivity up to 21000 $\mu S\ cm^{-1}$ and 20000 mg L^{-1} of total dispensed solids, up to 14000 mg of SO_4. Total organic carbon reaches 4 mg L^{-1} (Thiruvenkatachari et al. 2011). In the case of gold mines, the following wastewater contaminant levels are observed: 828 mg L^{-1} of Mg, 199 mg L^{-1} of Mn, 7742 mg L^{-1} of SO_4, and 21 mg L^{-1} of Zn (Shahedi et al. 2023). In the case of iron mines, the following wastewater composition was reported: Cu 0.27 mg L^{-1}, Fe 115.24 mg L^{-1}, Mn 0.96 mg L^{-1} and As 977 mg L^{-1} (Das and Nandi 2021).
- **The heavy industry** needs large quantities of energy. Such an energy source is a coke produced in coke plants that require large amounts of water (Sharma and Philip 2016). It is estimated that one tone of coke is generated from 0.3 to 4 m^3 of wastewater (Sharma and Philip 2016). Coke plant wastewater contains up to 10 mg L^{-1} of benzene, 1500–6000 mg L^{-1} of Chemical Oxygen Demand

(COD), 1000–2000 mg L^{-1} of Biological Oxygen Demand (BOD), 200 mg L^{-1} of total suspended solids, 150–2000 mg L^{-1} of phenols. Moreover, coke wastewater contains polyaromatic hydrocarbons, ammonia, and cyanide (Pal and Kumar 2014).

- **The metallurgical industry**: Heavy metals such as Cd, Cr, Cu, Ni, As, Pb, and Zn are the most hazardous water contaminations derived from this sector. Due to the high solubility in aquatic environments, living organisms can absorb and accumulate them (Gajbhiye and Wadnerwar 2021).
- **The textile industry**: The effluents from textile plants contain various contaminants such as salts, enzymes, surfactants, oxidizing and reducing agents. The major pollutants of textile wastewater include high suspended solids (SS), chemical oxygen demand, biological oxygen demand, heat, intense color, extreme pH, and other inorganic pollutants and nutrients (Jo et al. 2008). The generated wastewater is estimated to be 36 m^3 per ton of textiles (Mountassir et al. 2013). Chemical oxygen demand varies from 1.25 to 3.00 g L^{-1}, and Cl^- reached 400 mg L^{-1}. The most ubiquitous metal ions in described wastewater are iron 2.5 mg L^{-1}, lead and copper at level 1.4 mg L^{-1} and manganese 0.6 mg L^{-1}; pH reaches a value of 9–10 and turbidity 30 NTU (Zhang et al. 2021).
- **The brewing industry**: Advanced breweries are estimated to use 2.0–2.5 L of water per 1 L of produced beer. Whereas older breweries require higher amounts of water, reaching up to 7 L per 1 L of beer produced (Bak et al. 2001). Brewing wastewater contains cleaning agents and residues of technological processes, like wort, spent yeasts, beer residues, etc. The composition of brewing wastewaters is pH 6.9, COD 5650 mg L^{-1}, total nitrogen 43.77 mg L^{-1}, and total phosphorus 11.85 mg L^{-1} (Zibarev et al. 2022).
- **Meat industry**: Meat processing wastewater represents high biological and chemical oxygen demand, high fat content and high concentrations of dry matter, nitrogen and chlorides. COD varies from 2780 to 6720 mg L^{-1} of wastewater. BOD may be found in the range of 1200–3000 mg L^{-1}. Total nitrogen reaches up to 287 mg L^{-1} and phosphorus up to 70 mg L^{-1} (Sroka et al. 2004).
- **Pharmaceutical industry**: The primary sources of wastewater from this branch are production processes like chemical synthesis, fermentation, purification, extraction or formulation. Sludge composition from the pharmaceutical industry may vary according to the type of production or technologies used in the facility. Contaminants found in pharmaceutical wastewaters are residues of medicinal compounds, BOD up to 1800 mg L^{-1}, COD up to 28600 mg L^{-1}, ammonium nitrogen up to 164 mg L^{-1}, phosphate up to 47 mg L^{-1}, oil and grease up to 3964 mg L^{-1} (Rana et al. 2017).

16.3 Sustainable Bio-based Materials

Sustainable bio-based materials (SBMs) are materials composed of inorganic or organic matter of natural origin used in raw or modified forms. Below are briefly characterized the most typical ones used for wastewater treatment.

Inorganic Biomaterials

- **Activated carbon** (AC) is a highly porous material that undergoes specialized processing. It is primarily derived from raw materials like lignocellulosic biomass and coal. AC features a structure dominated by micropores, providing a large external surface area of up to 1100 $m^2\ g^{-1}$, pore volumes reaching 0.40 $cm^3\ g^{-1}$, bulk density, low ash content, and low conductivity (Isaeva et al. 2021). Other carbon-derived materials are **graphene** (Ali et al. 2019), **graphene oxide** (GO), and **reduced graphene oxide** (rGO) (Baig et al. 2019).
- **Silica**: Silicon dioxide (SiO_2) exhibits diverse structural configurations, ranging from amorphous to crystalline forms (e.g., quartz, cristobalite, tridymite), influencing its properties and making it a valuable, cost-effective adsorbent material (Diagboya and Dikio 2018).
- **Zeolites** are microporous aluminosilicate minerals with a crystalline structure composed of interconnected TO_4 tetrahedra (T = Si, Al, or other atoms). Their general formula is $M_x{}^{+}M_y{}^{2+}[Al_{(x+2y)}Si_{n-(x+2y)}O_{2n}]\ mH_2O$, where M^+ and M^{2+} represent exchangeable cations, and *n* and *m* are variable integers (Pérez-Botella et al. 2022). Natural zeolites are formed in geochemical processes. There are several types: clinoptilolite, chabazite, analcime erionite, faujasite, ferrierite, heulandite, laumontite, mordenite, natrolite, phillipsite, wairakite (Krstić 2021).
- **Clay minerals** are abundant, low cost, and renewable hydrated aluminosilicate minerals with a complex structure, typically consisting of silica tetrahedrons and aluminum oxide octahedrons arranged in layered, 2:1 (e.g. montmorillonite, illite, bentonite, vermiculite, attapulgite, sepiolite, and glauconite) or 1:1 interlayer structures (e.g. kaolinite, halloysite, or zeolite-lie frameworks (like zeolite) structures). Each mineral type's unique chemical formula and crystal structure influence its adsorption properties (Xie et al. 2024).

Organic Biomaterials

- **Cellulose** is the most abundant carbohydrate biopolymer in nature. It consists of β-D-glucopyranose units linked with β-1,4-glucosidic bonds and is organized into microfibrils (Chundawat et al. 2011). The physical form of cellulose is diverse due to the hydrogen bond system, which allows for the formation of different supramolecular semi-crystalline structures. Cellulose chains are rich in hydroxyl groups, which can be chemically modified using various agents (e.g., citric anhydride, maleic anhydride, *N*,*N*-carbonyldiimidazole, sodium metaperiodate, polyamines) (Wu et al. 2022).
- **Hemicellulose** is the second-most prevalent polysaccharide, representing approximately 30% of the total lignocellulosic biomass (Ibn Yaich et al. 2017). It is a

heterogeneous and branched polysaccharide comprising a variety of carbohydrate repeating units, including anhydrous neutral sugars (e.g. glucose, xylose, mannose, arabinose, galactose, and rhamnose) and uronic acids (e.g. glucuronic, galacturonic, and 4-*O*-methyl-D-glucuronic) (Wu et al. 2022).

- **Lignin** is a complex, heterogeneous biopolymer composed of three types of monomeric phenolic substructures: *p*-hydroxyphenyl (H), syringyl (S), and guaiacyl (G) units, that derive from three distinct aromatic alcohol precursors, specifically coumaryl, sinapyl, and coniferyl alcohol (Kai et al. 2016). The biopolymeric structure of lignin encompasses a range of functional groups, including hydroxyl, carboxylic, carbonyl, and methoxy groups, which impart the chemical and physical properties of lignin, such as reactivity, hydrophilicity, and functionality (Zhao et al. 2017).
- **Starch** is plants' most abundant storage polysaccharide, valued for its biocompatibility and degradability. Chemically, it is a mixture of two polymers: amylose and amylopectin. Amylose is a linear polysaccharide comprising 500–2000 glucose units linked by α-1-4 glycosidic bonds and constitutes 20–30% of starch. While amylopectin is a branched polysaccharide with chain lengths of 1,000,000 glucose units linked by *α*-1-6 glycosidic bonds, making up 70–80% of starch (Mohidem et al. 2023).
- **Pectins** are plant cell wall natural heteropolysaccharides composed mainly of *α*-1,4-D-galacturonic acid units, randomly interleaved by (1,2)-linked l-rhamnoside-chains and other neutral sugars (Wenpo et al. 2015). They possess three main structural domains: homogalacturonan, alternating with two types of highly branched rhamnogalacturonans regions RG-I and RG-II. However, there are also other structural classes of these polysaccharides, like xylogalacturonans, arabinogalactans, and arabinans (Lara-Espinoza et al. 2018).
- **Agarose** is a natural polysaccharide from marine microorganisms such as red algae known as agarophyte seaweeds (Renn 1997). Chemically, it is a neutral linear polymer of disaccharide repeating units of 1,4-linked 3,6-anhydro-*α*-l-galactose and 1,3-linked *β*-D-galactose molecules, which constitutes a major component of agar, including also minor agaropectin fraction (Goff et al. 2015).
- **Alginates** are a group of hydrophilic colloidal polysaccharides including alginic acid and its salts, naturally occurring in brown algae. Alginic acid is a linear unbranched form of a copolymer consisting of 1,4-linked *β*-D-mannuronic acid and 1,4-linked *α*-L-glucuronic acid residues. It forms water-soluble or insoluble salts with cations belonging to the first or second group of the periodic table, respectively. Alginates are the most frequently used polysaccharides in immobilization and encapsulation processes (Draget et al. 2002).
- **Chitosan** originates from chitin and is one of the most abundant natural polymers found in invertebrate animals and lower plants, presented in the outer skeletons of arthropods, crustaceans, and fungi (Muzzarelli and Muzzarelli 2005). Chitosan is a linear cationic copolymer consisting of 1,4 linked 2-amino-2-deoxy-*β*-D-glucan with a degree of acetylation varying from 0.5 to 1.0. Chemical crosslinking of chitosan with glutaraldehyde is the most common procedure for obtaining stable hydrogel materials (Gomez-Mascaraque et al. 2014).

- **Collagen**: A group of proteins with similar molecular and fibrous composition form a collagen family, in which 28 different types have been already identified (Ricard-Blum 2011). The typical chemical structure of collagen differs among the origin and consists of various aminoacids with different proportions. Nevertheless, in the polymer chain, the repeating fragments, including both glycine and proline −(Gly-X-Pro)− or hydroxyproline −(Gly-X-HPro)− and other aminoacids, are presented. The average amino acid composition includes 33% of glycine, 21% proline and, 10% hydroxyproline, 10% alanine, and the remaining 36% are various X aminoacids (Djabourov et al. 1988).
- **Gelatine** is a product of collagen denaturation, which can be proceeded by thermal, dilute alkali, or acid pre-treatment. It mainly consists of protein (85–92%) and smaller amounts of mineral salts and water (Schrieber and Gareis 2007). It consists of several reactive groups, such as amino, carboxyl, or hydroxy substituents, which can participate in a physical or chemical linking.
- **Biochar** is a carbonaceous material produced through the thermochemical conversion of biomass. It utilizes renewable biogenic feedstocks as its source material (Isaeva et al. 2021).

16.4 Application of Bio-based Materials in Sustainable Wastewater Treatment

Due to their pro-ecological nature, bio-based materials of natural origin are intensively studied for their application in wastewater treatment. Figure 16.2 depicts the most significant areas for their practical use. Adsorption, membrane separation, detection, monitoring, and degradation of hazardous compounds in water effluents have been considered and discussed in detail.

16.4.1 Bio-based Adsorbents

Adsorption is a fundamental process characterized by accumulating a specific substance, termed the adsorbate, at the interface of a two-phase system, which may include gas–solid, liquid–solid, gas–liquid, or liquid–liquid configurations. The phase exhibiting adsorption is called the adsorbent and may be either solid or liquid. Based on the interactions and forces involved, the phenomenon can be categorized into physical adsorption (physisorption) and chemical adsorption (chemisorption). In adsorption, a reduction in the adsorbate concentration within the bulk phase is accompanied by an increase at the adsorbent interface. Figure 16.3 presents the interactions between adsorbent and adsorbate in physisorption and chemisorption.

Physisorption is driven by weak intermolecular interactions that aim to minimize surface energy. Mechanisms of physisorption may include pore-filling, electrostatic attractions, hydrogen bonding, hydrophobic interactions, and various van der

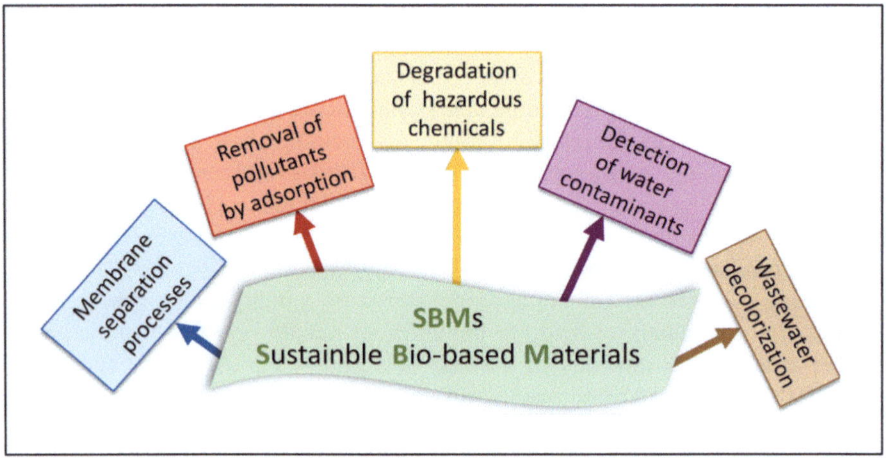

Fig. 16.2 Main application areas of sustainable bio-based materials for wastewater treatment

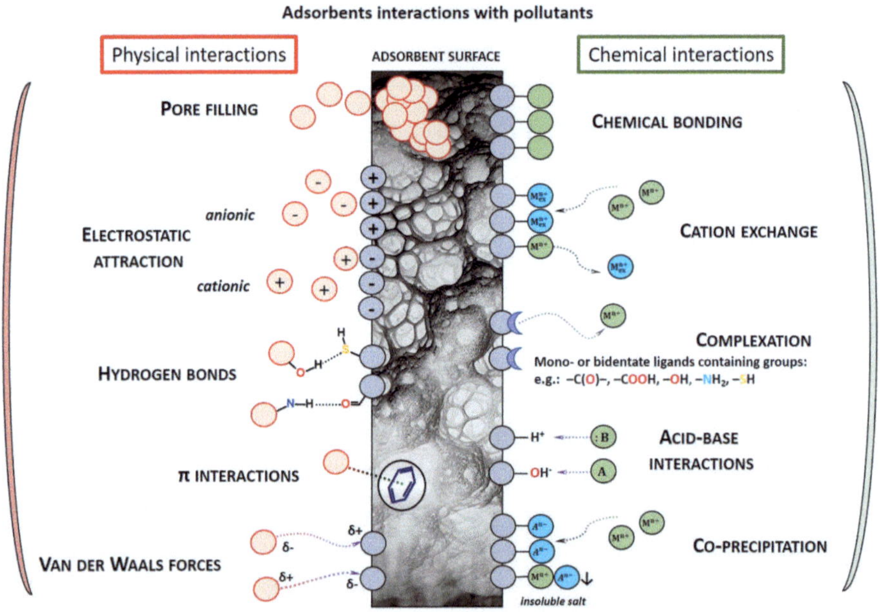

Fig. 16.3 Examples of physical and chemical interactions of pollutants with adsorbent surface. Description of the colors: grey spheres—specific groups on the adsorbent surface responsible for interaction with pollutants; red spheres—pollutants adsorbed with physical interactions; green spheres—pollutants adsorbed through chemical interactions; blue spheres—changeable metal cations; violet shapes—mono- or bidental ligands

Waals forces, including ion–dipole interactions, dipole–dipole interactions, induced dipole–dipole interactions, and π–π stacking interactions (Rosales et al. 2017). In physisorption, the mobility of the adsorbate molecules is relatively high, allowing them to migrate across the surface and form multiple layers. This process is generally reversible under low pressure and elevated temperatures. It enables adsorbed molecules to revert to their original fluid phase without undergoing property alterations, as no covalent bonds are formed. Furthermore, physisorption is inherently exothermic and achieves rapid equilibrium. Conversely, **chemisorption** entails the formation of strong, often covalent bonds between the adsorbate and specific active sites on the adsorbent surface. This interaction leads to substantial modifications in the electronic states of both the adsorbate and adsorbent and is often referred to as irreversible adsorption. Chemisorption is characterized by high heats of adsorption and specific site occupancy, typically resulting in monolayer coverage due to the spatial limitation of reactive sites. The process commonly necessitates activation energy for bond formation and is sensitive to temperature, with effective adsorption generally occurring at temperatures exceeding the critical point of the adsorbate. Unique spectroscopic techniques, including electron spin resonance and magnetic susceptibility measurements, provide insights into these strong, localized interactions (Lowell and Shields 1991).

Table 16.1 summarizes the most important, interrelated, and frequently determined parameters relevant to the thermodynamics of any sorption system, such as heat of adsorption, adsorption energy and activation energy (Inglezakis and Zorpas 2012), and binding energy (Shaji and Zachariah 2017; Costanzo et al. 2012). Together, they provide a comprehensive description of the basic energetics of the adsorption process, offering insights into how and why adsorption occurs and predicting its overall performance.

The enthalpy change, ΔH [kJ], represents the overall heat transfer occurring during the sorption process under isobaric conditions. In adsorption systems, the heat of adsorption, Q, can be described in two forms: total and differential. The total heat of adsorption refers to the average amount of energy released or absorbed per mole

Table 16.1 Thermodynamic typical parameters for physical and chemical sorption (Inglezakis and Zorpas 2012; Shaji and Zachariah 2017; Costanzo et al. 2012)

Parameter type	Adsorption type	
	Physisorption	Chemisorption
Reversibility	Reversible	Irreversible
Adsorption layer thickness	Multilayer	Monolayer
Enthalpy, ΔH [kJ]	$\Delta H < 0$ (exothermic)	$\Delta H > 0$ or $\Delta H < 0$ (endothermic/exothermic)
Energy bonding [meV]	10^1–10^2	10^3–10^4
Heat of adsorption, Q [kJ/mol]	<80	>80
Energy of adsorption [kJ/mol]	<8	>16
Activated energy [kJ/mol]	<40	>40

of adsorbate over a given range of surface coverage. Meanwhile, the differential heat of adsorption corresponds to the change in energy associated with the adsorption of an infinitesimally small additional amount of adsorbate at a given surface coverage. Under isobaric conditions, the differential heat of adsorption is directly related to the differential enthalpy change, so that $Q = \Delta H$. A negative value of ΔH indicates an exothermic adsorption process, while a positive value characterizes an endothermic process. The heat of adsorption (expressed in kJ/mol) reflects the strength of the interaction between the adsorbate and the adsorbent surface. Typically, higher values of heat of adsorption correspond to stronger chemical bonds (chemisorption), while lower values are associated with weaker physical interactions (physisorption). The heat of adsorption often decreases with increasing surface coverage, especially in heterogeneous systems, due to the gradual occupation of the most energetically favorable adsorption sites. In addition, the magnitude of the heat of sorption affects the ease with which adsorbed molecules can be subsequently desorbed. In physical sorption processes, sorption heat usually remains below 80 kJ/mol, while chemisorption often exceeds this threshold (Inglezakis and Zorpas 2012). Although closely related to the heat of sorption, the energy of sorption is a broader concept that encompasses the overall potential energy change when a molecule moves from the bulk phase to the surface of an adsorbent. This energy results from various intermolecular interactions, including van der Waals forces, hydrogen bonds and chemical bonds. The sorption energy is a critical parameter as it governs both the sorption capacity and selectivity of the sorption material, ultimately determining the stability of the sorbed molecules. Empirically, sorption energies below 8 kJ/mol are characteristic of physical adsorption, values between 8 and 16 kJ/mol correspond to ion exchange mechanisms, and values exceeding 16 kJ/mol are associated with chemical adsorption, see Table 16.1 (Inglezakis and Zorpas 2012; Shaji and Zachariah 2017; Costanzo et al. 2012). The activation energy for sorption represents the barrier that must be overcome for adsorption. It serves as an indicator of whether the process occurs spontaneously or requires external energy input. A low activation energy usually suggests rapid and spontaneous adsorption under ambient conditions. Furthermore, activation energy directly affects the kinetics of the process, regulating the rate at which equilibrium is reached.

Modifications of Adsorbents

Figure 16.4 presents the types of modifications for modified adsorbents, which exhibit enhanced properties that significantly improve their adsorptive capabilities.

Physical modifications, such as milling to increase surface area or freeze-drying to create porosity, make the biopolymers more effective by enhancing their capacity to capture pollutants. Similarly, composite formation—blending biopolymers with other materials—combines increased porosity with additional functional sites, enhancing adsorptive efficiency. Chemical modifications, on the other hand, allow targeted improvements by altering molecular structures or introducing specific functional groups that increase binding affinity for particular contaminants. Adding groups like amino or carboxyl, for example, enhances the biopolymer's interaction with heavy metals and dyes while cross-linking with stabilizing agents maintains the

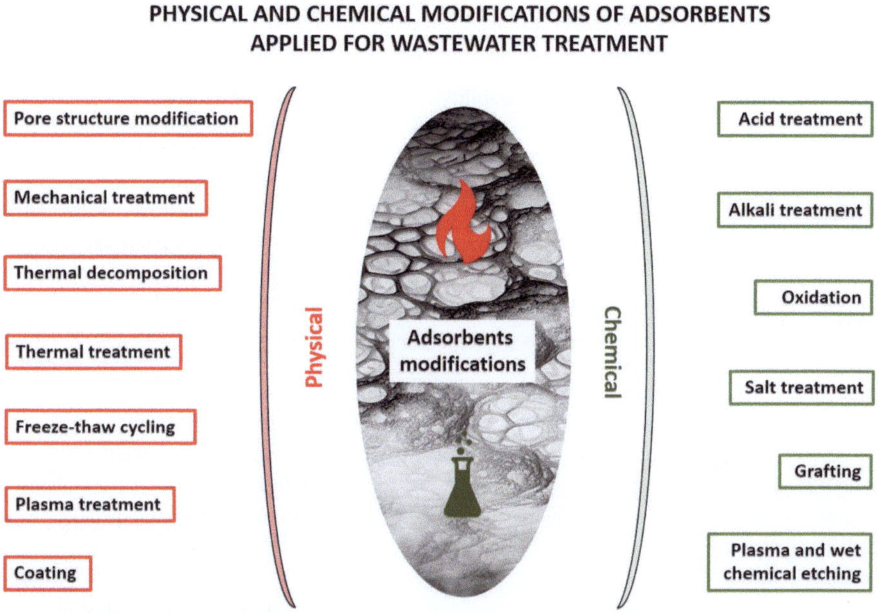

Fig. 16.4 Examples and brief characteristics of physical and chemical modifications of adsorbents applied in wastewater treatment

biopolymer's structural integrity during adsorption. Incorporating metal oxides also increases affinity for specific ions, further expanding the versatility of biopolymers in pollutant capture. Together, these modifications make biopolymers adaptable and efficient adsorbents capable of addressing a wide range of pollutants in environmental and industrial settings.

Required Features of Adsorbents

The design of an effective adsorption system for water treatment relies heavily on carefully selecting, developing, and characterizing the adsorbent material. Adsorbents are usually porous solids with highly developed surfaces insoluble in water and organic solvents. The ideal adsorbents must possess several critical qualities, including a large specific surface area, appropriate pore size distribution, and high mechanical and chemical stability. Additionally, they should be practical for use, reusable, and exhibit a significant affinity for the adsorbate, particularly at low initial concentrations of pollutants (Dotto and McKay 2020). These features are vital for ensuring the efficiency and sustainability of the overall adsorption process. The most essential features of adsorbents in wastewater treatment are included in Table 16.2.

Typical adsorbents are commonly utilized as pellets, grains, or granules, with grain sizes generally ranging from 0.1 to 2 mm. Natural adsorbents can be classified into inorganic and organic, further enhancing their properties and applications in water treatment processes.

Table 16.2 Key desirable adsorbent features for water treatment (Dotto and McKay 2020)

Feature category	Specific feature	Importance/rationale
Economic factors	Low cost and availability	Adsorbent accounts for 70% of operational costs; ease of procurement and transportation are crucial
Physical feature	Mechanical stability	Prevents high-pressure drops and preferential flow pathways in treatment columns
Chemical feature	Chemical stability	Resists degradation from varying water chemistries (pH, ionic strength, conductivity)
Adsorbent properties	High surface area and pore volume	Crucial for maximizing contaminant interaction
	Functional surface groups	Provide sites for contaminant binding
Performance metrics	High adsorption capacity	Minimizes adsorbent quantity needed, facilitates phase separation, and reduces plant size
	High efficiency	Ensures contaminant removal meets regulatory standards
	Fast kinetics	Achieves high capacity and efficiency quickly, reducing treatment time and plant size
Sustainability	Potential for regeneration and reuse	Significantly lowers operational costs by extending adsorbent lifespan

In essence, the efficiency of adsorption processes for removing emerging pollutants from aquatic environments is intricately linked to the adsorption mechanisms. It is heavily influenced by both the materials' physicochemical properties and the pollutants' chemical nature. A significant area of research is the development of biomaterials as adsorbents in wastewater treatment. Addressing the challenge of designing a multifunctional adsorbent, studies have focused on isolating natural materials from soil, rocks and plant, animal and microbial biomass. Thus, the various biomaterials investigated for wastewater pollutant removal can be broadly classified into inorganic categories (e.g., carbonaceous materials such as activated carbons, zeolites, and silica-based materials) and organic categories derived from renewable biomass (e.g., cellulose, chitosan, lignin). Comprehensive characterization and synthesis of these materials have been reported. While a diverse array of well-understood adsorbents exists, there is a continuous need to develop novel materials to satisfy the increasing performance requirements for effective water treatment. Characterization of adsorbents typically includes the evaluation of their efficacy against representative wastewater contaminants sourced from diverse anthropogenic activities. Such analyses are conducted using either raw wastewater samples that reflect the complex nature of real-world effluents or standard solutions with known concentrations of target pollutants, such as dyes, antibiotics, and heavy metal ions.

In the past decade, the literature has expanded significantly, detailing the application and optimization of biomaterials for pollutant adsorption. It is beyond the scope of this chapter to present all, or at least most, of the investigated biomaterials used in the adsorption of pollutants from water. Therefore, various adsorbents that are examples of unmodified biomaterial or characterized by high adsorption capacity while retaining the natural character of these materials will be presented using selected examples. Further details on the topics discussed are available in the cited original research and review articles. These provide detailed information, including the use of biopolymers in wastewater treatment as follows: for **activated carbon** (Pet et al. 2024; Muhammad et al. 2023; Sabzehmeidani et al. 2021), **graphene** (Chowdhury and Balasubramanian 2014; Parvathi et al. 2023; Hu et al. 2022), **silicas** (Diagboya and Dikio 2018; Cashin et al. 2018; Jal 2004), **zeolites** (Pérez-Botella et al. 2022; Velarde et al. 2023; Ugwu et al. 2022), **clays** (Xie et al. 2024; Orta et al. 2020; Khan et al. 2023) **alginates** (Qi et al. 2021; Radoor et al. 2024; Nawaz et al. 2023), **celluloses** (Nasrollahzadeh et al. 2021; Godage and Gionfriddo 2020; Boccia et al. 2023), **chitin/chitosan** (Godage and Gionfriddo 2020; Vakili et al. 2014; Russo et al. 2021), **pectins** (Płotka-Wasylka et al. 2024; Li et al. 2022a; Martínez-Sabando et al. 2023), **starch** (Abd El-Ghany et al. 2023; Haq et al. 2022; Ihsanullah et al. 2022), **proteins** (Nawaz et al. 2023; Nasrollahzadeh et al. 2021; Vinayagam et al. 2022), **biochar** (Biswal and Balasubramanian 2023; Jha et al. 2023; Qiu et al. 2022), and **lignins** (Fattahi et al. 2024; Barman et al. 2022; Wang et al. 2022).

16.4.1.1 Inorganic Natural Materials as Sorbents in Wastewater Treatment

Inorganic materials garnered significant interest because of their unique properties and the variety of carbon-based structures, including activated carbon, graphene, silicas, and zeolites (Muhammad et al. 2023; Boccia et al. 2023; Garg et al. 2023), which offer new possibilities across numerous subfields of chemistry, physics, and engineering The examples of inorganic adsorbents are included in Table 16.3. These materials offer diverse approaches for wastewater treatment, and combining different adsorbents can enhance their effectiveness, leveraging complementary features for improved pollutant removal. In the aqueous phase, molecules' adsorption mechanism on various adsorbents is primarily governed by weak van der Waals forces, induced dipole interactions, dipole–dipole forces, and hydrogen bonding between donor and acceptor species. Notably, hydrogen bonds, π–π interactions, covalent and electrostatic interactions, and the hydrophobic effect significantly contribute to adsorption.

Inorganic biopolymers such as activated carbon, graphene, silica, and zeolites all exhibit distinct adsorption characteristics. Their effectiveness is governed by the interplay of their structural properties, surface modifications, and the nature of the pollutants they target. Through ongoing research and innovation, these materials

Table 16.3 Examples of inorganic biomaterials applied as adsorbents for wastewater treatment

Raw material/modifications		Pollutant	Adsorption capacity, mg/g	References
Activated carbon	Plum kernels	Methylene blue	828	Sabzehmeidani et al. (2021)
	Baggase	Acid blue	391	
	Plum kernels	Acid blue 74	567	
	Olive seed chemically activated	Methylene blue	263.0	Saleem et al. (2019)
	Olive stone chemically activated (H_2SO_4 + NaOH reflux)	Safranin	526.3	
	Olive waste cake physically activated (H_2O/N_2) (60 min/ 850 °C)	Iodine	1495	
	Super-activated carbon prepared from food industry waste	Paracetamol	356.2	Vedenyapina et al. (2019)
	Macadamia nut shell chemically activated ($NaOH/N_2$)	Tetracycline	455.3	
	Cypress fruit	Pb(II) Cd(II) Co(II)	84.15 117.3 55.30	Al-Ma'abrehet al. (2024)
Graphene species	Graphene	Methylene blue	204.1	Liu et al. 2012b)
	Graphene oxide	Cu(II) Zn(II) Cd(II) Pb(II)	294 345 530 1119	Sitko et al. (2013)
		Methylene blue	1939	Zhang et al. (2011)
		Acridine orange	3333	Sun et al. (2012)
	Reduced graphene oxide, iron–iron oxide matrix	As(III) Cr(VI) Hg(II)	44.4 31.1 22.0	Bhunia et al. (2012)
	Reduced graphene oxide-iron oxide hybrid	Pb(II) 1-naphtol 1-naphthylamine	454.6 588.2 625.0	Yang et al. (2012)
Silicas	Large pore magnetic amino-silica	Cu(II) Pb(II) Cu(II) Cd(II)	628.3 880.6 628.3 492.4	Yuan et al. (2013)
	Guar-gum silica composite	Malachite green Safranin Pb(II) Cd(II)	781.2 281.7 645.2 709.2	Patra et al. (2017)

(continued)

Table 16.3 (continued)

Raw material/modifications		Pollutant	Adsorption capacity, mg/g	References
	Magnesium-silicate	Pb(II) Tetracycline UO_2^{2+}	557.9 381.3 482.8	Zhang et al. (2017)
Zeolites	Clinopilolite	Cd(II) Fe(III) Ni(II)	15.76 17.81 13.1	El-Azim and Mourad (2018)
	Zeolite X from kaolin	Cd(II)	82,0	Yuna (2016)
	Zeolite A from kaolin	Pb(II)	213	
	Cetylpyridinium bromide-modified zeolite	Tannic acid	110	Shi et al. (2018)
	N, N-dimethyl dehydroabietylamine oxide modified zeolite	Congo red	69.9	
Clays	Sepolite	Direct blue 85 Basic red 46	332 310	Santos and Boaventura (2016)
	Smectite	Acid brown 75 Basic yellow 28	8.33 76.92	Chaari et al. (2019)
	Montmorillonite modified with Fe_2O_3	Methylene blue	71.2	Cottet et al. (2014)
	Clay-alginate hybrid, Na-organobetonite and composite alginate (1:1)	Methylene blue	768	Belhouchat et al. (2017)
	ZnAl base clay	Indigo carmine	520	Bessaha et al. (2017)

can be further optimized to enhance their roles in environmental remediation and pollutant removal.

16.4.1.2 Organic Natural Materials as Sorbents in Wastewater Treatment

Organic biopolymers have become valuable adsorbents in wastewater treatment, recognized for their eco-friendly nature, biodegradability, and renewable sourcing. Derived from natural materials such as algae, plants, animal products, and microbial biomass, biopolymers like polysaccharides, proteins, and other natural polymers effectively adsorb inorganic and organic pollutants such as heavy metals and dyes or antibiotics. Their efficiency in adsorption arises from their porous structure, high surface area, and functional groups (like hydroxyl and amino groups), which interact readily with various contaminants. To further enhance these natural properties, biopolymers are widely modified to boost their adsorption performance, selectivity, and stability across diverse conditions.

This paragraph provides examples of organic biomaterials, included in Table 16.4, such as polysaccharides, proteins, lignins, and biochars used in processes for adsorption. Their role as eco-friendly alternatives to synthetic materials addresses the need for sustainable methods to remove pollutants, especially heavy metals and dyes, from wastewater (Nawaz et al. 2023; Doyo et al. 2023). Biopolymers like cellulose, chitosan, alginate, pectin, starch, proteins, and lignin are widely investigated for application in environmental remediation because they offer specific structural and functional characteristics that enable effective pollutant adsorption. These characteristics include their surface functional groups, porosity, molecular structures, and modifiability.

Organic-based wastewater adsorbents are valuable due to their inherent properties: high adsorption capacity, low cost, renewability, biodegradability, biocompatibility, and ease of modification. Both physical and chemical modifications significantly boost the biopolymers' adsorption capabilities, usually via increased surface area or enhanced binding through targeted functional group additions. The optimal biopolymer and modification approach is analyte-dependent.

16.4.2 Bio-based Membranes

Membrane technology is an integral component of wastewater treatment. In these processes, the membrane serves as a selective barrier between two phases, effectively separating water from contaminants. It is widely adopted as a leading strategy in water treatment due to its high efficiency and ease of implementation. Moreover, the broad diversity in membrane composition and physical structure allows for the selection of membrane types specifically tailored to target various water pollutants, enhancing the effectiveness of treatment (Obotey Ezugbe and Rathilal 2020).

While membranes made from synthetic polymers offer advantages such as good mechanical properties, controllable pore size and distribution, and high reproducibility, their use in wastewater treatment is hindered by high hydrophobicity, a propensity for fouling, high production costs, and poor environmental degradation. These challenges can be overcome by using bio-based materials in membrane production, including natural polymers, carbonaceous materials, and clay-based materials. In all cases, membranes can be produced entirely from bio-based materials or enhanced with natural bio-components, as in the case of hybrid membranes (Xue et al. 2024). Examples of membranes produced based on bio-based components applied for wastewater treatment are presented in Table 16.5.

New trends in membrane production using natural polymers involve the addition of nanoparticles, which, through electrospinning synthesis, enable the development of membranes with highly favorable properties for wastewater treatment applications, Fig. 16.5. The resulting membranes exhibit desirable characteristics such as high specific surface area, increased porosity, and robust mechanical strength (Kumar

Table 16.4 Examples of organic biomaterials applied as adsorbents for wastewater treatment

Biopolymer, modification/characteristic		Pollutant	Adsorption capacity, mg/g	References
Alginates	Calcium alginate, beds	CN^- (free) Ni(II) Cu(II) Zn(II)	1177 107.3 39.5 1.52	Pérez-Cid et al. (2020)
	Sodium alginate, tiosalicyclic acid Ca^{2+} cross-linking	Cu(II) Cd(II)	426.9 592.9	Alp Arıcı et al. (2020)
	Sodium alginate, cysteine functionalization	Pb(II)	770	Huamani-Palomino et al. (2019)
	Calcium alginate, thermo-plasma expanded graphite	Atrazine Bisphenol A 17-α ethinylestradiol Carbamazepine	0.186 0.208 0.292 0.290	Cuccarese et al. (2023)
	Alginate beds, impregnated graphene oxide	Au(III)	81.87	Saha et al. (2019)
	Sodium alginate aerogel beds, incorporated with L-cysteine-modified defective UiO-67	Pb(II) Cd(II) Cu(II)	661.2 296.2 326.4	Du et al. (2020)
	Calcium alginate, MnO_2 microcapsules	As(V) Cd(II)	6.5 63.6	Shim et al. (2019)
	Sodium alginate, cross-linked activated organo-bentonite composite	Methylene blue methyl orange	414 116	Belhouchat et al. (2017)
Chitin/ chitosan	Chitin from *Trypoxylus dichotomous,* hydrothermally treated deacetylated to chitosan	Pb(II) Cu(II) Fe(III); Levofloxacin Tetracycline HCl Sulfamethoxazole	934 462 270 26 22 67	Jiang et al. (2022)
	Chitosan, grafted polymerization of itaconic acid on chitosan to emulsion, crosslinking with glutaraldehyde	Hg(II) Pb(II)	870.1 1320	Ge et al. (2017)
	Chitosan aerogel, graphene oxide modification, glutaraldehyde cross-linking	Diesel oil	12,560	Guo et al. (2016)
	Chitosan hollow fibers, dry-wet spinning technique	Reactive blue 19	454.5	Mirmohseni et al. (2012)

(continued)

Table 16.4 (continued)

Biopolymer, modification/characteristic		Pollutant	Adsorption capacity, mg/g	References
	Chitosan megaporous scaffold, freeze-drying technique	Indigotine Red dye 40 Tartrazine Yellow 6 Food red 2	3032 3316 1078 788 1552	Esquerdo et al. (2014)
	Chitosan aerogel, composite based on citrus peel and bentonite	Cu(II)	861.6	Nie et al. (2023)
Cellulose	Cellulose filament with chitosan aerogel, citric acid cross-linking	Cu(II) Methylene blue	206 619	Varamesh et al. (2023)
	Cellulose fibers, acetylated	Pure crude oil Diesel oil Vacuum pump oil	55,360 29,320 31,500	Li et al. (2013)
	Cellulose, nanocrystalline form	Hydroxynaphtol blue Congo red	105.5 107.7	Putro et al. (2019)
	Porous cellulose monolith, freeze dying technique	Methylene blue	454	Salim et al. (2022)
	Nanocellulose, phosphorylated	Ag(I) Cu(II) Fe(III)	136 117 115	Liu et al. (2015)
Pectins	Pectin form pomelo pitch, unmodified	Acid blue 25	719.4	Shahrin et al. (2021)
	Low molecular weight pectin from seagrass, acid hydrolysis	Cd(II) Pb(II)	296.3 652.3	Khozhaenko et al. (2016)
	Pectin microgel particles, Ca^{2+} cross-linking	Methylene blue	284.09	Yu et al. (2018)
	Pectin-guar gum beads, Ca^{2+} cross-linking	Cd(II) Pb(II)	67.45 165.8	Jakóbik-Kolon et al. (2014)
	Pectin-cellulose microfibers, composite beds	Cd(II) Cu(II) Fe(II)	192.3 88.4 98.0	Lessa et al. (2020)
Starch	Corn starch, unmodified	Methylene blue	63.02	Amiri et al. 2023)
	Phosphate starch		73.17	
	Phosphate starch nanocrystals, phosphorylation, acid hydrolysis, mechanical treatment		88.53	

(continued)

Table 16.4 (continued)

Biopolymer, modification/characteristic		Pollutant	Adsorption capacity, mg/g	References
	Potato starch phosphate, microwave-ultrasound-assisted synthesis, cross-linking	Zn(II) Pb(II) Cd(II) Hg(II)	130.5 106.2 91.84 51.38	Bashir et al. (2020)
	Starch, cross-linking, etherifying	Reactive golden yellow SNE	208.8	Guo et al. (2019)
	Carboxymethystarch, composite material magnetite (Fe_2O_3) bentonite	Tetracycline	169.7	Shen et al. (2019)
Protein	Silk fibroin, unmodified	Fe(III)	12.82	Pilley et al. (2022)
	Mesoporous silica/gelatin aerogel, composite	Hg(II)	209	Herman et al. (2020)
	β-lactoglobulin nanofibrils, alkaline-thermal activation	Pb(II)	212	Zhang et al. (2020a)
	Porous lysozyme skeleton, amyloidal denaturation protocol	Congo red	993.6	Liang et al. (2022)
	Gelatine nancomposite beads, Fe_3O_4 nanoparticles and carbon nanotubes cross-linking	Methylene blue Direct red 80	465.5 380.7	Saber-Samandari et al. (2017)
Lignin	Soda lignin, alkaline extraction	Pb(II) Cd(II)	~7.5 ~7.0	Akpakpan et al. (2023)
	Succinylated soda lignin, alkaline extraction, esterification		~9.0 ~8.0	
	5-sulfosalicylic acid modified lignin, grafting, cross-linking	Malachite green Pb(II)	203.4 115.7	Du et al. (2024)
	Copper-doped activated lignin, activation, dopping	TeO_3^-, TeO_3^{2-}	248.4	Yao et al. (2022)
	Lignin-calcium, flocculation-sedimentation	Methylene blue	803.9	Dai et al. (2021)
Biochar	Biochar from muskmelon peel	Cu(II) Zn(II)	78.74 72.99	Biswal and Balasubramanian (2023)
	Biochar from rice husk; MgAl-layered double hydroxide-coated rice husk	Cd(II) Cu(II)	125.3 104.3	Li et al. (2022b)
	Biochar from litchi peel	Malachite green Congo red	2 468 404.4	Wu et al. (2020)
	Biochar from textile sludge	Cooking oil	121.1	Jha et al. (2023)

Table 16.5 Bio-based membraned utilized in wastewater treatment

Bio-based material	Membrane type	Water pollution	Improved properties of the membrane	References
Activated carbon	UF	Aromatic pollutants	Enhanced adsorption and separation properties, high stability and water flux, antifouling properties	Yang et al. (2020a)
Alginate	–	Methylene blue	High separation efficiency, selective adsorption	Wang et al. (2019a)
Cellulose	MF	Oil/water nanoemulsions	Hydrophilicity, high water flux	Hu et al. (2019)
Nanostructured cellulose	UF	Wastewater	High permeation flux, high separation efficiency, good flux recovery ratio, self-healing ability, hydrophilicity, antifouling properties	Yang et al. (2022)
Chitin	NF	Noble metal ions	Hydrophilicity and underwater oleophobic surface	Wang et al. (2019b)
Chitin, cellulose	–	Hg(II)	Easy regeneration	Tang et al. (2011)
Chitosan, bacterial cellulose	NF	Dyes	Antifouling properties for oil and protein	Zhijiang et al. (2018)
Chitosan	NF	Dyes, salts from textile water	High rejection rate and stability	Long et al. (2020)
Chitosan	MF	Oil droplets	High rejection rate, increased water flux, antifouling properties	Wan et al. (2022)
κ-carrageenan	NF	Emulsions, metal ions, organic dyes	High flux, antifouling properties, superoleophobicity	Prasannan et al. (2020)
Polylactic acid	UF	Organic compound	Antifouling properties	Khalil et al. (2022)

et al. 2023). The effectiveness of this method is supported by numerous studies, as extensively discussed in various review articles (Jahan and Zhang 2022; Abdulhamid and Muzamil 2023; Rocha et al. 2024). Furthermore, the natural polymers used in membrane production for wastewater treatment offer a high degree of modifiability through techniques such as crosslinking, grafting, acetylation, and quaternization, enabling enhanced adsorption properties, selectivity, affinity, and antifouling characteristics compared to traditional membranes (Saiyad et al. 2024).

Superior functional properties, such as high hydrophilicity, enhanced water permeability, adjustable surface charge, abundant surface functional groups, high porosity, and antifouling characteristics, characterize hybrid membranes enhanced

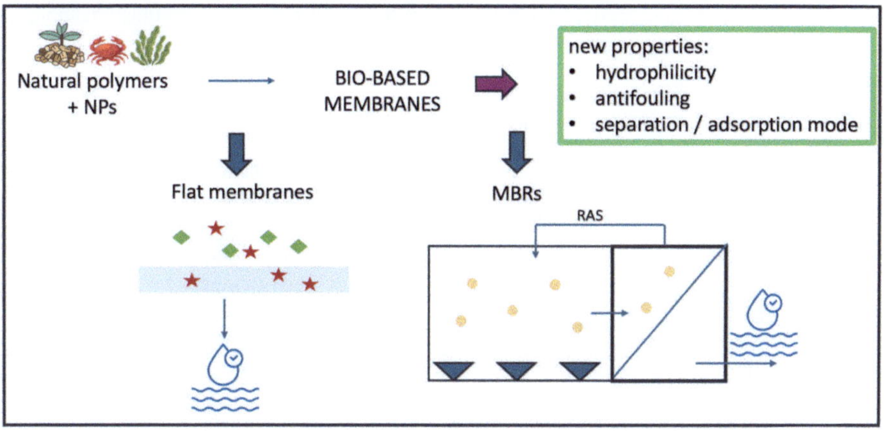

Fig. 16.5 Bio-based membranes for wastewater treatment, (RAS-returned activated sludge)

by biocomponents. Numerous review articles extensively document the advantages of incorporating bio-based additives in the production of wastewater treatment membranes (Xue et al. 2024; Yang et al. 2022).

Due to their abundance and unique properties, clay-based materials play a crucial role in specialized membrane production. Membranes produced with or enhanced by clay derivatives offer efficient, cost-effective solutions for wastewater treatment. These membranes have a high specific surface area, porous structure, high ion exchange capacity, excellent thermal stability, and low plasticity (Xue et al. 2024). Table 16.6 summarizes examples of commonly used clay-based materials for water remediation.

Another promising strategy involves hybrid membrane-adsorption systems, which enhance wastewater treatment by offering advantages such as excellent reusability, higher removal efficiency, faster kinetics, reduced membrane fouling, lower energy consumption, and decreased maintenance costs. Additionally, the versatility of operating in three modes-pre-adsorption, post-adsorption, and synergistic integration, enables the benefits of combined adsorption and membrane separation processes. This approach often incorporates natural polymers or bio-based nanomaterials and effectively removes heavy metals, dyes, pesticides, and pharmaceuticals from water. Examples of hybrid membrane-adsorption systems was discussed in the work of Davoodbeygi and co-workers (Davoodbeygi et al. 2023).

An alternative approach to membrane separation in water purification is the membrane bioreactor (MBR), which integrates biological processes with membrane filtration, Fig. 16.5. This solution enables efficient wastewater treatment through two system configurations: side-stream MBR and immersed MBR, which differ based on the placement of the membrane (Obotey Ezugbe and Rathilal 2020). Among the various types of membranes dedicated to MBR, polymers, ceramics and mixed matrix membranes (MMM), the last ones have gained great potential for wastewater

Table 16.6 Utilizing clay-based materials for the production of wastewater treatment membranes

Clay-based material	Main membrane component	Membrane type	Water pollution	Improved properties of membrane	References
Bentonite	Ceramic membrane made from bentonite	MF	Industrial wastewater	High mechanical properties, high retention rate	Bouazizi et al. (2016)
Bentonite	PVC	UF	Saline oily wastewater	Hydrophilicity, antifouling properties, high permeation rate	Ahmad et al. (2018)
Kaolinite	Ceramic membrane from Tunisian kaolin	UF	Textile wastewater	Good mechanical and chemical properties, high rejection rate	Bousbih et al. (2021)
Kaolinite	Natural phosphate and kaolinite	MF	Industrial wastewater	High permeability, porosity and flexural strength	Belgada et al. (2021)
Kaolinite	Ceramic membrane with perlite	MF	Dairy and textile wastewater	Regenerability, high flexural strength	Machtani Idrissi et al. (2023)
Zeolite	A-Al_2O_3 tubes	–	Desalination	High rejection, improving water flux	Wang et al. (2020a)
Zeolite	HFCM from natural zeolite	–	Ammonia	High rejection, high permeation flux	Adam et al. (2019)
Zeolite	PSF	UF	Heavy metals	Improved water hydraulic permeability	Yurekli (2016)
Potters' clay	Stone quarry dust, tea waste	MF	Methylene blue and congo red	Easy regeneration	Saikia et al. (2019)
Moroccan red clay	Natural phosphate	MF	Desalination, industrial wastewater	High mechanical stability	Mouiya et al. (2018)

PVC—poly(vinyl chloride), NPs—nanoparticles, CMC—carboxymethylated chitosan, HFCM—hollow fibre ceramic membrane, PSF—polysulfone

processing. They are characterized by high hydrophilicity, good chemical, mechanical, and thermal stability, and ease of synthesis (Abdullah et al. 2023). The efficiency of MBRs can be further enhanced by integrating traditional MBR processes with an additional separation unit, incorporating nanofiltration (NF) and reverse osmosis (RO) systems. This advanced configuration achieves high rejection rates for smaller dissolved pollutants, including humic-like substances, low-molecular-weight organics, heavy metals, and micropollutants (Kim et al. 2024), addressing one of the primary limitations of conventional MBR systems.

16.4.3 Bio-based Sensors

Sensors play a critical role in characterizing the quality of wastewater streams, ensuring their safe discharge and potential reuse. They are primarily focused on monitoring key parameters such as biological oxygen demand (BOD), total nitrogen (TN), and total phosphate (TP), as well as the detection of harmful compounds, including heavy metals, antibiotics, dyes, drugs, and pesticides (Burratti et al. 2024). Current technological solutions integrate smart technologies that enable real-time monitoring through soft-sensing techniques, utilizing Internet of Things (IoT) platforms, machine learning (ML) algorithms, and artificial intelligence (AI). Combined with bio-based materials, these advancements facilitate the development of high-performance sensors for effective water quality monitoring and management (Rando et al. 2024; Moretti et al. 2024).

The key component in sensor construction is the sensing element, connected to a transducer that converts physical or chemical changes into a measurable signal, as shown in Fig. 16.6. In biosensors, this sensing element is of biological origin (Fig. 16.6). The effectiveness and durability of sensors are closely linked to the physicochemical properties of the materials or carriers used in transducer production, which can be developed or enhanced using bio-based materials (Joshi et al. 2021).

The most widely used sensors for wastewater monitoring are optical and electrochemical sensors, both of which constructed using synthetic or natural components. Optical sensors offer advantages, including low production cost, ease of operation, and rapid detection. However, they suffer from limited sensitivity and specificity. In contrast, electrochemical sensors demonstrate high sensitivity, portability, and affordability, making them highly suitable for real-time water pollution monitoring, but they are limited by low specificity (Ateia et al. 2024).

Current trends in sensor construction focus on the utilization of polysaccharides, including natural and stimuli-responsive polymers, as well as biomass-derived carbon materials as sources of nanomaterials for the production of highly efficient transducers, Fig. 16.7. Nanomaterials due to their unique structural and physicochemical properties, such as enhanced mechanical strength, optical behavior, and electrical conductivity, which offer a high surface area-to-volume ratio, a critical factor in developing advanced sensing platforms. Integrating synthetic nanomaterials with bio-based components and green synthesis methods for producing nano-biomaterials from waste biomass, particularly lignocellulose, significantly enhances the potential for developing highly efficient sensors. This approach enables rapid, in-situ, and real-time detection of wastewater contaminants, providing improved sensitivity and selectivity for more accurate and sustainable monitoring solutions. Additionally, it

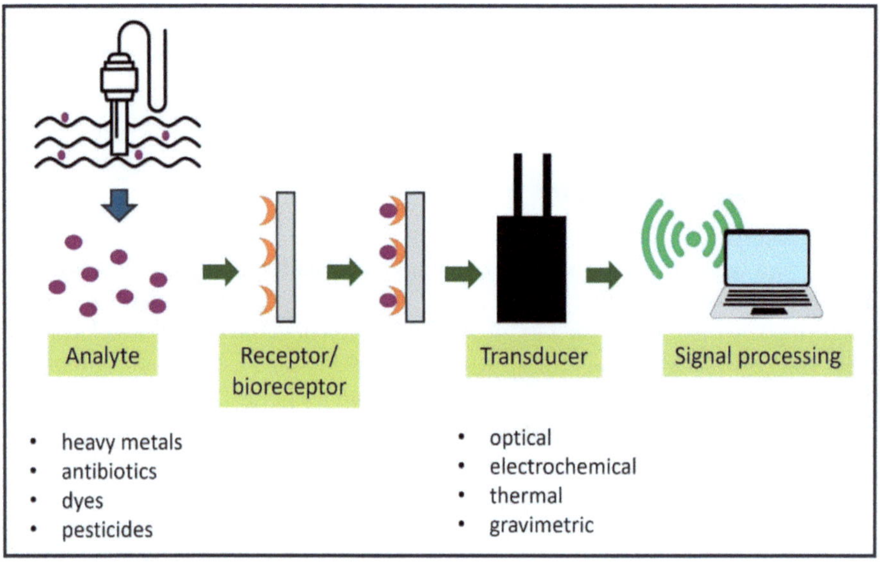

Fig. 16.6 Sensors/biosensors construction

minimizes the environmental impact of emerging technologies (Joshi et al. 2021; Aguilar-Pérez et al. 2020; Onfray and Thiam 2023).

Biomass-derived carbon nanomaterials, produced through pyrolysis or hydrothermal treatment, are particularly suitable for sensor technologies aimed at

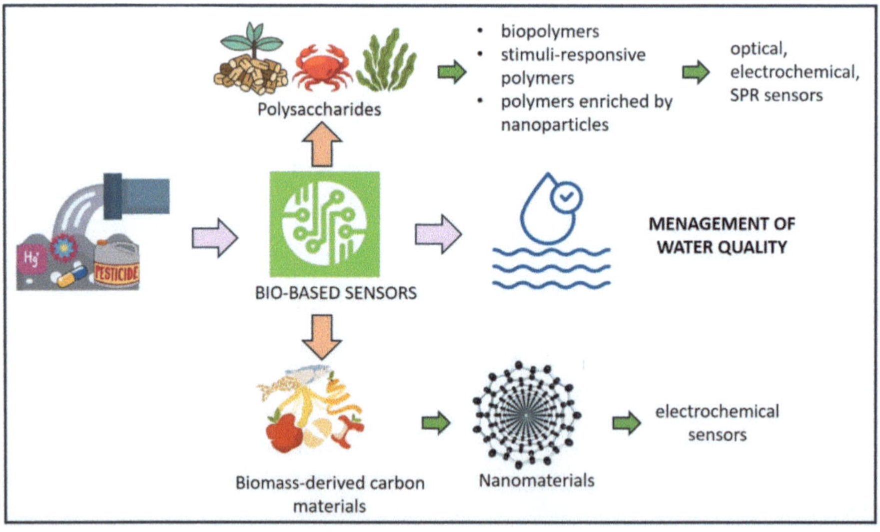

Fig. 16.7 Bio-based sensors in wastewater treatment

detecting phosphates, water hardness, pesticides, phenolic compounds, nitrogen species, and heavy metal ions. Their enhanced pore volumes and finely controlled microstructures enable these materials to outperform chemically synthesized alternatives (Mehmandoust et al. 2023). Examples of biomass-derived carbon material applications in wastewater treatment are shown in Table 16.7.

Polysaccharides (mainly chitosan, starch, cellulose, gelatine, dextran, agarose, alginate, and carrageenan), alongside nanomaterials and composites, have recently

Table 16.7 Electrochemical sensors based on biomass-derived carbon nanomaterials for wastewater treatment

Biomass-derived carbon nanomaterials	Biomass source	Water pollution	Detection limit	References
CNTs	Corn flour	Para-cresol	0.05 ppm	Zhao et al. (2021)
MCNPs	Corn powder	Pb(II) Cd(II)	1.72 pm 1.58 pm	Zeinu et al. (2016)
MNPs	*Areca nut* seeds	Nitrate	28.9 mA μM^{-1} cm^{-2}	Yallappa et al. (2018)
HAC	Banana stems	Nitrite	0.07 μm	Madhu et al. (2014a)
NDC	Bamboo fungi	Bisphenol A	1.068 μm	Xu et al. (2017)
N-CDs	Kiwi and sesame seeds	Nitrate	0.23 μm	Li et al. (2019)
N-NPC	Almond shells	Pb(II)	0.7 μg L^{-1}	Baikeli et al. (2019)
AC	Mango leaves	4-nitrophenol	0.16 μM	Madhu et al. (2014b)
AC	*Platanus* seeds	Pb(II) Cd(II)	0.13 μg L^{-1} 0.08 μg L^{-1}	Zhu et al. (2020)
CQD	Chia seeds	Hydrazine	39.7 μM	Sha et al. (2018)
OMCNs	Crab shell	Malachite green	0.05 μM	Yang et al. (2020b)
NBCF	Sweetgum fruit	Cd(II) Pb(II) Cu(II) Hg(II)	21.1 nM 7.72 nM 3.85 nM 1.68 nM	Dai et al. (2023)
Carbon paste	Coffee husks	Methylene blue	3 μM	Njanja et al. (2019)

CNTs—carbon nanotubes, MCNPs—mesoporous carbon nanoparticles, N-NPC—nitrogen-doped activated nanoporous carbon, HAC—heteroatom-enriched activated carbon, N-CDs—nitrogen Doped Carbon Dots, AC—activated carbon, NDC—nitrogen-doped carbon, CQD—carbon quantum dots, OMCNs—ordered mesoporous carbon nanofibers, NBCF—nano-black carbon film become widely used components in the transducers of sensors, including optical, electrical, labelled, and label-free types, proposed for detecting water pollutants (Adampourezare et al. 2024). These natural, bio-based sensing materials offer an appealing alternative to synthetic resins and materials whose use necessitates proper waste management. The diverse functional groups in polysaccharide structures are closely linked to their specific sensing performance and play a key role in wastewater sensing for detecting heavy metals, organic pollutants, pathogens, nutrients or environmental parameters. The interactions between polysaccharides and chemical analytes typically involve chelation, electrostatic interactions, and hydrogen bonding (Rando et al. 2024; Azeman et al. 2020).

Emerging technological strategies for sensor design incorporate cross-linking methodologies, wherein nanoparticles are embedded into hydrogel matrices to enhance the mechanical strength and flexibility of the hybrid materials employed as nano-bio-sensors (Sethi et al. 2023). These sensors can be classified based on transducer properties into electrochemical, optical, calorimetric, and mass-based biosensors (Aguilar-Pérez et al. 2020). Integrating nanoparticles with biopolymers or polysaccharides leverages the unique properties of each material, facilitating the creation of highly sensitive and selective sensors for the detection of water pollutants. Representative examples of these sensor applications are shown in Table 16.8.

A significant class of polysaccharides utilized in sensor development includes stimuli-responsive smart hydrogels, which enable both adsorption and detection of

Table 16.8 Examples of polysaccharide-nanoparticle sensors for the detection of water pollutants

Polysaccharide	Nanomaterial	Water pollution	Detection limit	Detection signal	References
Agarose	CuNPs	Pesticides, dimethoate	1.0 μg l^{-1}	Fluorescence	Kong et al. (2019)
Alginate	CDs	Fe(III)	1.42 μm	Fluorescence	Wang et al. (2018)
Carrageenan	AgNPs	Cu(II)	1.7 μm	Colorimetric	Wang et al. (2020b)
Cellulose	AuNCs	Hg(II)	0.09 μg/L	Fluorescence	Luo et al. (2021)
Chitosan	GO	Tetracycline	3.2×10^{-16} M	Electrochemical	Benvidi et al. (2022)
Dextran	MWCNT NGC	Cu(II)	0.63×10^{-6} M	Electrochemical	Zhao et al. (2019)
Gelatin	AgNPs	Hg(II)	2.37 μm	Colorimetric	Saenchoopa et al. (2021)
Starch	PTs/COF	Cd(II)	0.012 nm	Electrochemical	Li et al. (2023a)

AgNPs—silver nanoparticles, GO—graphene oxide, PTs—high affinity peptides, COF—covalent organic framework, MWCNT—multi-walled carbon nanotube, NGC—nano-gold composite, CDs—carbon dots, AuNCs—gold nanoclusters, CuNPs—copper nanoparticles water pollutants. They exhibit the ability to undergo conformational or structural changes in response to variations in environmental conditions or external stimuli such as temperature, pH, light, electric and magnetic fields, CO_2 concentration, or specific ions and molecules. The mechanism underlying their responsiveness is based on dynamic internal changes within the hydrogel matrix (Abousalman-Rezvani et al. 2022). Although the fabrication process of stimuli-responsive hydrogels is more complex than that of traditional sensors—requiring careful selection of polymers (responsive polymers, smart copolymers, or functional responsive groups) and cross-linking methods (chemical or physical)—the enhanced performance and functionality of these systems outweigh the mentioned challenges (Rando et al. 2024). Sensors based on stimuli-responsive hydrogels enable real-time data generation thanks to observable changes such as color shifts, swelling, or conductivity alterations, which provide crucial information for effective water quality management (Yang et al. 2024). Additionally, functionalized hydrogels exhibit selective adsorption of heavy metal ions through chelation, effectively removing these contaminants from the water stream and preventing their further accumulation (Li et al. 2023b). The use of stimuli-responsive hydrogels in wastewater treatment is well-documented in the literature. For nutrient contaminants such as nitrates and phosphates, the hydrogel response is typically pH-dependent, involving targeted ion binding through ionophores or interactions between nutrient ions and chromophores, which produces a colorimetric response (Rando et al. 2024). For dye detection, chitosan and PVA hydrogels containing GO nanosheets and Ce-ZnO nanoparticles respond to pH changes, enhancing sensitivity (Katowah et al. 2023). Additionally, fluorescent chitosan-gelatine hydrogels are effective for Cr^{6+} monitoring (Sethi et al. 2022).

Another valuable group of sensors in wastewater processing are Surface Plasmon Resonance (SPR) sensors, known for their high sensitivity, reliability, and cost-effectiveness compared to traditional optical sensors. In this approach, polysaccharides or nanoparticles derived from biomass function as effective sensing materials in SPR sensors, facilitating the detection of metal ions (Azeman et al. 2020). Examples of SPR sensors incorporating bio-based materials are summarized in Table 16.9.

16.4.4 Bio-based Materials with Immobilized Cells or Enzymes

One of the most promising strategies for the sustainable remediation of pollutants from wastewater is using biocatalysts immobilized in bio-based materials. They may be in the form of whole microbial cells or isolated enzyme molecules immobilized in a carrier of natural origin. Such systems offer a robust approach for the effective biodegradation of effluents containing a variety of hazardous chemicals, like pesticides, oils, azo dyes, phenolic compounds, polycyclic aromatic hydrocarbons (PAHs), or pharmaceutical residues (Fig. 16.8) (Zdarta et al. 2021; Najim et al. 2024;

Table 16.9 SPR sensors with bio-based materials for wastewater monitoring

Bio-based materials	Metallic layer	Water pollution	Detection limit	References
Alginate	Ag	Hg(II)	5.29 nm	Faghiri and Ghorbani (2019)
AuNPs from *Padina australis*	Au	Cu(II)	0.43 µm	Wu et al. (2023)
Carrageenan	Ag	NH_4^+	0.336 ppm	Bakar et al. (2024)
Carrageenan	Au	Pb(II)	1.354 ppm	Abdullah et al. (2018)
Chitosan	Au	Glyphosate	8 nm	Do et al. (2020)
Chitosan	Ag/ITO	Cd(II) Pb(II) Hg(II)	0.256 ppb 0.440 ppb 0.796 ppb	Verma and Gupta (2015)
CNCs	Au/V	Ni(II)	2–50 ppb	Omar et al. (2021)
CNCs	Au	Hg(II)	2 ppb	Ramdzan et al. (2021)
Starch	Fe/Ag	Hg(II)	1.48 nm	Balasurya et al. (2021)

CNCs—cellulose nanocrystals, ITO—indium tin oxide

Bouabidi et al. 2019). They can also serve as highly sensitive and selective biosensors for detecting a range of contaminants present in wastewater.

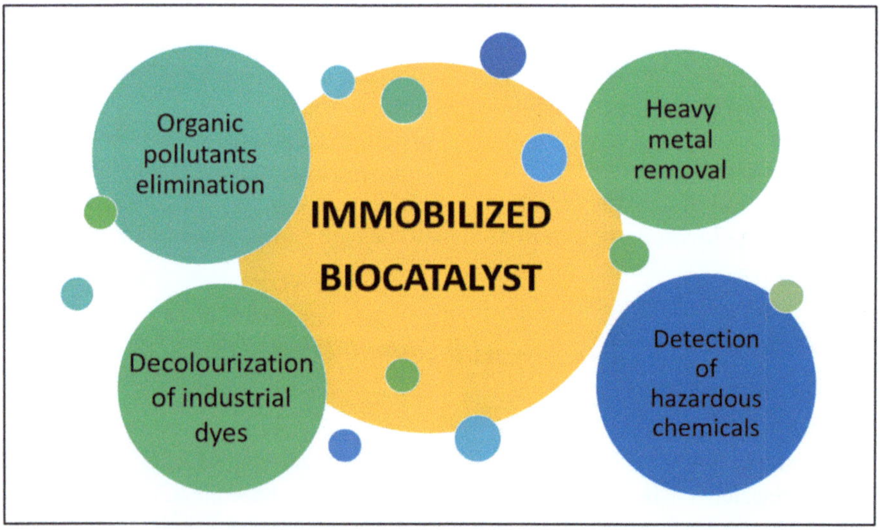

Fig. 16.8 Feasible applications of immobilised biocatalysts in wastewater treatment

Biocatalytic reactions offer a number of benefits over classical processes involving chemical catalysts. They are carried out under mild temperature and pH conditions and can be performed even at very low substrate concentrations. Furthermore, the selectivity and stereospecificity of biocatalysts are highly advantageous (Hauer 2020). This potential is also effectively explored in treating multi-component wastewater (Bouabidi et al. 2019; Hedar et al. 2023; Kołodziejczak-Radzimska et al. 2021). However, biocatalysts in their native form are often susceptible to adverse conditions in the reaction microenvironment, including elevated temperature, highly acidic or alkaline pH, salinity, and the presence of organic solvents. Therefore, it is crucial to improve their operational stability before practical use. This is most often achieved through immobilization, which is briefly based on the retention of biocatalyst molecules in a strictly defined region, typically limited by a solid barrier or support structure (Mohidem et al. 2023; Najim et al. 2024). This treatment allows for the enhancement of biocatalyst robustness against unfavorable environmental factors, facilitates their separation from the reaction medium and enables their reuse. There are numerous approaches for immobilization of enzymes and whole cells. The main ones are shown in Fig. 16.9.

Some are based on the physical retention of native enzyme molecules or whole cells in a strictly defined space, limited by insoluble porous barriers, like **entrapment** and **encapsulation**. Others involve the creation of various types of interactions between the biocatalyst and the carrier (Mohidem et al. 2023; Najim et al. 2024). In this regard, three main techniques can be distinguished:

(1) **adsorption**, in which weak interactions, including Van der Waals forces, hydrogen bonds, hydrophobic interactions, and electrostatic forces stabilize the immobilized molecules;

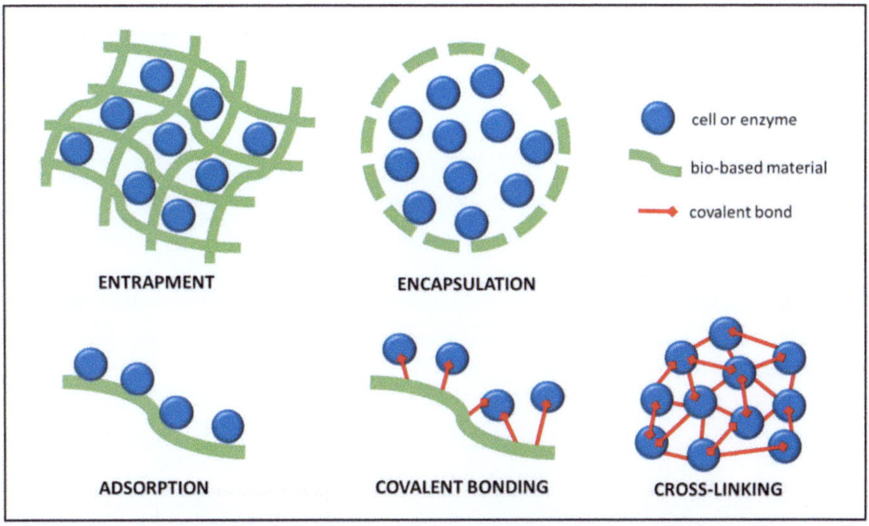

Fig. 16.9 Basic techniques used for enzyme and whole-cell immobilisation

(2) **covalent bonding**, which involves the formation of strong chemical bonds between the functional groups of the carrier and the biocatalyst;
(3) **cross-linking** is based on creating multiple covalent bonds directly between enzyme molecules or whole cells.

The aforementioned methods exhibit a range of characteristic features outlined in Table 16.10.

Consequently, the optimal choice is closely related to the intended use of a given biocatalyst. In the case of wastewater treatment, it is of the utmost importance that the immobilization method should be cost-effective, straightforward, and expeditious while also facilitating the creation of a highly active and long-lasting biocatalyst. Furthermore, using bio-based materials as biocatalyst carriers presents an additional advantage in this sector. They are biodegradable, environmentally friendly and represent a sustainable approach to wastewater treatment.

Table 16.10 Comparison of characteristic features for different immobilization techniques (Mohidem et al. 2023; Najim et al. 2024)

Feature	Immobilization technique				
	Entrapment	Encapsulation	Adsorption	Covalent bonding	Cross-linking
Methodology	Moderate	Moderate	Simple	Multistep and complicated	Moderate
Binding force	Moderate	Weak	Weak	Strong	Strong
Carrier regeneration	Impossible	Impossible	Possible	Impossible	–[a]
Rigidification of biocatalyst conformation	Low	Low	Low	High	High
Retention of activity	High	High	High	Moderate	Low
Biocatalyst stability	Moderate	Moderate	Moderate	High	High
Leakage of biocatalyst	Moderately possible	Moderately possible	Highly possible	Not noticed	Not noticed
Biocatalyst reusability	Moderate	Moderate	Low	High	High
Processing costs	Low	Low	Low	High	Moderate

[a] Carrier-free method

16.4.4.1 Bio-based Materials Containing Immobilized Cells for Wastewater Treatment

Living cells are usually immobilized using methods that are least invasive to their structure, i.e. by trapping, encapsulation, or adsorption. Thanks to this, they are protected against unfavorable environmental conditions and, at the same time, retain all their vital functions and activity properties. They can, therefore, be efficient micro-factories that successfully process pollutants present in wastewater. Examples of particular applications of such systems are provided in Table 16.11.

16.4.4.2 Bio-based Materials Containing Immobilized Enzymes for Wastewater Treatment

Enzymes for wastewater treatment are typically immobilized via entrapment often followed by an additional cross-linking step, or by creating covalent bonds with the bio-based matrix. Both approaches guarantee a significant increase in the stability and reusability of such biocatalysts under process conditions. Importantly, using bio-based materials with immobilized enzymes enables long-term biocatalytic wastewater processing with the expected efficiency and respect to the environment. Moreover, this minimizes the overall costs and idle time associated with batch replacement of enzymes in their native form.

The most powerful tools for the pro-ecological removal of contaminants from water are enzymes belonging to oxidoreductases (i.e. laccases, peroxidases, tyrosinases) and hydrolases (i.e. lipases, galactosidases) (Hedar et al. 2023; Lim and Mohamad 2022). Combined with bio-based materials, they become highly efficient machines that enable the processing of various types of pollutants in wastewater while maintaining sustainable development and circular economy principles. The particular examples of using such heterogeneous bio-based catalysts are listed in Table 16.12.

In addition to bioconversion reactions, enzymes immobilized using bio-based materials can be effectively used as biosensors for monitoring contaminants in wastewater. Such bioactive systems offer enhanced sensitivity and specificity, allowing for precisely detecting pollutants like heavy metals, pesticides, or organic compounds. They are invaluable tools for environmental monitoring, ensuring accurate, real-time data on water quality. Similarly to biocatalytic treatment, the largest share in the construction of biosensors is held by bio-based materials with immobilized enzymes from the oxidoreductase and hydrolase classes. These biocatalysts are most often immobilized by cross-linking or trapping them within natural-originated materials. Selected examples of such systems are listed in Table 16.13.

Table 16.11 Selected examples of the application of whole cells immobilized with bio-based materials in removal of contaminants from wastewater

Bio-based material	Cell species	Removed compound	Removal efficiency	References
Ba-alginate cellulose triacetate	*Klebsiella oxytoca*	Potassium cyanide	0.224 mM h^{-1} 0.192 mM h^{-1}	Chen et al. (2008)
Ca-alginate Agar Carrageenian	*Pseudomonas putida*	Sodium cyanide	98,0% 67.9% 59.8%	Chapatwala et al. (1998)
Ca-alginate	*Chlorella vulgaris*	Cu(II) Ni(II)	~100%	Mehta and Gaur (2001)
Ca-alginate	*Chlorella vulgaris*	Ammonium Phosphate	100% ~95%	Tam and Wong (2000)
Ca-alginate	*Scenedesmus intermedius*	Phosphorus Ammonium	35.26 mg L^{-1} 16.45 mg L^{-1} (per day)	Jiménez-Pérez et al. (2004)
Ca-alginate	*Trametes hirsuta*	Indigo carmine Phenol red	96% 69%	Domínguez et al. (2005)
Ca-alginate	*Funalia trogii*	Acid black 52	93.8%	Park et al. (2006)
Ca-alginate	*Rhodopseudomonas palustris*	Sulfadiazine	86%	Zeng et al. (2023)
Agar-alginate algal blocks	*Picochlorum maculatum*	Nitrite Ammonia Phosphate Nitrate	89.6% 98.5% 57.0% 46.4%	Dinesh Kumar et al. (2016)
Bacterial cellulose	*Pseudomonas moorei*	Paracetamol	~100%	Żur et al. (2020)
Carboxymethyl cellulose	*Chlamydomonas reinhardtii*	U(VI)	196.8 mg/g	Erkaya et al. (2014)
Cellulose acetate nanofibers	*Aeromonas eucrenophila, Clavibacter michiganensis Pseudomonas aeruginosa*	Methylene blue	~95%	San et al. (2014)
Chitosan	*Bacillus subtilis*	Cu(II)	100.7 mg/g	Liu et al. (2013)
Chitosan	*Chlorella vulgaris*	Nitrate	87.0%	Eroglu et al. (2012)
Chitosan	*Scenedesmus* sp.	Nitrate Phosphate	70.0% 94.0%	Fierro et al. (2008)
Tea waste biomass	*Bacillus sp.*	Cr(VI) Phenol	741.4 mg/g 7.761 mg/g	Gupta and Balomajumder (2015)

(continued)

Table 16.11 (continued)

Bio-based material	Cell species	Removed compound	Removal efficiency	References
Bamboo charcoal Wood charcoal	Degrading bacterial consortium	Nonylphenol	94% 70%	Lou et al. (2019)
Bamboo charcoal—Na-alginate blend	*Bacillus cereus*	Dimethomorph	96%	Zhang et al. (2020b)
Bentonite	*Pleurotus ostreatus*	Cd(II) Pb(II)	≥95% ≥95%	Kocaoba and Arısoy (2018)
Biochar (rice hull)	*Leclercia adecarboxylata*	Pb(II)	95%	Teng et al. (2020)

16.5 Current Trends and Future Perspectives

Despite significant advancements in wastewater treatment, the effective management of water resources remains a critical challenge that demands ongoing innovation. Continuous improvement is necessary to address the ever-evolving pressures of population growth and the intensifying impacts of climate change, both of which place increasing demands on water availability and sustainability.

Current trends in using biomaterials for wastewater treatment focus on enhancing adsorption efficiency and selectivity for specific pollutants, as well as improving the mechanical properties of bio-based materials. New strategies must also address the phenomenon where biodegradable materials, despite their high adsorption efficiency, can become secondary contaminants after pollutant removal, necessitating specific post-treatment processes (Fakhri et al. 2024).

Among others, the currently observed directions of technological advancements in wastewater processing based on bio-based materials include:

- Addition of nanoparticles to enhance adsorption properties and long-lasting utilization of bio-based materials. In this case, a combination of nanoparticles, which are characterized by their high surface-to-volume ratio and extraordinary mechanical capabilities, with biopolymers, which are biocompatible and renewable, results in the production of highly stable nanocomposite-based sorbents that exhibit significantly accelerated efficiency (Al-Gethami et al. 2024).
- Bio-based materials are used as photocatalyst matrices for antibiotics degradation. Currently, the main attention of light-driven degradation of antibiotics and other antimicrobial compounds is focused on the conjugation of semiconductor photocatalysts to different organic supports (e.g. biochar or various biopolymers). This approach enables the creation of novel heterogeneous photocatalytic systems, which are distinguished by improved operational stability and reusability that results in enhanced degradation efficiency of target compounds (Maldonado-Carmona et al. 2024).

Table 16.12 Selected examples of the application of enzymes immobilized with bio-based materials in removal of contaminants from wastewater

Bio-based material	Enzyme	Removed compound	Removal efficiency	References
Cu-alginate	Laccase	Textile dye effluent	65%	Sondhi et al. (2018)
Ca-alginate	Laccase	Bisphenol A	100%	Lassouane et al. (2019)
Ca-alginate	Lipase	Fats from food industry	50%	Jeganathan et al. (2006)
Ca-alginate	Horseradish peroxidase	Acid orange Acid blue	75% 69%	Fathollah et al. (2011)
Ca-alginate	Tyrosinase	Phenol	90%	El-Shora and El-Sharkawy (2020)
Ca and Cu-alginate, chitosan composite	Laccase	17α-ethinylestradiol	80.5%	Garcia et al. (2019)
Alginate-guar gum; Agarose-guar gum	Ginger peroxidase	Textile dye effluent	80% 69%	Ali and Husain (2018)
Biochar	Laccase	Carbamazepine	86%	Naghdi et al. (2017)
Bentonite	Laccase	Tetracycline	60%	Wen et al. (2019)
Cellulose	β-galactosidase	Lactose	>90%	Roy and Gupta (2003)
Bagasse cellulose	Keratinase	Melanoidin	85–90%	Zhang et al. (2019)
Green coconut fiber	Laccase	Textile dyes	Up to 90%	Cristóvão et al. (2012)
Cashew polysaccharide	Horseradish peroxidase	Phenol	100%	Silva et al. (2019)
Chitosan	Manganese peroxidase	Textile dye effluent	97%	Bilal et al. (2016)
Chitosan	Horseradish peroxidase	Remazol brilliant blue R Reactive black 5 Congo red Crystal violet	82.2% 97.8% 94.4% 87.4%	Bilal et al. (2017)
Chitosan	Lipase	Oil and grease	48%	Dumore and Mukhopadhyay (2012)
Chitosan Activated carbon	Tyrosinase	Phenol	100%; 100%	Mello et al. (2023)

(continued)

Table 16.12 (continued)

Bio-based material	Enzyme	Removed compound	Removal efficiency	References
Pectin	Lipase	Lipids from dairy effluents	69.2%	Das Neves et al. (2024)
Magnetite	Horseradish peroxidase	Chlorophenols mixture	97–100%	Tatsumi et al. (2000)
Zein nanofibers	Tyrosinase	Methyl orange Congo red Dye mixture	100% 100% 97%	Suryamathi et al. (2021)

Table 16.13 Selected examples of the application of enzymes immobilized with bio-based materials as biosensors for monitoring contaminants in wastewater

Bio-based material	Enzyme	Detected compound	Lower detection limit	References
Agarose-guar gum	Tyrosinase	Catechol	6.0 μM	Tembe et al. (2007)
Agarose-guar gum	Laccase	Phenol	40 μM	Kushwah and Bhadauria (2010)
Ca-alginate	Haloalkane dehalogenase	1,2-dichloroetan	111 μM	Campbell et al. (2006)
Gelatine	Tyrosinase	Catechol Phenol *p*-cresol	2.1×10^{-2} μM 1.5×10^{-2} μM 7.1×10^{-2} μM	Li et al. (2005)
Chitosan	Horseradish peroxidase	Catechol Resorcinol	450 μM 90 μM	Dabhade et al. (2020)
Chitosan	Alkaline phosphatase	2,4-D[a] 2,4,5-T[a]	2.26×10^{-3} μM 3.13×10^{-3} μM	Shyuan et al. (2008)
Chitosan	Glucose oxidase	Cr(VI)	9.6×10^{-1} μM	Dabhade et al. (2021)
Bovine serum albumin	Catalase	Pb(II)	1.8×10^{-5} μM	Elsebai et al. (2017)
Bovine serum albumin	Urease	Hg(II) Cu(II) Cd(II) Co(II) Pb(II) Sr(II)	1.0 μM 2.0 μM 5.0 μM 10 μM 20 μM 100 μM	Zhylyak et al. (1995)
Sugarcane biochar	Tyrosinase	Bisphenol A	3.18×10^{-3} μM	Liu et al. (2019)

[a] 2,4-D-2,4-dichlorophenoxyacetic acid; 2,4,5-T-2,4,5-trichlorophenoxyacetic acid

- Floating phytoremediation, which uses aquatic macrophytes to remove organic matter, nutrients, heavy metals, and pathogens from wastewater through bioaccumulation. Followed by converting the obtained biomass into bioenergy, animal feed, or fertilizer, which aligns this method with circular economy principles and enhances the sustainability of this approach (Sayanthan et al. 2024).
- 3D-printed bio-based materials for wastewater treatment, where biopolymers such as chitosan or alginate enable the production of filters, meshes, and membranes that effectively remove heavy metals, dyes, and oil contaminants. Moreover, these systems offer enhanced efficiency, reduced processing time, and lower costs, making them a promising alternative for advanced water treatment technologies (Roy Barman et al. 2023; Fijoł et al. 2022).
- Microbial biomass for the production of microbial fuel cells (MFCs), which presents an innovative solution combining wastewater treatment with power generation. Such systems can be effectively used to monitor water pollution. In this respect, MFC-based biosensors hold significant potential due to their ability to measure key parameters of water, such as biochemical oxygen demand (BOD), chemical oxygen demand (COD), dissolved oxygen (DO), and the presence of toxic compounds. This makes them a promising tool for efficient and real-time wastewater analysis (Sonawane et al. 2020).
- Focusing on the development of green technologies for the sustainable production of bio-based sorbents for wastewater treatment. The main goal of this approach is to obtain high-performance sorption biomaterials with the least complicated production route and the lowest financial outlays. Consequently, an increasing prevalence of the utilization of agricultural waste biomass in production technologies is currently observed. Among others, the use of fruit and vegetable peels in water depollution technology is very encouraging, given the near-zero cost of synthesizing these adsorbents (Boussaksou et al. 2025).
- Material innovations concerning the production of switchable biomaterials for wastewater treatment. In comparison to conventional matrices, stimuli-responsive ones can be regenerated and reused multiple times without significant loss of their functional properties. Moreover, switchable biomaterials often exhibit high specificity and efficiency in pollutant removal. The most advantageous aspect is that their physicochemical properties could be tailored by modifying their structural performance by incorporating various reactive groups. The ability to fine-tune these materials for preferential action to specific contaminants widely extends their practical applicability across a range of wastewater treatment sectors (Wang et al. 2025).

Although biomaterials have great potential in wastewater treatment, their use is also associated with some significant limitations. In this respect, the main challenges remain in enhancing scalability, recyclability, and multi-contaminant adaptability. In the future perspective research should focus on developing low-energy production technologies and maximizing functional properties and long-term stability of these materials under real-world wastewater treatment conditions.

References

Abd El-Ghany NA, Elella MHA, Abdallah HM, Mostafa MS, Samy M (2023) Recent advances in various starch formulation for wastewater purification via adsorption technique: a review. J Polym Environ 31:2792–2825. https://doi.org/10.1007/s10924-023-02798-x

Abdulhamid MA, Muzamil K (2023) Recent progress on electrospun nanofibrous polymer membranes for water and air purification: a review. Chemosphere 310:136886. https://doi.org/10.1016/j.chemosphere.2022.136886

Abdullah S, Azeman NH, Mobarak NN, Zan MSDA, Bakar AA (2018) Sensitivity enhancement of localized SPR sensor towards Pb(II) ion detection using natural bio-polymer based carrageenan. Optik 168:784–793. https://doi.org/10.1016/j.ijleo.2018.05.016

Abdullah R, Astira D, Widyanto AR, Dharma HNC, Hidayat ARP, Santoso L, Sulistiono DO, Rahmawati Z, Gunawan T, Jaafar J, Kusumawati Y, Othman MHD, Fansuri H (2023) Recent development of mixed matrix membrane as a membrane bioreactor for wastewater treatment: a review. Case Stud Chem Environ Eng 8:100485. https://doi.org/10.1016/j.cscee.2023.100485

Abousalman-Rezvani Z, Roghani-Mamaqani H, Riazi H, Abousalman-Rezvani O (2022) Water treatment using stimuli-responsive polymers. Polym Chem 13:5940–5964. https://doi.org/10.1039/D2PY00992G

Adam MR, Matsuura T, Othman MHD, Puteh MH, Pauzan MAB, Ismail AF, Mustafa AA, Rahman M, Jaafar J, Abdullah MS (2019) Feasibility study of the hybrid adsorptive hollow fibre ceramic membrane (HFCM) derived from natural zeolite for the removal of ammonia in wastewater. Process Saf Environ Protect 122:378–385). https://doi.org/10.1016/j.psep.2018.12.003

Adampourezare M, Nikzad B, Nasrollahzadeh S, Asadpour-Zeynali K, De La Guardia M, Ezzati Nazhad Dolatabadi J, Zhang F, Mahdi Jafari S (2024) Polysaccharide-based sensors and nanosensors: a review of recent progress and challenges. Microchem J 204:110944. https://doi.org/10.1016/j.microc.2024.110944

Aguilar-Pérez KM, Heya MS, Parra-Saldívar R, Iqbal HMN (2020) Nano-biomaterials in-focus as sensing/detection cues for environmental pollutants. Case Stud Chem Environ Eng 2:100055. https://doi.org/10.1016/j.cscee.2020.100055

Ahmad T, Guria C, Mandal A (2018) Synthesis, characterization and performance studies of mixed-matrix poly(vinyl chloride)-bentonite ultrafiltration membrane for the treatment of saline oily wastewater. Process Saf Environ Prot 116:703–717. https://doi.org/10.1016/j.psep.2018.03.033

Akpakpan A, Inam E, Itoro T (2023) Preparation, characterization and application of soda lignin and its ester derivatives as adsorbents in the adsorption of Pb^{2+} and Cd^{2+} from wastewater. J Mater Environ Sci 14:82–96

Al-Gethami W, Qamar MA, Shariq M, Alaghaz A-NMA, Farhan A, Areshi AA, Alnasir MH (2024) Emerging environmentally friendly bio-based nanocomposites for the efficient removal of dyes and micropollutants from wastewater by adsorption: a comprehensive review. RSC Adv 14:2804–2834. https://doi.org/10.1039/D3RA06501D

Ali M, Husain Q (2018) Guar gum blended alginate/agarose hydrogel as a promising support for the entrapment of peroxidase: stability and reusability studies for the treatment of textile effluent. Int J Biol Macromol 116:463–471. https://doi.org/10.1016/j.ijbiomac.2018.05.037

Ali I, Basheer AA, Mbianda XY, Burakov A, Galunin E, Burakova I, Mkrtchyan E, Tkachev A, Grachev V (2019) Graphene based adsorbents for remediation of noxious pollutants from wastewater. Environ Int 127:160–180. https://doi.org/10.1016/j.envint.2019.03.029

Al-Ma'abreh AM, Hmedat DA, Edris G, Hamed MA (2024) Removal of heavy metal ions (Pb2+, Co2+, and Cd2+) by activated carbon from cypress fruit: an investigation of kinetics, thermodynamics, and isotherms. J Chem 1–13. https://doi.org/10.1155/2024/1984821

Alp Arıcı T, Özcan AS, Özcan A (2020) Biosorption characteristics of Cu(II) and Cd(II) ions by modified alginate. J Polym Environ 28:3221–3234. https://doi.org/10.1007/s10924-020-01844-2

Amiri MJ, Raayatpisheh M, Radi M, Amiri S (2023) Preparation and characterization of biopolymer-based adsorbents and their application for methylene blue removal from wastewater. Sci Rep 13:17263. https://doi.org/10.1038/s41598-023-44613-6

Annu A, Mittal M, Tripathi S, Shin DK (2024) Biopolymeric nanocomposites for wastewater remediation: an overview on recent progress and challenges. Polymers 16:294. https://doi.org/10.3390/polym16020294

Ateia M, Wei H, Andreescu S (2024) Sensors for emerging water contaminants: overcoming road-blocks to innovation. Environ Sci Technol 58:2636–2651. https://doi.org/10.1021/acs.est.3c09889

Azeman NH, Arsad NA, Bakar AA (2020) Polysaccharides as the sensing material for metal ion detection-based optical sensor applications. Sensors 20:3924. https://doi.org/10.3390/s20143924

Barman P, Kadam R, Kumar A (2022) Lignin-based adsorbent for effective removal of toxic heavy metals from wastewater. Emergent Mater 5:923–943. https://doi.org/10.1007/s42247-021-00311-5

Baig N, Ihsanullah I, Sajid M, Saleh TA (2019) Graphene-based adsorbents for the removal of toxic organic pollutants: a review. J Environ Manag 244:370–382. https://doi.org/10.1016/j.jenvman.2019.05.047

Baikeli Y, Mamat X, Yalikun N, Wang Y, Qiao M, Li Y, Hu G (2019) Differential pulse voltammetry detection of Pb(II) using nitrogen-doped activated nanoporous carbon from almond shells. RSC Adv 9:23678–23685. https://doi.org/10.1039/C9RA03925B

Bak S, Osten E, Ekstam K, Harnulv G, Pajunen E, Prucha P, Rasi J (2001) European brewery convention: water in brewing, manual of good practice. In: European brewery convention, Nurnberg

Bakar MHA, Azeman NH, Mobarak NN, Nazri NAA, Mohd Daniyal WMEM, Zan MSD, Mahdi MA, Zain ARM, Gupta R, Abdullah F, Bakar AAA (2024) Real-time cationic sensing using plasmonic fiber optic sensor based phosphoryl-carrageenan. Sens Actuators A: Phys 377:115725. https://doi.org/10.1016/j.sna.2024.115725

Balasurya S, Syed A, Swedha M, Harini G, Elgorban AM, Zaghloul NS, Das A, Khan SS (2021) A novel SPR based Fe@Ag core–shell nanosphere entrapped on starch matrix an optical probe for sensing of mercury(II) ion: a nanomolar detection, wide pH range and real water sample application. Spectrochim Acta Part A: Mol Biomol Spectrosc 263:120204. https://doi.org/10.1016/j.saa.2021.120204

Bashir A, Manzoor T, Malik LA, Qureashi A, Pandith AH (2020) Enhanced and selective adsorption of Zn(II), Pb(II), Cd(II), and Hg(II) Ions by a Dumbbell- and flower-shaped potato starch phosphate polymer: a combined experimental and DFT calculation study. ACS Omega 5:4853–4867. https://doi.org/10.1021/acsomega.9b03607

Belgada A, Charik FZ, Achiou B, Ntambwe Kambuyi T, Alami Younssi S, Beniazza R, Dani A, Benhida R, Ouammou M (2021) Optimization of phosphate/kaolinite microfiltration membrane using Box-Behnken design for treatment of industrial wastewater. J Environ Chem Eng 9:104972. https://doi.org/10.1016/j.jece.2020.104972

Belhouchat N, Zaghouane-Boudiaf H, Viseras C (2017) Removal of anionic and cationic dyes from aqueous solution with activated organo-bentonite/sodium alginate encapsulated beads. Appl Clay Sci 135:9–15. https://doi.org/10.1016/j.clay.2016.08.031

Benvidi A, Nikmanesh M, Dehghan Tezerjani M (2022) Graphene oxide/chitosan based impedimetric aptasensor along with an ester linker for the detection of tetracycline. jns.kashanu.ac.ir 12. https://doi.org/10.22052/JNS.2022.01.020

Bessaha H, Bouraada M, De Ménorval LC (2017) Removal of an acid dye from water using calcined and uncalcined ZnAl-r anionic clay. Water Environ Res 89:783–790. https://doi.org/10.2175/106143017X14902968254809

Bhunia P, Kim G, Baik C, Lee H (2012) A strategically designed porous iron–iron oxide matrix on graphene for heavy metal adsorption. Chem Commun 48:9888. https://doi.org/10.1039/c2cc35120j

Bilal M, Asgher M, Iqbal M, Hu H, Zhang X (2016) Chitosan beads immobilized manganese peroxidase catalytic potential for detoxification and decolorization of textile effluent. Int J Biol Macromol 89:181–189. https://doi.org/10.1016/j.ijbiomac.2016.04.075

Bilal M, Iqbal HMN, Hu H, Wang W, Zhang X (2017) Enhanced bio-catalytic performance and dye degradation potential of chitosan-encapsulated horseradish peroxidase in a packed bed reactor system. Sci Total Environ 575:1352–1360. https://doi.org/10.1016/j.scitotenv.2016.09.215

Biswal BK, Balasubramanian R (2023) Use of biochar as a low-cost adsorbent for removal of heavy metals from water and wastewater: a review. J Environ Chem Eng 11:110986. https://doi.org/10.1016/j.jece.2023.110986

Boccia AC, Neagu M, Pulvirenti A (2023) Bio-based aerogels for the removal of heavy metal ions and oils from water: novel solutions for environmental remediation. Gels 10:32. https://doi.org/10.3390/gels10010032

Bouabidi ZB, El-Naas MH, Zhang Z (2019) Immobilization of microbial cells for the biotreatment of wastewater: a review. Environ Chem Lett 17:241–257. https://doi.org/10.1007/s10311-018-0795-7

Bouazizi A, Saja S, Achiou B, Ouammou M, Calvo JI, Aaddane A, Younssi SA (2016) Elaboration and characterization of a new flat ceramic MF membrane made from natural Moroccan bentonite. Application to treatment of industrial wastewater. Appl Clay Sci 132–133, 33–40. https://doi.org/10.1016/j.clay.2016.05.009

Bousbih S, Errais E, Darragi F, Duplay J, Trabelsi-Ayadi M, Daramola MO, Ben Amar R (2021) Treatment of textile wastewater using monolayered ultrafiltation ceramic membrane fabricated from natural kaolin clay. Environ Technol 42:3348–3359. https://doi.org/10.1080/09593330.2020.1729242

Boussaksou I, Aoulad El Hadj Ali Y, Azzouz A, Stitou M (2025) Recent trends in biosorption: the removal of emerging dye pollutants from aqueous medium.Euro-Mediterr J Environ Integr 10:161–193. https://doi.org/10.1007/s41207-024-00671-4

Burratti L, Sgreccia E, Bertelà F, Galiano F (2024) Metal nanostructures in polymeric matrices for optical detection and removal of heavy metal ions, pesticides and dyes from water. Chemosphere 362:142636. https://doi.org/10.1016/j.chemosphere.2024.142636

Campbell DW, Müller C, Reardon KF (2006) Development of a fiber optic enzymatic biosensor for 1,2-dichloroethane. Biotechnol Lett 28:883–887. https://doi.org/10.1007/s10529-006-9014-x

Cashin VB, Eldridge DS, Yu A, Zhao D (2018) Surface functionalization and manipulation of mesoporous silica adsorbents for improved removal of pollutants: a review. Environ Sci: Water Res Technol 4:110–128. https://doi.org/10.1039/C7EW00322F

Chaari I, Fakhfakh E, Medhioub M, Jamoussi F (2019) Comparative study on adsorption of cationic and anionic dyes by smectite rich natural clays. J Mol Struct 1179:672–677. https://doi.org/10.1016/j.molstruc.2018.11.039

Chapatwala KD, Babu GRV, Vijaya OK, Kumar KP, Wolfram JH (1998) Biodegradation of cyanides, cyanates and thiocyanates to ammonia and carbon dioxide by immobilized cells of Pseudomonas putida. J Ind Microbiol Biotechnol 20:28–33. https://doi.org/10.1038/sj.jim.2900469

Chen CY, Kao CM, Chen SC (2008) Application of Klebsiella oxytoca immobilized cells on the treatment of cyanide wastewater. Chemosphere 71:133–139. https://doi.org/10.1016/j.chemosphere.2007.10.058

Chen ZM, Xie W, Zhang HY (2012) Error of acid rain research, effect of atmosphere CO_2 ignored. AMR 610–613:381–384. https://doi.org/10.4028/www.scientific.net/AMR.610-613.381

Chowdhury S, Balasubramanian R (2014) Recent advances in the use of graphene-family nanoadsorbents for removal of toxic pollutants from wastewater. Adv Coll Interface Sci 204:35–56. https://doi.org/10.1016/j.cis.2013.12.005

Chundawat SPS, Beckham GT, Himmel ME, Dale BE (2011) Deconstruction of lignocellulosic biomass to fuels and chemicals. Annu Rev Chem Biomol Eng 2:121–145. https://doi.org/10.1146/annurev-chembioeng-061010-114205

Costanzo F, Silvestrelli PL, Ancilotto F (2012) Physisorption, diffusion, and chemisorption pathways of H2 molecule on graphene and on (2,2) carbon nanotube by first principles calculations. J Chem Theory Comput 8:1288–1294. https://doi.org/10.1021/ct300143a

Cottet L, Almeida CAP, Naidek N, Viante MF, Lopes MC, Debacher NA (2014) Adsorption characteristics of montmorillonite clay modified with iron oxide with respect to methylene blue in aqueous media. Appl Clay Sci 95:25–31. https://doi.org/10.1016/j.clay.2014.03.023

Cristóvão RO, Silvério SC, Tavares APM, Brígida AIS, Loureiro JM, Boaventura RAR, Macedo EA, Coelho MAZ (2012) Green coconut fiber: a novel carrier for the immobilization of commercial laccase by covalent attachment for textile dyes decolourization. World J Microbiol Biotechnol 28:2827–2838. https://doi.org/10.1007/s11274-012-1092-4

Cuccarese M, Van Hulle SWH, Mancini IM, Masi S, Caniani D (2023) Removal of organic micropollutants from water by adsorption on thermo-plasma expanded graphite encapsulated into calcium alginate. J Environ Health Sci Eng 21:497–512. https://doi.org/10.1007/s40201-023-00876-9

Dabhade A, Jayaraman S, Paramasivan B (2020) Colorimetric paper bioassay by horseradish peroxidase for the detection of catechol and resorcinol in aqueous samples. Prep Biochem Biotechnol 50:849–856. https://doi.org/10.1080/10826068.2020.1760883

Dabhade A, Jayaraman S, Paramasivan B (2021) Development of glucose oxidase-chitosan immobilized paper biosensor using screen-printed electrode for amperometric detection of Cr(VI) in water. 3 Biotech 11:183. https://doi.org/10.1007/s13205-021-02736-5

Dai K, Zhao G, Wang Z, Peng X, Wu J, Yang P, Li M, Tang C, Zhuang W, Ying H (2021) Novel mesoporous lignin-calcium for efficiently scavenging cationic dyes from dyestuff effluent. ACS Omega 6:816–826. https://doi.org/10.1021/acsomega.0c05401

Dai M, Xia J, Xue Z, Liu S, Pan X, Dong X, Wang J, Zhou G, Jiang D, Li Q (2023) Biomass-derived nano-black carbon film electrode for the electrochemical detection of hazardous ions. Microchem J 194:109267. https://doi.org/10.1016/j.microc.2023.109267

Das D, Nandi BK (2021) Treatment of iron ore beneficiation plant process water by electrocoagulation. Arab J Chem 14:102902. https://doi.org/10.1016/j.arabjc.2020.11.008

Das Neves AM, Visioli LJ, Enzweiler H, Paulino AT (2024) Lipase from Candida rugosa incorporated in pectin hydrogel via immobilization for hydrolysis of lipids in dairy effluents and production of fatty acids. J Water Process Eng 58:104821. https://doi.org/10.1016/j.jwpe.2024.104821

Davoodbeygi Y, Askari M, Salehi E, Kheirieh S (2023) A review on hybrid membrane-adsorption systems for intensified water and wastewater treatment: process configurations, separation targets, and materials applied. J Environ Manage 335:117577. https://doi.org/10.1016/j.jenvman.2023.117577

De Mello ACC, Da Silva FP, Gripa E, Salgado AM, Da Fonseca FV (2023) Phenol removal from wastewater using tyrosinase enzyme immobilized in granular activated carbon and activated chitosan beads. Water 15:3778. https://doi.org/10.3390/w15213778

Diagboya PNE, Dikio ED (2018) Silica-based mesoporous materials; emerging designer adsorbents for aqueous pollutants removal and water treatment. Microporous Mesoporous Mater 266:252–267. https://doi.org/10.1016/j.micromeso.2018.03.008

Dinesh Kumar S, Santhanam P, Park MS, Kim M-K (2016) Development and application of a novel immobilized marine microalgae biofilter system for the treatment of shrimp culture effluent. J Water Process Eng 13:137–142. https://doi.org/10.1016/j.jwpe.2016.08.014

Djabourov M, Leblond J, Papon P (1988) Gelation of aqueous gelatin solutions. I. Structural Investigation.. J De Physique 49:319–332. https://doi.org/10.1051/jphys:01988004902031900

Do MH, Dubreuil B, Peydecastaing J, Vaca-Medina G, Nhu-Trang T-T, Jaffrezic-Renault N, Behra P (2020) Chitosan-based nanocomposites for glyphosate detection using surface plasmon resonance sensor. Sensors 20:5942. https://doi.org/10.3390/s20205942

Domínguez A, Couto SR, Sanromán MÁ (2005) Dye decolorization by Trametes hirsuta immobilized into alginate beads. World J Microbiol Biotechnol 21:405–409. https://doi.org/10.1007/s11274-004-1763-x

Dotto GL, McKay G (2020) Current scenario and challenges in adsorption for water treatment. J Environ Chem Eng 8:103988. https://doi.org/10.1016/j.jece.2020.103988

Doyo AN, Kumar R, Barakat MA (2023) Recent advances in cellulose, chitosan, and alginate based biopolymeric composites for adsorption of heavy metals from wastewater. J Taiwan Inst Chem Eng 151:105095. https://doi.org/10.1016/j.jtice.2023.105095

Draget KI, Smidsrød O, Skjåk-Bræk G (2002) Alginates from algae. In: Steinbüchel A (ed) Biopolymers online. Wiley

Du L, Arauzo PJ, Meza Zavala MF, Cao Z, Olszewski MP, Kruse A (2020) Towards the properties of different biomass-derived proteins via various extraction methods. Molecules 25. https://doi.org/10.3390/molecules25030488

Du B, Wang Y, Zheng Q, Wang X, Chen X, Zhou J, Yang G, Sun R-C (2024) A novel modified lignin-based adsorbent for removal of malachite green and Pb^{2+} ions from wastewater. Sep Purif Technol 330:125495. https://doi.org/10.1016/j.seppur.2023.125495

Dumore NS, Mukhopadhyay M (2012) Removal of oil and grease using immobilized triacylglycerin lipase. Int Biodeterior Biodegradation 68:65–70. https://doi.org/10.1016/j.ibiod.2011.12.005

El Machtani Idrissi D, Elidrissi ZC, Achiou B, Ouammou M, Alami Younssi S (2023) Fabrication of low-cost kaolinite/perlite membrane for microfiltration of dairy and textile wastewaters. J Environ Chem Eng 11:109281. https://doi.org/10.1016/j.jece.2023.109281

El-Azim H, Mourad F (2018) Removal of heavy metals Cd (II), Fe (III) and Ni (II), from aqueous solutions by natural (Clinoptilolite) zeolites and application to industrial wastewater. AJEE 7:1–13. https://doi.org/10.9734/AJEE/2018/41004

Elgarahy AM, Eloffy MG, Guibal E, Alghamdi HM, Elwakeel KZ (2023) Use of biopolymers in wastewater treatment: a brief review of current trends and prospects. Chin J Chem Eng 64:292–320. https://doi.org/10.1016/j.cjche.2023.05.018

Elsebai B, Ghica ME, Abbas MN, Brett CMA (2017) Catalase based hydrogen peroxide biosensor for mercury determination by inhibition measurements. J Hazard Mater 340:344–350. https://doi.org/10.1016/j.jhazmat.2017.07.021

El-Shora HM, El-Sharkawy RM (2020) Tyrosinase from *Penicillium chrysogenum*: characterization and application in phenol removal from aqueous solution. J Gen Appl Microbiol 66:323–329. https://doi.org/10.2323/jgam.2020.01.002

Erkaya IA, Arica MY, Akbulut A, Bayramoglu G (2014) Biosorption of uranium(VI) by free and entrapped Chlamydomonas reinhardtii: kinetic, equilibrium and thermodynamic studies. J Radioanal Nucl Chem 299:1993–2003. https://doi.org/10.1007/s10967-014-2964-x

Eroglu E, Agarwal V, Bradshaw M, Chen X, Smith SM, Raston CL, Swaminathan Iyer K (2012) Nitrate removal from liquid effluents using microalgae immobilized on chitosan nanofiber mats. Green Chem 14:2682. https://doi.org/10.1039/c2gc35970g

Esquerdo VM, Cadaval TRS, Dotto GL, Pinto LAA (2014) Chitosan scaffold as an alternative adsorbent for the removal of hazardous food dyes from aqueous solutions. J Colloid Interface Sci 424:7–15. https://doi.org/10.1016/j.jcis.2014.02.028

Faghiri F, Ghorbani F (2019) Colorimetric and naked eye detection of trace Hg^{2+} ions in the environmental water samples based on plasmonic response of sodium alginate impregnated by silver nanoparticles. J Hazard Mater 374:329–340. https://doi.org/10.1016/j.jhazmat.2019.04.052

Fakhri V, Hamzehlouy A, Janmaleki Dehchani A, Moradi E, Tavakoli Dare M, Jafari A, Khonakdar HA (2024) Green solutions for blue waters: using biomaterials to purify water from microplastics and nanoplastics. J Water Process Eng 65:105854. https://doi.org/10.1016/j.jwpe.2024.105854

Fathollah G, Amir M, Simin N, Mohammad AF, Mahmood A (2011) Application of immobilized horseradish peroxidase for removal and detoxification of azo dye from aqueous solution. Res J Chem Environ 15:217–222

Fattahi N, Fattahi T, Kashif M, Ramazani A, Jung W-K (2024) Lignin: a valuable and promising bio-based absorbent for dye removal applications. Int J Biol Macromol 276:133763. https://doi.org/10.1016/j.ijbiomac.2024.133763

Fierro S, Del Pilar Sánchez-Saavedra M, Copalcúa C (2008) Nitrate and phosphate removal by chitosan immobilized Scenedesmus. Biores Technol 99:1274–1279. https://doi.org/10.1016/j.biortech.2007.02.043

Fijoł N, Aguilar-Sánchez A, Mathew AP (2022) 3D-printable biopolymer-based materials for water treatment: a review. Chem Eng J 430:132964. https://doi.org/10.1016/j.cej.2021.132964

Gajbhiye DR, Wadnerwar NN (2021) assessment of awareness regarding heavy metal toxicity in school children. JPRI 111–116. https://doi.org/10.9734/jpri/2021/v33i37B32028

Garcia LF, Lacerda MFAR, Thomaz DV, De Souza Golveia JC, Pereira MDGC, De Souza Gil E, Schimidt F, Santiago MF (2019) Optimization of laccase–alginate–chitosan-based matrix toward 17 α-ethinylestradiol removal. Preparat Biochem Biotechnol 49:375–383. https://doi.org/10.1080/10826068.2019.1573195

Garg S, Singh S, Shehata N, Sharma H, Samuel J, Khan NA, Ramamurthy PC, Singh J, Mubashir M, Bokhari A, Ooi DJ, Show PL (2023) Aerogels in wastewater treatment: a review. J Taiwan Inst Chem Eng 105299. https://doi.org/10.1016/j.jtice.2023.105299

Ge H, Hua T, Wang J (2017) Preparation and characterization of poly (itaconic acid)-grafted crosslinked chitosan nanoadsorbent for high uptake of Hg^{2+} and Pb^{2+}. Int J Biol Macromol 95:954–961. https://doi.org/10.1016/j.ijbiomac.2016.10.084

Godage NH, Gionfriddo E (2020) Use of natural sorbents as alternative and green extractive materials: a critical review. Anal Chim Acta 1125:187–200. https://doi.org/10.1016/j.aca.2020.05.045

Gomez-Mascaraque LG, Mendez JA, Fernandez-Gutierrez M, Vazquez B, San Roman J (2014) Oxidized dextrins as alternative crosslinking agents for polysaccharides: application to hydrogels of agarose-chitosan. Acta Biomater 10:798–811. https://doi.org/10.1016/j.actbio.2013.10.003

Guo X, Qu L, Zhu S, Tian M, Zhang X, Sun K, Tang X (2016) Preparation of three-dimensional chitosan-graphene oxide aerogel for residue oil removal. Water Environ Res 88:768–778. https://doi.org/10.2175/106143016X14609975747207

Guo J, Wang J, Zheng G, Jiang X (2019) Optimization of the removal of reactive golden yellow SNE dye by cross-linked cationic starch and its adsorption properties. J Eng Fibers Fabr 14:1558925019865260. https://doi.org/10.1177/1558925019865260

Gupta A, Balomajumder C (2015) Simultaneous removal of Cr(VI) and phenol from binary solution using Bacillus sp. immobilized onto tea waste biomass. J Water Process Eng 6:1–10. https://doi.org/10.1016/j.jwpe.2015.02.004

Haq F, Mehmood S, Haroon M, Kiran M, Waseem K, Aziz T, Farid A (2022) Role of starch based materials as a bio-sorbents for the removal of dyes and heavy metals from wastewater. J Polym Environ 30:1730–1748. https://doi.org/10.1007/s10924-021-02337-6

Hauer B (2020) Embracing nature's catalysts: a viewpoint on the future of biocatalysis. ACS Catal 10:8418–8427. https://doi.org/10.1021/acscatal.0c01708

Hedar M, Intisar A, Hussain N (2023) Immobilized enzyme systems for wastewater treatment. In: Advances in chemical pollution, environmental management and protection. Elsevier, pp 183–206

Herman P, Fábián I, Kalmár J (2020) Mesoporous silica-gelatin aerogels for the selective adsorption of aqueous Hg(II). ACS Appl. Nano Mater. 3:195–206. https://doi.org/10.1021/acsanm.9b01903

Hu M-X, Niu H-M, Chen X-L, Zhan H-B (2019) Natural cellulose microfiltration membranes for oil/water nanoemulsions separation. Colloids Surf A 564:142–151. https://doi.org/10.1016/j.colsurfa.2018.12.045

Hu H, Wen W, Ou JZ (2022) Construction of adsorbents with graphene and its derivatives for wastewater treatment: a review. Environ Sci Nano 9:3226–3276. https://doi.org/10.1039/D2EN00248E

Huamani-Palomino RG, Jacinto CR, Alarcón H, Mejía IM, López RC, Silva DDO, Cavalheiro ETG, Venâncio T, Dávalos JZ, Valderrama AC (2019) Chemical modification of alginate with cysteine and its application for the removal of Pb(II) from aqueous solutions. Int J Biol Macromol 129:1056–1068. https://doi.org/10.1016/j.ijbiomac.2018.09.096

Ibn Yaich A, Edlund U, Albertsson A-C (2017) Transfer of biomatrix/wood cell interactions to hemicellulose-based materials to control water interaction. Chem Rev 117:8177–8207. https://doi.org/10.1021/acs.chemrev.6b00841

Ihsanullah I, Bilal M, Jamal A (2022) Recent developments in the removal of dyes from water by starch-based adsorbents. Chem Rec 22:e202100312. https://doi.org/10.1002/tcr.202100312

Inglezakis VJ, Zorpas AA (2012) Heat of adsorption, adsorption energy and activation energy in adsorption and ion exchange systems. Desalin Water Treat 39:149–157. https://doi.org/10.1080/19443994.2012.669169

Isaeva VI, Vedenyapina MD, Kurmysheva AY, Weichgrebe D, Nair RR, Nguyen NPT, Kustov LM (2021) Modern carbon-based materials for adsorptive removal of organic and inorganic pollutants from water and wastewater. Molecules 26:6628. https://doi.org/10.3390/molecules26216628

Jahan I, Zhang L (2022) Natural polymer-based electrospun nanofibrous membranes for wastewater treatment: a review. J Polym Environ 30:1709–1729. https://doi.org/10.1007/s10924-021-02312-1

Jakóbik-Kolon A, Milewski AK, Mitko K, Lis A (2014) Preparation of pectin-based biosorbents for cadmium and lead ions removal. Sep Sci Technol 49:1679–1688. https://doi.org/10.1080/01496395.2014.906469

Jal P (2004) Chemical modification of silica surface by immobilization of functional groups for extractive concentration of metal ions. Talanta 62:1005–1028. https://doi.org/10.1016/j.talanta.2003.10.028

Jeganathan J, Bassi A, Nakhla G (2006) Pre-treatment of high oil and grease pet food industrial wastewaters using immobilized lipase hydrolyzation. J Hazard Mater 137:121–128. https://doi.org/10.1016/j.jhazmat.2005.11.106

Jha S, Gaur R, Shahabuddin S, Tyagi I (2023) Biochar as sustainable alternative and green adsorbent for the remediation of noxious pollutants: a comprehensive review. Toxics 11:117. https://doi.org/10.3390/toxics11020117

Jiang Q, Han Z, Li W, Ji T, Yuan Y, Zhang J, Zhao C, Cheng Z, Wang S (2022) Adsorption properties of heavy metals and antibiotics by chitosan from larvae and adult Trypoxylus dichotomus. Carbohydr Polym 276:118735. https://doi.org/10.1016/j.carbpol.2021.118735

Jiménez-Pérez MV, Sánchez-Castillo P, Romera O, Fernández-Moreno D, Pérez-Martínez C (2004) Growth and nutrient removal in free and immobilized planktonic green algae isolated from pig manure. Enzyme Microb Technol 34:392–398. https://doi.org/10.1016/j.enzmictec.2003.07.010

Jo H-J, Park E-J, Cho K, Kim E-H, Jung J (2008) Toxicity identification and reduction of wastewaters from a pigment manufacturing factory. Chemosphere 70:949–957. https://doi.org/10.1016/j.chemosphere.2007.08.018

Joshi P, Mishra R, Narayan RJ (2021) Biosensing applications of carbon-based materials. Curr Opin Biomed Eng 18:100274. https://doi.org/10.1016/j.cobme.2021.100274

Kai D, Tan MJ, Chee PL, Chua YK, Yap YL, Loh XJ (2016) Towards lignin-based functional materials in a sustainable world. Green Chem 18:1175–1200. https://doi.org/10.1039/C5GC02616D

Katowah DF, Ismail SH, Sadek AH, Rahman MM (2023) Ultrasensitive QCM sensor development for monitoring of methyl orange dye in aqueous phase based on novel cross-linked chitosan/PVA/GO/Ce-ZnO nanocomposite film. Mater Sci Eng B 297:116804. https://doi.org/10.1016/j.mseb.2023.116804

Khalil H, Hegab HM, Nassar L, Wadi VS, Naddeo V, Yousef AF, Banat F, Hasan SW (2022) Asymmetrical ultrafiltration membranes based on polylactic acid for the removal of organic substances from wastewater. J Water Process Eng 45:102510. https://doi.org/10.1016/j.jwpe.2021.102510

Khan S, Ajmal S, Hussain T, Rahman MU (2023) Clay-based materials for enhanced water treatment: adsorption mechanisms, challenges, and future directions. J Umm Al-Qura Univ Appll Sci. https://doi.org/10.1007/s43994-023-00083-0

Khozhaenko E, Kovalev V, Podkorytova E, Khotimchenko M (2016) Removal of the metal ions from aqueous solutions by nanoscaled low molecular pectin isolated from seagrass Phyllospadix iwatensis. Sci Total Environ 565:913–921. https://doi.org/10.1016/j.scitotenv.2016.01.108

Kim J, Wu B, Jeong S, Jeong S, Kim M (2024) Recent advances of membrane-based hybrid membrane bioreactors for wastewater reclamation. Front Membr Sci Technol 3:1361433. https://doi.org/10.3389/frmst.2024.1361433

Kocaoba S, Arısoy M (2018) Biosorption of cadmium(II) and lead(II) from aqueous solutions using *Pleurotus ostreatus* immobilized on bentonite. Sep Sci Technol 53:1703–1710. https://doi.org/10.1080/01496395.2018.1442477

Kołodziejczak-Radzimska A, Nghiem LD, Jesionowski T (2021) Functionalized materials as a versatile platform for enzyme immobilization in wastewater treatment. Curr Pollut Rep 7:263–276. https://doi.org/10.1007/s40726-021-00193-5

Kong D, Jin R, Wang T, Li H, Yan X, Su D, Wang C, Liu F, Sun P, Liu X, Gao Y, Ma J, Liang X, Lu G (2019) Fluorescent hydrogel test kit coordination with smartphone: robust performance for on-site dimethoate analysis. Biosens Bioelectron 145:111706. https://doi.org/10.1016/j.bios.2019.111706

Krstić V (2021) Role of zeolite adsorbent in water treatment. In: Handbook of nanomaterials for wastewater treatment. Elsevier, pp 417–481

Kumar S, Shandilya M, Uniyal P, Thakur S, Parihar N (2023) Efficacy of polymeric nanofibrous membranes for proficient wastewater treatment. Polym Bull 80:7145–7200. https://doi.org/10.1007/s00289-022-04417-6

Kushwah BS, Bhadauria S (2010) Development of biosensor for phenol detection using agarose–guar gum based laccases extracted from *Pleurotus ostreatus*. J Appl Polym Sci 115:1358–1365. https://doi.org/10.1002/app.31265

Lara-Espinoza C, Carvajal-Millán E, Balandrán-Quintana R, López-Franco Y, Rascón-Chu A (2018) Pectin and pectin-based composite materials: beyond food texture. Molecules 23:942. https://doi.org/10.3390/molecules23040942

Lassouane F, Aït-Amar H, Amrani S, Rodriguez-Couto S (2019) A promising laccase immobilization approach for Bisphenol A removal from aqueous solutions. Biores Technol 271:360–367. https://doi.org/10.1016/j.biortech.2018.09.129

Le Goff KJ, Gaillard C, Helbert W, Garnier C, Aubry T (2015) Rheological study of reinforcement of agarose hydrogels by cellulose nanowhiskers. Carbohyd Polym 116:117–123. https://doi.org/10.1016/j.carbpol.2014.04.085

Lessa EF, Medina AL, Ribeiro AS, Fajardo AR (2020) Removal of multi-metals from water using reusable pectin/cellulose microfibers composite beads. Arab J Chem 13:709–720. https://doi.org/10.1016/j.arabjc.2017.07.011

Li N, Xue M-H, Yao H, Zhu J-J (2005) Reagentless biosensor for phenolic compounds based on tyrosinase entrapped within gelatine film. Anal Bioanal Chem 383:1127–1132. https://doi.org/10.1007/s00216-005-0115-5

Li D, Zhu FZ, Li JY, Na P, Wang N (2013) Preparation and characterization of cellulose fibers from corn straw as natural oil sorbents. Ind Eng Chem Res 52:516–524. https://doi.org/10.1021/ie302288k

Li K, Xu J, Arsalan M, Cheng N, Sheng Q, Zheng J, Cao W, Yue T (2019) Nitrogen doped carbon dots derived from natural seeds and their application for electrochemical sensing. J Electrochem Soc 166:B56–B62. https://doi.org/10.1149/2.0501902jes

Li J, Yang Z, Ding T, Song Y-J, Li H-C, Li D, Chen S, Xu F (2022a) The role of surface functional groups of pectin and pectin-based materials on the adsorption of heavy metal ions and dyes. Carbohydr Polym 276:118789. https://doi.org/10.1016/j.carbpol.2021.118789

Li A, Zhang Y, Ge W, Zhang Y, Liu L, Qiu G (2022b) Removal of heavy metals from wastewaters with biochar pyrolyzed from MgAl-layered double hydroxide-coated rice husk: mechanism and application. Bioresour Technol 347:126425. https://doi.org/10.1016/j.biortech.2021.126425

Li J, Shu Y, Li C, Jiang Z (2023a) Highly catalytic nanoenzyme of covalent organic framework loaded starch-surface-enhanced Raman scattering/absorption bi-mode peptide as biosensor for ultratrace determination of cadmium. Front Nutr 9:1075296. https://doi.org/10.3389/fnut.2022.1075296

Li M, Shi Q, Song N, Xiao Y, Wang L, Chen Z, James TD (2023b) Current trends in the detection and removal of heavy metal ions using functional materials. Chem Soc Rev 52:5827–5860. https://doi.org/10.1039/D2CS00683A

Liang C, Qin S, Ai H, Li S, Du K (2022) Novel amyloid-like porous lysozyme skeletons as "green" superadsorbent presenting ultrahigh capacity and rapid sequestration towards hazardous Congo red. Chem Eng J 441:136005. https://doi.org/10.1016/j.cej.2022.136005

Lim J, Mohamad Z (2022) Enzymes immobilized polymeric supports for wastewater treatment application: a short review. Mater Today: Proc 65:2946–2952. https://doi.org/10.1016/j.matpr.2022.02.526

Liu SQ, Song BX, Zhang M, Wang WP (2012a) Current situation and outlook on biological treatment of wastewater in metallic mine. AMR 518–523:2460–2463. https://doi.org/10.4028/www.scientific.net/AMR.518-523.2460

Liu T, Li Y, Du Q, Sun J, Jiao Y, Yang G, Wang Z, Xia Y, Zhang W, Wang K, Zhu H, Wu D (2012b) Adsorption of methylene blue from aqueous solution by graphene. Colloids Surf B 90:197–203. https://doi.org/10.1016/j.colsurfb.2011.10.019

Liu Y, Liao T, He Z, Li T, Wang H, Hu X, Guo Y, He Y (2013) Biosorption of copper(II) from aqueous solution by Bacillus subtilis cells immobilized into chitosan beads. Trans Nonferrous Metals Soc China 23:1804–1814. https://doi.org/10.1016/S1003-6326(13)62664-3

Liu P, Borrell PF, Božič M, Kokol V, Oksman K, Mathew AP (2015) Nanocelluloses and their phosphorylated derivatives for selective adsorption of Ag^{+}, Cu^{2+} and Fe^{3+} from industrial effluents. J Hazard Mater 294:177–185. https://doi.org/10.1016/j.jhazmat.2015.04.001

Liu Y, Yao L, He L, Liu N, Piao Y (2019) Electrochemical enzyme biosensor bearing biochar nanoparticle as signal enhancer for Bisphenol A detection in water. Sensors 19:1619. https://doi.org/10.3390/s19071619

Long Q, Zhang Z, Qi G, Wang Z, Chen Y, Liu Z-Q (2020) Fabrication of chitosan nanofiltration membranes by the film casting strategy for effective removal of dyes/salts in textile wastewater. ACS Sustain Chem Eng 8:2512–2522. https://doi.org/10.1021/acssuschemeng.9b07026

Lou L, Huang Q, Lou Y, Lu J, Hu B, Lin Q (2019) Adsorption and degradation in the removal of nonylphenol from water by cells immobilized on biochar. Chemosphere 228:676–684. https://doi.org/10.1016/j.chemosphere.2019.04.151

Lowell S, Shields JE (1991) Chemisorption. In: Powder surface area and porosity. Springer Netherlands, Dordrecht, pp 210–216

Luo Q, Huang Y, Lei Z, Peng J, Xu D, Guo X, Wu Y (2021) Wood-derived nanocellulose hydrogel incorporating gold nanoclusters using in situ multistep reactions for efficient sorption and sensitive detection of mercury ion. Ind Crops Prod 173:114142. https://doi.org/10.1016/j.indcrop.2021.114142

Madhu R, Veeramani V, Chen S-M (2014a) Heteroatom-enriched and renewable banana-stem-derived porous carbon for the electrochemical determination of nitrite in various water samples. Sci Rep 4:4679. https://doi.org/10.1038/srep04679

Madhu R, Karuppiah C, Chen S-M, Veerakumar P, Liu S-B (2014b) Electrochemical detection of 4-nitrophenol based on biomass derived activated carbons. Anal Methods 6:5274. https://doi.org/10.1039/c4ay00795f

Maldonado-Carmona N, Piccirillo G, Godard J, Heuzé K, Genin E, Villandier N, Calvete MJF, Leroy-Lhez S (2024) Bio-based matrix photocatalysts for photodegradation of antibiotics. Photochem Photobiol Sci 23:587–627. https://doi.org/10.1007/s43630-024-00536-3

Martínez-Sabando J, Coin F, Melillo JH, Goyanes S, Cerveny S (2023) A review of pectin-based material for applications in water treatment. Materials 16:2207. https://doi.org/10.3390/ma16062207

Mehmandoust M, Li G, Erk N (2023) Biomass-derived carbon materials as an emerging platform for advanced electrochemical sensors: recent advances and future perspectives. Ind Eng Chem Res 62:4628–4635. https://doi.org/10.1021/acs.iecr.2c03058

Mehta SK, Gaur JP (2001) Removal of Ni and Cu from single and binary metalsolutions by free and immobilized Chlorella vulgaris. Eur J Protistol 37:261–271. https://doi.org/10.1078/0932-4739-00813

Mirmohseni A, Seyed Dorraji MS, Figoli A, Tasselli F (2012) Chitosan hollow fibers as effective biosorbent toward dye: preparation and modeling. Bioresour Technol 121:212–220. https://doi.org/10.1016/j.biortech.2012.06.067

Mohidem NA, Mohamad M, Rashid MU, Norizan MN, Hamzah F, Mat HB (2023) Recent advances in enzyme immobilisation strategies: an overview of techniques and composite carriers. J Compos Sci 7:488. https://doi.org/10.3390/jcs7120488

Moretti A, Ivan HL, Skvaril J (2024) A review of the state-of-the-art wastewater quality characterization and measurement technologies. Is the shift to real-time monitoring nowadays feasible? J Water Process Eng 60:105061. https://doi.org/10.1016/j.jwpe.2024.105061

Mouiya M, Abourriche A, Bouazizi A, Benhammou A, El Hafiane Y, Abouliatim Y, Nibou L, Oumam M, Ouammou M, Smith A, Hannache H (2018) Flat ceramic microfiltration membrane based on natural clay and Moroccan phosphate for desalination and industrial wastewater treatment. Desalination 427:42–50. https://doi.org/10.1016/j.desal.2017.11.005

Mountassir Y, Benyaich A, Rezrazi M, Berçot P, Gebrati L (2013) Wastewater effluent characteristics from Moroccan textile industry. Water Sci Technol 67:2791–2799. https://doi.org/10.2166/wst.2013.205

Muhammad S, Yahya EB, Abdul Khalil HPS, Marwan M, Albadn YM (2023) Recent advances in carbon and activated carbon nanostructured aerogels prepared from agricultural wastes for wastewater treatment applications. Agriculture 13:208. https://doi.org/10.3390/agriculture13010208

Muzzarelli RAA, Muzzarelli C (2005) Chitosan chemistry: relevance to the biomedical sciences. Adv Polym Sci 186:151–209. https://doi.org/10.1007/b136820

Naghdi M, Taheran M, Brar SK, Kermanshahi-pour A, Verma M, Surampalli RY (2017) Immobilized laccase on oxygen functionalized nanobiochars through mineral acids treatment for removal of carbamazepine. Sci Total Environ 584–585:393–401. https://doi.org/10.1016/j.scitotenv.2017.01.021

Najim AA, Radeef AY, Al-Doori I, Jabbar ZH (2024) Immobilization: the promising technique to protect and increase the efficiency of microorganisms to remove contaminants. J Chem Technol Biotechnol 99:1707–1733. https://doi.org/10.1002/jctb.7638

Nasrollahzadeh M, Sajjadi M, Iravani S, Varma RS (2021) Starch, cellulose, pectin, gum, alginate, chitin and chitosan derived (nano)materials for sustainable water treatment: a review. Carbohydr Polym 251:116986. https://doi.org/10.1016/j.carbpol.2020.116986

Nathanson J, Ambulkar A (n.d.) Wastewater treatment. https://www.britannica.com/technology/wastewater-treatment

Nawaz S, Tabassum A, Muslim S, Nasreen T, Baradoke A, Kim TH, Boczkaj G, Jesionowski T, Bilal M (2023) Effective assessment of biopolymer-based multifunctional sorbents for the remediation of environmentally hazardous contaminants from aqueous solutions. Chemosphere 329:138552. https://doi.org/10.1016/j.chemosphere.2023.138552

Nie J, Feng D, Shang J, Nasen B, Jiang T, Liu Y, Hou S (2023) Green composite aerogel based on citrus peel/chitosan/bentonite for sustainable removal Cu(II) from water matrices. Sci Rep 13:15443. https://doi.org/10.1038/s41598-023-42409-2

Njanja E, Mbokou SF, Pontie M, Nacef M, Tonle IK (2019) Comparative assessment of methylene blue biosorption using coffee husks and corn cobs: towards the elaboration of a lignocellulosic-based amperometric sensor. SN Appl Sci 1:513. https://doi.org/10.1007/s42452-019-0520-6

Obotey Ezugbe E, Rathilal S (2020) Membrane technologies in wastewater treatment: a review. Membranes 10:89. https://doi.org/10.3390/membranes10050089

Omar NAS, Fen YW, Ramli I, Azmi UZM, Hashim HS, Abdullah J, Mahdi MA (2021) Cellulose and vanadium plasmonic sensor to measure Ni^{2+} ions. Appl Sci 11:2963. https://doi.org/10.3390/app11072963

Onfray C, Thiam A (2023) Biomass-derived carbon-based electrodes for electrochemical sensing: a review. Micromachines 14:1688. https://doi.org/10.3390/mi14091688

Orta MDM, Martín J, Santos JL, Aparicio I, Medina-Carrasco S, Alonso E (2020) Biopolymer-clay nanocomposites as novel and ecofriendly adsorbents for environmental remediation. Appl Clay Sci 198:105838. https://doi.org/10.1016/j.clay.2020.105838

Pal P, Kumar R (2014) Treatment of coke wastewater: a critical review for developing sustainable management strategies. Sep Purif Rev 43:89–123. https://doi.org/10.1080/15422119.2012.717161

Pandey J (2020) Biopolymers and their application in wastewater treatment. In: Bharagava RN (ed) Emerging eco-friendly green technologies for wastewater treatment. Springer, Singapore, pp 245–266

Parashar N, Hait S (2023) Plastic rain—Atmospheric microplastics deposition in urban and peri-urban areas of Patna City, Bihar, India: distribution, characteristics, transport, and source analysis. J Hazard Mater 458:131883. https://doi.org/10.1016/j.jhazmat.2023.131883

Park C, Lee B, Han E-J, Lee J, Kim S (2006) Decolorization of acid black 52 by fungal immobilization. Enzyme Microb Technol 39:371–374. https://doi.org/10.1016/j.enzmictec.2005.11.045

Parvathi E, Dilraj N, Akshaya CV, Deepak NK (2023) A Review on graphene-based adsorbents for the remediation of toxic heavy metals from aqueous sources. Int J Environ Sci Technol 20:11645–11672. https://doi.org/10.1007/s13762-022-04587-w

Patra AS, Ghorai S, Sarkar D, Das R, Sarkar S, Pal S (2017) Anionically functionalized guar gum embedded with silica nanoparticles: an efficient nanocomposite adsorbent for rapid adsorptive removal of toxic cationic dyes and metal ions. Bioresour Technol 225:367–376. https://doi.org/10.1016/j.biortech.2016.11.093

Pérez-Botella E, Valencia S, Rey F (2022) Zeolites in adsorption processes: state of the art and future prospects. Chem Rev 122:17647–17695. https://doi.org/10.1021/acs.chemrev.2c00140

Pérez-Cid B, Calvar S, Moldes AB, Manuel Cruz J (2020) Effective removal of cyanide and heavy metals from an industrial electroplating stream using calcium alginate hydrogels. Molecules 25:5183. https://doi.org/10.3390/molecules25215183

Pet I, Sanad MN, Farouz M, ElFaham MM, El-Hussein A, El-sadek MSA, Althobiti RA, Ioanid A (2024) Review: recent developments in the implementation of activated carbon as heavy metal removal management. Water Conserv Sci Eng. 9:62. https://doi.org/10.1007/s41101-024-00287-3

Pilley S, Kaur H, Hippargi G, Gonde P, Rayalu S (2022) Silk fibroin: a promising bio-material for the treatment of heavy metal-contaminated water, adsorption isotherms, kinetics, and mechanism. Environ Sci Pollut Res 29:56606–56619. https://doi.org/10.1007/s11356-022-19833-4

Płotka-Wasylka J, Chabowska A, Pantanit S, Bunkoed O, Fares MY, Sajid M, Lambropoulou D, Kurowska-Susdorf A, Jatkowska N (2024) Natural/bio-based sorbents as greener extractive materials for endocrine disrupting compounds in samples of different matrix composition. TrAC Trends Anal Chem 176:117773. https://doi.org/10.1016/j.trac.2024.117773

Prasannan A, Udomsin J, Tsai H-C, Wang C-F, Lai J-Y (2020) Robust underwater superoleophobic membranes with bio-inspired carrageenan/laponite multilayers for the effective removal of emulsions, metal ions, and organic dyes from wastewater. Chem Eng J 391:123585. https://doi.org/10.1016/j.cej.2019.123585

Putro JN, Santoso SP, Soetaredjo FE, Ismadji S, Ju Y-H (2019) Nanocrystalline cellulose from waste paper: adsorbent for azo dyes removal. Environ Nanotechnol Monitor Manag 12:100260. https://doi.org/10.1016/j.enmm.2019.100260

Qi X, Tong X, Pan W, Zeng Q, You S, Shen J (2021) Recent advances in polysaccharide-based adsorbents for wastewater treatment. J Clean Prod 315:128221. https://doi.org/10.1016/j.jclepro.2021.128221

Qiu B, Shao Q, Shi J, Yang C, Chu H (2022) Application of biochar for the adsorption of organic pollutants from wastewater: modification strategies, mechanisms and challenges. Sep Purif Technol 300:121925. https://doi.org/10.1016/j.seppur.2022.121925

Radoor S, Karayil J, Jayakumar A, Kandel DR, Kim JT, Siengchin S, Lee J (2024) Recent advances in cellulose- and alginate-based hydrogels for water and wastewater treatment: a review. Carbohydr Polym 323:121339. https://doi.org/10.1016/j.carbpol.2023.121339

Ramdzan NS, Fen YW, Omar NA, Anas NA, Liew JY, Daniyal WM, Hashim HS (2021) Detection of mercury ion using surface plasmon resonance spectroscopy based on nanocrystalline cellulose/ poly(3,4-ethylenedioxythiophene) thin film. Measurement 182:109728. https://doi.org/10.1016/ j.measurement.2021.109728

Rana RS, Singh P, Kandari V, Singh R, Dobhal R, Gupta S (2017) A review on characterization and bioremediation of pharmaceutical industries' wastewater: an Indian perspective. Appl Water Sci 7:1–12. https://doi.org/10.1007/s13201-014-0225-3

Rando G, Scalone E, Sfameni S, Plutino MR (2024) Functional bio-based polymeric hydrogels for wastewater treatment: from remediation to sensing applications. Gels 10:498. https://doi.org/ 10.3390/gels10080498

Renn D (1997) Biotechnology and the red seaweed polysaccharide industry: status, needs and prospects. Trends Biotechnol 15:9–14. https://doi.org/10.1016/S0167-7799(96)10069-X

Ricard-Blum S (2011) The collagen family. Cold Spring Harb Perspect Biol 3:a004978–a004978. https://doi.org/10.1101/cshperspect.a004978

Rocha JM, Sousa RPCL, Fangueiro R, Ferreira DP (2024) The potential of electrospun membranes in the treatment of textile wastewater: a review. Polymers 16:801. https://doi.org/10.3390/pol ym16060801

Rock CM, Brassill N, Dery JL, Carr D, McLain JE, Bright KR, Gerba CP (2019) Review of water quality criteria for water reuse and risk-based implications for irrigated produce under the FDA Food Safety Modernization Act, produce safety rule. Environ Res 172:616–629. https://doi.org/ 10.1016/j.envres.2018.12.050

Rosales E, Meijide J, Pazos M, Sanromán MA (2017) Challenges and recent advances in biochar as low-cost biosorbent: from batch assays to continuous-flow systems. Bioresour Technol 246:176–192. https://doi.org/10.1016/j.biortech.2017.06.084

Roy Barman S, Gavit P, Chowdhury S, Chatterjee K, Nain A (2023) 3D-printed materials for wastewater treatment. JACS Au 3:2930–2947. https://doi.org/10.1021/jacsau.3c00409

Roy I, Gupta MN (2003) Lactose hydrolysis by Lactozym™ immobilized on cellulose beads in batch and fluidized bed modes. Process Biochem 39:325–332. https://doi.org/10.1016/S0032-9592(03)00086-4

Rozema E, VanderZaag A, Wood J, Drizo A, Zheng Y, Madani A, Gordon R (2016) Constructed wetlands for agricultural wastewater treatment in Northeastern North America: a review. Water 8:173. https://doi.org/10.3390/w8050173

Russo T, Fucile P, Giacometti R, Sannino F (2021) Sustainable removal of contaminants by biopolymers: a novel approach for wastewater treatment. Current state and future perspectives. Processes 9:719. https://doi.org/10.3390/pr9040719

Saber-Samandari S, Saber-Samandari S, Joneidi-Yekta H, Mohseni M (2017) Adsorption of anionic and cationic dyes from aqueous solution using gelatin-based magnetic nanocomposite beads comprising carboxylic acid functionalized carbon nanotube. Chem Eng J 308:1133–1144. https://doi.org/10.1016/j.cej.2016.10.017

Sabzehmeidani MM, Mahnaee S, Ghaedi M, Heidari H, Roy VAL (2021) Carbon based materials: a review of adsorbents for inorganic and organic compounds. Mater Adv 2:598–627. https://doi.org/10.1039/D0MA00087F

Saenchoopa A, Boonta W, Talodthaisong C, Srichaiyapol O, Patramanon R, Kulchat S (2021) Colorimetric detection of Hg(II) by γ-aminobutyric acid-silver nanoparticles in water and the assessment of antibacterial activities. Spectrochim Acta Part A: Mol Biomol Spectrosc 251:119433. https://doi.org/10.1016/j.saa.2021.119433

Saha S, Venkatesh M, Basu H, Pimple MV, Singhal RK (2019) Recovery of gold using graphene oxide/calcium alginate hydrogel beads from a scrap solid state detector. J Environ Chem Eng 7:103134. https://doi.org/10.1016/j.jece.2019.103134

Saikia J, Sarmah S, Bora JJ, Das B, Goswamee RL (2019) Preparation and characterization of low cost flat ceramic membranes from easily available potters' clay for dye separation. Bull Mater Sci 42:104. https://doi.org/10.1007/s12034-019-1767-7

Saiyad M, Shah N, Joshipura M, Dwivedi A, Pillai S (2024) Modified biopolymers in wastewater treatment: a review. Mater Today: Proc S2214785324000312. https://doi.org/10.1016/j.matpr.2024.01.031

Saleem J, Shahid UB, Hijab M, Mackey H, McKay G (2019) Production and applications of activated carbons as adsorbents from olive stones. Biomass Conv Bioref 9:775–802. https://doi.org/10.1007/s13399-019-00473-7

Salim MH, Kassab Z, Ablouh E, Sehaqui H, Aboulkas A, Bouhfid R, Qaiss AEK, El Achaby M (2022) Manufacturing of macroporous cellulose monolith from green macroalgae and its application for wastewater treatment. Int J Biol Macromol 200:182–192. https://doi.org/10.1016/j.ijbiomac.2021.12.153

Sallanko JT, Sarpola AT (2007) Dissolved phosphorus levels in fresh municipal wastewater. J Environ Sci Health Part A 42:1681–1683. https://doi.org/10.1080/10934520701518257

San NO, Celebioglu A, Tümtaş Y, Uyar T, Tekinay T (2014) Reusable bacteria immobilized electrospun nanofibrous webs for decolorization of methylene blue dye in wastewater treatment. RSC Adv 4:32249–32255. https://doi.org/10.1039/C4RA04250F

Santos SCR, Boaventura RAR (2016) Adsorption of cationic and anionic azo dyes on sepiolite clay: equilibrium and kinetic studies in batch mode. J Environ Chem Eng 4:1473–1483. https://doi.org/10.1016/j.jece.2016.02.009

Sayanthan S, Hasan HA, Abdullah SRS (2024) Floating aquatic macrophytes in wastewater treatment: toward a circular economy. Water 16:870. https://doi.org/10.3390/w16060870

Schrieber R, Gareis H (2007) Gelatine handbook: theory and industrial practice. Wiley

Sethi S, Medha, Singh G, Sharma R, Kaith BS, Sharma N, Khullar S (2022) Fluorescent hydrogel of chitosan and gelatin cross-linked with maleic acid for optical detection of heavy metals. J Appl Polym Sci 139:51941. https://doi.org/10.1002/app.51941

Sethi S, Singh A, Medha, Thakur S, Kaith BS, Khullar S (2023) Natural polymer-based nanocomposite hydrogels as environmental remediation devices. In: Shanker U, Hussain CM, Rani M (eds) Handbook of green and sustainable nanotechnology. Springer International Publishing, Cham, pp 407–441

Sha R, Jones SS, Vishnu N, Soundiraraju B, Badhulika S (2018) A novel biomass derived carbon quantum dots for highly sensitive and selective detection of hydrazine. Electroanalysis 30:2228–2232. https://doi.org/10.1002/elan.201800255

Shahedi A, Darban AK, Jamshidi-Zanjani A, Homaee M (2023) An overview of the application of electrocoagulation for mine wastewater treatment. Environ Monit Assess 195:522. https://doi.org/10.1007/s10661-023-11044-9

Shahrin EWE, Narudin NAH, Padmosoedarso KM, Kusrini E, Mahadi AH, Shahri NNM, Usman A (2021) Pectin derived from pomelo pith as a superior adsorbent to remove toxic Acid Blue 25 from aqueous solution. Carbohydr Polym Technol Appl 2:100116. https://doi.org/10.1016/j.carpta.2021.100116

Shaji A, Zachariah AK (2017) Surface area analysis of nanomaterials. In: Thermal and rheological measurement techniques for nanomaterials characterization. Elsevier, pp 197–231

Sharma NK, Philip L (2016) Combined biological and photocatalytic treatment of real coke oven wastewater. Chem Eng J 295:20–28. https://doi.org/10.1016/j.cej.2016.03.031

Shen Y, Abeynayake R, Sun X, Ran T, Li J, Chen L, Yang W (2019) Feed nutritional value of brewers ' spent grain residue resulting from protease aided protein removal. 1:1–10

Shi J, Yang Z, Dai H, Lu X, Peng L, Tan X, Shi L, Fahim R (2018) Preparation and application of modified zeolites as adsorbents in wastewater treatment. Water Sci Technol 2017:621–635. https://doi.org/10.2166/wst.2018.249

Shim J, Kumar M, Mukherjee S, Goswami R (2019) Sustainable removal of pernicious arsenic and cadmium by a novel composite of MnO_2 impregnated alginate beads: a cost-effective approach for wastewater treatment. J Environ Manag 234:8–20. https://doi.org/10.1016/j.jenvman.2018.12.084

Shyuan LK, Heng LY, Ahmad M, Aziz SA, Ishak Z (2008) Evaluation of pesticide and heavy metal toxicity using immobilized enzyme alkaline phosphatase with an electrochemical biosensor. Asian J Biochem 3:359–365. https://doi.org/10.3923/ajb.2008.359.365

Sievers JC, Wätzel T, Londong J, Kraft E (2016) Case study: characterization of source-separated blackwater and greywater in the ecological housing estate Lübeck "Flintenbreite" (Germany). Environ Earth Sci 75:1428. https://doi.org/10.1007/s12665-016-6232-y

Silva TME, Borges LL, Souza ERBE, Caramori SS (2019) Synthesis of immobilized biocatalysts for wastewater decontamination. Polímeros 29:e2019056. https://doi.org/10.1590/0104-1428.08918

Sitko R, Turek E, Zawisza B, Malicka E, Talik E, Heimann J, Gagor A, Feist B, Wrzalik R (2013) Adsorption of divalent metal ions from aqueous solutions using graphene oxide. Dalton Trans 42:5682. https://doi.org/10.1039/c3dt33097d

Sonawane JM, Ezugwu CI, Ghosh PC (2020) Microbial fuel cell-based biological oxygen demand sensors for monitoring wastewater: state-of-the-art and practical applications. ACS Sens 5:2297–2316. https://doi.org/10.1021/acssensors.0c01299

Sondhi S, Kaur R, Kaur S, Kaur PS (2018) Immobilization of laccase-ABTS system for the development of a continuous flow packed bed bioreactor for decolorization of textile effluent. Int J Biol Macromol 117:1093–1100. https://doi.org/10.1016/j.ijbiomac.2018.06.007

Soo A, Kim J, Shon HK (2024) Technologies for the wastewater circular economy—A review. Desalin Water Treat 317:100205. https://doi.org/10.1016/j.dwt.2024.100205

Sroka E, Kamiński W, Bohdziewicz J (2004) Biological treatment of meat industry wastewater. Desalination 162:85–91. https://doi.org/10.1016/S0011-9164(04)00030-X

Sun L, Yu H, Fugetsu B (2012) Graphene oxide adsorption enhanced by in situ reduction with sodium hydrosulfite to remove acridine orange from aqueous solution. J Hazard Mater 203–204:101–110. https://doi.org/10.1016/j.jhazmat.2011.11.097

Suryamathi M, Viswanathamurthi P, Vennila K, Palvannan T, Bertani R, Sgarbossa P (2021) Tyrosinase immobilized zein nanofibrous matrix as a green and recyclable material for biodegradation of azo dyes. Fibers Polym 22:2714–2725. https://doi.org/10.1007/s12221-021-0207-7

Tam NFY, Wong YS (2000) Effect of immobilized microalgal bead concentrations on wastewater nutrient removal. Environ Pollut 107:145–151. https://doi.org/10.1016/S0269-7491(99)00118-9

Taneez M, Österlund H, Lundy L, Viklander M (2023) Impacts of stormwater pipe materials and pipe repairs on stormwater quality: a review. Environ Sci Pollut Res 30:118065–118077. https://doi.org/10.1007/s11356-023-30508-6

Tang H, Chang C, Zhang L (2011) Efficient adsorption of Hg^{2+} ions on chitin/cellulose composite membranes prepared via environmentally friendly pathway. Chem Eng J 173:689–697. https://doi.org/10.1016/j.cej.2011.07.045

Tatsumi K, Wada S, Ichikawa H (2000) Removal of chlorophenols from wastewater by immobilized horseradish peroxidase. Biotechnol Bioeng 51:126–130. https://doi.org/10.1002/(SICI)1097-0290(19960705)51:1%3c126::AID-BIT15%3e3.0.CO;2-O

Tembe S, Inamdar S, Haram S, Karve M, D'Souza SF (2007) Electrochemical biosensor for catechol using agarose–guar gum entrapped tyrosinase. J Biotechnol 128:80–85. https://doi.org/10.1016/j.jbiotec.2006.09.020

Teng Z, Shao W, Zhang K, Yu F, Huo Y, Li M (2020) Enhanced passivation of lead with immobilized phosphate solubilizing bacteria beads loaded with biochar/ nanoscale zero valent iron composite. J Hazard Mater 384:121505. https://doi.org/10.1016/j.jhazmat.2019.121505

Thiruvenkatachari R, Younes M, Su S (2011) Coal mine site investigation of wastewater quality in Australia. Desalin Water Treat 32:357–364. https://doi.org/10.5004/dwt.2011.2722

Ugwu EI, Othmani A, Nnaji CC (2022) A review on zeolites as cost-effective adsorbents for removal of heavy metals from aqueous environment. Int J Environ Sci Technol 19:8061–8084. https://doi.org/10.1007/s13762-021-03560-3

UN Water red (2024) Water for prosperity and peace. UNESCO, Paris

Vakili M, Rafatullah M, Salamatinia B, Abdullah AZ, Ibrahim MH, Tan KB, Gholami Z, Amouzgar P (2014) Application of chitosan and its derivatives as adsorbents for dye removal from water and wastewater: a review. Carbohydr Polym 113:115–130. https://doi.org/10.1016/j.carbpol.2014.07.007

Varamesh A, Abraham BD, Wang H, Berton P, Zhao H, Gourlay K, Minhas G, Lu Q, Bryant SL, Hu J (2023) Multifunctional fully biobased aerogels for water remediation: applications for dye and heavy metal adsorption and oil/water separation. J Hazard Mater 457:131824. https://doi.org/10.1016/j.jhazmat.2023.131824

Vedenyapina MD, Kurmysheva AY, Rakishev AK, Kryazhev YG (2019) Activated carbon as sorbents for treatment of pharmaceutical wastewater (review). Solid Fuel Chem 53:382–394. https://doi.org/10.3103/S0361521919070061

Velarde L, Nabavi MS, Escalera E, Antti M-L, Akhtar F (2023) Adsorption of heavy metals on natural zeolites: a review. Chemosphere 328:138508. https://doi.org/10.1016/j.chemosphere.2023.138508

Verma R, Gupta BD (2015) Detection of heavy metal ions in contaminated water by surface plasmon resonance based optical fibre sensor using conducting polymer and chitosan. Food Chem 166:568–575. https://doi.org/10.1016/j.foodchem.2014.06.045

Vinayagam V, Murugan S, Kumaresan R, Narayanan M, Sillanpää M, Vo D-VN, Kushwaha OS (2022) Protein nanofibrils as versatile and sustainable adsorbents for an effective removal of heavy metals from wastewater: a review. Chemosphere 301:134635. https://doi.org/10.1016/j.chemosphere.2022.134635

Wan P, Yang X, Feng Q, Shi S, Deng B, Zhang L (2022) Biodegradable chitosan-based membranes for highly effective separation of emulsified oil/water. Environ Eng Sci 39:907–917. https://doi.org/10.1089/ees.2022.0254

Wang Y, Liang Z, Su Z, Zhang K, Ren J, Sun R, Wang X (2018) All-biomass fluorescent hydrogels based on biomass carbon dots and alginate/nanocellulose for biosensing. ACS Appl Bio Mater 1:1398–1407. https://doi.org/10.1021/acsabm.8b00348

Wang Q, Ju J, Tan Y, Hao L, Ma Y, Wu Y, Zhang H, Xia Y, Sui K (2019a) Controlled synthesis of sodium alginate electrospun nanofiber membranes for multi-occasion adsorption and separation of methylene blue. Carbohyd Polym 205:125–134. https://doi.org/10.1016/j.carbpol.2018.10.023

Wang Z, Li P, Fang Y, Yan L, Zhou W, Fan X, Liu H (2019b) One-step recovery of noble metal ions from oil/water emulsions by chitin nanofibrous membrane for further recycling utilization. Carbohydr Polym 223:115064. https://doi.org/10.1016/j.carbpol.2019.115064

Wang L, Yang J, Wang J, Raza W, Liu G, Lu J, Zhang Y (2020a) Microwave synthesis of NaA zeolite membranes on coarse macroporous α-Al_2O_3 tubes for desalination. Microporous Mesoporous Mater 306:110360. https://doi.org/10.1016/j.micromeso.2020.110360

Wang Y, Dong X, Zhao L, Xue Y, Zhao X, Li Q, Xia Y (2020b) Facile and green fabrication of carrageenan-silver nanoparticles for colorimetric determination of Cu^{2+} and S^{2-}. Nanomaterials 10:83. https://doi.org/10.3390/nano10010083

Wang T, Jiang M, Yu X, Niu N, Chen L (2022) Application of lignin adsorbent in wastewater treatment: a review. Sep Purif Technol 302:122116. https://doi.org/10.1016/j.seppur.2022.122116

Wang H, Chen X, Chen B, Yang M, Zhang B (2025) Switchable biomaterials for wastewater treatment: from material innovations to technological advancements. Chem Eng J 509:160928. https://doi.org/10.1016/j.cej.2025.160928

Wen X, Zeng Z, Du C, Huang D, Zeng G, Xiao R, Lai C, Xu P, Zhang C, Wan J, Hu L, Yin L, Zhou C, Deng R (2019) Immobilized laccase on bentonite-derived mesoporous materials for removal of tetracycline. Chemosphere 222:865–871. https://doi.org/10.1016/j.chemosphere.2019.02.020

Wenpo F, Gaofeng L, Shuying F, Yuanming Q, Keyong T (2015) Preparation and characterization of collagen–hydroxyapatite/pectin composite. Int J Biol Macromol 74:218–223. https://doi.org/10.1016/j.ijbiomac.2014.11.031

Wilfong M, Paolisso M, Patra D, Pavao-Zuckerman M, Leisnham PT (2023) Shifting paradigms in stormwater management—Hydrosocial relations and stormwater hydrocitizenship. J Environ Plan Policy Manag 25:429–442. https://doi.org/10.1080/1523908X.2023.2169262

Wu J, Yang J, Feng P, Huang G, Xu C, Lin B (2020) High-efficiency removal of dyes from wastewater by fully recycling litchi peel biochar. Chemosphere 246:125734. https://doi.org/10.1016/j.chemosphere.2019.125734

Wu S, Shi W, Li K, Cai J, Chen L (2022) Recent advances on sustainable bio-based materials for water treatment: fabrication, modification and application. J Environ Chem Eng 10:108921. https://doi.org/10.1016/j.jece.2022.108921

Wu L, Huang G, Xie T, Zhang A, Fu Y (2023) Green and ligand-free gold nanoparticles in Padina australis extract for colorimetric detection of Cu^{2+} in water. Colloids Surf A: Physicochem Eng Aspects 658:130773. https://doi.org/10.1016/j.colsurfa.2022.130773

Xiao W, Sun R, Hu S, Meng C, Xie B, Yi M, Wu Y (2023) Recent advances and future perspective on lignocellulose-based materials as adsorbents in diverse water treatment applications. Int J Biol Macromol 253:126984. https://doi.org/10.1016/j.ijbiomac.2023.126984

Xie S, Huang L, Su C, Yan J, Chen Z, Li M, Du M, Zhang H (2024) Application of clay minerals as adsorbents for removing heavy metals from the environment. Green Smart Mining Eng 1:249–261. https://doi.org/10.1016/j.gsme.2024.07.002

Xu Y, Lei W, Zhang Y, Fan H, Hao Q, Gao S (2017) Bamboo fungus-derived porous nitrogen-doped carbon for the fast, sensitive determination of bisphenol A. J Electrochem Soc 164:B3043–B3048. https://doi.org/10.1149/2.0021705jes

Xue Y, Kamali M, Al-Salem SM, Rossi B, Appels L, Dewil R (2024) Naturally derived materials to enhance the membrane properties in (waste)water treatment applications—Mechanisms, scale-up challenges and economic considerations. J Water Process Eng 57:104647. https://doi.org/10.1016/j.jwpe.2023.104647

Yallappa S, Shivakumar M, Nagashree KL, Dharmaprakash MS, Vinu A, Hegde G (2018) Electrochemical determination of nitrite using catalyst free mesoporous carbon nanoparticles from bio renewable *Areca nut* seeds. J Electrochem Soc 165:H614–H619. https://doi.org/10.1149/2.0561810jes

Yang X, Chen C, Li J, Zhao G, Ren X, Wang X (2012) Graphene oxide-iron oxide and reduced graphene oxide-iron oxide hybrid materials for the removal of organic and inorganic pollutants. RSC Adv 2:8821. https://doi.org/10.1039/c2ra20885g

Yang K, Pan T, Zhao Q, Chen C, Zhu X, Wang P, Chen B (2020a) Dual-function ultrafiltration membrane constructed from pure activated carbon particles via facile nanostructure reconstruction for high-efficient water purification. Carbon 168:254–263. https://doi.org/10.1016/j.carbon.2020.06.083

Yang B, Bin D, Tian T, Liu Y, Liu B (2020b) An ordered mesoporous carbon nanofiber array for the sensitive electrochemical detection of malachite green. ChemElectroChem 7:659–664. https://doi.org/10.1002/celc.201901689

Yang M, Lotfikatouli S, Chen Y, Li T, Ma H, Mao X, Hsiao BS (2022) Nanostructured all-cellulose membranes for efficient ultrafiltration of wastewater. J Membr Sci 650:120422. https://doi.org/10.1016/j.memsci.2022.120422

Yang S, Sarkar S, Xie X, Li D, Chen J (2024) Application of optical hydrogels in environmental sensing. Energy Environ Mater 7:e12646. https://doi.org/10.1002/eem2.12646

Yao G, Shao X, Qiu Z, Qiu F, Li Z, Zhang T (2022) Construction of lignin-based nano-adsorbents for efficient and selective recovery of tellurium (IV) from wastewater. Chemosphere 287:132058. https://doi.org/10.1016/j.chemosphere.2021.132058

Yu L, Jiang L, Wang S, Sun M, Li D, Du G (2018) Pectin microgel particles as high adsorption rate material for methylene blue: performance, equilibrium, kinetic, mechanism and regeneration studies. Int J Biol Macromol 112:383–389. https://doi.org/10.1016/j.ijbiomac.2018.01.193

Yuan Q, Li N, Chi Y, Geng W, Yan W, Zhao Y, Li X, Dong B (2013) Effect of large pore size of multifunctional mesoporous microsphere on removal of heavy metal ions. J Hazard Mater 254–255:157–165. https://doi.org/10.1016/j.jhazmat.2013.03.035

Yuna Z (2016) Review of the natural, modified, and synthetic zeolites for heavy metals removal from wastewater. Environ Eng Sci 33:443–454. https://doi.org/10.1089/ees.2015.0166

Yurekli Y (2016) Removal of heavy metals in wastewater by using zeolite nano-particles impregnated polysulfone membranes. J Hazard Mater 309:53–64. https://doi.org/10.1016/j.jhazmat.2016.01.064

Zdarta J, Jankowska K, Bachosz K, Degórska O, Kaźmierczak K, Nguyen LN, Nghiem LD, Jesionowski T (2021) Enhanced wastewater treatment by immobilized enzymes. Curr Pollut Rep 7:167–179. https://doi.org/10.1007/s40726-021-00183-7

Zeinu KM, Hou H, Liu B, Yuan X, Huang L, Zhu X, Hu J, Yang J, Liang S, Wu X (2016) A novel hollow sphere bismuth oxide doped mesoporous carbon nanocomposite material derived from sustainable biomass for picomolar electrochemical detection of lead and cadmium. J Mater Chem A 4:13967–13979. https://doi.org/10.1039/C6TA04881A

Zeng Q, Sun J, Bai X, Xu Z (2023) Immobilization of phototroph-derived extracellular polymer for simultaneous removal of antibiotics and heavy metals: a sustainable approach for advanced treatment of secondary effluent. J Clean Prod 396:136495. https://doi.org/10.1016/j.jclepro.2023.136495

Zhang W, Zhou C, Zhou W, Lei A, Zhang Q, Wan Q, Zou B (2011) Fast and considerable adsorption of methylene blue dye onto graphene oxide. Bull Environ Contam Toxicol 87:86–90. https://doi.org/10.1007/s00128-011-0304-1

Zhang S, Gao H, Li J, Huang Y, Alsaedi A, Hayat T, Xu X, Wang X (2017) Rice husks as a sustainable silica source for hierarchical flower-like metal silicate architectures assembled into ultrathin nanosheets for adsorption and catalysis. J Hazard Mater 321:92–102. https://doi.org/10.1016/j.jhazmat.2016.09.004

Zhang Z, Li D, Zhang X (2019) Enzymatic decolorization of melanoidins from molasses wastewater by immobilized keratinase. Bioresour Technol 280:165–172. https://doi.org/10.1016/j.biortech.2019.02.049

Zhang Q, Zhang S, Zhao Z, Liu M, Yin X, Zhou Y, Wu Y, Peng Q (2020a) Highly effective lead (II) removal by sustainable alkaline activated β-lactoglobulin nanofibrils from whey protein. J Clean Prod 255:120297. https://doi.org/10.1016/j.jclepro.2020.120297

Zhang C, Li J, Wu X, Long Y, An H, Pan X, Li M, Dong F, Zheng Y (2020b) Rapid degradation of dimethomorph in polluted water and soil by Bacillus cereus WL08 immobilized on bamboo charcoal–sodium alginate. J Hazard Mater 398:122806. https://doi.org/10.1016/j.jhazmat.2020.122806

Zhang L, Lei M, Feng T, Chang W, Ye A, Yi H, Yi C (2021) A case study of the wastewater treatment system modification in denim textile industry. J Text Inst 112:1666–1670. https://doi.org/10.1080/00405000.2020.1835155

Zhao W, Xiao L-P, Song G, Sun R-C, He L, Singh S, Simmons BA, Cheng G (2017) From lignin subunits to aggregates: insights into lignin solubilization. Green Chem 19:3272–3281. https://doi.org/10.1039/C7GC00944E

Zhao K, Veksha A, Ge L, Lisak G (2021) Near real-time analysis of para-cresol in wastewater with a laccase-carbon nanotube-based biosensor. Chemosphere 269:128699. https://doi.org/10.1016/j.chemosphere.2020.128699

Zhao X, Peng X, Zhao T, Yang L, Shu Y (2019) Preparation and electrochemical properties of multi-walled carbon nanotube/dextran/nano-gold composites. Nanosci Nanotechnol Lett 11:1404–1409. https://doi.org/10.1166/nnl.2019.3018

Zhijiang C, Ping X, Cong Z, Tingting Z, Jie G, Kongyin Z (2018) Preparation and characterization of a bi-layered nano-filtration membrane from a chitosan hydrogel and bacterial cellulose nanofiber for dye removal. Cellulose 25:5123–5137. https://doi.org/10.1007/s10570-018-1914-0

Zhu X, Liu B, Chen S, Wu L, Yang J, Liang S, Xiao K, Hu J, Hou H (2020) Ultrasensitive and simultaneous electrochemical determination of Pb^{2+} and Cd^{2+} based on biomass derived lotus root-like hierarchical porous carbon/bismuth composite. J Electrochem Soc 167:087505. https://doi.org/10.1149/1945-7111/ab82f7

Zhylyak GA, Dzyadevich SV, Korpan YI, Soldatkin AP, El'skaya AV (1995) Application of urease conductometric biosensor for heavy-metal ion determination. Sens Actuators B: Chem 24:145–148. https://doi.org/10.1016/0925-4005(95)85031-7

Zibarev NV, Politaeva NA, Andrianova MY (2022) Use of Chlorella sorokiniana (Chlorellaceae, Chlorellales) microalgae for purification of brewing-industry wastewaters. Biol Bull Russ Acad Sci 49:1776–1780. https://doi.org/10.1134/S1062359022100211

Żur J, Piński A, Michalska J, Hupert-Kocurek K, Nowak A, Wojcieszyńska D, Guzik U (2020) A whole-cell immobilization system on bacterial cellulose for the paracetamol-degrading Pseudomonas moorei KB4 strain. Int Biodeterior Biodegradation 149:104919. https://doi.org/10.1016/j.ibiod.2020.104919

Chapter 17
Advancements in the Sustainable Recycling of Polymer Composites

Soumyadeep Bera, P. S. Sangeeth, K. Parvanadev, Tomy Muringayil Joseph, and Sabu Thomas

Abstract Polymer composites are widely used across industries for their superior mechanical, thermal, and functional properties, yet their complex structures pose significant challenges for sustainable recycling and end-of-life management. This chapter explores recent advancements in the sustainable recycling of polymer composites, with emphasis on chemical, mechanical, and emerging green methodologies that improve recyclability without compromising performance. A literature-based comparative analysis of recycling pathways, supported by case studies on thermoplastic and thermoset systems and techno-environmental assessments, forms the methodological foundation. Key findings from recent studies (2020–2025) highlight the effectiveness of hybrid strategies such as catalyst-assisted solvolysis and enzymatic hydrolysis in enhancing monomer recovery and fibre retention. Compatibilization techniques—including PEG incorporation, maleic anhydride grafting, and nanofiller reinforcement—further improve recyclate quality. The discussion

S. Bera (✉)
International and Inter University Centre for Nanoscience and Nanotechnology (IIUCNN), Mahatma Gandhi University, Kottayam, Kerala, India
e-mail: soumyadeep.bera@my.jcu.edu.au; soumyadeepberart854@gmail.com

College of Science and Engineering, James Cook University, Townsville, QLD, Australia

P. S. Sangeeth · K. Parvanadev · S. Thomas
School of Energy Materials, Mahatma Gandhi University, Kottayam, Kerala, India
e-mail: sangeethps@mgu.ac.in

K. Parvanadev
e-mail: parvanadevk@mgu.ac.in

S. Thomas
e-mail: sabuthomas@mgu.ac.in

T. M. Joseph
Department of Polymer Technology, Faculty of Chemistry, Gdańsk University of Technology, Gdańsk, Poland
e-mail: tomy.muringayil@pg.edu.pl

S. Thomas
Trivandrum Engineering Science and Technology Research Park, Sreekariyam, Trivandrum, Kerala, India

L. Ravindran et al. (eds.), *Macro, Micro and Nanocomposites from Sustainable Sources*, Advanced Structured Materials 244, https://doi.org/10.1007/978-981-95-2469-3_17

addresses trade-offs between energy efficiency, product integrity, and process scalability, and evaluates the potential of green solvents and bio-based reinforcements in achieving circularity. While progress is notable, challenges in regulatory acceptance, economic feasibility, and industrial implementation remain. The chapter concludes with a forward-looking perspective on integrating innovative recycling technologies with sustainable materials design to close the loop on composite waste.

Keywords Composite recycling · Fiber-reinforced composites · End-of-life composites · Recyclability · Circular economy · Sustainable materials

17.1 Introduction

The rapid industrialization and urbanization of the modern world have significantly increased the consumption of materials, leading to a corresponding surge in waste generation. Recycling has emerged as a critical solution to address the environmental impacts of this trend, enabling the conservation of resources, reduction of greenhouse gas emissions, and minimization of landfill usage. Additionally, recycling approximately 20 million tons of plastics could prevent the depletion of about 50 million tons of natural resources (Gabisa et al. 2023). In today's era, marked by heightened awareness of climate change, resource scarcity, and pollution; recycling is no longer a choice but a necessity. It plays a pivotal role in transitioning from a linear to a circular economy, where materials are reused, repurposed, and reintegrated into the production cycle (García-Valderrama et al. 2024). Figure 17.1 presents a schematic overview of various eco-friendly, economical, and sustainable strategies for managing composite waste, highlighting emerging approaches that align with circular economy principles and environmental stewardship (Khalid et al. 2022).

Polymer composites, especially fibre-reinforced polymer composites (FRPCs), have revolutionized numerous industries including aerospace (Cha et al. 2025; Romero-Fierro et al. 2022) automotive (Khan et al. 2024; Skosana et al. 2024; Jagadeesh et al. 2022), marine (Baley et al. 2024; El Hawary et al. 2023), construction (Vijayan et al. 2023; Krauklis et al. 2021), and electronics (Dhanasekar et al. 2022; Li et al. 2022), due to their lightweight nature, high strength-to-weight ratio, chemical resistance, and design adaptability. These composites are typically composed of a polymer matrix (thermoplastic or thermoset) reinforced with fibres such as carbon or glass. However, their structural complexity poses significant challenges for end-of-life management. Thermosetting composites, in particular, are difficult to recycle because of their irreversible crosslinked network structures, which prevent remelting or reshaping. This leads to environmental accumulation, where most end-of-life composites are either landfilled or incinerated—both unsustainable and environmentally hazardous practices (Krauklis et al. 2021).

Recent data underscore the urgency of addressing composite waste. Studies project that global composite waste could reach several million tons annually by 2030, primarily driven by the increasing use of composite materials in renewable

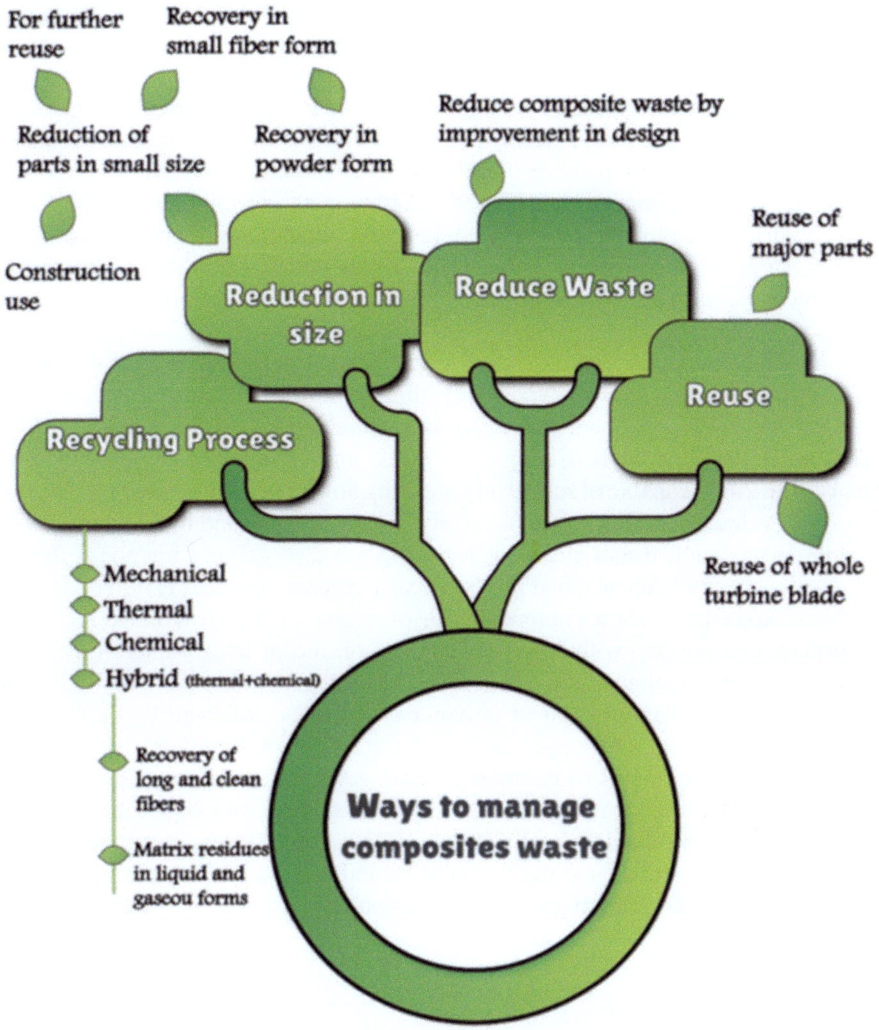

Fig. 17.1 Different ecofriendly, aconomical and sustainable development ways to manage composite wastes (Khalid et al. 2022). Reprinted from Khalid et al. (2022), with permission from Elsevier

energy sectors like wind turbines and electric vehicles. In the wind energy sector, the widespread adoption of composite materials for turbine blades has led to significant waste generation concerns. A study estimates that the usage of blade material waste is expected to grow from 1,000,000 tons in 2020 up to 2,000,000 tons in 2030, doubling in the current decade (Krauklis et al. 2021). Yet, current recycling rates remain alarmingly low, with less than 15% of composite waste being effectively recovered. Graphs and statistical reports from the European Composites Industry Association

(EuCIA) and other agencies show a stark gap between composite production and recycling infrastructure, emphasizing the need for targeted policy intervention and technological innovation.

Governments worldwide are beginning to respond to this challenge through regulatory amendments and sustainability mandates. The European Union, for instance, has imposed extended producer responsibility (EPR) directives and circular economy action plans that encourage recycling of complex materials (Joltreau 2022). Similarly, countries like Japan and the United States are funding research and pilot-scale recycling projects for composite materials. These policy shifts reflect increasing governmental concern and commitment to fostering circularity in advanced material use.

In parallel, academic and industrial research has intensified efforts to develop scalable and environmentally friendly recycling technologies. Recent literature highlights advancements in chemical recycling methods such as solvolysis (Okajima and Sako 2019), glycolysis (Chang et al. 2025) and supercritical fluid processing (Henry et al. 2016) capable of selectively breaking down polymer matrices and recovering high-value fibers. Innovative catalysts, green solvents, and hybrid approaches combining mechanical and chemical recycling are also gaining traction, offering improved recovery efficiency and product quality (Pérez et al. 2021).

This chapter provides a comprehensive overview of the sustainable recycling of polymer composites, with a particular focus on recent innovations in mechanical, chemical, thermal and biological recycling strategies. It begins by defining key terminologies and classification of polymer composites, followed by an in-depth discussion of recycling challenges and emerging solutions. Key data visualizations illustrate the current status of composite waste generation and recycling practices globally. Furthermore, it reviews legislative frameworks and highlights groundbreaking research efforts documented in recent literature. The chapter concludes by outlining future directions and the potential for industrial-scale integration of recycling technologies, contributing to a more sustainable and resource-efficient material lifecycle.

17.2 Recycling Techniques

17.2.1 Mechanical

Mechanical recycling is one of the primary techniques used for the recycling of polymer composites, particularly thermoset-based fibre-reinforced polymers (FRPs). This method involves size reduction through crushing, shredding, and milling of composite parts into smaller particles. These particles can then be further ground into powders or granules suitable for reuse, often as fillers in polymeric formulations or construction applications, such as artificial wood, asphalt, or cement (Conroy et al. 2006). Figure 17.2 gives an understanding of the mechanical recycling process

Fig. 17.2 Schematic illustration of mechanical recycling methods (Pietroluongo et al. 2020). Reprinted from Pietroluongo et al. (2020), with permission from Elsevier

and the stages involved during the process. The outputs from mechanical recycling typically fall into three categories: short fibres, fine powders with high resin content, and coarse recyclates with higher fiber content. Although the recyclates are generally low in value and struggle to compete with virgin materials, mechanical recycling remains the most economically viable option on an industrial scale, especially for glass fibre-reinforced thermosets (Pietroluongo et al. 2020).

However, mechanical recycling presents several limitations, primarily due to the degradation of both the polymer matrix and the reinforcing fibres. During repeated processing, fibre length significantly decreases, leading to reduced mechanical performance reported a progressive shortening of glass fibres and a decline in viscosity, thermal stability, and mechanical strength across multiple recycling cycles in PA66 composites, depicted in Fig. 17.3 (Pietroluongo et al. 2020). Similar findings were noted by Finnerty et al. (2023), where PLA-based composites reinforced with basalt and halloysite nanotubes (HNTs) exhibited increased glass transition temperatures and crystallinity with initial recycling, followed by degradation after several cycles. In particular, PLA/HNT composites maintained better fibre–matrix adhesion compared to PLA/BF, which showed visible voids and fibre pull-out under SEM analysis.

Fiber attrition also affects the recyclability of other thermoplastics, Le Duigou et al. (2008) demonstrated that flax fibres in PLLA biocomposites lost a substantial portion of their dimensions over six recycling cycles, leading to diminished

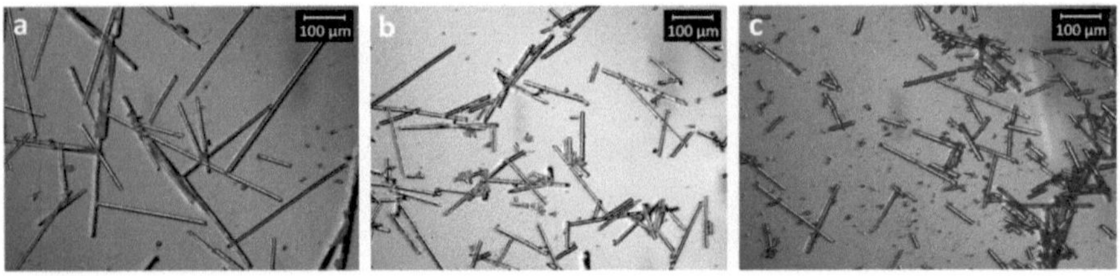

Fig. 17.3 Glass fibre evaluated by optical microscope at 100× for EoL fibres in (**a**) R fibres in (**b**) and RM1 fibres in (**c**) (Pietroluongo et al. 2020). Reprinted from, Pietroluongo et al. (2020), with permission from Elsevier

mechanical properties, which is shown in the SEM images in Fig. 17.4. Likewise, Chen et al. (2020b) found that TLCP/PP composites retained performance through fibril regeneration during recycling, whereas long glass fibre/PP composites suffered up to 30% tensile strength loss due to fibre shortening. The addition of maleic anhydride-grafted polypropylene (MAPP) improved interfacial adhesion in these systems without impairing recyclability (Chen et al. 2020b).

Howarth et al. (2014) analysed the energy intensity of mechanical recycling, describing a process where composite panels are milled using a high-speed drum with steel hammerhead inserts. Material removal parameters such as feed rate and width of cut were manipulated to model the energy demand under various conditions. They emphasized that mechanical recycling machines, despite fixed spindle speeds, allow control over feed rates to optimize energy use (Howarth et al. 2014).

Jubinville et al. (2023) showed that recycled HDPE used as a matrix for hemp hurd composites exhibited better mechanical strength and lower water absorption than virgin HDPE counterparts—likely due to crosslinking and chain cyclization confirmed via spectroscopy. In contrast, Almeras et al. (2004a, b) found that repeated

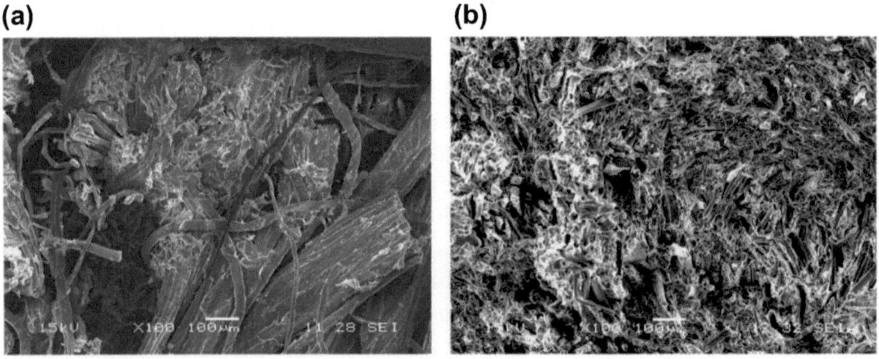

Fig. 17.4 SEM micrograph of fracture surface of BC-20% **a** after one injection, **b** after six injection cycles (Le Duigou et al. 2008). Reprinted from Le Duigou et al. (2008), with permission from Elsevier

reprocessing of intumescent ammonium polyphosphate (APP)/polyamide-6 (PA-6)/polypropylene (PP) blend led to degradation in fire retardancy and mechanical properties due to APP decomposition and matrix chain scission.

Economically, Hagnell and Åkermo (2019) emphasized that each recycling cycle results in fibre degradation and yield loss, progressively reducing the value of the recyclate. Eventually, recycled composites are relegated to low-grade uses, such as powder fillers, marking the end of their high-value lifecycle. Nevertheless, reintroducing reclaimed fibres into applications that tolerate shorter lengths can still yield structurally sound outcomes (Hagnell and Åkermo 2019). Figure 17.5 shows a pictorial representation of the flow of transfer from a structured aeronautical component to powder-filled thermoplastic component.

While, mechanical recycling is a cost-effective and scalable method for dealing with composite waste, it often results in property deterioration and lower-value products. The success of this method depends heavily on the nature of the fibre, matrix, and number of processing cycles, as well as application-specific tolerances for recycled material performance.

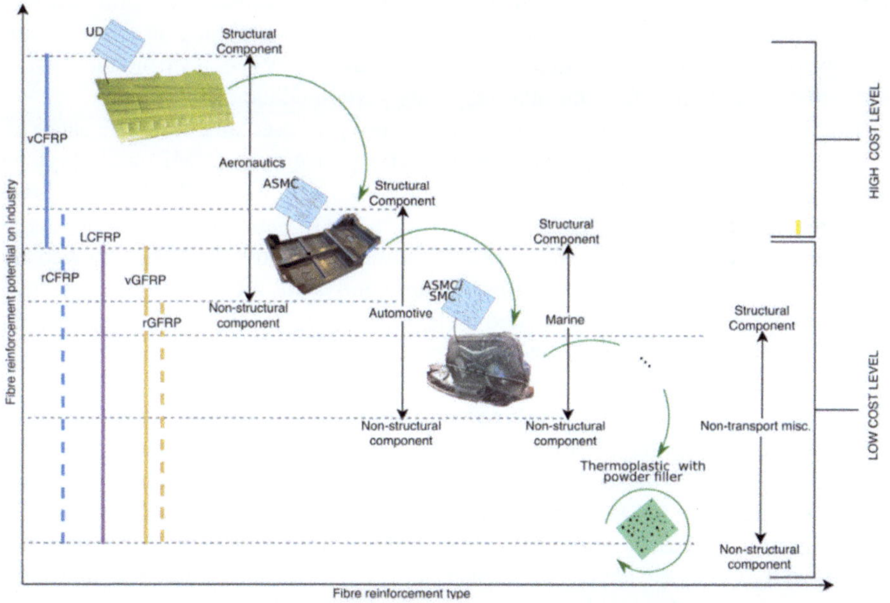

Fig. 17.5 Relevant application of researched recycled composites together with a conceptual closed loop material flow of transfer from a structured aeronautical component to powder-filled thermoplastic component (Hagnell and Åkermo 2019). Reprinted from Hagnell and Åkermo (2019), with permission from Elsevier

17.2.2 *Chemical*

17.2.2.1 Solvolysis

Solvolysis, a chemical recycling technique, is gaining significant attention for its ability to recover both valuable carbon fibres and useful chemical feedstocks from thermoset-based CFRP wastes. Unlike thermal or mechanical recycling, solvolysis enables depolymerization of the resin matrix under relatively mild conditions, often preserving the mechanical and surface integrity of the recovered fibres (Okajima and Sako 2019).

To improve the efficiency of solvolysis under mild conditions, researchers have explored various solvent and catalyst systems. For example, Zhao et al. (2022) developed an efficient solvolysis system employing a mixture of monoethanolamine (MEA) and potassium hydroxide (KOH) as the solvent–catalyst combination. Their process targeted anhydride-cured epoxy resins and operated at 160 °C for 60 min. This resulted in complete matrix decomposition, with a resin decomposition rate of approximately 99.3%. The recovered carbon fibres retained 96.5% of their original tensile strength and 95.2% of their modulus. Raman spectroscopy confirmed that the fibre surfaces were free of residues and retained their graphitic structure. Additionally, partial recovery of both MEA and degraded epoxy species from the solvent suggested the potential for a circular recycling system (Zhao et al. 2022).

Building on this direction, Kuang et al. (2018b) proposed an alternative catalytic solvolysis strategy using a blend of N-methyl-2-pyrrolidone (NMP) and ethylene glycol (EG), combined with 1,5,7-triazabicyclo[4.4.0]dec-5-ene (TBD) as an organic base catalyst. The reaction, conducted at 170 °C under ambient pressure, resulted in a 95% mass loss of the resin within 70 min. The recovered fibres showed no deterioration in mechanical properties. Significantly, the method proved effective even for CFRP panels with embedded metal electrodes, suggesting its versatility in handling complex composite waste (Kuang et al. 2018b).

Beyond optimizing solvent–catalyst systems, some researchers have focused on modifying the resin structure itself to enhance solvolytic breakdown. Shen et al. (2020) demonstrated that incorporating cleavable bonds—such as esters, acetals, or carbonates—into thermoset resin structures facilitates degradation under relatively mild conditions. For instance, thermosets with ester linkages degraded efficiently in 1 M sodium hydroxide within a few hours, and vanillin-based epoxies decomposed entirely in hydrochloric acid at 50 °C over 36 h. The inclusion of catalysts like TBD further accelerated the depolymerization process, highlighting a promising strategy for designing recyclable composites (Shen et al. 2020).

To better understand and optimize these processes, Luo et al. (2023) developed a theoretical model to simulate the diffusion and reaction kinetics of epoxy degradation during solvolysis. Their study highlighted the influence of fibre volume fraction (Vf) on solvent transport and reaction rate, noting that higher Vf slightly hindered solvent diffusivity. Nevertheless, the overall degradation was primarily governed by the solvent's intrinsic diffusivity and reaction kinetics. This model provides a valuable predictive tool for scaling up solvolysis processes in real-world applications (Luo et al. 2023).

The reuse potential of solvolytically recovered fibres has also been experimentally validated. Oliveux et al. (2017) demonstrated that discontinuous carbon fibres reclaimed from RTM6 resin-based laminates via solvolysis in a mixed organic solvent at 200 °C for 1 h could be reintegrated into new epoxy composites. Despite minor surface defects, the recycled fibres contributed to composites with mechanical performance comparable to the original material, underscoring their structural applicability (Oliveux et al. 2017).

In contrast, other studies have explored more aggressive solvolytic media to treat challenging waste streams. Feraboli et al. (2012) employed boiling sulfuric acid to recover carbon fibres from epoxy composites. While the fibre volume fraction of the recycled laminate (33%) was lower than that of the original (62%), the resulting composite retained 25%–30% of its original tensile strength. This method illustrates the viability of strong acid solvolysis, particularly when fibre quality is of secondary importance (Feraboli et al. 2012).

Expanding the potential of solvolysis, researchers have also turned to supercritical and subcritical fluid systems, which offer unique benefits due to their enhanced solvation power, low viscosity, and high diffusivity. These systems can operate without halogenated solvents, aligning with environmentally sustainable recycling goals (Oliveux et al. 2015).

One such approach was developed by Henry et al. (2016) who employed a semi-continuous flow reactor to process aerospace-grade CFRP using sub- and supercritical mixtures of water and ethanol. Operating at 200–350 °C, the method achieved over 98% resin removal while preserving the tensile integrity of the fibres. The hydrolysis-driven reaction was enhanced by the ethanol–water mixture, which also aided in the solubilization of degradation products (Henry et al. 2016). Additionally, Fig. 17.6 summarizes the products formed during the recycling of CFRPs using water/ethanol mixture.

A related system was explored by, Keith et al. (2019) who used a supercritical water–acetone mixture at 320 °C and 20 MPa to achieve over 95% resin recovery. Notably, the carbon fibres retained their mechanical strength. Energy analysis indicated that the method could reduce energy consumption by up to 91% compared to virgin carbon fibre production. However, the study also emphasized the importance of optimizing reaction time, as extended exposure led to partial fibre damage. Figure 17.7 represents the shrinking core model during the solvolysis technique (Keith et al. 2019).

Adding to this body of research, Okajima and Sako (2019) utilized supercritical acetone at 350 °C and 14 MPa to degrade CFRP waste. In just 60 min, 95.6% of the matrix resin was removed. The process preserved fibre quality and offered low post-processing environmental burden, due to acetone's recoverability and low toxicity (Okajima and Sako 2019).

Further advancing green solvolysis, Liu et al. (2020) implemented a closed-loop recycling system using supercritical n-butanol. At 330 °C and 6.5 MPa, the method achieved complete matrix degradation and recovered carbon fibres with more than

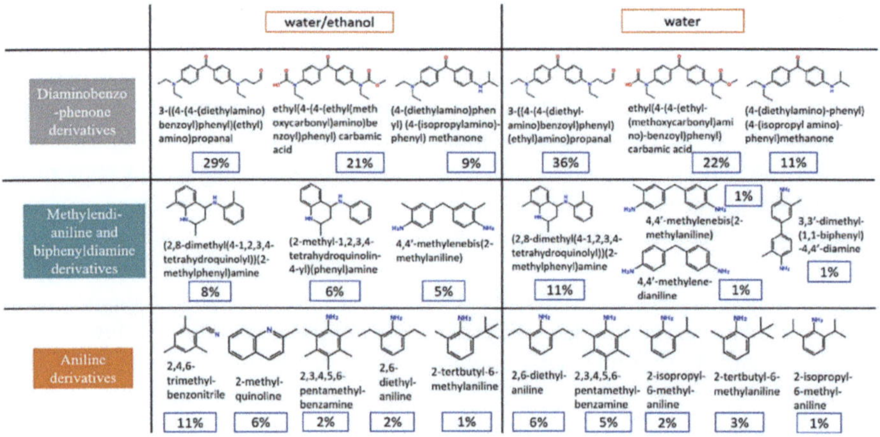

Fig. 17.6 Chemical structure and concentration of volatile and semi-volatile organic molecules formed during recycling of CFRPs using water/ethanol mixture (Henry et al. 2016). Reprinted from Henry et al. (2016), with permission from Elsevier

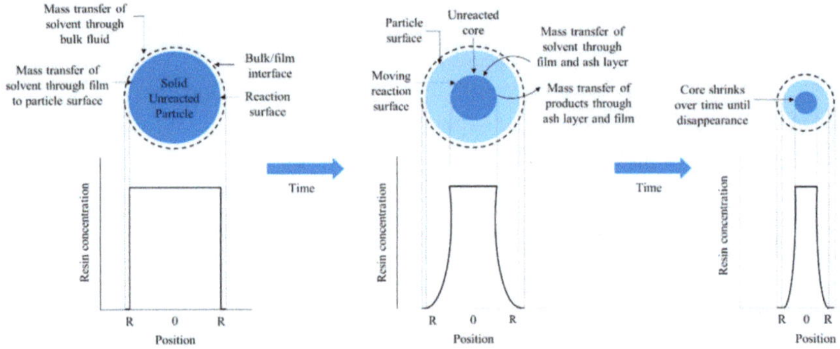

Fig. 17.7 Diagram of shrinking core model (Keith et al. 2019). Reprinted from Keith et al. (2019), with permission from Elsevier

95% of their original tensile strength. These fibres were successfully reused in epoxy- and polypropylene-based composites, with the latter even outperforming virgin fibre-based materials in tensile strength, suggesting value-added applications (Liu et al. 2020).

Glass fibre-reinforced plastics (GFRPs) have also been targeted using solvolysis. Mattsson et al. (2020) applied subcritical water with co-solvents like ethylene glycol, acetic acid, and formic acid to recycle GFRP waste from wind turbine blades. The process achieved resin degradation efficiencies between 74 and 85%, while also removing chlorine and nitrogen-based impurities, making it suitable for composites containing halogenated flame retardants (Mattsson et al. 2020).

In terms of acid-based solvolysis, Tortorici et al. assessed aqueous sulfuric acid at different concentrations for epoxy matrix removal. Solutions above 15 mol L^{-1} at 60 °C were effective in achieving near-complete resin removal, while lower concentrations left organic residues. Importantly, the carbon fibres retained their mechanical properties, supporting the method's feasibility for fibre recovery (Tortorici et al. 2024).

Beyond material recovery, solvolysis has also been extended toward chemical upcycling. Dang et al. (2002) proposed a nitric acid-based recycling route for bisphenol F-type epoxy resins, where the degraded fragments were reprocessed into a new thermoset. The resulting polymer exhibited improved mechanical and thermal characteristics, pointing toward the creation of higher-value materials from waste streams (Dang et al. 2002).

To accelerate the solvolysis process, Marinis et al. (2024) recently introduced a plasma-enhanced system combining nitrogen plasma with concentrated nitric acid. This technique significantly shortened processing time to 5.5 h and produced fibers with a tensile strength of 5.3 GPa. However, the method's high-power consumption and partial loss of acid present challenges that need to be resolved for scalable implementation (Marinis et al. 2024).

17.2.2.2 Hydrolysis

Hydrolysis has emerged as a promising strategy for the chemical recycling of polymer composites, offering a pathway to recover both the matrix and reinforcing fibres with varying degrees of efficiency and property retention. In the case of glass fibre-reinforced composites, Kao et al. (2012) investigated the mechanical properties of recycled glass fibres obtained through hydrolysis using sub-critical water to dissolve polyester resin. Their findings showed that while the tensile strength of recycled fibres decreased by approximately 40–70% compared to virgin fibres, the Young's modulus remained largely unchanged. Recycled fibres exhibited tensile strengths between 0.68 and 1.24 GPa and failure strains ranging from 0.91 to 1.61%. The efficiency of resin removal reached up to 95%, and it was found to be highly dependent on process parameters such as temperature (300–350 °C), time (5–30 min), and the presence of sodium hydroxide. SEM and EDX analyses revealed residual resin on the fibre surfaces, which impacted mechanical performance (Kao et al. 2012).

Hydrolysis has also been explored for recycling carbon fibre-reinforced composites. A notable study by Wang et al. (2021a) demonstrated an effective hydrolysis–oxidation synergistic catalytic method for recycling carbon fibres from unsaturated polyester resin (UPR) composites. Using an 80% hydrazine hydrate solution in the presence of sodium hydroxide at 80 °C, they achieved a degradation efficiency of 98.28% in just one hour. This was followed by a Fenton reaction to cleave remaining carbon–carbon bonds in the matrix. The recovered carbon fibres maintained their original surface morphology and achieved a tensile strength of 3.05 GPa—nearly matching virgin fibres (3.33 GPa)—highlighting the method's potential for high-quality fibre recovery. A related study employed the same strategy for epoxy-based

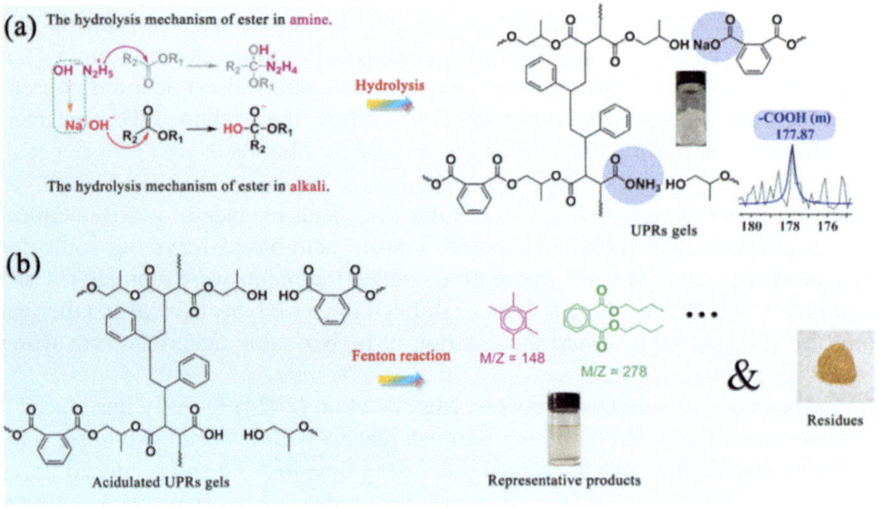

Fig. 17.8 The degradation of UPRs by hydrolysis-oxidation synergetic catalytical strategy (Wang et al. 2021a). Reprinted from Wang et al. (2021a), with permission from Elsevier

carbon fibre composites, achieving 95% degradation efficiency under a 4:1 molar ratio of hydrazine to sodium hydroxide at 80 °C within 2.5 h. The recycled carbon fibres retained 90% of their original tensile strength (2.8 GPa vs. 3.1 GPa), with SEM confirming their surface integrity and suitability for reuse. Figure 17.8 shows the schematic representation of the degradation of UPRs by hydrolysis-oxidation synergetic catalytical strategy (Wang et al. 2021a).

Thermal–chemical hybrid approaches have also been explored for carbon fibre recovery. Yao et al. (2022) implemented a two-step process involving thermal treatment at 400 °C for 1 h to remove 90% of the epoxy matrix, followed by chemical treatment in a 5% sodium hydroxide solution to enhance matrix removal. This method yielded carbon fibres with a tensile strength of 350 MPa and a modulus of 70 GPa—about 80% of the original values. The recycled fibres were reused to fabricate new CFRP composites, achieving a tensile strength of 250 MPa and a flexural strength of 300 MPa, thereby demonstrating practical reuse in structural applications. Another investigation applied a similar two-step method, confirming a total fibre recovery rate of 95% and validating the potential of these fibres in new composite formulations. The degradation products during of PBUP is shown in Fig. 17.9 (Yao et al. 2022).

Additionally, hydrolysis has been applied to the recycling of thermosetting resins such as urea–formaldehyde (UF). Liu et al. (2018) reported that hydrothermal hydrolysis under 140 °C for 2 h, using either 2 M hydrochloric acid or a 30% formaldehyde solution, resulted in up to 83% mass loss of cured UF resin. Chemical analyses, including ATR-FTIR and GC–MS, confirmed successful hydrolysis. The HCl process yielded ammonium chloride (NH_4Cl) as a byproduct, whereas the formaldehyde-based process enabled full resin decomposition and synthesis of a new UF resin incorporating about 6% of the recycled content. The newly formed resin retained similar

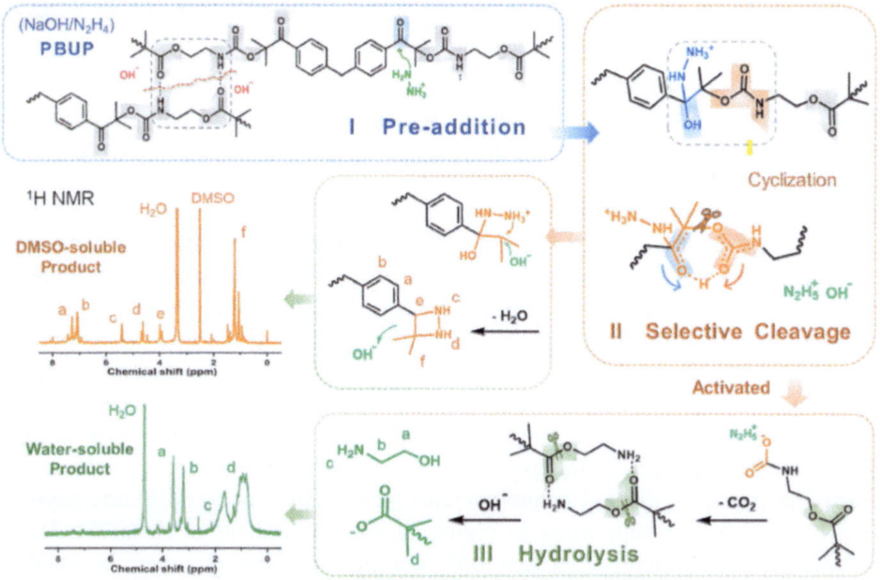

Fig. 17.9 Scheme of PBUP degradation and ^{1}H NMR spectra of degradation of products (Yao et al. 2022). Reprinted from Yao et al. (2022), with permission from Elsevier

chemical structures and bonding properties as conventional UF resin, supporting the feasibility of environmentally friendly and circular approaches to UF resin waste management (Liu et al. 2018).

These studies illustrate the flexibility and effectiveness of hydrolysis-based recycling methods across a range of fibre-reinforced and thermosetting polymer composites. By optimizing chemical conditions and combining with thermal or oxidative treatments, high recovery efficiencies and substantial preservation of mechanical properties can be achieved, advancing the circular economy in composite materials.

17.2.2.3 Glycolysis

Glycolysis has emerged as a highly promising and versatile chemical recycling method for polymer composites, enabling the recovery of valuable monomers, oligomers, polyols, and reinforcing fillers from both thermoplastic and thermoset systems. This process has gained attention for its ability to address complex and mixed waste streams, contributing to the development of sustainable materials and advancing the circular economy. A noteworthy example is the glycolytic depolymerization of interpenetrated polymer network (semi-IPN) foams composed of poly(vinyl chloride) (PVC) and crosslinked polyurea/isocyanurate. Calosi et al. (2022) demonstrated that under optimized conditions—165–175 °C for 20–30 min with dibutyltin dilaurate (DBTL) as a catalyst—more than 90% depolymerization of the crosslinked network was achieved. The recovered PVC showed approximately 90% purity, minimal degradation (evidenced by tetrahydrofuran solubility of 50–86%), and a glass transition temperature (Tg) similar to virgin PVC (85–87 °C). Additionally, the process produced a liquid phase rich in hydroxyl- and amine-terminated oligomers, which are suitable for synthesizing new polyurethane or polyurea-based materials (Calosi et al. 2022).

In the case of polyurethane (PU) recycling, glycolysis has demonstrated high efficiency and flexibility across various foam types and formulations. Chang et al. (2025) reported that glycolysis of PU foam using crude glycerol as a solvent achieved over 98% conversion at 210 °C with 2% catalyst loading. The regenerated polyols exhibited excellent physicochemical properties, including a hydroxyl number of 455 mg KOH/g and an acid number of 1.8 mg KOH/g. These polyols were effectively reused in new PU formulations, yielding foams with strong mechanical performance (compressive strength of 292 kPa), low thermal conductivity (0.035 W/m K), and good flame retardancy (limiting oxygen index of 27.8%). A techno-economic assessment of the process indicated an internal rate of return (IRR) of 30.64% and a payback period of 4.91 years, supporting its industrial viability (Chang et al. 2025). Simón et al. (2018) expanded this approach to PU composites containing thermoregulating microcapsules, achieving a 71% polyol recovery at 190 °C using crude glycerol and stannous octoate. The recovered polyols replaced up to 25% of virgin polyether polyol in new foam formulations without significantly compromising performance, underscoring the robustness of the glycolysis process even with advanced composite structures (Simón et al. 2018).

Epoxy thermosets, known for their high crosslink density and resistance to chemical degradation, have also been targeted for glycolysis-based recycling (Brivio et al. 2025). Figure 17.10 shows the block diagram of the hybrid PET depolymerization. Gersifi et al. (2003) investigated the depolymerization of diglycidyl ether of bisphenol-A (DGEBA) networks cured with polyamines using diethylene glycol (DEG) and titanium(IV) n-butoxide (TBT) at 245 °C. The process achieved complete matrix dissolution within 14 h and recovered phenolic-rich products, retaining over 75% of phenol groups. It also enabled the recovery of long glass fibres, maintaining their integrity for potential reuse in structural composites, thus enabling both organic and inorganic material recovery (Gersifi et al. 2003).

Glycolysis has also proven effective for recycling polyethylene terephthalate (PET) and polyester-based materials. Brivio et al. (2025) developed a hybrid glycolysis–methanolysis method that efficiently depolymerized PET from post-consumer textile waste, yielding dimethyl terephthalate (DMT) with 97.6% recovery and ≥99.1% purity. This process tolerated up to 10% non-PET contaminants and used a glycol-to-fibre ratio of 2 at 200 °C with 0.7 wt% catalyst loading, achieving over 99% PET conversion (Brivio et al. 2025). Complementary studies by, Wang et al. (2021b) showed that magnetic bimetallic $CoFe_2O_4$@ZIF-8/ZIF-67 catalysts enabled rapid glycolysis of PET and polybutylene terephthalate (PBT) at 200 °C, achieving an 84.3% yield of bis(2-hydroxyethyl) terephthalate (BHET) with high purity and catalyst reusability. Further advancements by Kim et al. (2024) and Salem et al.

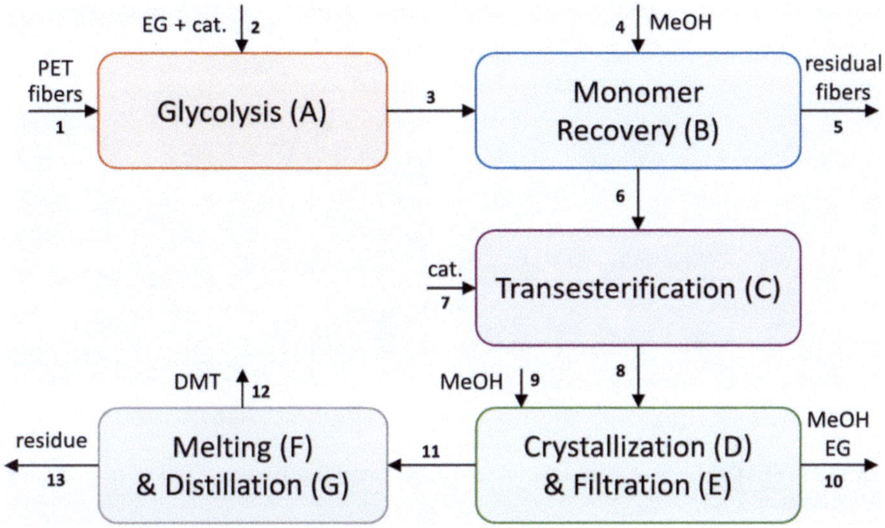

Fig. 17.10 Block diagram for proposed hybrid PET depolymerisation (Brivio et al. 2025). Reprinted from Brivio et al. (2025), with permission from Elsevier

(2025) highlighted the role of innovative catalytic systems—ranging from zinc-based compounds to co-catalysts such as dimethyl aniline—in enhancing reaction rates, product selectivity, and conversion efficiency under both conventional and microwave-assisted conditions. For instance, Salem et al. (2025) achieved full PET conversion and a 69.6% BHDET yield within just 1.5 h using a zinc acetate and dimethyl aniline co-catalyst system.

Microwave-assisted glycolysis has also proven to be a game-changer, significantly reducing reaction time and energy consumption. Donadini et al. (2024) reported a 94% reduction in reaction time and a 45% energy saving when recycling rigid PU foam under microwave irradiation. Using a PU/diethylene glycol mass ratio of 1.5 and potassium acetate as a catalyst, the process yielded polyols with low aromatic amine content (2.3 wt%) and performance characteristics suitable for replacing up to 30% of virgin polyol in new rigid foams (Donadini et al. 2024). Del Amo et al. (2021) also reported successful use of crude glycerol in PU glycolysis, attaining a 76.5% polyol purity and allowing up to 60% substitution of virgin polyol in new foams, while also recovering nanosilica fillers shown in Fig. 17.11, demonstrating the method's potential for multifunctional material recovery and the reaction mechanism is demonstrated by Fig. 17.12.

Glycolysis has additionally been utilized to synthesize new materials from recycled PET. Abdelaal et al. (2011) showed that propylene glycol (PG) outperforms other glycols such as diethylene glycol and triethylene glycol in producing hydroxyl-terminated oligomers suitable for unsaturated polyester (UP) resin synthesis. The resulting resins had molecular weights between 1787 and 2430 g/mol, with curing

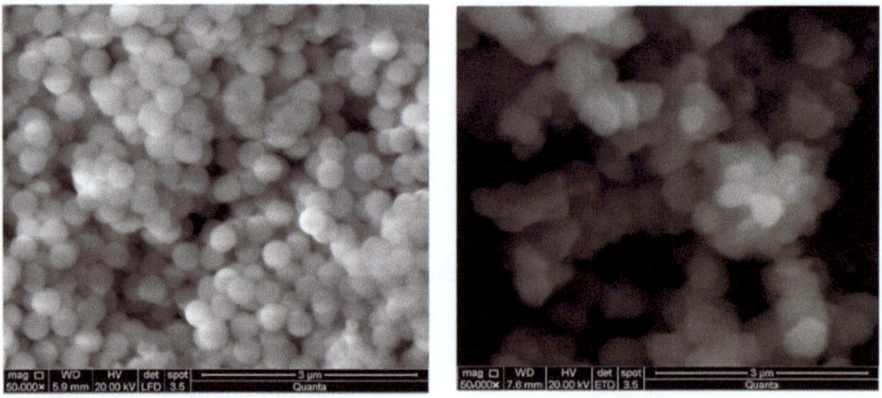

(a) (b)

Fig. 17.11 SEM pictures of fresh (**a**) and recovered nanosilica (**b**) with 50,000× magnification (Del Amo et al. 2021)

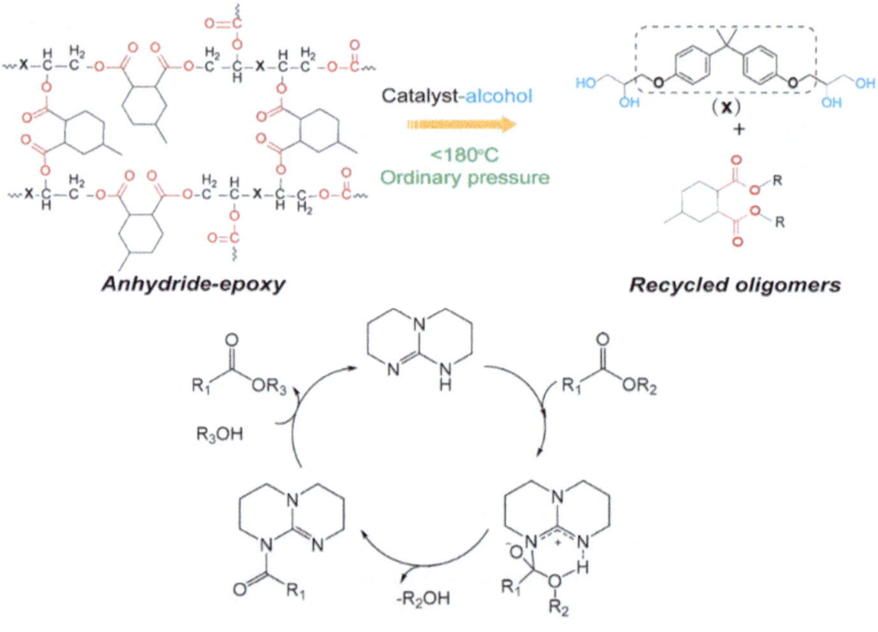

Fig. 17.12 The proposed dissolution mechanism of anhydride–epoxy via alcoholysis in a catalyst–alcohol solution at mild conditions (ordinary pressure below 180 C) (Del Amo et al. 2021)

behaviour influenced by glycol type and styrene content, allowing for customization based on performance needs (Abdelaal et al. 2011). Furthermore, Kirshanov et al. (2022, 2023) extended glycolysis applications to polyester tire cords, achieving complete PET depolymerization and a 46.07% devulcanization degree of the rubber matrix. Their kinetic modelling provided insights into the reaction pathways and molecular weight evolution, helping optimize glycolysis conditions for industrial applications (Kirshanov et al. 2022, 2023). Recent work by Anbarasu et al. (2024) and Leenders et al. (2025) further emphasized the scalability of glycolysis, employing nanocomposite catalysts and integrated processes for sequential recycling of poly-cotton textiles, respectively. These systems enabled simultaneous recovery of high-purity BHET and fermentable glucose, highlighting the feasibility of holistic material valorisation from complex, mixed waste streams (Anbarasu et al. 2024; Leenders et al. 2025).

Overall, these findings underscore the broad applicability and adaptability of glycolysis as a sustainable recycling strategy for polymer composites, particularly those based on PU, PET, and epoxy resins. The integration of novel catalytic systems, hybrid chemical routes, and energy-efficient technologies has significantly enhanced the performance and scalability of glycolysis. By enabling selective recovery of high-value materials even from contaminated or complex feedstocks, glycolysis offers a viable pathway toward circular material flows, reduced environmental burden, and economically feasible plastic waste management.

17.2.2.4 Other Chemical Lysis Techniques

Chemical lysis techniques, such as alcoholysis and methanolysis, have emerged as efficient methods for recycling various thermoset and thermoplastic polymer composites under relatively mild conditions. Alcoholysis has shown particular promise for depolymerizing crosslinked networks. Kuang et al. (2018a) demonstrated that anhydride-cured epoxy thermosets could be effectively dissolved via mild alcoholysis using 1,5,7-triazabicyclo[4,4,0]dec-5-ene (TBD) as a catalyst at temperatures below 180 °C. The process followed a surface erosion mechanism involving gel, swollen, and pure polymer layers, with ethylene glycol enhancing dissolution due to its high diffusivity and solubility (Kuang et al. 2018a). Similarly, Yuan et al. (2020) developed a novel poly(styrene-co-allyl sulfonic acid) (St-SAA) solid acid catalyst for the alcoholysis of dioscin, which outperformed conventional sulfuric acid in both yield (4.69%) and reusability, highlighting its potential as a sustainable and efficient alternative. Zou et al. (2022) explored microwave-assisted alcoholysis of silicone rubber waste using geraniol as a modifier, significantly improving mechanical and damping properties of the recovered elastomers. The addition of epoxidized natural rubber (ENR) enhanced interfacial compatibility, yielding a tensile strength of 4.8 MPa and elongation at break of 726%. For recycling PET-based composites, Ou et al. (2025) optimized the alcoholysis of PET blended with hybrid fibers using ethylene glycol and zinc acetate, producing glycolyzed oligomers suitable for synthesizing unsaturated polyester resin (UPR) with favorable mechanical properties (tensile strength of 57.1 MPa and bending strength of 71.9 MPa), demonstrating potential in high-performance construction materials.

Methanolysis has also been extensively studied for polycarbonate (PC) depolymerization. Bhogle and Pandit (2019) introduced an ultrasound-assisted methanolysis process at room temperature, which drastically reduced reaction time from 45 to 15 min and achieved BPA and DMC yields of 96.7% and 91.2%, respectively, under mild alkaline conditions (Bhogle and Pandit 2019). Complementing this, Liu et al. (2009) evaluated both methanolysis and hydrolysis of PC under moderate conditions. Using NaOH as a catalyst and tetrahydrofuran or 1,4-dioxane as solvents, they achieved over 95% BPA yield in methanolysis at 40 °C and 94% yield in hydrolysis at 100 °C. These studies collectively highlight the efficiency, scalability, and environmental viability of alcoholysis and methanolysis approaches for recycling polymer composites, paving the way for greener materials recovery and reuse strategies (Liu et al. 2009).

17.2.3 Thermal

Thermal recycling through pyrolysis is emerging as an effective and sustainable route for the valorization of polymer composites and plastic-based wastes. Pyrolysis involves thermochemical decomposition of polymers in the absence of oxygen, typically at temperatures ranging from 400 to 900 °C, producing valuable outputs such as pyrolysis oil, syngas, and char. These products serve as renewable energy sources and chemical feedstocks, making pyrolysis a key component in waste-to-energy strategies and circular economy models. Nanocomposites containing low silicate loadings (2–5%) can reduce heat release by 70–80% during combustion due to the formation of a protective ceramic char layer, thus offering safer thermal degradation pathways (Zanetti et al. 2002). Synthetic polymers possess high calorific values, often surpassing those of biomass and municipal solid waste, making them particularly suitable for energy recovery through pyrolysis. For example, the pyrolysis of waste tires and other rubbers has been well-documented, yielding bio-oil and syngas while minimizing environmental impact (Han et al. 2023; Menares et al. 2020). Similarly, slow batch pyrolysis of nitrile butadiene rubber (NBR) produced 39% oil with a heating value of 42.5 MJ/kg and demonstrated clear economic and environmental advantages over incineration, including net profits and significant reductions in greenhouse gas emissions shown in Table 17.1 (Pathiraja et al. 2025).

A range of polymer waste types have been successfully recycled using catalytic and advanced pyrolysis techniques. The pyrolysis of artificial marble waste, composed mainly of poly(methyl methacrylate) (PMMA) and aluminium hydroxide, yielded 66% methyl methacrylate (MMA) at 400 °C and converted the inorganic phase into γ-alumina (boehmite), enabling dual-material recovery (Tommaso et al. 2025). Waste polystyrene (PS), when subjected to catalytic pyrolysis with zeolite

Table 17.1 Properties of the NBR waste pyrolysis oil (Pathiraja et al. 2025)

S. no.	Description	Value
1	Density of pyrolysis oil (Kg/m^3)	972.50
2	Higher heating value (HHV) of pyrolysis oil (MJ/kg)	42.50
3	Flash point of pyrolysis oil (°C)	36.50
4	pH value of pyrolysis oil	8.75
5	Dynamic viscosity of pyrolysis oil at 40 °C (mPs s or CP)	25.44
6	Kinetic viscosity of pyrolysis oil at 40 °C (mm^2/s)	26.16

catalysts at 500 °C, produced 85% liquid yield with 70% aromatic hydrocarbons, demonstrating excellent selectivity and product value (Shen and Wu 2025). Figure 17.13 shows the overall representation of pyrolysis of epoxy resin. Likewise, polyvinyl chloride (PVC) presents challenges due to hydrochloric acid release; however, co-pyrolysis with calcium hydroxide ($Ca(OH)_2$) successfully reduced HCl emissions from 60 to 25% and enhanced aromatic yields by ~60%, while forming stable calcium salts in the char (Kuttiyathil et al. 2025).

Composite materials, including electronic and structural waste, have also been targeted. The pyrolysis of waste printed circuit boards (WPCBs) optimized via

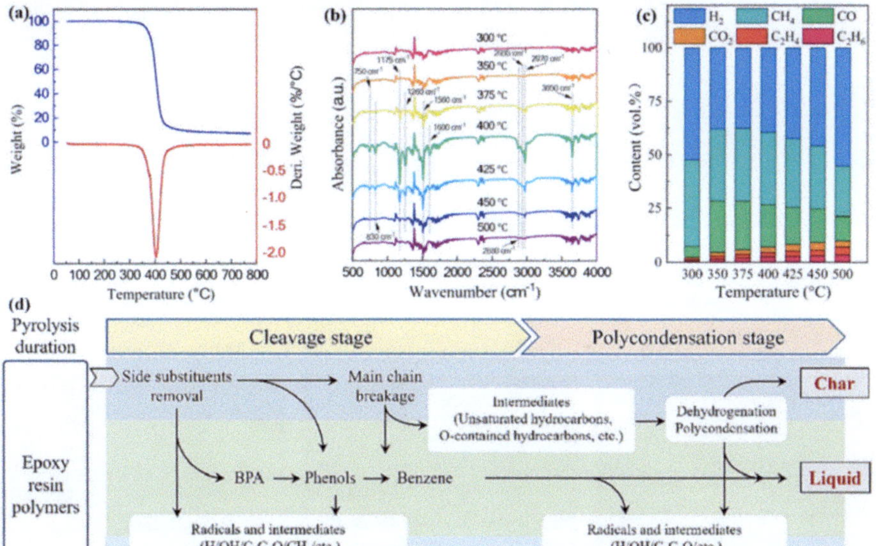

Fig. 17.13 Pyrolysis of epoxy resin: **a** TG/DTG curves, **b** FTIR spectra, **c** gas product composition, and **d** involved processes (Shen and Wu 2025). Reprinted from Shen and Wu (2025), with permission from Elsevier

response surface methodology achieved a high solid yield rich in metals and a liquid fraction with high calorific value (~35 MJ/kg), showcasing dual recovery potential (Boro and Tiwari 2025). In the wind energy sector, end-of-life wind turbine blades (WTBs) composed of carbon or glass fibre composites have been efficiently recycled. Transition-metal-assisted pyrolysis using zirconium chloride ($ZrCl_4$) enhanced resin decomposition from 52.13 to 75.59% and improved recovered glass fibre tensile strength by 34.74%, attributed to selective cleavage of C–O–C bonds and controlled graphitization (Lu et al. 2025). Figure 17.14 depicting the oxidation weight loss ratios and SEM images of the recovered fibre. Similarly, microwave-assisted catalytic pyrolysis of epoxy-based CFRPs using β-30 zeolite yielded up to 36.28% phenol, driven by bisphenol-A conversion and optimized microwave conditions, while balancing fibre integrity (Yao et al. 2025).

The evolution of carbon fibre properties through repeated pyrolysis is another area of importance. While repeated recycling via pyrolysis and partial oxidation increased fiber stiffness, tensile strength dropped to 55% after three cycles, indicating that material reuse must be matched to appropriate performance demands (Alguacil et al. 2025).

Pyrolysis has also been extended to wood-plastic composites (WPCs), where CO_2-mediated catalytic pyrolysis using a nickel-based catalyst increased syngas

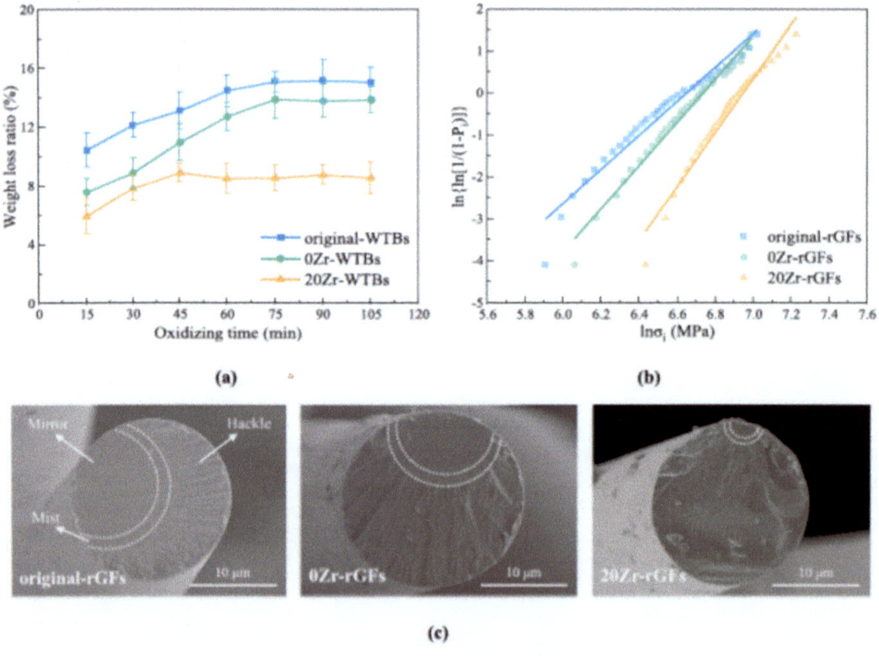

Fig. 17.14 Oxidation weight loss ratios of solid pyrolysis products at 500 °C (**a**), the distribution plots of Weibull tensile strength of the recovered glass fibres (**b**), and the tensile fracture surfaces of recovered glass fibres (**c**) (Lu et al. 2025). Reprinted from Lu et al. (2025), with permission from Elsevier

production by 14.2 times compared to nitrogen, and enhanced CO yields at lower onset temperatures, presenting a viable method for low-emission energy recovery (Kwon et al. 2025).

Further downstream, the hydroprocessing of mixed plastic pyrolysis oil (HPO) has shown promise as a diesel alternative. Compositional similarity to commercial diesel was as high as 98%, and emissions profiles for blends (30:70, 60:40, 90:10 HPO:Diesel) were 95% comparable to conventional fuels. HPO maintained high engine performance and lower brake-specific fuel consumption. Vedavalli et al. (2025) exemplifying the full-cycle utility of pyrolytic products in real-world applications. Overall, pyrolysis-based thermal recycling provides a robust, scalable, and increasingly optimized pathway for converting diverse polymer composite wastes into value-added fuels and chemicals, supporting both environmental sustainability and material circularity. Table 17.2 provides a detailed comparison of the pros and cons of the current recycling technique.

17.3 Environmental Challenges Associated with Current Recycling Techniques

Current recycling techniques for polymers are mechanical, chemical, and thermal, each present notable environmental challenges that limit their overall sustainability. Mechanical recycling, despite its widespread adoption, often results in material degradation due to repeated melting and reprocessing, which causes polymer chain scission. This leads to a reduction in mechanical properties and limits the recyclate's application to lower-performance products, a phenomenon known as downcycling (Newcomb 2016; Tseng et al. 2017). The process also requires highly sorted, clean input streams; contaminants such as food residues, inks, and incompatible polymers can significantly degrade product quality or render batches unusable (Titone et al. 2024). Furthermore, multi-layer or composite plastics—common in food packaging—pose a particular challenge, as the differing polymers in these materials are incompatible with mechanical recycling, necessitating their disposal through incineration or landfilling (Butenegro et al. 2021). In addition, the mechanical processing steps such as shredding, washing, and extrusion can release micro- and nano-sized plastic particles into wastewater systems, becoming a point source of microplastic pollution if not adequately managed (1-s2.0-S0269749122003281-main).

Chemical recycling offers a promising alternative by breaking down polymers into monomers or basic chemicals, making it theoretically possible to recover materials of near-virgin quality. However, this technique is also environmentally burdensome. Depending on the specific process—such as pyrolysis or solvolysis—chemical recycling can release volatile organic compounds (VOCs), carbon dioxide, and other pollutants unless advanced emission control systems are in place (Chanda 2021). Moreover, many plastics contain legacy additives like brominated flame retardants, bisphenol A (BPA), and phthalates that can persist through the recycling process and

Table 17.2 Pros and cons of existing recycling techniques

S. no.	Recycling technique	Pros and cons
1	Chemical recycling	*Pros* 1. Numerous studies have demonstrated that chemical recycling can significantly reduce the overall global warming potential (GWP) by substituting virgin plastic production. Credit allocation for avoided upstream emissions has shown reductions of up to 59% in total environmental impact (Al-Salem et al. 2014) 2. Chemical recycling processes effectively break down polymer chains into smaller constituents such as monomers, oligomers, and basic gases (e.g., CO_2, CO, CH_4, H_2). These intermediates can then be used to synthesize new polymers, enabling closed-loop recycling of plastics (Luo et al. 2024) 3. It allows for the treatment of plastic waste streams that are unsuitable for mechanical recycling such as multi-layered packaging, contaminated waste, and thermoset materials, transforming them into reusable petrochemical feedstocks (Luo et al. 2024) 4. Compared to mechanical recycling, chemical methods are capable of recovering more value-added products from plastic waste. In some assessments, this has been linked to a 60% reduction in environmental burden, in contrast to 25% for traditional mechanical recycling methods (Luo et al. 2024) 5. Theoretical models suggest that chemical upcycling can avoid as much as 2.9 kg of CO_2-equivalent per kg of PET waste. Similarly, other common plastics (LDPE, HDPE, PP, PS) show substantial emission savings when processed via solvolysis or pyrolysis (Luo et al. 2024) 6. Unlike mechanical approaches, which often result in degraded polymer properties, chemical recycling can yield materials with performance comparable to virgin plastics. This is critical for applications requiring high purity including food-grade packaging (Peti et al. 2025) *Cons* 1. There is evidence suggesting that a fraction of plastic processed through chemical recycling can escape as micro- or nano-sized particles into air or water. For instance, studies from UK facilities have reported emission rates of up to 13% (Singh and Walker 2024) 2. Many recycled plastics contain persistent hazardous chemicals, such as flame retardants, plasticizers, and stabilizers, which may not be fully removed during processing and can pose long-term health and environmental risks (Singh and Walker 2024) 3. Toxic substances leaching from recycled materials may pose occupational hazards to facility workers and downstream health risks to users of recycled products, especially where regulatory oversight is limited (Singh and Walker 2024) 4. Chemical recycling facilities are capital-intensive, requiring advanced reactors, purification units, safety systems, and emission controls, making their deployment financially demanding (Ragaert et al. 2017) 5. Certain plastic types, especially composites and multi-polymer blends, may not be easily or cost-effectively recycled chemically. This limits the scalability and efficiency of the process when dealing with heterogeneous plastic waste streams (Ragaert et al. 2017)

(continued)

Table 17.2 (continued)

S. no.	Recycling technique	Pros and cons
2	Mechanical recycling	*Pros* 1. Mechanical recycling typically produces lower carbon emissions compared to landfilling, incineration, or chemical recycling. This makes it an environmentally preferable option, especially for clean, homogeneous plastic waste (Akonda et al. 2012; Giorgini et al. 2015) 2. It involves relatively straightforward processes such as shredding, washing, melting, and re-extruding, which require minimal chemical usage. This simplicity reduces both energy requirements and operational costs (Howarth et al. 2014) 3. In the case of carbon fibre-reinforced composites (CFRPCs), mechanical recycling avoids the use of aggressive chemicals or heat, helping preserve the physical integrity of fibres. This enables reuse in secondary applications with minimal property loss (Durante et al. 2023) *Cons* 1. During processes such as grinding or extrusion, fiber length is reduced and surface quality deteriorates, leading to weakened tensile strength and lower performance in the final recycled product (Newcomb 2016; Tseng et al. 2017) 2. Adhesion between the recycled fibre and polymer matrix can be inadequate, which affects load transfer and mechanical performance. To restore functionality, surface modification of the fibres is often necessary (Butenegro et al. 2021)
3	Thermal recycling	*Pros* 1. Thermal processes such as pyrolysis and incineration can convert plastic waste into usable energy forms, including heat, electricity, and fuels. This makes it a viable option for managing non-recyclable or contaminated plastics (Nanda and Berruti 2021) 2. Thermal recycling infrastructure typically requires less initial investment and offers a faster return through energy generation, making it economically favourable in certain regions or waste management systems (Gałko and Sajdak 2022) 3. When properly optimized, pyrolysis processes can yield valuable hydrocarbon fractions that can be refined or reused in the chemical industry, effectively turning plastic waste into petrochemical feedstock substitutes (Lin and Yang 2007) 4. Thermal recycling can complement existing cascade waste treatment frameworks and, if regulated properly, reduce landfill reliance. However, systems must be engineered to avoid PVC or halogenated plastics that release harmful emissions (Gałko and Sajdak 2022) *Cons* 1. Unless advanced pollution controls are in place, thermal recycling can emit dioxins, furans, nitrogen oxides, and CO_2, posing serious environmental and health hazards during incineration or pyrolysis (Lewis and Gertsakis 2003) 2. Unlike chemical or mechanical recycling, thermal methods are destructive in nature, as polymers are broken down irreversibly. This eliminates the potential for closed-loop recycling (Gałko and Sajdak 2022) 3. The widespread adoption of thermal recycling is hindered by economic, regulatory, technical, and public perception challenges, especially in regions with strict air quality standards (Eze et al. 2021)

re-enter the environment or circulate in consumer products (Singh and Walker 2024). These substances pose potential toxicological risks to ecosystems and human health. Additionally, chemical recycling processes typically require high temperatures and the use of solvents or catalysts, leading to considerable energy consumption and associated greenhouse gas emissions—particularly when powered by non-renewable energy sources (Shuaib and Mativenga 2016).

Thermal recycling methods, such as incineration and pyrolysis, are often employed for plastic waste that is unsuitable for mechanical or chemical recycling. However, these methods are associated with substantial environmental drawbacks. The combustion of plastics, especially those containing halogens like polyvinyl chloride (PVC), can emit significant quantities of CO_2, nitrogen oxides (NO_x), sulphur oxides (SO_x), dioxins, and furans, contributing to air pollution and climate change (Lewis and Gertsakis 2003). In contrast to material recovery techniques, thermal recycling completely destroys the polymer, converting it to heat or energy, which limits circularity and represents a permanent loss of material value (Gałko and Sajdak 2022). Additionally, residues from these processes—such as fly ash and char—may contain toxic metals or persistent chemicals that pose risks of soil and groundwater contamination. The heterogeneity of input waste further complicates thermal recycling; mixed and contaminated plastic streams can lead to inconsistent combustion and emissions, making it difficult to maintain uniform environmental safeguards across facilities.

17.4 Recycling Different Types of Polymer Composites

17.4.1 Biodegradable Composites

Biodegradable polymers are environmentally friendly materials capable of breaking down through chemical and biological processes. These polymers are broadly categorized into natural and synthetic types, based on their source and degradation mechanisms (Doppalapudi et al. 2014). Natural biopolymers such as starch, cellulose, chitosan, and various proteins are derived from renewable resources and typically degrade enzymatically (Banerjee et al. 2014). On the other hand, synthetic biodegradable polymers—including polylactic acid (PLA), polyglycolic acid (PGA), polycaprolactone (PCL), and polybutylene succinate (PBS)—are petroleum-based and break down through hydrolysable or oxidizable chemical backbones (Vroman and Tighzert 2009).

Due to their biocompatibility and ability to degrade into non-toxic byproducts, biodegradable polymers have found widespread applications across various fields. In the packaging industry, they offer sustainable alternatives for food wrapping and containers (Siracusa et al. 2008). In agriculture, they serve roles such as controlled-release fertilizers and soil conditioning agents (Puoci et al. 2008). In the biomedical sector, these polymers are extensively used in drug delivery systems (Srivastava et al. 2016), tissue engineering (Gunatillake et al. 2003), and implantable devices

(Brem and Gabikian 2001), owing to their favourable degradation profiles and safety (Asghari et al. 2017). These biodegradable polymers are further classified as natural and synthetic polymers.

17.4.1.1 Natural

Natural biodegradable polymers—such as polysaccharides (e.g., cellulose, chitosan, hyaluronic acid, alginate) and proteins (e.g., collagen, gelatin, albumin)—are primarily derived from renewable plant sources like wheat, rice, and potatoes. These materials are highly valued for their biocompatibility, biodegradability, and sustainable origin. However, despite their advantages, natural biopolymers can suffer from limitations including batch-to-batch variability, potential antigenicity, and mechanical weaknesses (Doppalapudi et al. 2014).

Owing to their biological compatibility, these polymers have found widespread use in biomedical applications. They serve as effective carriers in drug delivery systems (Ngwuluka et al. 2014), support materials for tissue engineering scaffolds (Sell et al. 2010), agents for wound healing (Tavakoli et al. 2023), and components in surgical hemostats (Ghimire et al. 2021). Nonetheless, several challenges hinder their broader commercialization. These include relatively high production costs, insufficient industrial composting infrastructure, and low public demand and awareness (Rujnić-Sokele and Pilipović 2017). Despite these barriers, advancements in processing technologies and material innovation are expected to enhance their adoption and contribute significantly to reducing environmental pollution from conventional plastics (Awasthi et al. 2022).

Polysaccharides Composite

Polysaccharides are the most abundant class of naturally occurring polymers, composed of monosaccharide units linked by glycosidic bonds and classified as homo- or heteropolysaccharides based on their sugar composition. Sourced primarily from plants, algae, and microorganisms, common examples include cellulose, starch, chitin, chitosan, and alginate (Benalaya et al. 2024). These biopolymers are inherently biodegradable, renewable, and non-toxic, making them excellent candidates for sustainable material development. Their molecular architecture—rich in hydroxyl groups and capable of forming hierarchical structures—endows them with notable properties such as film-forming ability, mechanical strength, and compatibility with diverse chemical modifications (Valencia et al. 2019). Owing to these features, polysaccharides have found widespread applications in food packaging, biomedical engineering, pharmaceuticals, and environmental remediation. In recent years, the growing demand for green and circular materials has driven intensive research into transforming polysaccharides into advanced functional composites with enhanced recyclability, self-healing capabilities, and tailored multifunctionality.

Recent advancements in sustainable materials science have demonstrated significant progress in the development of recyclable, biodegradable, and high-performance cellulose- and polysaccharide-based composites. A notable contribution is the environmentally friendly, solvent-free, and catalyst-free fabrication of polysaccharide-vitrimer hybrid materials using chitin, chitosan, and sodium alginate without any chemical modification (Feng et al. 2023). These materials incorporate up to 40 wt% polysaccharides in a polyhydroxyurethane (PHU) vitrimer matrix, offering outstanding mechanical properties (tensile toughness of 13.7 MJ/m^3), thermal stability (degradation temperature ~457 °C), rapid self-healing (99% in 20 s at 80 °C), and high recyclability. The addition of carbon black or graphene further extends their functionality to flexible sensors and electromagnetic interference shielding applications.

Complementing this, Dong et al. (2025) developed recyclable nanocomposite films based on a PVA/starch blend reinforced with $Ti_3C_2T_x$ MXene nanoplatelets (2.5–10 wt%) via solution casting. These films exhibit significant mechanical improvements (up to 669% in Young's modulus and 292% in tensile strength), enhanced barrier properties (91% reduced water vapor permeability), and recyclability through redissolution and recasting without significant property degradation making them ideal for high-performance packaging (Dong et al. 2025).

Several innovations have utilized regenerated or recycled bacterial cellulose (rBC) and plant-derived cellulose for functional applications Hu et al. (2022) introduced magnetoelectric (ME) composites comprising rBC hydrogels, NdFeB magnetic particles, and copper coils. These biodegradable and recyclable composites are created through chemical and physical cross-linking and demonstrate effective electromechanical energy conversion, alongside the recyclability of both hydrogels and embedded components after degradation in water or enzymatic conditions (Hu et al. 2022). Similarly, Yuan et al. (2025) fabricated a biodegradable and recyclable piezoelectric macrofibre using imidazolium perchlorate crystallized in bacterial cellulose, showing high sensitivity (3.71 mV/kPa) and complete biodegradability within 9 h in enzymatic environments. Both studies highlight promising avenues for sustainable wearable electronics and flexible sensors (Yuan et al. 2025).

The development of all-cellulose composites (ACCs) has also seen considerable progress. Spörl et al. (2018) employed an ionic liquid, [EMIM][OAc], to dissolve cellulose and uniformly embed high-strength rayon fibres into a cellulose pulp matrix. The resultant ACCs possess strong mechanical properties and were successfully recycled up to three times by redissolving and remoulding, confirming their potential for closed-loop manufacturing in fibre-reinforced materials (Spörl et al. 2018).

Other efforts have explored eutectogel-based systems for self-healing and reusability. Wang et al. (2022) created a cellulose-based eutectogel composed of PVA, gelatin, and dialdehyde carboxymethyl cellulose (DCMC), crosslinked in a deep eutectic solvent (DES). This material boasts tensile stress of 1.25 MPa and 1400% elongation, alongside 80% self-healing efficiency under NIR irradiation. It maintains mechanical and adhesive properties even after five recycling cycles by reprocessing through freeze–thaw techniques (Wang et al. 2022).

In the realm of composite materials incorporating recycled cellulose, Platnieks et al. (2020) investigated woody-like composites formed from recycled cellulose (rCell) obtained from Tetra Pak waste and biodegradable PBS polymer. These composites exhibit enhanced hardness, modulus, and thermal stability, and biodegrade completely within 70 days in composting conditions, representing a practical solution for large-scale biodegradable packaging (Platnieks et al. 2020).

Another innovative approach by Su et al. (2023) used electrospinning and DES impregnation to develop a polysaccharide-based ultrathin membrane composed of PVA, Gleditsia sinensis polysaccharide, and glycerol. Post-treatment with a choline chloride-urea DES produced translucent membranes with high haze (92% at 550 nm), conductivity (1.32 S cm^{-1}), and recyclability via water dissolution in 5 s—suitable for optoelectronic and health monitoring applications (Su et al. 2023).

Recyclability has also been integrated into thermoset polymer matrices. Park et al. (2019) developed a healable and recyclable thermoset polyurethane composite reinforced with hydrophobized cellulose fibres via reversible Diels–Alder (DA) chemistry. The composites showed improved mechanical strength and retained performance after multiple healing and recycling steps, demonstrating potential for long-life structural materials (Park et al. 2019).

Lastly, Copenhaver et al. (2024) enhanced the recyclability of thermoplastics like PLA and PETg by incorporating spray-dried cellulose nanofibrils (CNFs) at 5–20 wt%. These composites maintained structural and mechanical integrity across up to seven thermomechanical recycling cycles. In particular, PETg composites with 20% CNFs retained performance equal to unreinforced PETg after three additional reuse cycles, underscoring CNFs' ability to enhance material circularity (Copenhaver et al. 2024).

17.4.1.2 Synthetic

Synthetic biodegradable polymers represent a highly adaptable group of materials, especially prominent in biomedical fields due to their tailorable properties, biodegradability, and biocompatibility (Tian et al. 2012). Despite their petroleum-based origins (Vroman and Tighzert 2009), many of these polymers—such as polylactic acid (PLA), polycaprolactone (PCL), and polyglycolic acid (PGA)—exhibit excellent biodegradability, making them suitable for a variety of sustainable applications (Al-Shalawi et al. 2023).

These polymers are extensively used not only in medicine, including drug delivery, tissue engineering, and implantable devices (Gunatillake et al. 2003), but also in sectors like food packaging (Swetha et al. 2023) and agriculture (Madhavan Nampoothiri et al. 2010). Their degradation typically occurs through hydrolysis of ester bonds, and copolymers like PLGA allow for controlled degradation rates, depending on the application. What makes these materials particularly attractive is their customizable mechanical behaviour and safe degradation into non-toxic products. However, they are not without limitations—some, such as PLA and PGA, can produce acidic degradation byproducts and have limited mechanical strength, especially under long-term load-bearing conditions (Gunatillake et al. 2003). Nevertheless, ongoing advances in polymer science continue to enhance their properties, broadening their utility in safer and more efficient medical and environmental applications.

PLA Composites

Polylactic acid (PLA) is a bio-based, biodegradable thermoplastic polyester synthesized from renewable resources such as corn starch or sugarcane. As an environmentally friendly alternative to petroleum-derived polymers, PLA has garnered significant attention in the pursuit of sustainable materials. It offers notable advantages including high stiffness, good mechanical strength, optical clarity, and ease of processing, making it suitable for various applications such as packaging, medical devices, automotive components, and 3D printing (Madhavan Nampoothiri et al. 2010; Faruk et al. 2012; Pang et al. 2010). However, PLA is not without its drawbacks. It suffers from brittleness, low thermal stability, and slow degradation under ambient conditions, which limits its functional scope (Qin et al. 2024).

To overcome these limitations, PLA is often reinforced with both organic and inorganic fillers to produce high-performance biocomposites. Common reinforcements include natural fibres like flax, jute, rice hulls, agave, and wood, as well as nanomaterials like nanoclays, carbon nanotubes, and graphene. These fillers enhance the mechanical, thermal, and barrier properties of PLA, while also improving biodegradability and sustainability (Majeed et al. 2013). The resulting PLA-based composites exhibit improved toughness, thermal resistance, and durability, making them viable for eco-friendly industrial and consumer applications.

Numerous studies have explored the recycling of PLA-based composites using mechanical, chemical, and emerging strategies. Mechanical recycling remains the most straightforward and industrially viable method, though it often results in polymer degradation and fibre attrition over multiple cycles. Le Duigou et al. (2008) examined the mechanical recycling of flax fibre-reinforced poly(L-lactide) (PLLA) biocomposites using injection molding over several cycles. Their results indicated a decrease in tensile strength and molecular weight due to thermo-mechanical degradation and fibre fragmentation, though the composites retained sufficient integrity for reuse over at least three cycles (Le Duigou et al. 2008).

Similar degradation behaviour was observed by Åkesson et al. (2016) in PLA composites reinforced with wood fibres and basalt fibres processed through multiple extrusion cycles. Mechanical properties such as tensile and impact strength decreased progressively with each cycle due to fiber degradation and matrix embrittlement. Despite these changes, structural integrity was preserved well enough to support reusability, especially in non-load-bearing applications (Åkesson et al. 2016). Another notable approach to enhancing the recyclability of PLA composites involves nanofillers. Salah et al. (2023) investigated the use of organically modified montmorillonite (OMMT) nanoclay in PLA matrices. Their results showed that the addition of OMMT improved both thermal and electromagnetic properties even after repeated recycling cycles, making the composites suitable for EMI shielding applications (Salah et al. 2023).

Natural fibre-based biocomposites also present a promising recycling route. Srebrenkoska et al. (2014) studied rice hull-reinforced PLA composites incorporating PLA-grafted-maleic anhydride (PLA-g-MA) as a compatibilizer. Although flexural modulus decreased by nearly 50% after recycling, the materials still performed comparably to conventional fibreboards, suggesting potential for eco-friendly construction applications (Srebrenkoska et al. 2014). Similarly, Ramos-Hernández et al. (2024) demonstrated that agave fibre-reinforced PLA biocomposites retained significant mechanical integrity after several mechanical recycling cycles. Their study emphasized the role of fiber-matrix interaction and processing conditions in ensuring long-term performance (Ramos-Hernández et al. 2024).

Beyond mechanical methods, advanced recycling strategies have also been developed to recover both matrix and fiber components more efficiently. Tian et al. (2017) introduced an innovative recycling method for continuous carbon fibre-reinforced PLA composites using 3D printing. The approach involved softening PLA with heat to recover intact carbon fibres, which could then be reused along with recycled PLA to produce new filaments. This process allowed for 100% recovery of carbon fibres and a 73% recovery rate for PLA, with remanufactured composites exhibiting improved flexural strength (Tian et al. 2017).

Chemical recycling has shown potential for breaking PLA down into its monomers, enabling high-purity recovery. Chen et al. (2023) developed a novel PLA composite containing a bio-based flame-retardant additive (PA-Fe-HBT) that not only enhanced thermal and UV stability but also facilitated PLA degradation in an alkaline medium catalysed by Fe^{3+} ions. The chemical recycling process yielded lactic acid with 88% purity, presenting a viable route for closed-loop material recovery (Chen et al. 2023).

In addition to mechanical and chemical approaches, biological recycling methods have also been explored. Singh et al. (2022) highlighted the use of enzymatic degradation under controlled environments as an emerging pathway for PLA recycling. Their study reviewed the impact of various recycling processes—including mechanical reprocessing, hydrolysis, and glycolysis—on structural and mechanical properties. It emphasized that optimizing fibre-matrix adhesion and processing conditions can mitigate performance loss during recycling (Singh et al. 2022). The integration of recycled materials from other waste streams has also been considered. Singh et al. (2022) further investigated PLA composites reinforced with chemically treated newspaper fibres, showcasing a method for combining waste paper and PLA into usable composites via melt blending and compression moulding. This approach highlights the dual benefit of reducing both paper and plastic waste while maintaining desirable mechanical properties (Singh et al. 2022).

Lastly, Al Zahmi et al. (2022) compared mechanical, chemical, and thermal recycling methods for carbon fibre-reinforced PLA composites. Their study found that although pyrolysis and solvent-based chemical recycling enabled higher fibre recovery and tensile strength retention (up to 98.9%), they also raised concerns over energy consumption and environmental impact. Mechanical recycling, despite some material degradation, remained the most practical and environmentally benign method, especially for short-term reuse applications (Al Zahmi et al. 2022).

PCL

Polycaprolactone (PCL) is a biodegradable and biocompatible aliphatic polyester synthesized through ring-opening polymerization of ε-caprolactone (Mandal and Shunmugam 2020; Bhattacharjee et al. 2021). Its low melting point (~60 °C), chemical resistance, and mechanical flexibility make it a versatile material for biomedical, environmental, and industrial applications (Backes et al. 2022; Jeong et al. 2018) PCL is particularly notable for its ease of processing via extrusion, injection moulding, and 3D printing.

One of the key attributes of PCL is its biodegradability, which primarily occurs through hydrolysis of its ester linkages. This degradation is enhanced by the presence of bacteria and fungi, making it suitable for composting and environmental degradation (Kumar Kalita et al. 2024; Li et al. 2000). Recent advances have accelerated this process through the incorporation of immobilized lipase enzymes into the PCL matrix. Kumar Kalita et al. (2024) reported that enzyme-embedded PCL achieved up to 90% mineralization in 145 days under composting conditions. Notably, the depolymerized products—mostly oligomers—could be repolymerized with a high conversion rate (~82%), enabling a closed-loop recycling pathway for PCL-based materials (Kumar Kalita et al. 2024).

While PCL is known for its flexibility and processability, its mechanical strength is generally lower than that of more rigid polymers like PLA or PET (Jeong et al. 2018). Additionally, its inherent hydrophobicity can limit applications involving aqueous environments. However, surface modification strategies—such as plasma treatment or chemical grafting with maleic anhydride—have proven effective in improving its wettability and interfacial adhesion with other materials (Khorramnezhad et al. 2021).

To overcome its mechanical limitations and expand its applications, PCL is often blended with natural polymers or reinforced with various fillers. Starch, a biodegradable and low-cost filler, has been widely used. Wu (2003) showed that blending PCL with starch, especially when compatibilized with maleic anhydride (PCL-g-MAH), enhanced tensile strength and phase compatibility while retaining good biodegradability (Wu 2003). Though the composites exhibited increased water resistance, this marginally slowed degradation. Similarly, Wahit et al. (2012) highlighted that incorporating natural fibres such as jute, flax, and kenaf into PCL matrices improved mechanical strength and biodegradation rates. However, repeated reprocessing could degrade the material's properties (Wahit et al. 2012).

Several studies have explored the use of recycled or waste-derived fillers with PCL. Luna et al. (2019) developed composites using wood powder (WP) and waste leather buff (WLB), which enhanced thermomechanical properties, reduced crystallinity, and increased stiffness. The addition of maleic anhydride-grafted PCL (PCL-MA) as a compatibilizer improved interfacial adhesion between the matrix and fillers. These composites retained their thermoplasticity and were suitable for repeated extrusion and moulding. Microcrystalline cellulose (MCC) was also shown to improve enzymatic degradation while preserving structural integrity (Luna et al. 2019).

Nanocomposite approaches have introduced novel reinforcements to PCL. Raghunath et al. (2009) used polyhedral oligomeric silsesquioxane (POSS) to reinforce PCL/polycarbonate polyurethane scaffolds, creating materials with balanced degradation and mechanical robustness for biomedical applications (Raghunath et al. 2009). Shokrollahi et al. (2014) reported that hydroxyapatite (HAp) nanoparticles, functionalized with supramolecular ureido-pyrimidinone (UPy) groups, significantly enhanced both mechanical properties and biodegradability. These results highlight the potential of nanofillers in improving performance without compromising sustainability (Shokrollahi et al. 2014).

In a more functional context, Luna et al. (2019) also explored electromagnetic interference (EMI) shielding using PCL blended with recycled α-Fe_2O_3 nanoparticles and oil palm fibers. The resulting composites showed improved microwave absorption and permittivity, demonstrating the feasibility of multifunctional sustainable materials (Luna et al. 2019).

Overall, PCL and its composites show strong promise for sustainable materials development. Its thermoplastic nature supports mechanical recycling, while its susceptibility to enzymatic and microbial degradation offers viable end-of-life options. Composites based on natural fibres, waste fillers, and nanomaterials expand the performance and application scope of PCL. However, ensuring long-term mechanical stability and optimizing degradation kinetics, especially under repeated use and recycling, remain key challenges for circular implementation.

PEG

Polyethylene glycol (PEG) is a hydrophilic, biocompatible, and flexible polyether extensively used across biomedical, pharmaceutical, environmental, and industrial applications owing to its excellent solubility in both aqueous and organic solvents, low toxicity, and versatile chemical structure (D'souza and Shegokar 2016; Chen et al. 2024). Composed of repeating ethylene glycol units [$-(CH_2CH_2O)_n-$], PEG—also known as Macrogol—offers tunable molecular weights, allowing for precise control of its physical properties and functions in various material systems. PEG can be synthesized via anionic polymerization of ethylene oxide using hydroxyl initiators like water, ethylene glycol, or diols, or by ring-opening polymerization of epoxyethane. Commercial PEGs are available in varying degrees of polymerization and functional group activation. The molecular weight of PEG ranges from 200 to several thousand, and its physical properties change with MW: it is a colourless, viscous liquid at MW < 600 and a waxy solid at MW > 800. It is water-soluble and hygroscopic at room temperature (D'souza and Shegokar 2016).

In polymer composites, PEG frequently serves as a soft segment, compatibilizer, or plasticizer, enhancing filler dispersion, interfacial adhesion, and phase compatibility. Its terminal hydroxyl groups enable chemical grafting or copolymerization with hydrophobic polymers such as polylactic acid (PLA), polycaprolactone (PCL), and polyurethanes, resulting in amphiphilic block copolymers with improved biodegradability, mechanical performance, and matrix interaction (Al Samad et al. 2016). PEG-based composites reinforced with nanoparticles (e.g., silica, hydroxyapatite, graphene oxide) or natural fibres demonstrate enhanced mechanical strength, controlled swelling, and tunable degradation rates, making them ideal for biomedical scaffolds, smart packaging, and drug delivery applications (Biondi et al. 2015). While PEG is not inherently biodegradable like PLA or PCL, its hydrophilic nature facilitates hydrolytic chain scission and enzyme accessibility when embedded within degradable matrices, thereby accelerating overall composite degradation (Hu et al. 2024). This positioned PEG not only as a functional matrix component but also as a multifunctional additive that improves the processability, compatibility, and environmental performance of advanced composite materials.

One study demonstrated the successful synthesis of laccase-immobilized hydrogels for dye wastewater treatment, utilizing dopamine-modified sodium alginate (SA) and PEG. These hydrogels retained 70.23% activity after 30 days of storage and 90.23% after six reuse cycles, showcasing excellent recyclability. The blend of natural bio-based polymers also improved biodegradability, resulting in 84.32% decolorization of Reactive Blue 19. This eco-friendly approach provides an effective solution for wastewater remediation (Su et al. 2024). Another study explored the development of multifunctional aerogel/fibre building insulation composites (AFCP-C) incorporating SiO_2 aerogel powder, α-alumina fibres, carbon fibre felt (CFF), and PEG. PEG acted as a dispersing agent, improving the distribution of aerogel particles in the fibre matrix, resulting in enhanced structure and thermal performance. This composite is fully recyclable, with biodegradable components, demonstrating sustainability and efficiency in building construction (Lv et al. 2024).

In the realm of biodegradable materials, a study developed PCL/PEG copolymer blends reinforced with cellulose nanofibers (CNF) for soft tissue engineering. The addition of CNF improved mechanical properties, and the scaffolds were shown to be biodegradable and compostable, supporting their potential for biomedical applications (Ghalia and Alhanish 2023). Recyclable solid–solid phase change materials (SSPCMs) were synthesized through reversible anhydride-alcohol crosslinking with PEG, offering excellent latent heat and recyclability without the need for reactive chemicals. This process enhances crystallinity and thermal performance, making it suitable for thermal energy storage applications (Wu et al. 2024).

PEG also plays a significant role in developing energy storage composites. In one study, a phase change material (PCM) was created by infusing PEG with lauric and capric acids into a porous wood structure. This composite exhibited significant energy storage capacity and excellent recyclability, demonstrating its potential for building applications (Chen et al. 2024). In smart material applications, the incorporation of PEG into epoxy resin films resulted in shape memory behaviors and improved recyclability. The films demonstrated high performance in terms of shape recovery, ductility, and thermal response, indicating their suitability for applications in aerospace, automotive, and medical fields (Hu et al. 2024). PEG has also been successfully applied to the chemical recycling of fibre-reinforced epoxy resin composites. In one study, a PEG/NaOH system facilitated the efficient decomposition of epoxy resin while preserving fiber integrity, making it possible to recycle high-value fibres for future use (Yang et al. 2014).

Overall, PEG continues to be a versatile and sustainable material, with applications spanning across biomedical, environmental, energy storage, and material recycling domains. It supports chemical recycling by facilitating efficient depolymerization of thermosets and mixed composites, enhances mechanical and processing performance in biodegradable blends, and promotes environmentally favourable degradation profiles. Furthermore, PEG's FDA approval, biocompatibility, and commercial availability continue to drive its adoption in biomedical devices, packaging, and eco-friendly materials. As the demand for recyclable and degradable polymers grows, PEG is poised to play an increasingly central role in the design of next-generation materials aligned with sustainability and circular economy principles.

17.4.2 Non-biodegradable Composites

Non-biodegradable polymers, such as polyethylene (PE), polypropylene (PP), polystyrene (PS), and polyvinyl chloride (PVC), are synthetic materials derived from petrochemicals (Dwivedi et al. 2019) and widely used in packaging (Zeng et al. 2024), agriculture (Sadeleer and Woodhouse 2024), and in medical field (Shastri 2003) due to their durability, chemical resistance, and low cost (Luckachan and Pillai 2011). However, their resistance to natural degradation causes them to persist in the environment for hundreds of years, leading to serious ecological issues, including landfill, accumulation, ocean pollution, and harm to wildlife (Taghavi et al. 2021). Unlike biodegradable polymers that break down into harmless byproducts, non-biodegradable plastics release toxic substances and microplastics as they degrade slowly (Luckachan and Pillai 2011). Traditional disposal methods like incineration can emit hazardous gases, and recycling is often limited by contamination and high processing costs (Bharadwaj et al. 2015). To address these challenges, alternative strategies such as pyrolysis for energy recovery (Dwivedi et al. 2019), development of biodegradable substitutes (Song et al. 2018), and improved recycling technologies are being explored.

17.4.2.1 Thermoplastic Composites

Polymers can be broadly classified into two categories based on their thermal response: thermoplastics and thermosets (Bîrca et al. 2019). This section focuses on thermoplastic polymers. The term "thermoplastic" is derived from "thermo" meaning heat and "plastic" meaning to flow. These polymers can be repeatedly softened and reshaped upon heating, making them ideal for recycling and reprocessing (Singh et al. 2021). Thermoplastics are typically composed of trans geometrically oriented polymer chains arranged in lamellae and spherulitic structures, which impart varying degrees of crystallinity. Compared to amorphous polymers, they exhibit a glass transition temperature above room temperature and have a sharp melting point (Lee and Knight 1970).

Common commodity thermoplastics include polypropylene (PP), polyethylene (PE), polyvinyl chloride (PVC), and polystyrene (PS), while engineering thermoplastics include acrylonitrile butadiene styrene (ABS) and polyethylene terephthalate (PET). Polyethylene, one of the most widely used thermoplastics, exists in different forms based on molecular structure: LDPE (low-density polyethylene) with long-chain branching for flexibility and impact resistance; HDPE (high-density polyethylene) with a linear structure for strength and chemical resistance; and LLDPE (linear low-density polyethylene), which balances strength and flexibility (Singh et al. 2021). Polypropylene is known for its lightweight, high melting point, and chemical resistance, although its brittleness at low temperatures can be mitigated through copolymerization or filler addition (Allen et al. 2025). PET, available in transparent amorphous or white semi-crystalline forms, offers good resistance to moisture and chemicals and maintains mechanical performance between −40 and 100 °C. Its strength is linked to intrinsic viscosity, indicating polymer chain length (Awaja and Pavel 2005).

Thermoplastics are vital across various industries due to their versatility, processability, and recyclability. For instance, PVC is used in pipes and seals, PE in films and containers, PP in automotive parts and textiles, PMMA in lenses, PS in appliances, nylon (PA) in gears and sportswear, PC in medical devices, POM in mechanical parts, CA in eyewear, and TPU in flexible footwear (Mărieş and Abrudan 2018). Despite their recyclability, only about 9% of the 300 million metric tons of plastic waste generated globally each year is recycled, while the rest is landfilled, incinerated, or lost to the environment, contributing significantly to pollution and climate change (Oladele et al. 2023). Figure 17.15 shows the polymer production from 2016 to 2020 (Mt) worldwide. In 2020, over 29 million metric tons of post-consumer plastic waste were collected in Europe; 23% ended in landfills, 40% in energy recovery, and just 34% was recycled. Packaging accounts for the largest share of plastic waste at 41%, followed by construction. Notably, nearly 30% of plastics produced globally serve packaging and chemical industries, growing at an annual rate of 12% (Azeko et al. 2016). From the 320 million tons of plastic produced in 2016, only 9–21% was recycled or incinerated, with the remainder degrading into microplastics in landfills or oceans (Tamrakar et al. 2023).

Azeko et al. (2016) examined physical recycling of LDPE into earth-based laterite/cement composites. LDPE dissolved in kerosene at 140 °C was quenched at 4 °C, filtered, and mixed with cement-laterite blends in varying proportions. At 20 vol% LDPE, the composite achieved its highest compressive strength (~4.1 MPa). Crack growth resistance improved with PE addition due to interactions between cracks and the microstructure (Azeko et al. 2016). Similarly, Stan et al. (2019) recycled LDPE/multi-walled carbon nanotube (MWCNT) composites, finding reduced

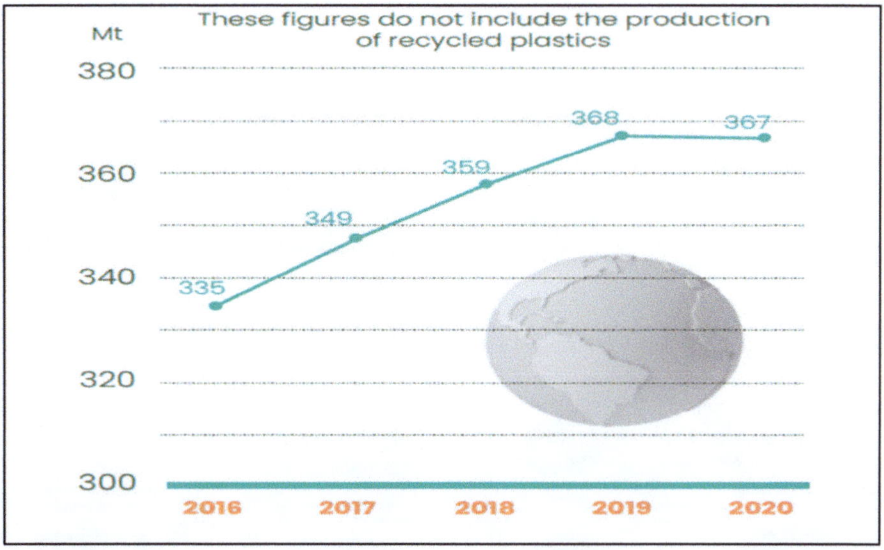

Fig. 17.15 Polymer production from 2016 to 2020 (Mt) worldwide. Plastics Europe. Plastics-the Facts. 2021. Available online: https://plasticseurope.org/knowledge-hub/ (accessed on 21 October 2021) (Oladele et al. 2023)

melt flow index and increased mechanical properties post-recycling, especially at 5 wt% MWCNT content, where conductive networks improved tensile and electrical performance (Stan et al. 2019).

Papaspyrides et al. (1994) used a dissolution/reprecipitation process to recycle LDPE greenhouse film. Dissolution in toluene at 85 °C followed by acetone precipitation yielded recycled LDPE with crystallinity levels slightly above the virgin polymer (37–38% vs. 34%), without notable loss in tensile strength (Papaspyrides et al. 1994). Figure 17.16 shows the recycling routes for PP-GF and PA6-CF based TPCLs and Fig. 17.17 shows the flow diagram for the experimental procedure. Meanwhile, Kiss et al. (2020) explored mechanical recycling of Tepex composite laminates (PP- and PA6-based, reinforced with GF or CF). Shredded laminates were hot compression molded into thermoplastic SMCs, which retained fibre volume fraction but lost 40–50% of flexural strength compared to virgin materials. However, PP-GF sandwich panels showed comparable flexural performance (~370 MPa strength, 18 GPa modulus) (Kiss et al. 2020).

Otheguy et al. (2009) recycled thermoplastic composite boat hulls via shredding and extrusion. Recycled materials blended with virgin PP were injection moulded into tensile bars. Paint contamination reduced tensile strength due to poorly bonded particles acting as stress concentrators, though some batches achieved impact strength comparable to pure GF/PP blends (Otheguy et al. 2009). Further studies by Papaspyrides et al. (1995) showed that ionomer and LDPE composites recycled through solvent-assisted fibre separation yielded improved tensile properties depending on washing steps. For ionomer composites, one wash preserved good

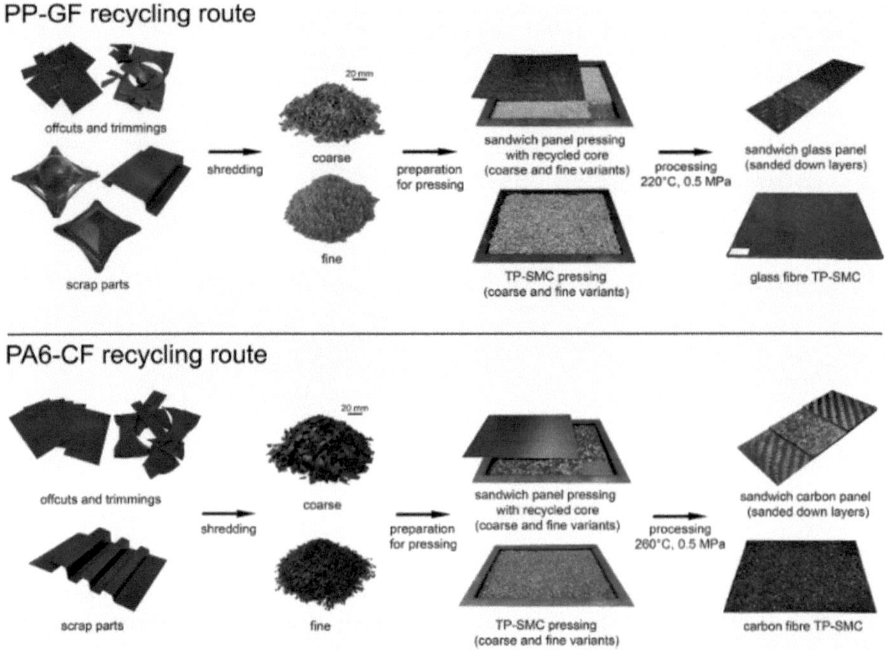

Fig. 17.16 Recycling routes for PP-GF and PA6-CF based TPCLs (Papaspyrides et al. 1994). Reprinted from Papaspyrides et al. (1994), with permission from Elsevier

fibre-polymer adhesion; for LDPE composites, three washes led to cleaner fibres and higher tensile strength, though reduced fibre length (Papaspyrides et al. 1995).

Srebrenkoska et al. (2008) studied PP composites reinforced with rice hulls (RH) and kenaf fibres (K), using melt mixing and reprocessing. After two recycling cycles, PP/K composites retained higher mechanical and thermal properties compared to PP/RH, with only a 5% drop in flexural strength and a 20% increase in modulus. TGA analysis showed improved thermal stability of PP/K composites, while SEM confirmed better fibre-matrix adhesion. These examples highlight how recycling strategies and fibre-matrix interactions significantly influence the performance of recycled thermoplastics (Srebrenkoska et al. 2008).

17.4.2.2 Thermosetting Composites

Thermosetting composites are created by combining a thermoset polymer resin—such as epoxy, polyester, phenolic, or vinyl ester—with reinforcing fibres like glass, carbon, aramid, or natural fibres, and then curing the mixture through heat, pressure, or chemical catalysts to form a rigid, cross-linked structure (Dorigato 2021). These materials are often chosen over thermoplastic composites due to their superior mechanical strength, dimensional stability, and heat resistance (Nishida et al. 2018).

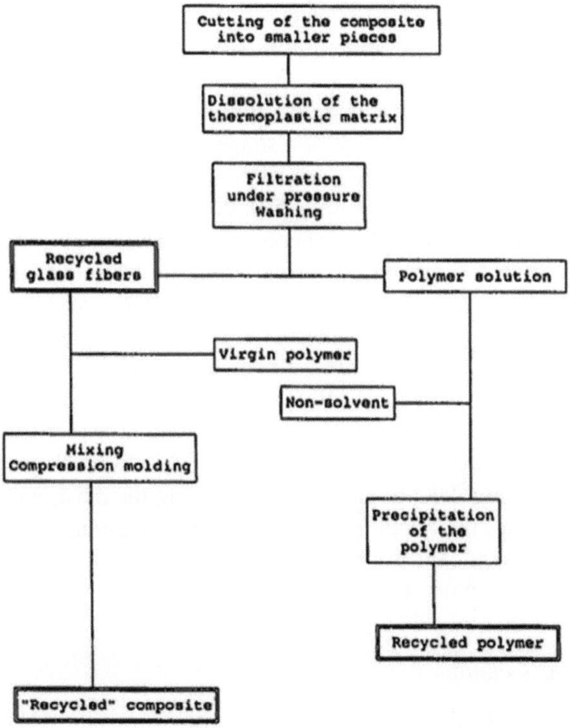

Fig. 17.17 Flow diagram for the experimental procedure (Papaspyrides et al. 1995). Reprinted from Papaspyrides et al. (1995), with permission from Elsevier

They are used in medical field (Madhav et al. 2019), aerospace (Hamerton and Kratz 2018), and in wind blade applications (Dorigato 2021; Murray et al. 2021). However, their irreversible curing process also makes recycling challenging, as they cannot be reprocessed like thermoplastics. Current recycling methods include mechanical grinding, pyrolysis, and chemical processes, but each has limitations in terms of material recovery and quality (Dorigato 2021).

Elastomers

Rubber, renowned for its unique viscoelastic properties (James and Guth 1943), exists in two primary forms: natural rubber (NR) and synthetic rubber. NR is predominantly composed of cis-1,4-polyisoprene and is extracted as latex—a white milky sap—from the *Hevea brasiliensis* tree, though alternative sources like guayule and Russian dandelion also exist (Joseph et al. 2022). For industrial use, NR is mainly harvested from *Hevea* latex. In contrast, synthetic rubbers such as polybutadiene, styrene-butadiene rubber (SBR), and ethylene-propylene-diene monomer (EPDM) are derived from petrochemicals (Sharma et al. 2016). Although ethylene and propylene are thermoplastics, their combination in EPDM disrupts crystallinity, conferring rubber-like behaviour. Rubber's elasticity stems from its flexible polymer chains, which are reinforced through vulcanization—typically with sulphur—creating crosslinks that enhance strength, durability, and shape recovery. NR can also crystallize under stress, increasing its toughness, and fillers like carbon black are often added to further improve performance (Moore 1950). Polymer characteristics are strongly influenced by molecular weight; NR inherently possesses high molecular weight, while synthetics allow for tunability during synthesis (Ali and Brocchini 2006). Glass transition temperature (Tg), a key factor in rubber's mechanical properties, can be adjusted using nanoparticles (Sharma et al. 2016). Rubber plays a vital role in a range of applications including tires, gloves, insulation, sensors, and protective materials (Venkatachalam et al. 2013; Rostami-Tapeh-esmaeil et al. 2021).

However, rubber's non-biodegradable nature poses significant recycling and environmental challenges (Asaro et al. 2020). Traditional disposal methods, such as incineration or landfilling, lead to pollution, prompting the development of alternative recycling techniques like supercritical CO_2 devulcanization, microwave-assisted processes, ultrasonic devulcanization, and mechano-chemical reclamation (Liu et al. 2016; Isayev et al. 2014; Guo et al. 2021).

Advance Recycling Techniques

Supercritical CO_2 devulcanization has emerged as a promising technique. Mangili et al. (2014) treated cryo-ground tire rubber (GTR, <0.4 mm) using diphenyl disulphide (DD) and supercritical CO_2 at 31.1–400 °C and 7.38–70 MPa, achieving a 52% devulcanization degree. The process primarily cleaved crosslinks while preserving the polymer backbone, as evidenced by increased sol fraction, lower molecular weight, and detection of radical intermediates like thiophenols (Mangili et al. 2014). Figure 17.18 shows the normalized gel fraction and FTIR graph of GTR and T-GTR. Similarly, Asaro et al. (2020) reported up to 90% devulcanization of sulphur-cured NR under 12 MPa at 160 °C using DD and $scCO_2$, with minimal main-chain degradation and a homogeneous microstructure, confirmed via FTIR, TGA, and SEM (Asaro et al. 2020). Figure 17.19 shows the SEM images of the untreated and treated NR.

Microwave-assisted devulcanization is another efficient strategy. Zanchet et al. (2012) recycled industrial SBR waste by exposing ground scraps to 900 W microwave radiation for up to 3 min, achieving selective sulphur bond cleavage and producing composites with virgin SBR. Liu et al. (2016) enhanced microwave devulcanization of waste rubber powders using ultrafine barium ferrite (BaM) as a microwave absorber. The process allowed partial devulcanization and formation of magnetic rubber composites, with improved dispersion and tunable magnetic properties (Zanchet et al. 2012).

Ultrasonic devulcanization has also shown promise. Isayev et al. (2014) utilized a twin-screw extruder equipped with a 40 kHz ultrasonic horn to process tire rubber particles, achieving significant bond cleavage and improved compatibility. Shabani and Jamshidi (2022) employed probe sonication in various media (toluene,

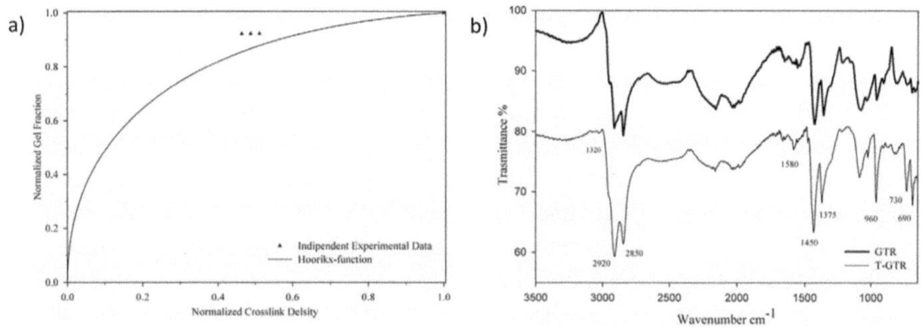

Fig. 17.18 Shows **a** Normalized gel fraction ([X]/[X0]) as function of normalized crosslink density (ξ/ξ0) compared to the Horikx function and **b** ATR spectra of GTR and T-GTR (Mangili et al. 2014). Reprinted from Mangili et al. (2014), with permission from Elsevier

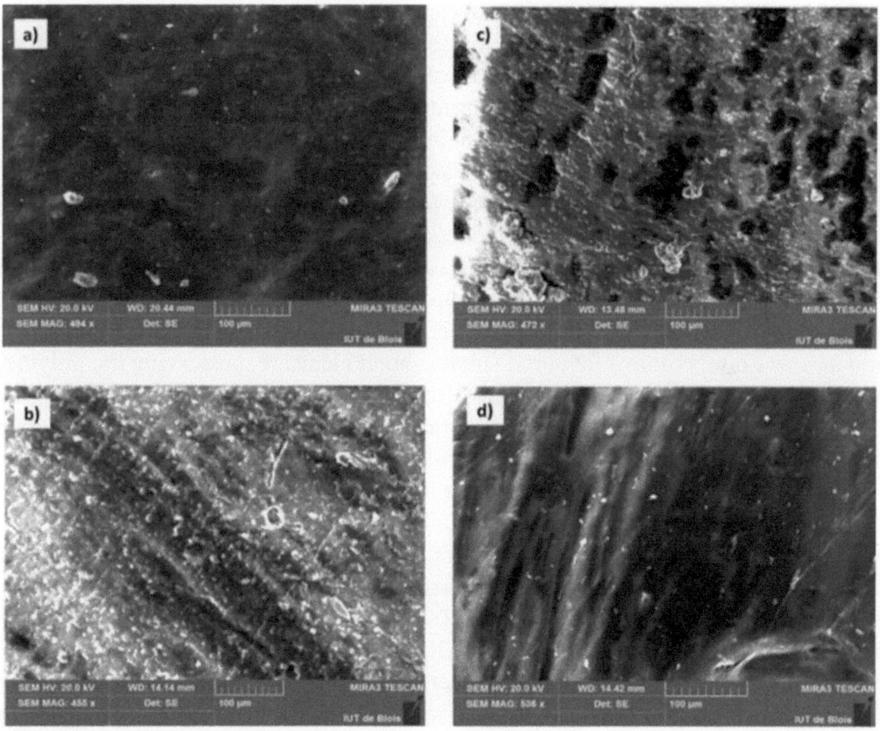

Fig. 17.19 SEM images of **a** virgin NR, **b** NR treated at 120 C, **c** NR treated at 140 C, and **d** NR treated at 160 C (Asaro et al. 2020). Reprinted from Asaro et al. (2020), with permission from Elsevier

oil, water), finding that toluene yielded the highest devulcanization (28%), which increased to 52% with the addition of VitaX. FTIR and TGA confirmed selective crosslink scission and preserved polymer integrity (Shabani and Jamshidi 2022).

Mechano-chemical approaches also offer viable routes. Ghosh et al. (2023) used silane-based devulcanization aids in an internal mixer at 140–160 °C to process vulcanized tire rubber. An alkenyl silane with an activator achieved 57.1% network breakdown while maintaining good mechanical properties and thermal stability. Overall, while each method presents distinct advantages, they also have limitations in scalability, efficiency, and material performance. Continued innovation in devulcanization processes is essential for advancing rubber recycling and sustainability (Ghosh et al. 2023).

Conventional Recycling Techniques

Thermo-mechanical and chemical devulcanization methods have shown promise in reclaiming waste rubber with improved selectivity and performance. Seghar et al. (2019) investigated the thermo-mechanical devulcanization of natural rubber (NR) using a twin-screw extruder, feeding ground rubber at 5 kg/h with a screw speed of 240 rpm and barrel temperatures ranging from 80 to 220 °C. The applied shear not only promoted devulcanization but also induced self-heating, which influenced the actual internal temperature. The study found that lower barrel temperatures (80–100 °C), in combination with shear-induced heating, achieved optimal devulcanization efficiency (~90%) with minimal energy input. Mooney viscosity and sol fraction analyses indicated that elevated temperatures led to greater polymer degradation rather than selective crosslink scission. Horikx plots confirmed that devulcanization at lower temperatures favoured sulphur bond cleavage, preserving the polymer backbone and enhancing recyclability (Seghar et al. 2019). Complementing this, Zhang et al. (2018) explored chemo-thermo-mechanical devulcanization of waste rubber powder (WRP) from heavy-duty truck tires using thiobisphenols as reclaiming agents. WRP was processed with varying thiobisphenol concentrations (0–3 g) and aromatic oil in an internal mixer at 180–200 °C, followed by compounding and moulding. Results showed increased sol content, reduced crosslink density, and lower Mooney viscosity with higher thiobisphenol loading and temperature. However, excessive devulcanization (at 3 g and 200 °C) compromised mechanical properties, while a moderate amount (0.5 g at 200 °C) offered the best balance between devulcanization and revulcanization performance. SEM and TGA analyses corroborated these findings, showing smoother surfaces and staged thermal degradation, respectively (Zhang et al. 2018).

In contrast, Guo et al. (2021) compared a low-temperature mechano-chemical devulcanization (LTMD) method to high-temperature atmospheric devulcanization (HTAD). HTAD relies on thermal-oxidative pyrolysis at 200–290 °C, resulting in non-selective bond cleavage and main chain degradation. In contrast, LTMD processes rubber at 40 °C with a devulcanizing agent and shear in an internal mixer at 60–120 °C, followed by extrusion at 50 °C to ensure uniformity. LTMD preserved 65–70% of the original mechanical strength, as confirmed by Horikx analysis and

TEM imaging, which showed smoother fracture surfaces and better filler dispersion. TGA results indicated superior thermal stability and lower oil content (4.5% vs. 12%) in LTMD-treated samples. Furthermore, LTMD avoided the release of toxic gases (e.g., H_2S, SO_2, CS_2), often associated with HTAD, and minimized recrosslinking. Together, these studies highlight the importance of optimizing processing parameters and additive concentrations to achieve selective crosslink scission while preserving polymer integrity, paving the way for more sustainable and high-performance rubber recycling strategies (Guo et al. 2021).

Resin

Thermosetting polymers, or thermosets, are high molecular weight crosslinked polymers extensively used in structural and engineering applications due to their superior mechanical properties, chemical resistance, and thermal stability. Unlike thermoplastics, thermosets cannot be re-melted or reshaped once cured, as their polymer chains form permanent covalent crosslinks during the curing process. This curing can be initiated by heat, UV radiation, electron beams, or chemical hardeners and catalysts. The irreversible nature of these crosslinks prevents thermosets from softening or deforming under heat, making them ideal for applications that require dimensional stability and resistance to harsh environments (Morici and Dintcheva 2022).

Several types of resins are classified as thermosets, each with distinct applications and properties. Phenolic resins, such as phenol–formaldehyde (PF), are among the earliest synthetic thermosets and are widely used in moulding compounds, electrical insulation, coatings, packaging, and fire-resistant and refractory materials due to their high thermal stability and flame resistance (Zhu et al. 2024). Unsaturated polyester resins are typically reinforced with glass fibres to form fibre-reinforced plastics (FRP), which are commonly found in marine applications (e.g., boats), tanks, motor housings, and sanitary fixtures like bathtubs (Nakagawa and Goto 2015). Epoxy resins, another important class, exhibit excellent mechanical strength, strong adhesion, and resistance to thermal and chemical degradation. These attributes make them indispensable in high-performance applications such as aerospace components, wind turbine blades, adhesives, protective coatings, and electronic encapsulation (Chen et al. 2020a; Jin et al. 2015). In high-tech sectors such as aerospace, military, automotive, and sporting goods, carbon fibre-reinforced polymers (CFRPs) are preferred for their high strength-to-weight ratios and durability (Yang et al. 2015).

However, one of the key drawbacks of thermosets is their resistance to conventional recycling processes due to the permanent crosslinked structure. Unlike thermoplastics, which can be remelted and reshaped, thermosets do not flow upon heating, making mechanical recycling ineffective and landfill disposal environmentally unsustainable. As a result, researchers have been developing innovative recycling strategies to overcome these limitations and promote circularity in thermoset use.

Several thermal and chemical methods have shown promise. Pyrolysis under limited oxygen conditions can break down the polymer matrix and remove the resin from composites like CFRP, while preserving the valuable reinforcing fibres (Yang et al. 2015). Chemical recycling approaches using acid pretreatment followed by alkaline glycolysis have demonstrated the ability to nearly completely remove epoxy resin from composites while preserving the mechanical properties of the recovered fibres (Jiang et al. 2017). Mixed solvent systems with embedded catalysts enable transesterification reactions at relatively low temperatures, facilitating efficient depolymerization of the matrix and simultaneous recovery of both fibers and resin components (Mu et al. 2022).

Alternative strategies also include mechanical reuse. For instance, mechanically shredded thermoset composites from decommissioned wind turbine blades can be integrated into new composite systems as fillers or reinforcements. However, challenges such as weak interfacial bonding between the matrix and the recycled filler limit mechanical performance (Beauson et al. 2016). Figure 17.20 shows the stress strain plot and SEM images of the fractured surface of SC composites. Supercritical solvolysis using acetone–water mixtures has shown potential in achieving high resin degradation rates with reduced energy consumption, although achieving complete fibre surface cleanliness remains an area for optimization (Keith et al. 2016). Figure 17.21 shows the SEM images of the washed and unwashed carbon fibres with acetone.

Bio-based and Dynamic Covalent Network (DCN) thermosets offer exciting prospects for sustainable design. Bio-based epoxy resins, such as those synthesized from furan derivatives, can degrade efficiently under mild oxidative or hydrolytic conditions, allowing recovery of valuable monomers like 2,5-furandicarboxylic acid (FDCA) with high purity (Chen et al. 2020a). Vitrimers, which incorporate dynamic covalent bonds like disulfides, enable reshaping, repair, and chemical recycling of thermoset composites through bond exchange reactions without significant loss in performance (Ruiz De Luzuriaga et al. 2016). Similarly, phenolic resins modified with dynamic boronic ester linkages have been shown to allow full recyclability of both resin and fiber under mild conditions, even at room temperature, with excellent recovery and reusability (Wang et al. 2018).

In summary, while traditional thermosets pose significant end-of-life challenges due to their permanent crosslinked structures, recent advances in recycling technologies—including thermal decomposition, chemical depolymerization, dynamic bond-based systems, and bio-based resin design—are paving the way toward a more sustainable, circular lifecycle for these high-performance materials. Continued innovation in this field is crucial to addressing the environmental impacts of thermoset waste, particularly in industries such as aerospace, wind energy, and automotive manufacturing, where such materials are used at scale.

17.5 Conclusion and Future Prospects

Addressing the environmental concerns linked to the disposal of polymer composites has become increasingly important due to their extensive use in industries such as aerospace, automotive, construction, and electronics. While mechanical recycling

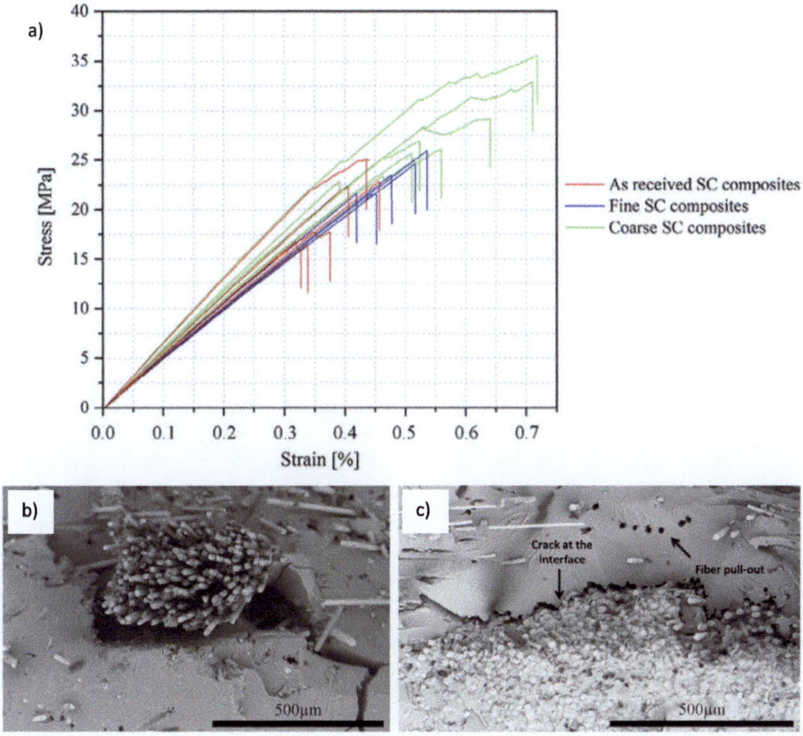

Fig. 17.20 **a** Stress–strain curves of the as received, fine and coarse SC composites with the highest fibre content on about 10 vol%. Whereas, **b** and **c** SEM images of fracture surface of SC composites (Beauson et al. 2016). Reprinted from Beauson et al. (2016), with permission from Elsevier

remains the most commonly adopted strategy because of its low cost and operational simplicity, it is often insufficient for composite materials. The mechanical treatment of composites typically leads to degradation of the fibre and polymer matrix, causing a decline in material performance and limiting their reuse in high-value applications—a process often referred to as downcycling.

In contrast, advanced recycling technologies are gaining momentum due to their ability to more effectively recover both polymer matrices and reinforcement fibres. Techniques such as chemical depolymerization (solvolysis), thermal decomposition (pyrolysis), and enzymatic degradation allow for greater selectivity in breaking down complex composite structures. Innovative methods like supercritical CO_2-assisted solvolysis have demonstrated environmentally friendly capabilities by avoiding harsh chemicals while promoting effective matrix breakdown under controlled pressure and temperature. Similarly, microwave-assisted recycling provides efficient, uniform heating that reduces processing times and energy input—making it a promising solution for thermoset composites.

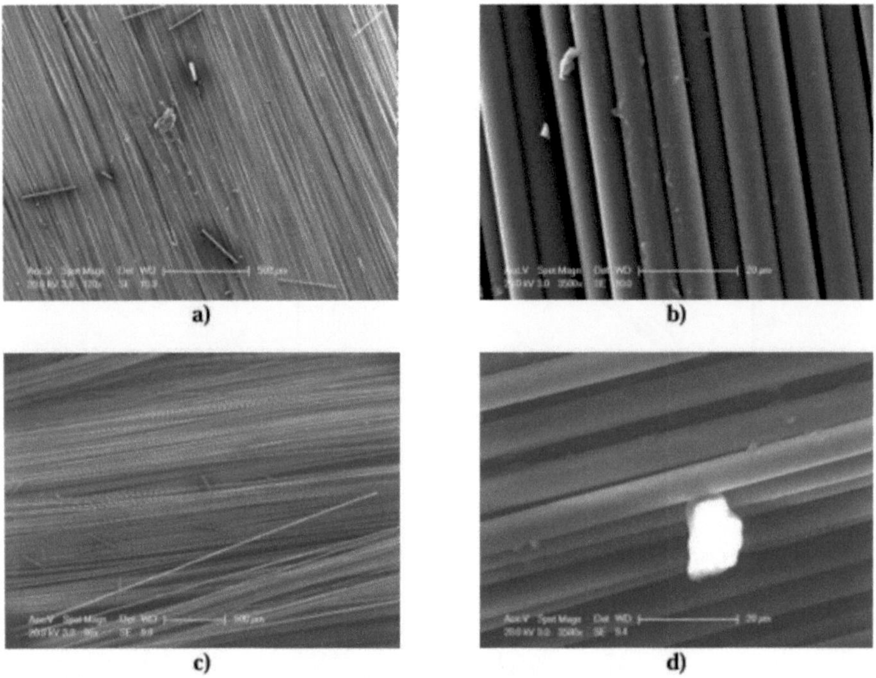

Fig. 17.21 ESEM images of: **a** and **b** unwashed carbon fibres recovered after solvolysis; **c** and **d** carbon fibres after slight wash with acetone (Keith et al. 2016)

Despite the promise of these emerging techniques, each presents specific technical and environmental drawbacks. Pyrolysis, for example, may release hazardous gases and carbon residues if not managed properly, while chemical recycling can involve toxic reagents, generate secondary waste, and require high energy consumption. Enzymatic degradation, though environmentally benign and substrate-specific, is typically limited by slow reaction rates and challenges in scaling. Moreover, most of these methods are still in development or pilot phases, and their integration into commercial recycling infrastructure remains limited.

From an environmental standpoint, the benefits of advanced recycling can be undermined by their resource intensity, byproduct management, and the potential emission of pollutants. As such, it is essential that future recycling technologies be evaluated using life-cycle assessment (LCA) tools to ensure that their environmental footprint is genuinely lower than current practices. The transition to green solvents, low-emission technologies, and renewable inputs is critical to reducing the overall impact of composite recycling.

Looking forward, the future of sustainable composite recycling will depend on the advancement of energy-efficient, scalable, and economically viable processes that can handle both thermoplastic and thermoset systems. Efforts should focus on closed-loop recycling, the use of renewable or recycled fillers, and the incorporation of eco-design principles from the material development stage. Real progress will require the removal of regulatory bottlenecks, the development of standardized recycling protocols, and stronger collaboration between researchers, industry stakeholders, and policymakers.

By aligning recycling innovations with the broader goals of a circular economy, polymer composite recycling has the potential to become a key strategy in sustainable materials management—one that not only reduces environmental burdens but also maximizes resource efficiency and product value retention over multiple life cycles.

References

Abdelaal MY, Sobahi TR, Makki MSI (2011) Chemical transformation of pet waste through glycolysis. Constr Build Mater 25(8):3267–3271. https://doi.org/10.1016/j.conbuildmat.2011.03.013

Åkesson D, Fazelinejad S, Skrifvars VV, Skrifvars M (2016) Mechanical recycling of polylactic acid composites reinforced with wood fibres by multiple extrusion and hydrothermal ageing. J Reinf Plast Compos 35(16):1248–1259. https://doi.org/10.1177/0731684416647507

Akonda MH, Lawrence CA, Weager BM (2012) Recycled carbon fibre-reinforced polypropylene thermoplastic composites. Compos Part A Appl Sci Manuf 43(1):79–86. https://doi.org/10.1016/j.compositesa.2011.09.014

Alguacil MC, Umeki K, You S, Joffe R (2025) Evolution of carbon fiber properties during repetitive recycling via pyrolysis and partial oxidation. Carbon Trends 18:100438. https://doi.org/10.1016/j.cartre.2024.100438

Ali, M., & Brocchini, S. (2006). Synthetic approaches to uniform polymers. Advanced drug delivery reviews, 58(15), 1671-1687.

Allen KA et al (2025) Understanding interfacial crystallization dynamics on carbon fiber reinforced polypropylene composite manufacturing. Compos B Eng 291. https://doi.org/10.1016/j.compositesb.2024.112027

Almeras X et al (2004a) Structure and morphology of an intumescent polypropylene blend. J Appl Polym Sci 93(1):402–411. https://doi.org/10.1002/app.20470

Almeras X et al (2004b) Artificial weathering and recycling effect on intumescent polypropylene-based blends. J Fire Sci 22(2):143–161. https://doi.org/10.1177/0734904104039609

Al-Salem SM, Evangelisti S, Lettieri P (2014) Life cycle assessment of alternative technologies for municipal solid waste and plastic solid waste management in the Greater London area. Chem Eng J 244:391–402. https://doi.org/10.1016/j.cej.2014.01.066

Al-Shalawi FD et al (2023) Biodegradable synthetic polymer in orthopaedic application: a review. Mater Today Proc 540–546. Elsevier Ltd. https://doi.org/10.1016/j.matpr.2022.12.254

Al Samad A et al (2016) PCL-PEG graft copolymers with tunable amphiphilicity as efficient drug delivery systems. J Mater Chem B 4(37):6228–6239. https://doi.org/10.1039/c6tb01841f

Al Zahmi S, Alhammadi S, Elhassan A, Ahmed W (2022) Carbon fiber/PLA recycled composite. Polymers (Basel) 14(11). https://doi.org/10.3390/polym14112194

Anbarasu M, Vinitha V, Preeyanghaa M, Neppolian B, Sivamurugan V (2024) Valorization of PES textile waste through glycolysis and aminolysis using bimetallic ZnO and carbon nitride nanocomposites. J Clust Sci 35(4):1105–1125. https://doi.org/10.1007/s10876-023-02534-4

Asaro L, Gratton M, Poirot N, Seghar S, Aït Hocine N (2020) Devulcanization of natural rubber industry waste in supercritical carbon dioxide combined with diphenyl disulphide. Waste Manag 118:647–654. https://doi.org/10.1016/j.wasman.2020.09.026

Asghari F, Samiei M, Adibkia K, Akbarzadeh A, Davaran S (2017) Biodegradable and biocompatible polymers for tissue engineering application: a review. Taylor and Francis Ltd. https://doi.org/10.3109/21691401.2016.1146731

Awaja F, Pavel D (2005) Recycling of PET. https://doi.org/10.1016/j.eurpolymj.2005.02.005

Awasthi SK et al (2022) A comprehensive review on recent advancements in biodegradation and sustainable management of biopolymers. Environ Pollut 307. https://doi.org/10.1016/j.envpol.2022.119600

Azeko ST, Mustapha K, Annan E, Odusanya OS, Soboyejo WO (2016) Recycling of polyethylene into strong and tough earth-based composite building materials. J Mater Civ Eng 28(2). https://doi.org/10.1061/(asce)mt.1943-5533.0001385

Backes EH et al (2022) Polycaprolactone usage in additive manufacturing strategies for tissue engineering applications: a review. Wiley. https://doi.org/10.1002/jbm.b.34997

Baley C et al (2024) Sustainable polymer composite marine structures: developments and challenges. Elsevier Ltd. https://doi.org/10.1016/j.pmatsci.2024.101307

Banerjee A, Chatterjee K, Madras G (2014) Enzymatic degradation of polymers: a brief review. Maney Publishing. https://doi.org/10.1179/1743284713Y.0000000503

Beauson J, Madsen B, Toncelli C, Brøndsted P, Ilsted Bech J (2016) Recycling of shredded composites from wind turbine blades in new thermoset polymer composites. Compos Part A Appl Sci Manuf 90:390–399. https://doi.org/10.1016/j.compositesa.2016.07.009

Benalaya I, Alves G, Lopes J, Silva LR (2024) A review of natural polysaccharides: sources, characteristics, properties, food, and pharmaceutical applications. Multidisciplinary Digital Publishing Institute (MDPI). https://doi.org/10.3390/ijms25021322

Bharadwaj A, Yadav D, Varshney S (2015) Non-biodegradable waste-its impact and safe disposal [Online]. https://www.researchgate.net/publication/301283637

Bhattacharjee J, Sarkar A, Panda TK (2021) Alkali and alkaline earth metal complexes as versatile catalysts for ring-opening polymerization of cyclic esters. Chem Rec 21(8):1898–1911. https://doi.org/10.1002/tcr.202100148

Bhogle CS, Pandit AB (2019) Ultrasound assisted methanolysis of polycarbonate at room temperature. Ultrason Sonochem 58. https://doi.org/10.1016/j.ultsonch.2019.104667

Biondi M, Borzacchiello A, Mayol L, Ambrosio L (2015) Nanoparticle-integrated hydrogels as multifunctional composite materials for biomedical applications. MDPI. https://doi.org/10.3390/gels1020162

Bîrca A, Gherasim O, Grumezescu V, Grumezescu AM (2019) Introduction in thermoplastic and thermosetting polymers. In: Materials for biomedical engineering: thermoset and thermoplastic polymers. Elsevier, pp 1–28. https://doi.org/10.1016/B978-0-12-816874-5.00001-3

Boro B, Tiwari P (2025) Pyrolysis of waste printed circuit board fragments: response surface methodology optimization and effect of temperature on product characteristics. Energy 319. https://doi.org/10.1016/j.energy.2025.134892

Brem H, Gabikian P (2001) Biodegradable polymer implants to treat brain tumors [Online]. www.elsevier.com/locate/jconrel

Brivio L, Sponchioni M, Innocenti P, Storti G, Moscatelli D (2025) Hybrid process for the depolymerization of polyethylene terephthalate (PET) from post-consumer mixed textile wastes. Chem Eng Sci 313:121702. https://doi.org/10.1016/j.ces.2025.121702

Butenegro JA, Bahrami M, Abenojar J, Martínez MÁ (2021) Recent progress in carbon fiber reinforced polymers recycling: a review of recycling methods and reuse of carbon fibers. MDPI. https://doi.org/10.3390/ma14216401

Calosi M et al (2022) Glycolysis of semi-interpenetrated polymer network foam based on poly(vinyl chloride) for recovery and reuse of the individual components. Waste Manag 153:229–239. https://doi.org/10.1016/j.wasman.2022.09.001

Cha JH et al (2025) Multidisciplinary space shield origami composite: incorporating cosmic radiation shielding, space debris impact protection, solar radiative heat shielding, and atomic oxygen erosion resistance. Compos B Eng 288. https://doi.org/10.1016/j.compositesb.2024.111876

Chanda M (2021) Chemical aspects of polymer recycling. KeAi Communications Co. https://doi.org/10.1016/j.aiepr.2021.06.002

Chang C, Jiang Y, Lin Y, Fu K, Xu C, Zhao S (2025) Glycolysis of waste polyurethane foam for preparing recycled polymer materials: optimization and techno-economic evaluation. Chem Eng J 507. https://doi.org/10.1016/j.cej.2025.160583

Chen X, Chen S, Xu Z, Zhang J, Miao M, Zhang D (2020a) Degradable and recyclable bio-based thermoset epoxy resins. Green Chem 22(13):4187–4198. https://doi.org/10.1039/d0gc01250e

Chen T, Mansfield CD, Ju L, Baird DG (2020b) The influence of mechanical recycling on the properties of thermotropic liquid crystalline polymer and long glass fiber reinforced polypropylene. Compos B Eng 200. https://doi.org/10.1016/j.compositesb.2020.108316

Chen J et al (2023) A new strategy for the preparation of polylactic acid composites with flame retardancy, UV resistance, degradation, and recycling performance. Chem Eng J 472. https://doi.org/10.1016/j.cej.2023.145000

Chen Y et al (2024) A recyclable energy storage wood composite with photothermal conversion properties for regulating building temperature. Adv Sustain Syst. https://doi.org/10.1002/adsu.202400627

Conroy A, Halliwell S, Reynolds T (2006) Composite recycling in the construction industry. Compos Part A Appl Sci Manuf 37(8):1216–1222. https://doi.org/10.1016/j.compositesa.2005.05.031

Copenhaver K et al (2024) Improving the recyclability of polymer composites with cellulose nanofibrils. J Polym Environ. https://doi.org/10.1007/s10924-024-03257-x

Dang W, Kubouchi M, Yamamoto S, Sembokuya H, Tsuda K (2002) An approach to chemical recycling of epoxy resin cured with amine using nitric acid [Online]. www.elsevier.com/locate/polymer

Del Amo J, Borreguero AM, Ramos MJ, Rodríguez JF (2021) Glycolysis of polyurethanes composites containing nanosilica. Polymers (Basel) 13(9). https://doi.org/10.3390/polym13091418

de Sadeleer I, Woodhouse A (2024) Environmental impact of biodegradable and non-biodegradable agricultural mulch film: a case study for Nordic conditions. Int J Life Cycle Assess 29(2):275–290. https://doi.org/10.1007/s11367-023-02253-y

De Tommaso J, Galli F, Nguyen TD, Zhuang Y, Dubois JL, Patience GS (2025) Waste artificial marble pyrolysis and hydrolysis. Waste Manag 195:129–144. https://doi.org/10.1016/j.wasman.2025.01.032

Donadini R, Boaretti C, Scopel L, Lorenzetti A, Modesti M (2024) Deamination of polyols from the glycolysis of polyurethane. Chem Eur J 30(3). https://doi.org/10.1002/chem.202301919

Durante M, Boccarusso L, De Fazio D, Formisano A, Langella A (2023) Investigation on the mechanical recycling of carbon fiber-reinforced polymers by peripheral down-milling. Polymers (Basel) 15(4). https://doi.org/10.3390/polym15040854

D'souza AA, Shegokar R (2016) Polyethylene glycol (PEG): a versatile polymer for pharmaceutical applications. Taylor and Francis Ltd. https://doi.org/10.1080/17425247.2016.1182485

Dhanasekar S et al (2022) Study of polymer matrix composites for electronics applications. J Nanomater 2022. https://doi.org/10.1155/2022/8605099

Dong M, Bilotti E, Zhang H, Papageorgiou DG (2025) Recyclable PVA/starch/$Ti_3C_2T_x$ MXene nanocomposite films with superior mechanical and barrier properties. Int J Biol Macromol 295. https://doi.org/10.1016/j.ijbiomac.2025.139545

Doppalapudi S, Jain A, Khan W, Domb AJ (2014) Biodegradable polymers-an overview. Wiley. https://doi.org/10.1002/pat.3305

Dorigato A (2021) Recycling of thermosetting composites for wind blade application. KeAi Communications Co. https://doi.org/10.1016/j.aiepr.2021.02.002

Dwivedi P, Mishra PK, Mondal MK, Srivastava N (2019) Non-biodegradable polymeric waste pyrolysis for energy recovery. Elsevier Ltd. https://doi.org/10.1016/j.heliyon.2019.e02198

El Gersifi K, Destais-Orvoën N, Durand G, Tersac G (2003) Glycolysis of epoxide-amine hardened networks. I. Diglycidylether/aliphatic amines model networks. Polymer (Guildf) 44(14):3795–3801. https://doi.org/10.1016/S0032-3861(03)00324-0

El Hawary O, Boccarusso L, Ansell MP, Durante M, Pinto F (2023) An overview of natural fiber composites for marine applications. MDPI. https://doi.org/10.3390/jmse11051076
Eze WU, Umunakwe R, Obasi HC, Ugbaja MI, Uche CC, Madufor IC (2021) Plastics waste management: a review of pyrolysis technology. Clean Technol Recycl 1(1):50–69. https://doi.org/10.3934/ctr.2021003
Faruk O, Bledzki AK, Fink HP, Sain M (2012) Biocomposites reinforced with natural fibers: 2000–2010. https://doi.org/10.1016/j.progpolymsci.2012.04.003
Feng Z, Zhao W, Jin L, Zhang J, Xue B, Ni Y (2023) Environmentally friendly strategy to access self-healable, reprocessable and recyclable chitin, chitosan, and sodium alginate based polysaccharide-vitrimer hybrid materials. Int J Biol Macromol 240. https://doi.org/10.1016/j.ijbiomac.2023.124531
Feraboli P, Kawakami H, Wade B, Gasco F, DeOto L, Masini A (2012) Recyclability and reutilization of carbon fiber fabric/epoxy composites. J Compos Mater 46(12):1459–1473. https://doi.org/10.1177/0021998311420604
Finnerty J et al (2023) Effect of mechanical recycling on the mechanical properties of PLA-based natural fiber-reinforced composites. J Compos Sci 7(4). https://doi.org/10.3390/jcs7040141
Gabisa EW, Ratanatamskul C, Gheewala SH (2023) Recycling of plastics as a strategy to reduce life cycle GHG emission, microplastics and resource depletion. Sustainability (Switzerland) 15(15). https://doi.org/10.3390/su151511529
Gałko G, Sajdak M (2022) Trends for the thermal degradation of polymeric materials: analysis of available techniques, issues, and opportunities. MDPI. https://doi.org/10.3390/app12189138
García-Valderrama T, Sánchez-Ortiz J, Pérez-González MDC, Puentes-Graña MDC (2024) Relationship between recycling, circular economy and eco-innovation in Europe. Application of a dynamic network data envelopment analysis model. Bus Strategy Environ. https://doi.org/10.1002/bse.3768
Ghalia MA, Alhanish A (2023) Mechanical and biodegradability of porous PCL/PEG copolymer-reinforced cellulose nanofibers for soft tissue engineering applications. Med Eng Phys 120. https://doi.org/10.1016/j.medengphy.2023.104055
Ghimire S et al (2021) Polymeric materials for hemostatic wound healing. MDPI. https://doi.org/10.3390/pharmaceutics13122127
Ghosh R et al (2023) New route of tire rubber devulcanization using silanes. Polymers (Basel) 15(13). https://doi.org/10.3390/polym15132848
Giorgini L et al (2015) Recovery of carbon fibers from cured and uncured carbon fiber reinforced composites wastes and their use as feedstock for a new composite production. Polym Compos 36(6):1084–1095. https://doi.org/10.1002/pc.23440
Gunatillake PA, Adhikari R, Gadegaard N (2003) Biodegradable synthetic polymers for tissue engineering. AO Research Institute Davos. https://doi.org/10.22203/eCM.v005a01
Guo L et al (2021) Rubber reclamation with high bond-breaking selectivity using a low-temperature mechano-chemical devulcanization method. J Clean Prod 279. https://doi.org/10.1016/j.jclepro.2020.123266
Hagnell MK, Åkermo M (2019) The economic and mechanical potential of closed loop material usage and recycling of fibre-reinforced composite materials. J Clean Prod 223:957–968. https://doi.org/10.1016/j.jclepro.2019.03.156
Hamerton I, Kratz J (2018) The use of thermosets in modern aerospace applications. In: Thermosets: structure, properties, and applications, 2nd edn. Elsevier, pp 303–340. https://doi.org/10.1016/B978-0-08-101021-1.00009-5
Han W, Han D, Chen H (2023) Pyrolysis of waste tires: a review. MDPI. https://doi.org/10.3390/polym15071604
Henry L, Schneller A, Doerfler J, Mueller WM, Aymonier C, Horn S (2016) Semi-continuous flow recycling method for carbon fibre reinforced thermoset polymers by near- and supercritical solvolysis. Polym Degrad Stab 133:264–274. https://doi.org/10.1016/j.polymdegradstab.2016.09.002

Howarth J, Mareddy SSR, Mativenga PT (2014) Energy intensity and environmental analysis of mechanical recycling of carbon fibre composite. J Clean Prod 81:46–50. https://doi.org/10.1016/j.jclepro.2014.06.023
Hu S et al (2022) Cellulose hydrogel-based biodegradable and recyclable magnetoelectric composites for electromechanical conversion. Carbohydr Polym 298. https://doi.org/10.1016/j.carbpol.2022.120115
Hu B et al (2024) Development of shape memory epoxy resin/polyethylene glycol film based on molecular dynamics simulation and performance investigation. J Appl Polym Sci. https://doi.org/10.1002/app.56153
Isayev AI, Liang T, Lewis TM (2014) Effect of particle size on ultrasonic devulcanization of tire rubber in twin-screw extruder. Rubber Chem Technol 87(1):86–102. https://doi.org/10.5254/RCT.13.87926
Jagadeesh P et al (2022) A comprehensive review on polymer composites in railway applications. Wiley. https://doi.org/10.1002/pc.26478
James HM, Guth E (1943) Theory of the elastic properties of rubber. J Chem Phys 11(10):455–481. https://doi.org/10.1063/1.1723785
Jeong H, Rho J, Shin JY, Lee DY, Hwang T, Kim KJ (2018) Mechanical properties and cytotoxicity of PLA/PCL films. Biomed Eng Lett 8(3):267–272. https://doi.org/10.1007/s13534-018-0065-4
Jiang J et al (2017) On the successful chemical recycling of carbon fiber/epoxy resin composites under the mild condition. Compos Sci Technol 151:243–251. https://doi.org/10.1016/j.compscitech.2017.08.007
Jin FL, Li X, Park SJ (2015) Synthesis and application of epoxy resins: a review. Korean Soc Ind Eng Chem. https://doi.org/10.1016/j.jiec.2015.03.026
Joltreau E (2022) Extended producer responsibility, packaging waste reduction and eco-design. Environ Resour Econ (Dordr) 83(3):527–578. https://doi.org/10.1007/s10640-022-00696-9
Joseph A et al (2022) Biodegradation of natural rubber by fungi and bacteria. Nat Environ Pollut Technol 21(3):1039–1048. https://doi.org/10.46488/NEPT.2022.v21i03.010
Jubinville D, Chen G, Mekonnen TH (2023) Simulated thermo-mechanical recycling of high-density polyethylene for the fabrication of hemp hurd plastic composites. Polym Degrad Stab 211. https://doi.org/10.1016/j.polymdegradstab.2023.110342
Kao CC, Ghita OR, Hallam KR, Heard PJ, Evans KE (2012) Mechanical studies of single glass fibres recycled from hydrolysis process using sub-critical water. Compos Part A Appl Sci Manuf 43(3):398–406. https://doi.org/10.1016/j.compositesa.2011.11.011
Keith MJ, Oliveux G, Leeke GA (2016) Optimisation of solvolysis for recycling carbon fibre reinforced composites
Keith MJ, Román-Ramírez LA, Leeke G, Ingram A (2019) Recycling a carbon fibre reinforced polymer with a supercritical acetone/water solvent mixture: comprehensive analysis of reaction kinetics. Polym Degrad Stab 161:225–234. https://doi.org/10.1016/j.polymdegradstab.2019.01.015
Khalid MY, Arif ZU, Ahmed W, Arshad H (2022) Recent trends in recycling and reusing techniques of different plastic polymers and their composite materials. Sustain Mater Technol 31:e00382. Elsevier B.V. https://doi.org/10.1016/j.susmat.2021.e00382
Khan F et al (2024) Advances of composite materials in automobile applications—a review. Elsevier B.V. https://doi.org/10.1016/j.jer.2024.02.017
Khorramnezhad M, Akbari B, Akbari M, Kharaziha M (2021) Effect of surface modification on physical and cellular properties of PCL thin film. Colloids Surf B Biointerfaces 200. https://doi.org/10.1016/j.colsurfb.2021.111582
Kim Y, Jang T, Hwang H, Sung Y, Kim BH (2024) Development of glycolysis catalysts for PET wastes including polyester textiles. Korean Fiber Society. https://doi.org/10.1007/s12221-024-00807-x
Kirshanov K, Toms R, Melnikov P, Gervald A (2022) Investigation of polyester tire cord glycolysis accompanied by rubber crumb devulcanization. Polymers (Basel) 14(4). https://doi.org/10.3390/polym14040684

Kirshanov KA, Toms RV, Balashov MS, Golubkov SS, Melnikov PV, Gervald AY (2023) Modeling of poly(ethylene terephthalate) homogeneous glycolysis kinetics. Polymers (Basel) 15(14). https://doi.org/10.3390/polym15143146

Kiss P et al (2020) In-house recycling of carbon- and glass fibre-reinforced thermoplastic composite laminate waste into high-performance sheet materials. Compos Part A Appl Sci Manuf 139. https://doi.org/10.1016/j.compositesa.2020.106110

Krauklis AE, Karl CW, Gagani AI, Jørgensen JK (2021) Composite material recycling technology—state-of-the-art and sustainable development for the 2020s. MDPI AG. https://doi.org/10.3390/jcs5010028

Kuang X, Shi Q, Zhou Y, Zhao Z, Wang T, Qi HJ (2018a) Dissolution of epoxy thermosets: via mild alcoholysis: the mechanism and kinetics study. RSC Adv 8(3):1493–1502. https://doi.org/10.1039/c7ra12787a

Kuang X, Zhou Y, Shi Q, Wang T, Qi HJ (2018b) Recycling of epoxy thermoset and composites via good solvent assisted and small molecules participated exchange reactions. ACS Sustain Chem Eng 6(7):9189–9197. https://doi.org/10.1021/acssuschemeng.8b01538

Kumar Kalita N, Hazarika D, Srivastava RK, Hakkarainen M (2024) Faster biodegradable and chemically recyclable polycaprolactone with embedded enzymes: revealing new insights into degradation kinetics. Chem Eng J 496. https://doi.org/10.1016/j.cej.2024.153982

Kuttiyathil MS, Ali L, Teoh WY, Altarawneh M (2025) Dechlorination of waste polyvinyl chloride (PVC) via its co-pyrolysis with $Ca(OH)_2$: a TG-IR-GCMS investigation. Case Stud Chem Environ Eng 11:101162. https://doi.org/10.1016/j.cscee.2025.101162

Kwon D, Kim JY, Song H, Lee J, Tsang YF, Kwon EE (2025) CO_2-mediated pyrolysis for syngas generation from wood-plastic composite waste. Energy 319. https://doi.org/10.1016/j.energy.2025.134917

Le Duigou A, Pillin I, Bourmaud A, Davies P, Baley C (2008) Effect of recycling on mechanical behaviour of biocompostable flax/poly(l-lactide) composites. Compos Part A Appl Sci Manuf 39(9):1471–1478. https://doi.org/10.1016/j.compositesa.2008.05.008

Lee WA, Knight GJ (1970) Ratio of the glass transition temperature to the melting point in polymers. Br Polym J 2(1):73–80

Leenders N et al (2025) Polycotton waste textile recycling by sequential hydrolysis and glycolysis. Nat Commun 16(1):738. https://doi.org/10.1038/s41467-025-55935-6

Lewis H, Gertsakis J (2003) Sustainability and the waste hierarchy a discussion paper on the waste management hierarchy and its relationship to sustainability. RMIT University, Melbourne

Li SM, Chen XH, Gross RA, Mccarthy SP (2000) Hydrolytic degradation of PCL/PEO copolymers in alkaline media. J Mater Sci Mater Med 11(4):227–233

Li L, Han L, Hu H, Zhang R (2022) A review on polymers and their composites for flexible electronics. Royal Society of Chemistry. https://doi.org/10.1039/d2ma00940d

Lin YH, Yang MH (2007) Catalytic pyrolysis of polyolefin waste into valuable hydrocarbons over reused catalyst from refinery FCC units. Appl Catal A Gen 328(2):132–139. https://doi.org/10.1016/j.apcata.2007.05.039

Liu FS, Li Z, Yu ST, Cui X, Xie CX, Ge XP (2009) Methanolysis and hydrolysis of polycarbonate under moderate conditions. J Polym Environ 17(3):208–211. https://doi.org/10.1007/s10924-009-0140-0

Liu J, Liu P, Zhang X, Lu P, Zhang X, Zhang M (2016) Fabrication of magnetic rubber composites by recycling waste rubber powders via a microwave-assisted in situ surface modification and semi-devulcanization process. Chem Eng J 295:73–79. https://doi.org/10.1016/j.cej.2016.03.025

Liu M, Wang Y, Wu Y, Wan H (2018) Hydrolysis and recycling of urea formaldehyde resin residues. J Hazard Mater 355:96–103. https://doi.org/10.1016/j.jhazmat.2018.05.019

Liu W, Huang H, Cheng H, Liu Z (2020) CFRP reclamation and remanufacturing based on a closed-loop recycling process for carbon fibers using supercritical N-butanol. Fibers Polym 21(3):604–618. https://doi.org/10.1007/s12221-020-9575-7

Lu Q et al (2025) Transition-metal-assisted pyrolysis to recover glass fibers from end-of-life wind turbine blades. J Anal Appl Pyrolysis 189:107081. https://doi.org/10.1016/j.jaap.2025.107081

Luckachan GE, Pillai CKS (2011) Biodegradable polymers—a review on recent trends and emerging perspectives. J Polym Environ 19(3):637–676. https://doi.org/10.1007/s10924-011-0317-1
Luna CBB, Siqueira DD, Da Silva Barbosa Ferreira E, Araújo EM, Wellen RMR (2019) Reactive compatilization of PCL/WP upon addition of PCL-MA. Smart option for recycling industry. Mater Res Express 6(12). https://doi.org/10.1088/2053-1591/ab55ad
Luo C, Chung C, Yu K (2023) A diffusion-reaction computational study to reveal the depolymerization mechanisms of epoxy composites for recycling. Mater Today Sustain 23. https://doi.org/10.1016/j.mtsust.2023.100452
Luo H, Tyrrell H, Bai J, Ibrahim Muazu R, Long X (2024) Fundamental, technical and environmental overviews of plastic chemical recycling. Royal Society of Chemistry. https://doi.org/10.1039/d4gc03127j
Lv Q et al (2024) Multifunctional and recyclable aerogel/fiber building insulation composites with sandwich structure. Constr Build Mater 423. https://doi.org/10.1016/j.conbuildmat.2024.135902
Madhav H, Singh N, Jaiswar G (2019) Thermoset, bioactive, metal-polymer composites for medical applications. In: Materials for biomedical engineering: thermoset and thermoplastic polymers. Elsevier, pp 105–143. https://doi.org/10.1016/B978-0-12-816874-5.00004-9
Madhavan Nampoothiri K, Nair NR, John RP (2010) An overview of the recent developments in polylactide (PLA) research. https://doi.org/10.1016/j.biortech.2010.05.092
Majeed K et al (2013) Potential materials for food packaging from nanoclay/natural fibres filled hybrid composites. https://doi.org/10.1016/j.matdes.2012.10.044
Mandal P, Shunmugam R (2020) Polycaprolactone: a biodegradable polymer with its application in the field of self-assembly study. Bellwether Publishing, Ltd. https://doi.org/10.1080/10601325.2020.1831392
Mangili I, Collina E, Anzano M, Pitea D, Lasagni M (2014) Characterization and supercritical CO_2 devulcanization of cryo-ground tire rubber: influence of devulcanization process on reclaimed material. Polym Degrad Stab 102(1):15–24. https://doi.org/10.1016/j.polymdegradstab.2014.02.017
Mărieş GRE, Abrudan AM (2018) Thermoplastic polymers in product design. IOP Conf Ser Mater Sci Eng. Institute of Physics Publishing. https://doi.org/10.1088/1757-899X/393/1/012118
Marinis D, Markatos D, Farsari E, Amanatides E, Mataras D, Pantelakis S (2024) A novel plasma-enhanced solvolysis as alternative for recycling composites. Polymers (Basel) 16(19). https://doi.org/10.3390/polym16192836
Mattsson C, André A, Juntikka M, Trnkle T, Sott R (2020) Chemical recycling of end-of-life wind turbine blades by solvolysis/HTL. IOP Conf Ser Mater Sci Eng. IOP Publishing Ltd. https://doi.org/10.1088/1757-899X/942/1/012013
Menares T, Herrera J, Romero R, Osorio P, Arteaga-Pérez LE (2020) Waste tires pyrolysis kinetics and reaction mechanisms explained by TGA and Py-GC/MS under kinetically-controlled regime. Waste Manag 102:21–29. https://doi.org/10.1016/j.wasman.2019.10.027
Moore J (1950) Some chemical and physical properties of rubber. Br J Appl Phys 1(1):6
Morici E, Dintcheva NT (2022) Recycling of thermoset materials and thermoset-based composites: challenge and opportunity. MDPI. https://doi.org/10.3390/polym14194153
Mu Q, An L, Hu Z, Kuang X (2022) Fast and sustainable recycling of epoxy and composites using mixed solvents. Polym Degrad Stab 199. https://doi.org/10.1016/j.polymdegradstab.2022.109895
Murray RE et al (2021) Structural validation of a thermoplastic composite wind turbine blade with comparison to a thermoset composite blade. Renew Energy 164:1100–1107. https://doi.org/10.1016/j.renene.2020.10.040
Nakagawa T, Goto M (2015) Recycling thermosetting polyester resin into functional polymer using subcritical water. Polym Degrad Stab 115:16–23. https://doi.org/10.1016/j.polymdegradstab.2015.02.005
Nanda S, Berruti F (2021) Thermochemical conversion of plastic waste to fuels: a review. Springer Science and Business Media Deutschland GmbH. https://doi.org/10.1007/s10311-020-01094-7

Newcomb BA (2016) Processing, structure, and properties of carbon fibers. Elsevier Ltd. https://doi.org/10.1016/j.compositesa.2016.10.018

Ngwuluka NC, Ochekpe NA, Aruoma OI (2014) Naturapolyceutics: the science of utilizing natural polymers for drug delivery. Mol Divers Preserv Int. https://doi.org/10.3390/polym6051312

Nishida H, Carvelli V, Fujii T, Okubo K (2018) Thermoplastic vs. thermoset epoxy carbon textile composites. IOP Conf Ser Mater Sci Eng. Institute of Physics Publishing. https://doi.org/10.1088/1757-899X/406/1/012043

Okajima I, Sako T (2019) Recycling fiber-reinforced plastic using supercritical acetone. Polym Degrad Stab 163:1–6. https://doi.org/10.1016/j.polymdegradstab.2019.02.018

Oladele IO et al (2023) Modern trends in recycling waste thermoplastics and their prospective applications: a review. MDPI. https://doi.org/10.3390/jcs7050198

Oliveux G, Dandy LO, Leeke GA (2015) Degradation of a model epoxy resin by solvolysis routes. Polym Degrad Stab 118:96–103. https://doi.org/10.1016/j.polymdegradstab.2015.04.016

Oliveux G, Bailleul JL, Gillet A, Mantaux O, Leeke GA (2017) Recovery and reuse of discontinuous carbon fibres by solvolysis: realignment and properties of remanufactured materials. Compos Sci Technol 139:99–108. https://doi.org/10.1016/j.compscitech.2016.11.001

Otheguy ME, Gibson AG, Findon E, Cripps RM, Ochoa Mendoza A, Aguinaco Castro MT (2009) Recycling of end-of-life thermoplastic composite boats. In: Plastics, rubber and composites, pp 406–411. https://doi.org/10.1179/146580109X12540995045642

Ou J, Dai Z, Kong Z, Yang R (2025) Acoholysis behavior of PET waste with hybrid fibers and their application in the synthesis of unsaturated resins. J Polym Res 32(4). https://doi.org/10.1007/s10965-025-04333-z

Pang X, Zhuang X, Tang Z, Chen X (2010) Polylactic acid (PLA): research, development and industrialization. https://doi.org/10.1002/biot.201000135

Papaspyrides CD, Poulakis JG, Varelides PC (1994) A model recycling process for low density polyethylene. Resour Conserv Recycl 12(3–4):177–184

Papaspyrides CD, Poulakis JG, Arvanitopoulos CD (1995) Recycling of glass fiber reinforced thermo-plastic composites. I. Ionomer and low density polyethylene based composites. Resour Conserv Recycl 14(2):91–101

Park KH, Shin C, Song YS, Lee HJ, Shin C, Kim Y (2019) Recyclable and mendable cellulose-reinforced composites crosslinked with Diels-Alder adducts. Polymers (Basel) 11(1). https://doi.org/10.3390/polym11010117

Pathiraja K, Nilmalgoda H, Wijethunga I, Koswattage K (2025) Sustainable management of nitrile butadiene rubber waste through pyrolysis. Sustainability (Switzerland) 17(3). https://doi.org/10.3390/su17030846

Pérez RL, Ayala CE, Opiri MM, Ezzir A, Li G, Warner IM (2021) Recycling thermoset epoxy resin using alkyl-methyl-imidazolium ionic liquids as green solvents. ACS Appl Polym Mater 3(11):5588–5595. https://doi.org/10.1021/acsapm.1c00896

Peti D, Dobránsky J, Michalík P (2025) Recent advances in polymer recycling: a review of chemical and biological processes for sustainable solutions. Multidisciplinary Digital Publishing Institute (MDPI). https://doi.org/10.3390/polym17050603

Pietroluongo M, Padovano E, Frache A, Badini C (2020) Mechanical recycling of an end-of-life automotive composite component. Sustain Mater Technol 23:e00143. Elsevier B.V. https://doi.org/10.1016/j.susmat.2019.e00143

Platnieks O, Barkane A, Ijudina N, Gaidukova G, Thakur VK, Gaidukovs S (2020) Sustainable tetra pak recycled cellulose/poly(butylene succinate) based woody-like composites for a circular economy. J Clean Prod 270. https://doi.org/10.1016/j.jclepro.2020.122321

Puoci F, Iemma F, Spizzirri G, Cirillo G, Curcio M, Picci N (2008) Polymer in agriculture: a review. Am J Agric Biol Sci 3(1):299–314

Qin W, Qiu Y, He H, Guo B, Li P (2024) Pyrogallic acid–compatibilized polylactic acid/thermoplastic starch blend produced via one-step twin-screw extrusion. Int J Biol Macromol 276. https://doi.org/10.1016/j.ijbiomac.2024.133758

Ragaert K, Delva L, Van Geem K (2017) Mechanical and chemical recycling of solid plastic waste. Elsevier Ltd. https://doi.org/10.1016/j.wasman.2017.07.044

Raghunath J et al (2009) Degradation studies on biodegradable nanocomposite based on polycaprolactone/polycarbonate (80:20%) polyhedral oligomeric silsesquioxane. J Biomed Mater Res A 91(3):834–844. https://doi.org/10.1002/jbm.a.32335

Ramos-Hernández T, Robledo-Ortíz JR, Martín del Campo AS, Rodrigue D, Cano A, Pérez-Fonseca AA (2024) Mechanical recycling of poly(lactic acid)/agave fiber biocomposites. J Reinf Plast Compos. https://doi.org/10.1177/07316844241253905

Romero-Fierro D, Bustamante-Torres M, Bravo-Plascencia F, Esquivel-Lozano A, Ruiz JC, Bucio E (2022) Recent trends in magnetic polymer nanocomposites for aerospace applications: a review. MDPI. https://doi.org/10.3390/polym14194084

Rostami-Tapeh-esmaeil E, Vahidifar A, Esmizadeh E, Rodrigue D (2021) Chemistry, processing, properties, and applications of rubber foams. MDPI AG. https://doi.org/10.3390/polym13101565

Ruiz De Luzuriaga A et al (2016) Epoxy resin with exchangeable disulfide crosslinks to obtain reprocessable, repairable and recyclable fiber-reinforced thermoset composites. Mater Horiz 3(3):241–247. https://doi.org/10.1039/c6mh00029k

Rujnić-Sokele M, Pilipović A (2017) Challenges and opportunities of biodegradable plastics: a mini review. SAGE Publications Ltd. https://doi.org/10.1177/0734242X16683272

Salah LS, Ouslimani N, Danlée Y, Beltrán FR, Huynen I, Ulagares de la Orden M (2023) Investigation of mechanical recycling effect on electromagnetic properties of polylactic acid (PLA)—nanoclay nanocomposites: towards a valorization of recycled PLA nanocomposites. Compos Part C Open Access 10. https://doi.org/10.1016/j.jcomc.2022.100339

Salem T, Elshahat M, Saafan AA, Kenawy ER (2025) Chemical recycling of poly(ethylene terephthalate) waste using a new co-catalytic system. Iranian Polym J (Engl Edn). https://doi.org/10.1007/s13726-024-01429-z

Seghar S, Asaro L, Rolland-Monnet M, Aït Hocine N (2019) Thermo-mechanical devulcanization and recycling of rubber industry waste. Resour Conserv Recycl 144:180–186. https://doi.org/10.1016/j.resconrec.2019.01.047

Sell SA, Wolfe PS, Garg K, McCool JM, Rodriguez IA, Bowlin GL (2010) The use of natural polymers in tissue engineering: a focus on electrospun extracellular matrix analogues. https://doi.org/10.3390/polym2040522

Shabani M, Jamshidi M (2022) Recycling NR/SBR waste using probe sonication as a new devulcanizing method; study on influencing parameters. RSC Adv 12(40):26264–26276

Sharma P, Roy S, Karimi-Varzaneh HA (2016) Validation of force fields of rubber through glass-transition temperature calculation by microsecond atomic-scale molecular dynamics simulation. J Phys Chem B 120(7):1367–1379. https://doi.org/10.1021/acs.jpcb.5b10789

Shastri VP (2003) Non-degradable biocompatible polymers in medicine: past, present and future

Shen Y, Wu Y (2025) Pyrolysis valorization of waste epoxy thermosets. J Anal Appl Pyrolysis 186:106908. https://doi.org/10.1016/j.jaap.2024.106908

Shen M, Cao H, Robertson ML (2020) Hydrolysis and solvolysis as benign routes for the end-of-life management of thermoset polymer waste. Annu Rev. https://doi.org/10.1146/annurev-chembioeng-120919-012253

Shokrollahi P, Mehmanchi M, Atai M, Omidian H, Shokrolahi F (2014) Effect of interface on mechanical properties and biodegradation of PCL HAp supramolecular nano-composites. J Mater Sci Mater Med 25(1):23–35. https://doi.org/10.1007/s10856-013-5039-6

Shuaib NA, Mativenga PT (2016) Effect of process parameters on mechanical recycling of glass fibre thermoset composites. Procedia CIRP 134–139. Elsevier B.V. https://doi.org/10.1016/j.procir.2016.03.206

Simón D, Rodríguez JF, Carmona M, Serrano A, Borreguero AM (2018) Glycolysis of advanced polyurethanes composites containing thermoregulating microcapsules. Chem Eng J 350:300–311. https://doi.org/10.1016/j.cej.2018.05.158

Singh N, Walker TR (2024) Plastic recycling: a panacea or environmental pollution problem. npj Mater Sustain 2(1). https://doi.org/10.1038/s44296-024-00024-w

Singh B et al (2021) Processing and recycling of thermoplastic polymers: current scenario and future challenges. Turk J Comput Math Educ (TURCOMAT) 12(2):2744–2753. https://doi.org/10.17762/turcomat.v12i2.2303

Singh R, Barwar A, Kumar A (2022) Investigations on primary and secondary recycling of PLA and its composite for biomedical and sensing applications. J Inst Eng (India): Ser C 103(4):821–836. https://doi.org/10.1007/s40032-022-00840-w

Siracusa V, Rocculi P, Romani S, Rosa MD (2008) Biodegradable polymers for food packaging: a review. https://doi.org/10.1016/j.tifs.2008.07.003

Skosana SJ, Khoathane C, Malwela T (2024) Driving towards sustainability: a review of natural fiber reinforced polymer composites for eco-friendly automotive light-weighting. SAGE Publications Ltd. https://doi.org/10.1177/08927057241254324

Song R, Murphy M, Li C, Ting K, Soo C, Zheng Z (2018) Current development of biodegradable polymeric materials for biomedical applications. Dove Medical Press Ltd. https://doi.org/10.2147/DDDT.S165440

Spörl JM et al (2018) Ionic liquid approach toward manufacture and full recycling of all-cellulose composites. Macromol Mater Eng 303(1). https://doi.org/10.1002/mame.201700335

Srebrenkoska V, Gaceva GB, Avella M, Errico ME, Gentile G (2008) Recycling of polypropylene-based eco-composites. Polym Int 57(11):1252–1257. https://doi.org/10.1002/pi.2470

Srebrenkoska V, Bogoeva-Gaceva G, Dimeski D (2014) Biocomposites based on polylactic acid and their thermal behavior after recycling. Maced J Chem Chem Eng 33(2):277–285

Srivastava A, Yadav T, Sharma S, Nayak A, Akanksha Kumari A, Mishra N (2016) Polymers in drug delivery. J Biosci Med (Irvine) 04(1):69–84. https://doi.org/10.4236/jbm.2016.41009

Stan F, Stanciu NV, Fetecau C, Sandu IL (2019) Mechanical recycling of low-density polyethylene/carbon nanotube composites and its effect on material properties. J Manuf Sci Eng 141(9). https://doi.org/10.1115/1.4044101

Su W et al (2023) Polysaccharides-based ultra-thin membrane with high haze, conductivity, and recyclability. Chem Eng J 472. https://doi.org/10.1016/j.cej.2023.144933

Su X et al (2024) Construction of immobilized laccase hydrogels via sodium alginate-dopamine/polyethylene glycol and its efficient degradation of dyeing wastewater. Int J Biol Macromol 279. https://doi.org/10.1016/j.ijbiomac.2024.134929

Swetha TA et al (2023) A comprehensive review on polylactic acid (PLA)—synthesis, processing and application in food packaging. Elsevier B.V. https://doi.org/10.1016/j.ijbiomac.2023.123715

Taghavi N, Udugama IA, Zhuang WQ, Baroutian S (2021) Challenges in biodegradation of non-degradable thermoplastic waste: from environmental impact to operational readiness. Elsevier Inc. https://doi.org/10.1016/j.biotechadv.2021.107731

Tamrakar S, Couvreur R, Mielewski D, Gillespie JW, Kiziltas A (2023) Effects of recycling and hygrothermal environment on mechanical properties of thermoplastic composites. Polym Degrad Stab 207. https://doi.org/10.1016/j.polymdegradstab.2022.110233

Tavakoli M, Labbaf S, Mirhaj M, Salehi S, Seifalian AM, Firuzeh M (2023) Natural polymers in wound healing: from academic studies to commercial products. Wiley. https://doi.org/10.1002/app.53910

Tian H, Tang Z, Zhuang X, Chen X, Jing X (2012) Biodegradable synthetic polymers: preparation, functionalization and biomedical application. Progress in Polymer Science 37(2):237–280

Tian X, Liu T, Wang Q, Dilmurat A, Li D, Ziegmann G (2017) Recycling and remanufacturing of 3D printed continuous carbon fiber reinforced PLA composites. J Clean Prod 142:1609–1618. https://doi.org/10.1016/j.jclepro.2016.11.139

Titone V, Botta L, Mistretta MC, La Mantia FP (2024) Influence of a biodegradable contaminant on the mechanical recycling of a low-density polyethylene sample. Polym Eng Sci 64(2):845–851. https://doi.org/10.1002/pen.26588

Tortorici D, Clemente R, Laurenzi S (2024) Solvolysis process for recycling carbon fibers from epoxy-based composites. Macromol Symp 413(4). https://doi.org/10.1002/masy.202400039
Tseng HC, Chang RY, Hsu CH (2017) Numerical prediction of fiber orientation and mechanical performance for short/long glass and carbon fiber-reinforced composites. Compos Sci Technol 144:51–56. https://doi.org/10.1016/j.compscitech.2017.02.020
Valencia GA, Zare EN, Makvandi P, Gutiérrez TJ (2019) Self-assembled carbohydrate polymers for food applications: a review. Blackwell Publishing Inc. https://doi.org/10.1111/1541-4337.12499
Vedavalli R et al (2025) Catalytic conversion of plastic waste into diesel fuel through pyrolysis and hydroprocessing. Process Saf Environ Prot 193:327–339. https://doi.org/10.1016/j.psep.2024.11.029
Venkatachalam P, Geetha N, Sangeetha P, Thulaseedharan A (2013) Natural rubber producing plants: an overview. Afr J Biotechnol 12(12):1297–1310. https://doi.org/10.5897/AJBX12.016
Vijayan DS, Sivasuriyan A, Devarajan P, Stefańska A, Wodzyński Ł, Koda E (2023) Carbon fibre-reinforced polymer (CFRP) composites in civil engineering application—a comprehensive review. MDPI. https://doi.org/10.3390/buildings13061509
Vroman I, Tighzert L (2009) Biodegradable polymers. https://doi.org/10.3390/ma2020307
Wahit MU, Akos NI, Laftah WA (2012) Influence of natural fibers on the mechanical properties and biodegradation of poly(lactic acid) and poly(μ-caprolactone) composites: a review. Polym Compos 33(7):1045–1053. https://doi.org/10.1002/pc.22249
Wang S, Xing X, Zhang X, Wang X, Jing X (2018) Room-temperature fully recyclable carbon fibre reinforced phenolic composites through dynamic covalent boronic ester bonds. J Mater Chem A Mater 6(23):10868–10878. https://doi.org/10.1039/c8ta01801d
Wang B et al (2021a) Recycling of carbon fibers from unsaturated polyester composites via a hydrolysis-oxidation synergistic catalytic strategy. Compos Sci Technol 203:108589. https://doi.org/10.1016/j.compscitech.2020.108589
Wang T, Shen C, Yu G, Chen X (2021b) Fabrication of magnetic bimetallic Co–Zn based zeolitic imidazolate frameworks composites as catalyst of glycolysis of mixed plastic. Fuel 304. https://doi.org/10.1016/j.fuel.2021.121397
Wang Y, Fu S, Lucia LA, Zhang H (2022) A cellulose-based self-healing composite eutectogel with reversibility and recyclability for multi-sensing. Compos Sci Technol 229. https://doi.org/10.1016/j.compscitech.2022.109696
Wu CS (2003) Physical properties and biodegradability of maleated-polycaprolactone/starch composite. Polym Degrad Stab 80(1):127–134. https://doi.org/10.1016/S0141-3910(02)00393-2
Wu Y et al (2024) Recyclable solid–solid phase change materials with superior latent heat via reversible anhydride-alcohol crosslinking for efficient thermal storage. Adv Mater 36(16). https://doi.org/10.1002/adma.202311717
Yang P, Zhou Q, Li XY, Yang KK, Wang YZ (2014) Chemical recycling of fiber-reinforced epoxy resin using a polyethylene glycol/NaOH system. J Reinf Plast Compos 33(22):2106–2114. https://doi.org/10.1177/0731684414555745
Yang J, Liu J, Liu W, Wang J, Tang T (2015) Recycling of carbon fibre reinforced epoxy resin composites under various oxygen concentrations in nitrogen-oxygen atmosphere. J Anal Appl Pyrolysis 112:253–261. https://doi.org/10.1016/j.jaap.2015.01.017
Yao Y, Zhao F, Wang B, Hu Z, Huang Y (2022) The designing of degradable unsaturated polyester based on selective cleavage activated hydrolysis and its application in recyclable carbon fiber composites. Compos Sci Technol 229:109692. https://doi.org/10.1016/j.compscitech.2022.109692
Yao Y et al (2025) Production of green phenol by microwave-assisted catalytic pyrolysis of epoxy-based carbon fiber reinforced plastic from wind turbine blades. J Energy Inst 119. https://doi.org/10.1016/j.joei.2025.102005
Yuan H et al (2020) Preparation of poly(styrene-co-allyl sulfonic acid) as a novel catalyst for the alcoholysis of dioscin. Polym Adv Technol 31(8):1776–1782. https://doi.org/10.1002/pat.4904

Yuan Z et al (2025) A biodegradable and recyclable piezoelectric macrofiber based on imidazole perchlorate crystallized in bacterial cellulose for mechanical sensing fabrics. Chem Eng J 509. https://doi.org/10.1016/j.cej.2025.161034

Zanchet A, Carli LN, Giovanela M, Brandalise RN, Crespo JS (2012) Use of styrene butadiene rubber industrial waste devulcanized by microwave in rubber composites for automotive application. Mater Des 39:437–443. https://doi.org/10.1016/j.matdes.2012.03.014

Zanetti M, Kashiwagi T, Falqui L, Camino G (2002) Cone calorimeter combustion and gasification studies of polymer layered silicate nanocomposites. Chem Mater 14(2):881–887. https://doi.org/10.1021/cm011236k

Zeng J et al (2024) Development of high-barrier composite films for sustainable reduction of non-biodegradable materials in food packaging application. Carbohydr Polym 330. https://doi.org/10.1016/j.carbpol.2024.121824

Zhang X, Saha P, Cao L, Li H, Kim J (2018) Devulcanization of waste rubber powder using thiobisphenols as novel reclaiming agent. Waste Manag 78:980–991. https://doi.org/10.1016/j.wasman.2018.07.016

Zhao Q, An L, Li C, Zhang L, Jiang J, Li Y (2022) Environment-friendly recycling of CFRP composites via gentle solvent system at atmospheric pressure. Compos Sci Technol 224. https://doi.org/10.1016/j.compscitech.2022.109461

Zhu B, Jiang X, Li S, Zhu M (2024) An overview of recycling phenolic resin. Multidisciplinary Digital Publishing Institute (MDPI). https://doi.org/10.3390/polym16091255

Zou Z et al (2022) Microwave-assisted alcoholysis of silicone rubber waste with geraniol for effective elastomer damping. Polym Eng Sci 62(5):1435–1443. https://doi.org/10.1002/pen.25933

1-s2.0-S0269749122003281-main

Chapter 18
Exploring the Durability and Longevity of Polymer Composites

Mohammadali Shirinbayan, Samia Nouira, and Joseph Fitoussi

Abstract Polymer composites are integral to industries such as aerospace, automotive, construction, and marine due to their exceptional mechanical properties, lightness, and adaptability. However, ensuring their long-term durability and reliability under complex service conditions—including mechanical loading, thermal cycling, and environmental exposure—remains a significant challenge. This chapter presents a novel multi-scale, multi-physics framework that integrates experimental investigations with advanced numerical modeling to elucidate the degradation mechanisms of polymer composites. Key findings include the identification of critical damage modes—such as matrix cracking, fiber-matrix debonding, delamination, and environmental degradation—and their interactions under realistic thermomechanical loading conditions. The chapter also presents state-of-the-art experimental techniques and modeling strategies that provide new insights into the durability and longevity of these materials. By addressing existing gaps in the understanding of polymer composite degradation, this work provides practical solutions to improve the structural integrity and lifetime of these materials in demanding industrial applications.

Keywords Polymer composites · Experimental and numerical dialogue · Damage mechanism · Mechanical properties

---

M. Shirinbayan (✉) · S. Nouira · J. Fitoussi
Arts et Metiers Institute of Technology, CNAM, CNRS, PIMM, Paris, France
e-mail: mohammadali.shirinbayan@ensam.eu

S. Nouira
e-mail: samia.nouira@univ-lorraine.fr

J. Fitoussi
e-mail: joseph.fitoussi@ensam.eu

S. Nouira
Université de Lorraine, Laboratoire LEM3 UMR 7239, 57045 Metz, France

L. Ravindran et al. (eds.), *Macro, Micro and Nanocomposites from Sustainable Sources*, Advanced Structured Materials 244, https://doi.org/10.1007/978-981-95-2469-3_18

18.1 Introduction

The performance and durability of composite materials are closely influenced by the interactions between their microstructural characteristics, manufacturing processes, applied mechanical loading, and environmental conditions. Unlike traditional homogeneous materials, the microstructure of composites is often established during fabrication, making it essential to understand how processing parameters affect material properties, structural behavior, and long-term durability.

Recent studies have highlighted the significant effects of various mechanical loadings—including quasi-static, dynamic, and cyclic—on local deformation, damage initiation, and fracture mechanisms in polymer matrix composites (Shirinbayan 2021; Shirinbayan et al. 2021a, b; Mortazavian and Fatemi 2015). Additionally, environmental factors such as temperature, humidity, and material aging can alter damage progression and service life (Fitoussi et al. 2022; Shreepannaga et al. 2022). As industrial applications increasingly demand lightweight, high-performance, and damage-tolerant materials, developing predictive approaches that account for these coupled effects remains a crucial challenge.

To accurately capture the behavior of composite materials under such conditions, experimental and numerical methodologies capable of covering a wide range of strain rates, from quasi-static (10^{-4} s^{-1}) to several hundred per second, are required (Shirinbayan et al. 2015, 2017a). However, rapid mechanical testing introduces difficulties such as wave propagation effects, which can cause spatio-temporal disturbances in measurement fields and affect result accuracy. Addressing these issues necessitates tailored testing procedures and advanced measurement techniques capable of providing reliable data during high-rate loading.

In addition to monotonic and dynamic loading, fatigue remains a major degradation mechanism in composite structures, especially under repeated stress cycles common in service environments (Imaddahen et al. 2020; Shirinbayan et al. 2019; Ayari et al. 2020a). Fatigue damage initiates at the microstructural level and gradually propagates, impacting macroscopic performance. A detailed understanding of damage thresholds, propagation rates, and the influence of microstructure is essential for reliable, long-term use in critical applications.

Structural health monitoring (SHM) methods, particularly those based on piezoelectric transducers capable of emitting and receiving Lamb Waves (LW), offer promising solutions for damage detection and durability assessment in composites (Worden et al. 2007; Yuan 2016; Sohn et al. 2002; Su et al. 2006; Su and Ye 2009; Mitra and Gopalakrishnan 2016). LW-based SHM systems enable continuous or periodic monitoring of damage development, including delamination, cracking, and fatigue-related defects, while providing valuable feedback on the effects of environmental conditions on structural integrity.

This chapter proposes a unified, multiscale methodology combining experimental characterization, numerical modeling, and SHM approaches to investigate damage mechanisms in composite materials under diverse mechanical and environmental conditions. The chapter is organized as follows: Sect. 18.2 reviews the relationships between manufacturing processes, microstructure, material properties, and structural performance. Section 18.3 presents an original methodology for characterizing damage mechanisms under quasi-static, dynamic, and cyclic loading.

18.2 Materials, Processes, and Induced Microstructures

Statistical analysis of microstructures and their influence on macroscopic mechanical behavior shows that mastery of the relationship between manufacturing process, microstructure, and mechanical properties is essential, especially when it comes to composite materials with discontinuous reinforcements. Indeed, the very wide variety of constituents (matrix, reinforcements, fillers, additives), elaboration processes (thermocompression, injection, etc.) and reinforcement architectures leads to a multitude of possible microstructures inducing complex and strongly coupled local phenomena that condition the mechanical properties of composites and the performance of their structures (Nikooharf et al. 2021; Shirinbayan et al. 2017b; Laribi et al. 2020).

It is important to note that organic matrix composite structures with discontinuous reinforcements cannot be considered homogeneous due to the nature of their forming processes. In addition, the heterogeneous material in the molten state flows during part development, which generally leads to a strong spatial variation of the microstructure, particularly in terms of reinforcement orientation. The spatial distribution of microstructures within a real part is therefore dependent on process parameters and part geometry. Variability in the volume fraction and orientation distribution of reinforcements has a significant impact on the material's mechanical properties in both the elastic phase, which is always anisotropic, and the nonlinear phase, which is often linked to the development of damage phenomena. In the case of materials with a thermoplastic matrix (Imaddahen et al. 2020; Nikooharf et al. 2021; Shirinbayan et al. 2024), this also affects the latter's nonlinearity and viscosity.

It can be reasonably deduced that the orientation of the reinforcements and the local volume fraction condition both influence the distribution of stress between the reinforcements and the matrix. The degree and type of anisotropy (Dumont et al. 2003) resulting from this variability in microstructure therefore also conditions elastic mechanical properties, as well as degradation, deformation, and fracture mechanisms. Consequently, in contrast to the general case of metallic materials or composites with continuous reinforcement (laminate type) (Feki et al. 2025), a single behavior law and a single limiting criterion are inadequate for describing the response of the actual structure. In theory, each distinct microstructure present within the part requires the development of a specific behavior law, which presents a significant challenge in the process of dimensioning the structures. As the spatial distribution of microstructures is contingent upon the selection of process parameters, it is imperative to possess a comprehensive understanding and exert control over the process-microstructure relationships. Furthermore, the development of predictive models that integrate microstructure as an input and account for the local phenomena it induces is of paramount importance. Consequently, a predictive approach, wherein the microstructure serves as an input data and induced nonlinear phenomena are taken into account, is indispensable.

18.2.1 Composite Materials and Associated Scales

Organic composites can be distinguished by their matrix (thermoset or thermoplastic) or by their reinforcement architecture (continuous or discontinuous) (Friedrich et al. 2005). They can be classified into three categories:

- Thermoset matrix composites with discontinuous reinforcements,
- Thermoplastic matrix composites with discontinuous reinforcements,
- Continuous-reinforced composites (long-fiber laminates).

Various techniques have been used to analyze their microstructures, including optical or scanning electron microscopy (SEM), X-ray microtomography (µtomoX), and an original ultrasonic methodology. For the purposes of quantifying micromechanical models, these microstructures were often quantified by statistical analysis of SEM or µtomoX images. Indeed, the volume fraction, orientation and slenderness distributions of reinforcements strongly condition the mechanical response of composites with discontinuous reinforcements (Shirinbayan et al. 2017c). These data are therefore essential for understanding and analyzing the influence of microstructure on the macroscopic behavior of these composites. They are also the input data for the multi-scale modeling developed in our research.

As scale transitions are fundamental to the study of composite materials (Friedrich et al. 2005), it is first necessary to define them. Indeed, terminology varies from one scientific community to another. Process experts, chemists, mechanics, etc., all use the same terminology. However, it is crucial to have a comprehensive understanding of the process-microstructure-properties relationship, particularly in the context of composites with discontinuous reinforcements. This part aims to present a definition of the different scales, followed by an introduction to an original ultrasonic characterization method developed for this study. Finally, a few examples of microstructural analysis will be provided to illustrate the diverse materials under investigation. This document defines three basic scales for analytical purposes:

- **Microstructure Scale (Micron Scale)**: This is the scale of the reinforcement, typically measuring approximately 10 µm. At this scale, the matrix and reinforcements can be distinguished. Quantification of the microstructure is typically based on observations made at this scale, including volume fraction and orientation distribution.
- **Representative Volume Element (RVE) Scale (Centimeter Scale)**: The RVE (Maragoni et al. 2018) is defined as the smallest statistically significant volume capable of encompassing the full range of heterogeneities present within the microstructure. Conversely, this volume must also be sufficiently substantial

to be regarded as homogeneously stressed. It is also important to note that microstructure data must always be treated statistically. Indeed, local variations in microstructure may exist even within the RVE. Finally, once the RVE has been defined, it can be considered as a homogeneous material subjected to homogeneous loads. This notion is particularly important when it comes to choosing the sample size for experimental characterization. Indeed, if the volume tested is too small, the dispersion of results will be too great. In some cases, a compromise will have to be found.

- **Structure Scale (Meter Scale)**: The structure scale represents the scale of the part. It is important to note that the structural scale differs from the RVE scale in that stress and strain fields are inhomogeneous. Additionally, microstructures can vary within a structure. Consequently, in contrast to the behavior observed in metals, multiple distinct RVEs can contribute to the composite structure.

In certain instances, it may be necessary to introduce scales smaller than the micron, particularly when the focus is on highly localized deformation and damage phenomena related to the polymer or polymer-reinforcement interfaces. Similarly, when the composite architecture and/or the phenomena observed necessitate it, an analysis at an intermediate scale between the RVE and the micron scale can be proposed. This is known as the mesoscopic scale. This is particularly useful for studying delamination in the case of laminated composites, which are also studied in this document.

18.2.2 Microstructure Variability Analysis

In the case of organic matrix composites with discontinuous reinforcements, the structure can never be considered spatially homogeneous. This is because, during shaping, the material flows, displacing the reinforcements and orienting them according to the part's manufacturing conditions and geometry. As a result, there is always a strong spatial variation in the microstructure from one part to another. This phenomenon occurs for the two main classes of manufacturing processes used in the automotive industry: thermoplastic injection molding and thermocompression.

By way of example, the microstructure of sheet molding compound (SMC) composites (polyester or vinylester matrix reinforced with glass fibers) used in automotive tailgates will be studied in detail (Ayari et al. 2020a; Tamboura et al. 2001; Abdessalem et al. 2021). SMC composites are used in a variety of automotive applications, including structural parts such as front and rear absorbers, bumpers and trim parts such as side panels, hoods, tailgates, rear quarter panels, floors and grilles. Figure 18.1 illustrates the results of a SEM observation of a standard SMC composite, which clearly demonstrates the extreme heterogeneity of these materials. Indeed, the material is composed of four distinct phases: resin, mineral fillers, fibrous reinforcements and porosities (Abdessalem et al. 2021). Additionally, there are zones of matrix that lack reinforcement. In the resin, the chalk particles and pores are distributed in a relatively uniform manner. Furthermore, SMCs can be considered as pseudo-layers, as strands of varying orientations are superimposed on one another at the local scale. This is clearly evident in the processing of images from X-ray microtomography analysis.

The microstructures of Advanced SMC (A-SMC) (Shirinbayan et al. 2015) and Low-density SMC (LD-SMC) (Shirinbayan et al. 2017c) (see Fig. 18.2) demonstrate comparable characteristics, with a higher density of reinforcements and fewer isolated matrix zones for A-SMC, and the presence of hollow glass spheres dispersed in the matrix zones for LD-SMC.

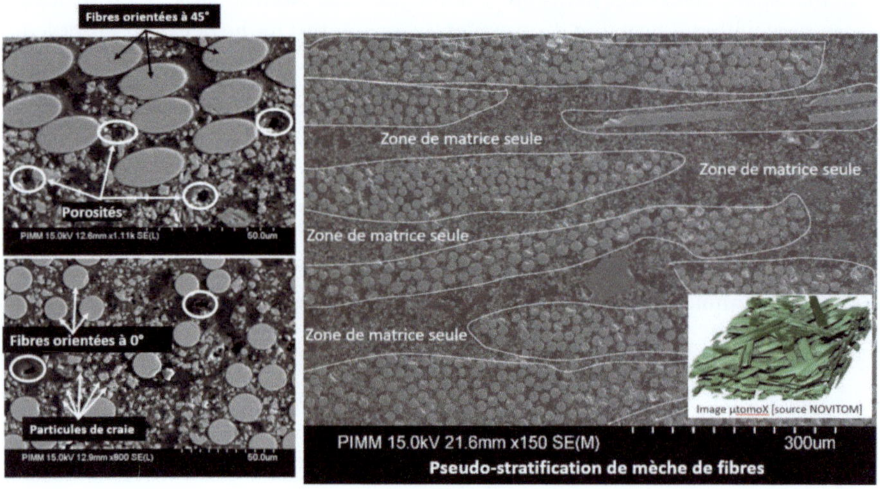

Fig. 18.1 Details of the microstructure of a standard SMC (Abdessalem et al. 2021)

Fig. 18.2 Microstructures of A-SMC and LD-SMC (Shirinbayan et al. 2015, 2017c)

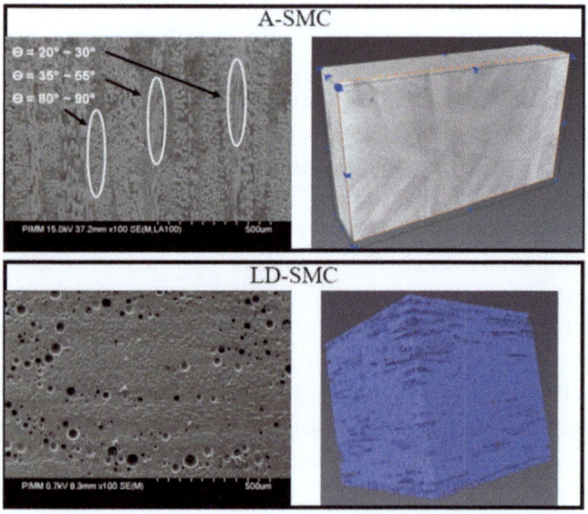

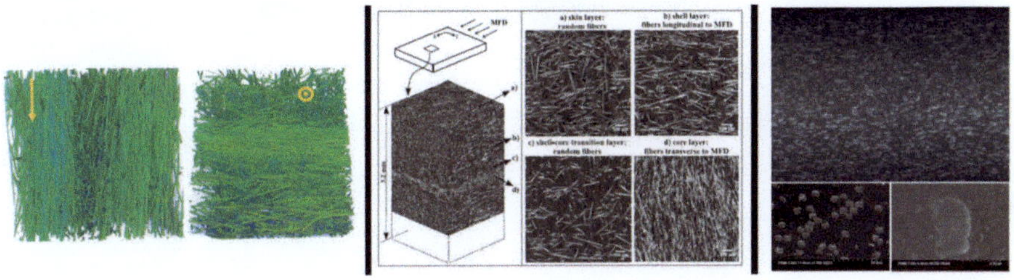

(a) (b) (c)

Fig. 18.3 Microstructures of thermoplastic matrix composites with discontinuous reinforcements

Other examples include thermoplastic matrix composites obtained by injection molding. Polypropylenes and polyamides reinforced with short fibers up to a few hundred microns in length are the most widely used in industry, particularly in the automotive sector. All composites feature the core-skin structure that is characteristic of injection-molded materials. The reinforcements are oriented in the direction of injection (MFD for Mold Flow Direction) in peripheral regions and perpendicular to the center. The relative thickness of these zones is dependent on the composition of the composite and the injection conditions (Meftah et al. 2018). Please refer to the accompanying illustrations of microstructures (Fig. 18.3). Figure 18.3c presents the microstructure of a 100% recycled carbon fiber-reinforced polypropylene composite. Note the non-circular shape of the carbon reinforcements.

18.2.3 Determination of Microstructure Mapping Using Ultrasonic Method

To rapidly determine the spatial distribution of microstructures within a real part, an original ultrasonic method has been developed (Laribi et al. 2020; Shirinbayan et al. 2017c). The process involves immersing a sample in water between two ultrasonic probes, with one acting as a transmitter and the other as a receiver (see Fig. 18.4a). This setup generates transverse waves, inducing shear along a plane defined by the normal to the sample and the wave propagation direction (see Fig. 18.4b, c). The sample is mounted on a precision rotation stage with $\pm 0.1^\circ$ accuracy, enabling continuous rotation around its normal axis. As the sample rotates, the transverse wave velocity is measured for each angular position. When fibers are aligned within the shear plane (Fig. 18.4b), wave velocity reaches its maximum, reflecting the higher proportion of sheared glass fibers. Conversely, when fibers are perpendicular to the shear plane (Fig. 18.4c), the wave velocity is at its minimum due to the reduced sheared fiber cross-section. The evolution of shear wave velocity as a function of

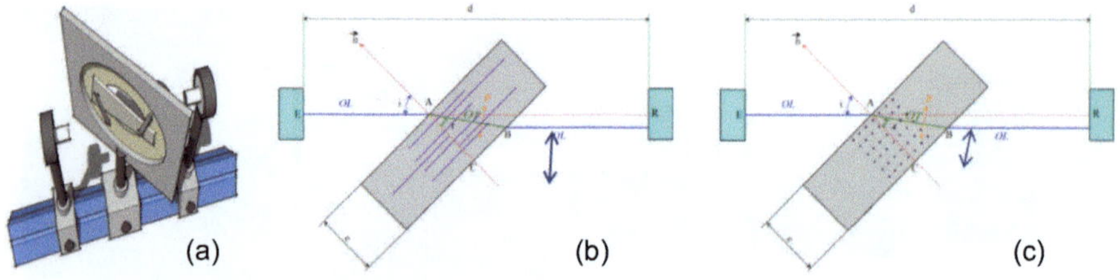

Fig. 18.4 Ultrasonic method for determining microstructure (Shirinbayan et al. 2017c)

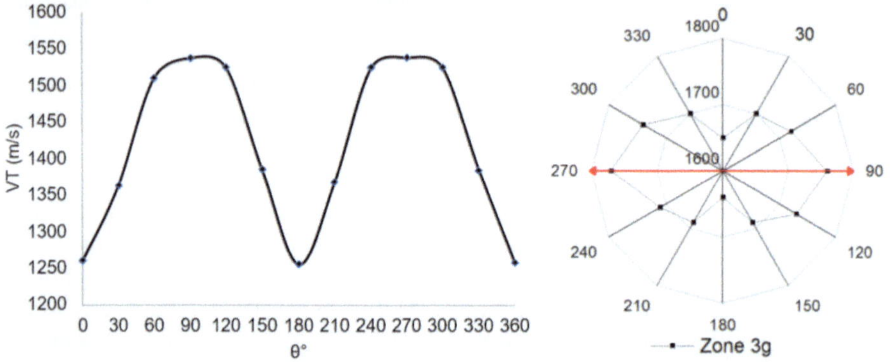

Fig. 18.5 Example of ultrasonic measurement results: sample strongly oriented at 90° (Shirinbayan et al. 2017c)

the rotation angle reveals a characteristic peak corresponding to the principal reinforcement direction. An example of this result on an SMC sample is presented in Fig. 18.5.

While this ultrasonic technique offers a rapid and effective means for mapping fiber orientation, it has several limitations. It provides a thickness-averaged evaluation of the microstructure, making it unsuitable for resolving through-thickness variations in parts thicker than a few millimeters. The angular accuracy of the rotary stage is $\pm 0.1°$, and the measurement uncertainty for transverse wave velocity remains within $\pm 1\%$ under controlled laboratory conditions. Moreover, the method assumes a planar fiber distribution and may be less reliable for detecting out-of-plane orientations or highly complex, multi-directional reinforcement architectures. It is particularly well-suited for thin-walled composite parts where fibers are predominantly in-plane, as is typically the case in automotive applications.

This characterization procedure is then applied to samples taken from real parts to generate microstructure maps. For example, Fig. 18.6 illustrates the microstructure mapping results obtained for the sheet molding compound (SMC) tailgate of the Peugeot 3008, as reported by Fitoussi et al. (2024). In this figure, the direction of the lines indicates the main fiber orientation, while their length represents the intensity of this orientation. Additionally, a birefringence parameter, K% (Shirinbayan

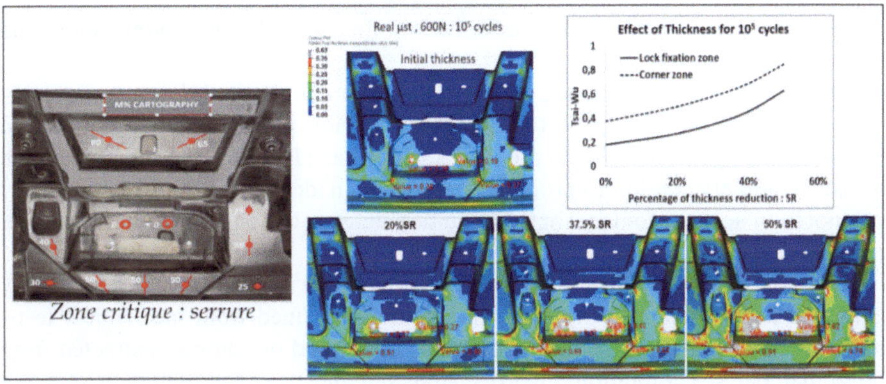

Fig. 18.6 Microstructure mapping of real composite parts optimizing the thickness of the Peugeot 3008 tailgate (Fitoussi et al. 2024).

et al. 2017c), derived from the amplitude of the transverse wave velocity curve (see Fig. 18.5), further characterizes the material. A K% value below 4 indicates a random orientation distribution, whereas a value exceeding 8 suggests a highly oriented material.

The microstructural data obtained through this technique serve as input for multi-scale numerical modeling aimed at predicting the mechanical behavior of composite parts. During these simulations, appropriate boundary conditions are applied to replicate the mechanical and geometrical constraints experienced by the actual components:

- Displacement boundary conditions are assigned based on the principal loading directions identified in service conditions.
- Symmetry boundary conditions are employed where applicable, reducing computational cost and reflecting the repetitive nature of fiber distributions.
- For local simulations within representative volume elements (RVEs), periodic boundary conditions are applied to ensure continuity of displacement and traction across the RVE boundaries, effectively simulating the behavior of an infinite medium.

These boundary conditions are essential for accurately reproducing the internal stress and strain distributions within the composite and for correctly interpreting the fiber orientation patterns revealed by the mapping process. Their integration ensures that both the microstructural maps and the associated mechanical property predictions remain reliable and representative of the component's actual service behavior.

The main strength of this multi-scale numerical approach lies in its ability to capture the influence of local fiber orientation distributions on the global mechanical performance of complex composite structures. By incorporating microstructural data directly into finite element (FE) models, it becomes possible to predict anisotropic stiffness, strength variations, and failure mechanisms throughout a component—providing valuable information for both design optimization and durability prediction.

Nevertheless, this methodology is subject to several limitations. It relies on simplifying assumptions such as perfect fiber-matrix adhesion, planar fiber distribution, and the use of homogenized material properties within each RVE. Furthermore, although periodic boundary conditions promote consistency in idealized microstructures, they may not fully account for local heterogeneities, defects, or non-ideal features present in actual parts.

To ensure accuracy and reliability, the numerical models are systematically calibrated and validated using experimental data obtained from mechanical tests, including monotonic tensile and fatigue tests performed on samples extracted from the actual parts. Calibration involves adjusting material parameters within the numerical model—such as Young's modulus, yield strength, and damage initiation criteria—to reproduce the experimentally measured global mechanical response. Validation is subsequently performed by comparing numerical predictions with additional independent experimental results not used during the calibration phase. This includes assessing the agreement between predicted and measured stress–strain curves, stiffness distributions, and failure locations. A good correlation between numerical and experimental results confirms the appropriateness of both the numerical models and the underlying assumptions in the multi-scale simulation framework.

18.3 Multi-scale Characterization of the Behavior, Damage, and Fracture of Composite Materials Subjected to Quasi-static, Dynamic, and Cyclic Loading

Composite materials are used in a variety of structures that are subjected to different thermomechanical loads, including quasi-static, fast dynamic, and cyclic. It is recognized that the macroscopic responses of these materials are influenced by the microstructure and the deformation and damage phenomena occurring at the local scale. To gain insight into the origin of the nonlinearities observed at the macroscopic scale and to develop the database required for the predictive numerical approaches, multi-scale experimental analyses should be conducted.

18.3.1 Mechanical Behavior of Composite Materials Under Dynamic Loading

The study and modeling of the mechanical behavior of materials and structures under dynamic loads (falls, impacts, crashes) represent a vast and complex field of research. For a long time, the study of the influence of rapid loads was almost exclusively dedicated to military applications. However, over the past twenty years, concerns about passenger safety in the transport sector have strongly contributed to the development of these studies within the framework of new challenges. Moreover, in the case of composite materials, automotive structures need to combine lightness to meet CO_2 emission reduction requirements with good energy absorption capacity in the event of impacts and accidents.

There has been a lack of exploration into the specific deformation, degradation and fracture mechanisms of materials subjected to dynamic loads at intermediate strain rates (up to a few hundred per second) due to a lack of suitable characterization tools. Indeed, studies have focused on either quasi-static behavior (up to 10^{-3}/s, accessible by standard testing machines) or on very fast dynamic behavior (over 1000/s, accessible by Hopkinson bar testing). It is in the slow (1–10 s^{-1}) and medium (10–1000 s^{-1}) dynamic regimes that materials become highly sensitive to strain rate. In the transport sector, for instance, materials are subjected to deformation speeds corresponding to the slow and medium regimes (up to a few hundred s^{-1}) during crashes. Therefore, the design of structures subjected to impacts and crashes, to ensure the protection of people, using numerical simulation by finite elements, generally requires behavior laws that translate the mechanical response of materials to deformation speeds covering the range from 1 to 200s^{-1}. It is therefore essential to develop and identify these laws on the basis of dynamic mechanical tests carried out in intermediate regimes. The mastery of such tests represents a significant challenge that has been identified by the scientific community (Jendli et al. 2005, 2009).

It is no longer sufficient to ignore inertia phenomena when designing structures that will be subjected to rapid loading. In addition to the sensitivity of materials to deformation speed, the loads experienced by a RVE contained in a real structure in the event of impact or crash can become highly complex. This is due to factors such as wave propagation, self-heating, and the spatio-temporal heterogeneity of fields. It is therefore essential to define clearly what is meant by characterization or "dynamic" behavior (Shirinbayan et al. 2015).

In terms of experimental characterization methods, it is essential to define the nature of the loading (quasi-static or dynamic) in relation to the relative values of the characteristic times of the loading and the material (or structure). It is important to note that if the characteristic loading time (rise time and duration) is significantly longer than the characteristic time required for the material (or structure) to deform, the latter may reach equilibrium after a phase during which the stress and strain fields are spatially variable within the sample or structure. In this case, the fields quickly become homogeneous (relative to the duration of the test). It is therefore possible to liken loading to quasi-static loading, despite the fact that kinetic energy cannot be ignored at the structural level. Conversely, if the characteristic time of the load is shorter than the characteristic time required for the material (or structure) to deform, the latter will never reach equilibrium and will undergo spatio-temporal variations in the fields throughout the duration of the load. It is important to consider the effects of inertia and wave propagation in such cases.

In accordance with the relevant technical standards, the term "dynamics" should be used only when a spatio-temporal variation in the fields is observed. However, the term "dynamic behavior" is often used to refer to the effect of strain rate. At the scale of the material of interest (REV), it is evident that the identification of behavior laws under rapid loading must be based on characterization tests during which only the effect of strain rate can be evaluated, without inertia or wave propagation effects interfering with the measurements. In other words, the material effect alone must be measured without being disturbed by the dynamic effect in the strict sense (Ahmed et al. 2021).

The analysis of the findings under dynamic loading demonstrates that there are certain similarities across the majority of composite materials, regardless of their architectural design (continuous or discontinuous reinforcements). The results obtained on SMC materials (Jendli et al. 2005) (Fig. 18.7) tested up to 150 s^{-1} are representative of the effects of strain rate on composite materials in general. Figure 18.7a illustrates the tensile curves obtained at varying strain rates. The results demonstrate that Young's modulus remains largely unaffected by an increase in the strain rate (Fig. 18.7b). Conversely, the damage threshold and failure stress increase significantly (Fig. 18.7c). Failure strain also rises with strain rate (Fig. 18.7d). Two distinct phases can be identified: an initial phase spanning from quasi-static speeds to approximately 1 s^{-1}, during which the rise in yield strength and stress at break is relatively modest. This is followed by a transition phase preceding an exponential surge in mechanical properties with rising strain rate.

From a macroscopic perspective, several key conclusions can be drawn about the impact of strain rate on composite materials. An increase in strain rate has the following effects: a higher damage threshold, an increase in break stress, and

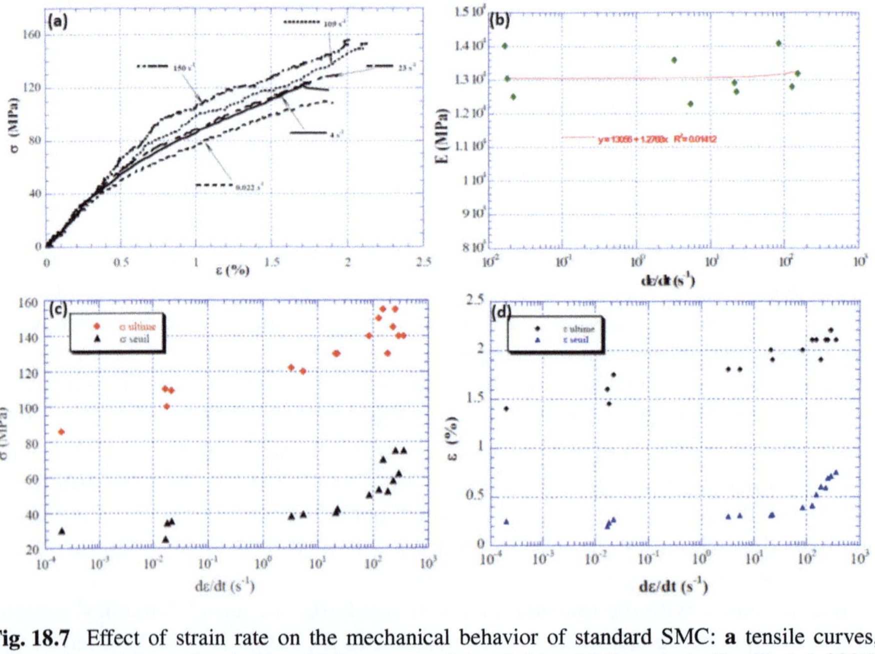

Fig. 18.7 Effect of strain rate on the mechanical behavior of standard SMC: **a** tensile curves, **b** Young's modulus, **c** threshold and failure stress, **d** threshold and failure strain (Jendli et al. 2005)

a reduction in damage kinetics. It is important to note that these three effects are generally applicable to all types of composite materials. It is notable that in the case of composites with lightly reinforced thermoplastic matrices, an increase in the strain rate has been observed to result in a reduction in failure strain. In this instance, it is the deformation of the matrix that determines the composite's overall response. However, the morphology of the polymers used in composites has a significant impact on their sensitivity to speed. For instance, the failure strain increases with speed for polyamide 6 and polycarbonate, while it decreases for polypropylene and polyamide 66 (Ahmed et al. 2021; Zhou and Mallick 2011). It is important to note that deformation phenomena are linked to macromolecular chain movements, which are more or less easy depending on the specific architecture of the polymer in question and the strain rate at which it is operating. Polyamide 6, for instance, exhibits high deformation rates at low speeds. It can be assumed that at this speed, certain chain movements are possible up to a reorientation of the macromolecular chains, which results in increased rigidity. At high speeds, these movements become impossible. The material's deformation mechanism changes. Similarly, an increase in speed erases the softening zone (threshold effect) in the case of polycarbonate, whereas it accentuates it in the case of polypropylene. It is possible to interpret these phenomena in terms of characteristic times once again. In fact, each deformation phenomenon is defined by a local movement with a characteristic time. As velocity increases, these movements may no longer be possible, as the characteristic time of the stress becomes too small. In such cases, the material will change its mechanism to accommodate the imposed deformation (Ahmed et al. 2021).

18.3.2 Visco-damage in Composites

The concepts of viscoelasticity and viscoplasticity are well established in the industry. However, based on the findings presented in the previous section, we would like to introduce the concept of visco-damage (Shirinbayan et al. 2015; Jendli et al. 2009) observed in composite materials. Visco-damage can be defined as the time- and strain rate-dependent evolution of internal damage mechanisms, such as matrix cracking, fiber/matrix debonding, and micro-delamination, which progressively degrade the material's stiffness and strength under mechanical loading. As previously demonstrated, the refined experimental approach enabled the observation of three crucial effects of strain rate typically seen in composite materials:

- An elevated damage threshold
- An augmented failure stress
- Reduced damage kinetics.

These effects were observed across all the composite materials under review. We can therefore conclude that the damage observed in composite materials is viscous in nature. To gain insight into the origin of visco-damage, a multi-scale analysis can be conducted using the interrupted test method in dynamic tests. This allows us

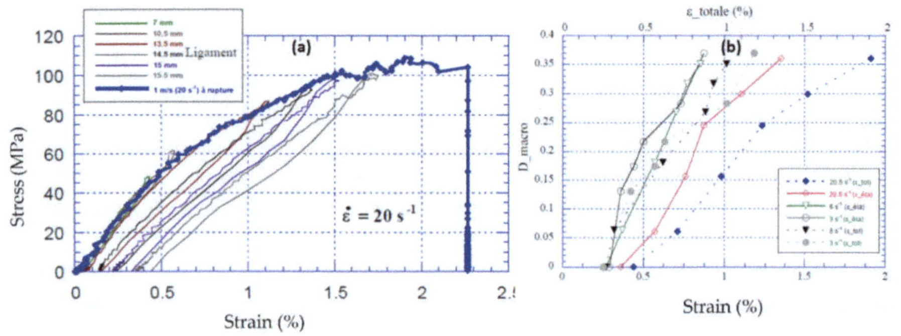

Fig. 18.8 Results of interrupted tests carried out on a standard SMC: **a** superimposed stress–strain curves and **b** evolution of the damage parameter (Jendli et al. 2009)

to analyze and quantify the origin of visco-damage in terms of damage threshold and kinetics. The objective of these tests is to produce pre-damaged samples at different levels. This enables us to carry out a qualitative and quantitative analysis of the evolution of damage at both microscopic and macroscopic scales after each loading (Shirinbayan et al. 2015; Jendli et al. 2009). This approach can be applied to different strain rates. From a macroscopic perspective, the impact of velocity on damage threshold and kinetics can be evaluated by examining the progression of the scalar damage parameter, as defined by Kachanov (1986):

$$D = 1 - \frac{E}{E_0}$$

where E and E_0 represent the Young's modulus of the damaged material (as measured during reloading) and of the original material, respectively. A typical result is shown in Fig. 18.8 (Jendli et al. 2009). This allows us to determine the damage threshold and kinetics (Fig. 18.8b).

As stated, the strain rate has a significant impact on the initiation and kinetics of damage. At the microscopic level, damage is primarily caused by decohesion at the fiber-matrix interface, which emerges at 20–30% of the load at failure and propagates diffusely. This mechanism is reflected in a local variable, d_{micro}, defined by the ratio between the rate of loosened fibers and the total fiber rate. The number of loosened reinforcements is estimated by means of SEM observations and image analysis carried out after each test interruption (Fig. 18.9) (Jendli et al. 2009). The interfacial decohesion process is initiated relatively early when the test is quasi-static (damage threshold strain: $\varepsilon = 0.24\%$). For strain rates of 8 s^{-1} and 20.5 s^{-1}, the first interface failures were observed at higher threshold strains of 0.26% and 0.43% respectively (Jendli et al. 2009). Furthermore, Fig. 18.9b demonstrates a paradoxical decrease in damage kinetics with increasing strain rate. The microscopic results thus confirm the macroscopic finding of delayed damage initiation and evolution, which is hampered by the "viscous" effect as strain rate increases.

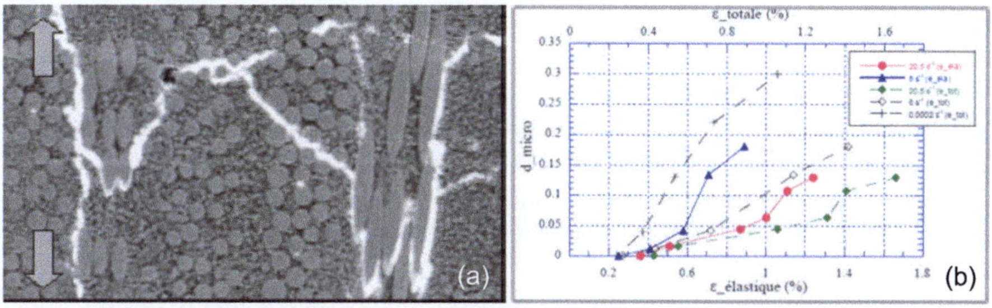

Fig. 18.9 Interrupted test results in SMC: **a** SEM microscopic observation: decohesion of the fiber-matrix interface ($d\varepsilon/dt = 25\ s^{-1}$) and **b** microscale evolution of damage at different strain rates (Jendli et al. 2009)

18.3.3 Characterization of Composite Materials Subjected to Cyclic Loading

The objective of reducing the weight of structures has led to the widespread adoption of discontinuous fiber-reinforced organic matrix composites, particularly in the automotive sector, where they are used in engine parts, radiators, heat exchangers, bodywork, flooring and other applications. The development process is relatively short, typically around 18 months in the automotive industry. It includes selecting the appropriate material for the application, sizing the structure, and validating it through experimentation. In this context, several iterations are often necessary. These iterations are long and costly, especially in terms of fatigue. It is therefore essential to rely on a predictive methodology that integrates the influence of the various variability parameters in relevant models. In addition, the predictive methodologies and tools proposed must be capable of integrating the influence of variable cyclic loading conditions. This requires a thorough understanding of the fatigue behavior of materials, and the ability to take into account the effects of microstructure and damage phenomena on service life. It is important to note that if the influence of cyclic loading is not properly managed, the use of these materials may be at risk. It is worth noting that many parts are subjected to repeated loading cycles, either continuously (such as engine parts) or intermittently (such as the tailgate, which is subjected to open-close cycles). These parts are typically operated at temperatures between −30 and 85 °C. Variations in amplitude or temperature may also be included in the functional specifications. Additionally, discontinuous-reinforced composites are highly sensitive to microstructural variability when subjected to cyclic loading. Furthermore, in polymer-based materials, the influence of frequency should not be overlooked, particularly the influence of self-heating, which can have a significant impact, especially for thermoplastic matrix materials. Wöhler curves are typically employed in the fatigue design of structural components. There are numerous variables that can influence the outcome, including the microstructure and the specific loading conditions (load orientation, load ratio R, frequency, temperature, etc.) (Imaddahen et al. 2020; Shirinbayan et al. 2019; Tamboura et al. 2001; Fitoussi et al. 2024; Zhou and

Mallick 2011). Figure 18.10 illustrates the outcomes of Wöhler curves established for various composite materials with discontinuous reinforcements subjected to different microstructures and fatigue conditions. The letters F or HO (fluted or highly oriented) indicate a material whose fibers have been strongly oriented in a preferred direction during processing. The letters NF or RO (non-fluted or randomly oriented) indicate a material whose reinforcements are randomly oriented in the plane.

In the case of SMC materials (Shirinbayan et al. 2019; Laribi et al. 2020), the fact that the reinforcements are strongly oriented in a particular direction does not significantly impact the service life of samples tested perpendicular to this direction. The Wöhler curves obtained in this case are almost similar to those obtained for materials with random fiber orientation (see Fig. 18.10a–d). In other words, the material's cyclic properties in a particular direction can be enhanced without undue effect on its properties in other directions, thanks to material creep during the thermocompression manufacturing process for SMC prepregs. This remains the case regardless of the working temperature. However, this is less evident in the case of thermoplastic

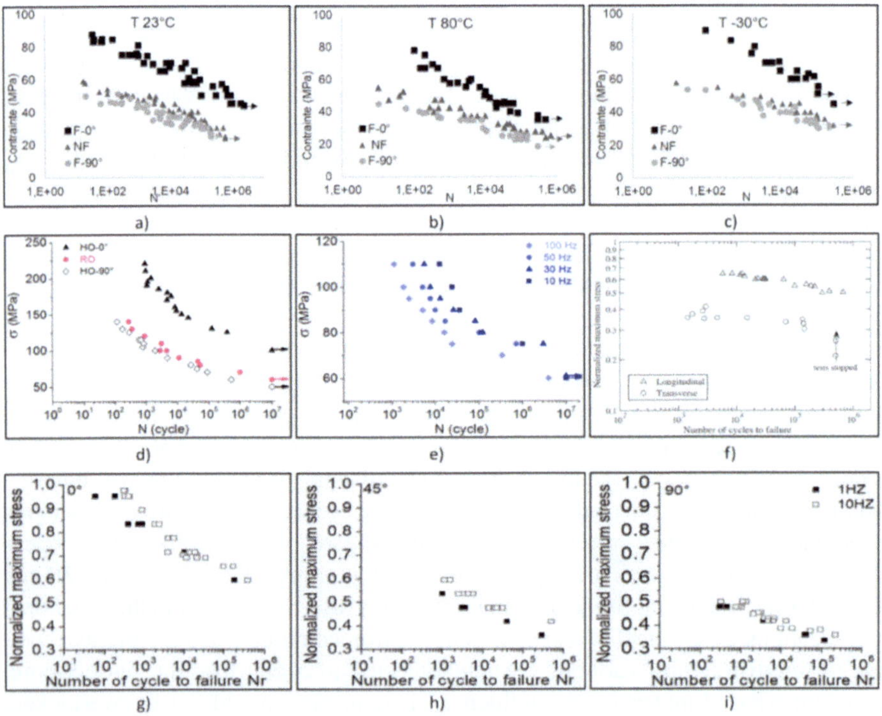

Fig. 18.10 Wöhler curves obtained for composite materials with discontinuous reinforcements, $R = 0.1$, **a–c** Standard SMC composite, effect of microstructure and temperature, **d** and **e** A-SMC composite, effect of microstructure (**d**) and frequency (**e**), **f** PA66GF30 composite, effect of microstructure, **g–i** PPGF30 composite, effect of microstructure and frequency (Imaddahen et al. 2020; Shirinbayan et al. 2019; Tamboura et al. 2001; Fitoussi et al. 2024; Zhou and Mallick 2011)

composites, where the orientation of the reinforcements in a given direction during injection molding has a significant impact on service life when the material is loaded perpendicular to the injection direction. This discrepancy is intrinsic to the specific microstructure. It is clear that in the case of SMC, the long fiber length associated with a pseudo-stratified architecture plays an important role as a "stopper" for cracks. This enables the delocalization of cracking from one local site to another and a damage diffusion phase that postpones the localization before final failure to a greater number of cycles. However, in the case of short-fiber injected composites, crack localization occurs rapidly despite the plasticity of the matrix. In all cases, the effect of frequency corresponds to a reduction in the service life of discontinuous-reinforced composites of varying magnitude, depending on frequency, and stress amplitude.

During fatigue tests (generally stress-controlled), it is recommended to measure the evolution of certain macroscopic variables, such as relative stiffness decrease, plastic deformation (or maximum deformation), and temperature rise (due to self-heating). This phenomenon is more pronounced in thermoplastic polymers, but also occurs in thermoset matrix materials. In particular, this effect is intensified as the frequency increases. The second phenomenon pertains to diffuse damage that develops on a local scale, particularly around and at the fiber-matrix interface.

In real-life applications, fatigue damage phenomena are often coupled with those resulting from monotonic quasi-static or fast dynamic loading. Furthermore, material degradation due to physico-chemical ageing often affects the durability of structures subjected to these same types of loading.

18.3.4 Synthesis of Damage Mechanisms in Composites

It has been established that when composite materials are subjected to thermomechanical stress, they undergo local damage mechanisms. Extensive research has been conducted into these processes through in-situ or interrupted testing, coupled with observation techniques using optical and scanning electron microscopy or X-ray micro-tomography. The direction of microcracking development is dependent on two key factors: the stress state and the microstructure parameters, particularly the orientation distribution of the reinforcements. One distinctive feature of composites with discontinuous reinforcements is diffuse damage (see Fig. 18.11) (Jendli et al. 2009; Fitoussi et al. 2013). It is important to note that the redistribution of stresses resulting from local microcracking can contribute to the delocalization of damage and, consequently, to its diffusion throughout the stressed volume, given the multitude of possible locations where damage can occur. Therefore, with each load increment, new microcracking sites are created without the initial cracks having propagated significantly. While damage phenomena typically manifest relatively early (20–30% of the load to failure), the systematic redistribution of stresses ultimately leads to diffuse damage and prolonged damage phases, resulting in extensive non-linear behavior.

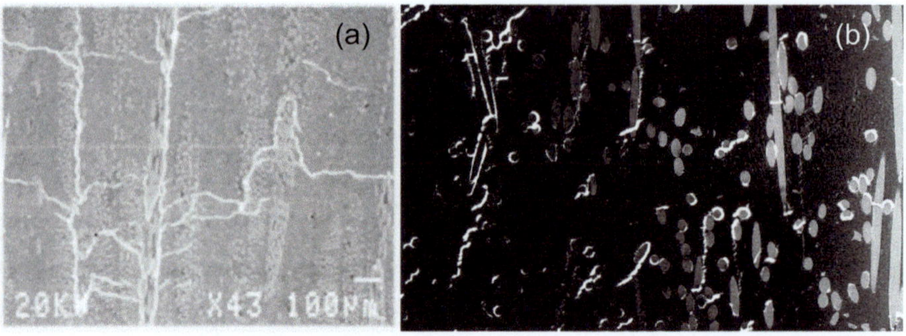

Fig. 18.11 Diffuse damage: **a** standard SMC composite and **b** PPGF30 composite (images taken close to fracture just before coalescence of microcracks into a macroscopic crack) (Jendli et al. 2009; Fitoussi et al. 2013)

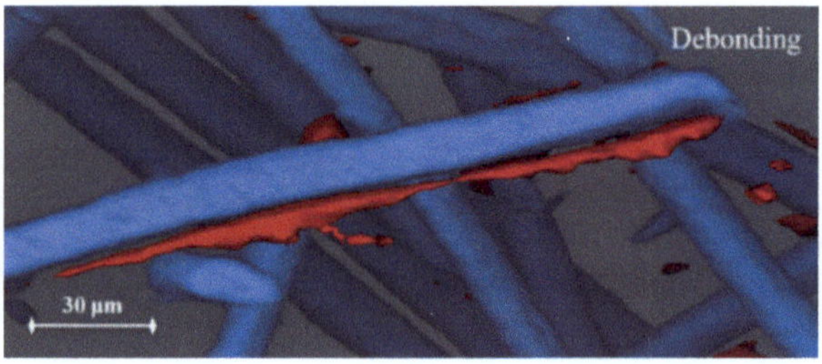

Fig. 18.12 Interface damage in a PA66-glass fiber composite (Arif et al. 2014a, b)

It is important to note that damage generally begins and is driven by micro-cracking at the fiber-matrix interface, which is often the weak point of these heterogeneous materials. This occurs particularly on reinforcements that are most disoriented with respect to the direction of the main stress (Figs. 18.12 and 18.13a) (Arif et al. 2014a, b).

It is important to note that the macroscopic stress applied is distributed locally, particularly at all points of the fiber-matrix interfaces. This occurs in the form of a combination of local normal stress and local shear stress (see Fig. 18.13b). To illustrate, a Mori and Tanaka-type homogenization calculation was carried out for a standard SMC composite (with random in-plane reinforcement orientation) subjected to a tensile stress of 1 MPa. This allowed us to plot the distribution of normal and shear stresses around the fiber-matrix interfaces (Fig. 18.13c, d). In this context, the letter "q" represents the orientation of the fiber with respect to the tensile direction, while the letter "F" identifies the specific interface point under consideration. Please note that "F" corresponds to an interface point contained in the plane, with "$F = 0°$" representing the interface point in question (Jendli et al. 2009; Fitoussi et al. 2013).

The calculations demonstrate that the maximum normal stress occurs at the interface points of fibers oriented at 90° to the tensile direction, while the maximum shear stress occurs for fibers oriented at 45°. In-situ tests conducted in a SEM chamber revealed that interfacial damage originated on the most disoriented fiber family with

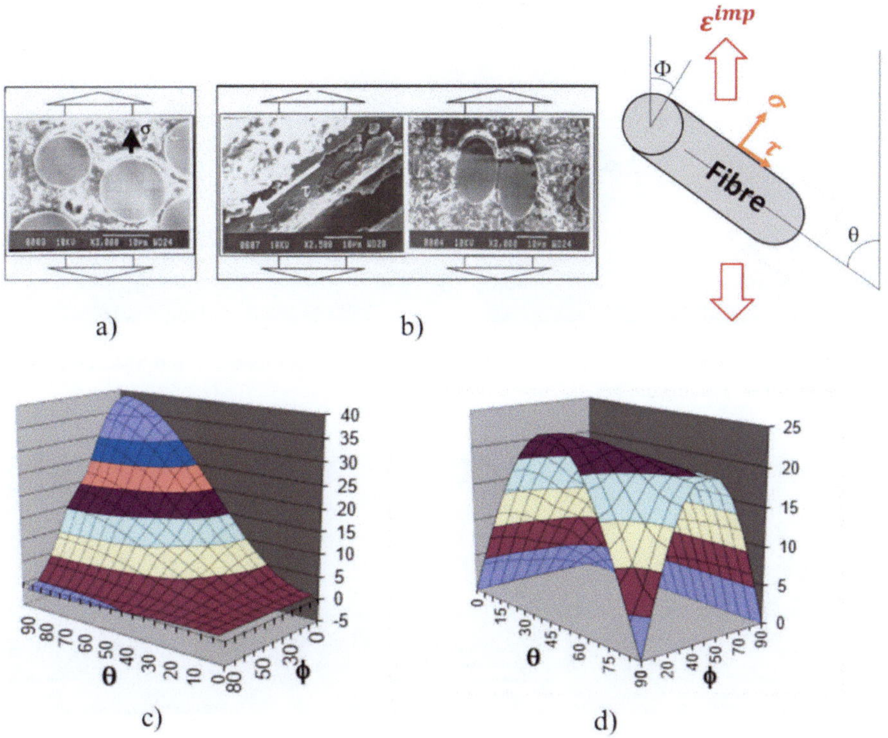

Fig. 18.13 Damage and local stress state at the fiber-matrix interface (standard SMC subjected to 1MPa tensile stress): **a** pure normal stress failure on a fiber oriented at 90° to the tensile stress, **b** normal stress failure coupled with shear stress on a fiber oriented at 45° to the tensile stress, **c** normal stress profile around the fiber-matrix interfaces, **d** shear stress profile around the fiber-matrix interfaces (Jendli et al. 2009; Fitoussi et al. 2013)

respect to the tensile direction and then progressively spread to the fiber families most oriented in the tensile direction. It is also important to note that, due to the local variation in microstructure, not all reinforcement interfaces of the same orientation are broken at the same time. There is a progressive fracture statistic for different interfacial sites of the same type.

In addition to interface damage, matrix cracking is typically a secondary mechanism in composites used in industry. It is commonly caused by interfacial cracks that split in the matrix at right angles to the main direction of loading, or in areas of high fiber depletion (see Fig. 18.14) (Imaddahen et al. 2020; Jendli et al. 2009).

In the case of thermoplastic matrix composites, it is possible to observe the phenomenon of matrix micro-cavitation and fibrillation (multiple cracks oriented perpendicular to the load direction, often generated by cavitation and favored between two damaged zones) in the matrix (see Fig. 18.15) (Arif et al. 2014a, b).

Fiber fragmentation (the third mechanism in competition) typically occurs in the final stages of loading before fracture. In the case of thermoplastics with highly

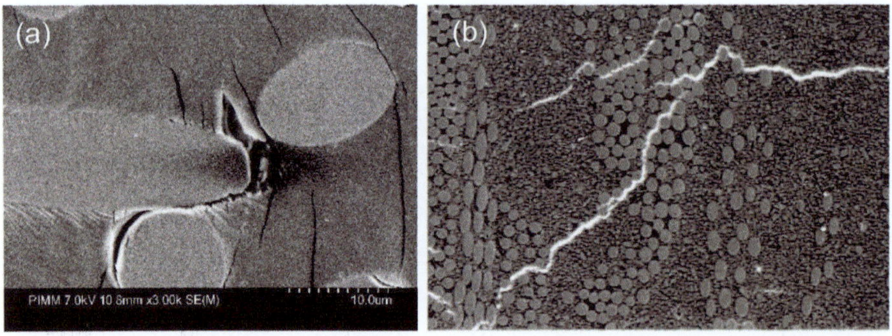

Fig. 18.14 Interfacial cracks bifurcating in the matrix: **a** PPGF composite and **b** SMC composite (Imaddahen et al. 2020; Jendli et al. 2009).

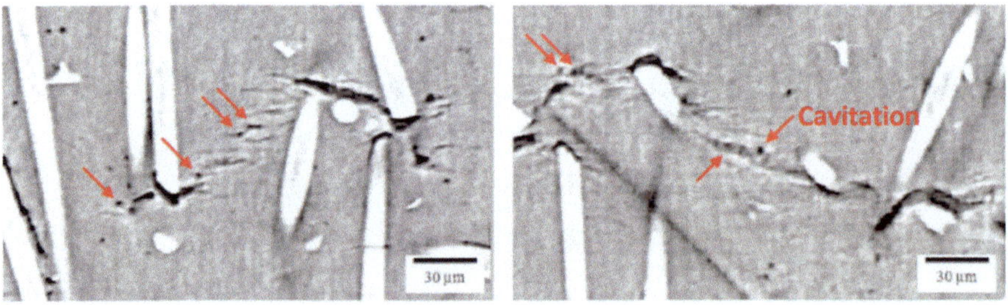

Fig. 18.15 Matrix damage by micro-cavitation and fibrillation in a PA66-glass fiber composite (Arif et al. 2014a, b)

oriented fibers in the tensile direction, fiber fragmentation may occur earlier. It is also possible to observe other local damage mechanisms, depending on the architecture or local phase behavior. For example, it is well known that in the case of thermoplastic matrix composites whose fibers are highly oriented in the main direction of the load, damage begins at the fiber tip (due to the lack of sizing caused by fiber breakage in the injection screw). The microcrack then propagates first in the direction of the load, and then spherically due to the local plasticity of the matrix (see Fig. 18.16) (Arif et al. 2014a, b).

It is also possible to observe effects linked to a particular architecture, as in the case of SMC materials. Indeed, local pseudo-stratification of the fiber strands leads to a phenomenon of local delamination between the strands, which we call pseudo-delamination (Shirinbayan et al. 2015) (Fig. 18.11a). Pseudo-delamination can be defined as a localized, internal separation between fiber bundles or clusters that mimics conventional interlaminar delamination, but occurs within a nominally homogeneous, non-laminated material. In certain loading conditions, crack propagation during pseudo-delamination plays a significant role in deformation and energy dissipation, particularly in the final stages before fracture.

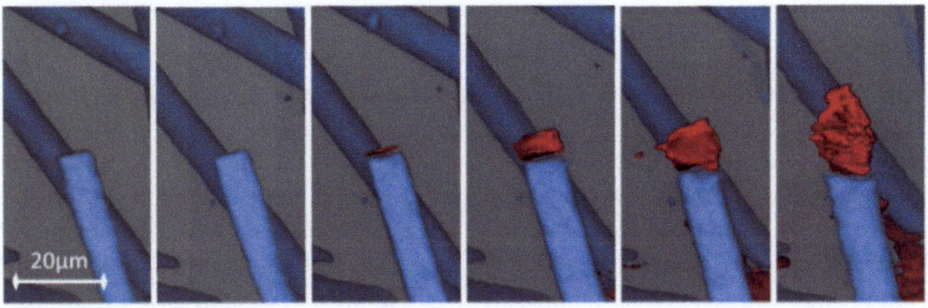

(a) $\varepsilon_1 = 0.0\%$ $\sigma_1 = 0.0$ MPa (b) $\varepsilon_2 = 1.1\%$ $\sigma_2 = 28.0$ MPa (c) $\varepsilon_3 = 2.1\%$ $\sigma_3 = 49.7$ MPa (d) $\varepsilon_4 = 3.2\%$ $\sigma_4 = 63.9$ MPa (e) $\varepsilon_5 = 4.2\%$ $\sigma_5 = 71.6$ MPa (f) $\varepsilon_6 = 5.3\%$ $\sigma_6 = 74.1$ MPa

Fig. 18.16 Glass fiber tip damage in a PA66-based composite (Arif et al. 2014a, b)

In the case of laminated materials (Nikbakt et al. 2018), the final phase of damage before failure is often characterized by crack propagation along the interfaces between plies. This damage phenomenon, which ultimately leads to composite failure, is known as delamination. The formation of free surfaces within the material is a significant factor in energy absorption during impact. It is crucial to analyze the impact of velocity on delamination for effective impact and crash design. To enhance the composite's capacity to absorb energy during a crash and/or optimize the ply stacking sequence, it is essential to consider the impact of loading speed on the property of the interface between plies. The capacity to absorb energy during delamination is typically quantified by the critical energy restitution rate (G_c). This is equivalent to the energy needed to separate two surfaces. While there has been a significant body of research conducted on mode I fracture, there has been comparatively less research conducted on mode II. It should be noted that in many applications, delamination failure involves both modes in a coupled manner. Furthermore, there is a lack of widespread characterization methods for material behavior in slow and medium dynamic regimes, which is reflected in the scarcity of experimental devices for delamination characterization in these regimes. It is often the case that design offices lack access to this information, which leads them to predict the failure of composites and assemblies at high stress speeds on the basis of properties obtained under quasi-static conditions. This generally results in over-dimensioning (Nikbakt et al. 2018).

A variety of experimental methodologies and modeling formalisms have been developed. These approaches are necessary to describe the response of materials on a macroscopic scale. However, they are not sufficient to understand and control the influence of loading speed on damage mechanisms at the ply scale (mesoscopic) or at the constituent scale (microscopic, such as the fiber-matrix interface). It is clear that the rate of crack advance is significantly influenced by the state of damage in the area surrounding the crack tip. This, in turn, is affected by the strain rate at the bottom of the crack and the anisotropy of the material. Therefore, it is essential to carry out a multi-scale analysis of damage and delamination failure phenomena.

The damage mechanisms in polymer composites are numerous, complex and highly coupled. However, our local analyses have shown that, for highly loaded industrial materials, damage at the fiber-matrix interface is typically the main cause of failure, accounting for at least 80% of the non-linear deformation phase. The damage thresholds and kinetics can be influenced by factors such as thermomechanical stress, the plastic deformation of the matrix, and the microstructure (in particular, the orientation distribution of reinforcements). It should be noted that the aforementioned damage scenarios are subject to further influence from external factors, including temperature, humidity, loading strain rate, and cycling. The type of loading, such as dynamic or fatigue, is of critical importance, and we also analyze the impact of interrupted fatigue sequences on residual dynamic properties. It is also important to consider the effects of ageing, particularly under varying environmental conditions, when evaluating the durability of composites.

It is important to note that local damage mechanisms often manifest as decreases in stiffness on a macroscopic scale. This structural-level degradation can be identified through a range of experimental procedures designed for quasi-static, dynamic, and cyclic loading. This helps to establish multi-scale relationships between local damage and overall material performance.

At the structural level, these localized damage mechanisms have significant performance implications for real-world applications. For example, fiber-matrix debonding and matrix cracking lead to progressive stiffness degradation, reducing the load carrying capacity and energy absorption potential of composite components under impact or crash conditions. Delamination and pseudo-delamination phenomena compromise structural integrity by facilitating crack propagation along weak planes, which can lead to catastrophic failure in automotive body parts such as tailgates or crash boxes. Fiber fragmentation, particularly in highly oriented fiber regions, reduces reinforcement efficiency, causing premature failure and limiting the service life of structural components. In applications subject to dynamic or cyclic loading, the accumulation of such damage mechanisms progressively reduces fatigue life, increases residual deformation and can induce sudden, brittle-like failure. Understanding these links between microscale damage and structural-level performance is therefore essential for accurate life prediction, optimized design and effective in-service monitoring of composite structures in demanding industrial environments.

Figure 18.17 brings together both experimental and numerical analyses related to the dynamic behavior of composite materials under high strain rates (case of an SMC tailgate), such as those encountered in vehicle crash tests. The results demonstrate that strain rate sensitivity is a key factor, with higher strain rates leading to increased stress levels. The load–displacement curves provide valuable insights into the failure behavior of the materials, including progressive damage mechanisms such as delamination or fiber failure. In addition, the finite element simulation offers a complementary approach to physical testing by predicting areas of high stress or potential failure. The primary objective is to gain a deeper understanding of how composite materials perform under high-speed impacts, which is a crucial aspect for industries such as automotive safety (Laribi et al. 2020; Jendli et al. 2005; Ayari et al. 2020b).

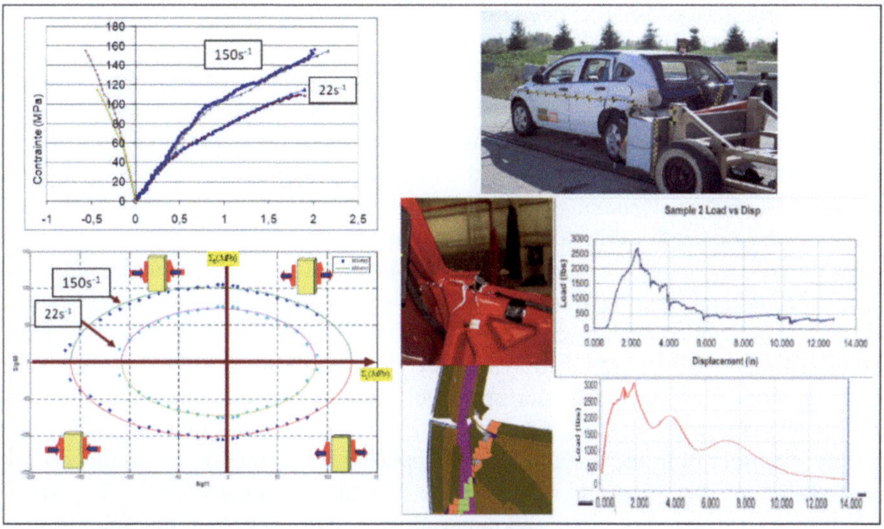

Fig. 18.17 Multi-scale modeling of the behavior of composite materials and structures, taking damage into account: influence of strain rate; case of an SMC tailgate (Laribi et al. 2020; Jendli et al. 2005; Ayari et al. 2020b)

The combination of local damage assessment with structural health monitoring (SHM) at the structural scale ensures a more comprehensive understanding of composite durability, enhancing both the reliability and longevity of the material. In practical terms, SHM systems have been successfully integrated into various industrial applications, such as monitoring fatigue damage in aircraft fuselage panels, delamination detection in wind turbine blades, or crack propagation tracking in automotive composite structures. These systems often use sensors like fiber optic sensors, piezoelectric transducers, and strain gauges that continuously monitor strain, acoustic emissions, or guided wave propagation.

Several industrial standards and practices govern the implementation of SHM systems in composite materials. For example, the ASTM D5229 standard outlines procedures for the accelerated aging and testing of polymer matrix composite materials used in structural applications, which is vital for SHM system calibration and validation. The ISO 13314:2011 standard provides guidelines on the determination of the fracture toughness of composite materials, which is crucial when assessing the integrity of structures through SHM. Furthermore, the SAE ARP 5786 standard focuses on fiber optic sensors for structural health monitoring in aerospace applications, ensuring the sensors meet performance requirements for detecting critical damage in composite structures.

Despite their effectiveness, SHM systems face challenges, such as sensor durability under harsh conditions, data management complexity, and the difficulty of accurately interpreting signals in anisotropic composite materials. Moreover, ensuring that the integration of sensors does not affect the structural integrity of lightweight composite parts is a significant concern.

Looking forward, future improvements in SHM technology may focus on the development of wireless, self-powered sensors, the use of artificial intelligence for real-time damage classification, and multi-parameter sensing networks that simultaneously track temperature, strain, and damage. These innovations will likely lead to more effective early damage detection, optimized maintenance protocols, and extended service life for critical composite components.

To provide a clearer overview of the various damage mechanisms identified in composite materials, a comparative summary is presented below. While the preceding sections have detailed each mechanism individually—from initiation causes to local manifestations—it is essential to synthesize this information to better grasp their collective impact on composite performance. This comparative table highlights the primary damage mechanisms, their typical initiation triggers, the most commonly employed observation techniques, and the consequences they have on the macroscopic behavior and reliability of composite structures. Such a synthesis is particularly useful for design engineers and materials scientists seeking to link microscopic damage phenomena to structural-level consequences, enabling more accurate durability predictions, damage-tolerant design strategies, and informed material selection for demanding industrial applications, as summarized Table 18.1.

Table 18.1 Damage mechanisms in composite materials

Damage mechanism	Cause/Initiation trigger	Observation methods	Composite performance
Fiber-matrix debonding	Stress concentrations at fiber-matrix interface, particularly on misaligned fibers	SEM, X-ray micro-CT, in-situ mechanical testing	Local stiffness reduction, promotes matrix cracking and delamination
Matrix cracking	Interfacial debonding, local matrix stress exceeding tensile strength	Optical microscopy, SEM, micro-CT	Decreases load transfer efficiency, enables crack coalescence
Delamination/pseudo-delamination	Localized stress concentrations between fiber bundles or plies	Ultrasonic C-scan, X-ray micro-CT, DIC, SEM	Loss of interlaminar strength, reduced energy absorption, crack propagation
Fiber fragmentation	High tensile stress, particularly in highly oriented fibers	SEM, micro-CT, in-situ tensile tests	Decreases reinforcing efficiency, reduces load-bearing capacity
Matrix micro-cavitation/fibrillation	Local plasticity and stress concentrations in matrix under tensile loading	SEM, optical microscopy, in-situ loading in SEM chamber	Promotes crack initiation and non-linear deformation behavior

18.4 Conclusion

Polymer composites have proven to be extremely useful in many high-performance applications due to their excellent mechanical properties, lightweight nature, and design flexibility. However, ensuring their long-term durability under complex thermomechanical and environmental conditions remains a significant challenge. This chapter has highlighted the key factors influencing the degradation of polymer composites, including mechanical loading, thermal effects, and environmental exposure. To address these challenges, several strategies can be implemented to enhance the durability and longevity of composites in industry-specific contexts. For example, in the aerospace sector, improving the fiber-matrix interface and utilizing advanced resin systems with higher thermal and oxidative resistance can significantly reduce the risk of delamination and material degradation. In automotive applications, multi-layered composite designs with optimized fiber orientation and the use of self-healing materials could help mitigate damage from repeated loading and harsh environmental exposure. Moreover, the integration of real-time monitoring systems such as structural health monitoring (SHM), combined with predictive maintenance, allows for early detection of damage, facilitating timely repairs and minimizing downtime. Additionally, the use of numerical simulations to optimize composite design and predict long-term behavior under various loading and environmental conditions can help manufacturers make informed decisions about material selection and component design. By leveraging these strategies, industries can improve the performance and reliability of composite materials, extending their service life and reducing maintenance costs. Collaborative efforts between material scientists, engineers, and industry practitioners will be essential to develop innovative solutions for enhancing the durability of composites in real-world applications.

References

Abdessalem A, Tamboura S, Fitoussi J, Daly HB, Tcharkhtchi A, Meraghni F (2021) Microstructure investigation of hydrothermal damage of aged SMC composites using micro-computed tomography and scanning electron microscopy. Eng Fail Anal 121:105177

Ahmed A et al (2021) A review on the tensile behavior of fiber-reinforced polymer composites under varying strain rates and temperatures. Constr Build Mater 294:123565. https://doi.org/10.1016/j.conbuildmat.2021.123565

Arif MF, Meraghni F, Chemisky Y, Despringre N, Robert G (2014a) In situ damage mechanisms investigation of PA66/GF30 composite: effect of relative humidity. Compos B Eng 58:487–495

Arif MF, Saintier N, Meraghni F, Fitoussi J, Chemisky Y, Robert G (2014b) Multiscale fatigue damage characterization in short glass fiber reinforced polyamide-66. Compos B Eng 61:55–65. https://doi.org/10.1016/j.compositesb.2014.01.019

Ayari H et al (2020a) Two hybrid approaches to fatigue modeling of advanced-sheet molding compounds (A-SMC) composite. Appl Compos Mater 27(1):19–36. https://doi.org/10.1007/s10443-019-09793-3

Ayari H et al (2020b) Micromechanical modelling of dynamic behavior of advanced sheet molding compound (A-SMC) composite. Appl Compos Mater 27(3):321–335. https://doi.org/10.1007/s10443-020-09811-9

Dumont P, Orgéas L, Le Corre S, Favier D (2003) Anisotropic viscous behavior of sheet molding compounds (SMC) during compression molding. Int J Plast 19(5):625–646. https://doi.org/10.1016/S0749-6419(01)00077-8

Feki I et al (2025) Multi-scale experimental investigation of porosity-induced damage effects in filament-wound carbon fiber reinforced epoxy composites used in hydrogen storage tanks. Polym Compos. https://doi.org/10.1002/pc.29121

Fitoussi J, Bocquet M, Meraghni F (2013) Effect of the matrix behavior on the damage of ethylene–propylene glass fiber reinforced composite subjected to high strain rate tension. Compos B Eng 45(1):1181–1191

Fitoussi J, Nikooharf MH, Kallel A, Shirinbayan M (2022) Mechanical properties and damage behavior of polypropylene composite (GF50-PP) plate fabricated by thermocompression process under high strain rate loading at room and cryogenic temperatures. Appl Compos Mater. https://doi.org/10.1007/s10443-022-10047-y

Fitoussi J et al (2024) Investigation of manufacturing process effects on microstructure and fatigue prediction in composite automotive tailgate design. Int J Adv Manuf Technol 130(9):4295–4310. https://doi.org/10.1007/s00170-024-12988-z

Friedrich K, Fakirov S, Zhang Z (2005) Polymer composites: from nano-to macro-scale. Springer Science & Business Media

Imaddahen MA, Shirinbayan M, Ayari H, Foucard M, Tcharkhtchi A, Fitoussi J (2020) Multi-scale analysis of short glass fiber-reinforced polypropylene under monotonic and fatigue loading. Polym Compos 41(11):4649–4662

Jendli Z, Fitoussi J, Meraghni F, Baptiste D (2005) Anisotropic strain rate effects on the fibre–matrix interface decohesion in sheet moulding compound composites. Compos Sci Technol 65(3–4):387–393

Jendli Z, Meraghni F, Fitoussi J, Baptiste D (2009) Multi-scales modelling of dynamic behaviour for discontinuous fibre SMC composites. Compos Sci Technol 69(1):97–103. https://doi.org/10.1016/j.compscitech.2007.10.047

Kachanov L (1986) Introduction to continuum damage mechanics. Springer Science & Business Media

Laribi MA et al (2020) Sheet molding compound automotive component reliability using a micromechanical damage approach. Appl Compos Mater 27(5):693–715

Maragoni L, Carraro PA, Quaresimin M (2018) Development, validation and analysis of an efficient micro-scale representative volume element for unidirectional composites. Compos Part A: Appl Sci Manuf 110:268–283. https://doi.org/10.1016/j.compositesa.2018.04.025

Meftah H, Tamboura S, Fitoussi J, Bendaly H, Tcharkhtchi A (2018) Characterization of a new fully recycled carbon fiber reinforced composite subjected to high strain rate tension. Appl Compos Mater 25(3):507–526

Mitra M, Gopalakrishnan S (2016) Guided wave based structural health monitoring: a review. Smart Mater Struct 25:053001

Mortazavian S, Fatemi A (2015) Fatigue behavior and modeling of short fiber reinforced polymer composites: a literature review. Int J Fatigue 70:297–321

Nikbakt S, Kamarian S, Shakeri M (2018) A review on optimization of composite structures. Part I: Laminated composites. Compos Struct 195:158–185. https://doi.org/10.1016/j.compstruct.2018.03.063

Nikooharf MH, Rezaei-Khamseh M, Shirinbayan M, Fitoussi J, Tcharkhtchi A (2021) Comparison of the physicochemical, rheological, and mechanical properties of core and surface of polypropylene composite (GF50-PP) plate fabricated by thermocompression process. Polym Compos

Shirinbayan M (2021) Multiscale damage analysis of the tension-tension fatigue behavior of a low-density sheet molding compound. J Appl Polym Sci 138(4):49721

Shirinbayan M, Fitoussi J, Meraghni F, Surowiec B, Bocquet M, Tcharkhtchi A (2015) High strain rate visco-damageable behavior of advanced sheet molding compound (A-SMC) under tension. Compos B Eng 82:30–41
Shirinbayan M, Fitoussi J, Bocquet M, Meraghni F, Surowiec B, Tcharkhtchi A (2017a) Multi-scale experimental investigation of the viscous nature of damage in advanced sheet molding compound (A-SMC) submitted to high strain rates. Compos B Eng 115:3–13
Shirinbayan M, Fitoussi J, Meraghni F, Surowiec B, Laribi M, Tcharkhtchi A (2017b) Coupled effect of loading frequency and amplitude on the fatigue behavior of advanced sheet molding compound (A-SMC). J Reinf Plast Compos 36(4):271–282
Shirinbayan M, Fitoussi J, Abbasnezhad N, Meraghni F, Surowiec B, Tcharkhtchi A (2017c) Mechanical characterization of a low density sheet molding compound (LD-SMC): multi-scale damage analysis and strain rate effect. Compos B Eng 131:8–20
Shirinbayan M, Fitoussi J, Meraghni F, Farzaneh S, Surowiec B, Tcharkhtchi A (2019) Effect of a post-fatigue damage on the residual dynamic behavior of advanced-SMC composites. Appl Compos Mater 26(5–6):1313–1331
Shirinbayan M, Rezaei-Khamseh M, Nikooharf MH, Tcharkhtchi A, Fitoussi J (2021a) Multi-scale analysis of mechanical properties and damage behavior of polypropylene composite (GF50-PP) plate at room and cryogenic temperatures. Compos Struct 114713. https://doi.org/10.1016/j.compstruct.2021.114713
Shirinbayan M, Rizi HB, Abbasnezhad N, Tcharkhtchi A, Fitoussi J (2021b) Tension, compression, and shear behavior of advanced sheet molding compound (A-SMC): multi-scale damage analysis and strain rate effect. Compos B Eng 225:109287
Shirinbayan M, Nouira S, Imaddahen M-A, Fitoussi J (2024) Microstructure-sensitive investigation on the plastic deformation and damage initiation of fiber-reinforced polypropylene composite. Compos Part B: Eng 286:111790. https://doi.org/10.1016/j.compositesb.2024.111790
Shreepannaga, Vijaya Kini M, Pai D (2022) The ageing effect on static and dynamic mechanical properties of fibre reinforced polymer composites under marine environment—a review. Mater Today Proc 52:689–696. https://doi.org/10.1016/j.matpr.2021.10.084
Sohn H, Farrar CR, Hemez FM, Czarnecki JJ (2002) A review of structural health review of structural health monitoring literature 1996–2001. Report number: LA-UR-02-2095, Research Org.: Los Alamos National Lab. (LANL), Los Alamos, NM
Su Z, Ye L (2009) Identification of damage using lamb waves. Springer, London
Su Z, Ye L, Lu Y (2006) Guided Lamb waves for identification of damage in composite structures: a review. J Sound Vib 295:753–780
Tamboura S, Sidhom H, Baptiste D, Fitoussi J (2001) Evaluation de la tenue en fatigue du composite SMC R42. Matériaux Tech 89(3–4):3–11
Worden K, Farrar CR, Manson G, Park G (2007) The fundamental axioms of structural health monitoring. Proc R Soc A Math Phys Eng Sci 463:1639–1664
Yuan FG (2016) Structural health monitoring (SHM) in aerospace structures. Woodhead Publishing is an imprint of Elsevier, Duxford
Zhou Y, Mallick P (2011) Fatigue performance of injection-molded short e-glass fiber reinforced polyamide-6, 6. II. Effects of melt temperature and hold pressure. Polym Compos 32(2):268–276

Chapter 19
Computational and Theoretical Modelling of Sustainable Polymers/Composites

Smitha Roy, Krishnan Aravind, and S. Mallya Sivamanjari

Abstract Sustainable polymer composites have attracted significant interest in recent years, not only due to its technological applications, but also because it combines 'sustainability', one among the important goals of research and technology (Inamuddin et al. in Sustainable polymer composites and nanocomposites. Springer, 2019). The properties of such composites are greatly influenced by the components incorporated within the system. Designing composites for specific applications thus requires careful selection of the components and specific understanding of the interactions influencing the characteristics of such systems; which can be deciphered using theoretical calculations (Shen et al. in Chem Rev 123:2242–2275, 2023). This chapter thus introduces briefly the computational methodology used in predicting the properties of such composite materials. Molecular dynamics simulations, one such method which studies the properties of bulk materials by allowing it to move for a specific time, have played a significant role in this direction in determining the mechanical and thermal properties of a series of polymeric systems (Zhou et al. in Adv Func Mater 32(14):2109881, 2022). First principal calculations, involving density functional theory, mark another notable method, involving quantum calculations for determining atomistic properties influencing the composite behaviour (Chen et al. in Comput Mater Sci 188:110229, 2021). Simulations including molecular dynamics, Monte Carlo and coarse grain model, etc. are also discussed in brief with examples (Wang and Gómez-Bombarelli in npj Comp Mater 5:1–9, 2019; Zhang et al. in Int J Geomech 21:04021048, 2021). Apart from this, the chapter also highlights the possibilists of utilizing computational approaches employed for other molecules

S. Roy
Union Christian College, Aluva, Kerala, India

K. Aravind (✉)
Saint Berchmans Autonomous College, Changanassery, Kottayam, Kerala, India
e-mail: aravind@sbcollege.ac.in

S. M. Sivamanjari
Sree Sankara College, Kalady, Kerala, India

School of Chemical Sciences, Mahatma Gandhi University, Kottayam, Kerala, India

L. Ravindran et al. (eds.), *Macro, Micro and Nanocomposites from Sustainable Sources*, Advanced Structured Materials 244, https://doi.org/10.1007/978-981-95-2469-3_19

including machine learning (artificial intelligence), for predicting the properties of sustainable polymeric materials (Han et al. in Constr Build Mater 244, 2020).

19.1 Introduction

Materials technology has evolved significantly since prehistoric times, from ancient ceramics to modern synthetic polymers, and more recent composite materials. However, their production and disposal contribute to ecological degradation, greenhouse gas emissions, habitat loss, pollution, biodiversity decline and depletion of resources. Novel material technologies can address these challenges by reducing the cost associated with manufacturing and waste generation, ultimately leading to more sustainable alternatives. Using computational tools is a crucial approach in developing sustainable materials considering material systems holistically across hierarchical scales, from nanoscale interactions to macroscale architectures (Inamuddin et al. 2019). The main advantage of such theoretical models enables a precise design of the required molecules by fine-tuning the underlining forces responsible for the required application. This green approach encourages the synthesis of the required system only, limiting the usage of chemicals. Accordingly, several applications have been highlighted in the literature about using these methods to tackle obstacles presented by sustainable material design. Quantum computational chemistry and molecular mechanics/dynamics are the two broad computational methodologies employed for theoretical studies. When the former deals with the wave function of electron, the latter uses Newtonian physics for determining the energy of the system. Quantum studies work well with small systems while polymers, DNA, RNA, proteins and similar large systems requires molecular mechanics (Scheme 19.1).

Quantum studies can be further divided in to may categories including ab initio method, semi empirical method, density functional theory method while molecular mechanics can be divided in to molecular dynamic simulations, Monte Carlo simulations, coarse grain simulations, etc. Recent studies also include a combination of the above two popularly known as QM-MM method. The above two methods and their combined studies falls under the broad category multiscale modelling. Both the methodologies have been used effectively to determine and predict molecular properties of polymer systems. Apart from the above methods, machine learning has revolutionized the theoretical prediction of molecular properties. Even though such studies for sustainable polymers are seldom reported, our book chapter briefly covers the methodology applicability of these methodologies which can be extended to the study of such systems.

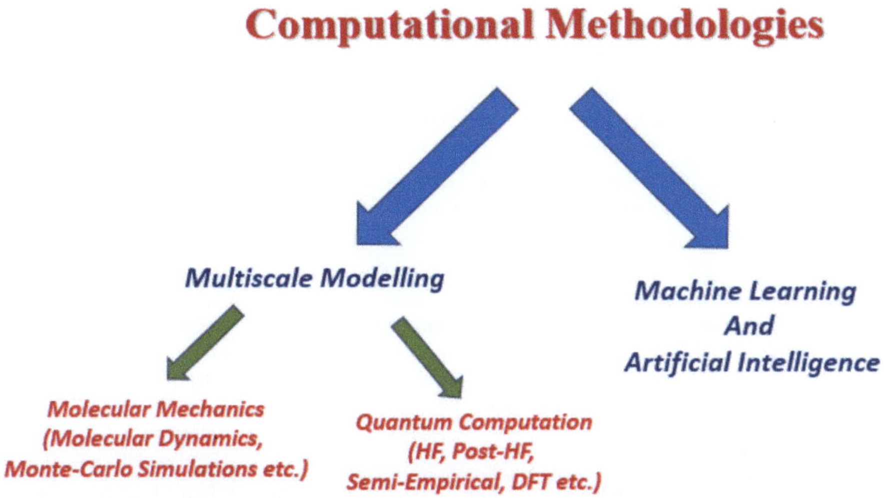

Scheme 19.1 Computational methodologies in general

19.2 Computational Methodologies

Materials technology has evolved significantly since prehistoric times, from ancient ceramics to modern synthetic polymers, and more recent composite materials. However, their production and disposal contribute to ecological degradation, greenhouse gas emissions, habitat loss, pollution, biodiversity decline and depletion of resources. Novel material technologies can address these challenges by reducing the energy costs associated with manufacturing and waste generation, ultimately leading to more sustainable alternatives. Using computational tools and materiomics is a crucial approach in developing sustainable materials considering material systems holistically across hierarchical scales, from nanoscale interactions to macroscale architectures. Accordingly, several applications have been highlighted in the literature about using these methods to tackle obstacles presented by sustainable material design. The various methodologies are as follows.

19.2.1 Multiscale Modelling

The fascinating world of multiscale mechanics can help engineer hierarchical materials with structures and properties similar to those of many biological systems. These materials can self-assemble, self-repair, adapt, evolve, and provide multiple functions controlled by external factors (Shandler et al. 2011). This involves both quantum mechanics and molecular mechanics. A general comparison of the various methods is given in the Table 19.1.

Table 19.1 Comparison of the various methods (Ramachandran et al. 2008)

Ab-initio	Semi empirical	Molecular mechanics	DFT
Counting all electrons	Ignoring some electrons (simplification)	Ignoring all electrons Only nuclei are taken into consideration	Considering only electron density
Limited to tens of atoms and best performance using a supercomputer	Limited to hundreds of atoms	Molecules containing thousands of atoms	Molecules containing more than hundred atoms
Can be applied to inorganics, organics, organo-metallics and molecular fragments (the catalytic components of an enzyme)	Can be applied to inorganics, organics, organo-metallics and small oligomers (peptide, nucleotide, saccharide)	Can be applied to inorganics, organics, oligonucleotides, peptides, saccharides, metallo-organics and inorganics	Can be applied to inorganics, organics, oligonucleotides, peptides, saccharides, metallo-organics and inorganics
Extended to a vacuum or implicit solvent environment	Extended to a vacuum or implicit solvent environment	Extended to a vacuum or implicit solvent environment	Extended to a vacuum or implicit solvent environment
Applicable to ground, transition, and excited states	Applicable to ground, transition, and excited states	Applicable to the ground state only. Thermodynamics and kinetics via molecular dynamics properties	Applicable to ground state. Is being modified to the excited state

19.2.1.1 Quantum Computation

Quantum methods are the first to be discussed in this category as these focuses on microscopic parts of the macromolecular system. Our group recently reported the use of an ionic system viz trimethylphenylammonium tribromide (TMPTB) with graphene (G) in improving the mechanical properties of unsaturated polyester resin (UPR) composites. A detailed computational study, a first of its kind, was carried out to determine the reason for increase in the tensile strength up to a value of 45% (Roy et al. 2025). DFT method threw light on the above observation, revealing the fact that a trimeric interaction between the ionic system, graphene and UPR was responsible for this exceptional increase. The interactions between the same was determined to be secondary forces in nature, further confirmed using QTAIM technique (quantum transmission for atoms in molecules method is one which defines bonding using topological analysis of electron density) which revealed the Laplacian of electron density at bond critical points falls under the category of secondary forces including hydrogen bonding, van der Waals forces, etc. The structure predicted by the theoretical calculation was observed in TEM analysis, ascertaining the importance and accuracy of computational techniques in deciphering interactions (Roy et al. 2025) (Fig. 19.1).

The same DFT method was used for sensing applications of chitin and chitosan in another work (Jeremić et al. 2018). When the above paper used B3LYP method with

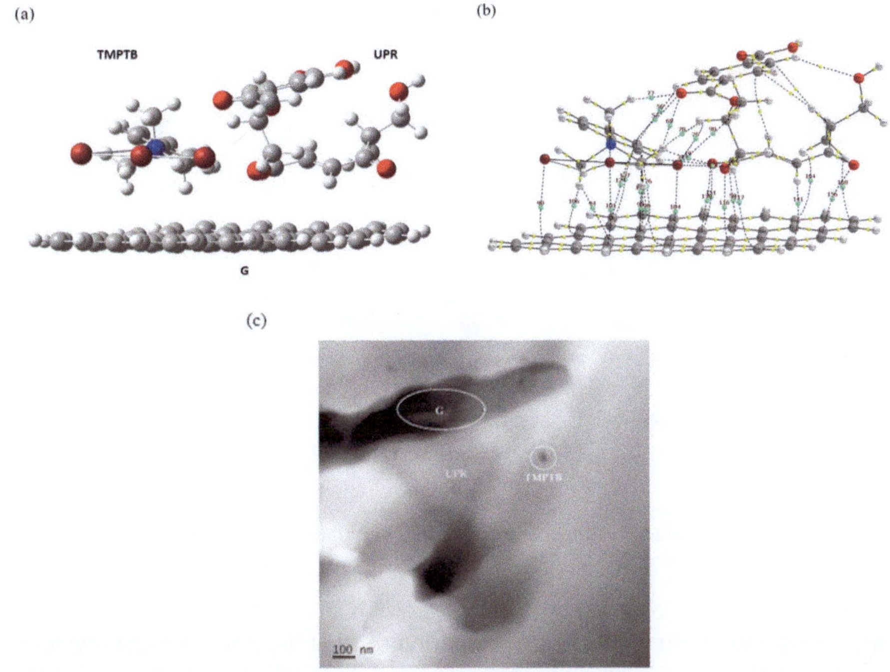

Fig. 19.1 **a** Geometry optimized computational prediction of trimitic interaction structure, **b** QTAIM analysis and **c** TEM image

6-31G(d) basis set for the calculation, the preset work used M06-2X/LanL2DZ level of theory. The input geometry for chitin and chitosan biopolymers were taken to be glucosamine and N-acetylglucosamine, respectively. Along with geometry optimization, a method used to bring the molecular structure to the ground state, the authors carried out enthalpies and Gibbs free energy calculations of the interactions between the biopolymers and metal ions namely Hg(I), Hg(II) Pb(I), and Pb(II). The authors claim that the obtained results may hopefully provide more insight into the metal ions adsorption properties of chitosan and chitin polymers in their environmental applications (Fig. 19.2).

The use of quantum methods is not only used to determine interaction alone. It can also be used to predict polymeric mechanical properties like Young's modulus, stress, strain, etc. An initial study pertaining to the above was carried out using polyethylene as a model compound (Suhai 1986). The tensile strength and microscopic longitudinal elastic modulus were calculated using first principle by the inclusion of perturbation theory to account for electron correlation.

Thomas Elder reported the quantum chemical determination of Young's Modulus of Lignin using first principle (Elder 2007). The study includes comparison of various quantum methods including semi empirical method with AM1 parametrization, ab initio Hartree–Fock method and DFT method using appropriate basis sets. The

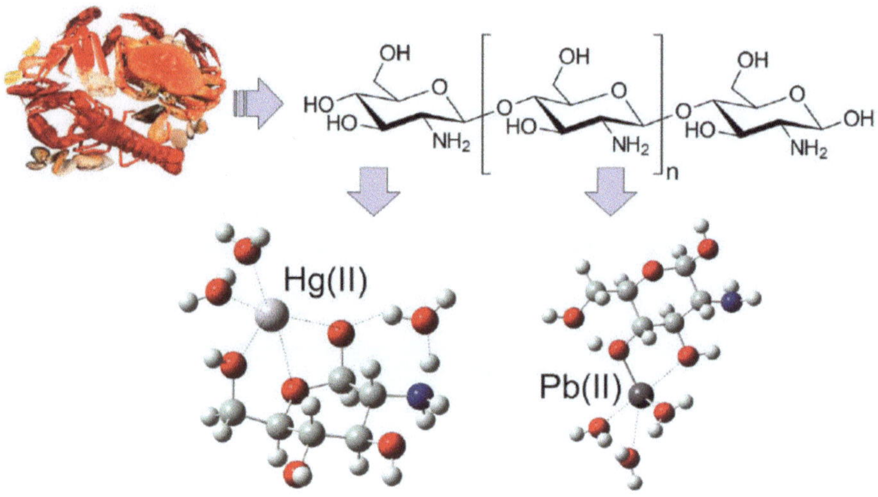

Fig. 19.2 Graphical abstract revealing the interaction of chitin with Hg (II) and Pb (II) (Reproduced from Jeremić et al. 2018)

procedure throws light on the age-old question of what happens when a polymer is stretched and hence is worth mentioning. The initial crystal structure, represented by Fig. 19.3, was geometry optimized using the above methods. The distance between C4 and C1′ representing the ends of the structure, was increased by a total of 5% in 10 increments, in such a way that the dimer is being strained in tension, optimizing all other variables at each step. A detailed comparison of the changes in the molecular descriptors including bond angle, bond lengths, dihedral angle, etc. were monitored and by the knowledge of the change in energy of the molecule and the increment in bond length of the end-to-end carbon atom, the Young's modulus was calculated. The authors report the accuracy of Hartree–Fock method in this regard. Cintrón et al. reported similar observations for cellulose using MM3 (molecular mechanics method) and quantum mechanics (Cintrón et al. 2011). A series of cellulose I_β models were used for the calculations and the authors were successful in deciphering the importance of intramolecular hydrogen bonding in molecular stiffness (for the first time) for disaccharide cellulose. Our group have carried out similar work to determine the mechanical characteristics of ionic liquid composites and ionic liquid graphene composites and were able to reproduce the young's modulus as per the experimental results (Matsumoto et al. 2018).

Molecular modelling methods of semiempirical tight-binding, DFT, and reaction rate calculations were involved in studying the fundamental mechanism involving the increase in thermal stability and mechanical performance of poly(styrene–butadiene–styrene) or SBS-modified asphalt on the addition of methylene diphenyl diisocyanate (MDI)-based additives. It was observed that MDI acts as a compatibilizing agent between asphaltenes and SBS, which enhances the phase stability of the MDI–SBS-modified asphalt (Ting et al. 2021). An extended semiempirical tight-binding

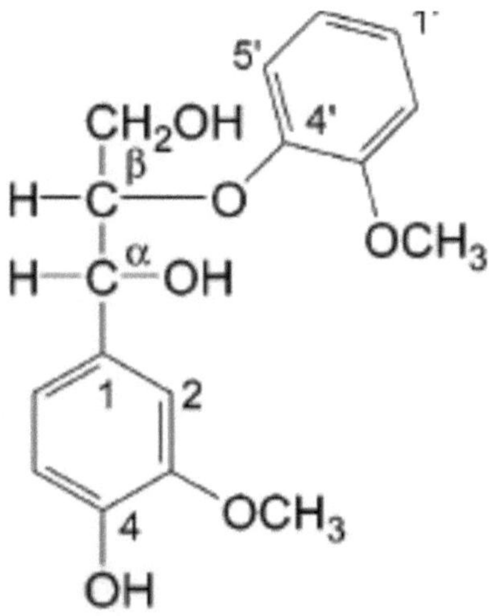

Fig. 19.3 Structure of Lignin used for young's modulus determination (Elder 2007)

quantum chemical method, called the GFN2-xTB method was well-suited for the fast calculation of molecular structures and non-covalent interaction energies, along with multipole electrostatics and density-dependent dispersion contributions. Compared to its predecessor GFN-xTB model, excellent computational performance and properties such as geometries, vibrational frequencies, and non-covalent interactions were mostly targeted. In addition, lower errors were observed for off-target properties such as barrier heights and molecular dipole moments (Bannwarth et al. 2019).

19.2.1.2 Molecular Mechanics

Apart from quantum studies, which are seldom reported in this regard, molecular mechanics has been extensively used to explore the properties of composites. Molecular methos like *molecular dynamics* method in which follows the trajectories of atoms and molecules over time, based on Newtonian mechanics. The evolution of the system is usually considered using 'ensembles', the selection of which depends on the kind of system and conditions one requires in accordance with his system. *Monte Carlo Simulations*, on the other hand, is a probabilistic method using random sampling to study the phase space of a system and determining its equilibrium states. *Coarse Grain model* is a more simplified form in which the atoms are grouped in to larger beads (particles); with the disadvantage of reduction in degrees of freedom, but with an advantage of simulation of bigger systems and larger simulation time (Leach 2001).

Frankland and coworkers reported the stress–strain behaviour of polymer–nanotube composites a single-walled carbon nanotube embedded in polyethylene from molecular dynamics simulation (Frankland et al. 2003). A periodic three-dimensional cell with nano tubes embedded inside the polymer. As shown in Fig. 19.4, the strain is applied in two directions ε11 and ε22, wherein, the volume of the box enclosing the polymeric matrix is increased in a regular manner and the system is allowed to attain equilibrium after each increase. The corresponding stress tensor is calculated as per the equation

$$\sigma = (\partial E / \partial e) / V$$

where σ is the stress tensor, E is the energy of the system, ε is the strain tensor, and V represents the total volume. They could effectively establish the fact that this composite showed higher stress in transverse direction compared to longitudinal direction. A similar work was reported by Zhu et al. for determining the influence of carbon nanotubes on the mechanical properties of Epon 862 composites (Zhu et al. 2007). Three systems, namely the Epon 862 matrix alone, a long and short CNT reinforced Epon composite were studied using the molecular dynamics simulations. Using the Epon 862 molecules and setting the Lennard–Jones potential for non-bonding interaction, an amorphous Epon 862 matrix with the size of 40.28 $A^{\circ 3}$ × 40.28 $A^{\circ 3}$ × 61.09 $A^{\circ 3}$ and density of 1 g/cm^3 was constructed at room temperature. NVT ensemble was then applied to the matrix and the molecules were allowed to minimize at the given conditions for 1 fs time frame. This was taken as the reference structure. to determine the stress–strain relations, procedure similar to the above-mentioned work was repeated by making a total of five increments with 2% increase in the dimensions of the box at each step, affirming uniform application of strain at each step. This was followed by the minimization atoms, i.e., the entries system was allowed to equilibrate within the new cell dimensions at the same ensemble NVT. The entire procedure was repeated after incorporating CNT within the MD box (Fig. 19.5). The authors could predict the Young's moduli for these materials effectively using the above method.

Besides determination of mechanical properties of polymers, Wen-Jong MA and Chin-Kun HU could effectively use molecular dynamic simulation for the determination of aggregation in polymeric systems with monomers connected by rigid bonds (Ma and Hu 2010). The studies were carried out in the environment with and without Lennard–Jones (L-J) fluid consisting of L-J molecules. To ensure zero angle

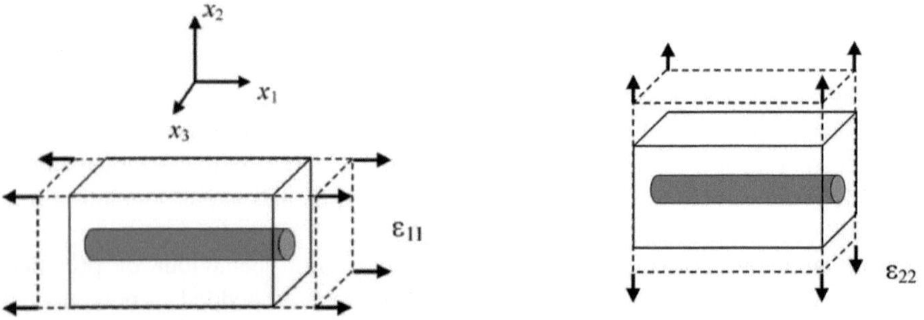

Fig. 19.4 Strains applied to the composites (Frankland et al. 2003)

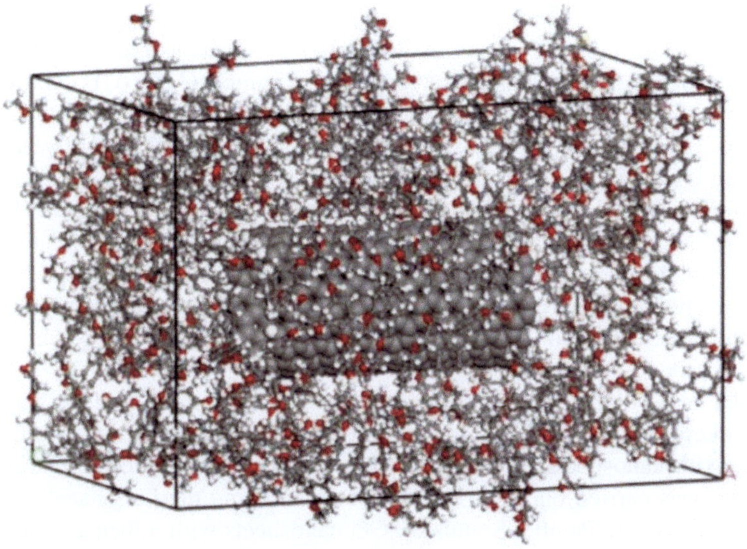

Fig. 19.5 Epon 862 molecules with CNT in MD box (Zhu et al. 2007)

bending phenomenon, the neighbouring monomer molecules were connected using rigid bonds. The authors also introduced fluid attractive linkers to study the aggregation. Factors like chain length of the polymer and effect of angle potentials were also explored in the above study. The snapshots of the various stages of polymer aggregation were recorded (Fig. 19.6). The work summarizes the fact that geometry constrains in the molecules regulate the driving forces required for the aggregation.

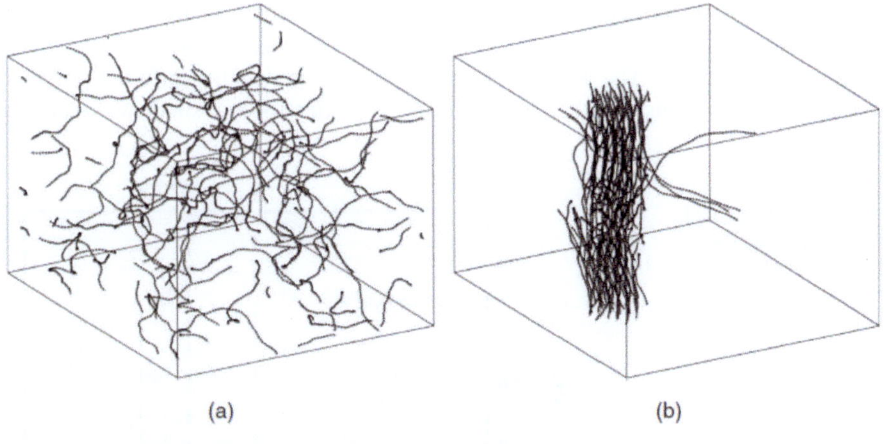

Fig. 19.6 Snapshots before (**a**) and after (**b**) the formation of aggregation (Ma and Hu 2010)

Similar to this work, Huan Tang and coworkers reported the aggregation process of graphene oxide in water (Tang et al. 2015). The authors constructed four different graphene oxide models, varying the oxygen content namely $C_{10}O_1(OH)_1(COOH)_{0.5}$, $C_{30}O_1(OH)_1(COOH)_{0.5}$, $C_{10}O_1(COOH)_{0.5}$, and $C_{10}O_1(OH)_1$ (termed OGO, RGO, GO-COOH, and GO-OH, respectively). It was found that RGO had a faster aggregation than OGO, and GO-OH aggregated easily than GO-COOH. The rate of the aggregation was also calculated and it was determined that hydrogen bonding, π–π stacking, electrostatic interaction, van der Waals attraction and collision of the solvent molecules played pivotal role in aggregation. Since graphene oxide serves as one of the potential ingredients in composites, the results of the work could throw light in the effective use of this material as a modifier; by replacing water with the polymer matrix.

Another important study featuring graphene oxide was reported by Xu Zhang and coworkers related to the determination of mechanical properties of multilayer graphene oxide nanosheets (Zhang et al. 2017). Molecular Dynamics simulation (MD simulation) was carried out with ReaxFF potential using the Large-scale Atomic/Molecular Massively Parallel Simulator. GO nanosheets with dimension of 4.7 nm $\times$ 4.7 nm and varying thicknesses, were chosen as the input geometry. The hydroxyl and epoxy functional groups randomly and uniformly distributed on the both sides of the basal plane of GO, based on the Lerf–Klinowski model with the ratio of epoxy and hydroxyl groups was set to be 1: 1 (Fig. 19.7).

From the study, young's modulus, fracture stress, and the fracture strain were found to be consistent with the experimental values of 175–425 GPa, 22–41 GPa, and 10–16%, respectively. The young's modulus was found to decrease with the increasing thickness of graphene oxide layer and the properties were found to degrade with increasing oxidation content. The results also provided mechanism of multilayer graphene oxide sheet fracture and hence the above work can be utilized to exploit graphene oxide nanosheets as structure reinforcing materials by tuning its structure.

A key study related to the effect of network architecture on mechanical properties of dynamically cross-linked polymer materials was performed by Zanjani and coworkers (2019). The work uses coarse grained models for different polymer network configurations and their mechanical properties using molecular dynamics

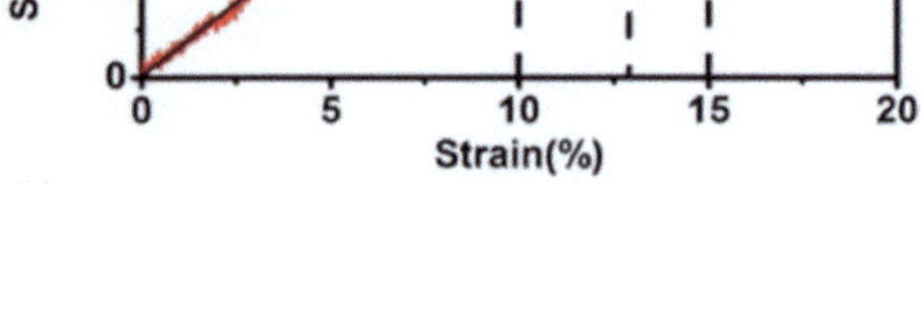

Fig. 19.7 Stress–strain curve of graphene oxide nanosheets with Young's Modulus fitted (Zhang et al. 2017)

(MD) simulations are predicted. Two kinds of network polymers, viz. IPN (interpenetrating networks) and SN (single networks), both based on hydroxy ethyl acrylate were used for the study. Both the systems with covalent cross-linker (such as furan-maleimide diacylate) and non-covalent cross-linker (2-ureido-4[1H]-pyrimidinone acrylate) were taken. MD simulations were carried using LAAMPS software. It was observed that the theoretical studies were consistent with the experimental values. The IPN could withstand stress higher stress compared to SN as expected. Additionally, the MD results demonstrate that the origin of this difference in mechanical behaviour between the two goes back to the difference in spatial degrees of freedom of the non-covalent cross-linkers, represented by non-bonded interactions within the two system types. The results of this work motivate future studies on the draft new systems with dual cross-linkers where one linkage exchanges rapidly providing dynamic character, while the other is a stimulus responsive dynamic covalent linkage that provides stability with dynamic exchange on-demand (Fig. 19.8).

Jeffrey M. Sanders et al. reported the use of MD simulations to measure the moisture uptake and plasticization effects of water on amorphous amylose starch model (Sanders et al. 2021). A new force field, namely OPLS3e was used for the present study to perform metadynamics calculations to create torsional free energy maps. The rate of water transportation and uptake was calculated mathematically and the activation energies of water uptake was calculated. The prediction of glass transition temperature pointed directly to the plasticization effect due to moisture content.

A very important application of polymeric sustainable composites is its use for drug delivery systems which was explored by Maleki and coworkers (2020). The

Fig. 19.8 Schematic demonstration of polymer networks with covalent and non-covalent of cross-linkers (Zanjani et al. 2019)

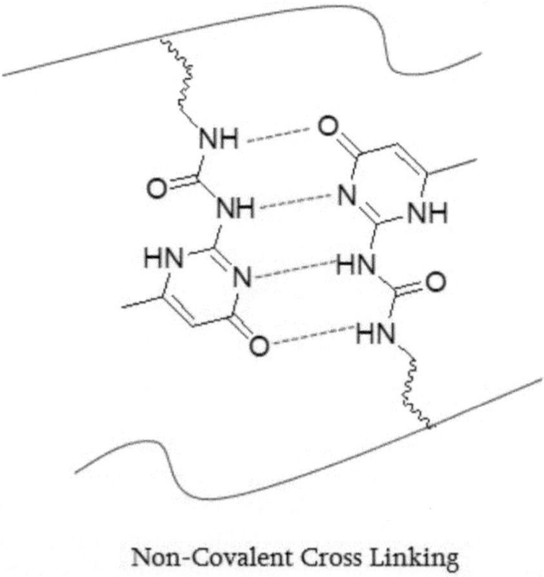

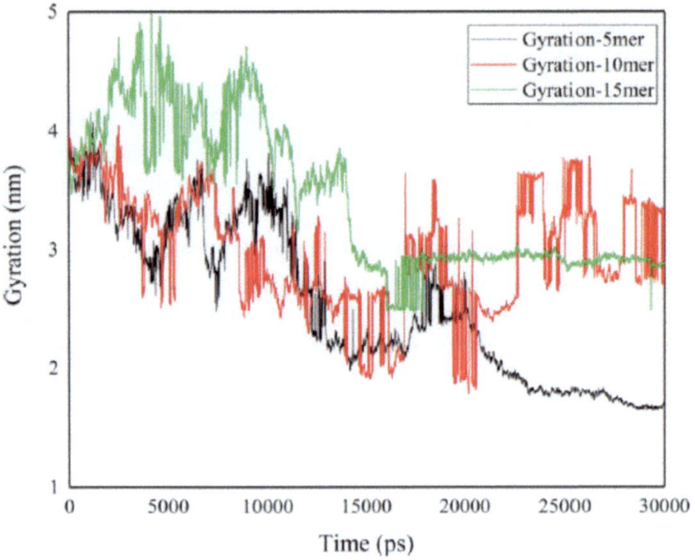

Fig. 19.9 Gyration radius of NIP (Maleki et al. 2020)

authors performed MD simulation of doxorubicin loading with N-isopropyl acrylamide carbon nanotube for drug delivery application. Three systems namely, NIP (polymer with various chain length), DOX (drug) and CNT, were modelled and MD simulations were performed using gromacs software. Radius of gyration, hydrogen bonds, van der Waals forces and electrostatic energies were monitored in the study. It was observed that shorter the polymer chain, more decrease of gyration radius was observed resulting in better interaction with drug and its delivery. The results also indicate that CNT and its derivatives can effectively act as plausible drug delivery systems (Fig. 19.9).

Nanoparticles are another kind of modifiers normally used in sustainable polymeric composites. The influence of nanoparticle-polymer interaction strength, nanoparticle volume fraction, features of nanoparticles, cross-linking density, temperature have been studies using the methods mentioned above ascertaining the importance of MD simulations in deciphering the mechanical properties of composites (Maleki et al. 2020).

Molecular mechanics (MM) has also been used to model small crystals of cellulose Iβ surrounded by water to find that the structured water layers around cellulose may pose a barrier to cellulase enzymes, leading to slower hydrolysis rates as observed experimentally (Matthews et al. 2006). Computational studies have been employed in biomedical applications including the binding interactions between breast cancer cells with high levels of cyclic arginylglycylaspartic acid (RGD) peptidomimetics and $\alpha_5\beta_1$ integrin (Guzzetti et al. 2017), a β-peptide targeting the transmembrane (TM) helices (Shandler et al. 2011), and the binding mechanism of RGD- and AGDV-peptides to the platelet integrin αIIbβ3 involved in platelet aggregation (Kononova et al. 2017). Such methods has also proven to be a powerful platform for predicting the mechanical properties and biological functions of silk (López Barreiro et al. 2019a). MD simulations successfully reveal that β-sheet nanocrystals confined to a few nanometres could achieve higher stiffness, strength and mechanical toughness than larger nanocrystals. It was found that nanoconfinement, which makes the most efficient use of hydrogen bonds, and the dissipative molecular stick–slip deformation significantly enhanced the mechanical properties of silk (Keten et al. 2010). In a study, conductive silk-based composites were developed from biobased carbon materials with MD modelling (López Barreiro et al. 2019b).

Understanding biomass interactions is critical for the bottom-up design of novel biocomposites and existing manufacturing processes (Zhou et al. 2022). The study of the structure and properties of plant biomass can help develop new hierarchical materials for the future (Zhou et al. 2021). A simulation framework determining the reaction kinetics and transport phenomena was developed to understand the solvolysis process of lignin, hemicellulose and cellulose involving whole biomass feedstocks, which is useful in emerging biorefinery strategies (Thornburg et al. 2020). In a similar study, a simple coarse grain model was employed to study the mesoscale mechanics of wood cell walls (Fig. 19.10) under axial strain, providing insights into fundamental structure–property relationships of wood. It was observed that the stick–slip deforming motion of hemicellulose on the application of fibril shear led to the softening of wood (Adler and Buehler 2013).

Cellulose pyrolysis was investigated with large-scale models by GPU-based ReaxFF MD simulation program (GMD-Reax). The reaction analysis tool (VARxMD) was employed to investigate the initial reaction mechanism of cellulose pyrolysis. The simulated evolution of the major pyrolysis products (glycolaldehyde, levoglucosan and water) agrees well with the Py-GC/MS experimental observations at a very close temperature range. The study found cellulose to be a good model fit

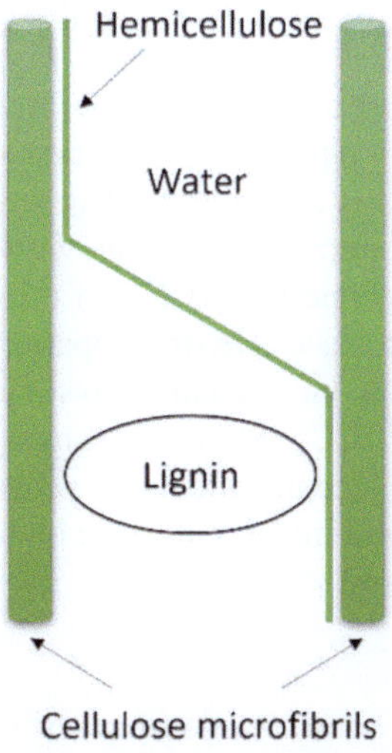

Fig. 19.10 Structure of cell wood (Adler and Buehler 2013)

to validate the ReaxFF force field in predicting the pyrolysis behaviour of complex molecular systems (Zheng et al. 2016).

The interactions between bioadhesives and fibres were elucidated via experimental bonding strength tests and MD simulations, revealing a nearly linear tensile strength-to-density correlation as a result of the hydrogen bonding content in the composite. In another study, the environmental-friendly scale and corrosion inhibition performance of polyaspartic acid (PASP), polyepoxysuccinic acid (PESA), oxidized starch (OS), and carboxymethyl cellulose (CMC) were investigated by molecular dynamics (MD) simulation and density functional theory (DFT) calculations using the COMPASS force field in Materials Studio 2018 (MS) program (Chen et al. 2021).

Multiscale modelling was also involved in the biomass pretreatment for biofuel production. A combination of three models including the microscale level (dealing with the chemical reaction), mesoscale level (dealing with diffusion in lignocellulosic material) and macroscale level (dealing with mixing and bulk diffusion) showed a high correlation with the experimental studies (Hosseini and Shah 2009). Designing and fabrication of new self-healing systems such as 'artificial leukocytes', polymer nanocomposites and polymer gels has also been achieved using such methods (Balazs 2007). A combination of computational and experimental approaches involved studying the impact of mechanical forces on the self-healing ability of dynamically cross-linked polymer systems. MD simulation results highlight the mechanism involved in non-covalent and dynamic covalent linkage responses to the external loading (Alwis Watuthanthrige et al. 2020).

Another important application is the investigation of the thermal properties of natural fibre (kenaf) reinforced composite materials. The model results agree favourably with the experimentally observed data. The method provides insights into the contribution of individual components to the thermal properties of the kenaf fibre (Guillou et al. 2018).

A hierarchical Bell model, providing a rigorous structure–property relationship was used to study the unfolding behaviour of α-helical (AH) and β-sheet (BS) domains of proteins (Fig. 19.11). It was observed that 3–4 parallel hydrogen bonds per turn favoured the protein's mechanical and thermodynamical stability, which was in agreement with the experimental findings (Ackbarow et al. 2007). The method was also able to give insights into the process of biomineralization including the interaction of silica with cellular membranes (Delle Piane et al. 2018), the role of citrate ions (CIT) in the regulation of polyaspartic acid (PASP) adsorption on the surface of hydroxyapatite (HAP) in the calcium phosphate solution (Zeng et al. 2021), the mechanism of lysozyme-mediated nucleation of amorphous calcium carbonate (ACC) (Rani and Saharay 2019), the aggregation and nucleation of calcium phosphate in bone tissue generation (Wang et al. 2021), study of 3D skeletal architecture and imaging in coral biomineralization (Scucchia et al. 2022; Reznikov et al. 2020), and so on.

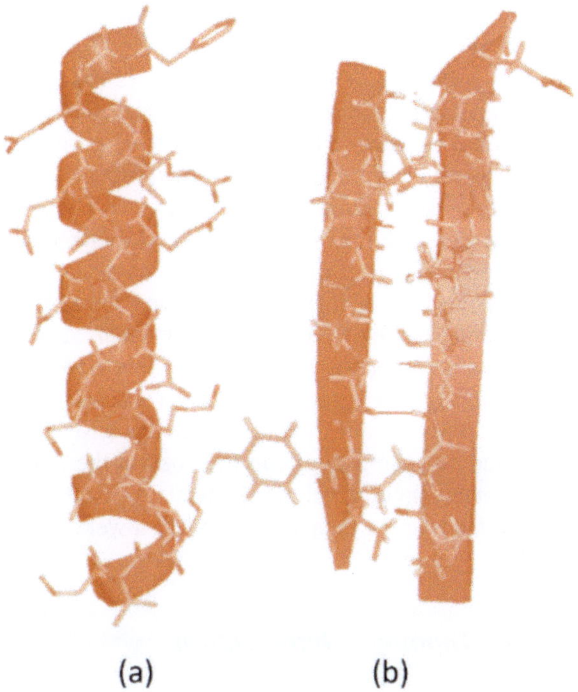

Fig. 19.11 **a** α-helical (AH) and **b** β-sheet (BS) domains of proteins (Ackbarow et al. 2007)

19.2.2 Machine Learning

An ML-based material design approach relies on generating a large dataset, expert preprocessing, and careful model selection to achieve optimal results. Recent advances in machine learning and deep learning (DL) algorithms can overcome old mechanics problems and unlock novel material design approaches (Guo et al. 2021). ML in computational chemistry has successfully established a statistical relationship between the chemical structure and potential energy without known facts about chemical bonding and other relevant interactions. One of its promising applications includes ML-based force fields (FFs) which can contribute to a better understanding of fundamental quantum chemical properties of systems and aim to bridge the gap between the accuracy of ab initio methods and the efficiency of classical FFs (Unke et al. 2021).

ML-based genetic algorithm and a deep learning framework were used to generate collagen sequences within specific melting temperature (Tm) values, which were later verified by experimental and computational methods. The model accurately predicts Tm values and the number of hydrogen bonds within collagen calculated with the molecular dynamics (MD) method was directly correlated to the experimental value (Khare et al. 2022).

In a study of polyhydroxyalkanoate (PHA)-based polymers, predictive modelling of their glass transition temperatures demonstrated rapid and accurate structure–property mappings of new PHA polymers based on properties such as topology, shape, and charge/polarity of specific chemical units forming the polymer backbone, which could be extended to a range of other properties (Pilania et al. 2019).

Tree-based classification models involving a series of 'if–then' rules were used to classify plastic wastes based on their metal contamination data to derive a general categorization rule for different plastic polymers. This ML-based technique could help understand plastic waste recycling and management practices (Chin et al. 2021). ML tools may be used in peptide-mediated biomineralization processes, an emerging technique in functional nanomaterial synthesis such as efficient catalysts, plasmonic materials and bimetallic nanoparticles. In a study, an enhanced hyperbox classifier was developed which predicts whether a given peptide sequence has a strong or weak binding affinity towards a gold surface (Janairo et al. 2020). The application of ML in the study of composites can help obtain the optimal design of such composites and accurately predict their mechanical properties, including toughness and strength, with limited training data and at a much lower computational cost (Gu et al. 2018).

19.2.3 Topology Optimization

The advancement in 3D printing technology has led to the development of topology optimization, a mathematical tool to spatially optimize the distribution of material within a defined domain (Rosinha et al. 2015). A study developed two-dimensional finite-element models of pure nickel with bimodal grain size distributions, featuring random and harmonic spatial arrangements of coarse and ultrafine-grain areas. While the bimodal random-structure material exhibited stress–strain heterogeneities at all scales, leading to premature global plastic instability; the bimodal harmonic-structure material demonstrated exceptional strength and ductility, surpassing expectations from the rule of mixtures (Shokry et al. 2021).

19.2.4 Neural Networks and Ologs

Neural machine translation is another modern approach to translating languages. In contrast to traditional statistical methods, it aims to build a unified neural network optimizing translation performance collectively. Recent translation models often employ an encoder-decoder architecture, where the encoder converts a source sentence into a fixed-length vector, and the decoder utilizes the vector to generate the translated output (Bahdanau 2014). The leading models for sequence transduction rely mainly on complex recurrent neural networks arranged in an encoder-decoder framework. A new streamlined architecture called the Transformer, entirely based on attention mechanisms, was found to eliminate the need for this recurrence, deliver superior translation quality, and be significantly quicker to train (Vaswani et al. 2017). A deep-learning-based inverse design model was constituted with a deep encoder–decoder architecture for targeted molecular design, to reconstruct new molecular structures with target properties. The model successfully generates light-absorbing molecules and host materials for phosphorescent organic light-emitting diodes (Kim et al. 2018).

Studies involving predicting stress, strain fields or fracture patterns could be easily accomplished at a lower cost using cycle-consistent adversarial and transformer neural networks compared to basic physics-based simulation methods such as molecular dynamics modelling. The process was also capable of forward and backward predictions, an approach closely related to inverse material design (Buehler and Buehler 2022). Field Perceiver, a general attention-based deep neural network model, was capable of effectively predicting physical field data and also involved in capturing size, orientation and geometry effects of crack problems for near- and far-field predictions (Buehler 2022).

Another study used a combined vector-quantized generative adversarial network and contrastive language-image pre-training neural networks to generate images. These were later translated into 3D architectures and 3D printed using fused deposition modelling into materials. These were further analysed using molecular dynamics modelling to understand their behaviour under extreme compressive loading (Hsu et al. 2022). Designing de novo protein systems was accomplished using ologs by drawing an analogy between an olog for the protein and an olog for a social network to describe their structural and functional properties (Spivak et al. 2011). In another interesting simulation approach, the foundation of classical music was extrapolated through a recursive method and translated into de novo protein folded into a 3D structure using deep learning. The resulting protein could be used as a basis for musical compositions (protein music). This integration of science and art can create new bioinspired materials (Wong et al. 2012; Milazzo et al. 2021) (Fig. 19.12).

19.3 Future Prospects

The above studies indicate the importance of computational techniques (including AI) serving as the key to the future of science and technology. Quantum studies, at the atomic level can be used effectively to determine the fundamental factors responsible for the observed properties, thereby helping us in fine-tuning the molecular structure to develop systems with desired properties. On the other hand, molecular mechanics can be further used to predict mechanical, thermal, electrical, etc. properties of the composite materials, thereby helping on to rationally pick the composite composition. When AI could serve as a comprehensive data base of molecular properties, machine learning can help enhance further prediction of molecular properties with greater accuracy.

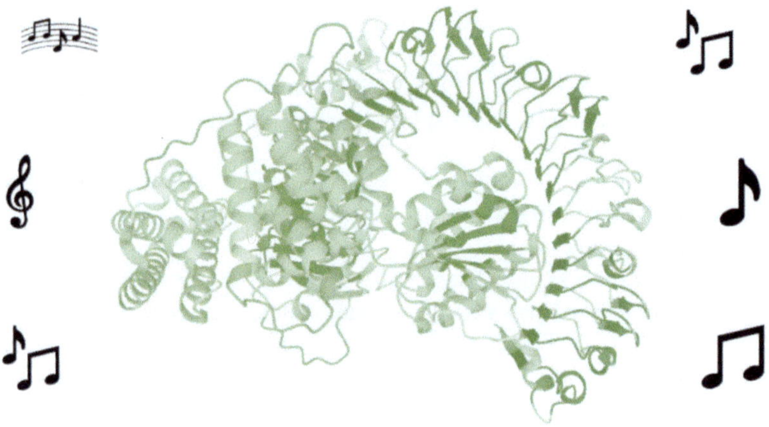

Fig. 19.12 Protein music-3D structure (Wong et al. 2012)

19.4 Conclusion

Computational tools like quantum mechanical studies, MD simulations, AI technology etc. are growing in such a rate that we it becomes impossible for one to ignore their contributions in the field of science and technology. From the above discussions, we can see that these tools have been effectively used to predict the properties pertaining to mechanical, thermal, drug delivery, aggregation and even art. These tools not only delve deep in to the scientific understanding of the system but also can be considered as 'greener', as it helps one to speculate the properties one requires in other words, it helps in accurately predicting the properties even before the system is made, thereby saving time and chemicals. This helps in conserving our resources and manage it in a proper way, leading to sustainability.

References

Ackbarow T, Chen X, Keten S, Buehler MJ (2007) Hierarchies, multiple energy barriers, and robustness govern the fracture mechanics of α-helical and β-sheet protein domains. Proc Natl Acad Sci 104(42):16410–16415

Adler DC, Buehler MJ (2013) Mesoscale mechanics of wood cell walls under axial strain. Soft Matter 9(29):7138–7144

Bahdanau D (2014) Neural machine translation by jointly learning to align and translate. arXiv: 1409.0473

Balazs AC (2007) Modeling self-healing materials. Mater Today 10(9):18–23

Bannwarth C, Ehlert S, Grimme S (2019) GFN2-xTB—an accurate and broadly parametrized self-consistent tight-binding quantum chemical method with multipole electrostatics and density-dependent dispersion contributions. J Chem Theory Comput 15(3):1652–1671

Buehler MJ (2022) FieldPerceiver: domain agnostic transformer model to predict multiscale physical fields and nonlinear material properties through neural ologs. Mater Today 57:9–25
Buehler EL, Buehler MJ (2022) End-to-end prediction of multimaterial stress fields and fracture patterns using cycle-consistent adversarial and transformer neural networks. Biomed Eng Adv 4:100038
Chen X, Chen Y, Cui J, Li Y, Liang Y, Cao G (2021) Molecular dynamics simulation and DFT calculation of "green" scale and corrosion inhibitor. Comput Mater Sci 188:110229
Chin HH, Varbanov PS, Klemeš JJ (2021) Plastic waste categorisation using machine learning methods-metals contaminations. In: 2021 6th international conference on smart and sustainable technologies (SpliTech), pp 1–6
Cintrón MS, Johnson GP, French AD (2011) Young's modulus calculations for cellulose Iβ by MM3 and quantum mechanics. Cellulose 18(3):505–516
De Alwis Watuthanthrige N, Ahammed B, Dolan MT, Fang Q, Wu J, Sparks JL, Zanjani MB, Konkolewicz D, Ye Z (2020) Accelerating dynamic exchange and self-healing using mechanical forces in crosslinked polymers. Mater Horiz 7(6):1581–1587
Delle Piane M, Potthoff S, Brinker CJ, Colombi Ciacchi L (2018) Molecular dynamics simulations of the silica-cell membrane interaction: insights on biomineralization and nanotoxicity. J Phys Chem C 122(37):21330–21343
Elder T (2007) Quantum chemical determination of Young's modulus of lignin. Calculations on a β-0–4′ model compound. Biomacromolecules 8(11):3619–3627
Frankland SJV, Harik VM, Odegard GM, Brenner DW, Gates TS (2003) The stress-strain behavior of polymer-nanotube composites from molecular dynamics simulation. Compos Sci Technol 63(11):1655–1661
Gu GX, Chen C-T, De BMJ (2018) Novo composite design based on machine learning algorithm. Extreme Mech Lett 18:19–28
Guillou J, Lavadiya DN, Munro T, Fronk T, Ban H (2018) From Lignocellulose to biocomposite: multi-level modelling and experimental investigation of the thermal properties of Kenaf fiber reinforced composites based on constituent materials. Appl Therm Eng 128:1372–1381
Guo K, Yang Z, Yu C-H, Buehler MJ (2021) Artificial intelligence and machine learning in design of mechanical materials. Mater Horiz 8(4):1153–1172
Guzzetti I, Civera M, Vasile F, Arosio D, Tringali C, Piarulli U, Gennari C, Pignataro L, Belvisi L, Potenza D (2017) Insights into the binding of cyclic RGD peptidomimetics to A5β1 integrin by using live-cell NMR and computational studies. ChemistryOpen 6(1):128–136
Han T, Siddique A, Khayat K, Huang J, Kumar A (2020) An ensemble machine learning approach for prediction and optimization of modulus of elasticity of recycled aggregate concrete. Constr Build Mater 244:118271
Hosseini SA, Shah N (2009) Multiscale modelling of biomass pretreatment for biofuels production. Chem Eng Res Des 87(9):1251–1260
Hsu Y-C, Yang Z, Buehler MJ (2022) Generative design, manufacturing, and molecular modeling of 3D architected materials based on natural language input. APL Mater 10(4):041107
Inamuddin, Thomas S, Mishra RK, Asiri AM (2019) Sustainable polymer composites and nanocomposites. Springer (2019)
Janairo JIB, Aviso KB, Promentilla MAB, Tan RR (2020) Enhanced hyperbox classifier model for nanomaterial discovery. AI 1(2):299–311
Jeremić S, Tran TH, Marković Z, Ngo TC, Dao DQ (2018) Insight into interaction properties between mercury and lead cations with chitosan and chitin: density functional theory studies. Comput Theor Chem 1138:99–106
Keten S, Xu Z, Ihle B, Buehler MJ (2010) Nanoconfinement controls stiffness, strength and mechanical toughness of β-sheet crystals in silk. Nat Mater 9(4):359–367
Khare E, Yu C-H, Gonzalez Obeso C, Milazzo M, Kaplan DL, Buehler MJ (2022) Discovering design principles of collagen molecular stability using a genetic algorithm, deep learning, and experimental validation. Proc Natl Acad Sci 119(40):e2209524119

Kim K, Kang S, Yoo J, Kwon Y, Nam Y, Lee D, Kim I, Choi Y-S, Jung Y, Kim S, Son W-J, Son J, Lee HS, Kim S, Shin J, Hwang S (2018) Deep-learning-based inverse design model for intelligent discovery of organic molecules. npj Comput Mater 4(1):67

Kononova O, Litvinov RI, Blokhin DS, Klochkov VV, Weisel JW, Bennett JS, Barsegov V (2017) Mechanistic basis for the binding of RGD- and AGDV-peptides to the platelet integrin αIIbβ3. Biochemistry 56(13):1932–1942

Leach AR (2001) Molecular modelling: principles and applications, 2nd edn. Pearson Education Ltd.

López Barreiro D, Yeo J, Tarakanova A, Martin-Martinez FJ, Buehler MJ (2019a) Multiscale modeling of silk and silk-based biomaterials—a review. Macromol Biosci 19(3):1800253

López Barreiro D, Martín-Moldes Z, Yeo J, Shen S, Hawker MJ, Martin-Martinez FJ, Kaplan DL, Buehler MJ (2019b) Conductive silk-based composites using biobased carbon materials. Adv Mater 31(44):1904720

Ma WJ, Hu CK (2010) Molecular dynamics approach to aggregation of polymer chains with monomers connected by rigid bonds. J Phys Soc Jpn 79(5)

Maleki R, Afrouzi HH, Hosseini M, Toghraie D, Rostami S (2020) Molecular dynamics simulation of doxorubicin loading with N-isopropyl acrylamide carbon nanotube in a drug delivery system. Comput Methods Programs Biomed 184

Matsumoto A, Iacob C, Noda T, Urakawa O, Runt J, Inoue T (2018) Introducing large counteranions enhances the elastic modulus of imidazolium-based polymerized ionic liquids. Macromolecules 51(11):4129–4142

Matthews JF, Skopec CE, Mason PE, Zuccato P, Torget RW, Sugiyama J, Himmel ME, Brady JW (2006) Computer simulation studies of microcrystalline cellulose Iβ. Carbohyd Res 341(1):138–152

Milazzo M, Anderson GI, Buehler MJ (2021) Bioinspired translation of classical music into de Novo protein structures using deep learning and molecular modeling. Bioinspir Biomim 17(1):015001

Pilania G, Iverson CN, Lookman T, Marrone BL (2019) Machine-learning-based predictive modeling of glass transition temperatures: a case of polyhydroxyalkanoate homopolymers and copolymers. J Chem Inf Model 59(12):5013–5025

Ramachandran KI, Deepa G, Namboori K (2008) Computational chemistry and molecular modeling: principles and applications. Springer

Rani RS, Saharay M (2019) Molecular dynamics simulation of protein-mediated biomineralization of amorphous calcium carbonate. RSC Adv 9(3):1653–1663

Reznikov N, Buss DJ, Provencher B, McKee MD, Piché N (2020) Deep learning for 3D imaging and image analysis in biomineralization research. J Struct Biol 212(1):107598

Rosinha IP, Gernaey KV, Woodley JM, Krühne U (2015) Topology optimization for biocatalytic microreactor configurations. In: Gernaey KV, Huusom JK, Gani R (eds) Computer aided chemical engineering, vol 37. Elsevier, pp 1463–1468

Roy S, Francis B, Menon S, Anas S, Krishnakumar KS, Krishnan A (2025) Effect of trimethylphenylammonium tribromide/carbonaceous nanosystems on mechanical and thermal properties of unsaturated polyester resin: experimental and computational perspective. Polymer (Guildf) 318:127982–127986

Sanders JM, Misra M, Mustard TJL, Giesen DJ, Zhang T, Shelley J, Halls MD (2021) Characterizing moisture uptake and plasticization effects of water on amorphous amylose starch models using molecular dynamics methods. Carbohydr Polym 252

Scucchia F, Sauer K, Zaslansky P, Mass T (2022) Artificial intelligence as a tool to study the 3D skeletal architecture in newly settled coral recruits: insights into the effects of ocean acidification on coral biomineralization. J Mar Sci Eng 10(3)

Shandler SJ, Korendovych IV, Moore DT, Smith-Dupont KB, Streu CN, Litvinov RI, Billings PC, Gai F, Bennett JS, DeGrado WF (2011) Computational design of a β-peptide that targets transmembrane helices. J Am Chem Soc 133(32):12378–12381

Shen SC, Khare E, Lee NA, Saad MK, Kaplan DL, Buehler MJ (2023) Computational design and manufacturing of sustainable materials through first-principles and materiomics. Chem Rev 123:2242–2275

Shokry A, Ahadi A, Ståhle P, Orlov D (2021) Improvement of structural efficiency in metals by the control of topological arrangements in ultrafine and coarse grains. Sci Rep 11(1):17445

Spivak DI, Giesa T, Wood E, Buehler MJ (2011) Category theoretic analysis of hierarchical protein materials and social networks. PLoS ONE 6(9):e23911

Suhai S (1986) Quantum mechanical calculation of the longitudinal elastic modulus and of deviations from Hooke's law in polyethylene. J Chem Phys 84(9):5071–5076

Tang H, Liu D, Zhao Y, Yang X, Lu J, Cui F (2015) Molecular dynamics study of the aggregation process of graphene oxide in water. J Phys Chem C 119(47):26712–26718

Thornburg NE, Pecha MB, Brandner DG, Reed ML, Vermaas JV, Michener WE, Katahira R, Vinzant TB, Foust TD, Donohoe BS, Román-Leshkov Y, Ciesielski PN, Beckham GT (2020) Mesoscale reaction-diffusion phenomena governing lignin-first biomass fractionation. Chemsuschem 13(17):4495–4509

Ting JH, Khare E, DeBellis A, Orr B, Jourdan JS, Martín-Martínez FJ, Jin K, Malonson BL, Buehler MJ (2021) Role of methylene diphenyl diisocyanate (MDI) additives on SBS-modified asphalt with improved thermal stability and mechanical performance. Energy Fuels 35(21):17629–17641

Unke OT, Chmiela S, Sauceda HE, Gastegger M, Poltavsky I, Schütt KT, Tkatchenko A, Müller K-R (2021) Machine learning force fields. Chem Rev 121(16):10142–10186

Vaswani A, Shazeer N, Parmar N, Uszkoreit J, Jones L, Gomez AN, Kaiser Ł, Polosukhin I (2017) Attention is all you need. In: Guyon I, Luxburg UV, Bengio S, Wallach H, Fergus R, Vishwanathan S, Garnett R (eds) Advances in neural information processing systems, vol 30. Curran Associates, Inc.

Wang W, Gómez-Bombarelli R (2019) Coarse-graining auto-encoders for molecular dynamics. npj Comp Mater 5:1–9

Wang K, Luo F, Wang L, Zhang B, Fan Y, Wang X, Xu D, Zhang X (2021) Biomineralization from the perspective of ion aggregation: calcium phosphate nucleation in the physiological environment. ACS Appl Mater Interfaces 13(41):49519–49534

Wong JY, McDonald J, Taylor-Pinney M, Spivak DI, Kaplan DL, Buehler MJ (2012) Materials by design: merging proteins and music. Nano Today 7(6):488–495

Zanjani MB, Zhang B, Ahammed B, Chamberlin JP, Chakma P, Konkolewicz D, Ye Z (2019) Computational investigation of the effect of network architecture on mechanical properties of dynamically cross-linked polymer materials. Macromol Theory Simul 28(4)

Zeng J, Yang S, Yu H, Xu Z, Quan X, Zhou J (2021) Simulation insight into the synergic role of citrate and polyaspartic peptide in biomineralization. Langmuir 37(11):3410–3419

Zhang X, Liu S, Liu H, Zhang J, Yang X (2017) Molecular dynamics simulation of the mechanical properties of multilayer graphene oxide nanosheets. RSC Adv 7(87):55005–55011

Zhang J, Li C, Wang S, Bai S (2021) Three-dimensional reconstruction of biomineralized sand and particle-flow-code numerical simulation of penetration resistance characteristics of biomineralization crust. Int J Geomech 21:04021048

Zheng M, Wang Z, Li X, Qiao X, Song W, Guo L (2016) Initial reaction mechanisms of cellulose pyrolysis revealed by ReaxFF molecular dynamics. Fuel 177:130–141

Zhou S, Jin K, Buehler MJ (2021) Understanding plant biomass via computational modeling. Adv Mater 33(28):2003206

Zhou S, Khan T, Jin K, Lee J, Buehler MJ (2022) Fundamental investigation of biomass interaction for green composites: experiments and molecular dynamics simulations. Adv Func Mater 32(14):2109881

Zhu R, Pan E, Roy AK (2007) Molecular dynamics study of the stress-strain behavior of carbon-nanotube reinforced Epon 862 composites. Mater Sci Eng A 447(1–2):51–57

Chapter 20
Regulatory and Policy Framework for Sustainable Polymer Composites

Girish Kumar Srivastava

Abstract This chapter analyzes the regulatory and policy mechanisms guiding the sustainable development and life cycle management of polymer composites. Regulatory frameworks (e.g., REACH, RoHS, and TSCA) establish mandatory legal requirements for chemical safety, environmental protection, and human health, while policy frameworks offer strategic guidance to encourage innovation and circularity. International standards and certifications, including ISO 14001, the Global Recycled Standard, and Cradle to Cradle, promote environmentally responsible practices and product transparency. Regional frameworks vary: the EU emphasizes circular economy principles through the Green Deal and Eco-design directives, while countries like Japan and India focus on green procurement and extended producer responsibility (EPR). The chapter highlights tools such as life cycle assessment (LCA), eco-design, and EPR, which are essential for evaluating environmental impact and facilitating sustainable material design. Despite growing support for bio-based and biodegradable composites, manufacturers face challenges related to cost, performance, and certification demands. Health and occupational safety regulations address risks from hazardous substances in polymer composites, with enforcement by agencies such as the EPA, OSHA, and EU-OSHA. Innovation is further supported through government funding, tax incentives, and public–private partnerships that advance material recycling and green technologies. Enforcement remains a challenge in global supply chains, where regulatory fragmentation and traceability limitations hinder compliance. Case studies from Patagonia, Covestro, and Japan illustrate effective regulatory integration. The chapter concludes by emphasizing future needs: stricter circularity standards, digital traceability, and international regulatory harmonization to ensure the sustainability of polymer composites at scale.

G. K. Srivastava (✉)
Departamento de Cirugía, Oftalmología, Otorrinolaringología y Fisioterapia, Facultad de Medicina, Universidad de Valladolid, Valladolid, Spain
e-mail: girishkumar.srivastava@uva.es

Instituto Universitario de Oftalmobiología Aplicada, Facultad de Medicina, Universidad de Valladolid, Valladolid, Spain

L. Ravindran et al. (eds.), *Macro, Micro and Nanocomposites from Sustainable Sources*, Advanced Structured Materials 244, https://doi.org/10.1007/978-981-95-2469-3_20

Fig. 20.1 Symbolic representation of interactions supporting the Regulatory and Policy Framework for Sustainable Polymer Composites, illustrating innovation, sustainability, and scientific advancement

20.1 Introduction to Regulatory and Policy Frameworks

20.1.1 Definition and Importance

Regulatory frameworks refer to the set of rules, laws, and guidelines established by governmental bodies or international organizations to govern the production, use, and disposal of materials, including polymer composites. These regulations ensure that industries operate within safe, ethical, and environmentally responsible boundaries, often setting mandatory standards for safety, environmental protection, and product quality. Policy frameworks, on the other hand, are broader strategic plans and principles set by governments or organizations to guide decision-making and encourage specific behaviors or outcomes, such as promoting research and development, fostering innovation, or incentivizing the use of eco-friendly materials. These frameworks are crucial for ensuring environmental sustainability, promoting safe and responsible use, driving innovation, facilitating market access, ensuring compliance, supporting the circular economy, protecting public health, aligning industry practices with global sustainability goals, encouraging transparency, and preventing environmental degradation. Thus, regulatory and policy frameworks are essential for ensuring that the development and use of polymer composites align with broader sustainability objectives. They provide the necessary structure to guide industries toward environmentally responsible practices, protect public health, and foster innovation in sustainable materials. Without these frameworks, the transition to sustainable polymer composites would be slow, uneven, and potentially harmful to both the environment and society.

20.1.2 Global Need for Sustainable Composites

The global demand for sustainable composites is rising due to increasing awareness of environmental issues, resource depletion, and the harmful effects of traditional materials (Development of sustainable biopolymer-based composites for lightweight applications from agricultural waste biomass: A review 2024; Prasad et al. 2024). Industries are under pressure to reduce carbon footprints, minimize waste, and use resources efficiently. Sustainable composites, made from renewable, recyclable, or biodegradable materials, offer a solution by combining performance with environmental benefits. This shift is driven by stricter regulations, consumer preference for green products, and the need to meet corporate sustainability goals. Regulatory frameworks are vital in this transition, setting standards, incentivizing innovation, and harmonizing global practices, ensuring sustainable composites are widely adopted across industries.

20.2 International Regulations and Standards

20.2.1 Key International Regulations

20.2.1.1 Registration, Evaluation, Authorisation, and Restriction of Chemicals (REACH)

REACH is a European Union regulation designed to protect human health and the environment from chemical risks. It requires manufacturers and importers to register chemical substances used in their products, including polymer composites, by submitting detailed information about their properties to the European Chemicals Agency (ECHA). Authorities then evaluate this information to assess the associated risks. Some substances may require authorization for use, while others may be restricted or banned under this regulation (REACH Regulation–European Commission, https://environment.ec.europa.eu/topics/waste-and-recycling/rohs-directive_en).

20.2.1.2 Restriction of Hazardous Substances (RoHS)

RoHS is an EU directive that restricts the use of specific hazardous materials in electrical and electronic equipment (EEE), promoting safe recycling and disposal. It limits the use of six hazardous substances, including lead and mercury, and applies to EEE that may include polymer composite components. Manufacturers must demonstrate compliance with this directive to market their products in the EU (RoHS Directive–European Commission, https://environment.ec.europa.eu/topics/waste-and-recycling/rohs-directive_en).

20.2.1.3 Waste Electrical and Electronic Equipment Directive (WEEE)

The WEEE Directive aims to reduce electronic waste and improve the environmental performance of electrical and electronic equipment throughout its life cycle. It encompasses a wide range of products, including household appliances and consumer electronics, many of which incorporate polymer composites. Manufacturers are responsible for financing the collection, treatment, recovery, and environmentally sound disposal of WEEE while adhering to specific recycling and recovery targets (Waste from Electrical and Electronic Equipment (WEEE)–European Commission, https://environment.ec.europa.eu/topics/waste-and-recycling/waste-electrical-and-electronic-equipment-weee_en).

20.2.1.4 ISO 14001 (Environmental Management Systems)

ISO 14001 is an internationally recognized standard for establishing an effective environmental management system (EMS) to enhance environmental performance. Organizations can use this framework to identify and manage environmental risks related to their operations, including the production and disposal of polymer composites. The standard encourages continuous improvement in environmental performance and compliance with legal requirements, promoting transparency and engagement with stakeholders (ISO 2024).

20.2.1.5 Toxic Substances Control Act (TSCA)

TSCA is a U.S. law that authorizes the Environmental Protection Agency (EPA) to regulate the introduction of new or existing chemicals to ensure they do not pose unreasonable risks to health or the environment. Manufacturers must submit a pre-manufacture notification (PMN) to the EPA before producing new chemical substances, which may include components used in polymer composites. The EPA evaluates existing chemicals to determine potential risks and can impose restrictions or bans on hazardous substances (O. US EPA 2024a).

20.2.1.6 California Proposition 65

Proposition 65 (also known as the Safe Drinking Water and Toxic Enforcement Act of 1986) requires businesses to provide warnings to Californians about significant exposures to chemicals that can cause cancer, birth defects, or other reproductive harm. The state maintains a list of harmful substances that may include chemicals found in polymer composites. Products containing these chemicals must carry warning labels if they expose consumers to levels above a certain threshold (Admin 2024).

20.2.1.7 EU Ecolabel

The EU Ecolabel is a voluntary certification scheme that promotes products and services with reduced environmental impact throughout their life cycle. This label sets specific environmental criteria for various product categories, including construction materials and consumer goods that may involve polymer composites. Products bearing the Ecolabel can appeal to environmentally conscious consumers and may provide a competitive advantage in the market (EU Ecolabel–Home, https://environment.ec.europa.eu/topics/circular-economy/eu-ecolabel_en).

20.2.1.8 REACH Annex XVII

Annex XVII of the REACH regulation lists substances subject to restrictions due to their significant risks to health and the environment. It includes restrictions on specific hazardous substances used in various applications, impacting manufacturers of polymer composites. Companies must ensure compliance with these restrictions when formulating and marketing their products (Lista de restricciones–ECHA, https://echa.europa.eu/es/substances-restricted-under-reach) (Table 20.1).

Table 20.1 Key international regulations on polymer composites

Regulation	Region	Purpose
Registration, Evaluation, Authorisation, and Restriction of Chemicals (REACH)	European Union (EU)	Protect human health and the environment from chemical risks
Restriction of Hazardous Substances (RoHS)	European Union (EU)	Restrict hazardous materials in electrical and electronic equipment (EEE), promoting safe recycling and disposal
Waste Electrical and Electronic Equipment Directive (WEEE)	European Union (EU)	Reduce electronic waste and improve the environmental performance of electrical and electronic equipment throughout its life cycle
ISO 14001 (Environmental Management Systems)	Global	Establish an effective environmental management system (EMS) to enhance environmental performance and manage operational risks
Toxic Substances Control Act (TSCA)	United States (US)	Regulate the introduction of new or existing chemicals to ensure they do not pose unreasonable risks to health or the environment
California Proposition 65	United States (US)	Protect against exposure to chemicals that cause cancer, birth defects, or other reproductive harm
EU Ecolabel	European Union (EU)	Promote products with reduced environmental impact throughout their life cycle, especially focusing on sustainable production practices
REACH Annex XVII	European Union (EU)	Restrict substances due to their significant risks to human health and the environment

20.2.2 *Global Standards and Certifications*

Global standards and certifications play a crucial role in ensuring the quality, safety, and sustainability of polymer composites. Key international organizations, such as the International Organization for Standardization (ISO) and the American Society for Testing and Materials (ASTM), establish standards that guide manufacturers in the development and evaluation of materials.

20.2.2.1 ISO Standards

ISO develops a wide range of standards that address various aspects of polymer composites, including manufacturing processes, material properties, and environmental impact. For example, ISO 9001 focuses on quality management systems, helping organizations ensure consistent quality in their products. Additionally, ISO 14001 provides a framework for effective environmental management, promoting practices that reduce waste and enhance sustainability throughout the production life cycle (ISO 2024).

20.2.2.2 ASTM Standards

ASTM International is known for its comprehensive testing standards, which are essential for assessing the performance and safety of polymer composites. Standards such as ASTM D638 for tensile properties and ASTM D790 for flexural properties help manufacturers evaluate the mechanical performance of their materials. By adhering to these standards, companies can ensure that their products meet industry benchmarks and regulatory requirements (ASTM International–Standards Worldwide 2024).

20.2.2.3 Certifications Related to Sustainability

In addition to safety and performance standards, various certifications focus on sustainability. For instance, the Global Recycled Standard (GRS) verifies recycled content and sustainable practices in production. Similarly, the Cradle to Cradle certification (Get Certified–Cradle to Cradle Products Innovation Institute 2024) assesses products based on their environmental impact and resource use throughout their life cycle. Achieving these certifications not only enhances marketability but also demonstrates a commitment to sustainability and responsible manufacturing.

By aligning with international standards and obtaining relevant certifications, manufacturers of polymer composites can enhance their credibility, meet regulatory requirements, and contribute to a more sustainable future. These frameworks foster

Table 20.2 Global standards and certifications for polymer composites

Standard/certification	Organization	Focus area
ISO 9001	International Organization for Standardization (ISO)	Quality management systems, ensuring consistent product quality
ISO 14001	International Organization for Standardization (ISO)	Environmental management systems, promoting sustainability and waste reduction throughout production life cycle
ASTM D638	ASTM international	Testing standards for tensile properties, essential for evaluating the performance of polymer composites
ASTM D790	ASTM international	Testing standards for flexural properties, helping assess the mechanical performance of polymer composites
Global Recycled Standard (GRS)	Textile exchange	Certification for verifying recycled content and sustainable practices in production
Cradle to cradle certification	Cradle to cradle products innovation institute	Certification assessing environmental impact and resource use throughout the product life cycle

consumer trust and promote best practices within the industry, ensuring that polymer composites are safe, effective, and environmentally responsible (Table 20.2).

20.3 Regional and National Regulatory Frameworks

Global regulatory frameworks for sustainable materials vary widely by region, shaped by distinct environmental policies, economic priorities, and regulatory approaches.

20.3.1 United States

In the United States, sustainable materials are governed by several federal agencies, most notably the Environmental Protection Agency (EPA) and the Federal Trade Commission (FTC).

20.3.1.1 EPA

It oversees regulations to control pollution and hazardous substances, playing a crucial role in setting standards for materials' environmental impact and waste management. The EPA enforces acts like the Toxic Substances Control Act (TSCA) and the Resource Conservation and Recovery Act (RCRA), which guide the sustainable disposal, recycling, and reuse of materials (O. US EPA 2024b).

20.3.1.2 FTC

It enforces "green" marketing claims under the Green Guides, which ensure that companies provide accurate information on the environmental benefits of their products, thus promoting transparency in sustainability claims (Federal Trade Commission|Protecting America's Consumers 2024).

20.3.2 European Union

The European Union has one of the most comprehensive sustainability regulatory frameworks, with initiatives aimed at achieving climate neutrality by 2050.

20.3.2.1 European Green Deal

It aims to transform the EU into a modern, resource-efficient economy by reducing greenhouse gas emissions, decoupling economic growth from resource use, and fostering a circular economy (The European Green Deal–European Commission 2024).

20.3.2.2 Circular Economy Action Plan

It introduces measures to promote product life cycle sustainability, including design, manufacturing, and disposal. It emphasizes the importance of recyclable materials, waste reduction, and eco-design in creating a circular economy (Circular economy action plan–European Commission 2024).

20.3.3 Asia–Pacific Region

The Asia–Pacific region is home to diverse approaches to sustainability, shaped by rapid industrialization and environmental challenges.

20.3.3.1 China

The Circular Economy Promotion Law and recent amendments to the Environmental Protection Law prioritize sustainable practices in manufacturing and waste reduction, while the China Standards 2035 strategy aims to enhance global competitiveness in green technology (The China Standards 2035 Strategy: Analyzing Recent Developments).

20.3.3.2 Japan

Japan's Act on Promoting Green Procurement mandates the use of eco-friendly products in government procurement, while its Basic Environment Law supports recycling, resource efficiency, and green innovation.

20.3.3.3 India

India's Plastic Waste Management Rules and Extended Producer Responsibility (EPR) guidelines encourage sustainable packaging and recycling practices to reduce plastic waste (Extended Producer Responsibility Guidelines Mandate Sustainable Plastic Packaging To Reduce Plastic Foot Print 2024).

20.3.4 Other Regions

Other regions, including Latin America and Africa, are gradually adopting sustainability-focused regulations.

20.3.4.1 Latin America

Countries like Brazil have introduced frameworks such as the National Solid Waste Policy, which emphasizes recycling and waste management (https://www.wiego.org/sites/default/files/resources/files/Pereira-Brazilian-Waste-Policy.pdf).

20.3.4.2 Africa

Some countries, including Kenya, have implemented strict bans on single-use plastics and promoted sustainable materials to reduce environmental impact (Table 20.3).

Table 20.3 Regional regulatory bodies and key focus areas for sustainable materials

Region	Regulatory body/policy	Key focus
United States	EPA, FTC	Pollution control, transparent green marketing
European Union	European green deal, circular economy plan	Climate neutrality, resource efficiency, circularity
Asia–Pacific	China (Circular Economy Law), Japan (Green Procurement), India (EPR)	Waste management, recycling, green procurement
Other regions	Latin America (Brazil's Waste Policy), Africa (Plastic Bans)	Plastic reduction, recycling, sustainable sourcing

20.4 Sustainability and Environmental Impact Assessments

Sustainability and environmental impact assessments are crucial for understanding the life cycle and long-term ecological effects of materials, including polymer composites. Key approaches such as Life Cycle Assessment (LCA), eco-design, and end-of-life management enable industries to minimize environmental footprints and align with regulatory frameworks that promote sustainable practices.

20.4.1 Life Cycle Assessment (LCA)

Life Cycle Assessment (LCA) is a comprehensive tool used to evaluate the environmental impact of a product throughout its life cycle, from raw material extraction to end-of-life disposal (Life Cycle Assessment, https://www.eea.europa.eu/help/glossary/eea-glossary/life-cycle-assessment).

20.4.1.1 Importance in Polymer Composites

For polymer composites, LCA is essential in assessing resource consumption, emissions, and waste at each stage. This process helps identify high-impact stages and opportunities for improvement in areas such as raw material selection, energy use, and recyclability.

20.4.1.2 Decision-Making

LCA provides data to support sustainable decision-making, helping manufacturers select eco-friendly materials, optimize production, and reduce overall carbon footprint.

20.4.2 *Eco-Design and Circular Economy*

Eco-design and circular economy principles focus on designing products with minimal environmental impact while promoting reuse, recycling, and resource efficiency.

20.4.2.1 Regulatory Support

Many regulatory frameworks, especially in the European Union, encourage or mandate eco-design practices to extend the life cycle of products and reduce waste. For instance, the EU's Circular Economy Action Plan promotes eco-design for greater durability, repairability, and recyclability of materials.

20.4.2.2 Impact on Polymer Composites

By applying eco-design principles, manufacturers can create polymer composites that are easier to disassemble, recycle, and repurpose, supporting a circular economy model and minimizing resource extraction.

20.4.3 *End-of-Life Management*

End-of-life management encompasses regulations and practices for handling products once they reach the end of their functional life. For polymer composites, this stage poses unique challenges due to their material complexity.

20.4.3.1 Recycling and Reuse

Regulations often mandate recycling or reuse programs to reduce landfill waste and conserve resources. Some regions require manufacturers to adopt Extended Producer Responsibility (EPR) programs, which hold them accountable for the end-of-life disposal of their products (https://pib.gov.in/pib.gov.in/Pressreleaseshare.aspx?PRID=1909909; Coalition (SPC) 2024).

20.4.3.2 Disposal

For polymer composites that cannot be recycled, sustainable disposal methods such as energy recovery are encouraged. Regulations, especially in developed

Table 20.4 Key sustainability and environmental impact assessment strategies for polymer composites

Strategy	Key focus	Environmental benefits
Life Cycle Assessment (LCA)	Evaluates environmental impact at each life cycle stage	Identifies high-impact areas and supports sustainable choices
Eco-design	Integrates sustainability into design	Promotes durability, recyclability, and resource efficiency
End-of-life management	Governs recycling, reuse, and disposal	Reduces landfill waste and encourages resource conservation

countries, restrict landfill disposal of certain composites to prevent environmental contamination (Table 20.4).

20.5 Regulations on Renewable and Biodegradable Composites

With growing environmental concerns, regulatory bodies worldwide are promoting the use of renewable and biodegradable composites to reduce reliance on fossil-based resources and minimize waste. However, while these regulations offer numerous environmental benefits, they also present challenges for manufacturers aiming to comply with strict guidelines and market demands.

20.5.1 Encouragement of Biodegradable Composites

Policies around the world are increasingly aimed at fostering the adoption of renewable and biodegradable materials across various industries.

20.5.1.1 Government Incentives and Regulations

Many countries offer financial incentives, tax breaks, and subsidies to manufacturers who incorporate biodegradable materials into their products. For instance, the European Union's Bioeconomy Strategy (https://research-and-innovation.ec.europa.eu/research-area/environment/bioeconomy/bioeconomy-strategy_en) and Circular Economy Action Plan (2024) promote the use of renewable resources and biodegradable materials to achieve sustainability goals.

20.5.1.2 Single-Use Plastics Bans

Various regions, such as the EU and parts of the Asia–Pacific, have implemented bans on certain single-use plastics, pushing industries toward biodegradable alternatives. This shift encourages the development of compostable and bio-based composites for packaging and consumer goods.

20.5.1.3 Green Labeling Programs

Programs like the U.S. Department of Agriculture's (USDA) BioPreferred Program label products (BioPreferred|Catalog, https://www.biopreferred.gov/BioPreferred/faces/catalog/Catalog.xhtml) made from renewable sources, helping consumers make eco-conscious choices and encouraging companies to meet these standards.

20.5.2 Challenges and Compliance

While policies promoting renewable and biodegradable composites are beneficial for the environment, manufacturers face several obstacles in meeting these regulatory demands.

20.5.2.1 Material Costs

Renewable and biodegradable materials often have higher production costs, which can affect profitability. Sourcing sustainable materials can be more expensive than conventional options, posing a barrier for small to medium enterprises.

20.5.2.2 Performance and Durability

Biodegradable composites may have different durability and performance characteristics than traditional materials. This can limit their application in high-performance sectors, such as automotive or construction, where specific material properties are required.

20.5.2.3 Compliance and Testing

Regulatory compliance involves rigorous testing to ensure biodegradability and sustainability claims, which can be time-consuming and costly. Certifications such as ASTM D6400 (for compostable plastics) (Standard Specification for Labeling of

Table 20.5 Key factors in the promotion and compliance of renewable and biodegradable composites

Aspect	Details	Implications for manufacturers
Government incentives	Subsidies, tax breaks for using biodegradable materials	Reduces costs, making eco-friendly materials more accessible
Single-use plastics bans	Bans on certain plastics drive demand for biodegradable options	Encourages innovation in biodegradable composite production
Compliance and certification	Standards such as ASTM D6400 and EN 13,432 require rigorous testing	Increases costs and time investment for market access

Plastics Designed to be Aerobically Composted in Municipal or Industrial Facilities, https://www.astm.org/d6400-21.html) and EN 13,432 (European standard for compostability) (https://docs.european-bioplastics.org/publications/bp/EUBP_BP_En_13432.pdf) are often required but can present technical and financial hurdles (Table 20.5).

20.6 Health and Safety Regulations

Health and safety regulations for polymer composites are crucial to ensuring both consumer safety and worker protection (A Systematic Review on Research Documentation of Ocular Medical Device Focused on Perfluorocarbon Liquid (PFCL) 2024; Suitability of ISO Standards for Medical Device Toxicity Measurement 2024; Srivastava et al. 2018). These regulations address issues related to toxicity, exposure, and workplace safety, which are especially important given the complex chemical makeup of many polymer composites and the potential health risks associated with their production and use.

20.6.1 Toxicity and Human Health

To protect consumers, regulatory agencies impose stringent guidelines regarding the toxicity and human health impacts of polymer composites used in consumer products.

20.6.1.1 Toxic Substances Control Act (TSCA) (U.S.)

Administered by the Environmental Protection Agency (EPA), TSCA regulates the production, importation, and use of chemicals in products, including polymers. It mandates toxicity testing to identify potential health hazards (O Us EPA 2024).

20.6.1.2 REACH Regulation (EU)

Under the Registration, Evaluation, Authorisation and Restriction of Chemicals (REACH), manufacturers must register and test polymer components for toxicity. This includes identifying substances that may cause harm through prolonged exposure or bioaccumulation (REACH Regulation–European Commission 2024).

20.6.1.3 Consumer Product Safety Commission (CPSC)

The CPSC enforces safety standards for materials used in consumer products, focusing on reducing harmful exposures, such as volatile organic compounds (VOCs), that can off-gas from composites and pose risks to human health (CPSC.gov, https://cpsc-d8-prod.ctacdev.com/).

These regulations require manufacturers to meet safety limits on toxic substances, ensuring that polymer composites do not release harmful chemicals that could negatively impact consumers' health.

20.6.2 Worker Safety

Worker safety regulations protect individuals involved in the production and handling of polymer composites, which can expose workers to potentially hazardous chemicals and materials.

20.6.2.1 Occupational Safety and Health Administration (OSHA) (U.S.)

OSHA sets standards for chemical handling, air quality, and protective equipment to safeguard workers in manufacturing environments. For polymer composites, OSHA regulates permissible exposure limits (PELs) for chemicals like styrene and formaldehyde, common in composite production (Home|Occupational Safety and Health Administration 2024).

20.6.2.2 European Agency for Safety and Health at Work (EU-OSHA)

In Europe, EU-OSHA provides guidelines for safe handling of composite materials, including requirements for ventilation, protective clothing, and exposure monitoring. Standards focus on reducing respiratory and dermal exposure to hazardous chemicals in the workplace (European Agency for Safety Health at Work–Information, Statistics, Legislation And Risk Assessment Tools 2024).

20.6.2.3 Globally Harmonized System of Classification and Labeling of Chemicals (GHS)

Adopted by many countries, the GHS standardizes chemical hazard communication, ensuring that workers worldwide have access to consistent information about chemical risks and safety precautions (About the GHS|UNECE, https://unece.org/about-ghs) (Table 20.6).

20.7 Innovation and Research Incentives

To drive the development of sustainable polymer composites, various governments and organizations offer incentives to support research and innovation in this field. These incentives encourage the exploration of environmentally friendly materials and processes, helping to meet global sustainability goals. Additionally, public–private partnerships play a significant role in advancing the adoption of sustainable composites by facilitating collaboration between government entities and private industries.

Table 20.6 Key health and safety regulations for polymer composites

Regulation/agency	Focus area	Key requirements for compliance
TSCA (EPA, U.S.)	Consumer toxicity and exposure limits	Requires testing of polymer toxicity and limits harmful substances
REACH (EU)	Toxicity in consumer products	Mandates registration/testing for chemicals in polymers
OSHA (U.S.)	Worker safety in production	Sets exposure limits, ventilation, and PPE requirements
EU-OSHA (EU)	Occupational safety standards	Enforces guidelines on protective gear and workplace exposure
GHS (Global)	Chemical hazard communication	Ensures labeling consistency for chemical safety information

20.7.1 Governmental Incentives

Governments worldwide recognize the environmental benefits of sustainable polymer composites and have established programs to stimulate research and development in this area.

20.7.1.1 Research Grants and Tax Incentives

Many governments provide research grants and tax credits for companies investing in sustainable materials. For instance, the U.S. Department of Energy (DOE) offers grants under the Energy Efficiency and Renewable Energy (EERE) program (Office of Energy Efficiency Renewable Energy 2024) to support innovations in bio-based polymers and recycling technologies.

20.7.1.2 EU's Horizon Europe Program

This initiative funds sustainable material research, including polymer composites, by providing financial support to companies and research institutions focusing on low-impact materials and circular economy innovations (Horizon Europe—European Commission 2024).

20.7.1.3 Green Bonds and Subsidies

Some governments issue green bonds or subsidies to fund projects aimed at reducing environmental impact, including projects related to sustainable polymer composites. These funds support projects from early research phases through to commercialization (Green bonds in Europe 2024).

These incentives reduce the financial burden on companies and research institutions, encouraging them to invest in the development of eco-friendly polymer composites.

20.7.2 Public–Private Partnerships

Public–private partnerships (PPPs) serve as a powerful tool to enhance research and innovation in sustainable polymer composites by bringing together resources and expertise from both the government and private sectors.

20.7.2.1 Collaborative R&D Projects

Programs like the National Institute of Standards and Technology's (NIST) Advanced Manufacturing Technology initiative in the U.S. encourage collaboration among companies, universities, and government agencies to advance sustainable manufacturing practices for polymer composites (https://www.nist.gov/oam/programs/advanced-manufacturing-technology-consortia-amtech-program).

20.7.2.2 EU Bio-based Industries Joint Undertaking (BBI JU)

This partnership between the European Union and the Bio-Based Industries Consortium (BIC) co-funds projects to develop bio-based materials, including sustainable composites, with a focus on commercial viability and scalability (Home|Circular Bio-based Europe Joint Undertaking (CBE JU), https://www.cbe.europa.eu/).

20.7.2.3 Pilot Projects and Testbeds

Many PPPs establish pilot projects and testbeds where companies can test sustainable polymer composites on a smaller scale before full commercialization. Such projects provide valuable data, help reduce costs, and mitigate risks.

These collaborations accelerate innovation by sharing resources, expertise, and risk, making it easier to overcome the technical and financial challenges of developing new sustainable materials (Table 20.7).

20.8 Compliance and Enforcement

Ensuring compliance with regulations for sustainable polymer composites involves ongoing monitoring and enforcement activities by government agencies and regulatory bodies. Compliance mechanisms, including audits and penalties, play a crucial role in maintaining adherence to environmental and safety standards. However,

Table 20.7 Key innovation and research incentives for sustainable polymer composites

Incentive type	Description	Examples
Government grants and tax credits	Financial support for sustainable R&D	DOE's EERE grants (U.S.), Horizon Europe (EU)
Green bonds and subsidies	Funding for eco-friendly projects	Green bonds for sustainable composite development
Public–private partnerships	Collaborative projects to reduce costs and share expertise	BBI JU (EU), NIST's Advanced Manufacturing (U.S.)

enforcing these regulations presents unique challenges, especially in complex, global supply chains where compliance varies across regions.

20.8.1 Monitoring and Enforcement

To monitor compliance, regulatory bodies employ a range of methods, from inspections to data reporting and independent audits.

20.8.1.1 Inspections and Audits

Agencies such as the Environmental Protection Agency (EPA) in the U.S. and the European Chemicals Agency (ECHA) (Inicio–ECHA, 2024) regularly inspect production facilities to ensure compliance with environmental and safety standards. These audits focus on production processes, emissions, waste management, and workplace safety.

20.8.1.2 Data Reporting Requirements

Regulations often mandate companies to submit data on emissions, chemical usage, and waste management practices. For example, the REACH regulation in the EU requires detailed reporting of chemicals used in manufacturing, ensuring transparency and traceability in polymer production.

20.8.1.3 Penalties for Non-compliance

Governments impose penalties for regulatory violations, which may include fines, production shutdowns, or legal actions. For instance, companies that fail to meet OSHA standards on worker safety may face substantial fines, while non-compliance with TSCA can result in product bans or recalls.

These monitoring and enforcement measures help ensure that companies adhere to sustainability and safety standards, fostering industry-wide accountability.

20.8.2 Challenges in Enforcement

Despite stringent regulatory frameworks, enforcing compliance with environmental and safety standards faces numerous challenges, particularly in global supply chains.

20.8.2.1 Complexity of Global Supply Chains

Products are often manufactured, assembled, and distributed across multiple countries, each with varying regulations. This makes it difficult to monitor compliance consistently throughout the supply chain, as standards may differ significantly between countries.

20.8.2.2 Resource Limitations

Regulatory bodies may lack the resources and personnel needed to effectively monitor every company and facility, especially in countries with limited infrastructure for regulatory enforcement. This can lead to gaps in compliance and unregulated activities.

20.8.2.3 Limited Traceability

Ensuring transparency and traceability of materials in polymer composites is challenging. Tracking sustainable and certified materials throughout the production process becomes increasingly complex in global supply chains, leading to potential misrepresentations of compliance.

These challenges highlight the need for coordinated international efforts and technological solutions, such as blockchain for supply chain transparency, to improve compliance and enforcement in the global market (Table 20.8).

Table 20.8 Key compliance and enforcement mechanisms for sustainable polymer composites

Mechanism	Description	Examples
Inspections and audits	On-site inspections and third-party audits	EPA facility audits (U.S.), ECHA inspections (EU)
Data reporting	Mandatory reporting on emissions and material use	REACH data submissions (EU), TSCA chemical reports (U.S.)
Penalties for violations	Fines, shutdowns, and legal consequences for non-compliance	OSHA fines for safety violations, product bans under TSCA

20.9 Case Studies

Case studies of successful implementation of sustainable polymer composite regulations offer valuable insights into best practices, challenges, and outcomes. These examples provide guidance for improving regulatory frameworks, enhancing sustainability, and promoting wider adoption of environmentally friendly materials in the industry.

20.9.1 Examples of Successful Implementation

Several companies and regions have led the way in adopting sustainable polymer composite regulations. These case studies highlight the benefits of regulatory compliance and innovation in sustainable materials.

20.9.1.1 Patagonia (U.S.)

This outdoor apparel brand is committed to using recycled and bio-based materials in its products. Patagonia collaborates with the Sustainable Apparel Coalition and complies with REACH and TSCA standards, ensuring that its polymer composites are both safe and environmentally friendly. Through initiatives like using recycled polyester and bio-based polymers, Patagonia has reduced its carbon footprint and set an industry standard for eco-conscious product design.

20.9.1.2 Covestro (Germany)

Covestro, a global supplier of polymer materials, has implemented closed-loop recycling for polycarbonates used in automotive and electronics applications. Working closely with the EU Circular Economy Action Plan, Covestro has developed innovative recycling methods to reprocess used materials into high-quality composites. This closed-loop system reduces waste and resource consumption, aligning with EU sustainability goals.

20.9.1.3 Japan's Green Purchasing Law

In Japan, government regulations mandate that public sector purchases prioritize eco-friendly products. This law has led to widespread adoption of sustainable polymer composites in public infrastructure projects, such as rail and construction, fostering demand for recycled and biodegradable materials. Through government-driven demand, Japanese manufacturers have developed durable, eco-friendly composites for public use, supporting national environmental targets (https://www.env.go.jp/content/000064788.pdf).

20.9.2 Lessons Learned

The success of these cases reveals important lessons that can guide future regulatory frameworks and industry practices.

20.9.2.1 Cross-Sector Collaboration

Collaboration among companies, regulators, and researchers proved crucial in all case studies. Patagonia's partnerships with industry coalitions and Covestro's work with EU initiatives show how cooperative efforts lead to more robust and effective solutions. Future regulatory frameworks can benefit from fostering partnerships that align industry goals with regulatory standards.

20.9.2.2 Market-Driven Compliance

Japan's Green Purchasing Law highlights the power of market demand in promoting sustainability. By requiring government agencies to purchase eco-friendly materials, Japan incentivized manufacturers to innovate. Similar demand-driven policies can stimulate broader industry adoption, creating a sustainable market for green composites.

20.9.2.3 Incentivizing Innovation and Recycling

Covestro's investment in closed-loop recycling for polymer composites underscores the importance of incentivizing innovative recycling technologies. Governments can encourage recycling innovation by providing grants, tax incentives, and other support to reduce waste in polymer composite industries (Table 20.9).

20.10 Future Trends and Challenges

As environmental concerns and technological advancements continue to shape the polymer composite industry, future regulatory frameworks will need to adapt and address emerging issues. Trends point toward stricter environmental standards and

Table 20.9 Key lessons from successful implementation of sustainable polymer composites

Lesson	Description	Case study examples
Cross-sector collaboration	Partnerships between companies, regulators, and researchers	Patagonia (U.S.), Covestro (Germany)
Market-driven compliance	Government-driven demand for sustainable materials	Japan's green purchasing law
Incentivizing recycling	Support for recycling technologies and closed-loop systems	Covestro's closed-loop recycling (Germany)

the need for a globally unified approach. However, achieving this global harmonization, along with overcoming technological and market challenges, will be crucial for the future of sustainable polymer composites.

20.10.1 Evolving Regulations

As technology advances and the urgency of environmental issues intensifies, regulatory frameworks are expected to evolve in several ways as follows.

20.10.1.1 Increased Emphasis on Circular Economy

Future regulations are likely to mandate stricter circular economy practices, such as extended producer responsibility (EPR) for manufacturers. Governments may require companies to design products that can be easily recycled or repurposed, reducing waste across industries.

20.10.1.2 Advanced Material Standards

With the rise of new sustainable materials, standards will evolve to cover innovations in bio-based and biodegradable polymers. These standards will likely include new criteria for safety, life cycle impact, and disposal.

20.10.1.3 Digital Tracking for Compliance

Blockchain and Internet of Things (IoT) technology will enable enhanced tracking of materials throughout the supply chain, improving transparency and traceability. Regulations may increasingly require digital tracking systems to ensure compliance, especially for recycled and sustainable composites.

By adapting regulations to cover these emerging trends, governments can better address sustainability goals and support new material technologies in reducing environmental impact.

20.10.2 Global Harmonization of Regulations

The global nature of the polymer composite supply chain underscores the need for harmonized regulatory standards across countries. However, achieving this harmonization poses challenges as follows.

20.10.2.1 Regulatory Disparities

Differences in regulatory rigor and standards among countries make it challenging to establish a single, unified framework. Countries with stringent regulations, like those in the EU, may face difficulties aligning with countries that have more relaxed or different environmental standards.

20.10.2.2 Economic and Technological Gaps

Developing nations often lack the resources and infrastructure to meet high regulatory standards, which may limit the feasibility of globally harmonized regulations.

20.10.2.3 International Collaboration

Harmonization would require extensive international cooperation through organizations like the United Nations Environment Programme (UNEP) and International Organization for Standardization (ISO). These organizations can play a key role in establishing common standards, but the process will require significant negotiation and compromise among participating nations.

Despite these challenges, a globally harmonized framework could simplify compliance, facilitate trade, and ensure that sustainable practices are upheld worldwide.

20.10.3 Technological and Market Challenges

Even with evolving regulations and potential harmonization, achieving widespread compliance and adoption of sustainable practices in the polymer composite industry will face challenges as follows.

20.10.3.1 High Costs of Sustainable Alternatives

Bio-based and biodegradable polymers often come at a premium compared to conventional materials. For widespread adoption, companies will need affordable, scalable solutions that make sustainable materials economically viable.

20.10.3.2 Technology Gaps in Recycling and Disposal

Efficient recycling and disposal solutions for polymer composites are still developing. Many sustainable composites are not yet fully recyclable or biodegradable, complicating end-of-life management. Investment in research to improve recycling and disposal technologies will be essential.

20.10.3.3 Consumer Demand and Market Readiness

Consumer demand for sustainable products varies globally. In regions where environmental concerns are less prioritized, companies may struggle to find markets for eco-friendly products. Market readiness is essential for supporting the shift toward sustainable practices.

These challenges highlight the need for continued innovation and investment in technology, as well as targeted educational campaigns to increase consumer awareness and demand for sustainable polymer composites (Table 20.10).

Table 20.10 Key future trends and challenges in sustainable polymer composite regulation

Future trend/challenge	Description	Examples
Evolving regulations	Stricter standards on circular economy and material tracking	EPR mandates, blockchain for supply chain transparency
Global harmonization of regulations	Need for unified international standards	UNEP and ISO collaboration on sustainability standards
Technological and market challenges	High costs, limited recycling solutions, and varying consumer demand	Investment in recycling tech, awareness campaigns

20.11 Conclusion

20.11.1 Summary of Key Points

In this chapter, the critical aspects of regulatory and policy frameworks shaping the sustainability of polymer composites are explored. It began by examining key international regulations, including REACH, RoHS, and WEEE, and their impact on the safety, health, and environmental footprint of polymer composites. It then presented the global standards and certifications such as ISO 9001, ASTM D638, and the Global Recycled Standard, highlighting their role in ensuring quality and performance while promoting sustainable manufacturing practices. Additionally, it discussed regulations concerning renewable and biodegradable composites, health and safety standards, innovation incentives, and the challenges of enforcement. Through case studies, successful implementations, and the lessons that can be drawn from them, future frameworks can be identified. The chapter concluded with an outlook on future trends and challenges, including the evolution of regulations and the need for global harmonization.

20.11.2 The Way Forward

The future of regulatory and policy frameworks in driving the sustainability of polymer composites looks promising, but several challenges remain. As technology advances and environmental concerns grow, regulatory frameworks must evolve to address emerging materials, manufacturing processes, and end-of-life management solutions. A key focus will be strengthening the integration of circular economy principles, such as designing for recyclability and reducing resource consumption. Governments and industry stakeholders must work together to create internationally harmonized regulations that ensure a level playing field and streamline compliance. Additionally, increasing emphasis on innovation incentives will encourage the development of new, more sustainable polymer composite materials. Ultimately, continued collaboration among regulatory bodies, manufacturers, and researchers will be crucial in advancing sustainability goals and ensuring the long-term viability of polymer composites in a circular economy.

References

«000064788.pdf». Accedido: 12 de noviembre de 2024. [En línea]. Disponible en: https://www.env.go.jp/content/000064788.pdf

About the GHS|UNECE. Accedido: 12 de noviembre de 2024. [En línea]. Disponible en: https://unece.org/about-ghs

Accedido: 12 de noviembre de 2024. [En línea]. Disponible en: https://docs.european-bioplastics.org/publications/bp/EUBP_BP_En_13432.pdf

Accedido: 12 de noviembre de 2024. [En línea]. Disponible en: https://www.wiego.org/sites/default/files/resources/files/Pereira-Brazilian-Waste-Policy.pdf

Admin O (2024) Proposition 65. OEHHA. Accedido: 12 de noviembre de 2024. [En línea]. Disponible en: https://oehha.ca.gov/proposition-65

Advanced Manufacturing Technology Consortia (AMTech) Program. NIST. Accedido: 12 de noviembre de 2024. [En línea]. Disponible en: https://www.nist.gov/oam/programs/advanced-manufacturing-technology-consortia-amtech-program

A Systematic Review on Research Documentation of Ocular Medical Device Focused on Perfluorocarbon Liquid (PFCL). Accedido: 12 de noviembre de 2024. [En línea]. Disponible en: https://www.allbiosolution.com/2022/11/a-systematic-review-on-research.html

ASTM International— Standards Worldwide (2024) Accedido: 12 de noviembre de 2024. [En línea]. Disponible en: https://www.astm.org/

"Bioeconomy strategy—European Commission". Accedido: 12 de noviembre de 2024. [En línea]. Disponible en: https://research-and-innovation.ec.europa.eu/research-area/environment/bioeconomy/bioeconomy-strategy_en

"BioPreferred|Catalog". Accedido: 12 de noviembre de 2024. [En línea]. Disponible en: https://www.biopreferred.gov/BioPreferred/faces/catalog/Catalog.xhtml

Coalition (SPC) SP (2024) Extended producer responsibility—SPC's guide. Accedido: 12 de noviembre de 2024. [En línea]. Disponible en: https://epr.sustainablepackaging.org

Circular economy action plan—European Commission. Accedido: 12 de noviembre de 2024. [En línea]. Disponible en: https://environment.ec.europa.eu/strategy/circular-economy-action-plan_en

CPSC.gov. U.S. Consumer Product Safety Commission. Accedido: 12 de noviembre de 2024. [En línea]. Disponible en: https://cpsc-d8-prod.ctacdev.com/

Development of sustainable biopolymer-based composites for lightweight applications from agricultural waste biomass: A review—ScienceDirect. Accedido: 12 de noviembre de 2024. [En línea]. Disponible en: https://www.sciencedirect.com/science/article/pii/S2542504823000337

EU Ecolabel–Home. Accedido: 12 de noviembre de 2024. [En línea]. Disponible en: https://environment.ec.europa.eu/topics/circular-economy/eu-ecolabel_en

European Agency for Safety & Health at Work—Information, statistics, legislation and risk assessment tools. Accedido: 12 de noviembre de 2024. [En línea]. Disponible en: https://osha.europa.eu/en

Extended Producer Responsibility guidelines mandate sustainable plastic packaging to reduce plastic foot print. Accedido: 12 de noviembre de 2024. [En línea]. Disponible en: https://pib.gov.in/pib.gov.in/Pressreleaseshare.aspx?PRID=1909909

Federal Trade Commission|Protecting America's Consumers. Accedido: 12 de noviembre de 2024. [En línea]. Disponible en: https://www.ftc.gov/

Get Certified—cradle to cradle products innovation institute. Accedido: 12 de noviembre de 2024. [En línea]. Disponible en: https://c2ccertified.org/get-certified

Green bonds in Europe. Accedido: 12 de noviembre de 2024. [En línea]. Disponible en: https://www.eea.europa.eu/en/analysis/indicators/green-bonds-8th-eap

Home|Circular Bio-based Europe Joint Undertaking (CBE JU). Accedido: 12 de noviembre de 2024. [En línea]. Disponible en: https://www.cbe.europa.eu/

Home|Occupational Safety and Health Administration. Accedido: 12 de noviembre de 2024. [En línea]. Disponible en: https://www.osha.gov/

Horizon Europe—European Commission. Accedido: 12 de noviembre de 2024. [En línea]. Disponible en: https://research-and-innovation.ec.europa.eu/funding/funding-opportunities/funding-programmes-and-open-calls/horizon-europe_en

Inicio—ECHA. Accedido: 12 de noviembre de 2024. [En línea]. Disponible en: https://echa.europa.eu/es/

ISO 14001:2015 (2024) ISO. Accedido: 12 de noviembre de 2024. [En línea]. Disponible en: https://www.iso.org/standard/60857.html
ISO–International Organization for Standardization, ISO (2024). Accedido: 12 de noviembre de 2024. [En línea]. Disponible en: https://www.iso.org/home.html
Life cycle assessment. European Environment Agency. Accedido: 12 de noviembre de 2024. [En línea]. Disponible en: https://www.eea.europa.eu/help/glossary/eea-glossary/life-cycle-assessment
Lista de restricciones—ECHA. Accedido: 12 de noviembre de 2024. [En línea]. Disponible en: https://echa.europa.eu/es/substances-restricted-under-reach
Office of Energy Efficiency & Renewable Energy. Energy.gov. Accedido: 12 de noviembre de 2024. [En línea]. Disponible en: https://www.energy.gov/eere/office-energy-efficiency-renewable-energy
O. US EPA (2024a) Summary of the toxic substances control act. Accedido: 12 de noviembre de 2024. [En línea]. Disponible en: https://www.epa.gov/laws-regulations/summary-toxic-substances-control-act
O. US EPA (2024b) U.S. Environmental Protection Agency. Accedido: 12 de noviembre de 2024. [En línea]. Disponible en: https://www.epa.gov/home
Prasad V, Alliyankal Vijayakumar A, Jose T, George SC (2004) A comprehensive review of sustainability in natural-fiber-reinforced polymers. Sustainability 16(3), Art. No. 3, ene. https://doi.org/10.3390/su16031223
REACH Regulation–European Commission. Accedido: 12 de noviembre de 2024. [En línea]. Disponible en: https://environment.ec.europa.eu/topics/chemicals/reach-regulation_en
RoHS Directive—European Commission. Accedido: 12 de noviembre de 2024. [En línea]. Disponible en: https://environment.ec.europa.eu/topics/waste-and-recycling/rohs-directive_en
Srivastava GK et al (2018) Comparison between direct contact and extract exposure methods for PFO cytotoxicity evaluation. Sci Rep 8(1):1425, ene. https://doi.org/10.1038/s41598-018-19428-5
Standard Specification for Labeling of Plastics Designed to be Aerobically Composted in Municipal or Industrial Facilities. Accedido: 12 de noviembre de 2024. [En línea]. Disponible en: https://www.astm.org/d6400-21.html
Suitability of ISO standards for medical device toxicity measurement. Accedido: 12 de noviembre de 2024. [En línea]. Disponible en: https://www.allbiosolution.com/2022/05/suitability-of-iso-standards-for.html
The China Standards 2035 Strategy: Analyzing recent developments. China Briefing News. Accedido: 12 de noviembre de 2024. [En línea]. Disponible en: https://www.china-briefing.com/news/china-standards-2035-strategy-recent-developments-and-their-implications-foreign-companies/
The European Green Deal—European Commission. Accedido: 12 de noviembre de 2024. [En línea]. Disponible en: https://commission.europa.eu/strategy-and-policy/priorities-2019-2024/european-green-deal_en
Waste from Electrical and Electronic Equipment (WEEE)—European Commission. Accedido: 12 de noviembre de 2024. [En línea]. Disponible en: https://environment.ec.europa.eu/topics/waste-and-recycling/waste-electrical-and-electronic-equipment-weee_en

Chapter 21
Exploring the Recyclability of Sustainable Polymer Composites

Ionut-Cristian Radu, Derniza-Elena Cozorici, Erika Blanzeanu, and Catalin Zaharia

Abstract Nowadays, the implementation of nanotechnology in various industry sectors is widely recognized and accepted as a promising tool due to its significant advantages. There are various types of material nanocomposites based on their main applications. Many classes of nanocomposites include polymers as matrix components or as fillers/reinforcements. These classes are collectively referred to as polymer nanocomposites. In addition to nanocomposites, micro- and macro-composites have been established as essential polymeric materials due to their wide applications over the decades. The book chapter outlines the major differences between nano-, micro-, and macro-composites regarding the internal structuring and sizes of their constitutive components, focusing on their main properties, applications, and recyclability. In terms of polymer composite applications and targeted properties, these materials can be viewed as polymers impregnated with inorganic or organic compounds. Essentially, the added inorganic or organic compound serves as a reinforcement or filler, enhancing the mechanical and thermal properties. However, nanotechnology extends the role of the added component to impart new specific properties such as barrier characteristics, antifouling, antimicrobial effects, conductivity, and more. The structure and composition of polymer composites influence the recycling management, guiding it toward appropriate procedures. Generally, the recycling process for composites is more complex compared to single polymeric materials, as separating the components of recyclable composites is more intricate, costly, and less

I.-C. Radu · D.-E. Cozorici · E. Blanzeanu · C. Zaharia (✉)
Faculty of Chemical Engineering and Biotechnology, Department of Bioresources and Polymer Science, Advanced Polymer Materials Group, National University of Science and Technology POLITEHNICA Bucharest, Bucharest, Romania
e-mail: catalin.zaharia@upb.ro

I.-C. Radu
e-mail: ionut_cristian.radu@upb.ro

D.-E. Cozorici
e-mail: derniza.cozorici@upb.ro

E. Blanzeanu
e-mail: erika.blanzeanu@upb.ro

L. Ravindran et al. (eds.), *Macro, Micro and Nanocomposites from Sustainable Sources*, Advanced Structured Materials 244, https://doi.org/10.1007/978-981-95-2469-3_21

economical. Because of this, many industrial polymeric composites are termed hard-to-recycle composites. Despite these challenges, manufacturers are driven to use and produce polymeric composites effectively due to their significant advantages. The book chapter recapitulates the major types of polymer matrices and possible polymer additives for the development and production of polymer composites. Therefore, the classification of polymers considers the polymer source: fully extracted from biomass (natural), biosynthesized via microbial fermentation, monomers extracted from biomass, natural polymers modified by synthetic procedures, and synthetic polymers derived from fuel sources. Furthermore, the book chapter addresses the inorganic and organic additives that form polymer composites. This approach is very important as it highlights the sustainability, reliability, and environmental friendliness of each type of polymer or additive, including rubbers, fibers, thermosets, bioplastics, bio-based, non-bio-based, biodegradable, and non-biodegradable materials. Each polymer class is presented in the context of its specific recyclable pathway and the complexity of recycling as a composite versus simple polymers, focusing on aspects such as separation and recovery. The general and newly developed recycling pathways are highlighted according to polymer and additive categories (from primary to quaternary/energy recovery): mechanical recycling, chemical recycling (partial or total monomer reconversion), pyrolysis, thermal treatment, enzyme-catalyzed processes, microbial degradation, and more. The final section provides valuable information on the importance of composite recycling and its contribution to the circular economy.

21.1 Introduction

Nanotechnology is now widely recognized and accepted as a promising tool in various industrial sectors, thanks to its significant and numerous advantages. There are various types of material nanocomposites, each tailored to specific applications. Many classes of nanocomposites include polymers as either the matrix component or the filler/reinforcement. These are collectively referred to as *polymer nanocomposites*. In addition to nanocomposites, micro- and macro-composites have long established polymeric materials as essential components in a wide range of applications over the past several decades. The first part of this book chapter highlights the key differences among nano-, micro-, and macro-composites, focusing on their internal structure and component size, with the aim of understanding their fundamental properties, applications, and recyclability. To fully understand these differences, it is essential to highlighting the history, evolution, and initial expectations of composites.

Composite materials have been developed since ancient times (4000–3000 B.C.), serving a wide range of applications, from construction to hunting and other daily uses. In those early times, most composite materials were not based on polymer matrices or polymer reinforcements; instead, they utilized whatever nature provided for exploitation, such as wood, glued wood, mud, clays, and bones. However, beginning in the 1800s, the chemical revolution significantly altered the development of composite materials. Advances in composition, coupled with the evolution of polymer science, paved the way for new possibilities in the creation of composite materials. The availability of new synthetic resins, capable of forming a solid state through liquid cross-molecular bonds (such as celluloid, melamine, Bakelite, phenolic-based resins, and unsaturated polyesters), marked a turning point into a new era. This advancement made it possible to create modern polymeric composites, utilizing a polymer-based matrix combined with an inorganic reinforcement or filler. In the early 30s, the new advancements in glass fibers and polymeric resins made possible the primary modern polymer composites as alternative for metal, wood, ceramic, or other materials. In the following decades, a plethora of polymers were discovered, significantly expanding the options for polymer-based matrices with new resins and thermoplastic polymers. Over time, the components and phases of polymer composite materials have evolved, with polymer-based reinforcements emerging in the form of various classes of polymeric fibers, including polyamides, polyesters, and polyolefins. This development further diversified the possibilities for designing advanced composite materials (Palanisamy et al. 2021; Tariq et al. 2024; Godara et al. 2021). In order to produce the high-quality polymer composites required, the manufacturing processes had to evolve at the same rate as the composition of the composites. While this book chapter does not aim to detail the evolution of manufacturing processes, it is important to note that the development of techniques such as vacuum bag molding, pultrusion, large-scale filament winding, compression molding, and spraying has significantly advanced polymer composites toward novel applications in aerospace, automotive, consumer products, and sporting goods. The next turning point was the development of carbon fibers in the early 60s, offering new alternatives for polymer composites with high stiffness, mechanical resistance, and low density. The decreasing cost of carbon fiber production is driving the evolution of polymer composites by making them more accessible for targeted high-tech application (Duraikkannu et al. 2021; Sathishkumar et al. 2014; Liu et al. 2015). By the middle of the 1900s and 2000s, polymer composites were already common materials due to cost-effective properties in every major industry (construction, electrical, electronics, aerospace, automotive, transportation, etc.) In general, polymer composites are classified into two categories based on their specific applications: high-performance composites and mass-consumption or low-performance composites. In essence, the previously mentioned polymer composites can be considered as either micro- or macro-composites, depending on the system's multiphase structure. In recent decades, as well as today, polymer composites have increasingly replaced other materials, even though polymers were once considered unsuitable. Undoubtedly, this trend will continue due to the inherent properties and sustainability potential of polymer composites. Moreover, new types of polymers, such as polyaryletherketone (PAEK) and polyetherimides (PEI), are continuously being developed and made available for important industrial applications, in alignment with human advancements (Hsissou et al. 2021). Polyaryletherketones (PAEKs) are semi-crystalline thermoplastic polymers distinguished by their exceptional mechanical properties and thermal stability. Their molecular structure features aromatic rings and a backbone consisting of alternating ether and ketone linkages (Sacks et al. 2024). The PAEK family comprises a range of polymers, including polyetherketone (PEK), polyetheretherketone (PEEK), polyetherketoneketone (PEKK), polyetheretherketoneketone (PEEKK), and polyetherketoneetherketoneketone (PEKEKK). The best-known PAEK polymers available on the market are polyetheretherketone (PEEK) and polyetherketoneketone (PEKK).

Polyetherimide (PEI) is an advanced thermoplastic characterized by the presence of both ether linkages and imide groups within its polymer backbone. As an amorphous high-performance polymer, PEI offers excellent thermal stability, good chemical resistance, inherent flame retardancy, and outstanding dimensional stability. It also demonstrates high tensile strength without reinforcement, very low smoke emission, and high hydrolytic stability. Due to its exceptional stability, PEI can be processed over a broader range of conditions compared to many other thermoplastics (Jiang et al. 2019; Ke et al. 2023; Xu et al. 2021).

Both PAEK and PEI are high-performance thermoplastic polymers characterized by high processing and transition temperatures. Specifically, PEEK exhibits a glass transition temperature of 143 °C and a melting point of 330 °C, while PEKK has a glass transition temperature of 160 °C and a melting point of 305 °C. PEI, an amorphous thermoplastic, has a higher glass transition temperature of 220 °C but lacks a defined melting point. PEEK, PEKK, and PEI are considered engineering polymers due to their exceptional mechanical strength, thermal stability, and chemical resistance. However, PEI is generally less chemically resistant than members of the PAEK family. Structurally, PEEK and PEKK are semi-crystalline, whereas PEI is amorphous (Alvaredo-Atienza et al. 2020; Jenkins 2001; Pedoto et al. 2022; https://lynxter.fr/en/blog/guide_en/material-data-sheet-pekk/).

Researchers and scientists have started also to handle environmental and recycling issues due to improved insights and knowledge in polymers and composites science. However, what is the driving force behind all these advancements that have directed vast amounts of energy, time, money, and the efforts of many researchers into the development of this new field? First, it is important to note that the initial and evolving expectations for polymer composites align with the driving force previously discussed. Polymer composites were designed to combine properties from different materials, achieving characteristics that neither the matrix nor the reinforcement could provide individually. A new material is developed with superior properties with respect to original materials. These superior characteristics mainly refer to higher stiffness and strength, lower expansion coefficient, higher fatigue resistance, superior thermal resistance or stress at high temperatures, superior corrosion resistance, ease of manufacture, low costs, etc. However, all of these new characteristics are closely linked to the type of agent incorporated into the polymer matrix (Rothon 2010). In the case of a relatively inert filler agent, the polymer composite exhibits a modified strength and thermal conductivity, higher hardness and cracking resistance, lower cost, lower expansion coefficient, improved processing, etc., according to American Society for Testing and Materials (ASTM-D-883). In contrast to these general features, a reinforcing agent must provide increased stiffness and strength, enhanced thermal resistance, the ability to withstand high temperatures, improved shock resistance, a higher elastic modulus, and greater fatigue resistance. The differences between reinforcing agents and filler agents lie in their chemistry, form, size, and constituent components. In general, reinforcing agents comprise fibers (fabric, textile, long fibers, short fibers, and meshes) and sphere, particles, flakes, irregulars, etc., with various dimensions (Fig. 21.1). According to continuous or discontinuous form, reinforcing agents lead to manufacturing high-performance or low-performance polymer composites. Filler agents are in general represented in forms of spheres, flakes, chopped, or irregular with micrometer dimensions. Thus, in addition to form and size, the chemical nature and degree of dispersion determine whether an agent is used as a filler or reinforcement (Aggarwal and Livigni 1977; Seymour 1976; Chauhan et al. 2020; Hsissou et al. 2020). The properties previously discussed are more specific to macro- and micro-composites, depending on the role of the added agent, whether as a filler or reinforcement. In the case of nanocomposites, the complexity and range of applications expand due to nano-scale dispersion and the unique physical laws governing nano-scale dimensions, such as quantum effects. Therefore, new specific applications are available based on inherent optical, electrical, and magnetic properties. When added agents exhibit nano-scale dimensions and are well-dispersed within the polymer matrix, quantum effects can emerge due to the proximity and interaction of the nano-agent particles. In such nanocomposites, the overall material can exhibit the specific properties of the embedded nanostructures, such as optical, electrical, and magnetic characteristics. This enables the design and development of quantum nanocomposites, superconductors, and conductors or semiconductors, which are driving significant advancements in next-generation smart materials and nanotechnology (Ghosh 2015; Kim et al. 2024).

Under these conditions, the additive agent comes into play, exhibiting a combined effect as both a reinforcing and filling material. These specialized applications must be achieved with low or very low additive concentrations (<5%) compared to those

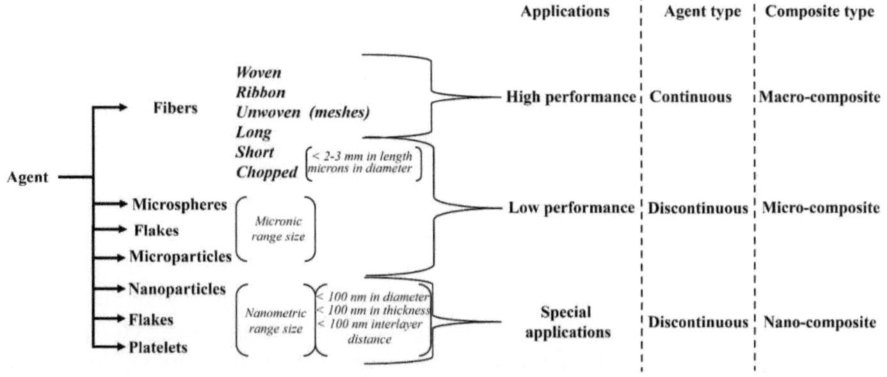

Fig. 21.1 Classification of reinforcing agents

used in micro- and macro-composites. In conclusion, the emergence of the novel polymer composite field did not result from the shortcomings of micro- and macro-composites, but rather from their limitations in meeting the specific demands of new applications aligned with advancing technological requirements.

21.2 Nano- Versus Micro- Versus Macro-composites

The major differences between composite types—nano, micro, and macro—are closely tied to the fundamental definitions of polymer composite. These differences not only relate to the primary properties and applications of the composites but also influence their sustainability and recycling approaches. Over time, polymer composites have been defined in various ways, all aiming to capture the key concept that these materials consist of multiple phases (at least two) with differing chemistry and either full or partial thermodynamic compatibility. One or more components, referred to as the discontinuous phase, are dispersed within another component, known as the continuous phase or matrix. In most cases, the polymeric component in the polymer composite is the matrix or continuous phase. A detailed discussion on the relationship between definitions, composition, properties, and applications will be addressed later in this book chapter. It is vital to understand the composite definition to highlight the relationship definition–internal structuring–composition. Thus, the current discussion aims to highlight the differences between micro- and macro-composites in terms of composition and internal structure. One major point regards the attention over the homogeneity approach as the composite phases have to be highly homogenized. However, homogeneity should be evaluated at a specific observation level within the context of composition. At the molecular level, the composite phases exhibit varying degrees of homogenization, depending on the chemical nature and the constituent components of each phase. In the case of a polymeric matrix, the constitutive components are macromolecules or agglomerates of macromolecules. The first category displays a nanometric range size while the second varies from hundreds of nanometers up to tens of microns. Despite having a homogeneous composition, the molecular symmetry of homopolymers is continuously affected by the chain length increase. This asymmetry is regarded as a natural consequence of the increasing molecular size in a linear configuration. In the case of chain composition heterogeneity (such as in proteins), the asymmetry becomes more complex, further increasing the overall complexity. Thus, it may lead to preferred interactions and spatial arrangement, resulting in further symmetry association by energy lowering. This approach can be realized through a higher degree of crystallization of multiple chains into various crystalline morphologies. Another characteristic of the increased size of macromolecules is the growing complexity and synergy among the components. The weak physical interactions among molecules at nano scale could collectively and cooperatively behave themselves at larger scales more predominant with respect to chemical bonds. Therefore, at the molecular level, the polymeric matrix contains its own constituent phases due to its intrinsic macromolecular nature, rather than being a continuous phase. Polymers display both crystalline and amorphous domains due to different orientation and ordering degree for the chain units (Lin et al. 2016; Zhang and Cheng 2020; Shimokawa et al. 2019). In this context, the definition of polymer composites becomes more complex when viewed at the molecular level, as each material reveals its own distinct phases. The complex nature of macromolecule-increased size makes possible a multitude variation of interactions with the agent constituent components: (i) In the case of nano-sized agents (nanofillers/reinforcing fillers), a high surface area-to-volume ratio specific to nanocomposite can be achieved only by interactions such as individual macromolecule—agent nanoparticle or nano-scale macromolecular aggregates/entanglement—agent nanoparticle. Depending on the processing methods, the dispersion of nanoparticles within the agent can be directed toward a higher degree of homogenization. In this context, it should be noted that achieving a precise interaction between individual macromolecules and agent nanoparticles is very challenging with the processing tools currently available in the industry. The weak inter- and intramolecular interactions manifest themselves collectively and cooperatively by keeping the macromolecules close together. In conclusion, it is not sufficient for only the polymer matrix or the agent to have nano-scale components in order to achieve the intimate interaction required to generate a high interfacial area. The intrinsic compatibility is also influenced by the phases present in the polymeric matrix: the amorphous phase versus the crystalline domains. The last one represents a highly ordered/compaction or the macromolecules by chain units' ability to adopt a lower energy conformation. In these domains, the interaction with agent nanoparticles can be hindered, as additives often act as a factor of disorder. There are also situations where the addition of a second additive may improve compatibility or special cases where agent nanoparticles behave as ordering agent/crystallization agent increasing the level of molecular organization. In cases where a nano-sized agent exhibits poor interaction with the polymeric matrix, the nanoparticles tend to form agglomerates or aggregates, ranging from hundreds of nanometers to several microns in size. These issues can be readily identified through morphological investigation methods and the evaluation of mechanical properties using both static and dynamic techniques. Probably the most significant insight that emerges from above mentions shows that not every nano-sized agent added into a polymeric matrix is able to generate nanocomposites, but the process is complex depending on the manufacturing methods or chemistry of each phase (Pandele et al. 2017, 2020; Tanasa et al. 2019, 2020; Vasile et al. 2020). Furthermore, nano-scale dispersions of agent nanoparticles can bring quantum effects; (ii) In the case of micro-sized agents (microparticles, microspheres, flakes, chopped fibers, etc.), the interaction with polymeric matrix is done at a micro-scale in general. These situations are specific for micro-composites where the additive acts more as a filler agent; (iii) Agents with one or two sizes in micro-scale domain (short fibers, chopped fibers, etc.). This situation is specific for low-performance micro-composites reinforced with inorganic or organic fibers. Micro-composites are regarded as heterogeneous materials at the micro-scale. There can be selectively identified phases or materials that are apparently homogeneous and interact with each other. The entire polymeric matrix is viewed as a continuous phase where the agent phase is discontinuously dispersed. The phases interface, type of polymeric matrix, and type of agent or processing methods control the typical micro-composite targeted properties. Thus, toughness, thermal conductivity, and cracking resistance are highly controlled by the interface being a matter of volume fraction. The interface that separates the composites' phases also represents the interaction surface between them. Viewed as a 2D surface, it is small with respect to agent microparticle size or distance between two microparticles. The agent surface modification or addition of an extra additive may improve the micro-composite characteristics, but it increases production costs, and it complicates the recycling process; (iv) Various reinforcing agents like long fibers, ribbons, or meshes produce high-performance macro-composites. To achieve superior mechanical properties, the matrix macromolecules must be effectively anchored to the fiber surface. Interfacial adhesion can be achieved through chemical reactions (using compatibilizers) or physical interactions that enhance the strength of the matrix. Thus, the strain resulting from the load can be transferred through the resin matrix from fiber to fiber. For the reinforcing stress to be transferred from one fiber to another, the fibers must be positioned close enough so that the area of influence or stress/strain lobes of each fiber overlap. When stress is applied to the composite, the fiber resists strain locally, causing a significantly higher stress to be induced in the fiber compared to the surrounding matrix. There is a minimum fiber population or loading level required for reinforcing to take place. Thus, the reinforcing agent content must be high, up to 70% of the composite volume in some cases. The increase of the elastic modulus and tensile strength of the composite is given by the continuous/long fibers, fabric, or textile: fibers can be disposed in various directions. This arrangement, along with the use of compatibilization agents, also increases production costs and presents an additional challenge for the recycling process (Chwał and Muc 2019; Picu and Keblinski 2003; Ali et al. 2021; Ilie and Hickel 2009).

21.3 Types of Polymeric Matrices and Specific Recyclability Pathways

The structure and composition of polymer composites guide recycling management toward the most suitable procedures. In general, the recycling process is more complex with respect to single polymeric materials because the component separation of the recyclable composite is far more complex, costly, and less economical. Because of that, many industrial polymeric composites are so-called *hard-to-recycle composites*. Even under these conditions, manufacturers are challenged to efficiently use and produce polymeric composites, driven by their key advantages. One major objective is to connect recyclability and sustainability efforts with polymers derived from renewable sources and those that decompose into environmentally friendly products. Consumers of polymer composites are increasingly expressing a clear preference for environmentally friendly materials that help reduce the volume of solid composite and polymer waste. These trends have emerged due to continuous regulatory changes and the rise of green movements in most developed and developing countries. It is crucial to summarizing several key terms that are closely linked to the concept of environmentally friendly materials and sustainability. Thus, terms such as biopolymer, bio-based polymer, compostable polymer, degradable or biodegradable polymer, renewable sources, and sustainable sources are overexpressed in the literature and may show multiple or overlapping meanings.

Biodegradability is a common term used to describe a polymer's ability to degrade under specific conditions, typically through the action of microorganisms such as molds, fungi, and bacteria, within a certain timeframe and environmental context. Degradability is also a broad term attributed to polymers that can disintegrate because of some particular processes such as physical degradation, and mechanical or chemical degradation. By this, biodegradability implies actions of some living organisms while degradability occurs under physical and chemical factors (Song et al. 2009; Innocenti et al. 2003; Samir et al. 2022; Niaounakis 2014).

Biopolymers are defined as polymers produced from natural sources (bio-based), biodegradable polymers, or having both properties. Because of that, biodegradable polymer and bio-based polymer have an interrelationship (Niaounakis 2014; Liu et al. 2023; Baranwal et al. 2022; Soroudi and Jakubowicz 2013; Rudin et al. 2013).

Bio-based polymers are those derived from renewable resources, meaning that the raw materials (monomers, oligomers, pre-formed polymers, etc.) are replaced or replenished through natural pathways at rates comparable to or faster than their consumption. Therefore, the carbon in these polymers must originate from renewable sources through natural processes (Niaounakis 2014; Edwards 2002; Golden et al. 2021; Tokiwa et al. 2009; Zia et al. 2021).

A renewable source is a source which produces plant and animal mass derived via photosynthesis from carbon dioxide recently fixed. All carbon from the raw materials is created through naturally recurrent processes. A sustainable polymer or polymer-based material must be produced from various resources (not only renewable) that can be maintained without affecting the environment, energy supply, or future generations and being connected with society's needs. In general, sustainable polymer emerges from renewable or recycled feedstock, or at the end of life can be recycled, biodegraded, or composted (Niaounakis 2014; Pellis et al. 2021; Mohanty et al. 2022; Hong and Chen 2019).

A compostable polymer must undergo biodegradation under biological conditions during the composting process, typically within several weeks, producing carbon dioxide, water, methane, and inorganic compounds at a consistent rate, without leaving any toxic residues or visible remains. There are numerous national and international regulation bodies that tried to summarize and clarify the usage of degradable, biodegradable, compostable materials so that situations where consumers are confused to be limited. Especially International Organization for Standardization (ISO) and ASTM released detailed definitions and testing procedures (Niaounakis 2014; Rudnik and Rudnik 2019; Gity Mir Mohamad and Sayaf 2015).

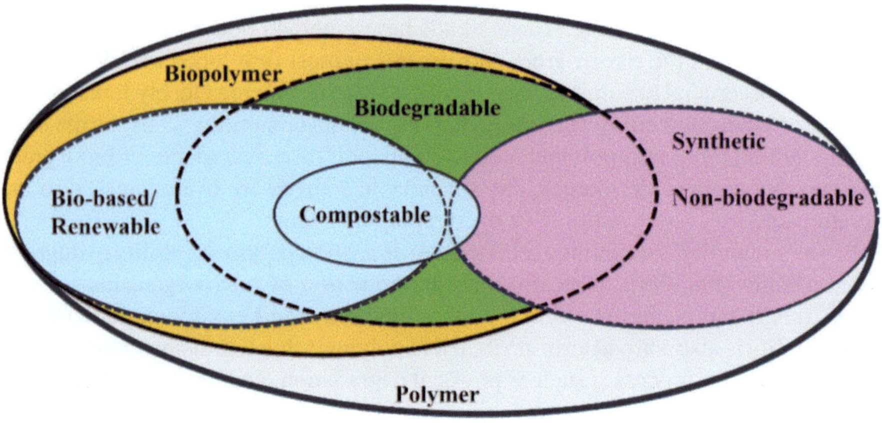

Fig. 21.2 Venn diagram of polymer classifications: biopolymers, biodegradable, compostable, and synthetic

In conclusion, a polymer can be non-biodegradable but still classified as a biopolymer if its monomer is produced from renewable sources that emit fewer greenhouse gases compared to traditional synthetic processes (Fig. 21.2).

Polymers can be degradable without being biodegradable by breaking down into small fragments or invisible particles. The degradation products are not assimilated by microorganisms to produce carbon dioxide, methane, and water. Furthermore, a polymer can be degradable and biodegradable without being compostable if the biodegradation rate is too slow or it does not fulfil several requirements. Even though the products of biodegradation are carbon dioxide, water, methane, and other compounds, compostable polymers must exhibit a high and consistent biodegradation rate while being compatible with the composting process, without compromising the quality of the compost. A sustainable polymer is not necessarily a renewable polymer, even if its recycling occurs through biodegradation, composting, or chemical or mechanical processes. Not all renewable polymers are sustainable if their production process is highly energy-intensive, the replenishment rate is too slow, or if they negatively impact the environment or future generations.

The book chapter will review the major types of polymer matrices used in the development and production of polymer composites, with a focus on sustainability and recyclability. The common classification of polymers as matrices for composites emphasizes three main categories: thermoplastics, thermosets (resins), and rubbers. However, this classification recapitulates the polymer intrinsic behavior and processability in concordance with the required properties for certain applications. The book chapter focuses more on recyclability potential and sustainability while the previous classification reveals poor insights over these approaches. Therefore, the classification of polymer types will consider their source and recycling potential: fully derived from biomass (pre-formed), biosynthesized through microbial fermentation, monomers extracted from biomass, natural polymers modified

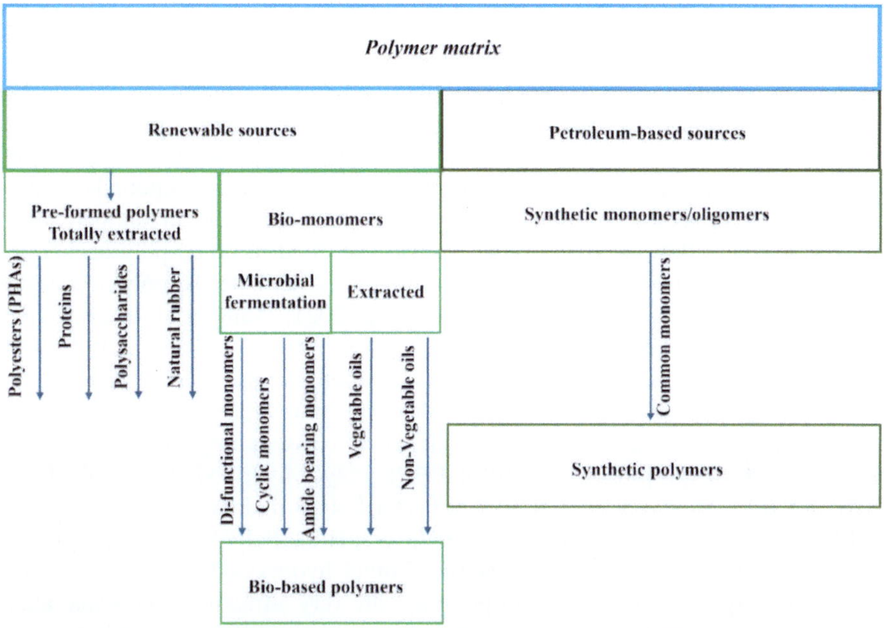

Fig. 21.3 Comprehensive classification of polymers: from source to degradability

through synthetic processes, and synthetic polymers produced from fossil fuels. Each polymer class is discussed in terms of its specific recycling pathways and the complexity involved in using them as composites versus simple polymers (e.g., separation, recovery). This approach is essential as it underscores the sustainability, reliability, and environmental impact of each polymer type (Fig. 21.3).

21.3.1 Pre-formed Natural Polymers

In this book chapter, fully extracted natural polymers will also be referred to as pre-formed natural polymers. These polymers are key structural and functional tools in the living world. They have a particular role in various cell walls (plants, fungi, etc.), brown algae, cells cytoplasm, and blood or are produced in certain situations (defense, hunting, reproduction, etc.). The main classes of extracted natural polymers are classified based on their chemical characteristics, including proteins, polysaccharides, polyesters, natural rubber, and polyphosphate. Proteins are large, complex macromolecules that play many major roles in the living organisms. They consist of one or more long, folded chains of amino acids (each amino acid is called a polypeptide). The amino acid composition and sequence of a protein are determined by the DNA sequence of the gene encoding that protein. The major protein types of high interest for composites include fibrous proteins such as fibroin, collagen, keratin, elastin, fibrin, and spidroin, as well as globular proteins like sericin, casein, and albumin (Hu et al. 2012; Mitra et al. 2023; Schultz 2017; Shen and Shen 2023). Polysaccharides are the most abundant naturally occurring polymers which are extracted from renewable sources. Polysaccharides can be classified based on their natural sources—such as plants, animals, microbes, and algae—or according to the type of repeating units they contain. Polysaccharide-based matrix for composites contains compounds such as cellulose, starch, pectin, chitosan, chitin, dextran, xanthan, pullulan, alginate, agar, and carrageenan. Polysaccharides can be also classified in homo-polysaccharides (starch, cellulose, bacterial cellulose, dextran, etc.) and hetero-polysaccharides (chitosan, alginate, agar, etc.) (Mohammed et al. 2021; Wang et al. 2023; Felicia et al. 1341). Polyesters are generally Poly(β-, γ-, δ-hydroxyalkanoate)s (PHAs) that are secreted as intracellular energy and carbon storage. The repeating units that make up polyesters are based on hydroxyalkanoic acids, which are components of various bacteria and higher animals. These classes comprise polymers such as Poly(3-hydroxypropionate (P3HP or PHP), Poly(3-hydroxybutyrate (P3HB or PHB), Poly(3-hydroxyvalerate) (P3HV or PHV), Poly(3-hydroxyhexanoate) (P3HH or PHH), Poly(3-hydroxyoctanoate) (P3HO or PHO), Poly(4-hydroxypropionate (P4HP), Poly(4-hydroxybutyrate) (P4HB), Poly(4-hydroxyvalerate) (P4HV), and various co-polymers. These types of polymers are very attractive as potential alternatives for petroleum-based polymers due to their biodegradability and processing potential. They can be processed by traditional techniques like injection molding, extrusion, or compressing molding (Niaounakis 2013; Samrot et al. 2021). It is important to note that not all biopolymers extracted from natural sources are easily biodegradable compared to synthetic polymers. Polysaccharides and proteins are biodegradable and compostable under specific conditions. PHAs are also biodegradable, but the biodegradation process varies depending on the type of polyester, and it may be more complex or require a longer period. However, composites with protein or polysaccharide matrices exhibit limited properties—such as thermal, mechanical, and barrier properties—compared to other polymer composites. Natural rubber, a non-biodegradable biopolymer, is commonly used in the production of polymer-based composites.

21.3.2 Bio-monomers Produced from Biomass via Microbial Fermentation or Extraction

In recent years, rising petroleum prices, reduced availability, and growing societal concern for adopting greener polymers have increased interest in producing bio-monomers from renewable sources. This is likely the fastest growing field in polymer and polymer composite science due to the immense opportunities it presents in addressing humanity's needs. The production of bio-monomers has emerged as a viable alternative to fuel-based monomers due to their lower greenhouse gas emissions. Beyond that, it also offers an alternative to polymers extracted from natural resources. The latter are limited in certain applications due to poor properties, the need for additives, or complex processing, compared to synthetic polymers. Thus, bio-monomers can be used to produce a wide variety of polymers with properties like synthetic polymers or to develop new types of polymers and polymer composites. Considering sustainability, these polymer-based materials closely touch the sustainability concept from the production point of view. One major challenge for these materials is developing efficient recycling methods to meet all the requirements of the sustainability concept (Hernández et al. 2014; Sharma et al. 2024). Some of these polymers rely also on bacterial activity for production but in different ways in comparison with previous classes. These biopolymers are not entirely synthetized from bacteria, but the raw material is produced via bacterial fermentation. In this case, the raw material may include the monomer/oligomer or the monomer precursor. Recent progress in metabolic engineering and catalysis has expanded fermentation for the production of many types of bio-monomers with direct industrial applications or with potential for production of new polymer-based materials. These bio-monomers are produced from vegetable mass conversion via bacterial fermentation using renewable feedstocks. Agriculture and the food industry, along with other sectors, generate large amounts of residues—such as cellulose, lignin, hemicellulose, and sugars—that can be sustainably used as feedstocks for biopolymer synthesis (Moradali and Rehm 2020; Kawaguchi et al. 2017; Minakawa et al. 2019). Other monomers or monomer precursors can be directly extracted from renewable plant-based resources without the need for bacterial fermentation. Generally, renewable sources are divided into vegetable oils (such as corn, castor, grape, canola, and soybean) and non-vegetable oils (such as lemon, orange, and mint). Bio-monomers are typically produced from these precursors through processes like epoxidation, maleinization, carbonation, or acrylation/methacrylation (Gaglieri et al. 2022; Kolář et al. 2023; Kaikade and Sabnis 2023).

21.3.3 Synthetic Polymers from Petroleum-Based Chemicals

These types of polymers were the first to be used in producing polymer composites and have long held the greatest interest in industry. This book chapter will not elaborate on these polymer classes, as there are already numerous papers, reviews, and books in the literature that provide detailed coverage of them. In this section, the main polymer classes will be recapitulated and compared to those previously discussed, focusing on their properties, degradation/biodegradation, and sustainability. The main classes of synthetic polymers are divided into thermoplastics (polyethylene terephthalate (PET), polyethylene (PE), polypropylene (PP), polystyrene (PS), polyvinyl chloride (PVC), polycarbonate (PC), polytetrafluoroethylene (PTFE), nylon, polylactic acid (PLA), etc.), thermosets (resins—epoxy, acrylic, unsaturated polyesters, phenolic, benzoxazine, urea-based, urethane-based), and elastomers (rubbers—polyisoprene, polybutadiene, butyl, nitril, and polychloroprene). Thermoplastics are defined as polymers that soften when heated, making them suitable for molding into various shapes while in their heated state. On cooling, thermoplastic becomes hard and rigid and retains the molded shape. They can also be reheated and reshaped multiple times, allowing for reversible molding into different forms. Thermosets are formed when two or more components chemically react under ambient conditions, or when triggered by radiation or heat, to create a highly cross-linked network. The formation of thermosets is an irreversible process. Elastomer is a rubber-like polymeric material with both elastic and viscous properties. Elastomers have the ability to return to their original shape after being stretched to significant extents. Due to the significant contribution of their viscous properties, elastomers are cross-linked (vulcanized) to promote a predominantly elastic behavior (Rasul and Anderson 2022). These petroleum-based plastics possess the essential thermomechanical properties required for the development of composites. More than that, the production cost for raw materials is low when compared to the synthetic monomers versus bio-monomers. Thus, synthetic polyethylene (PE) and polypropylene (PP) share similar properties with bio-based PE and PP, both being non-degradable, but the synthetic ones are notable for their lower production costs. In comparison, pre-formed polymers from renewable sources exhibit biodegradability. Bio-based monomers and extracted monomers are also derived from renewable sources, but the biodegradability of the resulting polymers remains a complex topic, depending on their specific chemical composition. On the other hand, synthetic polymers from petroleum sources are generally non-biodegradable, with a few exceptions, such as polycaprolactone (PCL), polybutylene succinate (PBS), and polybutylene adipate terephthalate (PBAT).

21.3.4 Types of Composite Additives and Specific Recycling Pathways

Furthermore, this book chapter will briefly discuss the inorganic and organic additives used in polymer composites, and how they complicate the recycling process. These additives include fibers, fillers, reinforcing fillers, plasticizers, lubricants, compatibilizers, coupling agents, stabilizers, and more. In this chapter, these agents are classified into synthetic agents and natural agents or natural polymers.

21.3.4.1 Synthetic Agents

Additives such as reinforcements, fillers, and reinforcing fillers are added to polymer to generate composites with superior properties as was previously described. There are numerous types of agents that can be added according to source, chemistry, shapes, or generated properties. Here, synthetic agents are classified as organic synthetic fibers and inorganic agents. Among synthetic polymer fibers, the most representative are carbon fibers, polyamide fibers (including aramid and aliphatic types), polyester fibers, and polyolefin fibers. Inorganic agents for composites include boron fibers, glass fibers, whiskers, glass particles, carbon black, graphite, carbonates, and oxides. Polymeric fibers share similar processing and recycling methods with their corresponding polymers, which are typically non-biodegradable. Cabon fibers, glass fibers, and other glass-based agents are recycled via specific treatment procedures such as mechanical recycling (shredding) or chemical recycling (vacuum pyrolysis and fluidized bed) (Karuppannan Gopalraj and Kärki 2020; Gonçalves et al. 2022; McConnell 2010; Tcherdyntsev 2021).

21.3.4.2 Natural Polymer Reinforcements/Fillers

These modified or grafted polymers are more suitable as reinforcing agents rather than as polymer matrices. The reinforcing agents are primarily based on cellulose fibers, due to their abundance across various industries. These natural fibers are basically renewable coming especially from agriculture, fashion, or food industry. Cellulose-based reinforcements can be added in forms of woven, non-woven, long fibers, short fibers, flakes, particles, or irregulars in a similar manner with synthetic reinforcements such as carbon fibers or glass fibers. Their composition consists of cellulose, hemicellulose, lignin, pectin, and low ash content. These reinforcements can be divided into several classes according to plant tissue and plant type in their production: bast, leaf, seed, fruit, and stalk. The hydrophilic nature of natural fibers can be an issue for the adhesion with composite matrix such as polyethylene, polypropylene, polylactic acid, acrylonitrile butadiene styrene graft copolymer, or polyvinyl chloride. In general, the preferred and most efficient grafting method is esterification considering that all polymeric components from fiber are rich in hydroxyl groups. There are also other grafting methods, but they are of limited interest in the field of composites. The most common reinforcements are jute, hemp, cotton, bamboo, rice straw, sisal, and kenaf (Kamarudin et al. 2022; Chandramohan and Presin Kumar 2017; Sharma et al. 2019; Baghaei and Skrifvars 2020). In addition to cellulose-based reinforcements, proteins and animal-derived products can also be used as natural reinforcements in composite development. Therefore, silk, wool, mohair, hair, and feathers are the second most abundant natural reinforcements after cellulose-based fibers (Mann et al. 2023; Murillo et al. 2024; Xueliang 2020). Natural reinforcements offer several key advantages over synthetic fibers from a sustainability perspective, including renewability and biodegradability. However, these composite-based materials currently exhibit lower performance compared to those with synthetic agents. Nevertheless, there are high expectations that sustainable composites will achieve their targeted applications in the near future, driven by advancements in technology and industrial processing methods.

21.3.4.3 Other Additives (Auxiliary Additives)

Special additives are generally required during polymer composite development for three main reasons: enhancing compatibility between phases, improving stability, and optimizing processability. Coupling agents and compatibilizers play a key role in the compatibility process. Coupling agents typically form chemical bonds between the polymeric matrix and the reinforcement, whereas compatibilizers rely on physical interactions between the phases. Coupling agents are additives that display chemical groups able to react with both phases. They are low molecular weight compounds: one side of the molecule reacts with the polymer matrix through grafting or condensation reactions, while the other side interacts with the reinforcement surface. They can either become part of the polymeric matrix and react with the reinforcing agent phase after mixing, or they can be bonded to the surface of the reinforcing agent and chemically target the polymeric matrix. Generally, coupling agents are used in high-performance or specialized composites, as they can increase production costs. Compatibilizers are typically macromolecular products (modified polymers) with a similar molecular structure: one part of the molecule is highly compatible with the polymeric matrix through physical interactions, such as long hydrocarbon chains, while the grafted segments interact with the reinforcing agent phase. The processability of polymer composites can be enhanced by the addition of lubricants and plasticizers, depending on the specific requirements of the polymeric matrix. In some cases, fillers improve processability by controlling or adjusting viscosity. Stabilizers slow down the degradation of polymers during processing and usage, thereby enhancing the durability of composites. Auxiliary additives are added in small quantities for reasons of efficiency and cost. Despite their low concentration, these additives can have a significant impact on the recycling process of polymer composites. During the recycling treatment, auxiliary additives complicate the process by additional steps or requirements. Furthermore, they can potentially migrate and undesirably lead to human exposure (Hahladakis et al. 2018; Marturano et al. 2017; Ambrogi et al. 2017; Colin et al. 2011).

For certain thermoplastic matrices, the use of auxiliary additives is essential due to their sensitivity to abiotic and biotic degradation factors. Common auxiliary additives include antioxidants as well as heat and light stabilizers, which are typically based on complex organic compounds such as hindered phenols, sulfur- and phosphorus-containing compounds, carbon black, and surface coatings like varnishes or paints. Primary and secondary antioxidants can be replaced by multi-functional antioxidants that have only recently become available. Due to their special molecular design, they combine primary and secondary antioxidant functions in one compound. Having several stabilizing functions combined in the same molecule, multi-functional antioxidants eliminate the need for co-stabilizers, such as phosphates and thioethers. Furthermore, natural-based oxygen and heat stabilizers have demonstrated effectiveness across various thermoplastic matrices. Polyolefins are particularly sensitive to light and oxygen; therefore, antioxidants such as α-tocopherol, a phenolic component of vitamin E, and flavonoids like quercetin and cyclodextrin are commonly used during processing (Al-Malaika and Issenhuth 2001). Tannins, a class of flavonoids, have also demonstrated antioxidant and UV-stabilizing efficiency in polyolefins, polyesters, and poly(vinyl alcohol), and can function as heat stabilizers in polyvinyl chloride (PVC) (Shnawa et al. 2016; Samper et al. 2013). Lubricants and plasticizers are used to soften polymers and improve their processability. External plasticizers are typically non-volatile phthalate esters, which are highly effective but raise significant concerns due to their potential toxicity to both humans and the environment. Bio-based plasticizers derived from agricultural biomass, such as epoxidized oils, succinates, citrates, glycerol, sugars, and cardanol, are increasingly being explored as alternatives to conventional plasticizers (Sun et al. 2024). Similarly, both internal and external lubricants are gradually being replaced by bio-based compounds sourced from plant oils or animal derivatives. Antimicrobials represent a class of additives that function as stabilizers against biotic agents such as bacteria, fungi, algae, molds, and even rodents. Due to growing concerns over the toxicity of arsenic-based compounds historically used in polymer processing and food-packaging applications, the industry has been compelled to develop safer alternatives. This has led to renewed interest in plant-derived antimicrobial agents. Various spices and herbs have been investigated for their natural antiseptic and medicinal properties, offering promising potential as bio-based antimicrobial additives (Sun et al. 2024; Marturano et al. 2023; Wang et al. 2024).

21.3.5 The General and Newly Developed Recycling Pathways for Polymer Composite

The current state of polymer composites, along with their impact on modern industry and recycling strategies, is further detailed. General and newly developed recycling strategies are highlighted for each industrial polymer composite category, as well as for emerging composites, in alignment with the concept of sustainability. Over time, polymers have driven significant advancements in industry and technology, ultimately improving the quality of human life. However, polymers have accumulated dangerously, spreading across the globe and posing a threat to environmental balance and human health. In this regard, humanity, industrial sector, and researchers needed to understand better what is "plastic," from what substances it is produced, and what are the specific characteristics and usage behavior. Various recycling strategies, alternatives to disposable plastics, and solutions for reducing petroleum-based production have been introduced to mitigate and eventually eradicate plastic pollution (Khare and Khare 2023; Law and Narayan 2022). The major recycling strategies are scattered from primary to quaternary/energy recovery methods: mechanical, chemical (partial/total monomer reconversion), pyrolysis, thermal treatment, enzyme-catalyzed procedures, microbial degradation, etc. All recycling methods are further described in alignment with each polymer composite discussed. Plastic pollution remains a pressing issue, far from being fully resolved. However, several clear sustainable strategies have begun to be implemented in the industrial sector, including mechanical recycling, energy recovery, reduced use of disposable plastics, and the development of bio-based polymers from renewable sources. Therefore, there is a grounded model that can be copied in the case of polymer composites too. This model must be adapted to the particular needs of polymer composites because of higher structural complexity (Ignatyev et al. 2014; Jung et al. 2023). First, it is crucial to recapitulating which the main aspects or approaches are that can make a polymer composite to be called sustainable. United Nations' statement of 1987 defined sustainability concept as "meeting the needs of the present without compromising the ability of future generations to meet their own needs" (Thomsen 2013). The definition is somehow broad, and it does not count general major aspects to be considered. Each situation, domain, or material has to be particularized in order to address the specific needs. In this regard, the light must be focused on composite manufacturing carbon footprint, sources of composite components and environmental impact, durability, recyclability, reusability, and life-cycle costs and impacts (Fig. 21.4).

Another major aspect to be discussed is the technological transfer from research level to small scale/pilot and finally to large-scale implementation. It is crucial for the research field to develop sustainable polymer composites with a high potential for commercial success, taking into account the relationship between eco-friendly materials, eco-friendly manufacturing, available resources, and cost impact. In particular, industrial companies aim to relay the profitability of certain manufacturing technology. Because of that, high financial resources can be seen as a barrier against recycling procedures considering that the recovered materials have low value with respect to primary raw materials. This economical sustainability concept manifests in the polymer industry, being also applied in the composite sector too. Therefore, it

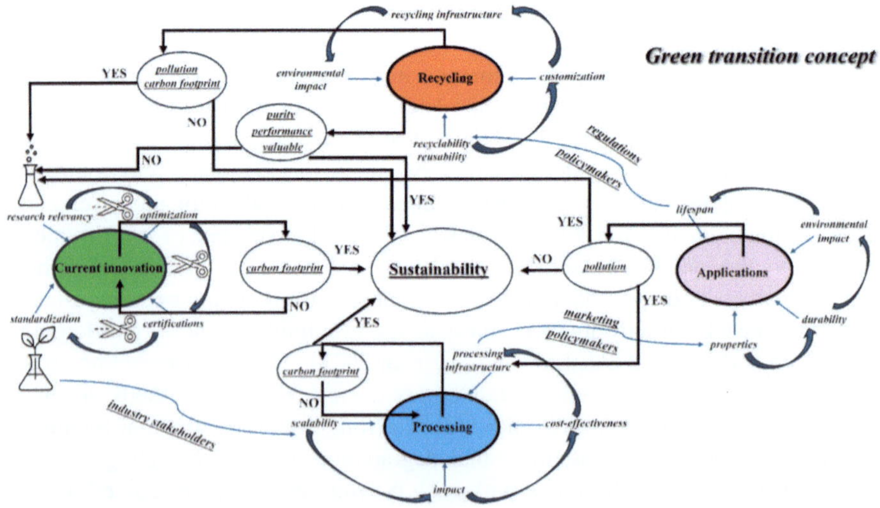

Fig. 21.4 Green transition in polymer composite innovation: recycling, processing, and applications

is highly important that the composition of composites is aligned with the requirements for economic sustainability as well. Otherwise, researchers deliver complex studies with high costs without pointing out the industry requirements (Hubbe 2023; Osburg et al. 2015; Pati 2022). A polymer composite may incorporate bio-based materials, but if the processing methods involve less environmentally friendly steps or the use of harmful solvents at high costs, its sustainability potential comes into question. Similarly, a polymer composite may contain bio-based materials, but if the renewable sources are limited and costly, its sustainability potential is also called into question. The idea of producing raw materials from renewable sources is very generous as was presented and defined in the previous section. However, the impact of using renewable sources must be divided according to different situations. In situations where raw materials are derived from natural resources or from industrial waste, several environmental impacts must be considered. In the case of using natural resources, these impacts may include carbon dioxide emissions, depletion of natural resources, ozone layer depletion, and water pollution caused by solvents, acids, or bases. Vegetal renewable sources may include deforestation or taking over agricultural lands. Furthermore, overall costs and carbon dioxide emissions can be higher due to additional steps and the resources required, such as large amounts of water, fertilizers, pesticides, fuel, machinery, and manual labor. In the case of raw materials derived from industrial waste, the concept and strategies applied are different. From an economic perspective, the overall costs are lower, as animal and plant waste have little value before being converted into bio-monomers, extracted monomers, preformed polymers, and similar products. However, there are several issues related to waste management that include proper waste distribution and access for the composite industry. Furthermore, animal and vegetal waste require pre-treatment steps and collection, transport, or storage (Papageorgiou 2018; Bansal and Rosenholm 2020; Sousa and Silvestre 2022; Bachmann et al. 2021; Rezić and Meštrović 2024).

More importantly, the benefits of polymer composite applications must be carefully balanced with their durability and recycling potential. The reinforcement or filler agent or auxiliary additive concentration must be fitted with the application and required properties. For composites where additives exceed a certain threshold, the properties may not improve and, in some cases, may even deteriorate. In such situations, the recycling process becomes more difficult, costs increase, and there are no added benefits in terms of properties or durability. Polymer nanocomposites and micro-composites often encounter such issues. This approach represents a life-cycle assessment, which seeks to recapitulate all environmental and social impacts. The goal of a life-cycle assessment is to provide a critical perspective on the current industrial state of polymer composites, which have a significant impact on society, and to suggest new research directions for recycling strategies, reuse methods, processing techniques, novel materials, or innovative applications (Hubbe 2023; Pati 2022; Papageorgiou 2018; Bansal and Rosenholm 2020; Sousa and Silvestre 2022; Bachmann et al. 2021; Rezić and Meštrović 2024).

Next, detailed aspects of polymer composites with a significant impact on the industrial sector will be described.

21.3.5.1 Fiber-Reinforced Polymer Composites (High- and Low-Performance Composites)

Polymer composites based on carbon and glass fibers are among the most used and studied in various industrial and research domains. Thus, the growing use of such composites has increased awareness about waste disposal. There is a notable paradox in that the demand for new composites continues to rise, despite the growing issue of composite accumulation (Meng et al. 2018; Ahmad et al. 2022; Andrzejewski et al. 2020). The annual global growth rate has increased by 12.5% in recent years, and the global market is expected to grow by 6.8% annually, reaching $33 billion by 2030, up from $22 billion. In this accumulation rate, thousands of tons of composite waste are accumulated every year. The primary industrial sectors that require new composites and generate significant amounts of composite waste annually are the aircraft and wind energy industries. The aerospace industry is mostly dominated by the carbon fiber composites while the wind sector mostly uses glass fiber composites. Both industries share a similar grow rate for each specific composite type.

The aircraft industry will generate around 23,000 tons per year of composite waste by 2035 without a proper recycling strategy. The automotive and construction industries were less affected by global events such as the pandemic and regional conflicts (Lefeuvre et al. 2017, 2019; Ozdemir et al. 2024). Industries such as electronics, sports, defense, transportation (rail transport), heavy industry (pressure vessels, rocket motor components, lightweight fuel storage), and consumer goods (toys, suitcases, leisure items) have shown fluctuating demand, but with an overall increasing trend (Zhang et al. 2023). Glass fibers are reinforcements produced by melting a mixture of silica sand, limestone, boric acid, and other components at temperatures exceeding 1200 °C. Compared to polymer fibers (aramid) and carbon fibers, glass fibers have lower strength but are less expensive. Glass fiber composites are among the most adaptable for industrial and domestic applications such as transportation (roadway, railways), aerospace, heavy industry (reservoirs, pipelines, pressure vessels, corrosion resistant materials), automotive, wind, or consumer goods applications. In 2022, the global market for glass fibers was valued at $19.69 billion, with a demand of 4,800 tons. It is expected to grow by 3.8–4.9% over the next 10 years. In 2023, glass fibers accounted for approximately 37.5% of the total global market. Compared to carbon and aramid fibers, glass fibers are less commonly used in high-cost industries such as aerospace and automotive, but are more prevalent in wind energy, transportation, and heavy industries (Paulet 2023; Duggal 2022; Bilisik et al. 2018; Shahari et al. 2021).

Aramide fibers are a type of synthetic polymer that contains aromatic rings into macromolecular structure. The most well-known aramid fiber is Kevlar that is mostly used in bulletproof vests, helmets, and other protective gear. Aramid fibers have excellent wear resistance, impact resistance, and abrasion, but are less stiff and strong upon heating. Aircraft industry is typical for aramid composites, where jet engine cowlings, leading edges of wings, and tail planes are manufactured. Probably, the most known application covers the military industries, such as ballistic layered composites and ballistic-rated body armor fabric. Another common application is the use of aramid fiber fabrics in combination with other reinforcement agents, such as carbon and glass fiber-based fabrics, to produce laminate lay-ups. In 2023, the global market for aramid fibers was valued at approximately $4.13 billion, and it is estimated to reach $4.49 billion in 2024. By 2034, the market is projected to exceed $10.23 billion, growing at a rate of 8.7% (Louwsma et al. 2021; Mikko 2020; Loos 2015; Precedence Research).

The supply deficit of virgin carbon fibers, aramid fibers, and glass fibers creates new opportunities for recycled fibers, which offer mechanical properties comparable to those of virgin materials. This recycling process must reduce the demand for new carbon fiber fabrication which is an energy-intensive consumer. This sustainable approach has a direct influence over the reduction of greenhouse gas emission and cost-effectiveness. The aim of the recycling strategy is to overview the composite composition by correlation recycling methods. In general, the polymer composite recycling methods are similar with polymer recycling methods, but with several particularities: mechanical treatment, chemical processing (solvolysis, depolymerization), and thermal processing (pyrolysis, fluidized bed) (Karuppannan Gopalraj and Kärki 2020). Each recycling treatment is highly influenced by the fiber type and matrix type. As discussed in the previous section, polymeric matrices can be thermoplastic, thermoset, or vulcanized elastomers. Each category has distinct characteristics and includes different types of polymers. Thermoplastics can be melted and reshaped a limited number of times, with each cycle affecting the integrity of the macromolecules and the overall properties of the polymer. Thermosets and rubber-based matrices cannot be melted or reshaped once formed. In the industrial sector, thermoset matrices are more widely used compared to thermoplastics or rubbers (Bilisik et al. 2018; Srivastava and Kumar 2022; Foruzanmehr et al. 2015). The main types of thermosets are epoxy, phenolic, vinyl ester, and polyester resins. Thus, the previously discussed resins are representative of glass fibers, aramid fibers, and carbon fibers. This is particularly important from a sustainability perspective, as none of these materials are biodegradable or compostable. Before processing, thermoset precursors have low viscosity that provide higher wetting properties with improved matrix–fiber interaction and easier processability. There is an increasing demand for thermoplastic matrix due to higher fractures and impact resistance or damage tolerance. Thermosets were not chosen solely for processing reasons; they also provide composites with superior mechanical properties, such as higher Young's modulus, strength, lower strain, and greater thermal stability. Over time, new types of thermoplastics were available for the industrial sector that reduced property gap with respect to thermosets. In this context, polymers such as the poly(aryletherketone) (PAEK) family, with the most notable being polyetheretherketone (PEEK) and polyetherketoneketone (PEKK), along with the polyetherimide (PEI) family, polytetrafluoroethylene (PTFE), and newer types of polyamides, polyesters, thermoplastic elastomers, and thermoplastic polyurethanes, have introduced characteristics that were not previously common in thermoplastic composites (Thiruchitrambalam et al. 2020; Pérez-Martín et al. 2021; Dou et al. 2023; Dickson et al. 2020; Singh et al. 2020). Therefore, the fiber recycling process is greatly influenced by the type of polymeric matrix. However, within the same matrix type, the recycling process tends to focus more on recovering the fibers rather than both the fibers and the matrix. The main reason is cost-related, as fiber manufacturing is more expensive and generates more greenhouse gases compared to the production of the polymeric matrix. Another important aspect to consider is the origin and condition of the composite material. Waste from the manufacturing process is easier to recycle because it has not undergone contamination or degradation. This category includes composite fiber offcuts (dry fibers), prepregs, and partially finished components, as well as fibers that can be recovered from the matrix (Ateeq 2023).

Mechanical recycling is the simplest technology used to break down polymer composites into small pieces (0.1–10 mm) through processes such as shredding and milling/grinding. These small pieces, known as recyclates, are sorted by content and size using cyclone machines and sieves. Recyclates produced through this process are classified as matrix-enriched (in powder form) and fiber-enriched (coarser in texture). The main goal is to reuse these recyclates as fillers or even as reinforcements in the production of new composites, particularly for the construction industry. It is crucial to incorporating virgin fibers to enhance the quality of the new composites, as recyclates tend to provide lower quality. However, this method is not suitable for producing new high-performance composites, as the reduced fiber size results in a loss of mechanical properties (Demski et al. 2024; Khalid et al. 2022). The main advantages of mechanical recycling are industrial scalability, low environmental impact, and economic viability of the process (cost-efficient). The main issues that have to be addressed are related to fiber mechanical properties because of the size reduction, the feedstock quality, and variation (Gonçalves et al. 2022). The chopped fibers in the best case are used as short reinforcement fibers for polymer composites with lower performance. However, the industry started to produce systems able to improve the efficiency of recyclates reutilization to reduce raw fiber consumption and environmental pollution. Industrial sector is constrained to work in line with sustainability concept and circular economy due to European directive 2000/53/EC toward reusing, remanufacturing, and recycling products from automotive and aerospace domains (La Rosa et al. 2016; Vieira et al. 2017). One of the industrial directions related to mechanical recycling is the peripheral down-milling process that can be a viable way for reaching economic and environmental benefits. Laminates made from unidirectional carbon fiber prepregs were processed by incorporating them into epoxy resins (Durante et al. 2023). Furthermore, the Tailorable Universal Feedstock for Forming (TuFF) technology is able to produce highly aligned short fiber preforms. The high fiber volume fraction parts of the new composites highlight mechanical properties comparable to continuous polymer fiber composites. The new processes are available for carbon fiber and extended to aramid fiber and glass fiber (https://hmroyal.com/blog/roy-tuff-aramid-fiber-blends/). A consortium of University of Delaware—Center for Composite Materials (UD-CCM), National Renewable Energy Laboratory (NREL), and Colorado State University (CSU) made a huge effort to develop and demonstrate a novel carbon fiber composites recycling process. The recycling process involves a separation step between the polymer matrix and fiber using various catalysts. The recovered short fibers must be cleaned by the polymer matrix via solvolysis and additional pyrolysis (Ozdemir et al. 2024; Giorgini et al. 2020; https://serdp-estcp.mil/projects/details/3a48a5ab-6999-4ea9-8dfc-afb66435fad5/wp21-1089-project-overview).

Chemical recycling: A more sustainable approach certainly involves focusing on recovering both phases: the fiber and the matrix. In mechanical recycling, the type of polymer matrix is less critical, as the process primarily aims to recover fibers as clean short fibers or recyclates. The polymer matrix is removed by solvolysis or pyrolysis, being considered an impurity for recovered fibers. The polymer matrix is highly important in the chemical recycling because this technique is focused on the polymer matrix and at the same time it is driven by the polymer matrix type: thermoplastic, thermoset, or rubbers (elastomers). Chemical recycling typically involves the use of chemical substances or processes to achieve recycling, either by acting as a reactant or solvent (through solvolysis or depolymerization). However, this approach can produce confusion with some thermolysis processes: solvolysis versus thermolysis. There are researchers who refer to thermolysis as a sub-category of chemical recycling. Thus, chemical recycling is a dissolution process (the physical processes of solubilizing a solid) or solvolysis processes. Both processes specifically involve bond-breaking through the action of a solvent medium (Branfoot et al. 2023). Some authors consider the thermolysis process a form of chemical recycling, as it involves the use of solvents under supercritical conditions, applying high temperatures in a thermal–chemical recycling approach. It is likely that, in the future, thermolysis will be regarded as a distinct treatment method, separate from both chemical and thermal recycling, due to emerging techniques and procedures that do not fit within the traditional definitions of these two categories. In this chapter, thermolysis is classified as a thermal recycling process, while solubilization at temperatures below 200 °C and under low pressure is considered a solvolysis process. Next, two main issues must be addressed: how the polymer matrix influences the recycling process, and which are the resulting products. In the case of a thermoplastic matrix, the entire macromolecule dissolves into the solution at lower temperatures and normal pressure. The dissolved polymer can be recovered by precipitation in a non-solvent, while solvent is typically recovered by distillation. In general, dissolution temperatures for thermoplastic must not exceed 200 °C for two reasons: the dissolution process is easier due to lack of crosslinking bridges; higher temperatures can negatively affect the macromolecules, especially at high pressure and long processes (thermal degradation). Even though some polymers are processed at temperatures above 200 °C, these processing steps are typically brief, lasting only seconds or minutes. Prolonged exposure, such as an hour during dissolution at temperatures above 200 °C, is detrimental to the integrity of the macromolecule. When dealing with a polymer matrix in the form of a network, the situation becomes far more complex. Special treatments, such as supercritical conditions, specialized solvents, and specific catalysts, must be applied to generate monomers, oligomers, linear polymers, or new compounds from a chemical recycling perspective. The classic treatment for the chemical recycling of thermosets is the solvolysis of the polymeric matrix or depolymerization mechanism by the insertion of reversible chemical bonds into the resin to produce sustainable resins. Solvolysis uses water, organic solvents including alcohols (ethanol, propanol, butanol, ethylene glycol), phenol, acetone, amines, and the mixture of any of these with water as a co-solvent. Furthermore, various acids (nitric, sulfuric, and acetic acids) and alkaline media are used for the process. The solvents used are hazardous to humans and not always environmentally friendly. The main issue of solvolysis is that it can only be used in recycling of polymer composites with chemically inert reinforcements (just carbon fiber-reinforced composites and glass fiber-reinforced composites). Furthermore, around 90% of the polymeric matrix can be removed via solvolysis (Khalid et al. 2022; Branfoot et al. 2023; Scaffaro et al. 2021). Polymeric fibers, such as aramids and polyesters, are unsuitable for solvolysis recycling due to the high risk of dissolution. The most commonly used thermosets and rubbers as polymer matrices for composites include amine-epoxy, anhydride-epoxy, unsaturated polyester (UP), PEEK (polyether ether ketone as a resin, not to be confused with thermoplastic PEEK), ethylene–propylene–diene monomer (EPDM), polypropylene (PP), styrene–butadiene rubber (SBR), polybutadiene rubber (BR), isobutylene isoprene rubber (IIR), and styrene–butadiene–styrene rubber (SBSR). There are various solvolysis procedures using solvents and specific catalysts applied to degrade network matrix. For resin-based matrix, the unsaturated polyester is the easiest to degrade and PEEK the hardest, while the epoxy categories are somewhere in between. Matrix tends to manifest in terms of the temperatures required to effect complete degradation according to different molecular structures. The thermoset solvolysis produced was able to generate liquid fractions that indicated some valuable small molecules may be recovered and valorized. There is a significant difference between thermosets and rubbers, even though both are structured as networks. Thermosets exhibit a higher degree of crosslinking, with short chains between crosslinking sites, while vulcanized rubbers are composed of macromolecular chains that crosslink during a specific vulcanization step. As a result, thermosets are expected to yield low molecular weight fractions during solvolysis, whereas rubbers produce linear macromolecules. In this context, devulcanization selectively breaks sulfur–sulfur (S–S) and carbon–sulfur (C–S) bonds without chemically altering the polymer backbone (Branfoot et al. 2023; Kabakçi et al. 2022).

In this chapter, the depolymerization process is deemed suitable for matrices with reversible or dynamic chemical bonds that break under specific conditions, such as temperature or the presence of a catalyst. The vitrimer class of polymers meets these criteria, offering dynamic bond exchange along with excellent mechanical properties, chemical resistance, and dimensional stability. Therefore, certain temperatures trigger the reversible chemical bonds of the resin to exchange dynamically. The resulting products can solve the problem of thermosets matrix recycling and may provide a closed loop for such polymer composites in the industrial sector.

The epoxy resin is the most used thermoset matrix for polymer composites. In recent years, various types of epoxy resins with reversible chemical bonds, such as ester, ether, disulfide, Schiff base, and acetal bonds, have been reported, depending on the monomer/oligomer and crosslinking agent used. As previously mentioned, bio-monomers from various sources exhibit ether and ester bonds, offering significant potential to produce polymer matrices in the industry. This last approach is very promising to achieve waste recycling and utilization in industry by transforming polymer composite waste into high value-added compounds (Tian et al. 2022; Montarnal et al. 2011).

In conclusion, chemical recycling works in line with sustainability concept and circular economy for certain situations where both phases, fiber and matrix, can be recovered and reused as valuable products. Recovered fiber requests a high purity for usage in new composite manufacturing. In comparison with mechanical recycling, long fibers can be recovered with target for new high-performance polymer composites. Polymer matrix must be converted into valuable products (monomers, oligomers, linear backbone) with a high purity and at a high content. Besides these approaches, other challenges need to be addressed in the context of sustainability: costs, available resources, scalability, etc. In this context, each type of chemical recycling must be carefully applied and adapted to the specific situation and composite type.

Thermal recycling: This type of recycling is primarily focused on fiber recovery, while the polymeric matrix is often sacrificed and treated as an impurity that needs to be removed. Thermal recycling is industrially scalable for thermoplastic and thermosets with application for various industrial sectors. There are different thermal treatment methods in air or inert media including classic pyrolysis, fluidized bed pyrolysis, microwave-assisted pyrolysis, and superheated steam pyrolysis/superheated steam thermolysis. An external heating is generally used to decompose the matrix phase, leaving fibers and inorganic fillers with a specific degree of purity. The matrix phase degradation products are oil and gases with different molecular weights. These degradation products can be further used as fuel in various situations (Oliveux et al. 2015; Job et al. 2016). Thermal treatment occurs at high temperatures above 400 °C (400–800 °C) that often can affect the fiber structure and further properties. Furthermore, the fiber purity is influenced by the thermal method as the pyrolysis carbon particles are left on the fiber surfaces, while persistent organic pollutants (gaseous fraction) such as dioxins and furans are released. These two presented facts limit the possibility of reusing fibers. Organic reinforcements such as aramid or polyester fibers are not suitable for thermal recycling considering that operating temperatures produce fiber destruction. However, thermal recycling has a great industrial application due to robustness for recycling contaminated polymer composites, high process efficiency, or low process complexity with respect to chemical recycling. By the new insights and advancements such as superheat steam methods, thermal recycling is a vital pillar able to efficiently recycle the composite polymers via avoiding landfilling, extending the products' lifespan and waste valorization (Jiang et al. 2024; Batista et al. 2024; Iglésias et al. 2023; Zhuravskii 2019; Shi 2011). Considering the thermal treatment, the polymer matrix recovery is not possible with respect to the chemical recycling. However, this approach should not be considered a minus from the sustainability point of view because thermal recycling has several superior facilities or particularities. This can be further adapted to industrial sustainability by energy consumption reduction, oil and gas better valorization,

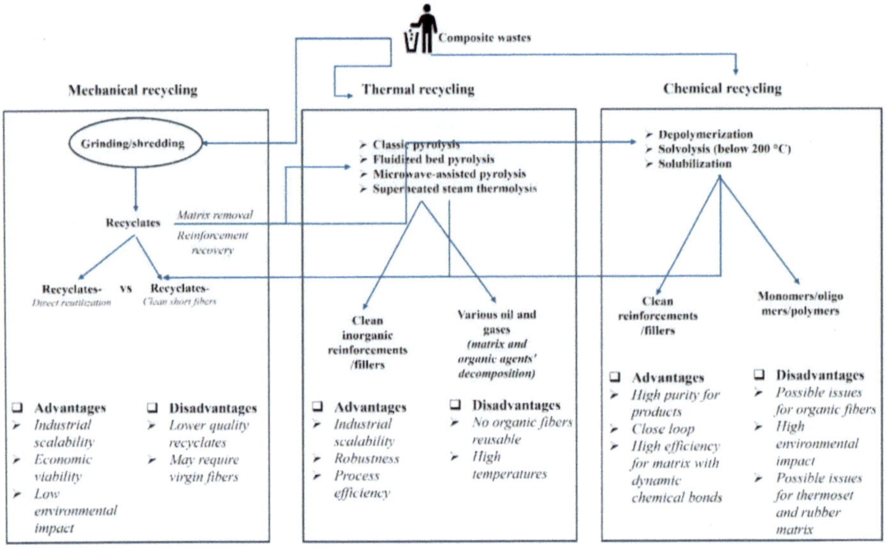

Fig. 21.5 Comparative overview of recycling methods: key differences, advantages, and disadvantages

and assessment of number of times carbon and glass fibers can be recycled by the method.

The key differences, along with the advantages and disadvantages of each discussed recycling method, are summarized in Fig. 21.5.

21.3.5.2 Thermoplastics Versus Thermosets: Key Differences and Separation Challenges

There is ongoing competition between thermoplastic and thermoset polymers in the manufacturing of polymeric composites. As previously discussed, each matrix type possesses distinct properties and characteristics. This section briefly outlines the main challenges associated with the separation of these materials during mechanical recycling, which remains the most widely implemented recycling method in industry. Unlike thermosets, thermoplastics can soften and be reprocessed upon heating. However, during mechanical recycling, polymer nanocomposites are not initially subjected to heat to remove the matrix. Both thermoplastic- and thermoset-based composites are processed similarly in the early stages, typically chopped and shredded into smaller fragments. At this point, short fibers can be more readily recovered and reintegrated in the case of thermoplastics. This is facilitated by the possibility of matrix solubilization, a process generally not applicable to thermosets due to their irreversible crosslinking. Furthermore, thermoplastic matrices allow for easier reintegration of short fibers into new composites through melting and reshaping, unlike thermosets, which undergo permanent hardening. Solvolysis and pyrolysis are applicable to both types of matrices, offering a route to recover valuable components. It is important to note that separating the reinforcement from the polymer matrix is only one aspect of composite sustainability. The manufacturing process itself plays a significant role in terms of energy consumption, processing time, and cost. Interestingly, the very characteristics that limit the recyclability of thermosets, such as irreversible curing, are the same features that make them favorable for high-performance applications in processing (Singh et al. 2018; Jin et al. 2019; Post et al. 2020).

21.3.5.3 Nanocomposites

Initially, several aspects of polymer nanocomposite applications are clarified. For macro- and micro-composites, industrial applications are well-established and widely recognized across various sectors, largely due to the processing of raw materials and the manufacturing of goods in factories. However, in the case of polymer nanocomposites, industrial applications are often not associated with traditional factory settings, as they tend to focus on novel, targeted, or niche applications. In the book chapter, the demonstrated applications of nanocomposites are referred to as commercial applications. Novel applications are also possible, particularly when compared to macro- and micro-composites, due to their unique structural and compositional characteristics (as previously described in the macro versus micro versus nanocomposite section). These include catalyst supports, actuators, sensors, high-performance gas barriers, specialized transport properties, structural reinforcements, nanofilms, supramolecular systems, adsorption, nano-lithographic patterned surfaces, and more (Hassan et al. 2021; Mahesh et al. 2023). The distinctive applications are closely correlated with the type of reinforcement or reinforcing filler, as well as the processing methods employed. One-, two-, and three-dimensional nanoscale agents impart specific characteristics to the entire nanocomposite. Advanced synthesis and processing techniques, such as solution mixing, ultrasonication, melt blending, electrospinning, and in situ polymerization, are utilized to achieve high levels of homogenization or to prevent exposure to high processing temperatures (Sundarram et al. 2015; Darwish et al. 2022). One of the earliest commercial applications emerged in the automotive sector, where Toyota introduced engine components made from nylon reinforced with montmorillonite clay (Kurauchi et al. 1991). Since then, other car manufacturers have developed nanocomposite components such as belt covers, rocker box covering, and body panels. The GM model Hummer also used trim, mid-bridge, canvas panel, and box protectors based on polypropylene reinforced with nanoclays. The manufactured components proved to be lighter and more fracture-proof in comparison with regular metal components. The packaging industry through beverage sector revealed nanocomposite with barrier properties (Hafeez 2022). Nanocellulose-derived composites showed special characteristics that have driven applications toward flame retardancy, thermal insulation, water treatment, or aerogels (Camani et al. 2022). There is a lack of long-term data and future projections regarding the use, safety, and environmental impact of polymer nanocomposites, as the technology behind nanomaterials is relatively new and commercial applications are still in their infancy. Long-term data is vital to offer a pertinent perspective over nanocomposite sustainability potential. Thus, the inherent environmental issues of nano-sized agents can compromise the expected nanocomposite benefits. The lifespan, durability, recyclability, reusability, cost-effectiveness, and environmental impact of nano-agents in nanocomposites are crucial missing links needed to provide a comprehensive assessment of their safety. Special additives or even nano-agents can migrate or be released from the nanocomposites and absorbed in the human body via ingestion, inhalation, or skin barriers during manufacturing, utilization, or end-of-life accumulation. The related problems must be addressed and studied to deliver a sustainability potential. Furthermore, the cost-effectiveness has a high sustainability relevance considering that common nano-agents such as nanocellulose, titanium dioxide, silver, gold, and carbon nanotubes require high costs and emit relevant greenhouse gases. Nano-agents such as clays, graphene, and graphite require less energy and lower costs compared to those mentioned earlier (Carroccio et al. 2022; Aysa et al. 2022; Yashas et al. 2021; Koo 2016). The sustainability approach for polymer matrices in nanocomposites is similar to that of micro and macro-composites, as the available polymeric matrices—thermoplastics, thermosets, and rubbers—are the same. However, the reinforcements (nano-agents) are specific and differ in their production processes when compared to traditional fibers. As previously mentioned, the production of nano-agents is costly, and furthermore, the processing methods for nanocomposites are more complex and expensive. The recycling of polymeric nanocomposites follows the same pathways as other composites, utilizing mechanical, chemical, and thermal recycling methods. However, there is no comprehensive data in the literature that clearly classifies or highlights the specific requirements of nanocomposites during recycling. This issue needs to be addressed, as the unique structural and compositional characteristics of nanomaterials may make them more sensitive or susceptible to certain recycling processes. Mechanical, thermal, or chemical recycling methods could potentially negatively affect nano-agents during grinding, exposure to high temperatures, or treatment with catalysts or solvents. Therefore, the recycling process must be tailored to the specific composition of each nanocomposite. Various established recycling treatments have been applied to nanocomposite waste, revealing structural, thermal, and crystalline properties comparable to those of virgin materials, even after multiple recycling cycles (Velásquez et al. 2024; Hou et al. 2023; Pathak and Yokozeki 2022). One major advantage of polymer nanocomposites is their potential to enhance sustainability, particularly when aligned with polymer recycling science. Nanocomposite matrices can be produced by recycling various polymer wastes. The resulting nanocomposites exhibit promising morphological, mechanical, rheological, and thermal properties, ensuring their potential for future commercial applications. However, the status of polymer nanocomposites derived from plastic waste, in terms of the materials used, processing methods, novel properties, sustainability, economics, safety, and potential applications, requires significant further research (Rezić and Meštrović 2024; Vazquez and Barbosa 2024; Rasilainen et al. 2024; Zare 2013).

21.4 Importance of Composite Recycling and Conclusions

The last part of the book chapter highlights valuable information regarding the importance of composite recycling and its contribution to the circular economy. The green transition toward sustainable development and processing of polymer composites comprises multifaceted challenges that claim integral comprehensive solutions. The main problem is that our global economic growth is mainly based on the overconsumption of natural resources without a sustainable foundation. The green transition serves as a driving force, compelling our research field, industrial recycling, and the economy to address gaps in research relevance, standardization, regulations, certifications, and recycling infrastructure, all of which are essential for achieving industrial sustainability. Overcoming these challenges requires substantial research and the development of innovative approaches to enhance current state-of-the-art methodologies for monitoring the quality of polymer composites during processing, usage, and recycling. This shift is achievable due to the multiple viable options available for selecting polymer sources, agent types, and processing methods, all of which influence carbon footprint, environmental impact, durability, recyclability, reusability, and life-cycle costs. However, the transition to green practices in polymer nanocomposite production and processing is more complex, largely due to the involvement of industry stakeholders. This transition demands a concerted effort from experts to tackle the challenges of scalability and cost-effectiveness, ensuring that continuous technological innovations are compatible with existing infrastructures. A collaborative evolution is needed between industrial recycling and processing infrastructure, alongside innovations, industry stakeholders, and policymakers, to fully harness the potential of green transitions in polymer composite production, paving the way for a sustainable and environmentally responsible future.

21.5 Emerging Trends and Future Research Directions in Composite Recycling

The future of polymeric composite materials depends on a concerted effort among experts in manufacturing, processing, recycling, materials design, and research. Processing and manufacturing methods must be adapted to accommodate a wide range of emerging polymeric materials for the development of high-value composite structures. These approaches should consider complex structural characteristics while ensuring low cost, rapid production, and efficient fabrication. Composite design must also account for factors such as polymer cost, carbon footprint, availability, mechanical and chemical properties, processability, and recyclability. The integration of machine learning and artificial intelligence is becoming increasingly vital for optimizing manufacturing processes and formulation strategies. This approach is going to revolutionize the field of composites by enabling the management of large datasets and accelerating the development of new, sustainable polymeric composites. In addition to the choice of material, processing and manufacturing methods play a critical role in defining and optimizing composite properties, service life, and recyclability.

Acknowledgements This work was funded by the EU's NextGenerationEU instrument through the National Recovery and Resilience Plan of Romania—Pillar III-C9-I8, managed by the Ministry of Research, Innovation and Digitalization, within the project entitled "Advanced 3D Printing for Designer Catalysts Applied to Biomass Conversion," contract no.760086/23.05.2023, code CF 43/14.11.2023.

References

Aggarwal SL, Livigni RA (1977) Structure and properties of microcomposites and macrocomposites from block polymers. Polym Eng Sci 17(8):498–505. https://doi.org/10.1002/pen.760170803

Ahmad J, Majdi A, Al-Fakih A, Deifalla AF, Althoey F, El Ouni MH et al (2022) Mechanical and durability performance of coconut fiber reinforced concrete: a state-of-the-art review. Materials 15(10):3601

Ali M, Al-Shukri A, Maghami MR, Gomes C (2021) Nano and bio-composites and their applications: a review. IOP Conf Ser Mater Sci Eng 1067:012093. https://doi.org/10.1088/1757-899X/1067/1/012093

Al-Malaika S, Issenhuth S (2001) The antioxidant role of vitamin E in polymers. IV. Reaction products of dl-α-tocopherol with lead dioxide and with polyolefins. Polymer 42(7):2915–2939. https://doi.org/10.1016/S0032-3861(00)00653-4

Alvaredo-Atienza Á, Chen L, San-Miguel V, Ridruejo Á, Fernández-Blázquez JP (2020) Fabrication and characterization of PEEK/PEI multilayer composites. Polymers 12(12):2765

Ambrogi V, Carfagna C, Cerruti P, Marturano V (2017) 4—Additives in polymers. In: Jasso-Gastinel CF, Kenny JM (eds) Modification of polymer properties. William Andrew Publishing, pp 87–108

Andrzejewski J, Gapiński B, Islam A, Szostak M (2020) The influence of the hybridization process on the mechanical and thermal properties of polyoxymethylene (POM) composites with the use of a novel sustainable reinforcing system based on biocarbon and basalt fiber (BC/BF). Materials 13(16):3496

Ateeq M (2023) A state of art review on recycling and remanufacturing of the carbon fiber from carbon fiber polymer composite. Compos Part C Open Access 12:100412. https://doi.org/10.1016/j.jcomc.2023.100412

Aysa NH, Shalan AE (2022) Green nanocomposites: magical solution for environmental pollution problems. In: Shalan AE, Hamdy Makhlouf AS, Lanceros-Méndez S (eds) Advances in nanocomposite materials for environmental and energy harvesting applications. Springer International Publishing, Cham, pp 389–417

Bachmann M, Kätelhön A, Winter B, Meys R, Müller LJ, Bardow A (2021) Renewable carbon feedstock for polymers: environmental benefits from synergistic use of biomass and CO_2. Faraday Discuss 230:227–246. https://doi.org/10.1039/D0FD00134A

Baghaei B, Skrifvars M (2020) All-cellulose composites: a review of recent studies on structure, properties and applications. Molecules 25(12):2836

Bansal KK, Rosenholm JM (2020) Synthetic polymers from renewable feedstocks: an alternative to fossil-based materials in biomedical applications. Ther Deliv 11(5):297–300. https://doi.org/10.4155/tde-2020-0033

Baranwal J, Barse B, Fais A, Delogu GL, Kumar A (2022) Biopolymer: a sustainable material for food and medical applications. Polymers 14(5):983

Batista LAPdS, Morgado GFdM, Rodrigues Brazil T, dos Anjos EGR, Guimarães A, Rezende MC et al (2024) Comparative analysis of thermal recycling approaches for carbon fiber recovery from CFRP waste. ACS Sustain Resour Manag 1(9):2108–2118. https://doi.org/10.1021/acssusresmgt.4c00201

Bilisik K, Kaya G, Özdemir H, Korkmaz M, Erdoğan Güler G (2018) Applications of glass fibers in 3D preform composites

Branfoot C, Folkvord H, Keith M, Leeke GA (2023) Recovery of chemical recyclates from fibre-reinforced composites: a review of progress. Polym Degrad Stab 215:110447. https://doi.org/10.1016/j.polymdegradstab.2023.110447

Camani PH, Lemes GD, Rosa DdS (2022) 4—Nanocellulose-based aerogels for various engineering applications. In: Sapuan SM, Norrrahim MNF, Ilyas RA, Soutis C (eds) Industrial applications of nanocellulose and its nanocomposites. Woodhead Publishing, pp 139–153

Carroccio SC, Scarfato P, Bruno E, Aprea P, Dintcheva NT, Filippone G (2022) Impact of nanoparticles on the environmental sustainability of polymer nanocomposites based on bioplastics or recycled plastics—a review of life-cycle assessment studies. J Clean Prod 335:130322. https://doi.org/10.1016/j.jclepro.2021.130322

Chandramohan D, Presin Kumar AJ (2017) Experimental data on the properties of natural fiber particle reinforced polymer composite material. Data Brief 13:460–468. https://doi.org/10.1016/j.dib.2017.06.020

Chauhan A, Singh A, Kumar D, Mishra K (2020) Properties of composite materials, pp 61–78

Chwał M, Muc A (2019) Design of reinforcement in nano- and microcomposites. Materials 12(9):1474

Colin X, Teyssedre G, Fois M (2011) Ageing and degradation of multiphase polymer systems, pp 797–841

Darwish MSA, Mostafa MH, Al-Harbi LM (2022) Polymeric nanocomposites for environmental and industrial applications. Int J Mol Sci 23(3):1023

Demski S, Misiak M, Majchrowicz K, Komorowska G, Lipkowski A, Stankiewicz K et al (2024) Mechanical recycling of CFRPs based on thermoplastic acrylic resin with the addition of carbon nanotubes. Sci Rep 14(1):11550. https://doi.org/10.1038/s41598-024-62594-y

Dickson AN, Abourayana HM, Dowling DP (2020) 3D printing of fibre-reinforced thermoplastic composites using fused filament fabrication—a review. Polymers 12(10):2188. https://doi.org/10.3390/polym12102188

Dou L, Yang B, Lan S, Liu Y, Liu Y, Nan C-W et al (2023) High-entropy-nanofibers enhanced polymer nanocomposites for high-performance energy storage. Adv Energy Mater 13(11):2203925. https://doi.org/10.1002/aenm.202203925

Duggal P (2022) Composites market size is evaluated 86.56 billion by 2021 and estimated to reach 150.28 billion by 2028

Duraikkannu SL, Castro-Muñoz R, Figoli A (2021) A review on phase-inversion technique-based polymer microsphere fabrication. Colloid Interface Sci Commun 40:100329. https://doi.org/10.1016/j.colcom.2020.100329

Durante M, Boccarusso L, De Fazio D, Formisano A, Langella A (2023) Investigation on the mechanical recycling of carbon fiber-reinforced polymers by peripheral down-milling. Polymers 15(4):854

Edwards W (2002) Farm Security and Rural Investment Act of 2002: commodity programs for crops. Iowa State University, Department of Economics, Staff General Research Papers

Felicia WXL, Rovina K, Nur'Aqilah MN, Vonnie JM, Erna KH, Misson M et al (2022) Recent advancements of polysaccharides to enhance quality and delay ripening of fresh produce: a review. Polymers 14(7):1341

Foruzanmehr MR, Elkoun S, Fam A, Robert M (2015) Degradation characteristics of new bio-resin based-fiber-reinforced polymers for external rehabilitation of structures. J Compos Mater 50. https://doi.org/10.1177/0021998315590262

Gaglieri C, Alarcon RT, de Moura A, Bannach G (2022) Vegetable oils as monomeric and polymeric materials: a graphical review. Curr Res Green Sustain Chem 5:100343. https://doi.org/10.1016/j.crgsc.2022.100343
Ghosh CK (2015) Quantum effect on properties of nanomaterials. In: Sengupta A, Sarkar CK (eds) Introduction to nano: basics to nanoscience and nanotechnology. Springer Berlin Heidelberg, Berlin, Heidelberg, pp 73–111
Giorgini L, Benelli T, Brancolini G, Mazzocchetti L (2020) Recycling of carbon fiber reinforced composite waste to close their life cycle in a cradle-to-cradle approach. Curr Opin Green Sustain Chem 26:100368. https://doi.org/10.1016/j.cogsc.2020.100368
Gity Mir Mohamad S, Sayaf M (2015) Compostable polymers and nanocomposites—a big chance for planet earth. In: Dimitris SA (ed) Recycling materials based on environmentally friendly techniques, Ch. 4. IntechOpen, Rijeka
Godara SS, Yadav A, Goswami B, Rana RS (2021) Review on history and characterization of polymer composite materials. Mater Today Proc 44:2674–2677. https://doi.org/10.1016/j.matpr.2020.12.680
Golden J, Daystar J, Handfield R, Pascual-González J (2021) The US biobased products industry: 2021
Gonçalves RM, Martinho A, Oliveira JP (2022) Recycling of reinforced glass fibers waste: current status. Materials 15(4):1596
Hafeez M (2022) Recent progress and overview of nanocomposites
Hahladakis JN, Velis CA, Weber R, Iacovidou E, Purnell P (2018) An overview of chemical additives present in plastics: migration, release, fate and environmental impact during their use, disposal and recycling. J Hazard Mater 344:179–199. https://doi.org/10.1016/j.jhazmat.2017.10.014
Hassan T, Salam A, Khan A, Khan SU, Khanzada H, Wasim M et al (2021) Functional nanocomposites and their potential applications: a review. J Polym Res 28(2):36. https://doi.org/10.1007/s10965-021-02408-1
Hernández N, Williams RC, Cochran EW (2014) The battle for the "green" polymer. Different approaches for biopolymer synthesis: bioadvantaged vs. bioreplacement. Org Biomol Chem 12(18):2834–2849. https://doi.org/10.1039/C3OB42339E
High-aligned short fiber (TuFF) technologies for sustainable composite repair. https://serdp-estcp.mil/projects/details/3a48a5ab-6999-4ea9-8dfc-afb66435fad5/wp21-1089-project-overview
Hong M, Chen EYX (2019) Future directions for sustainable polymers. Trends Chem 1(2):148–151. https://doi.org/10.1016/j.trechm.2019.03.004
Hou Y, Zhu G, Catt SO, Yin Y, Xu J, Blasco E et al (2023) Closed-loop recyclable silica-based nanocomposites with multifunctional properties and versatile processability. Adv Sci 10(35):2304147. https://doi.org/10.1002/advs.202304147
Hsissou R, Bekhta A, Dagdag O, El Bachiri A, Rafik M, Elharfi A (2020) Rheological properties of composite polymers and hybrid nanocomposites. Heliyon 6(6):e04187. https://doi.org/10.1016/j.heliyon.2020.e04187
Hsissou R, Seghiri R, Benzekri Z, Hilali M, Rafik M, Elharfi A (2021) Polymer composite materials: a comprehensive review. Compos Struct 262:113640. https://doi.org/10.1016/j.compstruct.2021.113640
https://hmroyal.com/blog/roy-tuff-aramid-fiber-blends/
https://lynxter.fr/en/blog/guide_en/material-data-sheet-pekk/
Hu X, Cebe P, Weiss AS, Omenetto F, Kaplan DL (2012) Protein-based composite materials. Mater Today 15(5):208–215. https://doi.org/10.1016/S1369-7021(12)70091-3
Hubbe MA (2023) Sustainable composites: a review with critical questions to guide future initiatives. Sustainability 15(14):11088
Iglésias M, Santos P, Lima TM, Leite L (2023) Thermal recycling of glass fibre composites: a circular economy approach. Sustainability 15(2):1396
Ignatyev IA, Thielemans W, Vander BB (2014) Recycling of polymers: a review. Chemsuschem 7(6):1579–1593. https://doi.org/10.1002/cssc.201300898

Ilie N, Hickel R (2009) Macro-, micro- and nano-mechanical investigations on silorane and methacrylate-based composites. Dent Mater 25(6):810–819. https://doi.org/10.1016/j.dental.2009.02.005

Innocenti FD (2003) Biodegradability and compostability. In: Chiellini E, Solaro R (eds) Biodegradable polymers and plastics. Springer, US, Boston, MA, pp 33–45

Jenkins MJ (2001) Crystallisation in miscible blends of PEEK and PEI. Polymer 42(5):1981–1986. https://doi.org/10.1016/S0032-3861(00)00438-9

Jiang S, Liao G, Xu D, Liu F, Li W, Cheng Y et al (2019) Mechanical properties analysis of polyetherimide parts fabricated by fused deposition modeling. High Perform Polym 31(1):97–106. https://doi.org/10.1177/0954008317752822

Jiang G, Lu Y, Weng S, Huang Z, Deng Z (2024) A review of the thermal recycling methods of carbon fibers from carbon fiber reinforced thermosetting resin composites. J Macromol Sci Part B 1–19. https://doi.org/10.1080/00222348.2024.2387407

Jin Y, Lei Z, Taynton P, Huang S, Zhang W (2019) Malleable and recyclable thermosets: the next generation of plastics. Matter 1(6):1456–1493. https://doi.org/10.1016/j.matt.2019.09.004

Job S, Mativenga P, Shuaib NA, Oliveux G, Leeke G, Pickering S (2016) Composites recycling—where are we now?

Jung H, Shin G, Kwak H, Hao LT, Jegal J, Kim HJ et al (2023) Review of polymer technologies for improving the recycling and upcycling efficiency of plastic waste. Chemosphere 320:138089. https://doi.org/10.1016/j.chemosphere.2023.138089

Kabakçi GC, Aslan O, Bayraktar E (2022) A review on analysis of reinforced recycled rubber composites. J Compos Sci 6(8):225

Kaikade DS, Sabnis AS (2023) Polyurethane foams from vegetable oil-based polyols: a review. Polym Bull 80(3):2239–2261. https://doi.org/10.1007/s00289-022-04155-9

Kamarudin SH, Mohd Basri MS, Rayung M, Abu F, Ahmad Sb, Norizan MN et al (2022) A review on natural fiber reinforced polymer composites (NFRPC) for sustainable industrial applications. Polymers 14(17):3698

Karuppannan Gopalraj S, Kärki T (2020) A review on the recycling of waste carbon fibre/glass fibre-reinforced composites: fibre recovery, properties and life-cycle analysis. SN Appl Sci 2(3):433. https://doi.org/10.1007/s42452-020-2195-4

Kawaguchi H, Ogino C, Kondo A (2017) Microbial conversion of biomass into bio-based polymers. Biores Technol 245:1664–1673. https://doi.org/10.1016/j.biortech.2017.06.135

Ke F, Yang X, Zhang M, Chen Y, Wang H (2023) Preparation and properties of polyetherimide fibers based on the synergistic effect of carbon nanotubes and ionic liquids. Polym Compos 44(12):8917–8927. https://doi.org/10.1002/pc.27747

Khalid MY, Arif ZU, Ahmed W, Arshad H (2022) Recent trends in recycling and reusing techniques of different plastic polymers and their composite materials. Sustain Mater Technol 31:e00382. https://doi.org/10.1016/j.susmat.2021.e00382

Khare R, Khare S (2023) Polymer and its effect on environment. J Indian Chem Soc 100(1):100821. https://doi.org/10.1016/j.jics.2022.100821

Kim D, Kim J, Chung S, Cho K (2024) Quantum-dot-induced energy filtering effect in organic thermoelectric nanocomposites. Adv Electron Mater 10(9):2300814. https://doi.org/10.1002/aelm.202300814

Kolář M, Machotová J, Hájek M, Honzíček J, Hájek T, Podzimek Š (2023) Application of vegetable oil-based monomers in the synthesis of acrylic latexes via emulsion polymerization. Coatings 13(2):262

Koo (2016) Environmental and health impacts for nanomaterials and polymer nanocomposites. In: Koo JH (ed) Fundamentals, properties, and applications of polymer nanocomposites. Cambridge University Press, Cambridge, pp 605–647

Kurauchi T, Okada A, Nomura T, Nishio T, Saegusa S, Deguchi R et al (1991) Nylon 6-clay hybrid—synthesis, properties and application to automotive timing belt cover. SAE Trans 100:571–577

La Rosa AD, Banatao DR, Pastine SJ, Latteri A, Cicala G (2016) Recycling treatment of carbon fibre/epoxy composites: materials recovery and characterization and environmental impacts through life cycle assessment. Compos B Eng 104:17–25. https://doi.org/10.1016/j.compositesb.2016.08.015

Law KL, Narayan R (2022) Reducing environmental plastic pollution by designing polymer materials for managed end-of-life. Nat Rev Mater 7(2):104–116. https://doi.org/10.1038/s41578-021-00382-0

Lefeuvre A, Garnier S, Jacquemin L, Pillain B, Sonnemann G (2017) Anticipating in-use stocks of carbon fiber reinforced polymers and related waste flows generated by the commercial aeronautical sector until 2050. Resour Conserv Recycl 125:264–272. https://doi.org/10.1016/j.resconrec.2017.06.023

Lefeuvre A, Garnier S, Jacquemin L, Pillain B, Sonnemann G (2019) Anticipating in-use stocks of carbon fibre reinforced polymers and related waste generated by the wind power sector until 2050. Resour Conserv Recycl 141:30–39. https://doi.org/10.1016/j.resconrec.2018.10.008

Lin C-C, Parrish E, Composto RJ (2016) Macromolecule and particle dynamics in confined media. Macromolecules 49(16):5755–5772. https://doi.org/10.1021/acs.macromol.6b00471

Liu Y, Zwingmann B, Schlaich M (2015) Carbon fiber reinforced polymer for cable structures—a review. Polymers 7(10):2078–2099

Liu L, Liang W, Zhang Y, Fu Q (2023) Nanoencapsulation in polymeric materials: weaving magical coats for microorganisms. Nano Today 52:101973. https://doi.org/10.1016/j.nantod.2023.101973

Loos M (2015) Chapter 2—Composites. In: Loos M (ed) Carbon nanotube reinforced composites. William Andrew Publishing, Oxford, pp 37–72

Louwsma J, Carvalho A, Lutz J-F, Joly S, Chan-Seng D (2021) Adsorption of phenylalanine-rich sequence-defined oligomers onto Kevlar fibers for fiber-reinforced polyolefin composite materials. Polymer 217:123465. https://doi.org/10.1016/j.polymer.2021.123465

Mahesh R, Vora K, Hanumanthaiah M, Shroff A, Kulkarni P, Makuteswaran S et al (2023) Removal of pollutants from wastewater using alumina based nanomaterials: a review. Korean J Chem Eng 40(9):2035–2045. https://doi.org/10.1007/s11814-023-1419-x

Mann GS, Azum N, Khan A, Rub MA, Hassan MI, Fatima K et al (2023) Green composites based on animal fiber and their applications for a sustainable future. Polymers 15(3):601

Marturano V, Cerruti P, Ambrogi V (2017) Polymer additives. Phys Sci Rev 2(6). https://doi.org/10.1515/psr-2016-0130

Marturano V, Marotta A, Salazar SA, Ambrogi V, Cerruti P (2023) Recent advances in bio-based functional additives for polymers. Prog Mater Sci 139:101186. https://doi.org/10.1016/j.pmatsci.2023.101186

McConnell VP (2010) Launching the carbon fibre recycling industry. Reinf Plast 54(2):33–37. https://doi.org/10.1016/S0034-3617(10)70063-1

Meng F, Olivetti E, Zhao Y, Chang J, McKechnie J (2018) Comparing life cycle energy and global warming potential of carbon fibre composite recycling technologies and waste management options. ACS Sustain Chem Eng 6:9854–9865. https://doi.org/10.1021/acssuschemeng.8b01026

Mikko K (2020) Advanced treatments of aramid fibers for composite laminates. In: Tri-Dung N (ed) Composite and nanocomposite materials, Ch. 3. IntechOpen, Rijeka

Minakawa H, Masuo S, Kaneko T, Takaya N (2019) Fermentation and purification of microbial monomer 4-amminocinnamic acid to produce ultra-high performance bioplastics. Process Biochem 77:100–105. https://doi.org/10.1016/j.procbio.2018.11.021

Mitra S, Bhattacharya P, Sen S, Patel TN (2023) Chapter 9—Protein-based composites and their applications. In: Inamuddin, Altalhi T, Alrooqi A (eds) Green sustainable process for chemical and environmental engineering and science. Elsevier, pp 205–250

Mohammed ASA, Naveed M, Jost N (2021) Polysaccharides; classification, chemical properties, and future perspective applications in fields of pharmacology and biological medicine (a review of current applications and upcoming potentialities). J Polym Environ 29(8):2359–2371. https://doi.org/10.1007/s10924-021-02052-2

Mohanty AK, Wu F, Mincheva R, Hakkarainen M, Raquez J-M, Mielewski DF et al (2022) Sustainable polymers. Nat Rev Methods Primers 2(1):46. https://doi.org/10.1038/s43586-022-00124-8

Montarnal D, Capelot M, Tournilhac F, Leibler L (2011) Silica-like malleable materials from permanent organic networks. Science 334(6058):965–968. https://doi.org/10.1126/science.1212648

Moradali MF, Rehm BHA (2020) Bacterial biopolymers: from pathogenesis to advanced materials. Nat Rev Microbiol 18(4):195–210. https://doi.org/10.1038/s41579-019-0313-3

Murillo M, Sánchez A, Gil A, Araya-Letelier G, Burbano-Garcia C, Silva YF (2024) Use of animal fiber-reinforcement in construction materials: a review. Case Stud Constr Mater 20:e02812. https://doi.org/10.1016/j.cscm.2023.e02812

Niaounakis M (2013) Biopolymers: reuse, recycling, and disposal. Biopolym Reuse Recycl Disposal 1–413

Niaounakis M (2014) Biopolymers: processing and products. Biopolym Process Products 1–601

Oliveux G, Dandy LO, Leeke GA (2015) Current status of recycling of fibre reinforced polymers: review of technologies, reuse and resulting properties. Prog Mater Sci 72:61–99. https://doi.org/10.1016/j.pmatsci.2015.01.004

Osburg V-S, Strack M, Toporowski W (2015) Consumer acceptance of wood-polymer composites: a conjoint analytical approach with a focus on innovative and environmentally concerned consumers. J Clean Prod 110. https://doi.org/10.1016/j.jclepro.2015.04.086

Ozdemir T, Deitzel JM, Crane R, Yarlagadda S, Blackwell C, Davis M et al (2024) Carbon fiber composites recycling technology enabled by the TuFF technology. Recycling 9(1):11

Palanisamy S, Mayandi K, Arumugaprabu V, Rajini N, Shanmugavel R (2021) History of composites and polymers, pp 1–21

Pandele AM, Andronescu C, Vasile E, Radu IC, Stanescu P, Iovu H (2017) Non-covalent functionalization of GO for improved mechanical performances of pectin composite films. Compos A Appl Sci Manuf 103:188–195. https://doi.org/10.1016/j.compositesa.2017.10.005

Pandele AM, Constantinescu A, Radu IC, Miculescu F, Ioan Voicu S, Ciocan LT (2020) Synthesis and characterization of PLA-micro-structured hydroxyapatite composite films. Materials 13(2):274

Papageorgiou GZ (2018) Thinking green: sustainable polymers from renewable resources. Polymers (Basel) 10(9). https://doi.org/10.3390/polym10090952

Pathak AK, Yokozeki T (2022) Recycled carbon nanofiber-polypropylene nanocomposite: a step towards sustainable structural material development. J Compos Sci 6(11):332

Pati S (2022) Life cycle assessment: future challenges

Paulet A (2023) Glass manufacturing market forecast & analysis report 2023

Pedoto G, Grandidier J-C, Gigliotti M, Vinet A (2022) Coupling between thermal ageing/degradation and creep behavior of PEKK and C/PEKK composites above the glass transition temperature. Compos A Appl Sci Manuf 153:106717. https://doi.org/10.1016/j.compositesa.2021.106717

Pellis A, Malinconico M, Guarneri A, Gardossi L (2021) Renewable polymers and plastics: performance beyond the green. New Biotechnol 60:146–158. https://doi.org/10.1016/j.nbt.2020.10.003

Pérez-Martín H, Mackenzie P, Baidak A, Ó Brádaigh CM, Ray D (2021) Crystallinity studies of PEKK and carbon fibre/PEKK composites: a review. Compos Part B: Eng 223:109127. https://doi.org/10.1016/j.compositesb.2021.109127

Picu C, Keblinski P (2003) Modeling of nanocomposites. Nanocompos Sci Technol 215–222

Post W, Arijana S, Rolf B, Karin M, Knoop RJI (2020) A review on the potential and limitations of recyclable thermosets for structural applications. Polym Rev 60(2):359–388. https://doi.org/10.1080/15583724.2019.1673406

Precedence Research. Aramid fiber market size, share, and trends 2024 to 2034. https://www.precedenceresearch.com/aramid-fiber-market

Rasilainen I, Lahtela V, Kärki T (2024) A review of plastic waste nanocomposites: assessment of features and applications. Discov Nano 19(1):112. https://doi.org/10.1186/s11671-024-04062-0

Rasul J, Anderson C (2022) Polymers

Rezić I, Meštrović E (2024) Challenges of green transition in polymer production: applications in zero energy innovations and hydrogen storage. Polymers 16(10):1310

Rothon R (2010) Functional fillers for plastics, 2nd edn, pp 115–140

Rudin A, Choi P (2013) Chapter 13—Biopolymers. In: Rudin A, Choi P (eds) The elements of polymer science & engineering, 3rd edn. Academic Press, Boston, pp 521–535

Rudnik E (2019) Chapter 2—Compostable polymer materials—definitions, structures and methods of preparation. In: Rudnik E (ed) Compostable polymer materials, 2nd edn. Elsevier, Boston, pp 11–48

Sacks G, Shah V, Yao L, Yan C, Shah D, Limeta L et al (2024) Polyaryletherketones: properties and applications in modern medicine. Biomed Technol 6:75–89. https://doi.org/10.1016/j.bmt.2023.11.002

Samir A, Ashour FH, Hakim AAA, Bassyouni M (2022) Recent advances in biodegradable polymers for sustainable applications. npj Mater Degrad 6(1):68. https://doi.org/10.1038/s41529-022-00277-7

Samper MD, Fages E, Fenollar O, Boronat T, Balart R (2013) The potential of flavonoids as natural antioxidants and UV light stabilizers for polypropylene. J Appl Polym Sci 129(4):1707–1716. https://doi.org/10.1002/app.38871

Samrot AV, Samanvitha SK, Shobana N, Renitta ER, Senthilkumar P, Kumar SS et al (2021) The synthesis, characterization and applications of polyhydroxyalkanoates (PHAs) and PHA-based nanoparticles. Polymers 13(19):3302

Sathishkumar T, Satheeshkumar S, Naveen J (2014) Glass fiber-reinforced polymer composites—a review. J Reinf Plast Compos 33(13):1258–1275. https://doi.org/10.1177/0731684414530790

Scaffaro R, Di Bartolo A, Dintcheva NT (2021) Matrix and filler recycling of carbon and glass fiber-reinforced polymer composites: a review. Polymers 13(21):3817

Schultz RM (2017) Proteins and protein structure☆. Reference module in life sciences. Elsevier

Seymour RB (1976) The role of fillers and reinforcements in plastics technology. Polym-Plast Technol Eng 7(1):49–79. https://doi.org/10.1080/03602557608063110

Shahari S, Fathullah M, Abdullah MMAB, Shayfull Z, Mia M, Budi Darmawan VE (2021) Recent developments in fire retardant glass fibre reinforced epoxy composite and geopolymer as a potential fire-retardant material: a review. Constr Build Mater 277:122246. https://doi.org/10.1016/j.conbuildmat.2021.122246

Sharma H, Komal U, Singh I, Misra J, Rakesh P (2019) Introduction to green composites, pp 1–13

Sharma M, Tellili N, Kacem I, Rouissi T (2024) Microbial biopolymers: from production to environmental applications—a review. Appl Sci 14(12):5081

Shen C-H (2023) Chapter 4—Gene expression: translation of the genetic code. In: Shen C-H (ed) Diagnostic molecular biology, 2nd edn. Academic Press, pp 91–125

Shi J (2011) Optimum decomposition conditions for glass fiber reinforced plastic recycling by superheated steam. Jpn J Appl Phys 50. https://doi.org/10.1143/JJAP.50.01AJ01

Shimokawa K, Ishihara K, Tezuka Y (2019) Topology meets polymers: introduction. In: Shimokawa K, Ishihara K, Tezuka Y (eds) Topology of polymers. Springer Japan, Tokyo, pp 1–5

Shnawa HA, Jahani Y, Khalaf MN, Taobi AH (2016) The potential of tannins as thermal co-stabilizer additive for polyvinyl chloride. J Therm Anal Calorim 123(2):1253–1261. https://doi.org/10.1007/s10973-015-5082-2

Singh AK, Prasad PB, Smita M, Kumar NS, Gupta MK (2018) Recent developments on epoxy-based thermally conductive adhesives (TCA): a review. Polym-Plast Technol Eng 57(9):903–934. https://doi.org/10.1080/03602559.2017.1354253

Singh J, Chawla K, Singh R (2020) Applications of thermoplastic polymers in 3D printing

Song JH, Murphy RJ, Narayan R, Davies GB (2009) Biodegradable and compostable alternatives to conventional plastics. Philos Trans R Soc Lond B Biol Sci 364(1526):2127–2139. https://doi.org/10.1098/rstb.2008.0289

Soroudi A, Jakubowicz I (2013) Recycling of bioplastics, their blends and biocomposites: a review. Eur Polymer J 49(10):2839–2858. https://doi.org/10.1016/j.eurpolymj.2013.07.025

Sousa AF, Silvestre AJD (2022) Plastics from renewable sources as green and sustainable alternatives. Curr Opin Green Sustain Chem 33:100557. https://doi.org/10.1016/j.cogsc.2021.100557

Srivastava JP, Kumar P (2022) Introduction to glass fiber-based composites and structures. Natural and synthetic fiber reinforced composites, pp 1–16

Sun S, Weng Y, Zhang C (2024) Recent advancements in bio-based plasticizers for polylactic acid (PLA): a review. Polym Test 140:108603. https://doi.org/10.1016/j.polymertesting.2024.108603

Sundarram S, Kim YH, Li W (2015) 4—Preparation and characterization of poly(ether imide) nanocomposites and nanocomposite foams. In: Mittal V (ed) Manufacturing of nanocomposites with engineering plastics. Woodhead Publishing, pp 61–85

Tanasa E, Zaharia C, Radu I-C, Surdu V-A, Vasile BS, Damian C-M et al (2019) Novel nanocomposites based on functionalized magnetic nanoparticles and polyacrylamide: preparation and complex characterization. Nanomaterials 9(10):1384

Tanasa E, Zaharia C, Hudita A, Radu I-C, Costache M, Galateanu B (2020) Impact of the magnetic field on 3T3-E1 preosteoblasts inside SMART silk fibroin-based scaffolds decorated with magnetic nanoparticles. Mater Sci Eng C 110:110714. https://doi.org/10.1016/j.msec.2020.110714

Tariq A, Ahmed SW, Bin Baharom M, Altaf K, Hussain G, Jamil A et al (2024) An investigation of thermal and hydrodynamic behaviour of compound fins under turbulent flow conditions: a numerical and experimental approach. Alex Eng J 89:149–171. https://doi.org/10.1016/j.aej.2024.01.051

Tcherdyntsev VV (2021) Reinforced polymer composites. Polymers 13(4):564

Thiruchitrambalam M, Bubesh Kumar D, Shanmugam D, Jawaid M (2020) A review on PEEK composites—manufacturing methods, properties and applications. Mater Today: Proc 33:1085–1092. https://doi.org/10.1016/j.matpr.2020.07.124

Thomsen C (2013) Sustainability (World Commission on environment and development definition), pp 2358–2363

Tian Z-s, Wang Y-q, Hou X-l (2022) Review of chemical recycling and reuse of carbon fiber reinforced epoxy resin composites. New Carbon Mater 37(6):1021–1041. https://doi.org/10.1016/S1872-5805(22)60652-8

Tokiwa Y, Calabia BP, Ugwu CU, Aiba S (2009) Biodegradability of plastics. Int J Mol Sci 10(9):3722–3742

Vasile E, Radu I-C, Galateanu B, Rapa M, Hudita A, Jianu D et al (2020) Novel nanocomposites based on bacterial polyester/LDH-SDS clay for stem cells delivery in modern wound healing management. Materials 13(20):4488

Vazquez YV, Barbosa SE (2004) Recycling of polymer nanocomposites. Plastic e-waste case study. Chem Phys Polym Nanocompos 759–774

Velásquez E, de Dicastillo CL, Rojas A, Garrido L, Pérez CJ, Lira M et al (2024) Multiple mechanical recycling of a post-industrial flexible polypropylene and its nanocomposite with clay: impact on properties for food packaging applications. Food Packag Shelf Life 43:101272. https://doi.org/10.1016/j.fpsl.2024.101272

Vieira DR, Vieira RK, Chang CM (2017) Strategy and management for the recycling of carbon fiber-reinforced polymers (CFRPs) in the aircraft industry: a critical review. Int J Sust Dev World 24(3):214–223. https://doi.org/10.1080/13504509.2016.1204371

Wang Z, Zhou X, Sheng L, Zhang D, Zheng X, Pan Y et al (2023) Effect of ultrasonic degradation on the structural feature, physicochemical property and bioactivity of plant and microbial polysaccharides: a review. Int J Biol Macromol 236:123924. https://doi.org/10.1016/j.ijbiomac.2023.123924

Wang Z, Liang G, Jiang S, Wang F, Li H, Li B et al (2024) Understanding the environmental impact and risks of organic additives in plastics: a call for sustained research and sustainable solutions. Emerg Contam 10(4):100388. https://doi.org/10.1016/j.emcon.2024.100388

Xu Z, Gehui L, Cao K, Guo D, Serrano J, Esker A et al (2021) Solvent-resistant self-crosslinked poly(ether imide). Macromolecules 54(7):3405–3412. https://doi.org/10.1021/acs.macromol.0c02860

Xueliang X (2020) Animal fibers. Handbook of fibrous materials, pp 37–74

Yashas SR, Shahmoradi B, Wantala K, Shivaraju HP (2021) Potentiality of polymer nanocomposites for sustainable environmental applications: a review of recent advances. Polymer 233:124184. https://doi.org/10.1016/j.polymer.2021.124184

Zare Y (2013) Recent progress on preparation and properties of nanocomposites from recycled polymers: a review. Waste Manag 33(3):598–604. https://doi.org/10.1016/j.wasman.2012.07.031

Zhang W-B, Cheng SZD (2020) Giant is different: size effects and the nature of macromolecules. Giant 1:100011. https://doi.org/10.1016/j.giant.2020.100011

Zhang J, Lin G, Vaidya U, Wang H (2023) Past, present and future prospective of global carbon fibre composite developments and applications. Compos B Eng 250:110463. https://doi.org/10.1016/j.compositesb.2022.110463

Zhuravskii GI (2019) Thermolysis of polymeric composite materials. J Eng Phys Thermophys 92(3):603–607. https://doi.org/10.1007/s10891-019-01968-w

Zia KM, Akram N, Tabasum S, Noreen A, Akbar MU (2021) 5—Processing of bio-based polymers. In: Zia KM, Akram N, Tabasum S, Noreen A, Akbar MU (eds) Processing technology for bio-based polymers. Elsevier, pp 151–189

Chapter 22
Going Green for Profit: Market Analysis and Economic Feasibility of Sustainable Polymer Composites

Nusarath Mohamed Kutty and Muhammed Nayeem Mullungal

Abstract Sustainable polymer composites (SPCs) combine environmental responsibility and industrial efficiency to revolutionize material science. These advanced materials use hemp, flax, and coir, bio-based resins like polylactic acid (PLA) and starch, and recycled polymers to reduce reliance on fossil fuels. SPCs can match or exceed the mechanical and thermal properties of traditional composites. Market developments further substantiate the commercial potential of SPCs. The global SPC market is projected to attain USD 86.43 billion by 2030, with a CAGR of 15.30%, driven by their use in the automotive, aerospace, and packaging sectors. Regulatory frameworks like the EU Green Deal and an increasing consumer inclination toward eco-friendly items are propelling this growth trajectory. Moreover, industry giants such as Ford and Boeing are diligently incorporating recycled carbon fibers and agro-industrial by-products, including wheat straw and coconut husks, into their supply chains to improve sustainability and cost-efficiency. Despite these gains, several issues remain in terms of mechanical performance and limited recycling infrastructure. Nevertheless, the utilization of SPCs across many industries is increasing. In the automobile industry, lightweight composites enhance electric car efficiency. Aerospace innovations which incorporate composites for more than 50% of its structure illustrate the capacity of SPCs to significantly decrease fuel usage. Biodegradable geopolymers and PLA-based materials are viable alternatives to traditional, non-degradable goods in the building and packaging industries. Sustainable polymer composites possess transformational potential for furthering ecological and industrial objectives.

N. M. Kutty
Department of Marketing, College of Business and Economics, Qatar University, Doha, Qatar

M. N. Mullungal (✉)
Environmental Science Program, Department of Biological and Environmental Sciences, College of Arts and Sciences, Qatar University, Doha, Qatar
e-mail: mmullungal@qu.edu.qa

L. Ravindran et al. (eds.), *Macro, Micro and Nanocomposites from Sustainable Sources*, Advanced Structured Materials 244, https://doi.org/10.1007/978-981-95-2469-3_22

22.1 Sustainable Polymer Composites

Sustainable polymer composites are innovative materials that incorporate environmentally friendly principles throughout their entire lifecycle, from production to disposal. These composites aim to reduce ecological impact by utilizing renewable resources, enhancing recyclability, and promoting biodegradability (Raja et al. 2024; Leite-Barbosa et al. 2024). By integrating natural residues such as sawdust and agro-industrial by-products into polymer matrices, they not only improve mechanical and thermal properties but also contribute to waste reduction and resource efficiency. This approach aligns with the principles of a circular economy, where materials are reused and repurposed to minimize environmental harm (Leite-Barbosa et al. 2024). The use of these materials has effectively replaced traditional materials such as metals in a variety of applications, resulting in a reduced weight while simultaneously increasing efficiency (Subadra and Griskevicius 2021).

Beyond material composition, sustainable polymer composites also focus on responsible end-of-life management. Fiber-reinforced polymer composites, for instance, emphasize recyclability, biodegradability, and product simplification to create a more sustainable lifecycle (Elser and Buchmeiser 2024). These characteristics make them suitable for various applications, from packaging to automotive and construction industries, where reducing environmental impact is a growing priority. There are other industries, including packaging and medicine. By integrating bio-based materials and improving disposal strategies, sustainable polymer composites play a vital role in advancing environmentally friendly material development. At present, this sector constitutes a significant source of revenue for certain economies, particularly those that have access to natural resources. This contributes to the development of environmentally conscious economies and strengthens the global community's efforts to achieve sustainability (Vazquez et al. 2021).

Sustainable polymer composites exhibit key characteristics that enhance both environmental responsibility and material performance. They are often composed of natural fibers and bio-based resins, such as polylactic acid (PLA), a biodegradable polymer derived from renewable sources (Elser and Buchmeiser 2024; Toprakci et al. 2024). A crucial aspect of their sustainability is recyclability, with advancements in processing techniques enabling the reuse of composite materials, thereby minimizing waste generation (Elser and Buchmeiser 2024; Shih et al. 2024). Additionally, emerging manufacturing methods, including 3D printing, facilitate the production of high-performance composites that maintain strength and durability while adhering to sustainable principles (Sehgal and Singh 2024; Xu et al. 2024). However, achieving an optimal balance between performance and environmental impact remains a challenge, necessitating continuous research and innovation to enhance the viability and applicability of these materials.

22.2 Type of Polymeric Materials Used and Their Applications

Sustainable polymer composites incorporate a diverse range of eco-friendly materials, including natural fibers, bio-based resins, and recycled materials, all of which contribute to reducing environmental impact while maintaining structural performance. Natural fibers, such as hemp, kenaf, and other agricultural residues, are widely used due to their biodegradability and mechanical strength (Bari et al. 2021). Specific materials like coconut mats and flax/hemp fabrics serve as insulation in geopolymer composites, enhancing both thermal efficiency and structural integrity (Przybek and Łach 2024). Additionally, wood flour and cellulose function as reinforcements in polyoxymethylene (POM) composites, improving mechanical properties while remaining biodegradable (Soćko and Andrzejewski 2024).

Biopolymers including polysaccharides (cellulose, starch) and proteins (collagen, silk fibroin) are sustainable (Table 22.1). Polyhydroxyalkanoates (PHAs) and polylactide (PLA) are also produced from renewable monomers and are also increasingly used to enhance the sustainability of composite materials (Cazan 2023). PLA, derived from renewable sources, is often plasticized to improve flexibility and can be enhanced with natural additives like perlite to strengthen mechanical properties (Toprakci et al. 2024). Modern polymer chemistry allows for the production of fossil-based polymers like polyethylene, PET, and polyurethanes from renewable feedstocks like biomass and CO_2. Recycled materials, including waste polypropylene (wPP), are integrated into composite structures, such as wPP combined with carbon-fiber-reinforced epoxy scrap, to create thermoplastic composites that support circular economy initiatives by repurposing industrial waste (Zolfaghari et al. 2024). Thus, some sustainable polymers are biodegradable (e.g., PLA in composting, PHA in nature, and chitosan in the body), whereas others are more lasting and require good recycling to lessen environmental impact.

Plant fibers are harvested from different plant parts, each with its unique properties and functions (Table 22.2). Bast or stem fibers are suitable for bags, textiles, and ropes due to their length and strength. In comparison, leaf fibers are coarse and weak. Carpets, rugs, and clothes are made using them. Fruit and seed fibers are lightweight and durable, making them ideal for brushes, carpets, and fabrics. Husks provide these fibers. Most grass fibers are used in cement composites. These fibers come from tall grasses. These fibers are found worldwide; however they are predominantly produced in tropical regions, especially Asia. Coir and jute are key commodities from India, Pakistan, and Bangladesh. China, along with other Asian nations, supplies wool and cotton.

Sustainable polymers can be used in many ways due to their versatility and degradability. Sustainable polymer composites made via blending, nanofiller inclusion, or active ingredient loading can be functionalized through chemical modification or improved manufacture. These materials are promising for packaging, health care,

Table 22.1 The source characteristics of various sustainable polymer composites

Material	Abundance	Sources	Applications	Properties
Cellulose	Most abundant in nature	Plants (e.g., wood, cotton)	Packaging, single-use items, cosmetics	Biodegradable, renewable, low cost
Starch	High availability	Plants (e.g., corn, potatoes, wheat)	Food tech, biodegradable packaging	Compostable, renewable
Pectin	Moderate	Fruits (e.g., apples, citrus peels)	Food tech (gelling agent), microbial growth promoter	Biodegradable, biocompatible
Chitosan	Limited	Crustacean shells (e.g., shrimp, crab), fungal sources	Biomedical, food tech, cosmetics	Biocompatible, antimicrobial
Alginate	Moderate	Phaeophyceae (brown seaweed)	Food tech, biomedical (wound dressing)	Biodegradable, chemical stability

Table 22.2 The major plant composition and their origin

Fiber type	Plant sources	Key countries
Bast/stem	Flax	India, Bangladesh, China, France
Leaf	Palm	Brazil, Philippines, Tanzania
Fruit/seed	Coconut	Indonesia, Sri Lanka, Mexico
Cotton/wool	Sheep + Boll	USA, India, Pakistan, Australia
Grass	Rice	Thailand, Vietnam, Myanmar

and the environment. The use of these sustainable materials demonstrates a significant shift toward environmentally responsible composite development, emphasizing both functionality and reduced ecological impact.

Sustainable polymer composites demonstrate promising mechanical properties, though they may not always match the performance of fully synthetic materials (Prasad et al. 2024). The incorporation of recycled plastics, ground tire rubber (GTR), and bio-based resources enhances tensile strength and impact resistance, contributing to a circular economy (Dasgupta et al. 2024; Singh and Singh 2024). These materials also offer significant thermal stability improvements, with studies showing that adding GTR to polyethylene composites increases melting temperatures, making them more durable in various applications (Dasgupta et al. 2024). Additionally, advanced manufacturing techniques, such as 4D printing and the use of bio-based additives, enhance design flexibility while reducing material waste and energy consumption (Olawumi et al. 2024).

Due to their multifunctional properties, sustainable polymer composites are increasingly utilized in industries such as automotive, aerospace, and construction (Islam et al. 2022). The adoption of sustainable processing methods, including 3D printing, has enabled the production of high-performance plastics with lower energy requirements, expanding their applicability across different sectors (Xu et al. 2024). By integrating innovative materials and processing techniques, sustainable polymer composites provide an environmentally friendly alternative to conventional materials while addressing key challenges such as waste reduction and resource efficiency.

22.3 Aligning Profitability with Sustainability

Aligning profitability with sustainability involves adopting eco-friendly practices that also make good business sense. Companies that switch to sustainable materials, like biodegradable plastics and recycled polymers, can lower waste management costs and comply with regulations more easily, saving money in the long run (Unni and Joseph 2024). Additionally, improving efficiency and reducing risks through sustainable strategies can lead to long-term financial benefits (Umana et al. 2024).

A circular economy approach helps businesses stay profitable by designing products for recyclability and reusability, reducing the need for new raw materials (Fagnani et al. 2022; Thomas et al. 2024). Innovations such as using bio-based ingredients and eco-friendly manufacturing methods maintain product quality while cutting environmental impact (Fagnani et al. 2022). Companies that engage stakeholders and adopt corporate social responsibility (CSR) initiatives also build a positive reputation, attracting customers and investors who support sustainability (Umana et al. 2024). By making smart choices that benefit both the planet and their bottom line, businesses can achieve long-term success while being environmentally responsible.

22.4 Key Market Insights

Sustainable polymer composites have become an essential focus in materials science due to the increasing demand for eco-friendly products across various industries. In particular, sectors such as automotive, aerospace, and construction are adopting these composites to meet sustainability goals while maintaining high performance. The market for sustainable polymer composites is expected to grow rapidly, with a projected compound annual growth rate (CAGR) of 15.30%, reaching USD 86.43 billion from 2022 to 2030 (Oh et al. 2024). This growth is attributed to the increasing consumer demand for environmentally friendly products, coupled with regulatory pressures aimed at reducing the carbon footprint and waste. According to a report by Zion Market Research (2024), the global sustainable polymers market was valued at approximately USD 8.27 billion in 2023 and is projected to reach USD 65.99 billion by 2032, reflecting a compound annual growth rate (CAGR) of 23.1% during the forecast period from 2024 to 2032 (Fig. 22.1).

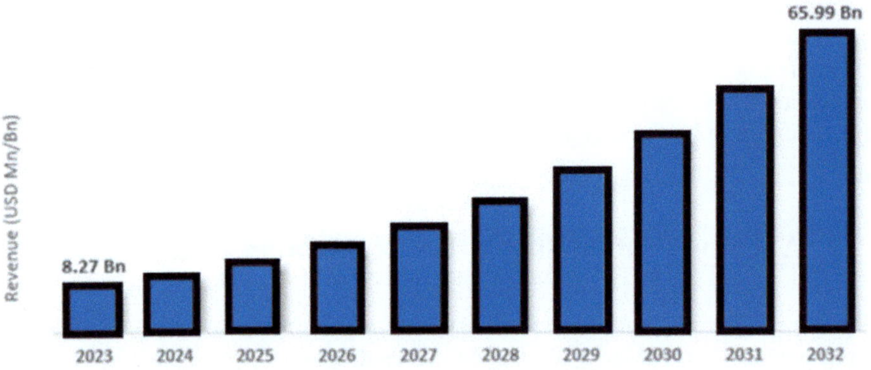

Fig. 22.1 Global sustainable polymers market projections for 2024–2032 in USD Billion, adopted from Zion Market Research, 2024

Projections indicate continued growth in the sustainable polymer composites market. IndustryARC forecasts the composites market to reach USD 158 billion by 2030, growing at a CAGR of 7.9% during 2024–2030, driven by the automotive industry's adoption of polymer composites (Composites Market Report—Forecast (2024–2030).

22.5 Current Market Trends in Sustainable Polymer Composites

The market for sustainable polymer composites is evolving rapidly, driven by various trends that are reshaping industries worldwide. Among these, the key trends include the following.

- **Sustainability Drivers**: Sustainability is one of the most significant drivers in the market, with national and international regulations, such as the European Union's Green Deal and the Circular Economy Action Plan, pushing industries toward adopting more eco-friendly materials. Additionally, increasing public concerns about climate change and the growing awareness of issues like microplastics are making sustainability a priority for both companies and consumers (Elser and Buchmeiser 2024). These trends highlight the need for industries to transition to materials that not only minimize environmental harm but also promote long-term sustainability.
- **End-of-Life Management**: One of the biggest challenges for fiber-reinforced polymer composites (FRPCs) is end-of-life management. Currently, a large percentage of these materials are landfilled, with only a small fraction being recycled or reused. For example, in the UK, 35% of carbon-fiber-reinforced polymer composites (CFRPCs) and 67% of glass-fiber-reinforced polymer composites

(GFRPCs) were landfilled in 2020. Therefore, improving recycling technologies and developing efficient end-of-life management processes are essential to reducing waste and enhancing the overall sustainability of FRPCs (Elser and Buchmeiser 2024).

- **Recyclability and Material Use**: Recyclability has become a central focus in the development of sustainable polymer composites. The incorporation of bio-based materials, such as natural fibers and biodegradable resins, is vital to enhancing the recyclability of composites. These materials not only reduce the reliance on fossil fuels but also help create products that are easier to recycle at the end of their lifecycle. Innovations in sustainable thermoset and thermoplastic matrices further contribute to the recyclability of these composites, ensuring that they are not only effective in their use but also easier to manage once their initial purpose has been fulfilled (Elser and Buchmeiser 2024).

22.6 Growth Drivers for Sustainable Polymer Composites

Several key growth drivers are fueling the market for sustainable polymer composites, particularly in industries that are increasingly focusing on environmental impacts. These drivers include the following.

- **Regulatory Support and Environmental Policies**: National and international regulations, particularly the EU's Green Deal and Circular Economy Action Plan, are powerful drivers of change in the sustainable polymer composites market. These regulations incentivize the use of sustainable materials by setting mandates for industries to reduce their carbon footprint and promote circular economy practices. Such regulatory frameworks are instrumental in encouraging the widespread adoption of more sustainable alternatives in manufacturing (Elser and Buchmeiser 2024).
- **Public Awareness and Environmental Concerns**: As the global public becomes more aware of environmental issues such as climate change and microplastic pollution, there is a growing demand for sustainable products. This trend is especially evident in industries like automotive, where companies are under increasing pressure to reduce emissions and improve the sustainability of their products. Consumer demand for eco-friendly products is pushing manufacturers to seek out alternative materials that are more sustainable and less harmful to the environment (Carvalho et al. 2024).
- **Technological Advancements**: Innovations in recycling technologies and the development of bio-based materials are significantly contributing to the growth of sustainable polymer composites. For example, new recycling methods such as enzymatic depolymerization are helping reduce the energy required for recycling processes, making it more economically viable to reuse composites. Additionally, advancements in materials science have led to the development of bio-based resins and natural fibers that enhance the performance and sustainability of polymer composites (Carvalho et al. 2024).

- **Market Demand for Sustainability**: Growing demand for sustainable products across industries such as transportation, construction, and electronics is a major driver of the sustainable polymer composites market. With industries aiming to reduce their environmental impact, there is increasing demand for alternative materials that align with sustainability goals. As these industries strive to meet consumer expectations for greener products, the adoption of sustainable polymer composites continues to rise (Aisyah et al. 2021).
- **Cost-Effectiveness**: As sustainability becomes a key focus for manufacturers, cost-effective alternatives to traditional materials have become essential. Natural fiber composites, for instance, are more affordable to produce than synthetic composites and offer a viable solution for reducing production costs. With their abundant availability and low production costs, natural fiber composites are particularly appealing for sectors like automotive and construction, where reducing costs while maintaining quality is crucial (Aisyah et al. 2021).
- **Shift Toward Bio-Based Materials**: The transition from fossil-based to bio-based feedstocks is another significant growth driver for sustainable polymer composites. Bio-based materials, such as those derived from lignin, cellulose, and other renewable resources, offer significant reductions in carbon emissions and resource depletion compared to traditional, fossil-based materials. This shift is becoming increasingly important in meeting sustainability targets and is aligning with consumer preferences for greener products (Carvalho et al. 2024).

22.7 Major Industries Adopting Sustainable Polymer Composites

Sustainable polymer composites are gaining traction across several key industries due to their ability to reduce environmental impact while maintaining performance. These materials, which often incorporate bio-based and recycled components, are revolutionizing traditional industries by offering eco-friendly alternatives to conventional materials. Below is an overview of how major industries, including automotive, aerospace, construction, and packaging, are adopting sustainable polymer composites to meet their sustainability goals.

22.7.1 Automotive Industry

The automotive industry is one of the largest sectors adopting sustainable polymer composites, driven primarily by the growing demand for electric vehicles (EVs) and urban air mobility (UAM) solutions. Sustainable composites such as biopolymers and natural fibers are favored for their lightweight properties, which contribute to improved fuel efficiency and reduced greenhouse gas emissions (Oh et al. 2024). These materials are essential for reducing the weight of vehicles, which directly translates into lower energy consumption and enhanced overall performance. With the automotive market's increasing shift toward greener alternatives, the demand for sustainable composites is set to rise, with projections indicating a compound annual growth rate (CAGR) of 15.30%, potentially reaching USD 86.43 billion by 2030 (Oh et al. 2024). This growth reflects the industry's commitment to sustainability and reducing its carbon footprint.

22.7.2 Aerospace Industry

In the aerospace industry, sustainable polymer composites are making significant strides as companies seek to minimize environmental impact while improving material performance. A key trend is the incorporation of recycled materials, particularly aerospace prepreg waste, into thermoplastic composites. This not only reduces waste but also ensures that materials meet the demanding mechanical requirements of aerospace applications (Butenegro et al. 2023). The aerospace sector, known for its stringent performance standards, is increasingly adopting composites that provide both environmental benefits and superior mechanical properties, making them suitable for high-performance applications. By integrating recycled content into these composites, the industry is able to reduce its environmental footprint without compromising on the quality and durability needed for aerospace components.

22.7.3 Construction Industry

The construction industry is increasingly adopting sustainable polymer composites due to their durability and eco-friendly properties. These materials are used in a wide variety of applications, from structural elements to insulation materials. The shift toward sustainable composites is driven by the growing recognition of the need for responsible resource management and the reduction of construction waste. Sustainable composites, often incorporating natural fibers and biopolymers, offer an attractive alternative to conventional materials that are less environmentally friendly (Khan et al. 2024). These composites help improve the energy efficiency of buildings and infrastructure, contributing to a reduction in overall carbon emissions from the construction sector. Additionally, the durability of these materials ensures that they last longer, reducing the need for frequent repairs or replacements.

22.7.4 Packaging Industry

The packaging industry is another key adopter of sustainable polymer composites, with a significant shift toward biodegradable alternatives to combat plastic waste.

Traditional plastic packaging, which poses serious environmental hazards, is being replaced with more sustainable options made from natural fibers and biopolymers (Khan et al. 2024; Maiti et. al. 2022). These green composites are not only biodegradable but also offer the required strength and flexibility for various packaging applications. The transition to biodegradable packaging materials is a crucial part of the global effort to reduce plastic pollution, and as consumer demand for sustainable products increases, the packaging industry is expected to see continued growth in the use of these eco-friendly composites. Innovations in material processing and design are further driving this shift, enabling packaging companies to meet sustainability goals without sacrificing functionality or performance.

22.8 Key Players and Market Dynamics

22.8.1 Automotive Industry

The automotive industry has been a leader in integrating sustainable polymer composites, particularly in light of global sustainability goals and the rise of electric vehicles (EVs). The use of Natural Fiber Reinforced Polymer Composites (NFRPCs) in vehicle manufacturing allows for reduced weight, which directly enhances fuel efficiency and decreases greenhouse gas emissions. Key automotive manufacturers have adopted these sustainable materials to align with their sustainability initiatives and meet evolving regulatory requirements regarding carbon emissions (Oh et al. 2024). Furthermore, the demand for green composites in the automotive industry is projected to grow significantly, with a compound annual growth rate (CAGR) of 15.30%, reaching USD 86.43 billion by 2030 (Skosana et al. 2024).

22.8.2 Material Suppliers

Companies specializing in biopolymers and natural fibers have become essential suppliers in the sustainable composites market. These suppliers focus on developing innovative materials that meet both environmental standards and performance expectations. This growing emphasis on biopolymers and natural fibers has allowed manufacturers to shift toward more sustainable production processes, which provide the dual benefit of reducing the ecological footprint while maintaining the desired material properties. As these materials become more widely available and cost-effective, the market for sustainable composites is expected to expand further.

22.8.3 Performance and Sustainability of Sustainable Composites

22.8.3.1 Mechanical Properties

One of the major challenges faced by NFRPCs is that, despite offering significant environmental advantages, they often struggle to match the mechanical performance of traditional composites. While these materials provide enhanced sustainability features, such as biodegradability and reduced environmental impact, their mechanical properties can fall short in demanding applications. To address this, ongoing research is focused on optimizing fiber-matrix interactions and exploring various treatment techniques to improve the strength and durability of these composites (Maiti et al. 2022). Continued innovation in this space will be key to expanding the applications of sustainable composites, especially in high-performance industries like automotive and aerospace (Skosana et al. 2024).

22.8.3.2 Recyclability and End-of-Life Management

The sustainability of NFRPCs is significantly enhanced by their recyclability and biodegradability. Effective end-of-life management is a crucial component of the sustainable composite lifecycle, reducing waste and promoting resource efficiency. While the industry has made significant progress in improving recycling technologies, there are still challenges associated with recycling complex composite structures. However, innovations in material formulations and recycling methods are making NFRPCs increasingly viable in terms of environmental impact reduction (Elser and Buchmeiser 2024). The focus on recyclability and circularity in material production aligns with global sustainability goals, fostering more sustainable manufacturing practices (Maiti et al. 2022).

22.8.4 Economic Feasibility of Sustainable Composites

22.8.4.1 Material Costs

The cost comparison between sustainable and conventional polymer composites reveals significant differences, both in material costs and long-term environmental savings. Sustainable composites, such as those made with vanillin-based polymers, typically have higher material costs due to the use of renewable resources and advanced manufacturing techniques. For example, some sustainable nanocomposites can cost around \$25 per kilogram (Sehgal et al. 2024). In contrast, traditional composites like carbon fiber-reinforced polymers (CFRPs) tend to be less expensive initially, but they contribute to higher environmental costs over time, such as increased carbon footprints and waste disposal issues (Zamani et al. 2024).

Interestingly, sustainable composites using bio-based fillers can often be cheaper than their conventional counterparts, like silica-based or petroleum-based fillers. This is due to their lower density and biodegradable properties, which reduce costs in production and disposal (Dadkhah and Messori 2024). Despite the higher upfront costs, the long-term savings from recyclability and a smaller environmental footprint are increasingly recognized as valuable.

22.8.4.2 Processing Costs

The processing costs of sustainable composites are often higher than conventional composites. This is due to the specialized techniques required for recycling and manufacturing, such as the use of compression molding in vanillin-based CFRPs (Zamani et al. 2024). Conventional composites benefit from established, cost-effective manufacturing processes, but they lack the environmental benefits offered by sustainable alternatives. As recycling technologies improve and become more cost-effective, processing costs for sustainable composites are expected to decrease, making them more competitive in the long run (Sánchez et al. 2024).

22.8.4.3 Environmental Impact

Sustainable composites are renowned for their lower carbon footprints and higher recyclability rates compared to traditional materials. For example, certain sustainable nanocomposites have a carbon footprint of just 8 kg CO_2 per kilogram (Sehgal et al. 2024). In comparison, conventional composites, while potentially cheaper, often contribute significantly to environmental degradation and waste. The environmental impact of conventional composites is far-reaching, from the extraction of petroleum-based raw materials to the disposal of non-recyclable waste. The shift toward sustainable composites can mitigate these environmental challenges, supporting the industry's transition to more responsible practices (Maiti et al. 2022).

Example Ford's Use of Sustainable Polymer Composites

Ford has been at the forefront of incorporating sustainable polymer composites into its vehicle manufacturing processes. The company has partnered with materials scientists and suppliers to transition away from traditional petroleum-based plastics, replacing them with biodegradable and renewable composites. This move not only supports sustainability but also aligns with Ford's broader goals of reducing its environmental impact (Bekemeyer 2021).

Examples of Ford's Sustainable Composites:

- **Soy-Based Polyurethane Foam**: Ford uses soy-based polyurethane foam in car seats and headrests, which has significantly reduced the use of petroleum in manufacturing. This innovation leads to a reduction of over 5 million pounds of petroleum consumption annually, showcasing Ford's commitment to environmental sustainability.
- **Wheat Straw-Reinforced Plastic**: First introduced in the 2010 Ford Flex, wheat straw-reinforced plastic replaces traditional talc-filled plastic. This switch reduces the weight of the vehicle and contributes to improved sustainability by utilizing renewable resources.
- **Kenaf Fiber-Reinforced Door Panels**: In models like the Ford Escape, door panels made with kenaf fiber-reinforced composites help reduce vehicle weight by 25% compared to conventional plastics. This reduction in weight contributes directly to better fuel efficiency, supporting the company's sustainability goals.
- **Coconut Fiber Composites**: Ford also utilizes coconut fiber composites in various interior parts, such as trunk liners and mats, replacing synthetic fillers that would otherwise be made from petroleum-based materials. This helps further reduce the use of non-renewable resources.
- **Recycled Carbon Fiber Composites**: In collaboration with Boeing, Ford has integrated recycled carbon fiber into vehicle structures, providing lightweight and sustainable options for vehicle components. This innovation plays a crucial role in reducing the overall carbon footprint of their manufacturing process.

22.8.4.4 Impact and Benefits

- **Lighter Vehicles—Better Fuel Efficiency**: The use of these sustainable composites results in lighter vehicles, which directly contribute to improved fuel efficiency. This benefit is critical in reducing the carbon footprint of Ford's vehicles while maintaining high performance.
- **Lower Carbon Footprint—Sustainable Manufacturing**: By substituting petroleum-based materials with renewable and biodegradable composites, Ford is actively lowering its carbon footprint. This shift toward sustainable materials aligns with the company's broader environmental strategy and commitment to sustainable manufacturing practices.
- **Circular Economy Approach—Use of Renewable Resources**: Ford's use of natural fibers and bio-based materials promotes a circular economy by incorporating renewable resources into the manufacturing process. This helps close the loop on material usage, ensuring that the lifecycle of materials used in Ford's vehicles contributes to environmental sustainability.

22.8.5 Industry Expansion

The adoption of sustainable polymer composites is not limited to the automotive sector alone. Other industries, such as aerospace, construction, and consumer goods, are also embracing the potential of bio-based and recycled composites for more sustainable production practices. Companies like **Airbus**, **Adidas**, and **Unilever** are incorporating these materials into their products and manufacturing processes. Airbus, for instance, uses bio-based composites in its aircraft structures, while Adidas has adopted sustainable materials for its clothing lines, and Unilever focuses on using recycled plastics in its consumer goods packaging. This widespread adoption across industries highlights the growing importance and versatility of sustainable polymer composites in shaping the future of manufacturing and product development.

Boeing 787 Dreamliner: Pioneering Sustainable Aerospace Materials

The Boeing 787 Dreamliner incorporates approximately 50% composite materials by weight, significantly reducing the aircraft's overall mass. This weight reduction leads to improved fuel efficiency and lower greenhouse gas emissions, contributing to more sustainable air travel. The use of CFRP composites also enhances the aircraft's durability and reduces maintenance requirements.

In addition to using advanced composites in aircraft construction, Boeing has implemented recycling initiatives to promote sustainability. For instance, the company has tested recycled carbon fiber composite materials in interior components, such as moisture-absorbing floor panels in self-disinfecting lavatories. These efforts not only reduce waste but also support the development of a circular economy in aerospace manufacturing. By integrating sustainable polymer composites into its aircraft design and manufacturing processes, Boeing demonstrates a commitment to environmental responsibility and innovation in the aerospace industry.

22.9 Conclusion

Sustainable polymer composites (SPCs) are becoming an ecological need and a strategic economic advantage. Their creation was a milestone for innovation, legislation, and industry collaboration. SPCs are becoming more competitive with traditional materials because to bio-based resins like PLA and cutting-edge recycling methods like enzymatic depolymerization. Positive policy frameworks like the EU Circular Economy Action Plan are altering market incentives and rewarding sustainable practices, accelerating this innovation.

Additionally, cross-industry collaboration is crucial to growing SPC usage. Ford and Boeing's carbon fiber recycling and agro-waste reuse partnerships show how cooperative research and development may save production costs and improve environmental performance. These efforts reflect a larger industrial shift toward circularity and sustainability for long-term profitability. SPCs must overcome critical obstacles to reach their revolutionary potential. The performance-cost trade-offs and lack of standardized infrastructure for recycling fiber-reinforced composites limit their use. Hybrid composites that mix natural fibers with high-performance components like nanocellulose will require persistent research and development to overcome these hurdles.

SPCs will support green industrialization as these firms speed their net-zero emissions transformation. Their track shows that environmental care and economic growth are mutually reinforcing. To maximize SPCs' potential, stakeholders must commit to innovation, policy integration, and systemic circularity, proving that sustainable materials are viable and crucial to manufacturing and the world. To enable scalable, circular supply chains and maximize SPC market effect, it is crucial to prioritizing R&D in high-performance hybrid composites (e.g., natural fiber/nanocellulose blends) and advocating for globally standardized recycling methods.

References

Aisyah HA et al (2021) A comprehensive review on advanced sustainable woven natural fibre polymer composites. Polymers 13(3):471

Bari E et al (2021) Current strategies for the production of sustainable biopolymer composites. Polymers 13(17):2878

Beena Unni A, Muringayil Joseph T (2024) Enhancing polymer sustainability: eco-conscious strategies. Polymers 16(13):1769

Bekemeyer J (2021) Ford and MSU collaborate on plant-based composite for car parts. https://www.dbusiness.com/tech-mobility-news/ford-and-msu-collaborate-on-plant-based-composite-for-car-parts/

Butenegro JA et al (2023) Novel sustainable composites incorporating a biobased thermoplastic matrix and recycled aerospace prepreg waste: development and characterization. Polymers 15(16):3447

Carvalho D et al (2024) Advancing sustainability in the automotive industry: bioprepregs and fully bio-based composites. Compos Part C Open Access 100459

Cazan C (2023) Advances in sustainable polymeric materials. MDPI-Multidisciplinary Digital Publishing Institute, 540 pp

Dadkhah M, Messori MA (2024) Comprehensive overview of conventional and bio-based fillers for rubber formulations sustainability. Mater Today Sustain 100886

Dasgupta A, Sen B, Dutta P (2024) Development of a sustainable polymer composite: enhancing properties of waste plastic with ground tire rubber reinforcement. Phys Scr 99(9):095013

Elser I, Buchmeiser MR (2024) Toward Sustainable fiber-reinforced polymer composites. Macromol Mater Eng 309(9):2400013

Fagnani DE et al (2022) Sustainable green polymerizations and end-of-life treatment of polymers

IndustryARC (2024) Composites market report—forecast (2024–2030). https://www.industryarc.com/Report/15225/composites-market.html

Islam MZ et al (2022) Green composites from natural fibers and biopolymers: a review on processing, properties, and applications. J Reinf Plast Compos 41(13–14):526–557

Khan UASrivastava A, Asthana N (2024) Harnessing green composites for energy efficiency in the era of industry 4.0. In: Engineering materials for efficient energy storage and conversion. IGI Global, pp 195–222

Leite-Barbosa O et al (2024) Polymer composites reinforced with residues from Amazonian agro-extractivism and timber industries: a sustainable approach to enhancing material properties and promoting bioeconomy. Polymers 16(23):3282

Maiti S et al (2022) Sustainable fiber-reinforced composites: a review. Adv Sustain Syst 6(11):2200258
Oh E et al (2024) Sustainable green composite materials in the next-generation mobility industry: review and prospective
Olawumi MA, Omigbodun FT, Oladapo BI (2024) Integration of sustainable and net-zero concepts in shape-memory polymer composites to enhance environmental performance. Biomimetics 9(9):530
Prasad V et al (2024) A comprehensive review of sustainability in natural-fiber-reinforced polymers. Sustainability 16(3):1223
Przybek A, Łach M (2024) Research on the physical properties of an eco-friendly layered geopolymer composite. Materials 17(19):4937
Raja SKS et al (2024) Optimization of sustainable polymer composites for surface metamorphosis in FDM processes. Appl Chem Eng 7(4). https://doi.org/10.59429/ace.v7i4.5531
Sánchez JG et al (2024) Sustainable composite manufacturing from non-expiring carbon fiber/epoxy prepregs based on a vitrimeric matrix. J Manuf Process 119:902–910
Sehgal R, Singh BJ (2024) Sustainable approaches to improve the characteristics of polymer-based composites. Green Compos Manuf Sustain Approach 19:133
Sehgal A et al (2024) Polymer matrix nanocomposites for lightweight sustainable automotive parts. E3S Web Conf 547:03013
Shih YF et al (2024) Application of sustainable wood-plastic composites in energy-efficient construction. Buildings 14(4):958
Singh P, Singh DN (2024) Polymer composites from waste plastics and reclaimed rubber from tires for sustainable development. J Appl Polym Sci 141(46):e56244
Skosana S, Khoathane C, Malwela T (2024) Driving towards sustainability: a review of natural fiber reinforced polymer composites for eco-friendly automotive light-weighting. J Thermoplast Compos Mater 38. https://doi.org/10.1177/08927057241254324
Soćko A, Andrzejewski J (2024) The development of sustainable polyoxymethylene (POM)-based composites by the introduction of natural fillers and melt blending with poly(lactic acid)-PLA. J Compos Sci 8(8):315
Subadra SP, Griskevicius P (2021) Sustainability of polymer composites and its critical role in revolutionising wind power for green future. Sustain Technol Green Econ 1(1):1–7. https://doi.org/10.21595/stge.2021.21974
Thomas J et al (2024) Navigating the labyrinth of polymer sustainability in the context of carbon footprint. Coatings 14(6):774
Toprakci HAK, Savcı DY, Toprakçı O (2024) Sustainable biopolymer composites from perlite, plasticized and unplasticized poly(lactic acid). Int J Environ Geoinf 11(3):60–68
Umana AU, Garba BMP, Audu AJ (2024) Sustainable business development in resource-intensive industries: balancing profitability and environmental compliance. Int J Multidiscip Res Updates 8(2):64–78
Vázquez-Núñez E et al (2021) Green composites and their contribution toward sustainability: a review. Polym Polym Compos 29(9 Suppl):S1588–S1608
Xu G, Hou L, Wu P (2024) Sustainable plastics with high performance and convenient processibility. Adv Sci 11(35):2405301
Zamani P et al (2024) Assessing sustainability and green chemistry in synthesis of a vanillin-based vitrimer at scale: enabling sustainable manufacturing of recyclable carbon fiber composites. Compos Part A Appl Sci Manuf 179:108016
Zion Market Research Report (2024) Sustainable polymers market by type and application: forecast 2024–2032. https://www.zionmarketresearch.com/report/sustainable-polymers-market
Zolfaghari E, Infurna G, Alessi S, Dispenza C, Dintcheva NT (2024) Sustainable composites from waste polypropylene added with thermoset composite waste or recovered carbon fibres. Polymers 16(20):2922. https://doi.org/10.3390/polym16202922